Series on Number Theory and Its Applications* ISSN 1793-3161

Published

Vol. 5 Algebraic Geometry and Its Applications
edited by J. Chaumine, J. Hirschfeld & R. Rolland

Vol. 6 Number Theory: Dreaming in Dreams
edited by T. Aoki, S. Kanemitsu & J.-Y. Liu

Vol. 7 Geometry and Analysis of Automorphic Forms of Several Variables
Proceedings of the International Symposium in Honor of Takayuki Oda on the Occasion of His 60th Birthday
edited by Yoshinori Hamahata, Takashi Ichikawa, Atsushi Murase & Takashi Sugano

Vol. 8 Number Theory: Arithmetic in Shangri-La
Proceedings of the 6th China–Japan Seminar
edited by S. Kanemitsu, H.-Z. Li & J.-Y. Liu

Vol. 9 Neurons: A Mathematical Ignition
by Masayoshi Hata

Vol. 10 Contributions to the Theory of Zeta-Functions: The Modular Relation Supremacy
by S. Kanemitsu & H. Tsukada

Vol. 11 Number Theory: Plowing and Starring Through High Wave Forms
Proceedings of the 7th China–Japan Seminar
edited by Masanobu Kaneko, Shigeru Kanemitsu & Jianya Liu

Vol. 12 Multiple Zeta Functions, Multiple Polylogarithms and Their Special Values
by Jianqiang Zhao

*For the complete list of the published titles in this series, please visit:
www.worldscientific.com/series/sntia

MULTIPLE ZETA FUNCTIONS, MULTIPLE POLYLOGARITHMS AND THEIR SPECIAL VALUES

Series on Number Theory and Its Applications Vol. 12

MULTIPLE ZETA FUNCTIONS, MULTIPLE POLYLOGARITHMS AND THEIR SPECIAL VALUES

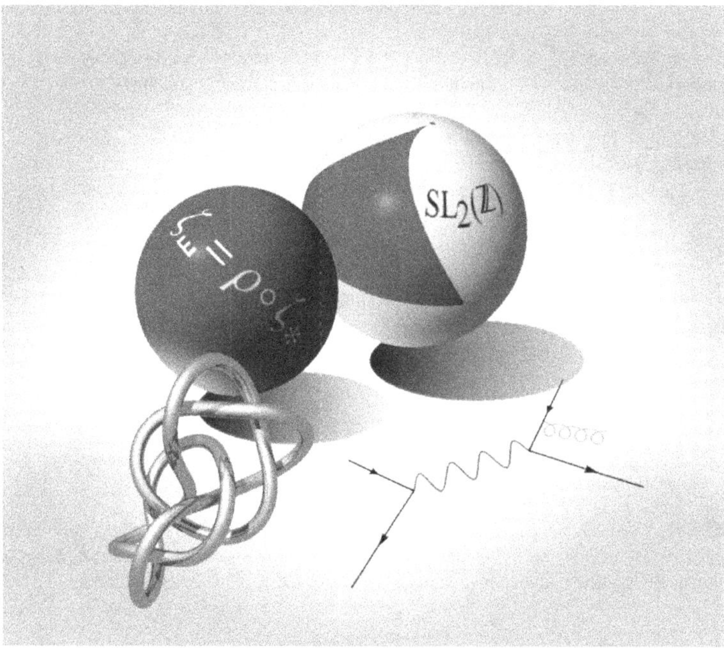

Jianqiang Zhao

World Scientific

NEW JERSEY • LONDON • SINGAPORE • BEIJING • SHANGHAI • HONG KONG • TAIPEI • CHENNAI • TOKYO

Published by

World Scientific Publishing Co. Pte. Ltd.
5 Toh Tuck Link, Singapore 596224
USA office: 27 Warren Street, Suite 401-402, Hackensack, NJ 07601
UK office: 57 Shelton Street, Covent Garden, London WC2H 9HE

Library of Congress Cataloging-in-Publication Data
Names: Zhao, Jianqiang.
Title: Multiple zeta functions, multiple polylogarithms, and their special values / by Jianqiang Zhao (Georgia Southern University, USA).
Description: New Jersey : World Scientific, 2016. | Series: Series on number theory and its applications ; volume 12 | Includes bibliographical references and index.
Identifiers: LCCN 2015043387 | ISBN 9789814689397 (hardcover : alk. paper)
Subjects: LCSH: Functions, Zeta. | Logarithmic functions.
Classification: LCC QA351 .Z43 2016 | DDC 515/.56--dc23
LC record available at http://lccn.loc.gov/2015043387

British Library Cataloguing-in-Publication Data
A catalogue record for this book is available from the British Library.

谨以此书献给岚、盈、心雅

To Lan, Annie and Emma

Contents

Preface

Since my postdoc years at UPenn, I have taken many notes from various papers to help myself keep abreast with the current research of the multiple zeta values and multiple polylogarithms. As the notes became thicker and thicker, I felt a strong urge to organize them into a book form so that not only my students and I but also other people interested in these objects would benefit from them. More recently, these objects have become more and more important due to their newly discovered close relations to numerous other branches of mathematics and even mathematical physics, so that such a textbook seems more appropriate.

Originally I planned to include the most basic topics as well as some more advanced theories such as the p-adic multiple zeta values and the motivic fundamental groups. But as more details were written out, I found it very difficulty to maintain the book to an ideal length while keeping all of them together. Consequently, I must choose to stay on the more elementary side and insert only several optional sections (marked by *) for the more inquisitive reader.

During the multi-year long writing process, I have benefited enormously from multiple visits to the Max Planck Institute for Mathematics, IHES, Taiwan National University, the Academia Sinica, and the Morningside Center of Mathematics in Beijing. This project was also supported partially by NSF grant DMS1162116.

Many people have contributed to this work in various ways. Herbert Gangl was responsible for prompting me to prepare the first outline of this book in early 2010. He was also extremely helpful in providing with me many valuable opinions from the content of the book to the pictures on the cover. Thanks are also due to my students John Cutie and Margaret Sundberg, who suggested a lot of grammatical improvements. I want to express my heart-felt gratitude to Henrik Bachmann, Johannes Blümlein, David Bradley, David Broadhurst, Francis Brown, Tianxin Cai, Ching-Li Chai, Chieh-Yu Chang, Marc Conrad, Pierre Deligne, Kurusch Ebrahimi-Fard, Hidekazu Furusho, Herbert Gangl, Alexander Goncharov, David Goss, Li Guo, Richard Hain, Khodabakhsh and Tatiana Hessami Pilehrood, Hoang Ngoc Ming, Mike Hoffman, Masanobu Kaneko, Maxim Kontsevich, Ulf Kühn, Peng Lei, W.-C. Winnie Li, Zhonghua Li, Erin Linebarger, Kohji Matsumoto, Dominique Manchon, Megan McCoy, Erik Panzer, Johannes Singer, Ismael Soudères, Roberto Tauraso, Danesh Thakur, Kevin Thielen, Jens Vollinga, Jos Vermaseren, Liuquan Wang, Fei Xu, Seidai Yasuda, Haiping Yuan, Don Zagier, Doron Zeilberger, Bin Zhang, Xia Zhou and Wadim Zudilin, for their valuable discussions, fruitful collaborations, and constant encouragement.

I also want to thank Prof. Keqin Feng, Prof. Mike Rosen, and Prof. Jing Yu for their patience, insightful guidance, fatherly love, and immense help during the difficult times.

Finally, I want to thank my wife for her unfailing love over the years, to my daughters for the joys they bring me, and to my parents for their sacrifice for me. This book would not be possible without their continuous support.

Jianqiang Zhao

Introduction

The Riemann zeta function $\zeta(s)$ is one of the most fascinating objects in mathematics. It is a well-known historical fact that L. Euler had studied $\zeta(s)$ as a real function long before B. Riemann did. In particular, he was able to evaluate the function at positive even numbers up to 26 using Bernoulli numbers. He also established a few very nice relations among the double zeta values defined by the nested double sums

$$\zeta(s,t) := \sum_{m>n\geq 1} \frac{1}{m^s n^t}, \quad s,t \in \mathbb{N},$$

which are said to have weight $s+t$. As a matter of fact, Euler worked with the double zeta *star* values $\zeta^\star(s,t)$ which are defined by replacing $m > n$ by $m \geq n$ in the above. Among his many amazing discoveries one stands out as the beautiful identity $\zeta(2,1) = \zeta(3)$, simple to state but not so easy to prove. If you have not seen it before, perhaps you should try it by yourself.

Around 1990, more than 200 years after Euler, this subject was revitalized and generalized to the multiple zeta values (MZVs) by M. Hoffman and D. Zagier independently and almost simultaneously. These values provide the multiple variable version of zeta and double zeta values defined by the iterated sums

$$\zeta(s_1,\ldots,s_d) := \sum_{k_1>\cdots>k_d>0} \frac{1}{k_1^{s_1}\cdots k_d^{s_d}}. \tag{0.1}$$

It is important to note that another common way to define these in the literature is to use the increasing order on the indices.

Since their debut MZVs have appeared unexpectedly in many diverse areas of mathematics and physics. Most prominently, they have shown up in the computation of some infinite classes of the Feynman integrals and knot invariants. Further, F. Brown proved in 2012 in [102] that all the periods of the category of mixed Tate motives unramified over $\mathbb{Z}$ are $\mathbb{Q}[\frac{1}{2\pi i}]$-linear combinations of the MZVs, revealing their fundamental nature in algebraic geometry.

History often repeats itself. Just like Riemann considered $\zeta(s)$ as a complex function years after Euler's work, multiple zeta functions $\zeta(s_1, \ldots, s_d)$, where $s_1, \ldots, s_d$ are complex variables, were investigated almost a decade after Hoffman and Zagier's seminal works. We will visit this analytic theory via the Euler–MacLaurin summation formula in Chap. 1.

In the mid 1990s, mathematical physicists discovered that one needs not only the MZVs to express some Feynman integrals but also the special values of the multiple polylogarithms (MPLs) at the sixth root of unity. Here, MPLs are multiple variable generalizations of the classical polylogarithms, which will be considered in Chap. 2. Besides the analytic properties, we also look at the mixed Hodge structures associated with MPLs and, as a by-product, sketch a way to define their single-valued versions.

The next four chapters form the central part of this book. The reader will find many detailed results concerning the MZVs, proved using a variety of methods, such as the theory of hypergeometric series, Hopf algebras, and Lie algebras. Since these tools are rarely found in the standard graduate curriculum, we list a few key results and properties we need in the appendices found at the back of the book.

One of the most fruitful ideas to study the rational structure of the MZVs is to consider so-called double shuffle relations. If two MZVs are multiplied, we can use either their infinite series representations or their iterated integral representations first observed by M. Kontsevich. The former leads to the quasi-shuffle structure while the latter yields the shuffle one. Comparing the two ways of computing the same product, one obtains so-called finite double shuffle relations. However, even in the simplest case, these relations are insufficient. For example, Euler's identity $\zeta(2,1) = \zeta(3)$ cannot be proved this way simply because the smallest weight of a product of the MZVs is 4. To prove this identity, we are naturally led to the regularizations of the MZVs in two different ways corresponding to the two different product structures. We present this beautiful work of K. Ihara, M. Kaneko and D. Zagier [308] in Chap. 3.

A surprising application of the MZV theory is that they can be used to express the Drinfeld associator which plays a key role in V. Drinfeld's work on the deformation of some Hopf algebras related to the quantum Yang–Baxter equations [172, 173]. The first half of Chap. 4 centers around this key connection, which provides us with an invaluable framework to study the symmetrized MZVs in Chap. 6. The second half introduces a single variable cousin of MPLs, whose single-valued version will be crucial in evaluating several classes of the Feynman integrals in Chap. 15.

Chapter 5 is a sort of compendium of various identities involving the MZVs whose proofs require many different methods. Chapter 6, on the contrary, is rather short and concentrates on only one special variant of MZVs constructed by a kind of cyclic symmetry. These symmetrized MZVs will be linked to the finite version of the MZVs to be considered in Chap. 8.

When the infinite sums in Eq. (0.1) are truncated to finite partial sums, one obtains so-called multiple harmonic sums (MHSs), which will be treated in Chap. 7 together with their alternating version. As these sums are obviously rational numbers we are naturally led to examine their Van Hamme type congruence properties [292, 539, 619, 625]. Only recently do we realize that there should be close relations between the MZVs and their finite versions which are defined essentially by embedding the MHSs into a kind of adèle structure. In particular, this surprising connection motivates the study of the double shuffle relations among the finite MZVs and finite Euler sums. Note that *a priori* we have only one way to multiply the MHSs by their series definition and no integral expression is available. However, due to a beautiful conjecture of Kaneko and Zagier, we can now use some symmetrized versions of the MZVs to link to finite MZVs thereby hinting at a kind of double shuffle relations among them. This will be done in Chap. 8.

For many interesting objects in number theory we often can and do consider their q-analogs. We will carry this out for the MHSs in Chap. 9. Many results contained in this chapter imply the corresponding ones for q-MZVs when we take the number of terms in the finite sums to infinity. When we further let $q \to 1^-$ we arrive at statements for ordinary MZVs. And we can also extend this framework to their star version, which is the content of Chap. 10.

In the next two chapters we continue to deal with q-analogs, first for the multiple zeta functions and then for their special values (q-MZVs). It transpires that there are quite a few different ways to define q-MZVs, which are often affiliated, although sometimes implicit at the first glance. Our attention here is mostly devoted to the double shuffle structures among these q-MZVs.

We return to MPLs in Chaps. 13 and 14 where we shall consider the special values of these functions at the Nth roots of unity. We call these numbers colored MZVs of level N. Unlike MZVs, the double shuffle relations only provide some of the $\mathbb{Q}$-linear relations between colored MZVs.

Chaps. 13 explains how to obtain more relations from algebraic and geometric considerations. Chaps. 14 gives more detailed structural results for lower level colored MZVs.

In the last chapter we provide an attractive and dynamic link recently found between the MZVs/MPLs and the mathematical physics: the study of the Feynman integrals. This is one of the most active research areas in mathematical physics in the past 20 years. We consider two valuable techniques in these computations. The first is the graphical functions of O. Schnetz [505] which are used to evaluate the boxed ladder graphs, the wheel graphs, and the zig-zag graphs by applying the theory of single-valued MPLs and single-valued MZVs developed in Chap. 4. The second is to use the Schwinger parametric form of the Feynman integrals to transform these to projective integrals, which is particularly effective in interpreting results concerning Feynman integrals in terms of algebraic geometric objects such as motives and the cosmic Galois group originally conjectured by P. Cartier and currently being developed by F. Brown.

Our primary goal in writing this book is to provide a comprehensive and accessible introduction to the theory of the MZVs and MPLs for advanced undergraduate students, graduate students and researchers who are interested in learning this beautiful subject. Due to the rapid development of these subjects in recent years as well as space limitation, I have refrained from wading into a few advanced topics such as the motivic theory [102,164], the p-adic theory [48,49,216,218,224,319], the transcendence theory [35,105,642], the relations between MPLs to cluster algebras [243,469], and the generalizations to elliptic MPLs [39, 57, 94, 95, 106, 200, 424], to name just a few. The omission does not mean these are negligible. On the contrary, to have a thorough understanding of the MZVs/MPLs these are often indispensable tools. For example, there is a close connection between the p-adic MZVs and the finite MZVs which provides a very promising approach to proving some of the most fundamental conjectures according to the recent works of D. Jarossay [318,319] and S. Yasuda.

The results contained in this book are due to so many people that it is impossible to attribute each individually. However, I have tried to compile them in the last section of each chapter as some historical notes. Further, some relevant research works are often stated without proofs in these sections. Hopefully, the interested readers can explore further and find their own next research project by themselves.

To facilitate the study, I also provide a few exercise problems at the end of each chapter. Most of these problems are straightforward but some are probably difficult. The serious readers are encouraged to try at least some of them, especially those statements that are cited in the main body of the book.

The subject of the MZVs/MPLs is often very computationally intensive and has been a hot bed for experimental mathematics for a couple of decades [32, 66]. We thus provide quite a few computer programming related problems scattered throughout the book and sketch a few crucial computer pseudo-codes in an appendix. Also included in the appendices are a few tables of useful numerical data and the answers to a few selected exercise problems.

Many people have read the manuscript. Although a lot of inaccuracies have been caught and corrected during the editing process, there will likely be some more errors, for which I am solely responsible. If you notice any please send an email to me: zhaoj@ihes.fr.

Chapter 1

Multiple Zeta Functions

The multiple zeta functions are nested generalizations of the Riemann zeta function. In this chapter we start by summarizing briefly the most important results of the Riemann zeta function. Then we bring in one of our major players in this book — the multiple zeta function — and outline its main properties and problems. Some results will be generalized to higher levels in Chap. 13, but often with more advanced and different proofs. So, we still provide complete proofs of these results in this chapter.

1.1 Riemann Zeta Function

There are numerous accounts in the literature about the Riemann zeta function defined as follows:

$$\zeta(s) := \sum_{n=1}^{\infty} \frac{1}{n^s}, \qquad \operatorname{Re}(s) > 1.$$

As motivations for further generalizations, we list some of its most important properties below.

- **Euler product formula.** For $\operatorname{Re}(s) > 1$,

$$\zeta(s) = \prod_{p:\text{prime}} \left(1 - p^{-s}\right)^{-1}.$$

- **The functional equation.** The Riemann zeta function can be analytically continued to a meromorphic function on the whole complex s-plane, which is holomorphic everywhere except for a simple pole at $s = 1$ with residue 1. Moreover, it satisfies

$$\zeta(s) = 2^s \pi^{s-1} \sin\left(\frac{\pi s}{2}\right) \Gamma(1-s)\, \zeta(1-s). \tag{1.1}$$

- **Special values.** For any positive even number $2n$,

$$\zeta(2n) = -\frac{1}{2} \cdot \frac{B_{2n}}{(2n)!}(2\pi i)^{2n}, \tag{1.2}$$

where B_{2n} are Bernoulli numbers defined by the generating function

$$\frac{t}{e^t - 1} = \sum_{n=0}^{\infty} B_n \frac{t^n}{n!}. \tag{1.3}$$

For negative integers $-n < 0$, one has

$$\zeta(-n) = -\frac{B_{n+1}}{n+1}. \tag{1.4}$$

- **Generating function of the special values.** Let $\psi(x) := \frac{\Gamma'(x)}{\Gamma(x)}$ be the digamma function defined for all x unequal to a negative integer. Then for all $|x| < 1$

$$\psi(1 - x) = -\sum_{k \geq 2} \zeta(k) x^{k-1} - \gamma, \tag{1.5}$$

where $\gamma \approx 0.57721566\ldots$ is Euler's constant. See, for example, [6, § 6.3]. Moreover, ψ satisfies the reflection, recurrence and duplication relations

$$\begin{aligned} \psi(x) &= \psi(1-x) - \pi\cot(\pi x) = \psi(x+1) - \frac{1}{x}, \\ \psi(2x) &= \frac{1}{2}\psi(x) + \frac{1}{2}\psi(x + 1/2) + \log 2, \end{aligned} \tag{1.6}$$

and it has an important integral expression

$$\psi(x+1) = -\gamma + \int_0^1 \frac{1 - t^x}{1 - t} \mathrm{d}t, \tag{1.7}$$

- **Zeros, the critical line, and the Riemann hypothesis.** From Eq. (1.4) we see that $\zeta(s)$ has zeros at negative even integers (called trivial zeros). In his landmark 1859 paper [483] Riemann conjectured that all non-trivial zeros of $\zeta(s)$ must lie on the critical line $\mathrm{Re}(s) = 1/2$. This is the famous "Riemann hypothesis".

1.2 Multiple Zeta Functions

The multiple zeta functions are multiple variable generalizations of the Riemann zeta function.

Definition 1.2.1. For fixed positive integer d and d-tuple of complex variables $\mathbf{s} = (s_1, \dots, s_d)$, the *multiple zeta function* is defined by

$$\zeta(\mathbf{s}) := \sum_{k_1 > \cdots > k_d > 0} k_1^{-s_1} \cdots k_d^{-s_d}. \tag{1.8}$$

To guarantee convergence, $\mathbf{s}$ must satisfy $\mathrm{Re}(s_1 + \cdots + s_j) > j$ for all $j = 1, \dots, d$ where $\mathrm{Re}(s)$ is the real part of a complex number s (see Exercise 1.4). The number d is called the *depth*, denoted by $\mathrm{dp}(\mathbf{s})$. When all the variables are positive integers we call $|\mathbf{s}| := s_1 + \cdots + s_d$ the *weight*. Furthermore, the *multiple zeta star function* is defined by

$$\zeta^\star(\mathbf{s}) := \sum_{k_1 \geq \cdots \geq k_d \geq 1} k_1^{-s_1} \cdots k_d^{-s_d}. \tag{1.9}$$

Remark 1.2.2. (i) In the literature the order $0 < k_1 < \cdots < k_d$ is used sometimes, which gives rise to $\zeta(s_d, \dots, s_1)$ in our notation.

(ii) It is not hard to see that $\zeta^\star(\mathbf{s})$ can be expressed using the multiple zeta functions of different depths:

$$\zeta^\star(s_1, \dots, s_d) = \sum \zeta(s_1 \circ \cdots \circ s_d) \tag{1.10}$$

where $\circ$ is either "+" or ",". So we will consider only the multiple zeta functions in the rest of this chapter.

Similar to the Riemann zeta function one can study both the algebraic theory of the special values of the multiple zeta functions at positive integers, called the *multiple zeta values* and the analytic theory of the multiple zeta functions. In this chapter we will consider mainly the analytic side and postpone the treatment of the multiple zeta values to Chaps. 3–6.

1.3 Analytic Continuation

Since the multiple zeta function of depth d in Eq. (1.8) is defined only for complex arguments with certain restrictions it is essential to continue it analytically to the whole d-dimensional complex space in order to study special values at arbitrary integer arguments.

Recall that the Bernoulli polynomials $B_k(x)$ are defined by the generating function

$$\frac{te^{xt}}{e^t - 1} = \sum_{k=0}^{\infty} B_k(x) \frac{t^k}{k!}$$

and the "periodic Bernoulli polynomials" are defined by

$$\widetilde{B}_k(x) := B_k(\{x\}), \quad x \geq 1,$$

where $\{x\}$ is the fractional part of x. Then we have (see [571, Ch. IX, Misc. Ex. 12])

$$\widetilde{B}_k(x) = -k! \sum_{n\in\mathbb{Z}\setminus\{0\}} \frac{e^{2\pi i n x}}{(2\pi i n)^k}. \tag{1.11}$$

Recall that the Bernoulli numbers satisfy $B_k = B_k(1)$ if $k \ge 2$ while $B_0 = 1$ and $B_1 = -1/2$.

Lemma 1.3.1. *For every positive integer $M \ge 2$ and $x > 1$, we have*

$$\left|\widetilde{B}_M(x)\right| \le \frac{4M!}{(2\pi)^M}.$$

Proof. It follows from the fact that $\zeta(M) \le \zeta(2) = \pi^2/6 < 2$ for $M \ge 2$. □

Let's recall the classical Euler–Maclaurin summation formula [571, 7.21]. Let $f(x)$ be any (complex-valued) C^∞ function on $[1,\infty)$ and let m and M be two positive integers with M **even**. Then we have

$$\begin{aligned}\sum_{n=1}^{m} f(n) = &\int_1^m f(x)\,\mathrm{d}x + \frac{1}{2}(f(1)+f(m)) + \sum_{r=2}^{M} \frac{B_r}{r!}\left(f^{(r-1)}(m) - f^{(r-1)}(1)\right)\\ &- \frac{1}{M!}\int_1^m \widetilde{B}_M(x) f^{(M)}(x)\,\mathrm{d}x.\end{aligned} \tag{1.12}$$

Suppose $\mathrm{Re}(s) > 1$, $f(x) = 1/x^s$. Taking $m = \infty$ and $m = k$ respectively in Eq. (1.12) and then taking the difference we get:

$$\begin{aligned}\sum_{n>k} \frac{1}{n^s} &= \sum_{n=1}^{\infty} \frac{1}{n^s} - \sum_{n=1}^{k} \frac{1}{n^s}\\ &= \int_k^\infty f(x)\,\mathrm{d}x - \frac{1}{2}f(k) - \sum_{r=2}^{M} \frac{B_r}{r!} f^{(r-1)}(k) - \frac{1}{M!}\int_k^\infty \widetilde{B}_M(x) f^{(M)}(x)\,\mathrm{d}x\\ &= \frac{1}{(s-1)k^{s-1}} - \frac{1}{2k^s} + \sum_{r=2}^{M} \frac{B_r}{r!}\frac{(s)_{r-1}}{k^{s+r-1}} - \frac{(s)_M}{M!}\int_k^\infty \frac{\widetilde{B}_M(x)}{x^{s+M}}\,\mathrm{d}x,\end{aligned}$$

where

$$(s)_M = \frac{\Gamma(s+M)}{\Gamma(s)} = \begin{cases} 1, & \text{if } M = 0;\\ s(s+1)\cdots(s+M-1), & \text{if } M > 0,\end{cases} \tag{1.13}$$

is the Pochhammer symbol. Here we have used the fact that $B_k = 0$ if k is odd ≥ 3. Setting $(s)_0 = 1$ and $(s)_{-1} = 1/(s-1)$, one has

$$\sum_{n>k} \frac{1}{n^s} = \sum_{r=0}^{M} \frac{B_r}{r!}\frac{(s)_{r-1}}{k^{s+r-1}} - \frac{(s)_M}{M!}\int_k^\infty \frac{\widetilde{B}_M(x)}{x^{s+M}}\,\mathrm{d}x. \tag{1.14}$$

Theorem 1.3.2. *Let d be a positive integer. Define*

$$\mathsf{S}_d = \left\{ (s_1, \dots, s_d) \in \mathbb{C}^d \,\middle|\, \begin{array}{l} s_1 = 1, s_1 + s_2 = 2, 1, -2m \ \forall m \in \mathbb{Z}_{\geq 0}, \\ \text{and } s_1 + \cdots + s_j \in \mathbb{Z}_{\leq j} \ \forall j \geq 3. \end{array} \right\}. \quad (1.15)$$

(i) *If $d = 1$ then we have*

$$\zeta(s) = \frac{1}{s-1} + \frac{1}{2} + \sum_{r=1}^{M} \frac{B_{r+1}}{(r+1)!}(s)_r - \frac{(s)_{M+1}}{(M+1)!} \int_1^\infty \widetilde{B}_{M+1}(x) x^{-s-M-1} \, \mathrm{d}x. \quad (1.16)$$

(ii) *If $d \geq 2$ then for all $\mathbf{s} = (s_1, \dots, s_d) \in \mathbb{C}^d \setminus \mathsf{S}_d$ and positive integer $M > d + |\mathrm{Re}(s_1)| + \cdots + |\mathrm{Re}(s_d)|$ we have*

$$\zeta(\mathbf{s}) = \sum_{r=0}^{M} \frac{B_r}{r!}(s_1)_{r-1} \cdot \zeta(s_1 + s_2 + r - 1, s_3, \dots, s_d) - \frac{(s_1)_M}{M!} \sum_{k_2 > \cdots > k_d > 0} \frac{1}{k_2^{s_2} \cdots k_d^{s_d}} \int_{k_2}^\infty \frac{\widetilde{B}_M(x)}{x^{s_1+M}} \, \mathrm{d}x. \quad (1.17)$$

These provide the analytic continuation of the multiple zeta function $\zeta(\mathbf{s})$ to a meromorphic function on $\mathbb{C}^d$ with simple poles given by S_d.

Proof. We prove the theorem by induction on the depth d. When $d = 1$, taking $k = 1$ and replacing M by $M + 1$ in Eq. (1.14) we get Eq. (1.16). Since we may choose M arbitrary large, this gives the analytic continuation of $\zeta(s)$ to the whole s-plane.

Now we assume $d \geq 2$. In Eq. (1.8) we replace s_1, k_1, k_2 by s, n and k, respectively. Then Eq. (1.17) follows directly from Eq. (1.14) if we can show that the series in Eq. (1.17) converges. Lemma 1.3.1 implies

$$\sum_{k_2 > \cdots > k_d > 0} \left| \frac{1}{k_2^{s_2} \cdots k_d^{s_d}} \int_{k_2}^\infty \frac{\widetilde{B}_M(x)}{x^{s_1+M}} \, \mathrm{d}x \right| \leq \frac{4M!}{(2\pi)^M (M-1+\mathrm{Re}(s_1))} \sum_{k_2 > \cdots > k_d > 0} \frac{1}{k_2^{M-1+\mathrm{Re}(s_1)+\mathrm{Re}(s_2)} k_3^{\mathrm{Re}(s_3)} \cdots k_d^{\mathrm{Re}(s_d)}}$$

which, by induction, converges absolutely whenever $M > d + |\mathrm{Re}(s_1)| + \cdots + |\mathrm{Re}(s_d)|$.

This completes the proof of the theorem by induction. □

1.4 Other Multiple Zeta Functions

In the 2002 Arizona Winter School, Deligne designed a project for some participating graduate students to find a good candidate in the étale theory corresponding to the multiple zeta function defined by Eq. (1.8), called the p-adic multiple zeta functions. Subsequently Furusho successfully defined p-adic multiple zeta *values* by generalizing Coleman's construction of p-adic polylogarithms and explained his results with the Tannakian formalism in [216–218]. Moreover, Chatzistamatiou [127] showed that these values are p-adic integers.

Over global fields, one can generalize the multiple zeta functions to a kind of multiple Dedekind zeta functions over arbitrary number field F. Let O_F be the ring of integers of F and $N(I) = |O_F : I|$ the norm for an integral ideal I in F. Let d be a positive integer and $F_1, \dots, F_d$ be d number fields. R. Masri [416] defined the multiple Dedekind zeta function over $F_1, \dots, F_d$ as (after slightly change of ordering of the indices)

$$\zeta_d(\{F_i\}_{i=1}^d; s_1, \dots, s_d) := \sum_{\substack{\mathfrak{n}_1 \in O_{F_1}, \dots, \mathfrak{n}_d \in O_{F_d} \\ N(\mathfrak{n}_1) > \dots > N(\mathfrak{n}_d)}} N(\mathfrak{n}_1)^{-s_1} \cdots N(\mathfrak{n}_d)^{-s_d}.$$

By Riemann–Roch he proves that there are positive numbers $\varepsilon_1, \dots, \varepsilon_d$ such that the series $\zeta_d(\{F_i\}_{i=1}^d; s_1, \dots, s_d)$ is absolutely convergent and analytic in the region

$$\text{Re}(s_1 + \dots + s_j) > j + \varepsilon_1 + \dots + \varepsilon_j, \quad \forall j = 1, 2, \dots, d.$$

In [611] we consider the case where $F_1 = \dots = F_d = F$ and define $\zeta_F(s_1, \dots, s_d) = \zeta_d(\{F\}_{i=1}^d; s_1, \dots, s_d)$. In [300] Horozov approached this problem from a different point of view. In particular, his version of the multiple Dedekind zeta functions has analytic continuations and their special values can be expressed using iterated integrals.

It would be very interesting to obtain some numerical data concerning the following generalization of Zagier's conjecture concerning Dedekind zeta functions (cf. Conjecture 2.2.1).

Problem 1.4.1. For large enough positive integers $n_1, \dots, n_d$ such that $\zeta_F(n_1, \dots, n_d)$ converges, is there an expression of $\zeta_F(n_1, \dots, n_d)$ in terms of a determinant with entries given by $\mathscr{L}i_{n_1,\dots,n_d}$ evaluated at F-rational points up to some factors determined only by the number field F (such as the discriminant, the number of real and complex embeddings, etc.)? Here $\mathscr{L}i_{n_1,\dots,n_d}$ is the single-valued version of the multiple polylogarithms to be defined in Sec. 2.9.

In another direction, Akiyama and Ishikawa [8] defined the multiple Hurwitz zeta function by

$$\zeta(s_1,\dots,s_d;\theta_1,\dots,\theta_d) = \sum_{k_1>\cdots>k_d>0} (k_1+\theta_1)^{-s_1}\cdots(k_d+\theta_d)^{-s_d}, \quad (1.18)$$

where $0 \le \theta_j < 1$ for $j = 1,\dots,d$ and derive its analytic continuation by using a modified Euler–Maclaurin summation formula. Murty and Sinha obtained the analytic continuation in [432] by using the binomial series expansion and Hartog's Theorem in complex analysis. All of the authors above considered also the cognate multiple L-series

$$L(s_1,\dots,s_d;\chi_1,\dots,\chi_d) := \sum_{k_1>\cdots>k_d\ge 1} \frac{\chi_1(k_1)\dots\chi_d(k_d)}{k_1^{s_1}\cdots k_d^{s_d}}, \quad (1.19)$$

where χ_j's are Dirichlet characters. These values are essentially the colored multiple zeta values which we will study in Chap. 13.

Kelliher and Masri [345] realized the analytic continuation of Eq. (1.18) with an extension of the generalized function approach first adopted in [605] and further determined some residues explicitly. M. Hoang Ngoc *et al.* [286] also studied these functions and used a certain shuffle product structure to produce an algorithm for computing the special values of the multiple polylogarithms (see Chap. 2 for definition).

Generalizing the ideas of Wu, Matsumoto and *et al.*, Zhao and Zhou (see [418, 423, 635]) defined the *Mordell–Tornheim L-functions* by

$$L_{\mathrm{MT}}(\mathbf{s};\boldsymbol{\chi}) := \sum_{k_1=1}^{\infty}\cdots\sum_{k_d=1}^{\infty} \frac{\chi_1(k_1)\dots\chi_d(k_d)\chi_{d+1}(k_1+\cdots+k_d)}{k_1^{s_1}\cdots k_d^{s_d}(k_1+\cdots+k_d)^{s_{d+1}}}. \quad (1.20)$$

They derived the analytic continuation by utilizing so-called colored Mordell–Tornheim functions. When all the characters in Eq. (1.20) are principal these functions are nothing but the traditional Mordell–Tornheim zeta functions

$$\zeta_{\mathrm{MT}}(s_1,\dots,s_{d+1}) := \sum_{k_1=1}^{\infty}\cdots\sum_{k_d=1}^{\infty} \frac{1}{k_1^{s_1}\cdots k_d^{s_d}(k_1+\cdots+k_d)^{s_{d+1}}}. \quad (1.21)$$

Zhou *et al.* provided some signed q-analogs of Eq. (1.21) in [638], Okamoto studied a parametrized analogue in [450] while Tsumura and Okamoto considered its alternating analogues in [551] and [453], respectively.

For another type of the multiple zeta functions, Matsumoto (see [421]) replaces the factor $k_j + \theta_j$ in the denominator of Eq. (1.18) by linear polynomials of $k_1,\dots,k_j$. More generally, while pursuing his Ph.D. Essouabri considered the multiple zeta functions of the form

$$\sum_{\mathbf{k}=(k_1,\dots,k_d)\in\mathbb{N}^d} P_1(\mathbf{k})^{-s_1}\cdots P_d(\mathbf{k})^{-s_d}$$

where $P_j(\mathbf{k})$'s are polynomials of d variables. In his thesis [206, 207] Essouabri established the analytic continuation in the case $d = 1$ and stated that his method could be applied to arbitrary depth for suitable P_j's including all the linear polynomials. Later, Essouabri *et al.* [208] extended this to more general types of the multiple zeta functions involving vectorial functions in the denominators.

One of the most interesting multiple zeta functions originates from physics. Around 1990, related to his work on topological quantum field theory, Witten [573] found the volumes of the moduli spaces of representations of the fundamental groups of two dimensional surfaces are some special values of a new type of zeta functions attached to complex semisimple Lie algebras $\mathfrak{g}$ at positive integers. Inspired by this result, Zagier [598] defined these complex functions as follows:

$$\zeta_W(s;\mathfrak{g}) := \sum_{\rho} \frac{1}{(\dim \rho)^s},$$

where ρ runs over all finite dimensional irreducible representations of $\mathfrak{g}$. By physical considerations Witten showed that for any positive integer n

$$\zeta_W(2n;\mathfrak{g}) = c(2n;\mathfrak{g})\pi^{2nr},$$

where $c(2n;\mathfrak{g}) \in \mathbb{Q}$ and r is the number of positive roots of $\mathfrak{g}$. Such formulas are now called Witten volume formulas. Many results along this direction can be found in [621, 624, 626, 636].

Matsumoto and his collaborators [354–357, 422] generalized Witten's zeta functions to the multiple variable setting over an arbitrary semi-simple Lie algebra $\mathfrak{g}$. Let Δ_+ be the set of positive roots of $\mathfrak{g}$, $\{\alpha_1, \dots, \alpha_d\}$ the fundamental roots with coroots $\{\alpha_1^\vee, \dots, \alpha_d^\vee\}$, and $\{\lambda_1, \dots, \lambda_d\}$ the fundamental weights such that $\langle \alpha_i^\vee, \lambda_j \rangle = \delta_{i,j}$ (the Kronecker symbol). Then the multiple Witten zeta function over $\mathfrak{g}$ is defined by

$$\zeta_{\mathfrak{g}}(\{s_\alpha\}_{\alpha\in\Delta_+}) := \sum_{k_1,\dots,k_d=1}^{\infty} \prod_{\alpha\in\Delta_+} \langle \alpha^\vee, k_1\lambda_1 + \cdots + k_d\lambda_d \rangle^{-s_\alpha}, \tag{1.22}$$

which satisfies (see [354])

$$\zeta_W(s;\mathfrak{g}) = \zeta_{\mathfrak{g}}(\{s\}^d) \prod_{\alpha\in\Delta_+} \langle \alpha^\vee, \lambda_1 + \cdots + \lambda_d \rangle,$$

where $\{s\}^d$ means the variable s repeats itself d times. One of the most exciting results is a statement on their functional relations produced by the Weyl group symmetry in the underlying Lie algebra structure (see [352, 357] for details). For other relevant works, see [451, 452].

It is well known that there is a huge supply of analogies between the number fields and the function fields over finite fields. Carlitz and Weil worked on these in the early years of twentieth century, continued by many researchers recently (see [255, 545] for a lot of fascinating results along this line). In particular, Thakur devoted a section of his book [545, § 5.10] to the study of various possible multiple zeta values over the rational function field $\mathbb{F}_q[t]$ over the finite field $\mathbb{F}_q$ of q elements. For global function field K over $\mathbb{F}_q$, we now state a result of Masri [417]. For a divisor D of K, let $\deg(D)$ be its degree and $|D| = q^{\deg(D)}$ be its norm. Let $\mathfrak{D}_K^+$ be the sub semi-group of effective divisors of K. Masri defines

$$Z_d\left(K; s_1, \ldots, s_d\right) := \sum_{\substack{(D_1,\ldots,D_d)\in\mathfrak{D}_K^+\times\cdots\times\mathfrak{D}_K^+ \\ 0\le\deg(D_1)\le\cdots\le\deg(D_d)}} \prod_{k=1}^{d} |D_k|^{-s_k},$$

and shows that it has a meromorphic continuation to all of $\mathbb{C}^d$ and is a rational function in each of $q^{-s_1}, \ldots, q^{-s_d}$ with a specified denominator. Further, $Z_d\left(K; s_1, \ldots, s_d\right)$ has all possible simple poles on the linear subvarieties

$$s_k + \cdots + s_d = 0, 1, \ldots, d-k+1, \quad k = 1, \ldots, d.$$

It has been discovered that Thakur's multiple zeta value over the function fields enjoy some similar identities to those over $\mathbb{Q}$ (see [11, 125, 129, 130, 379–382]). Due to the lack of integral representation of the function field multiple zeta values, one loses the important family of so-called double shuffle relations (see Chap. 5). However, on the transcendence of these values Yu's classical results in [588] and the new breakthrough by Chang [124] goes far beyond what we can do over the number fields.

1.5 Historical Notes

Despite its name the Riemann zeta function was studied first by Euler before the mid eighteenth century (see, for example, [211]) although he considered it as a function of a real variable. It is Riemann who treated it as a complex function for the first time in his famous 1859 article [483].

The genesis of the multiple zeta functions can be traced back to a series of correspondences between Leonhard Euler and Christian Goldbach (see [323]). On the Christmas Eve of 1742, with different notation Goldbach wrote down some special cases of the following infinite sum on a letter to Euler:

$$\sum_{a\ge b\ge 1} \frac{1}{a^m b^n},$$

where m and n are positive integers. In modern notation this is $\zeta(m,n)+\zeta(m+n)$ where $\zeta(m,n)$ is called a double zeta value. Later, Euler discovered so-called decomposition formula for the double zeta values:

$$2\zeta(n,1)=n\zeta(n+1)-\sum_{i=1}^{n-2}\zeta(n-i)\zeta(i+1), \tag{1.23}$$

for all positive integer $n\geq 2$.

There are two independent sources of the revival of the multiple zeta functions in recent years. From late 1980s to early 1990s, motivated by some problems of Moen (see [297]) Hoffman initiated the study of the multiple zeta values from the viewpoint of their underlying algebraic structures in a series of papers (see [288, 289]). During the same time period, Zagier was attracted to the multiple zeta functions by their mysterious connections to Witten's zeta functions, Vassiliev knot invariants and the theory of mixed Tate motives over $\mathbb{Z}$. After many discussions with Drinfeld, Kontsevich and Goncharov, Zagier included section 9 in his 1994 paper [598] which became one of the most influential papers in the modern study of the multiple zeta functions.

However, the analytic properties of the double zeta function were considered much earlier than the general multiple variable case, similar to the situation of Euler's study of the double zeta values at positive integers. The analytic continuation of the double zeta function was first realized by Atkinson [23] using the Poisson summation formula in the 1940s, and by Apostol and Vu in [20] independently much later. For arbitrary depth, there are many different approaches: the generalized functions or distributions of [462, 605], the Euler–Maclaurin summation formula of [10, 160], the Mellin–Barnes integral [419, 420], the multiple Hurwitz–Lerch zeta functions [351], and the binomial series expansion and Hartog's Theorem [432]. In Sec. 1.3 we gave a polished proof of the second approach using the Euler–Maclaurin summation formula outlined first in [615].

Most of the current research on multiple zeta functions is on their special values. However, there are some important results concerning their analytic properties. For example, Matsumoto showed a functional equation for the double zeta functions (see [142, 420]) and Nakamura and Pankowski discovered the existence of nontrivial zeros of general multiple zeta functions in [435, 437] although their precise locations are still beyond our reach. For a good survey of the analytic theory of these functions and further generalizations, Matsumoto's paper [418] is a good reference for results up to around 2001.

There are some investigations of the behavior of the multiple zeta functions at non-positive integers. At depth greater than 1, strictly speaking, different ways to define these values may lead to different results. This phenomenon already occurs at depth two, see Exercise 1.6. In [9, 10], Akiyama *et al.* provided several possible definitions. Komori [351] and Sasaki [497, 498] also considered this problem from different perspectives. By Connes–Kreimer's algebraic approach with Rota–Baxter algebras, Guo and Zhang [264] first defined renormalized MZVs at non-positive integers satisfying stuffle relations. Manchon and Paycha [414] further provided a way to renormalize all MZVs at singular points. By choosing appropriate finite "linear" combination of multiple zeta-functions with some argument shifts, Furusho et al. [225] managed to desingularize the multiple zeta functions, which yields a rigorous definition of multiple zeta functions at non-positive integers. In [184], we showed that all the different ways to renormalize MZVs satisfying the stuffle relations form a set on which a certain transfer group acts freely and transitively.

Using Raabe's identity, Sadaoui [490] linked a certain multiple integral and a classical inversion argument to obtain an analytic continuation of the multiple zeta function defined at negative integer arguments. This process has been used to derive some closed expressions for these values by Moll *et al.* [427].

In the literature there exist a few functions named as "multiple zeta functions" but are defined completely differently from Eq. (1.8). For example, Kurokawa's multiple zeta function considered in [367] is a tensor product of the general zeta functions that are defined by Euler products and are meromorphic of finite order with functional equations over the entire space. In particular, the tensor product of two zeta functions is defined by the Hadamard product with zeros and poles given by the sum of those of the two zeta functions. So Kurokawa's multiple zeta function differs drastically from ours since we do not have Euler product expansion associated with our version (see [436] for a proof and [491] for a related work on multiple zeta-star functions). In [473] one can also find the multiple zeta functions of the free Abelian groups.

A caveat in reading papers on multiple zeta functions is that sometimes the word "multiple" is used to refer to multiple summation indices even though the function is a single variable function. For instances, the p-adic multiple zeta function of [535] and the Barnes multiple zeta-function of [353] are both single variable functions.

Exercises

1.1. Show that $B_{2n+1} = 0$ for all $n > 0$ by the definition Eq. (1.3).

1.2. Prove Eq. (1.4) using the functional equation given by Eq. (1.1) and Eq. (1.2).

1.3. Prove the relation between multiple zeta star function and the multiple zeta function given by Eq. (1.10).

1.4. Show that Eq. (1.8) converges if and only if $\mathrm{Re}(s_1 + \cdots + s_j) > j$ for all $j = 1, \ldots, d$.

1.5. Complete the proof of Thm. 1.3.2 by induction.

1.6. Carry out the analytic continuation of the double zeta function $\zeta(s_1, s_2)$ explicitly. Then compare the two different limits

$$\lim_{s_1\to 0}\lim_{s_2\to 0}\zeta(s_1, s_2) \quad \text{and} \quad \lim_{s_2\to 0}\lim_{s_1\to 0}\zeta(s_1, s_2)$$

at $(s_1, s_2) = (0, 0)$. Are they the same? Why?

Chapter 2

Multiple Polylogarithms (MPLs)

In this chapter we consider single- and multi-variable generalizations of the natural logarithm, called the polylogarithms (polylogs) and the multiple polylogarithms (multiple polylogs or MPLs), respectively.

2.1 Chen's Iterated Integrals

We briefly recall the theory of K.-T. Chen's iterated integrals in this section. It is an indispensable tool in the subsequent study of the analytic properties of MPLs. For $r > 1$, define inductively

$$\int_a^b f_1(t)\,\mathrm{d}t \cdots f_r(t)\,\mathrm{d}t = \int_a^b f_1(\tau)\,\mathrm{d}\tau \cdots f_{r-1}(\tau)\left(\int_a^\tau f_r(t)\,\mathrm{d}t\right)\mathrm{d}\tau.$$

When $r = 0$ one sets the integral to be 1 as a convention. More generally, let $w_1, \dots, w_r$ be some 1-forms (repetition allowed) on a manifold M and let $\alpha : [0,1] \to M$ be a piecewise smooth path. Write $\alpha^* w_i = f_i(t)\mathrm{d}t$ and define the iterated integral

$$\int_\alpha w_1 \cdots w_r := \int_0^1 f_1(t)\,\mathrm{d}t \cdots f_r(t)\,\mathrm{d}t.$$

Remark 2.1.1. Our order of the 1-forms in the iteration here is opposite to K.-T. Chen's original order.

The following results are crucial in the application of Chen's theory of the iterated path integrals. However, the proofs are beyond the scope of this book so we recommend the interested reader to read K.-T. Chen's original papers.

Lemma 2.1.2. *Let w_i $(i \geq 1)$ be $\mathbb{C}$-valued 1-forms on a manifold M. Then*

(i) *The value of $\int_\alpha w_1 \cdots w_r$ is independent of the parametrization of α.*

(ii) *If* $\alpha, \beta : [0,1] \longrightarrow M$ *are composable paths* (i.e., $\alpha(1) = \beta(0)$), *then*

$$\int_{\beta\alpha} w_1 \cdots w_r = \sum_{j=0}^{r} \int_{\beta} w_1 \cdots w_j \int_{\alpha} w_{j+1} \cdots w_r,$$

where we set $\int_\alpha \phi_1 \cdots \phi_m = 1$ *if* $m = 0$ *and* $\beta\alpha$ *denotes the composition of* α *and* β.

(iii) *For every path* $\alpha : [0,1] \longrightarrow M$, *let* $\alpha^{-1}(t) = \alpha(1-t)$. *Then*

$$\int_{\alpha^{-1}} w_1 \cdots w_r = (-1)^r \int_{\alpha} w_r \cdots w_1.$$

(iv) *Let* $\mathfrak{S}_n$ *be the symmetry group of* n *letters. For every path* α,

$$\int_\alpha w_1 \cdots w_r \int_\alpha w_{r+1} \cdots w_{r+s} = \sum_{\substack{\sigma \in \mathfrak{S}_{r+s} \\ \sigma(1)<\cdots<\sigma(r) \\ \sigma(r+1)<\cdots<\sigma(r+s)}} \int_\alpha w_{\sigma^{-1}(1)} \cdots w_{\sigma^{-1}(r+s)}.$$

Proof. (i) can be derived from the theorem on p. 361 of [132]. (ii) and (iii) are formulas (1.6.1) and (1.6.2) of [132] respectively. Ree [481] discovered the shuffle relation (iv) which appeared as (1.5.1) in [132]. □

Lemma 2.1.3. *If* $w_i^{(j)}$ *are closed 1-forms for* $1 \le i \le r$ *and* $1 \le j \le n$ *such that* $\sum_j w_1^{(j)} \wedge w_2^{(j)} = \sum_j w_2^{(j)} \wedge w_3^{(j)} = \cdots = \sum_j w_{r-1}^{(j)} \wedge w_r^{(j)} = 0$ *then* $\sum_j \int_\alpha w_1^{(j)} w_2^{(j)} \cdots w_r^{(j)}$ *only depends on the homotopy class of* α.

Proof. The case $r \le 2$ is proved on pp. 366-368 of [132]. The general case follows from a similar argument. □

2.2 Classical Polylogarithms

For positive integer $n > 1$, the classical polylog is defined by

$$\mathrm{Li}_n(x) := \sum_{k=1}^{\infty} \frac{x^k}{k^n}, \qquad |x| \le 1.$$

Clearly the Riemann zeta value $\zeta(n) = \mathrm{Li}_n(1)$ if $n \ge 2$. If $n = 1$ one sees easily that

$$\mathrm{Li}_1(x) = \sum_{k=1}^{\infty} \frac{x^k}{k} = -\log(1-x), \qquad |x| < 1.$$

Note that one can express $\mathrm{Li}_n(x)$ using the following iterated integral

$$\mathrm{Li}_n(x) = \int_0^x \left(\frac{dt}{t}\right)^{n-1} \frac{dt}{1-t},$$

from which one obtains an analytic continuation of $\mathrm{Li}_n(x)$ to $\mathbb{C}$ as a multi-valued function with a singularity at 1. Since multi-valued functions are hard to grasp there have been several attempts to find their single-valued cousins. In [392] Lewin defined the following polylog of a real variable $x < 1$:

$$L_n(x) = \sum_{j=0}^{n} \frac{(-\log|x|)^j}{j!} \mathrm{Li}_{n-j}(x) + \frac{(-\log|x|)^{n-1}}{n!} \log|1-x|, \tag{2.1}$$

which can be extended to $x \in \mathbb{R}$ by setting $L_n(x) = (-1)^{n-1} L_n(1/x)$ for $x \geq 1$. He further hypothesized that $L_n(x)$ should satisfy "clean" functional equations and proved those equations found by Kummer with $n < 5$. Another version of single-valued polylog is defined by Ramakrishnan (see [479, 594]) for complex variable x with $|x| \leq 1$

$$D_n(x) := \begin{cases} \mathrm{Re} \\ \mathrm{Im} \end{cases} \sum_{j=0}^{n} \frac{(-\log|x|)^j}{j!} \mathrm{Li}_{n-j}(x) - \frac{(-\log|x|)^n}{2 \cdot n!} \quad \begin{array}{l} \text{if } n \text{ is odd;} \\ \text{if } n \text{ is even.} \end{array} \tag{2.2}$$

For $|x| \geq 1$ one can set $D_n(x) = (-1)^{n-1} D_n(1/x)$. Note that this function is real analytic. But it seems that Zagier's version of single-valued polylogs $\mathscr{L}i_n$ in [596] behaves the most regularly in the sense that the functional equations of $\mathscr{L}i_n$ do not involve lower order polylogs, at least for $n \leq 3$. It is defined on the one dimensional projective complex space $\mathbb{CP}^1$ for any positive integer $n \geq 2$ by

$$\mathscr{L}i_n(x) := \begin{cases} \mathrm{Re} \\ \mathrm{Im} \end{cases} \sum_{j=0}^{n-1} \frac{2^j B_j}{j!} \log^j |x| \, \mathrm{Li}_{n-j}(x) \quad \begin{array}{l} \text{if } n \text{ is odd;} \\ \text{if } n \text{ is even,} \end{array} \tag{2.3}$$

where B_j is the jth Bernoulli number. These functions are all single-valued and real-analytic by straightforward monodromy computations. In particular $\mathscr{L}i_n(0) = \mathscr{L}i_n(\infty) = \mathscr{L}i_{2n}(1) = 0$, and $\mathscr{L}i_{2n+1}(1) = \zeta(2n+1)$.

We now turn briefly to the discussion of polylog functional equations. In high school, students already know that $\log|x| + \log|y| = \log|xy|$. For dilogarithms (dilogs), one has the following identity (see [600, p. 9]): if $1/x + 1/y + 1/z = 1$ then (see Exercise 2.1)

$$\mathrm{Li}_2(x) + \mathrm{Li}_2(y) + \mathrm{Li}_2(z) = \frac{1}{2}\left[\mathrm{Li}_2\left(-\frac{xy}{z}\right) + \mathrm{Li}_2\left(-\frac{yz}{x}\right) + \mathrm{Li}_2\left(-\frac{zx}{y}\right)\right]. \tag{2.4}$$

In fact, this is a consequence of so-called five-term relation discovered by Abel [1], which says, in terms of the single-valued Bloch–Wigner dilog function $\mathscr{L}i_2(x)$, that for all $x, y \neq 0, 1$ and $x \neq y$

$$\mathscr{L}i_2(x) - \mathscr{L}i_2(y) + \mathscr{L}i_2\left(\frac{y}{x}\right) - \mathscr{L}i_2\left(\frac{1-x^{-1}}{1-y^{-1}}\right) + \mathscr{L}i_2\left(\frac{1-x}{1-y}\right) = 0. \tag{2.5}$$

For a more symmetric form, see Exercise 2.3.

Similar result holds for the trilog $\mathscr{L}i_3(x)$: for all x, y, z, we have the 22-term relations

$$\mathscr{L}i_3(-xyz) + \bigoplus_{\mathrm{Cyc}(x,y,z)} \Big\{ \mathscr{L}i_3(zx - x + 1) + \mathscr{L}i_3\big(\frac{zx - x + 1}{zx}\big)$$
$$- \mathscr{L}i_3\big(\frac{zx - x + 1}{z}\big) + \mathscr{L}i_3\big(\frac{x(yz - z + 1)}{-(zx - x + 1)}\big) + \mathscr{L}i_3\big(\frac{yz - z + 1}{y(zx - x + 1)}\big)$$
$$+ \mathscr{L}i_3(z) - \mathscr{L}i_3\big(\frac{yz - z + 1}{yz(zx - x + 1)}\big)\Big\} = 3\,\mathscr{L}i_3(1). \tag{2.6}$$

Here $\bigoplus_{\mathrm{Cyc}(x,y,z)} f(x, y, z) := f(x, y, z) + f(y, z, x) + f(z, x, y)$. This was discovered first by Goncharov [245] when he confirmed the following Zagier's Conjecture for $\zeta_F(3)$ where $\zeta_F(s)$ is the Dedekind zeta function over a number field F. Zagier verified his conjecture for $\zeta_F(2)$ in [593].

Conjecture 2.2.1. (Zagier's Conjecture) *Let F be a number field. Let $\sigma_1, \dots, \sigma_{r_1} : F \to \mathbb{R}$ be the real embeddings and $\sigma_{1+r_1} = \overline{\sigma_{r_1+1+r_2}}, \dots, \sigma_{r_1+r_2} = \overline{\sigma_{r_1+2r_2}} : F \to \mathbb{C}$ be the complex embeddings. Then there exist $y_1, \dots, y_{r(n+1)} \in \mathbb{Q}[F \setminus \{0, 1\}]$ such that*

$$\zeta_F(n) = \begin{cases} \pi^{nr_2} D_F^{-1/2} \det\Big[\mathscr{L}i_n\big(\sigma_i(y_j)\big)\Big]_{1\le i,j\le r_1+r_2}, & \text{if } n \text{ is odd;} \\ \pi^{n(r_1+r_2)} D_F^{-1/2} \det\Big[\mathscr{L}i_n\big(\sigma_{r_1+i}(y_j)\big)\Big]_{1\le i,j\le r_2}, & \text{if } n \text{ is even,} \end{cases}$$

where D_F is the discriminant of F.

2.3 Definition of MPLs

The multiple polylogs are iterated generalizations of polylogs to the multi-variable situation.

Definition 2.3.1. Let d, $s_1, \dots, s_d$ be positive integers, and let $\mathbf{x} = (x_1, \dots, x_d) \in \mathbb{C}^d$ be a complex vector variable such that $|\prod_{j=1}^{\ell} x_j| < 1$ for all $\ell = 1, \dots, d$. The multiple polylog (MPL) function of depth d is defined by

$$\mathrm{Li}_{s_1,\dots,s_d}(\mathbf{x}) := \sum_{k_1>k_2>\cdots>k_d\ge1} \frac{x_1^{k_1} x_2^{k_2} \cdots x_d^{k_d}}{k_1^{s_1} k_2^{s_2} \cdots k_d^{s_d}}. \tag{2.7}$$

As in the definition of the multiple zeta functions, the order of indices in Eq. (2.7) is sometimes reversed in research papers, even by the author himself. Note also that occasionally the term "multiple polylog" means the single variable function $\mathrm{Li}_{s_1,s_2\dots,s_d}(t,1,1,\dots,1)$ in the literature, which will in fact be needed in this book later (see Sec. 4.2 and Sec. 14.1).

Now by setting $a_i=\prod_{j=1}^{i}x_j^{-1}$ it is not hard to check (see Exercise 2.4) that

$$\mathrm{Li}_{s_1,\dots,s_d}(x_1,\dots,x_d)=\int_0^1\left(\frac{\mathrm{d}t}{t}\right)^{s_1-1}\frac{\mathrm{d}t}{a_1-t}\cdots\left(\frac{\mathrm{d}t}{t}\right)^{s_d-1}\frac{\mathrm{d}t}{a_d-t}. \tag{2.8}$$

This is an iterated path integral in the sense of Chen (see [132, 133]) whose path lies in $\mathbb{C}$. Thus one can easily enlarge its domain of definition to some open subset of $\mathbb{C}^d$. However, it is difficult to study the monodromy of the MPLs by this integral expression.

2.3.1 *Divergence Control:* $\mathsf{DivLog}_{\mathbb{N}}$ *and* DivLog_D

Let $D=\{z\in\mathbb{C}:|z|<1\}$ be the open unit disk and $\mathsf{H}(D)$ be the set of holomorphic functions on D.

Definition 2.3.2. Let $\mathbb{C}^{\mathbb{N}}$ be the set of infinite sequences in $\mathbb{C}$. Let $\mathsf{DivLog}_{\mathbb{N}}\subset\mathbb{C}^{\mathbb{N}}$ such that $\mathbf{S}=(S_1,S_2,\dots)\in\mathsf{DivLog}_{\mathbb{N}}$ if and only if there is an asymptotic expansion

$$S_n=\mathrm{As}_{\mathbf{S}}(\log n)+O(\log^\alpha(n)/n)\quad\text{as } n\to\infty,$$

where $\mathrm{As}_{\mathbf{S}}(t)\in\mathbb{C}[t]$ and $\alpha\in\mathbb{N}$.

Example 2.3.3. In applications we often think $\mathbf{S}$ as coming from some partial sum sequence of an infinite series. For instance, it is well known that the partial sums of harmonic series, namely, the harmonic numbers $H_n=1+\frac{1}{2}+\cdots+\frac{1}{n}$ have the asymptotic formula

$$H_n=\log n+\gamma+O(\log(n)/n)\quad\text{as } n\to\infty,$$

where $\gamma\approx 0.57721566\dots$ is Euler's constant. Hence $\mathrm{As}_{\mathbf{H}}(t)=t+\gamma$ for $\mathbf{H}=(H_1,H_2,\dots)$. Similarly, for any $n\in\mathbb{N}_0$, we have $\mathrm{As}_{\mathsf{a}(n)}(t)=t^n$ where $\mathsf{a}(n)=((H_1-\gamma)^n,(H_2-\gamma)^n,\dots)$. Thus $\mathrm{As}:\mathsf{DivLog}_{\mathbb{N}}\to\mathbb{C}[t]$ is surjective.

Definition 2.3.4. Let DivLog_D be the set of functions $f\in\mathsf{H}(D)$ whose asymptotic expansion, as $z\to 1$ in $D\cap\mathbb{R}$, is given by

$$f(z)=\mathrm{As}_f(-\log(1-z))+O\Big(\big|(1-z)\log^\alpha(1-z)\big|\Big),$$

where $\mathrm{As}_f(t)\in\mathbb{C}[t]$ and $\alpha\in\mathbb{N}_0$.

Example 2.3.5. For any $n \in \mathbb{N}_0$, if we take

$$f(z) = (\mathrm{Li}_1(z))^n = (-\log(1-z))^n$$

then we see that $\mathrm{As}_{\mathrm{Li}_1^n}(t) = t^n$. Thus $\mathrm{As} : \mathsf{DivLog}_D \to \mathbb{C}[t]$ is surjective.

Remark 2.3.6. Define a ring structure on $\mathbb{C}^{\mathbb{N}}$ by the componentwise addition and multiplication. Then it is not hard to see that $\mathsf{DivLog}_{\mathbb{N}}$ becomes a sub-ring of $\mathbb{C}^{\mathbb{N}}$. It is also clear that DivLog_D is a sub-ring of $\mathsf{H}(D)$.

2.3.2 *Relations Between* $\mathsf{DivLog}_{\mathbb{N}}$ *and* DivLog_D

For any sequence $\mathbf{S} = (S_1, S_2, \dots)$, we define the formal power series associated with it as follows:

$$\mathrm{Fp}(\mathbf{S}) := \sum_{n\geq 1} u_n z^n \tag{2.9}$$

where $S_n = u_1 + \cdots + u_n$.

Lemma 2.3.7. *If* $\mathbf{S} \in \mathsf{DivLog}_{\mathbb{N}}$ *and* $\mathrm{As}_{\mathbf{S}} = 0$ *then the associated power series* $f = \mathrm{Fp}(\mathbf{S}) \in \mathsf{DivLog}_D$ *and* $\mathrm{As}_f = 0$.

Proof. By the definition, there is $\alpha \in \mathbb{N}$ such that

$$S_n = u_1 + \cdots + u_n = O(\log^\alpha(n)/n) \quad \text{as } n \to \infty.$$

By Abel's transformation we see that

$$f(z) = \sum_{n\geq 1} u_n z^n = \sum_{M\geq 1} (z^M - z^{M+1}) \sum_{n=1}^{M} u_n = (1-z) \sum_{M\geq 1} z^M O(\log^\alpha(M)/M).$$

We claim that

$$\sum_{M\geq 1} z^M O(\log^\alpha(M)/M) = O\big(|\log^{\alpha+1}(1-z)|\big). \tag{2.10}$$

To establish this, we break the infinite sum into two parts, under the assumption that $z < 1$ but is very close to 1. First, by the Integral Test

$$\sum_{0<M<1/(1-z)} z^M \frac{\log^\alpha(M)}{M} \leq \int_1^{1/(1-z)} \frac{\log^\alpha(t)}{t} \mathrm{d}t = \frac{(-\log(1-z))^{\alpha+1}}{\alpha+1}.$$

On the other hand, if $M \geq 1/(1-z)$, then $\log^\alpha(M)/M \leq (-\log(1-z))^\alpha(1-z)$ since the function $\log^\alpha(t)/t$ is decreasing for all $t > e^\alpha$. Thus

$$\begin{aligned}\sum_{M\geq 1/(1-z)} z^M \frac{\log^\alpha(M)}{M} &\leq (-\log(1-z))^\alpha (1-z) \sum_{M\geq 1/(1-z)} z^M \\ &\leq z^{1/(1-z)} (-\log(1-z))^\alpha.\end{aligned}$$

Since $\lim\limits_{z\to 1^-} z^{1/(1-z)} = \lim\limits_{x\to 0^+} (1-x)^{1/x} = 1/e$ the claim Eq. (2.10) follows immediately. This finishes the proof of the lemma. □

By Lemma 2.3.7, we see that there exists a comparison map cp making the following diagram commutate:

$$\begin{array}{ccc} \mathsf{DivLog}_{\mathbb{N}} & \xrightarrow{\quad \mathrm{As} \quad} & \mathbb{C}[t] \\ {\scriptstyle \mathrm{Fp}}\big\downarrow & & \big\downarrow{\scriptstyle \mathrm{cp}} \\ \mathsf{DivLog}_{D} & \xrightarrow[\quad \mathrm{As} \quad]{} & \mathbb{C}[t] \end{array}$$

The map cp will be constructed in the proof of Prop. 13.3.28.

Proposition 2.3.8. *Let $s \in \mathbb{N}$ and $z \in \overline{D}$, where $\overline{D}$ is the closed unit disk. Let $\mathbf{S} = (S_1, S_2, \dots) \in \mathsf{DivLog}_{\mathbb{N}}$ and define the sequence $\mathbf{T} = (T_1, T_2, \dots)$ by*

$$T_M := \sum_{n=1}^{M-1} \frac{S_n}{n^s} z^n \quad \forall M \in \mathbb{N}.$$

Then $\mathbf{T} \in \mathsf{DivLog}_{\mathbb{N}}$. Moreover, $\mathbf{T}$ converges if $(s, z) \neq (1, 1)$. If $(s, z) = (1, 1)$ then $\frac{\mathrm{d}}{\mathrm{d}t} \mathrm{As}_{\mathbf{T}}(t) = \mathrm{As}_{\mathbf{S}}(t)$.

Proof. By linearity, we may assume without loss of generality that $\mathrm{As}_{\mathbf{S}}(t) = t^k$ for some $k \in \mathbb{N}_0$, *i.e.*, $S_n = \log^k(n) + O(\log^\alpha(n)/n)$ for some constant α.

If $s \geq 2$ then the major term of $\sum_{n \geq M} T_n$ is dominated by $\sum_{n \geq M} \log^k(n)/n^2$. Set

$$I_k := \int_M^\infty \log^k(x)/x^2 \, \mathrm{d}x.$$

Integration by parts yields easily that $I_k = k \log^{k-1}(M)/M + kI_{k-1}$ with $I_0 = 1/M$. Thus $\mathbf{T}$ converges absolutely by the Integral Test.

If $s = 1$ but $z \neq 1$, then we set $u_1 = -S_1$ and $u_n = \frac{S_{n-1}}{n-1} - \frac{S_n}{n}$ for all $n \geq 2$. By Abel's transformation,

$$\sum_{n=1}^\infty \frac{S_n}{n} z^n = -\sum_{n=1}^\infty z^n \sum_{m=1}^n u_m = -\sum_{m=1}^\infty u_m \sum_{n=m}^\infty z^m = \frac{1}{z-1} \sum_{m=1}^\infty u_m z^m. \tag{2.11}$$

Observe that $u_m > 0$ for sufficiently large m as $\log^k(t)/t$ is decreasing when $t > e^k$. Thus the series in Eq. (2.11) converges absolutely since $|z| \leq 1$ and $\sum u_m$ converges (see Exercise 2.2).

If $(s, z) = (1, 1)$, then the big-O term contributes to $O(\log^\alpha(n)/n^2)$ which can be handled as above. So we only need to consider the leading term $\log^k(n)/n$. Using the decomposition $\log(n+1) = \log(n) + 1/n + O(1/n^2)$ we see that

$$\frac{\log^k(n)}{n} = \frac{\log^{k+1}(n+1) - \log^{k+1}(n)}{k+1} + O(\log^k/n^2).$$

Summing up we thus get the major term for T_M as $\log^{k+1}(M)/(k+1)$. Hence $\mathrm{As}_{\mathbf{T}}(t) = t^{k+1}/(k+1)$ whose derivative is t^k.

This finishes the proof of the proposition. □

We have a quantitative version of the case $s = 1, z \neq 1$ of Prop. 2.3.8.

Lemma 2.3.9. *For all $k \geq 0$, we have*

$$\sum_{m=1}^{\infty} \frac{\log^k m}{m} z^{m-1} = O\left(\log^{k+1}\left(\frac{1}{1-z}\right)\right) \quad \text{as } z \to 1 \text{ in } D \cap \mathbb{R}.$$

Proof. We proceed with induction on k. If $k = 0$ then the left-hand side is $z^{-1}\log\left(\frac{1}{1-z}\right) = O\Big(\log\left(\frac{1}{1-z}\right)\Big)$ as $z \to 1$ in $D \cap \mathbb{R}$. Assume the lemma holds for $k = l \geq 1$ for some l. Then we have

$$\log^{l+1} m = (l+1)\int_1^m \frac{\log^l x}{x}\,\mathrm{d}x \leq 2(l+1)\sum_{n=1}^{m} \frac{\log^l n}{n}. \tag{2.12}$$

Here we have used the fact that the function $f(x) = \log^l x/x$ is increasing on $(1, e^l)$ and then decreasing on (e^l, ∞). Hence if $m < e^l$ then Eq. (2.12) following from a simple application of integral test technique (and the 2 is unnecessary). If $m > e^l$ then suppose $a < e^l < a+1$ for some integer $a < m$. We see that the area below $f(x)$, $1 \leq x \leq m$, can be covered by the following three kinds of rectangles:

(1) $a-1$ rectangles, each based on $[i-1, i]$ with height $f(i)$, $i = 2, \cdots, a$;

(2) $m-a-1$ rectangles, each based on $[i, i+1]$ with height $f(i)$, $i = a+1, \ldots, m-1$;

(3) one rectangle based on $[a, a+1]$ with height $f(e^l)$.

Note that the $m-1$ nonzero terms in $\sum_{n=1}^m \frac{\log^l n}{n}$ give the first $m-2$ rectangular areas in (1) and (2) and the last term $\frac{\log^l m}{m}$ is not used. The reason is the value l/e^l is larger than all the values of the $f(x)$ at the division points $x = 2, \cdots, m$. So we need to add the one rectangle in (3) by doubling the sum. In fact we need only the two terms with $n = 2$ and 3 since $f(e^l) = l/e^l < \log^l 2/2 + \log^l 3/3$ for all $l \geq 1$. From Eq. (2.12) we get

$$\begin{aligned}\sum_{m=1}^{\infty} \frac{\log^{l+1} m}{m} z^{m-1} &\leq 2(l+1) \sum_{m=1}^{\infty} \frac{z^{m-1}}{m} \sum_{n=1}^{m} \frac{\log^l n}{n} \\ &= 2(l+1) \sum_{n=1}^{\infty} \frac{\log^l n}{n} z^{n-1} \sum_{r=1}^{\infty} \frac{z^{r-1}}{r+n-1} \\ &< 2(l+1) \left(\frac{1}{t} \log \left(\frac{1}{1-z} \right) \right) \sum_{n=1}^{\infty} \frac{\log^l n}{n} z^{n-1}.\end{aligned}$$

Thus the lemma follows from the induction assumption. □

For $M \in \mathbb{N}$, we may define the Mth partial sum of a MPL as

$$\mathrm{Li}^{(M)}_{s_1,\ldots,s_d}(z_1, \ldots, z_d) := \sum_{M \geq k_1 > \cdots > k_d > 0} \frac{z_1^{k_1} \cdots z_d^{k_d}}{k_1^{s_1} \cdots k_d^{s_d}}. \tag{2.13}$$

Corollary 2.3.10. *For all* $\mathbf{s} = (s_1, \ldots, s_d) \in \mathbb{N}^d$ *and* $\mathbf{z} = (z_1, \ldots, z_d) \in \overline{D}^d$, *the sequence* $\mathrm{Li}(\mathbf{s}; \mathbf{z}) := \big(\mathrm{Li}^{(M)}_{\mathbf{s}}(\mathbf{z})\big)_{M \geq 1} \in \mathsf{DivLog}_{\mathbb{N}}$. *It converges if and only if* $(s_1, z_1) \neq (1, 1)$.

Proof. This follows from Prop. 2.3.8 and an induction on the depth d. □

2.3.3 *Application of Iterated Integrals*

Let $a_1, \ldots, a_k \in \mathbb{C}$ and fix a path $\gamma : [0, 1] \to \mathbb{C}$. Define

$$\begin{aligned}I_\gamma(a_1, \ldots, a_k) := \int_0^1 (\gamma^* \omega_{a_1})(t_1) \int_0^{t_1} (\gamma^* \omega_{a_2})(t_2) \cdots \\ \int_0^{t_{k-2}} (\gamma^* \omega_{a_{k-1}})(t_{k-1}) \int_0^{t_{k-1}} (\gamma^* \omega_{a_k})(t_k),\end{aligned} \tag{2.14}$$

where, for all $z \in \mathbb{C}$

$$\omega_z(t) = \begin{cases} \dfrac{dt}{z^{-1} - t}, & \text{if } z \neq 0; \\ \dfrac{dt}{t}, & \text{if } z = 0. \end{cases}$$

By Lemma 2.1.3 we see that if the image of $\gamma \in \mathbb{C} \setminus \{a_1^{-1}, \ldots, a_d^{-1}\}$ then I_γ is well defined and depends only on the homotopy class of γ.

By convention, we set $I_\gamma = 0$ if $k = 0$. Also, for any $(\alpha, \beta) \in \mathbb{C}^2$, we denote by $[\alpha, \beta]$ the straight path $t \mapsto \alpha + (\beta - \alpha)t$, $t \in [0, 1]$.

Proposition 2.3.11. *Let $d \in \mathbb{N}$, $\mathbf{s} = (s_1, \dots, s_d) \in \mathbb{N}^d$ and $z_1, \dots, z_d \in \overline{D} \setminus \{0\}$. Set $z'_k = \prod_{j=1}^k z_j$ for $k = 1, \dots, d$. Then for all $z \in D$,*

$$\mathrm{Li}_{\mathbf{s}}(zz_1, z_2, \dots, z_d) = I_{[0,z]}(\{0\}^{s_1-1}, z'_1, \{0\}^{s_2-1}, z'_2, \dots, \{0\}^{s_d-1}, z'_d).$$

Proof. This is essentially Eq. (2.8). See Exercise 2.4. □

2.4 Analytic Continuation of Multiple Logarithms

In this section we describe a method of defining the analytic continuation of multiple logarithms by Chen's iterated integrals. This can be generalized to arbitrary MPLs but the notation is too cumbersome so we leave it to the interested reader (see [612]).

For any positive integer n, we set $\mathfrak{L}_n(\mathbf{x}) = \mathrm{Li}_{\underbrace{1,\dots,1}_{n \text{ times}}}(\mathbf{x})$ and define the index set

$$\mathcal{I}_n := \{0,1\}^n = \{\mathbf{i} = (i_1, \dots, i_n) : \ i_t = 0 \text{ or } 1 \ \forall t = 1, \dots, n\}. \tag{2.15}$$

Put $\mathbf{0} = (0, \cdots, 0)$ and $\mathbf{1} = (1, \dots, 1)$. The weight function on the indices $|(i_1, \dots, i_n)| = \sum_{t=1}^n i_t$ is the number of nonzero components.

Lemma 2.4.1. *Let $\mathbf{x} = (x_1, \dots, x_n)$ be an ordered set of complex variables. Let*

$$\mathbf{x}_j = (x_1, \dots, x_{j-1}, x_j x_{j+1}, x_{j+2}, \dots, x_n)$$

for $j = 1, \dots, n-1$ and $\mathbf{x}_n = (x_1, \dots, x_{n-1})$. Then the total differential

$$\begin{aligned} \mathrm{d}\mathfrak{L}_n(\mathbf{x}) &= \sum_{j=1}^n \mathrm{d}_j \mathfrak{L}_n(\mathbf{x}) \\ &= \sum_{j=1}^n \left(\frac{\mathfrak{L}_{n-1}(\mathbf{x}_j)}{1 - x_j} \mathrm{d}x_j + \frac{\mathfrak{L}_{n-1}(\mathbf{x}_{j-1})}{x_j(x_j - 1)} \mathrm{d}x_j \right) && (2.16) \\ &= \sum_{j=1}^n \mathfrak{L}_{n-1}(\mathbf{x}_j) \mathrm{d}\log\left(\frac{1 - x_{j+1}^{-1}}{1 - x_j} \right), && (2.17) \end{aligned}$$

where $x_{n+1} = \infty$, $\mathfrak{L}_0(\mathbf{x}_1) = 1$ and when $t = 1$ the second term in the sum of Eq. (2.16) does not appear.

Proof. Suppose $j > 1$ (the case $j = 1$ is simpler). Then by the definition

$$\begin{aligned} \mathrm{d}_j \mathfrak{L}_n(\mathbf{x}) &= \sum_{k_1 > \cdots > \widehat{k_j} > \cdots > k_n} \frac{x_1^{k_1} \cdots \widehat{x_j^{k_j}} \cdots x_n^{k_n}}{k_1 \cdots \widehat{k_j} \cdots k_n} \sum_{k_j = k_{j+1}+1}^{k_{j-1}-1} x_j^{k_j - 1} \, \mathrm{d}x_j \\ &= \sum_{k_1 > \cdots > \widehat{k_j} > \cdots > k_n} \frac{x_1^{k_1} \cdots \widehat{x_j^{k_j}} \cdots x_n^{k_n}}{k_1 \cdots \widehat{k_j} \cdots k_n} \cdot \frac{x_j^{k_{j-1}-1} - x_j^{k_{j+1}}}{x_j - 1} \mathrm{d}x_j \\ &= \frac{\mathfrak{L}_{n-1}(\mathbf{x}_j)}{1 - x_j} \mathrm{d}x_j + \frac{\mathfrak{L}_{n-1}(\mathbf{x}_{j-1})}{x_j(x_j - 1)} \mathrm{d}x_j, \end{aligned}$$

where we have set $k_{n+1} = 0$. This proves Eq. (2.16). Equation (2.17) now follows immediately (see Exercise 2.5). □

Suppose $\mathbf{i} = (i_1, \ldots, i_n)$ has weight k and let $1 \le \tau_1 < \cdots < \tau_k \le n$ such that $i_{\tau_1} = \cdots = i_{\tau_k} = 1$. We define

$$\mathbf{x}(\mathbf{i}) = \mathbf{y} = (y_1, \ldots, y_k), \text{ where } y_m = \prod_{\alpha = 1 + \tau_{m-1}}^{\tau_m} x_\alpha, \quad 1 \le m \le k \tag{2.18}$$

with $\tau_0 = 0$. Set $w_0(\mathbf{x}) = 0$ and

$$w_t(\mathbf{x}) := \mathrm{d}\log\Big(\frac{1 - x_{t+1}^{-1}}{1 - x_t}\Big), \qquad \text{for } \ 1 \le t \le n.$$

We now define a partial order $\prec$ on $\mathcal{J}_n$ as follows: Given $\mathbf{i} = (i_1, \ldots, i_n)$ and $\mathbf{j} = (j_1, \ldots, j_n)$, $\mathbf{j} \prec \mathbf{i}$ (or, equivalently, $\mathbf{i} \succ \mathbf{j}$) if $j_t \le i_t$ for every $1 \le t \le n$ and $j_s < i_s$ for some s.

Theorem 2.4.2. *Let $\mathbf{x} = (x_1, \ldots, x_n)$ be an ordered set of complex variables. The multiple logarithm $\mathfrak{L}_n(\mathbf{x})$ is a multi-valued holomorphic function on*

$$\mathcal{S}'_n = \mathbb{C}^n \setminus \left\{ (x_1, \ldots, x_n) : \prod_{1 \le j \le n} (1 - x_j) \prod_{1 \le j < k \le n} (1 - x_j \cdots x_k) = 0 \right\} \tag{2.19}$$

and can be expressed by

$$\mathfrak{L}_n(\mathbf{x}) = \sum_{\substack{0 \ne \mathbf{j}_1 \prec \cdots \prec \mathbf{j}_n, \\ \mathbf{j}_1, \ldots, \mathbf{j}_n \in \mathcal{J}_n}} \int_{\mathbf{0}}^{\mathbf{x}} w_{\mathbf{j}_n - \mathbf{j}_{n-1}}(\mathbf{x}(\mathbf{j}_n)) \cdots w_{\mathbf{j}_2 - \mathbf{j}_1}(\mathbf{x}(\mathbf{j}_2)) w_1(\mathbf{x}(\mathbf{j}_1)), \tag{2.20}$$

where the subtraction $- : \mathcal{J}_n \to \mathbb{Z}$ is a binary operation such that

$$(i_1, \ldots, i_n) - (j_1, \ldots, j_n) = \begin{cases} s', & \text{if } i_s = j_s + 1 \text{ and } i_\ell = j_\ell \quad \forall \ell \ne s; \\ 0, & \text{otherwise,} \end{cases}$$

where j_s is the s'th nonzero component of $(j_1, \ldots, j_n)$.

Proof. The proof is straightforward so we leave it as an exercise (see Exercise 2.6). Note that $\mathbf{j}_1 - \mathbf{0}$ is always equal to 1. □

It is clear that every nonzero term in the sum of Eq. (2.20) is provided by some vector index $(\mathbf{j}_1, \dots, \mathbf{j}_n) \in \mathscr{I}_n^n$ with $|\mathbf{j}_\ell| = \ell$ for all $\ell = 1, \dots, n$, which we call *admissible* for convenience.

Example 2.4.3. When $n = 1$,

$$\mathrm{Li}_1(x) = \int_0^x \mathrm{d}\log\Big(\frac{1}{1-x}\Big) = -\log(1-x).$$

When $n = 2$, $\mathfrak{I}_2 = \{(0,0),(0,1),(1,0),(1,1)\}$ and there are two admissible elements in $\mathfrak{I}_2^2$: $((1,0),(1,1))$ and $((0,1),(1,1))$. Let $\mathbf{x} = (x, y)$ then

$$\mathbf{x}(1,0) = x, \quad \mathbf{x}(0,1) = xy, \quad \mathbf{x}(1,1) = (x,y).$$

Thus

$$\begin{aligned}
\mathrm{Li}_{1,1}(x,y) &= \int_{\mathbf{0}}^{\mathbf{x}} w_2(\mathbf{x}) w_1(\mathbf{x}(1,0)) + w_1(\mathbf{x}) w_1(\mathbf{x}(0,1)) \\
&= \int_{(0,0)}^{(x,y)} \frac{\mathrm{d}y}{1-y}\frac{\mathrm{d}x}{1-x} + \Big(\frac{\mathrm{d}x}{1-x} + \frac{\mathrm{d}y}{y(y-1)}\Big)\frac{\mathrm{d}(xy)}{1-xy}.
\end{aligned}$$

The iterated path integrals are understood in the sense of Chen. For example, if $\alpha : [0,1] \to \mathbb{C}^2$ is the path from $(0,0)$ to (x_0, y_0) then

$$\int_{(0,0)}^{(x_0,y_0)} \frac{\mathrm{d}y}{1-y}\frac{\mathrm{d}x}{1-x} = \int_0^1 \Big(\int_0^t \frac{\mathrm{d}x \circ \alpha(s)}{1 - x\circ\alpha(s)}\Big) \frac{\mathrm{d}y\circ\alpha(t)}{1 - y\circ\alpha(t)},$$

where $x \circ \alpha(s)$ (resp. $y \circ \alpha(s)$) is the x- (resp. y-) coordinate of $\alpha(s)$. When $n = 3$ let $\mathbf{x} = (x, y, z)$. Then

$$\begin{aligned}
\mathbf{x}(1,0,0) =& x, \quad \mathbf{x}(0,1,0) = xy, \quad \mathbf{x}(0,0,1) = xyz, \quad \mathbf{x}(1,1,0) = (x,y), \\
\mathbf{x}(1,0,1) =& (x, yz), \quad \mathbf{x}(0,1,1) = (xy, z), \quad \mathbf{x}(1,1,1) = (x,y,z).
\end{aligned}$$

Thus

$$\begin{aligned}
&\mathrm{Li}_{1,1,1}(x,y,z) \\
&= \int_{(0,0,0)}^{(x,y,z)} \frac{\mathrm{d}z}{1-z}\frac{\mathrm{d}y}{1-y}\frac{\mathrm{d}x}{1-x} + \frac{\mathrm{d}z}{1-z}\Big(\frac{\mathrm{d}x}{1-x} + \frac{\mathrm{d}y}{y(y-1)}\Big)\frac{\mathrm{d}(xy)}{1-xy} \\
&\quad + \Big(\frac{\mathrm{d}x}{1-x} + \frac{\mathrm{d}y}{y(y-1)}\Big)\frac{\mathrm{d}z}{1-z}\frac{\mathrm{d}(xy)}{1-xy} + \Big(\frac{\mathrm{d}y}{1-y} + \frac{\mathrm{d}z}{z(z-1)}\Big)\frac{\mathrm{d}(yz)}{1-yz}\frac{\mathrm{d}x}{1-x} \\
&\quad + \Big(\frac{\mathrm{d}y}{1-y} + \frac{\mathrm{d}z}{z(z-1)}\Big)\Big(\frac{\mathrm{d}x}{1-x} + \frac{\mathrm{d}(yz)}{yz(yz-1)}\Big)\frac{\mathrm{d}(xyz)}{1-xyz} \\
&\quad + \Big(\frac{\mathrm{d}x}{1-x} + \frac{\mathrm{d}y}{y(y-1)}\Big)\Big(\frac{\mathrm{d}(xy)}{1-xy} + \frac{\mathrm{d}z}{z(z-1)}\Big)\frac{\mathrm{d}(xyz)}{1-xyz}.
\end{aligned}$$

2.5 Monodromy of Multiple Logarithms

The main goal of this section is to define the variation matrix associated with the multiple logarithms by utilizing the vector index set $\mathfrak{I}$ as defined in Eq. (2.15), together with the partial order $\prec$.

Write $\mathbf{x} = (x_1, \ldots, x_n)$ as before. We know the multiple logarithm $\mathfrak{L}_n(\mathbf{x})$ is related to $\mathfrak{L}_m(\mathbf{x})$ for $m < n$. This can be seen easily, for instance, when we take the derivatives as given by Lemma 2.4.1.

Definition 2.5.1. Suppose $\mathbf{i} \in \mathcal{I}_n$ has weight k. We define the $\mathbf{i}$-th *retraction* map $\rho_\mathbf{i} : \mathcal{I}_n \to \mathfrak{I}_k$ as follows.

(1) If $\mathbf{i} \not\succ \mathbf{j}$ then $\rho_\mathbf{i}(\mathbf{j}) = (0, \ldots, 0)$
(2) Suppose $\mathbf{i} \succ \mathbf{j}$ and $\mathbf{j}$ has weight l. Suppose further that the 1's occur at the positions $\tau_1, \ldots, \tau_k$ in $\mathbf{i}$ and $t_1, \ldots, t_l$ in $\mathbf{j}$, respectively. Then we can put $t_r = \tau_{\alpha_r}$ for $1 \le r \le l$. Set the entry of $\rho_\mathbf{i}(\mathbf{j})$ to be 1 if it is at the α_r-th $(1 \le r \le l)$ component and 0 otherwise. In other words, $\rho_\mathbf{i}(\mathbf{j})$ is obtained from $\mathbf{j}$ by deleting all of its entries whose corresponding entries of $\mathbf{i}$ are equal to 0.

For instance, $\rho_{(1101)}\big((1100)\big) = (110) \in \mathfrak{I}_3$ and $\rho_{(1101)}\big((0100)\big) = (010)$. In general, $\rho_\mathbf{i}(\mathbf{i}) = \mathbf{1}_k$ for all $\mathbf{i}$ with $|\mathbf{i}| = k$.

We are now ready to define the entries of the $2^n \times 2^n$ variation matrix $\mathcal{M}_{[n]}(\mathbf{x}) := \mathcal{M}_{\underbrace{1,\ldots,1}_{n \text{ times}}}(\mathbf{x})$ associated with the n-tuple logarithm $\mathfrak{L}_n(\mathbf{x})$.

Definition 2.5.2. Fix a point $\mathbf{x} \in \mathcal{S}_n$ where

$$\mathcal{S}_n = \mathbb{C}^n \setminus \left\{ (x_1, \ldots, x_n) : \prod_{1 \le j \le n} x_j(1 - x_j) \prod_{1 \le j < k \le n} (1 - x_j \cdots x_k) = 0 \right\}. \tag{2.21}$$

For $1 \le s \le n$, write $a_s = a_s(\mathbf{x}) = (x_1 \cdots x_s)^{-1}$ and

$$\theta_s = \theta_s(\mathbf{x}) = \frac{\mathrm{d}t}{t - a_s} = \frac{\mathrm{d}t}{t - a_s(\mathbf{x})}.$$

(i) If $\mathbf{j} \not\prec \mathbf{i}$ then we define the $(\mathbf{i}, \mathbf{j})$-th entry of $\mathcal{M}_{[n]}(\mathbf{x})$ to be 0.
(ii) Suppose $\mathbf{i} \succ \mathbf{j}$ as given in Definition 2.5.1(2). By Eq. (2.18)

$$\mathbf{x}(\mathbf{i}) = \mathbf{y} = (y_1, \ldots, y_k), \quad y_m = \prod_{\alpha = 1 + \tau_{m-1}}^{\tau_m} x_\alpha = \frac{a_{\tau_{m-1}}(\mathbf{x})}{a_{\tau_m}(\mathbf{x})}, \quad 1 \le m \le k,$$

with $\tau_0 = 0$ and $a_0(\mathbf{x}) = 1$. Set $t_0 = \alpha_0 = 0$, $t_{l+1} = n+1$, $\alpha_{l+1} = k+1$, $a_{n+1}(\mathbf{x}) = a_{k+1}(\mathbf{y}) = 0$. Define the $(\mathbf{i},\mathbf{j})$-th entry of $\mathcal{M}_{[n]}(\mathbf{x})$ as $(2\pi i)^l E_{\mathbf{i},\mathbf{j}}(\mathbf{x})$ where

$$E_{\mathbf{i},\mathbf{j}}(\mathbf{x}) = \gamma^k_{\rho_{\mathbf{i}}(\mathbf{j})}(\mathbf{y}) := (-1)^{k-l}\prod_{r=0}^{l}\int_{a_{\alpha_{r+1}}(\mathbf{y})}^{a_{\alpha_r}(\mathbf{y})} \theta_{\alpha_r+1}(\mathbf{y})\cdots\theta_{\alpha_{r+1}-1}(\mathbf{y})$$
$$= (-1)^{k-l}\prod_{r=0}^{l}\int_{p_r} \theta_{\tau_{\alpha_r+1}}(\mathbf{x})\cdots\theta_{\tau_{\alpha_{r+1}-1}}(\mathbf{x}) \quad (2.22)$$

Here the $l+1$ paths $p_0,\dots,p_l$ for the $l+1$ integrals are *independent of* $\mathbf{i}$ where p_r is any fixed contractible path from $a_{t_{r+1}}(\mathbf{x})$ to $a_{t_r}(\mathbf{x})$ in the punctured complex plane $\mathbb{C}\setminus\bigcup_{t_r<s<t_{r+1}}\{a_s(\mathbf{x})\}$, and the integral $\int_{p_r} = 1$ if $\alpha_r + 1 = \alpha_{r+1}$. We get the second equality by observing that

$$a_m(\mathbf{y}) = (y_1\cdots y_m)^{-1} = a_{\tau_m}(\mathbf{x}) \Longrightarrow a_{\alpha_r}(\mathbf{y}) = a_{\tau_{\alpha_r}}(\mathbf{x}) = a_{t_r}(\mathbf{x}).$$

Proposition 2.5.3. *Suppose* $\mathbf{i}$ *and* $\mathbf{j}$ *are given as in Definition* 2.5.2(ii). *As multi-valued functions*

$$E_{\mathbf{i},\mathbf{j}}(\mathbf{x})$$
$$= \prod_{r=0}^{l}\mathfrak{L}_{\alpha_{r+1}-\alpha_r-1}\left(\frac{a_{t_r}(\mathbf{x})-a_{t_{r+1}}(\mathbf{x})}{a_{\tau_{\alpha_r+1}}(\mathbf{x})-a_{t_{r+1}}(\mathbf{x})},\dots,\frac{a_{\tau_{\alpha_{r+1}-2}}(\mathbf{x})-a_{t_{r+1}}(\mathbf{x})}{a_{\tau_{\alpha_{r+1}-1}}(\mathbf{x})-a_{t_{r+1}}(\mathbf{x})}\right) \quad (2.23)$$
$$= \mathfrak{L}_{k-\alpha_l}\left(x_{1+t_l}\cdots x_{\tau_{\alpha_l+1}},\dots,x_{1+\tau_{k-1}}\cdots x_{\tau_k}\right)$$
$$\times\prod_{r=0}^{l-1}\mathfrak{L}_{\alpha_{r+1}-\alpha_r-1}\left(\frac{1-x_{1+t_r}\cdots x_{t_{r+1}}}{1-x_{1+\tau_{\alpha_r+1}}\cdots x_{t_{r+1}}},\dots,\frac{1-x_{1+\tau_{\alpha_{r+1}-2}}\cdots x_{t_{r+1}}}{1-x_{1+\tau_{\alpha_{r+1}-1}}\cdots x_{t_{r+1}}}\right). \quad (2.24)$$

Here $\mathfrak{L}_0 = 1$ *and* $a_0 = 0$.

Proof. By a direct calculation using substitution we get

$$(-1)^{\alpha_{r+1}-\alpha_r-1}\int_{p_r}\theta_{\tau_{\alpha_r+1}}(\mathbf{x})\cdots\theta_{\tau_{\alpha_{r+1}-1}}(\mathbf{x})$$
$$= \mathfrak{L}_{\alpha_{r+1}-\alpha_r-1}\left(\frac{a_{t_r}(\mathbf{x})-a_{t_{r+1}}(\mathbf{x})}{a_{\tau_{\alpha_r+1}}(\mathbf{x})-a_{t_{r+1}}(\mathbf{x})},\dots,\frac{a_{\tau_{\alpha_{r+1}-2}}(\mathbf{x})-a_{t_{r+1}}(\mathbf{x})}{a_{\tau_{\alpha_{r+1}-1}}(\mathbf{x})-a_{t_{r+1}}(\mathbf{x})}\right).$$

The proposition follows immediately. $\square$

Example 2.5.4. In the last row of $\mathcal{M}_{[n]}(\mathbf{x})$ one has

$$E_{1,\mathbf{j}}(\mathbf{x})=\gamma^n_{\mathbf{j}}(\mathbf{x})=\prod_{r=0}^{l}\mathfrak{L}_{t_{r+1}-t_r-1}\Big(\frac{a_{t_r}-a_{t_{r+1}}}{a_{t_r+1}-a_{t_{r+1}}},\dots,\frac{a_{t_{r+1}-2}-a_{t_{r+1}}}{a_{t_{r+1}-1}-a_{t_{r+1}}}\Big)$$

$$=\prod_{r=0}^{l}\mathfrak{L}_{t_{r+1}-t_r-1}\Big(\frac{1-x_{1+t_r}\cdots x_{t_{r+1}}}{1-x_{2+t_r}\cdots x_{t_{r+1}}},\dots,\frac{1-x_{t_{r+1}-1}x_{t_{r+1}}}{1-x_{t_{r+1}}}\Big) \quad (2.25)$$

where $\mathfrak{L}_0=1$ and $x_{n+1}=\infty$. In particular, $E_{1,\mathbf{0}}=\gamma^n_{\mathbf{0}}(\mathbf{x})=\mathfrak{L}_n(\mathbf{x})$ and $E_{1,\mathbf{1}}=\gamma^n_{\mathbf{1}}(\mathbf{x})=1$.

For any $n\ge 2$, let $L_n=L_n(\mathbf{x})=[C_{\mathbf{0}}\ \dots\ C_{\mathbf{1}}]$ be the matrix with 2^n columns $C_{\mathbf{j}}$ ($\mathbf{j}\in\mathcal{I}_n$) where

$$C_{\mathbf{j}}=\sum_{\mathbf{i}\succ\mathbf{j}}\gamma^{|\mathbf{i}|}_{\rho_{\mathbf{i}}(\mathbf{j})}\big(\mathbf{x}(\mathbf{i})\big)e_{\mathbf{i}}$$

where $e_{\mathbf{i}}$ is the **i**-th standard unit column vector. In listing the columns we have used a complete order "$<$" on $\mathcal{I}_n$: If $|\mathbf{i}|<|\mathbf{j}|$ then $\mathbf{i}<\mathbf{j}$ (or, equivalently, $\mathbf{j}>\mathbf{i}$). If $|\mathbf{i}|=|\mathbf{j}|$ then the usual lexicographic order from left to right is in force with $1<0$ (not $0<1$!).

We now fix a standard basis $\{e_{\mathbf{i}}:\mathbf{i}\in\mathcal{I}_n\}$ of $\mathbb{C}^{2^n}$ consisting of column vectors. Suppose $|\mathbf{i}|=k$. It follows from definition that the **i**-th row is

$$R_{\mathbf{i}}:=\sum_{\mathbf{j}\prec\mathbf{i}}(2\pi i)^{|\mathbf{j}|}\gamma^k_{\rho_{\mathbf{i}}(\mathbf{j})}\big(\mathbf{x}(\mathbf{i})\big)e^T_{\mathbf{j}}=(2\pi i)^k e^T_{\mathbf{i}}+\sum_{\mathbf{j}\precneqq\mathbf{i}}(2\pi i)^{|\mathbf{j}|}\gamma^k_{\rho_{\mathbf{i}}(\mathbf{j})}\big(\mathbf{x}(\mathbf{i})\big)e^T_{\mathbf{j}} \quad (2.26)$$

where $e^T_{\mathbf{j}}$ are now row vectors. Note that $\gamma^k_{\rho_{\mathbf{i}}(\mathbf{i})}=\gamma^k_{\mathbf{1}_k}=1$ by the definition. It is clear that the first entry (*i.e.*, $\mathbf{j}=\mathbf{0}$) of this row is $\mathfrak{L}_k\big(\mathbf{x}(\mathbf{i})\big)$.

Example 2.5.5. Using the above definition we can easily get the variation matrix associated with the double logarithm:

$$\mathcal{M}_{1,1}(x,y)=\begin{bmatrix}1&&&\\ \mathrm{Li}_1(x)&2\pi i&&\\ \mathrm{Li}_1(xy)&&2\pi i&\\ \mathrm{Li}_{1,1}(x,y)&2\pi i\,\mathrm{Li}_1(y)&2\pi i(\mathrm{Li}_1(x)-\mathrm{Li}_1(y)-\log y)&(2\pi i)^2\end{bmatrix}.$$

For any $k=0,\dots,n$, let us call the minor of $\mathcal{M}_{[n]}(\mathbf{x})$ consisting of rows beginning with k-tuple logarithms (corresponding to $\mathbf{i}=(\{1\}^k,0,\dots,0)$) and ending just before the $(k+1)$-tuple logarithms *the kth block*. It has $\binom{n}{k}$ rows with row indices $|\mathbf{i}|=k$.

Lemma 2.5.6. *The matrix $\mathcal{M}_{[n]}(\mathbf{x})$ is a lower triangular matrix. Moreover, the columns with $|\mathbf{j}|=k$ of the kth block of $\mathcal{M}_{[n]}(\mathbf{x})$ is $(2\pi i)^k$ times the identity matrix of rank $\binom{n}{k}$.*

Proof. The lemma follows directly from Eq. (2.26) because if $\mathbf{j} \precneqq \mathbf{i}$ then $\mathbf{j} < \mathbf{i}$. □

Lemma 2.5.7. *The* $\mathbf{j}$*-th column of* $\mathcal{M}_{[n]}(\mathbf{x})$ *is*

$$(2\pi i)^{|\mathbf{j}|} C_{\mathbf{j}} = (2\pi i)^{|\mathbf{j}|} \sum_{\mathbf{i} \succ \mathbf{j}} \gamma^{|\mathbf{i}|}_{\rho_{\mathbf{i}}(\mathbf{j})}(\mathbf{x}(\mathbf{i})) e_{\mathbf{i}}$$

where $\mathbf{x}(\mathbf{i})$ *is defined by Eq.* (2.18) *depending on* $\mathbf{i}$.

Proof. Use Eq. (2.26). □

Proposition 2.5.8. *The columns of* $\mathcal{M}_{[n]}(\mathbf{x})$ *form the set of the fundamental solutions of the following system of differential equations*

$$\begin{cases} \mathrm{d}X_{\mathbf{0}} = 0, \\ \mathrm{d}X_{\mathbf{i}} = \displaystyle\sum_{|\mathbf{k}|=|\mathbf{i}|-1,\, \mathbf{k} \prec \mathbf{i}} X_{\mathbf{k}}\, \mathrm{d}\gamma^{|\mathbf{i}|}_{\rho_{\mathbf{i}}(\mathbf{k})}(\mathbf{x}(\mathbf{i})) \quad \text{for all } 1 \le |\mathbf{i}| \le n \end{cases} \tag{2.27}$$

where $\mathbf{x}(\mathbf{i})$ *is determined as in Eq.* (2.18).

Proof. We justify the statement by induction on n. It is easy to see the proposition is valid for $n = 1$ and $n = 2$. Suppose $n \ge 3$ and assume the proposition is true for $\le n - 1$. Let us now consider the $\mathbf{j}$-th column as expressed in Lemma 2.5.7. The cases $|\mathbf{i}| = 1$ or $\mathbf{j} > \mathbf{i}$ are obvious. Suppose

(1) $1 < |\mathbf{i}| < n$ and $\mathbf{j} \le \mathbf{i}$. There are two cases. (i) $\mathbf{j} \not\prec \mathbf{i}$. This is trivial because each term of both sides is zero. (ii) $\mathbf{j} \prec \mathbf{i}$. Then there is a t such that $i_t = j_t = 0$. We denote $\mathbf{i}' \in \mathfrak{I}_{n-1}$ the corresponding index after deleting the i_t-th component. By induction

$$\sum_{|\mathbf{k}'|=|\mathbf{i}'|-1,\, \mathbf{j}' \prec \mathbf{k}' \prec \mathbf{i}'} \gamma^{|\mathbf{k}'|}_{\rho_{\mathbf{k}'}(\mathbf{j}')}(\mathbf{x}'(\mathbf{k}'))\, \mathrm{d}\gamma^{|\mathbf{i}'|}_{\rho_{\mathbf{i}'}(\mathbf{k}')}(\mathbf{x}'(\mathbf{i}')) = \mathrm{d}\gamma^{|\mathbf{i}'|}_{\rho_{\mathbf{i}'}(\mathbf{j}')}(\mathbf{x}'(\mathbf{i}'))$$

where we set $\mathbf{x}' = (x_1, \dots, x_{i_t-1}, x_{i_t} x_{i_t+1}, x_{i_t+2}, \dots, x_n)$. Since $|\mathbf{i}'| = |\mathbf{i}|$ and $|\mathbf{k}'| = |\mathbf{k}|$ we can get the desired equation by inserting 0 before the i_t-th components of $\mathbf{i}'$, $\mathbf{j}'$ and $\mathbf{k}'$.

(2) $\mathbf{i} = \mathbf{1}$ and $|\mathbf{j}| = l$. We need to show

$$\mathrm{d}\gamma^n_{\mathbf{j}}(\mathbf{x}) = \sum_{|\mathbf{k}|=n-1,\, \mathbf{j} \prec \mathbf{k}} \gamma^{n-1}_{\rho_{\mathbf{k}}(\mathbf{j})}(\mathbf{x}(\mathbf{k}))\, \mathrm{d}\gamma^n_{\mathbf{k}}(\mathbf{x}). \tag{2.28}$$

This is trivial when $l = n$. The case $l = 0$ follows from Eq. (2.17) of Lemma 2.4.1:

$$\mathrm{d}\mathfrak{L}_n(\mathbf{x}) = \sum_{t=1}^{n} \mathfrak{L}_{n-1}(x_1, \dots, x_{t-1}, x_t x_{t+1}, x_{t+2}, \dots, x_n)\, \mathrm{d}\log \frac{1 - x^{-1}_{t+1}}{1 - x_t}.$$

So we may assume $0 < l < n$, $j_{t_1} = \cdots = j_{t_l} = 1$ and $j_t = 0$ for all other indices t. By the definition Eq. (2.25) we have

$$\mathrm{d}\gamma^n_{\mathbf{j}}(\mathbf{x}) = \sum_{r=0}^{l} \sum_{t_r < s < t_{r+1}} \gamma^{n-1}_{\rho_{\mathbf{v}_s}(\mathbf{j})}\big(\mathbf{x}(\mathbf{v}_s)\big)\,\mathrm{d}\gamma^n_{\mathbf{v}_s}(\mathbf{x}),$$

where $t_0 = 0$, $t_{l+1} = n+1$ and

$$\begin{array}{l} \mathbf{v}_s = (1,\ldots,\underset{\substack{\uparrow\\ t_r\text{-th place}}}{1},\ldots\ldots\ldots,1,\underset{\substack{\uparrow\\ s\text{-th place}}}{0},1,\ldots\ldots\ldots,\underset{\substack{\uparrow\\ t_{r+1}\text{-th place}}}{1},\ldots,1). \end{array}$$

Under the retraction map $\rho_{\mathbf{v}_s}$ the numbering of the indices changes as follows: $t \rightsquigarrow t$ if $t < s$ and $t \rightsquigarrow t-1$ if $t > s$. We also have

$$a_t\big(\mathbf{x}(\mathbf{v}_s)\big) = \begin{cases} a_t(\mathbf{x}) & \text{if } t < s, \\ a_{t+1}(\mathbf{x}) & \text{if } t > s. \end{cases}$$

Hence for each s such that $t_r < s < t_{r+1}$ the integral expression of $\gamma^{n-1}_{\rho_{\mathbf{k}}(\mathbf{j})}\big(\mathbf{x}(\mathbf{k})\big)$ is unchanged under $\rho_{\mathbf{k}}$ ($\mathbf{j} \prec \mathbf{k}$) except the $\mathbf{v}_s$-term. Equation (2.28) now follows immediately from the Leibniz rule and therefore the proposition is proved. □

Theorem 2.5.9. *Let* $\mathscr{M}_{[n]}(\mathbf{x}) = \big[E_{\mathbf{i},\mathbf{j}}(\mathbf{x})\big]_{\mathbf{i},\mathbf{j}\in\mathcal{I}_n}$ *where* $E_{\mathbf{i},\mathbf{j}}(\mathbf{x})$ *is defined by Eq.* (2.22). *Let* $1 \le j \le k \le n$ *and* $q_{jk} \in \pi_1(S_n, \mathbf{x})$ *(resp.* $1 \le j < n$ *and* q_{j0}*) enclose* $\mathscr{D}_{jk} = \{x_j \ldots x_k = 1\}$, *(resp.* $\mathscr{D}_{j0} = \{x_j = 0\}$*) only once but no other irreducible component of* D_n *such that* $\int_{q_{jk}} \mathrm{d}\log(1 - x_j \ldots x_k) = 2\pi i$ *(resp.* $\int_{q_{j0}} \mathrm{d}\log x_j = 2\pi i$*). Then*

$$M(q_{j0}) = I + \big[n_{\mathbf{i},\mathbf{j}}\big]_{\mathbf{i},\mathbf{j}\in\mathcal{I}_n}, \quad M(q_{jk}) = I + \big[m_{\mathbf{i},\mathbf{j}}\big]_{\mathbf{i},\mathbf{j}\in\mathcal{I}_n}$$

where I *is the identity matrix of rank* 2^n,

$$n_{\mathbf{i},\mathbf{j}} = \begin{cases} -1 & \textit{if } \mathbf{i} = \mathbf{j} + \mathbf{u}_{s+1} \textit{ and } t_r \le n - j \le s \le t_{r+1} - 2,\ r \ge 1, \\ 0 & \textit{otherwise}, \end{cases} \tag{2.29}$$

and

$$m_{\mathbf{i},\mathbf{j}} = \begin{cases} 1 & \textit{if } \mathbf{i} = \mathbf{j} + \mathbf{u}_{n-k+1},\ t_r = j \le n-k \le t_{r+1} - 2,\ r \ge 1, \\ -1 & \textit{if } \mathbf{i} = \mathbf{j} + \mathbf{u}_i,\ t_r + 1 \le i \le n - k = t_{r+1} - 1,\ r \ge 0, \\ 0 & \textit{otherwise}. \end{cases} \tag{2.30}$$

Here $\mathbf{i}$ *and* $\mathbf{j}$ *satisfy the condition in Definition* 2.5.2(2) *in the case of* $m_{\mathbf{i},\mathbf{j}} = \pm 1$ *and* $n_{\mathbf{i},\mathbf{j}} = -1$.

Proof. If we analytically continue every integral entry of $\mathcal{M}_{[n]}(\mathbf{x})$ along the same loop $q \in \pi_1(S_n, \mathbf{x})$ we will denote this action of q by $\Theta(q)$. By the definition it is clear that if $\mathbf{i} \not\succ \mathbf{j}$ then $\Theta(q)E_{\mathbf{i},\mathbf{j}}(\mathbf{x}) = E_{\mathbf{i},\mathbf{j}}(\mathbf{x})$ which is either 0 or 1. Thus we are only concerned with $E_{\mathbf{i},\mathbf{j}}$ with $\mathbf{i} \succ \mathbf{j}$.

We now fix some $\mathbf{j}$. If $|\mathbf{j}| = n$ then clearly $(\Theta(q) - I)C_1 = [0, \ldots, 0]^T$ for any loop q. This proves the proposition for $|\mathbf{j}| = n$. We now assume $|\mathbf{j}| < n$. Let $\mathbf{i}$ and $\mathbf{j}$ be given as in Definition 2.5.2(2). Then the theorem follows from the explicit expression Eq. (2.24) using the monodromy computation contained in Thm. 4.3, Props. 5.4 and 5.5 of [612]. Since this computation is straightforward we leave the details to the interested reader. □

Corollary 2.5.10. *The monodromy representation of* $\mathcal{M}_{[n]}(\mathbf{x})$

$$\rho_{\mathbf{x}} : \pi_1(S_n, \mathbf{x}) \longrightarrow \mathrm{GL}_{2^n}(\mathbb{Z})$$

is unipotent.

Proof. This is clear. □

2.6 Variation Matrix for MPLs

In this section we provide the procedure to construct the variation matrix $\mathcal{M}_{\mathbf{n}}(\mathbf{x})$ associated with the general MPL $\mathrm{Li}_{\mathbf{n}}(\mathbf{x})$.

Let $\mathbf{n} = (n_1, \ldots, n_d)$ and $n = |\mathbf{n}|$. Let $\mathbf{x} = (x_1, \ldots, x_d)$ and $a_j = \prod_{k=1}^{j} x_k$ for all $j = 1, \ldots, d$. We will describe precisely how to construct the entries in the last row of $\mathcal{M}_{\mathbf{n}}(\mathbf{x})$ (say, with column number c) together with the first column entry with row number also equal to c. The other rows can follow the same pattern since every row starts with some MPL by merging some of the variables in $\mathbf{x}$ (merging means multiplying some consecutive variables). The procedure always produces the correct column entries but only produces the possible candidates for row entries (called *pseudo* row entries). But the key point is that all weight $n-1$ pseudo row entries are genuine ones. And, fortunately, most of the time the pseudo row entries of lower weight are correct, too. So the most efficient way to construct the variation matrix is to write down all these pseudo entries first. Iteratively, to determine whether a certain row entry of weight k ($k < n-1$) is genuine or not one should apply the monodromy properties of the columns corresponding to those weight $k+1$ (genuine) entries to try to modify this pseudo entry if necessay. We will carry out this procedure explicitly in all the weight four double polylog cases in Sec. 2.6.2.

2.6.1 General Procedure Leading to Pseudo Row Entries

Let $\mathbf{n}$, $\mathbf{x}$ and a_j's be the same as above. Set $a_0 = 1$ and rename $(1, \{0\}^{n_1-1}, a_1, \dots, \{0\}^{n_d-1}, a_d, 0)$ as $(b_0, b_1, \dots, b_n, b_{n+1})$. We now choose $i_0 < i_1 < \cdots < i_\ell < i_{\ell+1}$ subject to the following conditions:

(i) $i_0 = 0$, b_0 is the leading 1, $i_{\ell+1} = n+1$ and b_{n+1} is the last 0.
(ii) For all $1 \le j \le l$, either $i_j + 1 = i_{j+1}$ or one of b_{i_j} and $b_{i_{j+1}}$ is nonzero.
(iii) For fixed $1 \le j \le d$, when choosing the 0's between a_{j-1} and a_j (if there are any) always start from the first 0 (denoted by b_{i_s}) immediately after a_{j-1} and choose consecutively until, say, b_{i_t}.

Set $b'_{i_t} = b_{i_t}$ except in case (iii) if b_{i_t} is not the last 0 immediately before a_j then we define $b'_{i_t} = a_{j-1}$ (one may think of a collapse of these consecutive 0's onto a_{j-1} if not all of the 0's between a_{j-1} and a_j are chosen). For each such choice of $1 \le i_1 < \cdots < i_\ell \le n$ satisfying (i)-(iii) above, we set the corresponding column entry of $\mathcal{M}_{\mathbf{n}}(\mathbf{x})$ as

$$(-1)^{\#\{j: b_{i_j} \ne 0\}} \int_0^1 \frac{dt}{t - b_{i_1}} \cdots \frac{dt}{t - b_{i_\ell}} \tag{2.31}$$

and the corresponding row entry as

$$I(b_0; b_1, \dots, b_{i_1-1}; b_{i_1}) I(b'_{i_1}; b_{i_1+1}, \dots, b_{i_2-1}; b_{i_2}) \cdots I(b'_{i_\ell}; b_{i_\ell+1}, \dots, b_n; b_{n+1}), \tag{2.32}$$

where

$$I(b'_k; b_{k+1}, \dots, b_{m-1}; b_m) = \begin{cases} 1, & \text{if } m = k+1; \\ \operatorname{sgn}(\mathbf{b}_{k,m}) \displaystyle\int_{b_m}^{b'_k} \frac{dt}{t - b_{k+1}} \cdots \frac{dt}{t - b_{m-1}}, & \text{if } m > k+1, \end{cases}$$

where $\operatorname{sgn}(\mathbf{b}_{k,m}) = 1$ if $b_j = 0$ for all $k < j < m$, $\operatorname{sgn}(\mathbf{b}_{k,m}) = (-1)^{\#\{j: b_j \ne 0, k<j<m\}}$ if $b_k = 0$, and $\operatorname{sgn}(\mathbf{b}_{k,m}) = (-1)^{\#\{j: b_j \ne b_m, k<j<m\}}$ otherwise. As a check, the value in Eq. (2.32) should always be nonzero. Note that in the above iterated integral $I(b'_k; \cdots; b_m)$ we have the two boundary values reversed. Sometimes, different choices of subsets of b_j's may give rise to the same column entry, then we need to combine all the corresponding pseudo row entries into one entry by adding them up (for instance, see Example 2.6.2 for $\mathcal{M}_{2,2}(x, y)$). A general choice looks like the following:

$$(1, \{0\}^{m_1-1}, a_{i_1}, \dots, \{0\}^{m_j-1}, a_{i_j}, \dots, \{0\}^{m_\ell-1}, a_{i_\ell}, 0)$$

where $1 \le i_1 < \cdots < i_\ell \le d$ and the block of zeros $\{0\}^{m_j-1}$ (for $m_j > 1$) must be chosen from a block of consecutive 0's between a_r

and a_{r+1} (starting from the 0 immediately after a_r) for some r satisfying $0 \le i_{j-1} \le r < i_j \le d$. The corresponding column entry is then given by $\mathrm{Li}_{m_1,\dots,m_\ell}(1/a_{i_1}, a_{i_1}/a_{i_2}, \dots, a_{i_{\ell-1}}/a_{i_\ell})$. We order these entries by the following complete ordering: (i) $\mathbf{m}_1 < \mathbf{m}_2$ if $|\mathbf{m}_1| < |\mathbf{m}_2|$, and (ii) if $|\mathbf{m}_1| = |\mathbf{m}_2|$ then we order by the indices $\mathbf{i}$ using the lexicographic order with $1 < 2 < \cdots$. For example, $\mathrm{Li}_2(x)$ (with $(i_1) = (1)$) comes before $\mathrm{Li}_{1,1}(x, y)$ (with $(i_1, i_2) = (1, 1)$) which in turn comes before $\mathrm{Li}_2(xy)$ (with $(i_1) = (2)$).

2.6.2 *Some Examples in Weight Four*

In this section we use the notation $\ell_n(x) = \log^n(x)/n!$ and set $a_0 = 1$, $a_1 = 1/x$ and $a_2 = 1/(xy)$. For each choice of $1 \le i_1 < \cdots < i_\ell \le n$ given in the last subsection, we write $\alpha \otimes \beta$ to mean the corresponding column entry α and pseudo row entry β.

Example 2.6.1. For the double polylog $\mathrm{Li}_{1,3}(x, y)$, we have

$$\mathrm{Li}_{1,3}(x, y) = I(1; a_1, 0, 0, a_2; 0).$$

Here are the possible choices:

(1) $\emptyset$: $1 \otimes \mathrm{Li}_{1,3}(x, y)$
(2) $\{a_1\}$: $-I(1; a_1; 0) \otimes -I(a_1; 0, 0, a_2; 0) = \mathrm{Li}_1(x) \otimes \mathrm{Li}_3(y)$
(3) $\{a_2\}$: $-I(1; a_2; 0) \otimes -I(1; a_1, 0, 0; a_2) = \mathrm{Li}_1(xy) \otimes e(x, y)$
(4) $\{a_1, a_2\}$: $I(1; a_1, a_2; 0) \otimes I(a_1; 0, 0; a_2) = \mathrm{Li}_{1,1}(x, y) \otimes \ell_2(y)$
(5) $\{0, a_2\}$: $-I(1; 0, a_2; 0) \otimes -I(1; a_1; 0)I(a_1; 0; a_2) = \mathrm{Li}_2(xy) \otimes \beta_5 = \mathrm{Li}_2(xy) \otimes \mathrm{Li}_1(x)\ell_1(y)$ (There is a collapse of 0. And this produces the only pseudo row entry in this case.)
(6) $\{a_1, 0, a_2\}$: $I(1; a_1, 0, a_2; 0) \otimes I(a_1; 0; a_2) = \mathrm{Li}_{1,2}(x, y) \otimes \ell_1(y)$ (There is a collapse of 0.)
(7) $\{0, 0, a_2\}$: $I(1; 0, 0, a_2; 0) \otimes -I(1; a_1; 0) = \mathrm{Li}_3(xy) \otimes \mathrm{Li}_1(x)$
(8) $\{0, a_1, 0, a_2\}$: $\mathrm{Li}_{1,3}(x, y) \otimes 1$

Note that variation matrix $\mathscr{M}_3(xy)$ is well known (see [267]) and $\mathscr{M}_{1,2}(x, y)$ is given in Exercise 2.10. Thus the variation matrix $\mathscr{M}_{1,3}(x, y)$ associated with $\mathrm{Li}_{1,3}(x,y)$ over S_2 is

$$\begin{bmatrix} 1 \\ \mathrm{Li}_1(x) & 1 \\ \mathrm{Li}_1(xy) & 0 & 1 \\ \mathrm{Li}_{1,1}(x,y) & \mathrm{Li}_1(y) & h(x,y) & 1 \\ \mathrm{Li}_2(xy) & 0 & \ell_1(xy) & 0 & 1 \\ \mathrm{Li}_{1,2}(x,y) & \mathrm{Li}_2(y) & f(x,y) & \ell_1(y) & \mathrm{Li}_1(x) & 1 \\ \mathrm{Li}_3(xy) & 0 & \ell_2(xy) & 0 & \ell_1(xy) & 0 & 1 \\ \mathrm{Li}_{1,3}(x,y) & \mathrm{Li}_3(y) & e(x,y) & \ell_2(y) & \beta_5 \rightsquigarrow r_5 & \ell_1(y) & \mathrm{Li}_1(x) & 1 \end{bmatrix} \tau_{1,3}(2\pi i)$$

where

$$\begin{aligned} h(x,y) &= -\,I(1;a_1;a_2) = -I\Big(1;\frac{a_1-a_2}{1-a_2};0\Big) \\ &= \mathrm{Li}_1\left(\frac{1-xy}{1-y}\right) = \mathrm{Li}_1(x) - \mathrm{Li}_1(y^{-1}), \\ f(x,y) &= I(1;a_1,0;a_2) = I\Big(1;\frac{a_1-a_2}{1-a_2},\frac{-a_2}{1-a_2};0\Big) \\ &= \mathrm{Li}_{1,1}\left(\frac{1-xy}{1-y},1-y\right) = \mathrm{Li}_2(y^{-1}) - \mathrm{Li}_2(x) + \ell_1(xy)\,\mathrm{Li}_1(x), \\ e(x,y) &= -\,I(1;a_1,0,0;a_2) \\ &= -\,I\Big(1;\frac{a_1-a_2}{1-a_2},\frac{-a_2}{1-a_2},\frac{-a_2}{1-a_2};0\Big) = \mathrm{Li}_{1,1,1}\left(\frac{1-xy}{1-y},1-y,1\right) \\ &= \mathrm{Li}_3(x) - \mathrm{Li}_3(y^{-1}) - \ell_1(xy)\,\mathrm{Li}_2(x) + \ell_2(xy)\,\mathrm{Li}_1(x) \end{aligned}$$

by using Exercise 2.14 and Eq. (2.48). Here $\beta_5 \rightsquigarrow r_5$ means that we need to replace the pseudo row entry β_5 by

$$r_5 = \ell_1(xy)\,\mathrm{Li}_1(x) - \mathrm{Li}_2(x).$$

This can be found by taking the monodromy of $e(x,y)$ about the four divisors $\{x=0\}$, $\{y=0\}$, $\{x=1\}$, and $\{y=1\}$. We do not need to consider the column on top of $\mathrm{Li}_3(y)$ as it will contribute nothing because of the 0 on its 5th row. Notice the diagonal entry 1 lying on top of β_5 is on the 5th row, or in other words, the column number of β_5 is 5.

Example 2.6.2. For $\mathrm{Li}_{2,2}(x,y)$, we have

$$\mathrm{Li}_{2,2}(x,y) = I(1;0,a_1,0,a_2;0) = I(1;0_1,a_1,0_2,a_2;0),$$

where, in order to distinguish the two interior 0's, we have used the set $\{0_1,a_1,0_2,a_2\}$. Here are the possible choices:

(1) $\emptyset$: $1 \otimes \mathrm{Li}_{2,2}(x,y)$

(2) $\{a_1\}$: $-I(1;a_1;0) \otimes -I(1;0_1;a_1)I(a_1;0_2,a_2;0) = \mathrm{Li}_1(x) \otimes \ell_1(x)\,\mathrm{Li}_2(y)$
(3) $\{a_2\}$: $-I(1;a_2;0) \otimes -I(1;0_1,a_1,0_2;a_2) = \mathrm{Li}_1(xy) \otimes k(x,y)$
(4) $\{0_1,a_1\}$: $-I(1;0_1,a_1;0) \otimes -I(a_1;0_2,a_2;0) = \mathrm{Li}_2(x) \otimes \mathrm{Li}_2(y)$
(5) $\{0_1,a_2\}$: $-I(1;0_1,a_2;0) \otimes -I(0_1;a_1,0_2;a_2) = \mathrm{Li}_2(xy) \otimes \mathrm{Li}_2(y^{-1})$ (Note there is no collapse of 0's! Row entry $\mathrm{Li}_2(y^{-1})$ is to be combined with $\mathrm{Li}_2(x)$ from the next case.)
(6) $\{0_2,a_2\}$: $-I(1;0_2,a_2;0) \otimes -I(1;0_1,a_1;0_2) = \mathrm{Li}_2(xy) \otimes \mathrm{Li}_2(x)$
(7) $\{a_1,a_2\}$: $I(1;a_1,a_2;0) \otimes I(1;0_1;a_1)I(a_1;0_2;a_2) = \mathrm{Li}_{1,1}(x,y) \otimes \ell_1(x)\ell_1(y)$
(8) $\{a_1,0_2,a_2\}$: $I(1;a_1,0_2,a_2;0) \otimes I(1;0_1;a_1) = \mathrm{Li}_{1,2}(x,y) \otimes \ell_1(x)$
(9) $\{0_1,a_1,a_2\}$: $I(1;0_1,a_1,a_2;0) \otimes I(a_1;0_2;a_2) = \mathrm{Li}_{2,1}(x,y) \otimes \ell_1(y)$
(10) $\{0,a_1,0,a_2\}$: $\mathrm{Li}_{2,2}(x,y) \otimes 1$

Thus the variation matrix $\mathscr{M}_{2,2}(x,y)$ associated with $\mathrm{Li}_{2,2}(x,y)$ over S_2 is

$$\begin{bmatrix}
1 \\
\mathrm{Li}_1(x) & 1 \\
\mathrm{Li}_1(xy) & 0 & 1 \\
\mathrm{Li}_2(x) & \ell_1(x) & 0 & 1 \\
\mathrm{Li}_{1,1}(x,y) & \mathrm{Li}_1(y) & h(x,y) & 0 & 1 \\
\mathrm{Li}_2(xy) & 0 & \ell_1(xy) & 0 & 0 & 1 \\
\mathrm{Li}_{2,1}(x,y) & \ell_1(x)\,\mathrm{Li}_1(y) & g(x,y) & \mathrm{Li}_1(y) & \ell_1(x) & \alpha(y) & 1 \\
\mathrm{Li}_{1,2}(x,y) & \mathrm{Li}_2(y) & f(x,y) & 0 & \ell_1(y) & \mathrm{Li}_1(x) & 0 & 1 \\
\mathrm{Li}_{2,2}(x,y) & \ell_1(x)\,\mathrm{Li}_2(y) & k(x,y) & \mathrm{Li}_2(y) & \ell_1(x)\ell_1(y) & \begin{matrix}\mathrm{Li}_2(y^{-1}) \\ +\,\mathrm{Li}_2(x)\end{matrix} & \ell_1(y) & \ell_1(x) & 1
\end{bmatrix} \tau'$$

where $\tau' = \tau_{2,2}(2\pi i)$ and

$$\alpha(y) = -\,\mathrm{Li}_1(y^{-1}) = -\,\mathrm{Li}_1(y) - \ell_1(y),$$

$$\begin{aligned}
g(x,y) &= I(1;0_1,a_1;a_2) = I\Big(1;\frac{-a_2}{1-a_2},\frac{a_1-a_2}{1-a_2};0\Big) \\
&= \mathrm{Li}_{1,1}\left(1-xy,\frac{1}{1-y}\right) = \mathrm{Li}_2(x) - \mathrm{Li}_2(y^{-1}) - \ell_1(xy)\,\mathrm{Li}_1(y^{-1})
\end{aligned}$$

by using Eq. (2.44), and

$$\begin{aligned}
k(x,y) &= -I(1;0_1,a_1,0_2;a_2) = -I\Big(1;\frac{-a_2}{1-a_2},\frac{a_1-a_2}{1-a_2},\frac{-a_2}{1-a_2};0\Big) \\
&= \mathrm{Li}_{1,1,1}\left(1-xy,\frac{1}{1-y},1-y\right) \\
&= 2\,\mathrm{Li}_3(x) - \ell_1(x)\,\mathrm{Li}_2(x) - 2\,\mathrm{Li}_3(y^{-1}) + \ell_1(y)\,\mathrm{Li}_2\left(\frac{1-xy}{x(1-y)}\right) \\
&\quad - \ell_1(y)\,\mathrm{Li}_2(y^{-1}) - \ell_1(xy)\Big(\mathrm{Li}_2(y^{-1}) - \ell_1(y)\,\mathrm{Li}_1(y^{-1})\Big)
\end{aligned}$$

$$- \ell_1(y)\operatorname{Li}_2\left(1 - \frac{1}{xy}\right) - \ell_1(y)\operatorname{Li}_2\left(\frac{1-xy}{1-y}\right)$$

by using Exercise 2.14, Eq. (2.48) and the following identities:

$$\operatorname{Li}_3(y^{-1}) - \operatorname{Li}_3(y) = \ell_3(y), \quad \operatorname{Li}_2(y^{-1}) + \operatorname{Li}_2(y) = 2\operatorname{Li}_2(1) - \ell_2(y). \tag{2.33}$$

Notice in particular that all the pseudo entries are genuine in this case.

Example 2.6.3. For $\operatorname{Li}_{3,1}(x, y)$, we have

$$\operatorname{Li}_{3,1}(x, y) = I(a_0; 0, 0, a_1, a_2; 0) = I(1; 0, 0, a_1, a_2; 0).$$

The following are possible choices of the subsets of $\{0, 0, a_1, a_2\} = \{0_1, 0_2, a_1, a_2\}$:

(1) $\emptyset$: $1 \otimes \operatorname{Li}_{3,1}(x, y)$
(2) $\{a_1\}$: $-I(1; a_1; 0) \otimes -I(1; 0, 0; a_1)I(a_1; a_2; 0) = \operatorname{Li}_1(x) \otimes \ell_2(x)\operatorname{Li}_1(y)$
(3) $\{a_2\}$: $-I(1; a_2; 0) \otimes -I(1; 0, 0, a_1; a_2) = \operatorname{Li}_1(xy) \otimes l(x, y)$
(4) $\{0_1, a_1\}$: $-I(1; 0_1, a_1; 0) \otimes -I(1; 0_2; a_1)I(a_1; a_2; 0) = \operatorname{Li}_2(x) \otimes \ell_1(x)\operatorname{Li}_1(y)$ (collapse of 0_1 onto $a_0 = 1$)
(5) $\{0_1, a_2\}$: $-I(1; 0_1, a_2; 0) \otimes I(1; 0_2, a_1; a_2) = \operatorname{Li}_2(xy) \otimes g(x, y)$ (collapse of 0_1. This gives the only pseudo row entry in this case.)
(6) $\{a_1, a_2\}$: $I(1; a_1, a_2; 0) \otimes I(1; 0, 0; a_1) = \operatorname{Li}_{1,1}(x, y) \otimes \ell_2(x)$
(7) $\{0_1, a_1, a_2\}$: $I(1; 0_1, a_1, a_2; 0) \otimes I(1; 0_2; a_1) = \operatorname{Li}_{2,1}(x, y) \otimes \ell_1(x)$ (collapse of 0_1)
(8) $\{0, 0, a_1\}$: $-I(1; 0, 0, a_1; 0) \otimes -I(a_1; a_2; 0) = \operatorname{Li}_3(xy) \otimes \operatorname{Li}_1(y)$
(9) $\{0, 0, a_2\}$: $-I(1; 0, 0, a_2; 0) \otimes -I(0; a_1; a_2)$
(10) $\{0, 0, a_1, a_2\}$: $\operatorname{Li}_{3,1}(x, y) \otimes 1$

Thus the variation matrix $\mathscr{M}_{3,1}(x, y)$ associated with $\operatorname{Li}_{3,1}(x, y)$ over S_2 is

$$\begin{bmatrix}
1 \\
\operatorname{Li}_1(x) & 1 \\
\operatorname{Li}_1(xy) & 0 & 1 \\
\operatorname{Li}_2(x) & \ell_1(x) & 0 & 1 \\
\operatorname{Li}_{1,1}(x,y) & \operatorname{Li}_1(y) & h(x,y) & 0 & 1 \\
\operatorname{Li}_2(xy) & 0 & \ell_1(xy) & 0 & 0 & 1 \\
\operatorname{Li}_3(x) & \ell_2(x) & 0 & \ell_1(x) & 0 & 0 & 1 \\
\operatorname{Li}_{2,1}(x,y) & \ell_1(x)\operatorname{Li}_1(y) & g(x,y) & \operatorname{Li}_1(y) & \ell_1(x) & \alpha(y) & 0 & 1 \\
\operatorname{Li}_3(xy) & 0 & \ell_2(xy) & 0 & 0 & \ell_1(xy) & 0 & 0 & 1 \\
\operatorname{Li}_{3,1}(x,y) & \ell_2(x)\operatorname{Li}_1(y) & l(x,y) & \ell_1(x)\operatorname{Li}_1(y) & \ell_2(x) & g \rightsquigarrow \tilde{g} & \operatorname{Li}_1(y) & \ell_1(x) & \alpha(y) & 1
\end{bmatrix}\tilde{\tau}$$

where $\tilde{\tau} = \tau_{3,1}(2\pi i)$, $\tilde{g}(x,y) = g(x,y) - \mathrm{Li}_2(x) = -\mathrm{Li}_2(y^{-1}) - \ell_1(xy)\,\mathrm{Li}_1(y^{-1})$ and

$$\begin{aligned} l(x,y) &= -\,I(1;0,0,a_1;a_2) = -I\Big(1;\frac{-a_2}{1-a_2},\frac{-a_2}{1-a_2},\frac{a_1-a_2}{1-a_2};0\Big) \\ &= \mathrm{Li}_{1,1,1}\left(1-xy,1,\frac{1}{1-y}\right) \\ &= \mathrm{Li}_3(x) - \mathrm{Li}_3(y^{-1}) - \ell_1(xy)\,\mathrm{Li}_2(y^{-1}) - \ell_2(xy)\,\mathrm{Li}_1(y^{-1}) \end{aligned}$$

by using Exercise 2.14, Eqs. (2.48), (2.33) and the following identities:

$$\begin{aligned} \mathrm{Li}_3(1-y^{-1}) + \mathrm{Li}_3(1-y) + \mathrm{Li}_3(y) &= \mathrm{Li}_3(1) + \ell_2(y)\,\mathrm{Li}_1(y) + \ell_3(y), \\ \mathrm{Li}_2(1-y^{-1}) + \mathrm{Li}_2(1-y) &= -\,\ell_2(y). \end{aligned}$$

We find $\tilde{g}(x,y)$ by considering the monodromy of $l(x,y)$. Note that the weight three entry $\ell_2(x)\,\mathrm{Li}_1(y)$ does not affect $\tilde{g}(x,y)$ at all.

2.6.3 *A General Result*

Usually it is tedious to write down explicitly the variation matrix associated with any given MPL $\mathrm{Li}_{m_1,\dots,m_n}(\mathbf{x})$. However, the following general statement is true according to Deligne and Goncharov [164] (see next section for relevant definitions):

> *The MPL* $\mathrm{Li}_{n_1,\dots,n_d}(\mathbf{x})$ *underlies a good unipotent graded-polarizable variation of mixed Hodge–Tate structures* $(V_{n_1,\dots,n_d}, W_\bullet, F^\bullet)$ *over*
>
> $$\mathbb{C}^d \setminus \Big\{\prod_{i=1}^{d} x_i(1-x_i) \prod_{1\le i<j\le d} (1-x_i\dots x_j) = 0\Big\}$$
>
> *with the weight-graded quotients* gr^W_{-2k} *being given by* c_k *copies of the Tate structure* $\mathbb{Z}(k)$ *which are nonzero only for* $0 \le k \le n := n_1 + \dots + n_d$.

Here c_k is the number of different ways to pick ordered $(k+2)$-tuples $(b_{i_0},\dots,b_{i_{k+1}})$ from the ordered numbers $(b_0,\dots,b_{n+1})$ in the following tableau

$$\left|\, b_0 \,\right| \cdots \left|\, b_{n+1} \,\right| = \left|\, a_0 \,\right| \underbrace{0 \,\Big|\, \cdots \,\Big|\, 0}_{n_1-1 \text{ times}} \left|\, a_1 \,\right| \cdots\cdots \left|\, a_{d-1} \,\right| \underbrace{0 \,\Big|\, \cdots \,\Big|\, 0}_{n_d-1 \text{ times}} \left|\, a_d \,\right| 0 \,\Big| \tag{2.34}$$

such that all three conditions (i)-(iii) above are satisfied. It is apparent that

$$c_k \ge d_k(n_1,\dots,n_d) := \sum_{\substack{k_1+\dots+k_d=k \\ 0\le k_j<n_j\ \forall j}} 1.$$

Each term in the sum corresponds to the following choice: for every $j = 1, \dots, n$, choose k_j 0's immediately after a_{j-1}. This was established using the theory of motives [54, 273, 390, 561], so we omit its proof in this book due to space limitation.

Example 2.6.4. By the definition, we always have $c_0 = c_n = 1$. When $n_1 = \cdots = n_d = 1$ tableau Eq. (2.34) becomes

$$\Big| \, b_0 \, \Big| \, \cdots \, \Big| \, b_{n+1} \, \Big| = \Big| \, \underbrace{1 \, \Big| \, \cdots \, \Big| \, 1}_{n+1 \text{ times}} \, \Big| \, 0 \, \Big|$$

Hence c_k is the number of ways to choose k elements from the set $\{b_1, \dots, b_n\}$, *i.e.*, $c_k = \binom{n}{k}$.

For ease of statement, let us put a box $\square$ on a number whenever we choose it.

Example 2.6.5. This example will help you to do Exercise 2.9 and 2.10. First let's consider $\mathrm{Li}_{2,1}$. There are six nontrivial ways to put boxes on $\big| \, 1 \, \big| \, a_1 \, \big| \, a_2 \, \big| \, 0 \, \big| \, 0 \, \big|$:

$$(1) \ \Big|\boxed{1}\Big| \, 0 \, \Big|\boxed{a_1}\Big| \, a_2 \, \Big|\boxed{0}\Big| \qquad (2) \ \Big|\boxed{1}\Big| \, 0 \, \Big| \, a_1 \, \Big|\boxed{a_2}\Big|\boxed{0}\Big|$$

$$(3) \ \Big|\boxed{1}\Big|\boxed{0}\Big|\boxed{a_1}\Big| \, a_2 \, \Big|\boxed{0}\Big| \qquad (4) \ \Big|\boxed{1}\Big| \, 0 \, \Big|\boxed{a_1}\Big|\boxed{a_2}\Big|\boxed{0}\Big|$$

$$(5) \ \Big|\boxed{1}\Big|\boxed{0}\Big| \, a_1 \, \Big|\boxed{a_2}\Big|\boxed{0}\Big| \qquad (6) \ \Big|\boxed{1}\Big|\boxed{0}\Big|\boxed{a_1}\Big|\boxed{a_2}\Big|\boxed{0}\Big|$$

Together with the trivial choice $\Big|\boxed{1}\Big| \, 0 \, \Big| \, a_1 \, \Big| \, a_2 \, \Big|\boxed{0}\Big|$ we get $c_0 = c_3 = 1$, $c_1 = 2$ and $c_2 = 3$.

However, for $\mathrm{Li}_{1,2}$ we have altogether only six ways to do this:

$$(1) \ \Big|\boxed{1}\Big| \, a_1 \, \Big| \, 0 \, \Big| \, a_2 \, \Big|\boxed{0}\Big| \qquad (2) \ \Big|\boxed{1}\Big|\boxed{a_1}\Big| \, 0 \, \Big| \, a_2 \, \Big|\boxed{0}\Big|$$

$$(3) \ \Big|\boxed{1}\Big| \, a_1 \, \Big| \, 0 \, \Big|\boxed{a_2}\Big|\boxed{0}\Big| \qquad (4) \ \Big|\boxed{1}\Big|\boxed{a_1}\Big| \, 0 \, \Big|\boxed{a_2}\Big|\boxed{0}\Big|$$

$$(5) \ \Big|\boxed{1}\Big| \, a_1 \, \Big|\boxed{0}\Big|\boxed{a_2}\Big|\boxed{0}\Big| \qquad (6) \ \Big|\boxed{1}\Big|\boxed{a_1}\Big|\boxed{0}\Big|\boxed{a_2}\Big|\boxed{0}\Big|$$

Thus $c_0 = c_3 = 1$, $c_1 = c_2 = 2$.

2.7 Variations of Mixed Hodge Structures of Multiple Polylogarithms

2.7.1 *Definition: A Review*

In this section we briefly review the theory of variations of mixed Hodge structures (for which we use the abbreviation MHS **only in this chapter**). See [122] for another short introduction to variations of Hodge structures.

A pure ($\mathbb{Z}$-)Hodge structure (HS) of weight k consists of a finitely generated abelian group $H(\mathbb{Z})$ and a decreasing Hodge filtration $F^\bullet$ on $H(\mathbb{C}) := H(\mathbb{Z}) \otimes_{\mathbb{Z}} \mathbb{C}$ such that $H(\mathbb{C}) = F^p \oplus \overline{F^{k-p+1}}$ for all integers p. Here the "bar" is the complex conjugation on the second factor of the tensor product. A special example is the Tate structure $\mathbb{Z}(-k)$ of weight $2k$ consisting of $H(\mathbb{Z}) = \mathbb{Z}$ and the filtration $F^p = 0$ for $p > k$ and $F^p = H(\mathbb{C})$ for $p \le k$. If we replace $\mathbb{Z}$ by $\mathbb{Q}$ in the above then we get a pure ($\mathbb{Q}$-)HS of weight k.

A MHS consists of a finitely generated abelian group $H(\mathbb{Z})$ and two filtrations: an increasing weight filtration $W_\bullet$ on $H(\mathbb{Q}) := H(\mathbb{Z}) \otimes_{\mathbb{Z}} \mathbb{Q}$ and a decreasing filtration $F^\bullet$ on $H(\mathbb{C})$, which are compatible in the following sense. On each graded piece of the weight filtration $\mathrm{gr}_k^W = W_k/W_{k-1}$ the induced Hodge filtration determined by

$$F^p(\mathrm{gr}_k^W)(\mathbb{C}) = {(F^p \cap W_k(\mathbb{C}) + W_{k-1}(\mathbb{C}))}\Big/{W_{k-1}(\mathbb{C})}$$

is a pure Hodge structure of weight k where $W_k(\mathbb{C}) := W_k \otimes_{\mathbb{Z}} \mathbb{C}$. If all the pure Hodge structures induced as above are direct sums of Tate structures then we say the MHS is a Tate structure. For a mixed Hodge–Tate structure, we can put a framing as in [45, § 1.3.4, § 1.4].

Following Steenbrink and Zucker (see Definitions 3.1, 3.2 and 3.4 in [526]) we have

Definition 2.7.1. A *variation of HS of weight* k defined over $\mathbb{Q}$ and a complex manifold S is a collection of data $(\mathbb{V}_{\mathbb{Q}}, F^\bullet)$ where

(a) $\mathbb{V}_{\mathbb{Q}}$ is a locally constant sheaf (local system) of $\mathbb{Q}$-vector spaces on S;

(b) $F^\bullet$ is a decreasing filtration by holomorphic subbundles of the locally free sheaf $\mathcal{V} = \mathcal{O}_S \otimes_{\mathbb{Q}} \mathbb{V}_{\mathbb{Q}}$;

(c) At each $s \in S$, $F^\bullet$ induces the Hodge filtration $F_s^\bullet$ of a Hodge structure of weight k on the fiber $\mathcal{V}_s$ of $\mathcal{V}$ such that

(i) whenever $p + q = k$ one has $\mathcal{V}_s = F_s^p \oplus \overline{F_s^{q+1}}$, where the "bar" denotes the complex conjugation,

(ii) equivalently, one has $\mathcal{V}_s = \bigoplus_{p+q=k} H_s^{p,q}$ where $H_s^{p,q} = F_s^p \cap \overline{F_s^q}$.

(d) (Griffiths transversality) Under the connection ∇ over $\mathcal{V}$,

$$\nabla F^p \subset \Omega^1_S \otimes_{\mathcal{O}_S} F^{p+1} \qquad \text{for all } p.$$

Definition 2.7.2. *A polarization over* $\mathbb{Q}$ of a variation of Hodge structure of weight k over $\mathbb{Q}$ is a non-degenerate and flat bilinear pairing:

$$\beta : \mathbb{V}_{\mathbb{Q}} \times \mathbb{V}_{\mathbb{Q}} \longrightarrow \mathbb{Q},$$

such that β is $(-1)^k$-symmetric, and the Hermitian form $\beta_s(C_s v, \bar{w})$ is positive on each fiber. Here C_s denotes the Weil operator with respect to F_s, namely the direct sum of multiplication by i^{p-1} on $H^{p,q}_s$. A variation is called *polarizable* (over $\mathbb{Q}$) if it admits a polarization (over $\mathbb{Q}$).

Definition 2.7.3. A *variation of MHS* defined over $\mathbb{Q}$ and a complex manifold S is a collection of data $(\mathbb{V}_{\mathbb{Q}}, W_\bullet, F^\bullet)$ where

(a) $\mathbb{V}_{\mathbb{Q}}$ is a local system of $\mathbb{Q}$-vector spaces on S,

(b) $W_\bullet$ is an increasing filtration of the $\mathbb{V}_{\mathbb{Q}}$ by local subsystems,

(c) $F^\bullet$ is a decreasing filtration by holomorphic subbundles of $\mathcal{V} = \mathcal{O}_S \otimes_{\mathbb{Q}} \mathbb{V}_{\mathbb{Q}}$.

(d) $\nabla F^p \subset \Omega^1_S \otimes_{\mathcal{O}_S} F^{p+1}$ for all p.

(e) The data

$$\left(\operatorname{gr}^W_k \mathbb{V}_{\mathbb{Q}}, F^\bullet\left(\mathcal{O}_S \otimes_E W_k / \mathcal{O}_S \otimes_E W_{k-1}\right)\right) \tag{2.35}$$

is a variation of HS of weight k defined over $\mathbb{Q}$; or equivalently, on the fiber over $s \in S$, (V_s, W_s, F_s) is a MHS defined over $\mathbb{Q}$.

(f) If the induced collection of variations of HS Eq. (2.35) are all polarizable then the MHS is called graded-polarizable.

Remark 2.7.4. By extension of scalars in $\mathbb{V}_{\mathbb{Q}}$ one can define $\mathbb{V}_{\mathbb{F}}$ for any field $\mathbb{F}$ such that $\mathbb{Q} \subset \mathbb{F} \subset \mathbb{R}$.

Giving a local system $\mathbb{V}_{\mathbb{Q}}$ is equivalent to specifying its monodromy representation

$$\rho_x : \pi_1(S, x) \longrightarrow \operatorname{Aut}_{\mathbb{Q}} \mathcal{V}_x.$$

A variation is called *unipotent* if this representation is unipotent. From [270, Prop. 1.3] we know that a variation of MHS $(\mathbb{V}_{\mathbb{Q}}, W_\bullet, F^\bullet)$ is unipotent if and only if each of the variations of Hodge structure $\operatorname{gr}^W_k \mathbb{V}_{\mathbb{Q}}$ is constant.

In general, the behavior of a variation of MHS over a non-compact base S at "infinity" is very hard to control. In [526] Steenbrink and Zucker consider the case when S is a curve and define the admissibility condition at infinity. For higher dimensional S, Kashiwara, M. Saito, and others define a variation over S to be admissible if its restriction to every curve is admissible in the sense of Steenbrink–Zucker.

However, the behavior of *unipotent* variations of MHS at infinity can be controlled rather easily. We have the classical result of Deligne (see [158, Prop. 5.2]) which defines the *canonical extension* $\tilde{\mathcal{V}}$ of $\mathcal{V}$.

Theorem 2.7.5. *Let $\tilde{S}$ be a normalization of S. Let $(\mathbb{V}_{\mathbb{Q}}, W_\bullet, F^\bullet)$ be a unipotent variation of MHS over S. Then*
(a) There is a unique extension $\tilde{\mathcal{V}}$ of $\mathcal{V}$ over $\tilde{S}$ satisfying the following equivalent conditions:

> *(i) Inside every section of $\tilde{\mathcal{V}}$, every flat section of $\mathcal{V}$ increases at most at the rate of $O(\log^k \|x\|)$ (k large enough) on every compact set of $D = \tilde{S} - S$.*
>
> *(ii) Similarly, every flat section of $\mathcal{V}^\vee$ (the dual) increases at most at the rate of $O(\log^k \|x\|)$ (k large enough).*

(b) The combination of the two conditions (i) and (ii) is equivalent to the combination of the following two conditions:

> *(iii) In terms of any local basis of $\tilde{\mathcal{V}}$ the connection matrix $\boldsymbol{\omega}$ of $\mathcal{V}$ has at most logarithmic singularities along D.*
>
> *(iv) The residue of $\boldsymbol{\omega}$ along any irreducible component of D is nilpotent.*

We will verify conditions (iii) and (iv) by Prop. 2.5.8 for the multiple logarithm variations of MHS. They are unipotent variations by Thm. 2.5.9.

Definition 2.7.6. Let $\tilde{S}$ be a compactification of S. Then a unipotent variation of MHS $(\mathbb{V}_{\mathbb{Q}}, W_\bullet, F^\bullet)$ over S is said to be *good* if it satisfies the following conditions at infinity:

> (1) the Hodge filtration bundles $F^\bullet$ extend over $\tilde{S}$ to sub-bundles $\tilde{F}^\bullet$ of the canonical extension $\tilde{\mathcal{V}}$ of $\mathcal{V}$ such that they induce the corresponding thing for each pure subquotient $\mathrm{gr}^W_k \mathbb{V}_{\mathbb{Q}}$,
>
> (2) for the nilpotent logarithm N_j of a local monodromy transformation about a component D_j of D, the weight filtration of N_j relative to $W_\bullet$ exists.

A slightly different definition first appeared in [270, 271] with the extra assumption that $D = \tilde{S} - S$ is a normal crossing divisor. In these papers, Hain and Zucker classified good unipotent variations of MHS on algebraic manifolds. With constant pure weight subquotients these variations behave well at infinity.

2.7.2 *Mixed Hodge Structures of Multiple Logarithms*

Fix an embedding $\mathbb{C}^n \hookrightarrow \mathbb{CP}^n$. Let $\mathscr{D}_n = D_n \cup (\mathbb{CP}^n \setminus \mathbb{C}^n)$. Let $M_{2^n}(\mathbb{C})$ be the set of $2^n \times 2^n$ matrices over $\mathbb{C}$. Put

$$\boldsymbol{\omega} = \Big(c_{\mathbf{i},\mathbf{j}}\Big)_{\mathbf{i},\mathbf{j}\in\mathcal{I}_n} \in H^0(\mathbb{CP}^n, \Omega^1_{\mathbb{CP}^n}(\log(\mathscr{D}_n))) \otimes M_{2^n}(\mathbb{C}) \tag{2.36}$$

where

$$c_{\mathbf{i},\mathbf{j}} = \begin{cases} \mathrm{d}\gamma^{|\mathbf{i}|}_{\rho_{\mathbf{i}}(\mathbf{j})}(\mathbf{x}(\mathbf{i})) & \text{if } |\mathbf{j}| = |\mathbf{i}| - 1,\ \mathbf{j} \prec \mathbf{i}, \\ 0 & \text{otherwise.} \end{cases}$$

It is not difficult to check that all of the 1-forms in $\boldsymbol{\omega}$ have logarithmic singularity on $\mathscr{D}_n$. Moreover, because $\mathcal{M}_{[n]}(\mathbf{x})$ is invertible and $\boldsymbol{\omega}$ is closed we get

$$\mathrm{d}\boldsymbol{\omega} = 0, \quad \boldsymbol{\omega}\wedge\boldsymbol{\omega} = 0. \tag{2.37}$$

This shows that $\boldsymbol{\omega}$ is integrable.

Now we define a meromorphic connection ∇ on the trivial bundle

$$\mathbb{CP}^n \times \mathbb{C}^{2^n} \longrightarrow \mathbb{CP}^n \tag{2.38}$$

by

$$\nabla f = \mathrm{d}f - \boldsymbol{\omega} f$$

where $f : S_n \to \mathbb{C}^{2^n}$ is a section. This connection has regular singularities along $\mathscr{D}_n$ because $\boldsymbol{\omega}$ is integrable by Eq. (2.37) and all the 1-forms in $\boldsymbol{\omega}$ are logarithmic in any compactification of S_n. By the explicit construction of $\boldsymbol{\omega}$ we see immediately that the conditions (iii) and (iv) of Thm. 2.7.5 are satisfied. Proposition 2.5.8 further implies that the columns $(2\pi i)^{|\mathbf{j}|}C_{\mathbf{j}}(\mathbf{x})$ of $\mathcal{M}_{[n]}(\mathbf{x})$ satisfy $\nabla f = 0$ and are therefore flat sections of Eq. (2.38). Even though they are multi-valued, their $\mathbb{Z}$-linear span is well defined thanks to Thm. 2.5.9. Hence $V_{[n]}(\mathbf{x})$ forms a local system over S_n.

Definition 2.7.7. The local system $V_{[n]}(\mathbf{x})$ is called the *n-tuple logarithm local system.*

To define the MHS on $V_{[n]}$ we can define the weight filtration by putting $W_{2k+1} = W_{2k}$ and

$$W_{-2k}V_{[n]}(\mathbf{x}) := \langle (2\pi i)^{|\mathbf{i}|}C_{\mathbf{i}} : \ |\mathbf{i}| \geq k\rangle_{\mathbb{Q}}$$

which is the $\mathbb{Q}$ vector space with basis $\{(2\pi i)^{|\mathbf{i}|}C_{\mathbf{i}} : \ |\mathbf{i}| \geq k\}$. In particular, $W_{-2k}V_{[n]}(\mathbf{x}) = 0$ if $k > n$ and $W_{2k}V_{[n]}(\mathbf{x}) = V_{[n]}(\mathbf{x})$ if $k \geq 0$. By regarding

$e_{\mathbf{i}}$'s as column vectors one can define the Hodge filtration on $V_{[n]}(\mathbf{x}) \otimes \mathbb{C} = V_{[n],\mathbb{C}}$ as follows:

$$F^{-k} V_{[n],\mathbb{C}} := \langle e_{\mathbf{i}} : \ |\mathbf{i}| \leq k \rangle_{\mathbb{C}}.$$

So in particular, $F^{-k} V_{[n],\mathbb{C}} = 0$ for $k < 0$ and $F^{-k} V_{[n],\mathbb{C}} = V_{[n],\mathbb{C}}$ for $k \geq n$.

By induction on n and using Lemma 2.5.6 it is easy to show that

$$F^{-p} \cap W_{-2k} V_{[n],\mathbb{C}} = \begin{cases} 0 & \text{if } p \leq k-1 \\ \langle (2\pi i)^{|\mathbf{i}|} e_{\mathbf{i}} : \ k \leq |\mathbf{i}| \leq p \rangle & \text{if } k \leq p \leq n \, . \\ \langle (2\pi i)^{|\mathbf{i}|} e_{\mathbf{i}} : \ k \leq |\mathbf{i}| \leq n \rangle & \text{if } p \geq n \end{cases}$$

This implies that

$$F^{-p} \operatorname{gr}^W_{-2k} V_{[n],\mathbb{C}} = \begin{cases} 0 & \text{if } p \leq k-1 \\ W_{-2k} V_{[n],\mathbb{C}} / W_{-2k-1} V_{[n],\mathbb{C}} & \text{if } p \geq k. \end{cases}$$

In other words, $F^q \operatorname{gr}^W_{-2k} V_{[n],\mathbb{C}} = 0$ for $q \geq -k+1$ and $F^q \operatorname{gr}^W_{-2k} V_{[n],\mathbb{C}} = \operatorname{gr}^W_{-2k} V_{[n],\mathbb{C}}$ for $q \leq -k$. This means that the Hodge filtration induces a pure HS of weight $-2k$ on each weight graded piece. Furthermore, it is not hard to see by checking the powers of $2\pi i$ appearing on the diagonal of $\mathcal{M}_{[n]}(\mathbf{x})$ that this induced structure on $\operatorname{gr}^W_{-2k} V_{[n],\mathbb{C}}$ is isomorphic to the direct sum of $\binom{n}{k}$ copies of the Tate structure $\mathbb{Z}(k)$ by Lemma 2.5.6.

2.8* Limit Mixed Hodge Structures of Multiple Logarithms

Let the monodromy of $\mathcal{M}_{[n]}(\mathbf{x})$ at any subvariety $\mathcal{D}$ of $\mathbb{CP}^n$ be given by the matrix $T_{\mathcal{D}}$ and the local monodromy logarithm by $N_{\mathcal{D}} = \log T_{\mathcal{D}}/(2\pi i)$. Note that $T_{\mathcal{D}}$ is unipotent so $N_{\mathcal{D}}$ is well defined.

Now let us recall the construction of the unipotent variations of limit MHS at the "infinity" with *normal crossing*. Let S be a complex manifold of dimension d. Suppose that S is embedded in $\tilde{S}$, via the mapping j, such that $D = \tilde{S} - S$ is a divisor with normal crossings. Let $\mathbb{V}$ be any unipotent local system of complex vector spaces on S, and $\mathcal{V}$ the corresponding vector bundle. According to Thm. 2.7.5 by Deligne there is a canonical extension $\tilde{\mathcal{V}}$ of $\mathcal{V}$ over $\tilde{S}$. Moreover, when the local monodromy is nilpotent $\tilde{\mathcal{V}}$ is a subsheaf of $j_*\mathcal{V}$. The local picture of $S \subset \tilde{S}$ is $(\Delta^*)^r \times \Delta^{d-r} \subset \Delta^d$ where Δ is the unit disk and Δ^* is the punctured one. We let $t_1, \dots, t_r$ denote the variables on $(\Delta^*)^r$, and $N_1, \dots, N_r$ the (commuting) local nilpotent logarithms of the associated monodromy transformations of the fibre. For $z_1, \dots, z_r$ in the upper half-plane, the universal covering mapping for $(\Delta^*)^r$ is given by

$$t_j = \exp(2\pi i z_j), \quad j = 1, \cdots, r.$$

Let $v_1, \ldots, v_m$ be a basis of the multi-valued sections of $\mathbb{V}$ over $(\Delta^*)^r \times \Delta^{d-r}$, the formula

$$[\tilde{v}_1, \ldots, \tilde{v}_m] = [v_1, \ldots, v_m] \exp\left(-\sum_{j=1}^{r} 2\pi i z_j N_j\right) = [v_1, \ldots, v_m] \prod_{j=1}^{r} t_j^{-N_j}$$

determines a basis of the sections of $\mathscr{V}$ over Δ^d and these provide, by the definition, the generators of $\tilde{\mathscr{V}}$ over Δ^d.

In our situation, although the divisor D_n is not normal crossing Thm. 2.7.5 is still valid because the image of the global holomorphic logarithmic forms in the complex of smooth forms on S is independent of the normal crossings compactification (see [269, Prop. (3.2)]). In fact, the forms we are considering lie in the subcomplex generated by 1-forms of the type df/f where f is a rational function. Such forms are automatically logarithmic in any compactification and therefore our connection is automatically regular. Hence the admissibility and the existence of the limit MHS is an automatic consequence of the admissibility of our variations restricted to every curve in S_n. Moreover, the pullback to $\tilde{S}_n$ of our trivial bundle Eq. (2.38) restricted to S_n is exactly Deligne's canonical extension of Eq. (2.38), and the pullbacks of the subbundles $F^\bullet$ and $W_\bullet$ are the correct extended Hodge and weight subbundles. Therefore we have

Theorem 2.8.1. *The n-tuple logarithm underlies a good unipotent graded-polarizable variation of mixed Hodge–Tate structures $(V_{[n]}, W_\bullet, F^\bullet)$ over*

$$S_n = \mathbb{C}^n \setminus \Big\{ \prod_{1\le j\le n} x_j(1-x_j) \prod_{1\le i<j\le n} (1 - x_i \ldots x_j) = 0 \Big\}$$

with the weight-graded quotients gr^W_{-2k} being given by $\binom{n}{k}$ copies of the Tate structure $\mathbb{Z}(k)$.

Proof. It is clear that all the odd graded weight quotients are zero so that we can let the polarizations on the weight graded quotients gr^W_{-2k} be the ones that give each vector $2\pi i e_{\mathbf{j}}$ ($|\mathbf{j}| = k$) length 1. Then everything is clear except the Griffiths transversality condition. But this condition is also satisfied because $\mathrm{d}C_{\mathbf{j}} = \boldsymbol{\omega} C_{\mathbf{j}}$ for every $\mathbf{j} \in \mathscr{I}_n$ by Prop. 2.5.8 since all the entries of $\boldsymbol{\omega}$ on and above the main diagonal vanish. □

If we want to determine the limit MHS of multiple logarithms explicitly we can still apply the techniques used in the normal crossing case. We will carry this out only for the depth two and three cases. The general picture is similar but much more complicated.

2.8.1 *Limit Mixed Hodge Structures of Double Polylogs*

First we look at the double logarithm variation of MHS. We have

$$\mathcal{M}_{1,1}(x,y)=\begin{bmatrix}1 & & & \\ \mathfrak{L}_1(x) & 2\pi i & & \\ \mathfrak{L}_1(xy) & 0 & 2\pi i & \\ \mathfrak{L}_2(x,y) & 2\pi i\mathfrak{L}_1(y) & 2\pi i h(x,y) & (2\pi i)^2\end{bmatrix}$$

where $h(x,y)=\mathfrak{L}_1(x)-\mathfrak{L}_1(y)-\log y$.
(i) Let us first try to extend the MHS to the divisor $\mathcal{D}_{20}=\{y=0\}$ along the tangent vector $\partial/\partial y$. We have

$$T_{\{y=0\}}=\begin{bmatrix}1 & & & \\ 0 & 1 & & \\ 0 & 0 & 1 & \\ 0 & 0 & -1 & 1\end{bmatrix},\quad N_{\{x=0\}}=\frac{\log T_{\{x=0\}}}{2\pi i}=\begin{bmatrix}0 & & & \\ 0 & 0 & & \\ 0 & 0 & 0 & \\ 0 & 0 & -\frac{1}{2\pi i} & 0\end{bmatrix}.$$

Let $\mathcal{M}_{1,1}(x,y)=[C_0(x,y)\ \cdots C_3(x,y)]$. Define

$$\begin{aligned}[s_0\ s_1\ s_2\ s_3]&=\lim_{t\to 0}\mathcal{M}_{1,1}(x,t)\begin{bmatrix}1 & & & \\ 0 & 1 & & \\ 0 & 0 & 1 & \\ 0 & 0 & \log t/(2\pi i) & 1\end{bmatrix}\\ &=\begin{bmatrix}1 & & & \\ \mathfrak{L}_1(y) & 2\pi i & & \\ 0 & 0 & 2\pi i & \\ 0 & 0 & 2\pi i\mathfrak{L}_1(x) & (2\pi i)^2\end{bmatrix}.\end{aligned}$$

Let $V_{\mathbb{Q},\{y=0\}}$ be the $\mathbb{Q}$-linear span of s_0,s_1,s_2,s_3, and $V_{\mathbb{C},\{y=0\}}=\mathbb{C}\otimes V_{\mathbb{Q},\{y=0\}}$. Let $\{e_j: j=0,\cdots,3\}$ be the standard basis of $\mathbb{C}^4$ where the only nonzero entry of e_j is at the $(j+1)$st component. Then the limit MHS on $\{(x,y): y=0, x\neq 1\}$ along $\partial/\partial y$ are given by

$$((V_{\mathbb{Q},\{y=0\}},W_\bullet),(V_{\mathbb{C},\{y=0\}},F^\bullet))$$

where for $k=0,\dots,3$

$$W_{-2k}V_{\mathbb{Q},\{x=0\}}=\langle s_k,\dots,s_3\rangle,\ W_{-2k}=W_{-2k+1}\tag{2.39}$$

and

$$F^{-k}V_{\mathbb{C},\{x=0\}}=\langle e_0,\dots,e_k\rangle.\tag{2.40}$$

(ii) A similar calculation shows that along the tangent vector $\partial/\partial y$ the limit MHS on the divisor $\mathcal{D}_{22} = \{(x,1) : x \neq 1\}$ is the $\mathbb{Q}$-linear span of $s_0, \ldots, s_3$ where

$$[s_0\ s_1\ s_2\ s_3] = \begin{bmatrix} 1 & & & \\ \mathfrak{L}_1(x) & 2\pi i & & \\ \mathfrak{L}_1(x) & 0 & 2\pi i & \\ \mathfrak{L}_2(x,1) & 0 & 2\pi i \mathfrak{L}_1(x) & (2\pi i)^2 \end{bmatrix}.$$

It is easy to see by differentiation that $\mathfrak{L}_2(x,1) = \big(\mathfrak{L}_1(x)\big)^2/2$.
(iii) The extension of MHS to $\mathcal{D}_{11} = \{(1,y) : y \neq 0,1\}$ along the tangent vector $\partial/\partial x$ is given by the $\mathbb{Q}$-linear span of $s_0, \ldots, s_3$ where

$$[s_0\ s_1\ s_2\ s_3] = \begin{bmatrix} 1 & & & \\ 0 & 2\pi i & & \\ -\operatorname{Li}_1(\frac{y}{y-1}) & 0 & 2\pi i & \\ \operatorname{Li}_2(\frac{y}{y-1}) & 2\pi i \operatorname{Li}_1(y) & -2\pi i \log \frac{y}{y-1} & (2\pi i)^2 \end{bmatrix}.$$

(iv) Limit MHS on $\mathcal{D}_{12} = \{(x, 1/x) : x \neq 0,1\}$ along the tangent vector $\partial/\partial y$ is given by the $\mathbb{Q}$-linear span of $s_0, \ldots, s_3$ where

$$[s_0\ s_1\ s_2\ s_3] = \begin{bmatrix} 1 & & & \\ -\operatorname{Li}_1(\frac{x}{x-1}) & 2\pi i & & \\ 0 & 0 & 2\pi i & \\ -\operatorname{Li}_2(\frac{x}{x-1}) & 2\pi i \log \frac{x}{x-1} & 0 & (2\pi i)^2 \end{bmatrix}.$$

(v) $\mathcal{D}_{20} \cap \mathcal{D}_{11} = (1,0)$. From (i) we see that there are limit MHS on the open set $\mathcal{D}_{20} \setminus \{(1,0)\}$ of $\mathcal{D}_{20}$. We now can easily extend these MHS to $(1,0)$ along the vector $\partial/\partial x$ and find the limit MHS at $(1,0)$ to be the $\mathbb{Q}$-linear span of $s_0, \cdots, s_3$ where

$$[s_0\ s_1\ s_2\ s_3] = \begin{bmatrix} 1 & & & \\ 0 & 2\pi i & & \\ 0 & 0 & 2\pi i & \\ 0 & 0 & 0 & (2\pi i)^2 \end{bmatrix}.$$

If we start from (iii) and then extend the MHS to $(1,0)$ along the tangent vector $\partial/\partial x$ we will get the same limit MHS.
(vi) $\mathcal{D}_{22} \cap \mathcal{D}_{12} = \mathcal{D}_{12} \cap \mathcal{D}_{11} = \mathcal{D}_{11} \cap \mathcal{D}_{22} = (1,1)$. We can start from either case (ii) or (iii) or (iv). Extending the limit MHS of case (ii) we see immediately that the along the tangent vector $\partial/\partial y$ the limit MHS at $(1,1)$ is given by the $\mathbb{Q}$-linear span of

$$[s_0\ s_1\ s_2\ s_3] = \begin{bmatrix} 1 & & & \\ 0 & 2\pi i & & \\ 0 & 0 & 2\pi i & \\ E_{4,1} & 0 & 0 & (2\pi i)^2 \end{bmatrix}. \tag{2.41}$$

If we extend the limit MHS of case (iii) to $(1,1)$ along the tangent vector $\partial/\partial x$ we find that only the lower left corner entry is different from the above. Instead of 0 it is given by

$$E_{4,1} = \lim_{y\to 1}\left(\mathrm{Li}_2\left(\frac{y}{y-1}\right) + \frac{\log^2(1-y)}{2} + \log y\,\mathrm{Li}_1(y)\right) = -\mathrm{Li}_2(1) = -\frac{\pi^2}{6},$$

since for all $t \neq 0$

$$\mathrm{Li}_2(1-t) + \mathrm{Li}_2(1-1/t) + \frac{1}{2}\log^2 t = 0. \tag{2.42}$$

But if we take $s_0' = s_0 - s_3/48$ we obtain the same basis as in Eq. (2.41). Exactly the same phenomenon occurs if we start from case (iv) and then use tangent vector $\partial/\partial y$.

If we extend the limit MHS of (iv) to the point $(1,1)$ along the tangent vector $\partial/\partial x$ then we find that

$$E_{4,1} = \lim_{x\to 1} -\mathrm{Li}_2\left(\frac{x}{x-1}\right) - \frac{1}{2}\log^2(1-x) = \mathrm{Li}_2(1) = \frac{\pi^2}{6}$$

by taking $t = 1-x$ in Eq. (2.42). Now if we let $s_0' = s_0 + \frac{1}{48}s_3$ then we get the same basis as in Eq. (2.41). This phenomenon happens in higher logarithm cases too.

In Exercise 2.11 we will see another example of limit MHS. These examples suggest the following conjecture.

Conjecture 2.8.2. *The variations of mixed Hodge–Tate structures associated with every MPL can be produced as the variations of the limit mixed Hodge–Tate structures of some multiple logarithm.*

2.9 Single-Valued MPLs of Multi-Variables

Using the monodromy of MPLs obtained from Sec. 2.5 we can construct the single-valued MPLs. Although this procedure can be carried out in general the computation in each particular case will involve very complicated notations. So we will treat only several lower weight cases in detail in order to illustrate the main idea.

2.9.1 *General Procedure of Construction*

In this subsection we first describe the general procedure to define the single-valued real analytic version of $\mathrm{Li}_{m_1,\dots,m_n}(\mathbf{x})$, denoted by $\mathscr{L}i_{s_1,\dots,s_n}(\mathbf{x})$, provided the variation matrix $\mathscr{M}_{s_1,\dots,s_n}(\mathbf{x})$ is already found.

We will work with the multiple logarithms below, but the idea is exactly the same for the general case.

Let $\mathcal{M}_{[n]} = \mathcal{M}_{[n]}(\mathbf{x}) = L_n(\mathbf{x})\tau_n(2\pi i)$ where

$$\tau_n(\lambda) = \mathrm{diag}\big[\lambda^{|\mathbf{j}|}\big]_{\mathbf{j}\in\mathcal{J}_n}.$$

Define the matrix

$$B_{[n]} = \tau_n(i)\mathcal{M}_{[n]}\overline{\mathcal{M}}_{[n]}^{-1}\tau_n(i)$$

where $\overline{\mathcal{M}}_{[n]}$ is the complex conjugation of $\mathcal{M}_{[n]}$. By the monodromy computation we see that $B_{[n]}$ is a single-valued matrix function defined over S_n because all the monodromy automorphisms are given by real matrices. Now that both $\mathcal{M}_{[n]}$ and $\overline{\mathcal{M}}_{[n]}^{-1}$ are lower triangular matrices, so is $B_{[n]}$. Moreover its diagonal is given by $\tau_n(i)\tau_n(2\pi i)\overline{\tau_n(2\pi i)}^{-1}\tau_n(i) = I$, the identity matrix. Hence we may assume $B_{[n]} = I + N$ where N a nilpotent matrix. We see that $\log B_{[n]}$ is well defined and since $\overline{B}_{[n]} = B_{[n]}^{-1}$ it satisfies

$$\overline{\log B_{[n]}} = -\log B_{[n]},$$

namely, $\log B_{[n]}$ is a pure imaginary matrix. Then we define $\mathcal{L}i_{\{1\}^n}(\mathbf{x})$ by multiplying $i^{2[n/2]-1}/2$ on the lower left corner entry of $\log B_{[n]}$, which is a single-valued real analytic version of the multiple logarithm $\mathrm{Li}_{\{1\}^n}(\mathbf{x})$.

Remark 2.9.1. Our method is slightly different from that of [44]. In fact, in the polylog case the matrix $B_{[n]}$ constructed as above is the conjugate of the one in [44] by $\tau(i)$.

2.9.2 *Single-Valued Double Logarithms*

We have seen from Example 2.5.5 that $\mathcal{M}_{1,1}(x,y) = L_{1,1}(x,y)\tau_{1,1}(2\pi i)$, where

$$L_{1,1}(x,y) = \begin{bmatrix} 1 & & & \\ \mathrm{Li}_1(x) & 1 & & \\ \mathrm{Li}_1(xy) & & 1 & \\ \mathrm{Li}_{1,1}(x,y) & \mathrm{Li}_1(y) & \log\frac{y-1}{y(1-x)} & 1 \end{bmatrix} \text{ and } \tau_{1,1}(\lambda) = \begin{bmatrix} 1 & & & \\ & \lambda & & \\ & & \lambda & \\ & & & \lambda^2 \end{bmatrix}.$$

Let $B_{1,1}(x,y) = \tau_{1,1}(i)L_{1,1}(x,y)\tau_{1,1}(-1)\overline{L_{1,1}(x,y)}^{-1}\tau_{1,1}(i)$. Then $B_{1,1}(x,y)$ is unipotent and single-valued. An easy calculation shows:

$$\log B_{1,1}(x,y) = \begin{bmatrix} 0 & & & \\ -2i\log|1-x| & 0 & & \\ -2i\log|1-xy| & & 0 & \\ -2i\,\mathcal{L}i_{1,1}(x,y) & -2i\log|1-y| & 2i\log\left|\frac{y-1}{y(1-x)}\right| & 0 \end{bmatrix}$$

where

$$\mathscr{L}i_{1,1}(x,y) = \operatorname{Im}\big(\operatorname{Li}_{1,1}(x,y)\big) + \arg(1-xy)\log\left|\frac{y-1}{y(1-x)}\right| \\ - \arg(1-x)\log|1-y| \quad (2.43)$$

is a single-valued real analytic version of $\operatorname{Li}_{1,1}(x,y)$.

By differentiation it is easy to check that (see Exercise 2.13)

$$\operatorname{Li}_{1,1}(x,y) = \operatorname{Li}_2\Big(\frac{xy-x}{1-x}\Big) - \operatorname{Li}_2\Big(\frac{x}{x-1}\Big) - \operatorname{Li}_2(xy). \quad (2.44)$$

So by using the single-valued Bloch–Wigner dilog function

$$\mathscr{L}i_2(z) = \operatorname{Im}\big(\operatorname{Li}_2(z)\big) + \arg(1-z)\log|z| \quad (2.45)$$

we can also recover Eq. (2.43) as

$$\mathscr{L}i_{1,1}(x,y) = \mathscr{L}i_2\Big(\frac{xy-x}{1-x}\Big) - \mathscr{L}i_2\Big(\frac{x}{x-1}\Big) - \mathscr{L}i_2(xy). \quad (2.46)$$

Further, using the five term relation we find

$$\mathscr{L}i_{1,1}(x,y) + \mathscr{L}i_{1,1}(y,x) + \mathscr{L}i_2(xy) = 0.$$

One can check quickly that $\mathscr{L}i_{1,1}$ satisfies the functional equation

$$\mathscr{L}i_{1,1}(x,y) = -\mathscr{L}i_{1,1}\Big(\frac{x}{x-1}, 1-y\Big)$$

by using the functional equations $\mathscr{L}i_2(x) = -\mathscr{L}i_2(1-x) = -\mathscr{L}i_2(1/x)$.

2.9.3 Single-Valued Double Polylogs $\mathscr{L}i_{1,2}$ and $\mathscr{L}i_{2,1}$

By [595] a single-valued version of $\operatorname{Li}_3(x)$ can be defined as

$$\mathscr{L}i_3(z) = \operatorname{Re}\big(\operatorname{Li}_3(z)\big) - \log|z|\operatorname{Re}\big(\operatorname{Li}_2(z)\big) - \frac{1}{3}(\log|z|)^2\log|1-z|. \quad (2.47)$$

We now consider $\operatorname{Li}_{2,1}(x,y)$ and $\operatorname{Li}_{1,2}(x,y)$. By the procedure outlined in Sec. 2.9.1 we find that a single-valued version of $\operatorname{Li}_{2,1}(x,y)$ is (see Exercise 2.9):

$$\begin{aligned}\mathscr{L}i_{2,1}(x,y) &= \operatorname{Re}\operatorname{Li}_{2,1}(x,y) - \arg(1-xy)\big[\mathscr{L}i_2(x) + \mathscr{L}i_2(y)\big] \\ &\quad + \log|1-y|\operatorname{Re}\operatorname{Li}_2(x) - \log|x|\operatorname{Re}\operatorname{Li}_{1,1}(x,y) - \log\left|\frac{1-y}{y}\right|\operatorname{Re}\operatorname{Li}_2(xy) \\ &\quad + \frac{1}{3}\log|x|\Big[2\log|1-x|\log|1-y| + \log|1-xy|\log|y(1-x)|\Big] \\ &\quad - \frac{1}{3}\log|x^2y|\log|1-xy|\log\left|\frac{1-y}{y}\right|.\end{aligned}$$

A single-valued version of $\mathrm{Li}_{1,2}(x,y)$ is (see Exercise 2.10):

$$\begin{aligned}\mathscr{L}i_{1,2}(x,y) &= \mathrm{Re}\,\mathrm{Li}_{1,2}(x,y) + \arg(1-xy)\big[\mathscr{L}i_2(x)+\mathscr{L}i_2(y)\big]\\ &\quad - \arg(1-x)\,\mathscr{L}i_2(y) + \log|1-x|\,\mathrm{Re}\,\mathrm{Li}_2(xy) - \log|y|\,\mathrm{Re}\,\mathrm{Li}_{1,1}(x,y)\\ &\quad + \frac{1}{3}\log|y|\Big[\log|1-x|\log|1-y| + \log|1-xy|\log\Big|\frac{y(1-x)}{1-y}\Big|\Big]\\ &\quad + \frac{1}{3}\log|1-x|\log|xy|\log|1-xy|.\end{aligned}$$

Using the single-valued versions of and trilog $\mathscr{L}i_3(z)$ given by Eq. (2.47) we can express $\mathscr{L}i_{1,2}(x,y)$ as

$$\begin{aligned}\mathscr{L}i_{1,2}(x,y) &= \mathscr{L}i_3\Big(\frac{y-xy}{1-xy}\Big) - \mathscr{L}i_3\Big(\frac{1-x}{1-xy}\Big)\\ &\quad + \mathscr{L}i_3(1-xy) - \mathscr{L}i_3(y) + \mathscr{L}i_3(1-x) - \mathscr{L}i_3(1).\end{aligned}$$

This follows from the relation first discovered by Zagier after Goncharov's conviction that such an identity should exist:

$$\begin{aligned}\mathrm{Li}_{1,2}(x,y) &= \mathrm{Li}_3\Big(\frac{y-xy}{1-xy}\Big) - \mathrm{Li}_3\Big(\frac{1-x}{1-xy}\Big) - \log\Big(\frac{1-x}{1-xy}\Big)\mathrm{Li}_2(y)\\ &\quad + \mathrm{Li}_3(1-xy) - \mathrm{Li}_3(y) + \mathrm{Li}_3(1-x) - \mathrm{Li}_3(1)\\ &\quad - \log(1-xy)\big(\mathrm{Li}_2(1)+\mathrm{Li}_2(1-x)\big) + \frac{1}{2}\log(y)\log^2(1-xy).\end{aligned} \tag{2.48}$$

The geometric origin of this identity can be found in [607, Remark 3.18]. By straightforward computation one can find the following interesting formula:

$$\mathscr{L}i_{1,2}(x,y) + \mathscr{L}i_{2,1}(y,x) + \mathscr{L}i_3(xy) = 2\log|y|\,\mathrm{Re}(\mathrm{Li}_1(y)\,\mathrm{Li}_1(x)).$$

One should compare this with

$$\mathrm{Li}_{1,2}(x,y) + \mathrm{Li}_{2,1}(y,x) + \mathrm{Li}_3(xy) = \mathrm{Li}_2(y)\,\mathrm{Li}_1(x).$$

Finally, one can also express the triple logarithm $\mathscr{L}i_{1,1,1}(x,y,z)$ by using the trilog $\mathscr{L}i_3(z)$ (see Exercise 2.14). Thus in weight three we essentially need only $\mathscr{L}i_3(z)$. But it may be a little surprising that similar identities in higher weight cases do not exist in general from geometric considerations (see [607, p. 146]). For example, $\mathscr{L}i_{2,2}(x,y)$ cannot be expressed by only tetralogarithm $\mathscr{L}i_4$.

2.10 Aomoto Polylogs

Usually it is much easier to grasp a concept when geometric figures are attached to it. This is certainly true when we study MPLs. Quite a few deep results have been obtained up to weight three especially because in these lower weights we can draw pictures and visualize/discover various relations between them. Moving up to higher dimensions it is natural to consider so-called Aomoto polylogs.

Definition 2.10.1. Let F be an arbitrary field. Fix a positive integer n. By a *simplex* in the projective spaces $\mathbb{P}^n_F$ we mean an ordered set of hyperplanes $L = (L_0, \dots, L_n)$. It is *nondegenerate* if the intersection of all the hyperplanes L_i is empty. A *face* of L is any nonempty intersection of the hyperplanes. A pair of simplices is *admissible* if they do not have any common faces of the same dimension. It is a *generic pair* if all the faces of the two simplices are in the general positions.

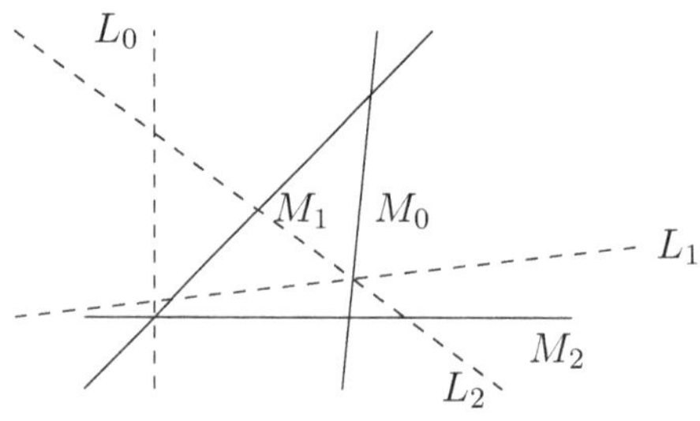

Figure 2.1: An admissible non-generic pair of triangles.

By projective invariance to be defined below, given a nondegenerate simplex L we may choose the coordinate system $[t_0, \dots, t_n]$ in $\mathbb{P}^n_F$ such that $L_i = \{t_i = 0\}$ for $0 \le i \le n$. If $F = \mathbb{C}$ then we define the canonical differential form associated with L as

$$\omega_L = \frac{\mathrm{d}t_1}{t_1} \wedge \frac{\mathrm{d}t_2}{t_2} \wedge \dots \wedge \frac{\mathrm{d}t_n}{t_n}.$$

The Aomoto n-logarithm is a multi-valued function on the configurations of nondegenerate admissible pairs of simplices $(L; M)$ in $\mathbb{CP}^n$ defined as follows:

$$\Lambda_n(L; M) = \int_{\Delta_M} \omega_L,$$

where Δ_M is the n-cycle representing a generator of the relative homology $H_n(\mathbb{CP}^n, \cup_{j=0}^n M_j; \mathbb{Z})$.

We now define the *double scissors congruence groups* $A_n(F)$ whose defining relations should reflect (conjecturally all of) the functional equations of Aomoto polylogs. Furthermore, the associated graded object should form a Hopf algebra which provides a bridge to the study of motivic polylogic complexes.

Definition 2.10.2. Define $A_0(F) = \mathbb{Z}$. If $n > 0$ then $A_n(F)$ is the abelian group generated by admissible pairs of n-simplices $(L;\, M)$ subject to the following relations:

(R1) *Nondegeneracy.* $(L;\, M) = 0$ if and only if L or M is degenerate.

(R2) *Trivial intersection.* Suppose $L_0, \ldots, L_n$ and $M_0, \ldots, M_n$ are hyperplanes in $\mathbb{P}_F^{n+1}$. If $N = L_i$ or $N = M_i$ for some i then

$$(N|L;\, M) := \big((L_0 \cap N, \ldots, L_n \cap N);\, (M_0 \cap N \ldots, M_n \cap N)\big) = 0.$$

(R3) *Skew symmetry.* For every permutation σ of $\{0, \ldots, n\}$,

$$(\sigma L;\, M) = (L; \sigma M) = \operatorname{sgn}(\sigma)(L;\, M),$$

where $\sigma(L_0, \ldots, L_n) = (L_{\sigma(0)}, \ldots, L_{\sigma(n)})$.

(R4) *Additivity in L and M.* For any hyperplanes $L_0, \ldots, L_{n+1}$ and n-simplex M in $\mathbb{P}_F^n$,

$$\sum_{j=0}^{n+1} (-1)^j ((L_0, \ldots, \widehat{L_j}, \ldots, L_{n+1});\, M) = 0$$

if every pair $((L_0, \ldots, \widehat{L_j}, \ldots, L_{n+1});\, M)$ is admissible. A similar relation is satisfied by any n-simplex L and hyperplanes $M_0, \ldots, M_{n+1}$.

(R5) *Projective invariance.* For every $g \in \mathrm{PGL}_{n+1}(F)$,

$$(gL;\, gM) = (L;\, M).$$

Remark 2.10.3. The additivity relation (R4) can be visualized in dimension two by the following picture:

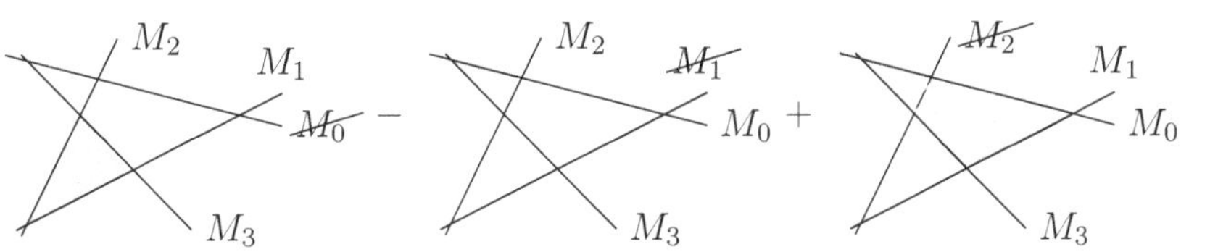

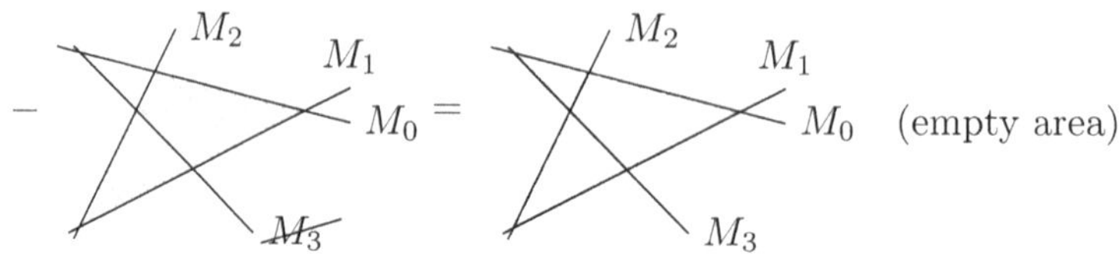

Example 2.10.4. We now consider the simplest case: $n = 1$. By projective invariance (R5) we may assume without loss of generality that $L = (\infty, 0)$ and $M = (1, a)$ for some $a \neq 0, 1, \infty$. Then the first de Rham cohomology $H^1_{\text{dR}}(\mathbb{CP}^1 \setminus L) = H^1(\mathbb{C}^*)$ is generated by $[dz/z]$ while the relative singular homology $H_1(\mathbb{CP}^1, M)$ is generated by $[\Delta_M]$ where Δ_M is the line segment from 1 to a. So we get the period (see [361])

$$\int_1^a \frac{\mathrm{d}z}{z} = \log a.$$

Example 2.10.5. Let $n = 2$. Take $L_1 = \{x = 0\}$, $L_2 = \{y = 0\}$ and $L_0 = \{z = 0\}$. Let $M_0 = \{x = z\}$, $M_1 = \{y = tz\}$, and $M_2 = \{x + y = z\}$. We have the following picture in Fig. 2.2. Hence the period is given by

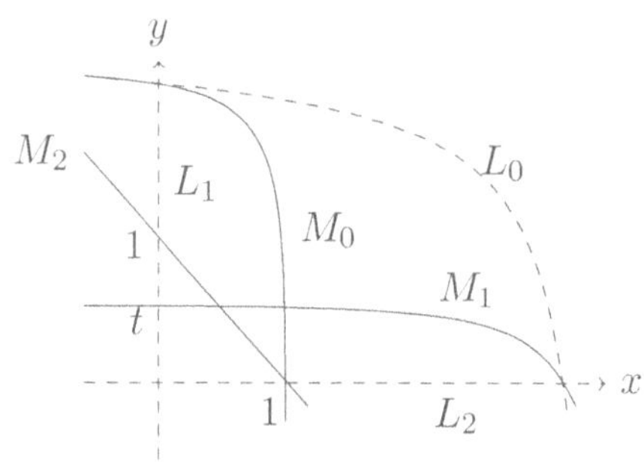

Figure 2.2: A real picture for dilog.

$$\int_{0<1-x<y<t} \frac{\mathrm{d}x}{x}\frac{\mathrm{d}y}{y} = -\int_{0<x<y<t} \frac{\mathrm{d}y}{y}\frac{\mathrm{d}x}{1-x} = -\operatorname{Li}_2(t).$$

Definition 2.10.6. The n-simplex L whose faces are $L_i = \{t_i = 0\}$ $(0 \leq i \leq n)$ is called the *standard (coordinate) simplex*. For an arbitrary field F, we denote by $\Lambda_{n_1,\dots,n_d}$ the subgroup of A_n $(n = n_1 + \dots + n_d)$ generated by all *MPL pairs* $\Lambda_{n_1,\dots,n_d}(x_1, \dots, x_d)$ corresponding to the MPLs (all possible $x_1, \dots, x_d$). Specifically, we can represent $\Lambda_{n_1,\dots,n_d}(x_1, \dots, x_d)$ by $(L; M)$

where L is the standard simplex in $\mathbb{P}^n_F$ and M is determined as follows: Take $a_j = 1/(x_1 \dots x_j)$ and the vertex facing M_0 to be

$$m_0 = [z_0, \dots, z_n] = [\underbrace{0, \dots, 0}_{n_1-1 \text{ times}}, -a_1, \underbrace{0, \dots, 0}_{n_2-1 \text{ times}}, -a_2, \dots, \underbrace{0, \dots, 0}_{n_d-1 \text{ times}}, -a_d, 1].$$

The vertex m_j $(1 \le j \le n)$ facing M_j is given by

$$m_i = [z_1 + 1, \dots, z_j + 1, z_{j+1}, \dots, z_n, 1].$$

If $F = \mathbb{C}$ and $M = \Lambda_{n_1,\dots,n_d}(x_1, \dots, x_d)$ a real simplex then the integral

$$\int_{\Delta_M} \frac{\mathrm{d}t_1}{t_1} \wedge \cdots \wedge \frac{\mathrm{d}t_n}{t_n} = (-1)^d \operatorname{Li}_{n_1,\dots,n_d}(x_1, \dots, x_d), \tag{2.49}$$

where Δ_M is the convex real span by the faces of M. This is the origin of the name *MPL pair.*

The key to the study of the double scissors congruence groups A_n is to equip the amalgamation $A_\bullet = \bigoplus_{n\ge 0} A_n$ with a suitable Hopf algebra structure. In the lower weight cases (≤ 4) this can be done explicitly using some geometric combinatorial argument (see [607,610]) but in general this is still unclear. The product μ is easy to define but the coproduct is highly nontrivial. On the generic part the coproduct is defined in [606] (see Exercise2.17). If this can be done on A_n for arbitrary n then we have the following surprising conjecture.

Conjecture 2.10.7. *The restricted coproduct provides a complex*

$$A_{>0} \longrightarrow A_{>0} \otimes A_{>0} \longrightarrow A_{>0} \otimes A_{>0} \otimes A_{>0} \longrightarrow \cdots$$

whose graded n-piece

$$A_n \longrightarrow \bigoplus_{k=1}^{n-1} A_k \otimes A_{n-k} \longrightarrow \cdots$$

gives

$$H^i_{(n)}(A_\bullet, \mathbb{Q}) \cong \operatorname{gr}^\gamma_n K_{2n-i}(F)_\mathbb{Q}$$

where $\operatorname{gr}^\gamma_n K_{2n-i}(F)_\mathbb{Q}$ *is the* γ*-filtration of the* K*-groups with* $\mathbb{Q}$*-coefficients.*

In fact, according to the Tannakian formalism the category $\mathsf{MTM}(F)$ of mixed Tate motives over a field F is supposed to be equivalent to the category of graded modules over a certain graded commutative Hopf algebra $\mathcal{A}_\bullet$ (see [46] and [248, Chap. 3]). Therefore the Ext groups in the category $\mathsf{MTM}(F)$ of mixed Tate motives over $\operatorname{Spec}(F)$ are isomorphic to the cohomology groups of the Hopf algebra $\mathcal{A}_\bullet$. Beilinson *et al.* conjectured that $A_\bullet$ is isomorphic to $\mathscr{A}_\bullet$ and therefore the groups $A_n \otimes \mathbb{Q}$ should have a Hopf algebra structure over $\mathbb{Q}$. This is the primary motivation to study the groups A_n. Further, by Beilinson's conjecture, there is a negatively graded Lie algebra $L_\bullet(F)$ over $\mathbb{Q}$ such that

$$\mathrm{Ext}^i_{\mathsf{MTM}(F)}(\mathbb{Q}(0), \mathbb{Q}(n)) \cong H^i_{(n)}(L_\bullet(F)) \cong \mathrm{gr}^n_\gamma K_{2n-i}(F) \otimes \mathbb{Q}.$$

This is why we have Conjecture 2.10.7.

One hopes to construct the motivic Lie algebra $L_\bullet(F)$ explicitly, at least when the weight is not too large. To this end, let $\mathscr{B}_n(F) = \mathbb{Z}[\mathbb{P}^1_F]/\mathscr{R}_n(F)$ where the subgroup $\mathscr{R}_n(F)$ reflects (conjecturally all of) the functional equations of the single-valued polylog $\mathscr{L}i_n(z)$ defined by Eq. (2.3). A lot of evidence in [244] shows that $\mathscr{B}_n(F) \otimes \mathbb{Q}$ (for $n = 1, 2, 3$) is dual to the motivic Lie algebra $L_{-n}(F)$ (the dual is between the ind and pro $\mathbb{Q}$-vector spaces). It follows from this line of thought that the following conjecture should be true (see [245]).

Conjecture 2.10.8. *Let* $\Pi_n = \bigoplus_{j=1}^{n-1} \mu(A_j \otimes A_{n-j})$ *be the subgroup of prisms of* A_n *then*

$$(A_n/\Pi_n) \otimes \mathbb{Q} \cong L_{-n}(F)^\vee$$

and for $n = 1, 2, 3$

$$(A_n/\Pi_n) \otimes \mathbb{Q} \cong L_{-n}(F)^\vee \cong \mathscr{B}_n(F) \otimes \mathbb{Q}.$$

The first isomorphism means that the dual to the Hopf algebra $A_\bullet \otimes \mathbb{Q}$ is isomorphic to the universal enveloping algebra of the Lie algebra $L_\bullet(F)$. This conjecture is trivial when $n = 1$ since all are isomorphic to $F^\times \otimes \mathbb{Q}$ by the cross ratio (see Exercise 2.3). For $n = 2$, the first isomorphism is settled in [47] by taking $L_{-2}(F)^\vee$ to be the Bloch group $B_2(F)$ which is the quotient group of $\mathbb{Z}[\mathbb{P}^1_F]$ modulo the five term relations.[1] It is known that for number fields $\mathscr{B}_2(F) \cong B_2(F)$ modulo torsion (see [45, 47]).

Similar to $B_2(F)$ the group $B_3(F)$ is defined by Goncharov [245] as the quotient of $\mathbb{Z}[\mathbb{P}^1_F]$ by the subgroup generated by the generic seven-term relations of some generalized cross ratio r_3 and the Kummer–Spence relations (see [607, Sec. 2.4] for details). The main result of [607] is that modulo torsion one has

$$A_3(F)/\Pi_3(F) \cong B_3(F).$$

[1]In the literature the Bloch group sometimes refers to the subgroup of $B_2(F)$ which is isomorphic to the indecomposable part of $K_3(F)$ modulo torsions.

2.11 Historical Notes

It is well known that the logarithm was first invented by John Napier in 1614. Although the original intention was to simplify the multiplication and the division to the level of addition and subtraction, logarithms have found many other applications in modern sciences.

From the early days of the modern physics people knew logarithms and $\pi^2 = 6\zeta(2)$ are indispensable for computations in quantum electrodynamics (see, for example, [276, 463, 507]). The trilogs also found its way in perturbative quantum field theory in the late 1940s (see [339, 340][2]). Polylogarithms of larger weights emerged gradually throughout the later half of twentieth century, although they sometimes appeared in disguised forms (see, for example, [37,325,528]). The expressions are often very complicated even for very simple diagrams. For example, on p. 39 of [187] Eden *et al.* pointed out that there are 192 dilog terms in Wu's calculation (see [579]) on a certain fourth order single loop with all scalar particles. It was discovered later that the *multiple* zeta values and the Euler sums appear in the counterterms of quantum field theory in higher loop electron magnetic moments (see [90, 376]). In recent years, MPLs are linked to the study of conformal field theory [214,347], amplitudes [96, 177, 253, 546] and Feynman graphs [64, 101, 108, 403, 505] in physics as well as Hopf and Lie algebras, combinatorics (double shuffle relations) [284,286], algebraic geometry [230,232,244,245,365,524,607,608,614], hyperbolic volumes [440,593], modular forms [233], Mahler measures [372] and cluster algebras [213, 242] in mathematics.

The first serious study of classical polylogs seemed to be carried out by Jonquière [321]. Starting from 1950s Leonard Lewin, an engineer, systematically studied polylogs by generalizing a lot results for the dilog in [391,392]. Apparently he chose the notation *Li* because of "polylogarithm integral". Later on, with his collaborators, he found a few very interesting ladder relations such as the following of order 12 [5]: let $\phi = (\sqrt{5}-1)/2$ then

$$\frac{\mathrm{Li}_n(\phi^{12})}{12^{n-1}} - \frac{3}{2}\frac{\mathrm{Li}_n(\phi^6)}{6^{n-1}} - \frac{\mathrm{Li}_n(\phi^4)}{4^{n-1}} + \frac{11}{48}\frac{\mathrm{Li}_n(\phi^2)}{2^{n-1}} = \frac{13}{48}\frac{\log^n \phi}{n!} - \frac{\zeta(2)}{48}\frac{\log^{n-2}\phi}{(n-2)!}$$
$$+ \frac{19\zeta(4)}{1728}\frac{\log^{n-4}\phi}{(n-4)!} + \frac{67\zeta(5)}{6912}\frac{\log^{n-5}\phi}{(n-5)!}, \quad 1 \le n \le 5.$$

A few higher order relations have been discovered numerically since then

[2] Unfortunately the paper contains an error which was corrected only around 1957 in [470, 520].

(see, for example, [33, 144]). For weight 6 and 7, Gangl [229] obtains the first proven ladder relations.

In a series of papers spanning over twenty years K.-T. Chen developed his far-reaching theory of the iterated integrals in [131–133] which was unfortunately overlooked for several decades. Since the 1990s, especially because of its applications in the multiple zeta functions and MPLs, its importance has been gradually recognized by more and more mathematicians. Richard Hain, one of K.-T. Chen's Ph.D. students, has played a pivotal role in this process. His survey article [267] and lecture notes [268] are good sources to learn Chen's theory.

During his study of monodromy properties of $\mathrm{Li}_n(x)$ in [479] Ramakrishnan implicitly came upon the functions defined in Eq. (2.2) (written down explicitly by Zagier [594] and by Wojtkowiak [574] in a modified form). When $n = 2$ this coincides with the Bloch–Wigner dilog which was brought to prominence in Bloch's works [53, 55]. Note that the function defined by Eq. (2.3) also generalizes the Bloch–Wigner function.

Around year 1991/92 Herbert Gangl investigated various functional equations of polylogs $\mathrm{Li}_n(x)$ for $n \leq 7$ (see [227,228,442]). Although there is a lot of information available for these functional equations when n is small (see [575]) the most general functional equation of $\mathrm{Li}_n(x)$ is still a mystery whose complete solution will provide key information in the construction of some motivic complexes in [244, 245].

In [230] Gangl and Müller-Stach discovered an algebraic cycle version of the five-term relation Eq. (2.5) satisfied by the dilog in some higher Chow group. They also found a certain Kummer relation in the algebraic cycle-version for the trilog. This has been enhanced to Goncharov's 22-term relation for the trilog Eq. (2.6) by the author [614].

We should point out that in the literature, there is an alternative approach to the polylogs, treating both α and s in $\mathrm{Li}_\alpha(s)$ as complex variables, for example, see [152, 321, 474].

Turning to the MPLs, Kummer, Poincaré, Lappo-Danilevsky (see [366, 377, 378, 475], respectively) and K.-T. Chen all examined them in various disguised forms from different points of view ranging from the differential equations to the iterated integrals. In [281, 283], Hoang Ngoc also presented certain parametrized Dirichlet functions in the iterated integral form of which the MPLs are special cases. In the mid 1990s, Goncharov utilized the MPLs to solve successfully some difficult problems in number theory and algebraic geometry. Sometimes, the term "hyperlogarithm" (for e.g., see [346]) is used instead of MPL, but this is not standard since for other people (see [475]) hyperlogarithm means a different function. Moreover, physicists often refer the MPLs as Goncharov's polylogs. In [72], a certain variant of the MPL is also called a multidimensional polylog. Further, Zlobin [640] considered some special values of MPLs, too.

Back in the early 1980s Deligne noticed that the dilog function Li_2 gives rise to a good unipotent variation of mixed Hodge–Tate structures (see [163]). In his survey paper [267] Hain gave a detailed account of the polylog local systems as variations of mixed Hodge structure building on Ramakrishnan's explicit computation of the monodromy of polylogs in [479]. The monodromy computation also yields the single-valued variant $\mathscr{L}i_n(z)$ of the polylogs (see [44, 54]). These functions in turn have significant applications in arithmetic algebraic geometry such as Zagier's Conjecture 2.2.1. On the other hand, as pointed out by Goncharov [247], "higher cyclotomy theory" should study the MPL motives at roots of unity, not only those of the polylogs. This is one of the reasons why it is crucial to consider the variations of mixed Hodge structures associated with the MPLs.

The idea in Sec. 2.9 follows closely that of Beilinson and Deligne's paper [44] as given in [54]. By this approach one can easily construct the single-valued version of any MPL as long as one can determine explicitly its associated matrix of variations.

In order to understand the motivic origin of MPLs it is beneficial to work with Aomoto polylogs defined first in [17] which are geometrically more intuitive. Our treatment here is quite superficial and avoids a lot of technicalities but hopefully still gives the reader a glance of what can possibly be done with this powerful concept. In particular, the double scissors congruence groups are first introduced in [46] and then elaborated upon by Beilinson *et al.* in the two companion papers [45, 47]. They were able to come up with the weight two motivic complex modulo torsion using the geometric combinatorial properties satisfied by the double scissors congruence group of pairs of triangles in the projective plane. This has been generalized to dimension three in [607]. In the weight four case, Gangl [227, 228] has made some important progress recently building on his previous results. Note that in their original definition, Beilinson *et al.* imposed another type of relations called *Intersection Additivity*. However, in [606] we demonstrated that these relations follow from the others. Finally, Dupont elucidated the coproduct structure of double scissors congruence groups in [178].

There are several other types of MPLs. To study so-called multiple zeta star values (see Chap. 10) it is useful to consider the following variation of one variable MPLs. Let $\mathbf{s} = (s_1, \dots, s_d) \in \mathbb{N}^d$ and define

$$\mathrm{Le}_{\mathbf{s}}(z) = \sum_{k_1 \geq \cdots \geq k_d \geq 1} \frac{z^{k_1}}{k_1^{s_1} \cdots k_d^{s_d}}, \qquad |z| < 1. \tag{2.50}$$

In Chap. 4 we will construct the single-variable multiple polylogs (called "generalized polylogs" in [553]). In the case of the classical polylogs, it is possible to lift it to a two variable function: for $s, x \in \mathbb{C}$ with $|x| < 1$

$$\mathrm{Li}_s(x) = \sum_{k=1}^{\infty} \frac{x^k}{k^s}.$$

Clearly when $x \to 1$ one recovers the Riemann zeta function. One can also consider this function at general s, or at negative integers or even half integers (see [152, 321, 548]). In fact, we will need this version later (see Definition 12.8.4). There are several other analogs such as the infinitesimal polylogs [121], the p-adic polylogs [48,576,577], the Grassmannian polylogs [252], and Kontsevich's finite polylogs in [48, 360] which are related to the classical polylogs via a certain "derivation map". See [196] for details. In a series of papers [369–371] Lagarias and Li considered the Lerch zeta functions and discovered that for positive integer m the Lerch transcendent $\sum_{n\geq 0} z^n/(n+c)^m$ is a one-parameter deformation of $\mathrm{Li}_m(z)$ depending on $c \in \mathbb{C}$, whose associated $(m+1)$-dimensional monodromy representation is a non-algebraic deformation of the monodromy of $\mathrm{Li}_m(z)$. For general treatment of MPLs from the algebro-geometric point of view, see [44,45,47, 230–232, 246, 249, 252, 569, 572, 607, 610, 611]. For general surveys of earlier results of polylogs, see [267, 391–393, 442]. For more recent accounts up to year 2002, see [72, 76, 117, 564, 608].

Exercises

2.1. Verify Eq. (2.4) by differentiation.

2.2. Show that the infinite series

$$\sum_{n \geq 2} \left(\frac{\log^k(n-1)}{n-1} - \frac{\log^k(n)}{n} \right)$$

converges for all $k \in \mathbb{N}_0$.

2.3. Define the cross-ratio

$$r(a, b, c, d) = \frac{(a-c)(b-d)}{(a-d)(b-c)}$$

for all distinct $a, b, c, d \in \mathbb{CP}^1$. Show that for any five distinct points $x_0, \ldots, x_4 \in \mathbb{CP}^1$ we have the five term relation of dilog

$$\sum_{i=0}^{4} (-1)^i \mathscr{L}i_2(r(x_0, \ldots, \widehat{x}_i, \ldots, x_4)) = 0,$$

where $\widehat{x}_i$ means x_i is removed.

2.4. Prove Eq. (2.8) by geometric series expansion.

2.5. Derive Eq. (2.17) from Eq. (2.16) by breaking it into two sums and making an index substitution.

2.6. Prove Thm. 2.4.2 by induction on n using Lemma 2.4.1.

2.7. Derive Eq. (2.24) from Eq. (2.23) in Prop. 2.5.3, using substitutions and the integral expression of MPLs given by Eq. (2.8).

2.8. Verify the variation matrix for the double logarithm in Example 2.5.5 by Definition 2.5.2.

2.9. Check that the variation matrix $\mathscr{M}_{2,1}(x, y)$ associated with $\mathrm{Li}_{2,1}(x, y)$ over S_2 is

$$\begin{bmatrix} 1 & 0 & 0 & 0 & 0 & 0 & 0 \\ \mathrm{Li}_1(x) & 1 & 0 & 0 & 0 & 0 & 0 \\ \mathrm{Li}_1(xy) & 0 & 1 & 0 & 0 & 0 & 0 \\ \mathrm{Li}_2(x) & \log(x) & 0 & 1 & 0 & 0 & 0 \\ \mathrm{Li}_{1,1}(x,y) & \mathrm{Li}_1(y) & \mathrm{Li}_1\left(\frac{1-xy}{1-y}\right) & 0 & 1 & 0 & 0 \\ \mathrm{Li}_2(xy) & 0 & \log(xy) & 0 & 0 & 1 & 0 \\ \mathrm{Li}_{2,1}(x,y) & \mathrm{Li}_1(y)\log(x) & g(x,y) & \mathrm{Li}_1(y) & \log x & -\mathrm{Li}_1(y^{-1}) & 1 \end{bmatrix} \tau_{2,1}(2\pi i)$$

where

$$g(x, y) = -\int_{a_2}^{1} \frac{dt}{t} \frac{dt}{t - a_1} = \mathrm{Li}_2(x) - \mathrm{Li}_2(y^{-1}) - \log(xy)\,\mathrm{Li}_1(y^{-1}).$$

The columns of $\mathscr{M}_{2,1}(x, y)$ form the fundamental solutions of the differential equation over S_2

$$d\lambda = \begin{bmatrix} 0 & 0 & 0 & 0 & 0 & 0 \\ d\,\mathrm{Li}_1(x) & 0 & 0 & 0 & 0 & 0 \\ d\,\mathrm{Li}_1(xy) & 0 & 0 & 0 & 0 & 0 \\ 0 & d\,\mathrm{Li}_1(y) & d\,\mathrm{Li}_1\left(\frac{1-xy}{1-y}\right) & 0 & 0 & 0 \\ 0 & d\log(x) & 0 & 0 & 0 & 0 \\ 0 & 0 & d\log(xy) & 0 & 0 & 0 \\ 0 & 0 & 0 & d\log(x) & d\,\mathrm{Li}_1(y) & -d\,\mathrm{Li}_1(y^{-1}) & 0 \end{bmatrix} \lambda$$

Describe the weight and Hodge filtration in detail.

2.10. Check that the variation matrix $\mathcal{M}_{1,2}(x, y)$ associated with $\mathrm{Li}_{1,2}(x, y)$ over S_2 is

$$\begin{bmatrix} 1 & 0 & 0 & 0 & 0 & 0 \\ \mathrm{Li}_1(x) & 1 & 0 & 0 & 0 & 0 \\ \mathrm{Li}_1(xy) & 0 & 1 & 0 & 0 & 0 \\ \mathrm{Li}_{1,1}(x,y) & \mathrm{Li}_1(y) & \mathrm{Li}_1\left(\frac{1-xy}{1-y}\right) & 1 & 0 & 0 \\ \mathrm{Li}_2(xy) & 0 & \log(xy) & 0 & 1 & 0 \\ \mathrm{Li}_{1,2}(x,y) & \mathrm{Li}_2(y) & f(x,y) & \log y & \mathrm{Li}_1(x) & 1 \end{bmatrix} \tau_{1,2}(2\pi i)$$

where

$$f(x, y) = -\int_{a_2}^{1} \frac{dt}{t-a_1}\frac{dt}{t} = \mathrm{Li}_2(y^{-1}) - \mathrm{Li}_2(x) + \log(xy)\,\mathrm{Li}_1(x).$$

The columns of $\mathcal{M}_{1,2}(x, y)$ form the fundamental solutions of the differential equation over S_2

$$d\lambda = \begin{bmatrix} 0 & 0 & 0 & 0 & 0 & 0 \\ d\,\mathrm{Li}_1(x) & 0 & 0 & 0 & 0 & 0 \\ d\,\mathrm{Li}_1(xy) & 0 & 0 & 0 & 0 & 0 \\ 0 & d\,\mathrm{Li}_1(y) & d\,\mathrm{Li}_1\left(\frac{1-xy}{1-y}\right) & 0 & 0 & 0 \\ 0 & 0 & d\log(xy) & 0 & 0 & 0 \\ 0 & 0 & 0 & d\log y & d\,\mathrm{Li}_1(x) & 0 \end{bmatrix} \lambda$$

Describe the weight and Hodge filtration in detail.

2.11. Compute the limit MHS of the triple logarithm $\mathfrak{L}_3(x, y, z)$ on $\mathcal{D}_{11} = \{x = 1\}$ over $\{(1, y, z) : yz(1-y)(1-z)(1-yz) \neq 0\}$ along the vector $\partial/\partial x$. Show that the variation matrix is essentially $\mathcal{M}_{2,1}\left(\frac{y}{y-1}, \frac{z(y-1)}{yz-1}\right)$. This gives some evidence for Conjecture 2.8.2.

2.12. Prove the functional equation expressed by Eq. (2.42) for the dilog by differentiation.

2.13. Verify the identity (2.44) by differentiation. Then use it to compute the integral (see (2.97) on p. 812 of [507])

$$\int_0^1 \left(\frac{\log((1+x)/2)}{1-x} - \frac{\log((1-x)/2)}{1+x}\right)\frac{dx}{x}$$

by first expanding the integrand using power series and then expressing it using double logarithms, dilogs and products of logarithms.

2.14. Prove the identity

$$\begin{aligned}\mathrm{Li}_{1,1,1}(x,y,z) &= \mathrm{Li}_{1,2}\Big(\frac{1}{1-y},\frac{z(1-y)}{z-1}\Big)-\mathrm{Li}_{1,2}\Big(\frac{1-yx}{1-y},\frac{z(1-y)}{z-1}\Big)\\ &\quad -\log(1-x)\,\mathrm{Li}_2\Big(\frac{z}{z-1}\Big)-\log(1-z)\,\mathrm{Li}_{1,1}(x,y)\end{aligned}$$

by differentiation. Then use Eq. (2.48) to show that

$$\begin{aligned}\mathscr{L}i_{1,1,1}(x,y,z) &= \mathscr{L}i_3\Big(\frac{(y-1)(1-xyz)}{y(1-x)(1-z)}\Big)+\mathscr{L}i_3\Big(\frac{y}{y-1}\Big)+\mathscr{L}i_3(yz)\\ &\quad -\mathscr{L}i_3\Big(\frac{1-xyz}{1-z}\Big)-\mathscr{L}i_3\Big(\frac{1-xyz}{yz(1-x)}\Big)-\mathscr{L}i_3\Big(\frac{y-xy}{y-1}\Big)\\ &\quad -\mathscr{L}i_3\Big(\frac{y-yz}{y-1}\Big)+\mathscr{L}i_3(1-z)\end{aligned}$$

modulo products of logarithms and dilogs.

2.15. Prove Eq. (2.49) for $d=2$ and $d=3$.

2.16. How can one decompose the pair in Figure 2.3 into two dilog pairs? **Hint:** The dilog pairs (L, M) are determined by the following geometric combinatorial properties: there are exactly two edges of M each of which contains a vertex of L while exactly one vertex of M facing one of the above two edges lies on an edge of L.

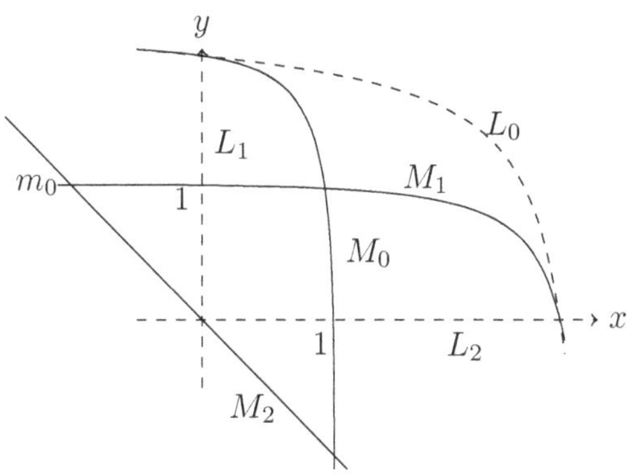

Figure 2.3: Decompose this pair of simplices into two dilog pairs.

2.17. Let $A_0^0 = A_0$ and A_n^0 $(n>0)$ the subgroup of A_n whose elements can be represented by admissible pairs of simplices in generic position. For any index sets $I=\{i_1<\cdots<i_k\}\subseteq\{1,\ldots,n\}$ and $J=\{j_1<\cdots<j_{n-k}\}\subseteq$

$\{1,\ldots,n\}$, we define $\operatorname{sgn}(I;J)=\operatorname{sgn}(I,\bar{I})\cdot\operatorname{sgn}(J,\bar{J})$, where $\operatorname{sgn}(I,\bar{I})$ is the sign of the permutation $(I,\bar{I})\to\{1,\cdots,n\}$ (similarly for $\operatorname{sgn}(J,\bar{J})$) and $\bar{I}$ (resp. $\bar{J}$) is the set complementary to I (resp. J) in $\{1,\cdots,n\}$.

Define the coproduct map on A_0 as follows:

$$\begin{aligned}\nu: A_0 &\longrightarrow A_0\otimes A_0\\ 1 &\longmapsto 1\otimes 1\end{aligned}$$

and when $n>0$,

$$\begin{aligned}\nu:\quad A_n^0 &\longrightarrow \bigoplus\nolimits_{k=0}^n A_{n-k}^0\otimes A_k^0\\ [L;\,M] &\longmapsto \sum_{k=0}^n \nu_{n-k,k}\big([L;\,M]\big),\end{aligned}$$

where $\nu_{n,0}=\mathrm{id}\otimes 1_F$, $\nu_{0,n}=1_F\otimes\mathrm{id}$ and for $0<k<n$:

$$\begin{aligned}\nu_{n-k,k}:\quad A_n^0 &\longrightarrow A_{n-k}^0\otimes A_k^0\\ [L;\,M] &\longmapsto \sum_{|I|=k}\operatorname{sgn}(I)\nu_{n-k,k}^{(I)}\big([L;\,M]\big)\end{aligned}$$

where

$$\nu_{n-k,k}^{(I)}\big([L;\,M]\big)=\sum_{|J|=n-k}\operatorname{sgn}(J)[L_I|L_0,L_{\bar{I}};M_0,M_J]\otimes[M_J|L_0,L_I;M_0,M_{\bar{J}}].$$

Note that both factors in each term are admissible since L and M are in generic position. We similarly define $\nu_{n-k,k}^{(J)}$ for any fixed index set J and define

$$\nu_{n-k,k}=\sum_{|J|=n-k}\operatorname{sgn}(J)\nu_{n-k,k}^{(J)}.$$

Prove that the coproduct ν is well defined, *i.e.*, it is compatible with the five axioms (R1)-(R5) in Definition 2.10.2.

Chapter 3

Multiple Zeta Values (MZVs)

Most of the on-going studies on multiple zeta functions concentrate on the algebraic aspect of the multiple zeta values (MZVs for abbreviation), which are the convergent special values of the multiple zeta functions at positive integers. Sometimes MZVs are also called multi-zeta values, or poly-zeta values, or Euler–Zagier sums.

Throughout this chapter we will use s_j's to denote integer variables.

3.1 Shuffle Product and Euler's Decomposition Formula

Kontsevich first observed that MZVs can be represented by the iterated integrals. Put

$$\mathrm{x}_0 = \frac{\mathrm{d}t}{t}, \qquad \mathrm{x}_1 = \frac{\mathrm{d}t}{t-1}.$$

Setting all the variables $x_1 = \cdots = x_d = 1$ in Eq. (2.8) we obtain an iterated integral expression of the MZVs

$$\zeta(s_1, \ldots, s_d) = (-1)^d \int_0^1 \mathrm{x}_0^{s_1-1}\mathrm{x}_1 \cdots \mathrm{x}_0^{s_d-1}\mathrm{x}_1. \tag{3.1}$$

Using this we can prove the following Euler's decomposition formula quite easily.

Theorem 3.1.1. *Let $m, n \geq 2$ be two positive integers. Then*

$$\zeta(m)\zeta(n) = \sum_{\substack{i\geq 2, j\geq 1\\ i+j=m+n}} \left[\binom{i-1}{m-1} + \binom{j-1}{n-1}\right]\zeta(i,j). \tag{3.2}$$

Proof. Set $b_1 = b_2 = \mathrm{x}_1$. By Eq. (3.1) and using Lemma 2.1.2(iv)

$$\zeta(m)\zeta(n) = \int_0^1 \mathrm{x}_0^{m-1} b_1 \int_0^1 \mathrm{x}_0^{n-1} b_2$$

$$
\begin{aligned}
&= \sum_{k=0}^{n-1} \binom{m-1+k}{k} \int_0^1 \mathrm{x}_0^{m-1+k} b_1 \mathrm{x}_0^{n-1-k} b_2 \\
&\quad + \sum_{k=0}^{m-1} \binom{n-1+k}{k} \int_0^1 \mathrm{x}_0^{n-1+k} b_2 \mathrm{x}_0^{m-1-k} b_1 \\
&= \sum_{k=0}^{n-1} \binom{m-1+k}{k} \zeta(m+k, n-k) \\
&\quad + \sum_{k=0}^{m-1} \binom{n-1+k}{k} \zeta(n+k, m-k)
\end{aligned}
$$

which is equivalent to Eq. (3.2). □

3.2 Hoffman's Algebraic Setup

One of the key ideas to study the $\mathbb{Q}$-linear relations between the MZVs is to formalize them using the theory of words. In order to realize fully the product structures we actually need two algebras equipped with two different multiplications whose underlying set of words are closely intertwined.

3.2.1 *Shuffle Relations*

With the aid of Chen's theory of the iterated integrals one can define a shuffle algebra reflecting the multiplication rule of two MZVs using these integrals. Specifically, as we have seen in the proof of Euler's decomposition formula in the preceding section, we can apply Chen's formula (*i.e.*, Lemma 2.1.2(iv)) for the product of two iterated integrals and define some noncommutative polynomial algebra $\mathfrak{A} = \mathbb{Q}\langle \mathrm{x}_0, \mathrm{x}_1 \rangle$ so that the MZVs can be viewed as values of a homomorphism (in fact, two different homomorphisms) from a sub-algebra of $\mathfrak{A}$ to the real number field $\mathbb{R}$. As an immediate consequence, the product of two MZVs can be expressed as a $\mathbb{Z}$-linear combination of the MZVs, which is called a *shuffle relation.*

Definition 3.2.1. Define the alphabet $\mathsf{X} = \{\mathrm{x}_0, \mathrm{x}_1\}$ and let X^* be the set of words over X (*i.e.*, monomials in the letters in X) including the *empty word* 1. Let $\mathfrak{A} = \mathbb{Q}\langle \mathsf{X} \rangle$ be the (weight) graded noncommutative polynomial $\mathbb{Q}$-algebra of words generated over the alphabet X. The number of letters in the word $\mathbf{w}$, denoted by $|\mathbf{w}|$, is called the *weight*[1] of $\mathbf{w}$, and the number

[1]In the study of words we usually use the term "length". But here we avoid using this term because in the study of the MZVs this may cause confusion as some authors use it to mean depth.

of x_1's, the *depth*.

Definition 3.2.2. Let $\mathfrak{A}^0$ be the subalgebra of $\mathfrak{A}$ generated by words not beginning with x_1 and not ending with x_0. The words in $\mathfrak{A}^0$ are called *admissible words*. Further, we define $\mathfrak{A}^1 = \mathbb{Q}1 \oplus \mathfrak{A}\mathsf{x}_1$ to be the subalgebra of $\mathfrak{A}$, which is a noncommutative polynomial algebra on generators $\mathsf{z}_k = \mathsf{x}_0^{k-1}\mathsf{x}_1$ for all $k \geq 1$.

Clearly, there is a many-to-one correspondence between the MZVs and admissible words and the definition of weight and depth is consistent. For example, Euler already proved $\zeta(3) = \zeta(2,1)$. So the depth of a MZV depends on its expression, however, the weight shouldn't by a folklore conjecture.

Definition 3.2.3. Let $\mathfrak{A}_{Ш}$ be the algebra of $\mathfrak{A}$ together with the multiplication defined by shuffle product $Ш$: for any letters $x, y \in \mathsf{X}$ and words $\mathbf{u}, \mathbf{v} \in \mathsf{X}^*$ we define recursively

$$1 \,Ш\, \mathbf{u} = \mathbf{u} \,Ш\, 1 = \mathbf{u}, \qquad x\mathbf{u} \,Ш\, y\mathbf{v} = x(\mathbf{u} \,Ш\, y\mathbf{v}) + y(x\mathbf{u} \,Ш\, \mathbf{v}).$$

Here we write $\mathsf{x}_0\mathbf{u} \,Ш\, \mathsf{x}_1\mathbf{v}$ rather than $(\mathsf{x}_0\mathbf{u}) \,Ш\, (\mathsf{x}_1\mathbf{v})$ to minimize the number of parentheses. Namely, concatenation has higher operational priority than $Ш$.

Proposition 3.2.4. *We have*

(i) $\mathfrak{A}_{Ш}$ *is a commutative and associative* $\mathbb{Q}$*-algebra.*

(ii) *For any admissible word* $\mathbf{w}$ *of depth* d, *let* $\zeta(\mathbf{w}) = (-1)^d \int_0^1 \mathbf{w}$. *Extend the map to a* $\mathbb{Q}$*-linear map* $\zeta_{Ш} : \mathfrak{A}^0_{Ш} \longrightarrow \mathbb{R}$. *Then* $\zeta_{Ш}$ *is an algebra homomorphism.*

Proof. (i) We demonstrate only the commutativity and leave the associativity to the interested reader. For commutativity, we only need to show that for all words $\mathbf{u}$ and $\mathbf{v}$

$$\mathbf{u} \,Ш\, \mathbf{v} = \mathbf{v} \,Ш\, \mathbf{u}. \tag{3.3}$$

We prove this by induction on $|\mathbf{u}| + |\mathbf{v}|$. If either $\mathbf{u}$ or $\mathbf{v}$ is 1 then Eq. (3.3) is clear by the definition. Now assume Eq. (3.3) holds for any two words if the sum of their weights is $< \ell \in \mathbb{N}$. Let $|\mathbf{u}| + |\mathbf{v}| = \ell$ and suppose without loss of generality $\mathbf{u} = x\mathbf{u}'$, and $\mathbf{v} = y\mathbf{v}'$ where $x, y \in \{\mathsf{x}_0, \mathsf{x}_1\}$. Then by the definition

$$\mathbf{u} \,Ш\, \mathbf{v} = x(\mathbf{u}' \,Ш\, y\mathbf{v}') + y(x\mathbf{u}' \,Ш\, \mathbf{v}') = x(y\mathbf{v}' \,Ш\, \mathbf{u}') + y(\mathbf{v}' \,Ш\, x\mathbf{u}')$$

by inductive assumption. But by the definition the last expression is equal to $\mathbf{v} ш \mathbf{u}$. Thus Eq. (3.3) is proved and therefore $\mathfrak{A}_ш$ is commutative.

(ii) We need to establish that for any $\mathbf{u}, \mathbf{v} \in \mathfrak{A}^0_ш$ we have

$$\zeta_ш(\mathbf{u} ш \mathbf{v}) = \zeta_ш(\mathbf{u})\zeta_ш(\mathbf{v}).$$

To avoid ambiguity let's denote by ш the shuffle operation defined by Definition 3.2.3 and ш′ the shuffle operation defined by

$$(l_1 \cdots l_r) ш' (l_{r+1} \cdots l_{r+s}) := \sum_\sigma l_{\sigma(1)} \cdots l_{\sigma(r+s)}, \tag{3.4}$$

where the letter $l_j \in \{\mathrm{x}_0, \mathrm{x}_1\}$ for all $j = 1, \ldots, r+s$, and σ ranges over all shuffles of type (r, s), *i.e.*, permutations σ of $r+s$ letters with $\sigma^{-1}(1) < \cdots < \sigma^{-1}(r)$ and $\sigma^{-1}(r+1) < \cdots < \sigma^{-1}(r+s)$. We now prove that

$$\mathbf{u} ш' \mathbf{v} = \mathbf{u} ш \mathbf{v} \quad \forall \mathbf{u}, \mathbf{v} \in \mathfrak{A} \tag{3.5}$$

by induction on $|\mathbf{u}| + |\mathbf{v}|$. Clearly when $\mathbf{u}, \mathbf{v} \in \mathfrak{A}$ then $|\mathbf{u}| + |\mathbf{v}| \geq 0$. If $|\mathbf{u}| + |\mathbf{v}| \leq 1$ then Eq. (3.5) is trivial. Now assume Eq. (3.5) holds for $|\mathbf{u}| + |\mathbf{v}| = k \geq 1$ and let $U, V \in \mathfrak{A}$ with $|U| + |V| = k+1$. If either $U = 1$ or $V = 1$ then Eq. (3.5) is clear. Otherwise we may assume $U = x\mathbf{u}, V = y\mathbf{v} \in \mathfrak{A}$ for some $x, y \in \{\mathrm{x}_0, \mathrm{x}_1\}$. By the definition and induction assumption

$$U ш V = x(\mathbf{u} ш y\mathbf{v}) + y(x\mathbf{u} ш \mathbf{v}) = x(\mathbf{u} ш' y\mathbf{v}) + y(x\mathbf{u} ш' \mathbf{v}) = U ш' V \tag{3.6}$$

because, according to Eq. (3.4), we can partition the terms in $U ш' V$ into two disjoint sets depending on whether the leading letter is x_0 or x_1.

The proposition now follows immediately from Eqs. (3.1), (3.5) and Lemma 2.1.2(iv). □

We now provide an interesting result which will be used later.

Proposition 3.2.5. *Let p, q and j be nonnegative integers such that $\min(p,q) \geq j$. Let T^j_{p+q} denote the sum of all those words occurring in $(\mathrm{x}_0\mathrm{x}_1)^p ш (\mathrm{x}_0\mathrm{x}_1)^q$ that contain the subword x_0^2 exactly j times. Then we have*

$$(\mathrm{x}_0\mathrm{x}_1)^p ш (\mathrm{x}_0\mathrm{x}_1)^q = \sum_{j=0}^{\min(p,q)} \binom{p+q-2j}{p-j} \cdot 4^j T^j_{p+q}.$$

Proof. Let $\mathbf{w}$ be any word term in the sum T^j_{p+q}. We now count the multiplicity of $\mathbf{w}$ in $(\mathrm{x}_0\mathrm{x}_1)^p ш (\mathrm{x}_0\mathrm{x}_1)^q$, namely, the number of ways the letters of $\mathbf{w}$ can be colored in either blue (from $(\mathrm{x}_0\mathrm{x}_1)^p$) or red (from $(\mathrm{x}_0\mathrm{x}_1)^q$) following the shuffle product rule.

By the definition there are $p+q$ $\mathtt{x}_0$'s in $\mathbf{w}$, of which $2j$ $\mathtt{x}_0$'s are contained in factors $\mathtt{x}_0^2$ and $p+q-2j$ "single" $\mathtt{x}_0$'s have only $\mathtt{x}_1$'s as neighbors. This implies there are exactly j factors $\mathtt{x}_1^2$ (see Exercise 3.2). For each factor $\mathtt{x}_0^2$, one $\mathtt{x}_0$ is red and the other is blue because of the shuffle product rule. Similar pattern holds for factors $\mathtt{x}_1^2$. These contribute a factor of $2^j \cdot 2^j = 4^j$ to the multiplicity of $\mathbf{w}$. On the other hand, exactly $p-j$ of the "single" $\mathtt{x}_0$'s are colored blue, contributing a factor of $\binom{p+q-2j}{p-j}$ to the multiplicity of $\mathbf{w}$. We can now complete the proof of the proposition by observing that the coloring of the "single" $\mathtt{x}_1$'s is now determined uniquely. □

Proposition 3.2.6. *For any positive integers $n \geq m$, we have*

$$\sum_{k=1-n}^{n} (-1)^k k^{2m-1} \left[(\mathtt{x}_0\mathtt{x}_1)^{n-k} ш (\mathtt{x}_0\mathtt{x}_1)^{n-1+k} \right] = \sum_{j=1}^{m} 4^{n-j} T_{2n-1}^{n-j} \cdot a_{2m-1,j}, \tag{3.7}$$

$$\sum_{k=-n}^{n} (-1)^k k^{2m} \left[(\mathtt{x}_0\mathtt{x}_1)^{n-k} ш (\mathtt{x}_0\mathtt{x}_1)^{n+k} \right] = \sum_{j=1}^{m} 4^{n-j} T_{2n}^{n-j} \cdot a_{2m,j}, \tag{3.8}$$

where for $j = 1, \dots, m$ we set

$$a_{2m-1,j} = \sum_{k=1-j}^{j} (-1)^k k^{2m-1} \binom{2j-1}{j-k}, \quad a_{2m,j} = \sum_{k=-j}^{j} (-1)^k k^{2m} \binom{2j}{j-k}. \tag{3.9}$$

Proof. By Prop. 3.2.5 the left-hand side of Eq. (3.7) can be written as

$$\left(\sum_{k=1-n}^{0} \sum_{j=0}^{n-1+k} + \sum_{k=1}^{n} \sum_{j=0}^{n-k} \right) (-1)^k k^{2m-1} \binom{2n-1-2j}{n-k-j} \cdot 4^j T_{2n-1}^j$$

which after reordering is equal to

$$\sum_{j=0}^{n-1} 4^j T_{2n-1}^j \cdot \sum_{k=1-n+j}^{n-j} (-1)^k k^{2m-1} \binom{2n-1-2j}{n-k-j}.$$

Then from Eq. (3.9) (with $n-j$ in the place of j)

$$\begin{aligned} &\sum_{k=1-n}^{n} (-1)^k k^{2m-1} \left[(\mathtt{x}_0\mathtt{x}_1)^{n-k} ш (\mathtt{x}_0\mathtt{x}_1)^{n-1+k} \right] \\ &\qquad = \sum_{j=n-m}^{n-1} 4^{n-j} T_{2n-1}^j \cdot a_{2m-1,n-j}. \end{aligned}$$

The proof of Eq. (3.8) is similar and is left to the interested reader. □

3.2.2 *Stuffle Relations*

On the other hand, it is easy to see that

$$\zeta(2)\zeta(3) = \zeta(2,3) + \zeta(3,2) + \zeta(5).$$

This can be generalized to other MZVs without too much difficulty. Relations such as this are called the *stuffle relations*. Indeed, one can even extract some formal algebra structures from MZVs.

Definition 3.2.7. Let $\mathfrak{A}_*$ be the algebra of $\mathfrak{A}$ together with the multiplication defined by the stuffle product $*$: for any nonnegative integers m, n and any words $\mathbf{u}, \mathbf{v}$ in $\mathfrak{A}$ we define:

$$1 * \mathbf{u} = \mathbf{u} * 1 = \mathbf{u},$$
$$\mathrm{x}_0^m * \mathbf{u} = \mathbf{u} * \mathrm{x}_0^m = \mathbf{u}\mathrm{x}_0^m,$$
$$\mathrm{x}_0^m \mathrm{x}_1 \mathbf{u} * \mathrm{x}_0^n \mathrm{x}_1 \mathbf{v} = \mathrm{x}_0^m \mathrm{x}_1 \big(\mathbf{u} * \mathrm{x}_0^n \mathrm{x}_1 \mathbf{v}\big) + \mathrm{x}_0^n \mathrm{x}_1 \big(\mathrm{x}_0^m \mathrm{x}_1 \mathbf{u} * \mathbf{v}\big) + \mathrm{x}_0^{m+n+1} \mathrm{x}_1 (\mathbf{u} * \mathbf{v}).$$

Here, again, concatenation has higher operational priority than $*$.

Similar to Prop. 3.2.4, one can prove the following fact for the stuffle product.

Proposition 3.2.8. *We have*

(i) $\mathfrak{A}_*$ *is a commutative and associative algebra.*
(ii) *For any admissible word* $\mathbf{w} \in \mathfrak{A}^0$ *of depth* d, *let* $\zeta(\mathbf{w}) = (-1)^d \int_0^1 \mathbf{w}$. *We then can extend the map to a* $\mathbb{Q}$*-linear map* $\zeta_* : \mathfrak{A}_*^0 \longrightarrow \mathbb{R}$. *Then* ζ_* *is an algebra homomorphism.*

Proof. Similar to Prop. 3.2.4, this can be proved by induction. We leave the details to the interested reader. □

3.3 Double Shuffle Relations

By comparing Prop. 3.2.4 and Prop. 3.2.8 we set

Definition 3.3.1. The *finite double shuffle relations* (finite DBSFs) of the MZVs are the $\mathbb{Q}$-linear relations generated by

$$\zeta(\mathbf{u} * \mathbf{v} - \mathbf{u} ш \mathbf{v})$$

for all admissible words $\mathbf{u}$ and $\mathbf{v}$.

Example 3.3.2. The simplest finite DBSF is obtained by computing $\zeta(2)^2$ in two different ways. First,

$$\begin{aligned}\zeta(2)^2 &= \int_0^1 \mathrm{x}_0\mathrm{x}_1 \cdot \int_0^1 \mathrm{x}_0\mathrm{x}_1 = \int_0^1 \mathrm{x}_0\mathrm{x}_1 ш \mathrm{x}_0\mathrm{x}_1 \\ &= 2\int_0^1 \mathrm{x}_0\mathrm{x}_1\mathrm{x}_0\mathrm{x}_1 + 4\int_0^1 \mathrm{x}_0\mathrm{x}_0\mathrm{x}_1\mathrm{x}_1 = 2\zeta(2,2) + 4\zeta(3,1).\end{aligned}$$

On the other hand, by the stuffle relation

$$\zeta(2)^2 = 2\zeta(2,2) + \zeta(4).$$

Therefore we get the finite DBSF

$$\zeta(4) = 4\zeta(3,1).$$

3.3.1 *Lyndon Words*

Note that it is impossible to prove the famous identity $\zeta(2,1) = \zeta(3)$ by only the finite DBSFs because the smallest weight of the MZVs is two given by $\zeta(2)$ and thus $\zeta(3)$ cannot be a product of two MZVs. To solve this puzzle, we need the *regularized* DBSFs. The key idea is contained in the next theorem. Before stating it we need some results concerning so-called Lyndon words.

Definition 3.3.3. Let A be any alphabet with an complete order $<$ on the set of words A^*. A nonempty word $\mathbf{w} \in A^*$ is called *Lyndon* if $\mathbf{w} < \mathbf{v}$ for any nontrivial factorization $\mathbf{w} = \mathbf{uv}$.

Hence, to define Lyndon words in $\mathfrak{A} = \mathbb{Q}\langle \mathsf{X} \rangle$ we first need to put an order on it. If $\mathbf{u}, \mathbf{v}$ are words with $\mathbf{v} \neq 1$, we let $\mathbf{u} < \mathbf{uv}$. For any words $\mathbf{w}_1, \mathbf{w}_2, \mathbf{w}_3$, set $\mathbf{w}_1\mathrm{x}_0\mathbf{w}_2 < \mathbf{w}_1\mathrm{x}_1\mathbf{w}_3$.

Example 3.3.4. It is straightforward to see that x_0 is the smallest London word while there is no largest word. Lyndon words can have the form $\mathbf{z}_{s_1} \cdots \mathbf{z}_{s_d} := \mathrm{x}_0^{s_1-1}\mathrm{x}_1 \cdots \mathrm{x}_0^{s_d-1}\mathrm{x}_1$ with $s_j < s_1$ for all $j > 1$. But $\mathrm{x}_0^2\mathrm{x}_1\mathrm{x}_0^2\mathrm{x}_1\mathrm{x}_0\mathrm{x}_1$ is also a Lyndon word.

Lemma-Definition 3.3.5. *Every word* $\mathbf{w} \in \mathfrak{A}$ *has a unique factorization as non-increasing Lyndon words, called the Lyndon factorization of* $\mathbf{w}$.

Proof. Write $\mathbf{w} = \mathbf{z}_{s_1} \cdots \mathbf{z}_{s_d}\mathrm{x}_0^m$ for some positive integers $s_1, \ldots, s_d$ and some nonnegative integer m. We will define a subsequence $\{i_1, \ldots, i_k\}$ of $\{1, \ldots, d\}$ such that for each $j = 1, \ldots, k$ we have $s_{i_j} > s_\ell$ for all $\ell \in$

$\{i_j+1,\ldots,i_{j+1}-1\}$ and $s_{i_j} \le s_{i_{j+1}}$, where $i_{k+1}=d+1$. Clearly $\mathbf{w} = \mathbf{w}_1\cdots\mathbf{w}_k\mathtt{x}_0^m$ where

$$\mathbf{w}_j = \mathbf{z}_{s_{i_j}}\cdots\mathbf{z}_{s_{i_{j+1}-1}}$$

is a Lyndon word for every j. However, it is not always true that $\mathbf{w}_j \ge \mathbf{w}_{j+1}$ for all j. Problem may occur if $s_{i_j} = s_{i_{j+1}}$. So we need to modify the factorization by possibly concatenating some of the $\mathbf{w}_j$'s as follows.

For simplicity, we write $I_j(\mathbf{z}_{t_1}\cdots\mathbf{z}_{t_l}) = t_j$ for each composition $(t_1,\ldots,t_l)\in\mathbb{N}^l$. Continue with the notation in the preceding paragraph, if $I_1(\mathbf{w}_{j-1}) > I_1(\mathbf{w}_j) = I_1(\mathbf{w}_{j+1}) = \cdots = I_1(\mathbf{w}_\ell) > I_1(\mathbf{w}_{\ell+1})$ then we compare $I_j(\mathbf{w}_\alpha)$ and $I_j(\mathbf{w}_{\alpha+1})$ for all $\alpha=1,\ldots,\ell-1$ and for increasing $j=2,3,\cdots$, until we find either of the two:

(i) $I_{j-1}(\mathbf{w}_\alpha) = I_{j-1}(\mathbf{w}_{\alpha+1}) \ne 0$ but $I_j(\mathbf{w}_\alpha)\ne I_j(\mathbf{w}_{\alpha+1})$. If $I_j(\mathbf{w}_\alpha) > I_j(w_{\alpha+1}) > 0$ or $I_j(\mathbf{w}_\alpha) = 0 < I_j(\mathbf{w}_{\alpha+1})$ then we concatenate $\mathbf{w}_\alpha$ and $\mathbf{w}_{\alpha+1}$, otherwise we don't. Here $I_j(\mathbf{w}_\alpha)=0$ means $\mathbf{w}_\alpha$ has depth $j-1$ only.

(ii) $\mathbf{w}_\alpha = \mathbf{w}_{\alpha+1}$. Then we don't concatenate them.

Now it is not hard to see we do have non-increasing Lyndon words in the factorization. □

Example 3.3.6. We have the following Lyndon factorizations

$$(\mathtt{x}_1)(\mathtt{x}_0),\quad (\mathtt{x}_1)(\mathtt{x}_0\mathtt{x}_1)(\mathtt{x}_0^3\mathtt{x}_1\mathtt{x}_0^3\mathtt{x}_1\mathtt{x}_0\mathtt{x}_1)(\mathtt{x}_0),\quad (\mathtt{x}_0^2\mathtt{x}_1)(\mathtt{x}_0^3\mathtt{x}_1\mathtt{x}_0^2\mathtt{x}_1\mathtt{x}_0^3\mathtt{x}_1\mathtt{x}_0\mathtt{x}_1),$$
$$(\mathtt{x}_0^3\mathtt{x}_1\mathtt{x}_0\mathtt{x}_1)(\mathtt{x}_0^3\mathtt{x}_1\mathtt{x}_0^2\mathtt{x}_1),\quad (\mathtt{x}_0^3\mathtt{x}_1\mathtt{x}_0\mathtt{x}_1)(\mathtt{x}_0^3\mathtt{x}_1).$$

Our next goal is to show that every word $\mathbf{w}\in\mathfrak{A}^1_* = (\mathfrak{A}^1,*)$ has a unique expression as a $\mathbb{Q}$-linear combination of the stuffle products of non-increasing Lyndon words.

Lemma 3.3.7. *For* $1\ne\mathbf{w}\in\mathfrak{A}$, *let* $\mathrm{LW}(\mathbf{w})$ *be the largest word appearing in* $\mathbf{w}$. *Let* $\mathbf{w},\mathbf{u},\mathbf{v}\in\mathfrak{A}$ *such that* $\mathbf{u}<\mathbf{v}$. *Then* $\mathrm{LW}(\mathbf{u}*\mathbf{w}) < \mathrm{LW}(\mathbf{v}*\mathbf{w})$

Proof. Use induction on the weight sum $|\mathbf{u}|+|\mathbf{v}|$. □

Lemma 3.3.8. *Let* $\mathbf{u},\mathbf{v}\in\mathfrak{A}$ *such that* $\mathbf{u}$ *is a Lyndon word and* $\mathbf{u}\ge\mathbf{v}$. *Then* $\mathrm{LW}(\mathbf{u}*\mathbf{v}) = \mathbf{uv}$.

Proof. Since $\mathtt{x}_0$ commutes with every word under stuffle we may assume that $\mathbf{u}=\mathbf{z}_{s_1}\cdots\mathbf{z}_{s_d}$ and $\mathbf{v}=\mathbf{z}_{t_1}\cdots\mathbf{z}_{t_\ell}\in\mathfrak{A}^1$ without loss of generality. Recall that in $\mathbf{u}*\mathbf{v}$ there are two kinds of words, one is obtained by pure shuffle, while the other involves stuffing. Assume $s_1=t_1,\ldots,s_i=t_i$ and, $0<s_{i+1}<t_{i+1}$ or $t_{i+1}=0$ since $\mathbf{u}\ge\mathbf{v}$. Then in each term of the shuffle,

except for $\mathbf{uv}$, $\mathbf{z}_{t_1}$ appears between $\mathbf{z}_{s_{j-1}}$ and $\mathbf{z}_{s_j}$ for some $1 \le j \le d$ ($\mathbf{z}_{s_0}$ is vacuous). If $j > i$ then clearly $s_j < s_1 \le t_1$ since $\mathbf{u}$ is Lyndon word and thus this term is smaller than $\mathbf{uv}$. If $j \le i$ then all of $\mathbf{z}_{s_j}$ up to $\mathbf{z}_{s_i}$ in this shuffle term are pushed to the right and thus producing a smaller term than $\mathbf{uv}$ since $\mathbf{uv}$'s $\mathbf{z}_{s_{i+1}}$-term corresponds to some $\mathbf{z}_{t_k}$ or some $\mathbf{z}_{s_k}$ with $k \le i$. Finally, if some stuffing is involved, then $t_1, \dots, t_i$ cannot become smaller resulting a stuffing term smaller than $\mathbf{uv}$. This shows that $\mathrm{LW}(\mathbf{u} * \mathbf{v}) = \mathbf{uv}$ and the lemma is proved. □

Lemma 3.3.9. *Let $\mathbf{u}, \mathbf{v} \in \mathfrak{A}^1$ such that $\mathbf{u}$ is a Lyndon word and $\mathbf{u} \ge \mathbf{v}$. Then for any nonnegative integer n we have $\mathrm{LW}(\mathbf{u} * \mathbf{u}^n\mathbf{v}) = \mathbf{u}^{n+1}\mathbf{v}$.*

Proof. We now use induction on n. Lemma 3.3.8 takes care of the case $n = 0$. By the same consideration as in Lemma 3.3.8 we see that $\mathrm{LW}(\mathbf{u} * \mathbf{u}^n\mathbf{v})$ can only come from either $\mathbf{u}(\mathbf{u}^n\mathbf{v}) = \mathbf{u}^{n+1}\mathbf{v}$ or $\mathbf{u}(\mathbf{u} * \mathbf{u}^{n-1}\mathbf{v})$. By induction we are done. □

Lemma 3.3.10. *Let $\mathbf{w}_1 \ge \mathbf{w}_2 \ge \cdots \ge \mathbf{w}_k$ be a sequence of Lyndon words. Then*

$$\mathrm{LW}(\mathbf{w}_1 * \mathbf{w}_2 * \cdots * \mathbf{w}_k) = \mathbf{w}_1\mathbf{w}_2 \cdots \mathbf{w}_k.$$

Proof. We proceed by induction on k. The case $k = 1$ is trivial. Suppose the theorem is true if the number of Lyndon words is less than k. We then have two cases.

1) If $\mathbf{w}_1 > \mathbf{w}_2$ then $\mathbf{w}_1 > \mathbf{w}_2 \cdots \mathbf{w}_k$ since $\mathbf{w}_j$'s are all Lyndon words. By Lemmas 3.3.7, Lemma 3.3.8 and the induction assumption we get

$$\mathrm{LW}(\mathbf{w}_1 * \mathbf{w}_2 * \cdots * \mathbf{w}_k) = \mathrm{LW}(\mathbf{w}_1 * \mathbf{w}_2 \cdots \mathbf{w}_k) = \mathbf{w}_1\mathbf{w}_2 \cdots \mathbf{w}_k.$$

2) If $\mathbf{w}_1 = \mathbf{w}_2$ then $\mathbf{w}_2 \cdots \mathbf{w}_k$ has the form $\mathbf{w}_1^n\mathbf{v}$ for $\mathbf{v} < \mathbf{w}_1$, and the theorem follows from Lemmas 3.3.7, Lemma 3.3.9 and the induction assumption. □

Theorem 3.3.11. *As a commutative algebra, $\mathfrak{A}_*$ is the free polynomial algebra on the Lyndon words.*

Proof. Let $\mathbf{w} \in \mathfrak{A}$ and let $\mathbf{w} = \mathbf{w}_1 \cdots \mathbf{w}_k$ be its Lyndon factorization guaranteed by Lemma 3.3.5. Then by Lemma 3.3.10 we know there is some $c \in \mathbb{Q}$ such that $\mathrm{LW}(\mathbf{w} - c\mathbf{w}_1 * \cdots * \mathbf{w}_k)$ is smaller than $\mathbf{w}$. Since there are only finitely many words having the same degree as that of $\mathbf{w}$, by repeating this we see that $\mathbf{w}$ can be written as a $\mathbb{Q}$-linear combination of the stuffle products of non-increasing Lyndon words. Clearly this expression is unique.

Let $\mathfrak{L}$ be the polynomial algebra with one generator for each Lyndon word. The above shows that $\mathfrak{L}$ maps onto $\mathfrak{A}$. Let $\mathfrak{A}_n$ be the $\mathbb{Q}$ vector space generated by all weight n words in $\mathfrak{A}$. We see easily that the Poincaré series of $\mathfrak{A}$ is

$$\sum_{n=0}^{\infty}(\dim_{\mathbb{Q}} \mathfrak{A}_n)t^n = \sum_{n=0}^{\infty} 2^n t^n = \frac{1}{1-2t}.$$

On the other hand, another equivalent definition of a Lyndon word is the following: it is a primitive word (*i.e.*, not a true power of another word) that is minimal in its conjugacy class (*i.e.*, the set of words it produces by moving a right factor to the left). Now if a word $\mathbf{w}$ of weight n can be written as $\mathbf{w} = \mathbf{u}^m$ with $\mathbf{u}$ primitive of weight d then $n = dm$ and the number of conjugates of $\mathbf{w}$ is exactly d. Since every weight n word is produced in this way we get

$$2^n = \sum_{d|n} dN(d). \tag{3.10}$$

where $N(n)$ is the number of Lyndon words of weight n. Clearly the Poincaré series of $\mathfrak{L}$ is

$$f(t) := \prod_{n=1}^{\infty}(1-t^n)^{-N(n)}.$$

The coefficient of t^m in the logarithmic derivative $(\log f(t))'$ is

$$\sum_{n|(m+1)} nN(n) = 2^{m+1}$$

by Eq. (3.10). Hence $f(t) = 1/(1-2t)$ and therefore $\mathfrak{L} \cong \mathfrak{A}$. □

Theorem 3.3.12. *As a commutative algebra, $\mathfrak{A}_{Ш}$ is the free polynomial algebra on the Lyndon words.*

Proof. The proof is similar to that of Thm. 3.3.11 and can be found in [482, Thm. 6.1], so we leave it to the interested reader. □

Remark 3.3.13. By Möbius inversion formula (see Exercise 3.10) the number of Lyndon words of weight n is given by

$$N(n) = \frac{1}{n}\sum_{d|n} 2^{n/d}\mu(d), \tag{3.11}$$

where the Möbius function

$$\mu(d) = \begin{cases} 1, & \text{if } d = 1; \\ 0, & \text{if } d \text{ has square factor} > 1; \\ (-1)^m, & \text{if } d \text{ is the product of } m \text{ distint primes.} \end{cases} \tag{3.12}$$

Equation (3.11) can be generalized easily to words built on a larger alphabet. One only need to replace 2 by the number of letters in the alphabet.

Theorem 3.3.14. *The evaluation maps $\zeta_{ш}$ and ζ_* can be extended to the algebra homomorphisms:*

$$\zeta_{ш} := \zeta_{ш}(-;T) : \mathfrak{A}^1_{ш} \longrightarrow \mathbb{R}[T], \quad \zeta_* := \zeta_*(-;T) : \mathfrak{A}^1_* \longrightarrow \mathbb{R}[T],$$

where

$$\mathfrak{A}^1_{ш} = (\mathfrak{A}^1, ш), \quad \mathfrak{A}^1_* = (\mathfrak{A}^1, *), \text{ and } \zeta_{ш}(\mathrm{x}_1;T) = \zeta_*(\mathrm{x}_1;T) = -T.$$

These provide two ways of regularizations of the MZVs.

Remark 3.3.15. We will see that for any non-admissible word $\mathbf{w} \in \mathfrak{A}^1$, both $\zeta_{ш}(\mathbf{w})$ and $\zeta_*(\mathbf{w})$ are polynomials in T with all coefficients except for the leading one are given by convergent MZVs. To signify this we sometimes write $\zeta_{ш}(\mathbf{w};T)$ and $\zeta_*(\mathbf{w};T)$, which are called regularized MZVs.

Proof. By Thm. 3.3.11, as commutative algebras,

- $\mathfrak{A}^1_*$ is the free polynomial algebra on the Lyndon words (with one word x_0 removed),
- $\mathfrak{A}^0_*$ is the free polynomial algebra on the Lyndon words (with two words x_0 and x_1 removed).

Thus

$$\mathfrak{A}^1_* \cong \mathfrak{A}^0_*[\mathrm{x}_1]. \tag{3.13}$$

So we can naturally extend $\zeta_* : \mathfrak{A}^0_* \longrightarrow \mathbb{R}$ to $\zeta_* : \mathfrak{A}^1_* \longrightarrow \mathbb{R}[T]$ by setting $\zeta_*(\mathrm{x}_1) = -T$. Similarly, by Thm. 3.3.12, as commutative algebras,

- $\mathfrak{A}^1_{ш}$ is the free polynomial algebra on the Lyndon words (with one word x_0 removed),
- $\mathfrak{A}^0_{ш}$ is the free polynomial algebra on the Lyndon words (with two words x_0 and x_1 removed).

Thus

$$\mathfrak{A}^1_{ш} \cong \mathfrak{A}^0_{ш}[\mathrm{x}_1]. \tag{3.14}$$

So we can naturally extend $\zeta_{ш} : \mathfrak{A}^0_{ш} \longrightarrow \mathbb{R}$ to $\zeta_{ш} : \mathfrak{A}^1_{ш} \longrightarrow \mathbb{R}[T]$ by setting $\zeta_{ш}(\mathrm{x}_1) = -T$. □

Example 3.3.16. For example,

$$\begin{aligned}
\mathtt{x}_0^2\mathtt{x}_1\mathtt{x}_0^3\mathtt{x}_1 &= \mathtt{x}_0^3\mathtt{x}_1 * \mathtt{x}_0^2\mathtt{x}_1 - \mathtt{x}_0^3\mathtt{x}_1\mathtt{x}_0^2\mathtt{x}_1 + \mathtt{x}_0^6\mathtt{x}_1,\\
\mathtt{x}_0^2\mathtt{x}_1\mathtt{x}_0^2\mathtt{x}_1 &= \frac{1}{2}(\mathtt{x}_0^2\mathtt{x}_1 * \mathtt{x}_0^2\mathtt{x}_1 + \mathtt{x}_0^5\mathtt{x}_1),\\
\mathtt{x}_1^2\mathtt{x}_0 &= \frac{1}{2}\mathtt{x}_1 \sqcup\!\sqcup\, \mathtt{x}_1 \sqcup\!\sqcup\, \mathtt{x}_0 - \mathtt{x}_1 \sqcup\!\sqcup\, \mathtt{x}_0\mathtt{x}_1 + \mathtt{x}_0\mathtt{x}_1^2.
\end{aligned}$$

3.3.2 *Regularization in Two Ways*

Before proving the main theorem for the regularized DBSFs we need a few preliminary results describing how to regularize the divergent series and the divergent iterated integral definition, and how to relate these two kinds of regularizations.

Lemma 3.3.17. *Define an $\mathbb{R}$-module $\mathbb{R}$-linear automorphism ρ of $\mathbb{R}[T]$ by*

$$\rho(\exp(Tu)) = \exp\left(\sum_{n=2}^{\infty} \frac{(-1)^n}{n}\zeta(n)u^n\right)\exp(Tu). \tag{3.15}$$

Let $P(T) \in \mathbb{R}[T]$ and $Q(T) = \rho(P(T))$. Then

$$\sum_{m=1}^{\infty} P(\log m + \gamma)t^{m-1} = \frac{1}{1-t}Q\left(\log\frac{1}{1-t}\right) + O\left(\log^{\deg P}\left(\frac{1}{1-t}\right)\right)$$

as $t \to 1^+$.

Proof. By linearity we only need to prove the identity for $P(T) = (T-\gamma)^l$ and $Q(T) = \rho((T-\gamma)^l)$. First, Weierstrass defined $\Gamma(x)$ by the following:

$$\frac{1}{\Gamma(u+1)} = \exp(\gamma u)\prod_{k=1}^{\infty}\left(1+\frac{u}{k}\right)\exp\left(-\frac{u}{k}\right).$$

Hence

$$-\log\Gamma(u+1) = \gamma u + \sum_{k=1}^{\infty}\left(\log\left(1+\frac{u}{k}\right) - \frac{u}{k}\right) = \gamma u + \sum_{k=1}^{\infty}\sum_{n=2}^{\infty}\frac{(-1)^{n-1}}{n}\frac{u^n}{k^n}. \tag{3.16}$$

Rearranging the sum we get

$$\Lambda(u) := \exp\left(\sum_{n=2}^{\infty}\frac{(-1)^n}{n}\zeta(n)u^n\right) = e^{\gamma u}\Gamma(1+u) \qquad \forall |u| < 1. \tag{3.17}$$

For a power series $g(u) = \sum_{n\ge 0} a_n u^n$ let $\mathrm{Coeff}_l(g) = a_l$. Then by linearity

$$Q(T) = \rho((T-\gamma)^l) = l!\,\mathrm{Coeff}_l\,\rho(\exp((T-\gamma)u))$$

$$
\begin{aligned}
&= l!\,\mathrm{Coeff}_l\left[\Lambda(u)\exp((T-\gamma)u)\right]\\
&= \frac{\mathrm{d}^l}{\mathrm{d}u^l}\left[\Lambda(u)\exp((T-\gamma)u)\right]_{u=0}\\
&= \frac{\mathrm{d}^l}{\mathrm{d}u^l}\left[\Gamma(1+u)\exp(Tu)\right]_{u=0}
\end{aligned}
$$

Thus

$$
\begin{aligned}
\frac{1}{1-t}Q\left(\log\frac{1}{1-t}\right) &= \frac{\mathrm{d}^l}{\mathrm{d}u^l}\left[\frac{\Gamma(1+u)}{(1-t)^{1+u}}\right]_{u=0}\\
&= \frac{\mathrm{d}^l}{\mathrm{d}u^l}\left[\sum_{m=1}^{\infty}\frac{\Gamma(1+u)(1+u)(2+u)\cdots(m-1+u)}{(m-1)!}t^{m-1}\right]_{u=0}\\
&= \frac{\mathrm{d}^l}{\mathrm{d}u^l}\left[\sum_{m=1}^{\infty}\frac{\Gamma(m+u)}{\Gamma(m)}t^{m-1}\right]_{u=0}\\
&= \sum_{m=1}^{\infty}\frac{\Gamma^{(l)}(m)}{\Gamma(m)}t^{m-1}.
\end{aligned}
$$

Using the following expression (from which one can derive the Stirling's formula)

$$
\log\Gamma(x) = \left(x-\frac{1}{2}\right)\log x - x + C - \int_0^{\infty}\frac{t-[t]-1/2}{x+t}\,\mathrm{d}t
$$

we get

$$
\Gamma'(x) = \Gamma(x)\log x + \Gamma(x)O(x^{-1}) \quad \text{as } x\to\infty.
$$

By the Leibnitz rule and the complete induction on l, we have

$$
\frac{\Gamma^{(l)}(m)}{\Gamma(m)} = \log^l m + O\left(\frac{\log^{l-1} m}{m}\right) \quad \text{as } m\to\infty,
$$

for all $l \geq 0$. By Lemma 2.3.9 we get, as $t \to 1^+$,

$$
\begin{aligned}
\sum_{m=1}^{\infty}\frac{\Gamma^{(l)}(m)}{\Gamma(m)}t^{m-1} &= \sum_{m=1}^{\infty}(\log^l m)t^{m-1} + O\left(\log^l\frac{1}{1-t}\right)\\
&= \sum_{m=1}^{\infty}P(\log m+\gamma)t^{m-1} + O\left(\log^l\frac{1}{1-t}\right),
\end{aligned}
$$

as desired. □

For $s \in \mathbb{N}$, define $\mathbf{z}_s = \mathrm{x}_0^{s-1}\mathrm{x}_1$. For any composition $\mathbf{s} = (s_1, \ldots, s_d) \in \mathbb{N}^d$, we set $\mathbf{z}_{\mathbf{s}} := \mathbf{z}_{s_1}\cdots\mathbf{z}_{s_d}$ and

$$
\zeta_*(\mathbf{s};T) := (-1)^d\zeta_*(\mathbf{z}_{\mathbf{s}}), \quad \zeta_{\sqcup\!\sqcup}(\mathbf{s};T) := (-1)^d\zeta_{\sqcup\!\sqcup}(\mathbf{z}_{\mathbf{s}}).
$$

Remark 3.3.18. We will always use $\zeta_{\sqcup\!\sqcup}(\mathbf{s};T)$ and $\zeta_*(\mathbf{s};T)$ when we consider them as polynomials of T, while we will write $\zeta_{\sqcup\!\sqcup}(\mathbf{s}) := \zeta_{\sqcup\!\sqcup}(\mathbf{s};0)$ and $\zeta_*(\mathbf{s}) := \zeta_*(\mathbf{s};0)$ for the constant terms only.

For $M > 0$, we define the partial sum of the MZVs

$$H_M(\mathbf{s}) := \sum_{M \ge k_1 > \cdots > k_d > 0} k_1^{-s_1} \cdots k_d^{-s_d}. \tag{3.18}$$

It is easy to see that

$$(-1)^{d+\ell} H_M(\mathbf{s}) H_M(\mathbf{t}) = H_M(\mathbf{z_s} * \mathbf{z_t})$$

where H_M is the $\mathbb{Q}$-linear map $\mathfrak{A}_*^1 \longrightarrow \mathbb{R}$ such that $H_M(\mathbf{z_s}) = (-1)^d H_M(\mathbf{s})$.

Lemma 3.3.19. *Put $J = \mathrm{dp}(\mathbf{s})$. Then as $M \to \infty$ we have*

$$H_M(\mathbf{s}) = \zeta_*(\mathbf{s}; \log M + \gamma) + O(M^{-1} \log^J M).$$

Moreover, if $\mathbf{s} = (\{1\}^\ell, \mathbf{s}')$ for admissible $\mathbf{s}'$ then the leading term of $\zeta_*(\mathbf{s};T)$ is $(1/\ell!)\zeta(\mathbf{s}')T^\ell$, where $\zeta(\emptyset) = 1$.

Proof. We first prove that for all $d \ge 1$ and $j \le d$ we have

$$A_{j,d}(M) = \sum_{k_1 > \cdots > k_j > M \ge k_{j-1} > \cdots > k_d > 0} \frac{1}{k_1^2 k_2 \cdots k_d} = O(M^{-1} \log^d M). \tag{3.19}$$

Indeed, if $d = 1$ we have $j = 1$ and

$$A_{1,1}(M) = \sum_{k>M} \frac{1}{k^2} < \frac{1}{M} \sum_{k>M} \frac{1}{k} = O(M^{-1} \log M).$$

Now by induction

$$\begin{aligned}
&\sum_{k_1 > \cdots > k_j > M \ge k_{j-1} > \cdots > k_d > 0} \frac{1}{k_1^2 k_2 \cdots k_d} \\
&\ll \sum_{k_1 > \cdots > k_j > M} \frac{1}{k_1^2 k_2 \cdots k_j} \sum_{M \ge k_{j-1} > \cdots > k_{d-1} > 0} \frac{\log k_{d-1}}{k_{j+1} \cdots k_{d-1}} + O\left(\frac{\log^{d-1} M}{M}\right) \\
&\ll \cdots \\
&\ll (\log^{d-j} M) \sum_{k_1 > M} \frac{\log^{j-1} k_1}{k_1^2} + O\left(\frac{\log^{d-1} M}{M}\right) \\
&\ll \frac{\log^{d-j} M}{M} \int_M^\infty \frac{\log^{j-1} x}{x} \, \mathrm{d}x + O\left(\frac{\log^{d-1} M}{M}\right) \ll \frac{\log^{d-1} M}{M}.
\end{aligned}$$

This proves Eq. (3.19).

We now proceed by induction on the number ℓ of leading 1's in $\mathbf{s}$. If $\ell = 0$ we have $s_1 > 1$.

$$\zeta(\mathbf{s}) - H_M(\mathbf{s}) = \sum_{j=1}^{d} \sum_{k_1>\cdots>k_j>M\geq k_{j-1}>\cdots>k_d>0} k_1^{-s_1}\cdots k_d^{-s_d}$$
$$\leq \sum_{j=1}^{d} \sum_{k_1>\cdots>k_j>M\geq k_{j-1}>\cdots>k_d>0} \frac{1}{k_1^2 k_2\cdots k_d} = O\left(\frac{\log^{d-1} M}{M}\right)$$

by Eq. (3.19). This proves the case $\ell = 0$ with $\zeta_*(\mathbf{s}; T) = \zeta(\mathbf{s})$ a constant. Suppose the lemma is true when the number of leading 1's in $\mathbf{s}$ is less than $\ell \geq 1$. Suppose further that $\mathbf{s} = (\{1\}^{\ell-1}, t_1, \ldots, t_d)$ with $t_1 > 1$ for some $d \geq 0$ (when $d = 0$ no t_j's will appear). Then

$$-H_M(\mathrm{x}_1) = H_M(1) = \sum_{k=1}^{M} \frac{1}{k} = \log M + \gamma + O(M^{-1}) \quad \text{as } M \to \infty,$$

where γ is Euler's constant. Using the stuffle relation we have

$$\ell H_M(1,\mathbf{s}) = H_M(1)H_M(\mathbf{s}) - \sum_{j=1}^{d} H_M(\{1\}^{\ell-1}, t_1, \ldots, t_j, 1, t_{j+1}, \ldots, t_d)$$
$$- \sum_{j=1}^{\ell-1} H_M(\{1\}^{j-1}, 2, \{1\}^{\ell-j-1}, t_1, \ldots, t_d)$$
$$- \sum_{j=1}^{d} H_M(\{1\}^{\ell-1}, t_1, \ldots, t_{j-1}, t_j + 1, t_{j+1}, \ldots, t_d).$$

By the induction assumption

$$\ell H_M(1,\mathbf{s})$$
$$= (\log M+\gamma)\zeta_*(\mathbf{s}; \log M+\gamma) + O\left(\frac{\log^{J+1} M}{M}\right)$$
$$- \sum_{j=1}^{d} \zeta_*(\{1\}^{\ell-1}, t_1, \ldots, t_j, 1, t_{j+1}, \ldots, t_d); \log M+\gamma) + O\left(\frac{\log^{J+1} M}{M}\right)$$
$$- \sum_{j=1}^{\ell-1} \zeta_*(\{1\}^{j-1}, 2, \{1\}^{\ell-j-1}, t_1, \ldots, t_d; \log M+\gamma) + O\left(\frac{\log^{J} M}{M}\right)$$
$$- \sum_{j=1}^{d} \zeta_*(\{1\}^{\ell-1}, t_1, \ldots, t_{j-1}, t_j+1, t_{j+1}, \ldots, t_d; \log M+\gamma) + O\left(\frac{\log^{J} M}{M}\right)$$
$$= \ell\zeta_*(1, \mathbf{s}; \log M+\gamma) + O(M^{-1}\log^{J+1} M),$$

since $\zeta_*(1;T)\zeta_*(\mathbf{s};T) = T\zeta_*(\mathbf{s};T)$ can be computed by the stuffle, which leads to the expression of $\zeta_*(1,\mathbf{s};T)$. In particular, the leading term of $\zeta_*(1,\mathbf{s};T)$ is equal to $\frac{T}{\ell}$ times the leading term of $\zeta_*(\mathbf{s};T)$. Thus the lemma follows from the inductive assumption. □

Now for $|z| < 1$ we define the single-variable MPL by

$$\mathrm{L}_{\mathbf{s}}(z) := \sum_{k_1>\cdots>k_d\geq 1} \frac{z^{k_1}}{k_1^{s_1}\cdots k_d^{s_d}} = (-1)^d \int_0^z \mathrm{x}_0^{s_1-1}\mathrm{x}_1\cdots\mathrm{x}_0^{s_d-1}\mathrm{x}_1, \tag{3.20}$$

where $\mathrm{x}_0 = \frac{\mathrm{d}x}{x}$ and $\mathrm{x}_1 = \frac{\mathrm{d}x}{x-1}$.

Lemma 3.3.20. *Put* $J = \mathrm{dp}(\mathbf{s})$. *Then as* $t \to 1^+$

$$\mathrm{L}_{\mathbf{s}}(t) = \zeta_{\sqcup\!\sqcup}\left(\mathbf{s}; \log\frac{1}{1-t}\right) + O\left((1-t)\log^J\left(\frac{1}{1-t}\right)\right). \tag{3.21}$$

Proof. We prove the lemma by induction on the number ℓ of leading 1's in $\mathbf{s}$. First assume $\ell = 0$ and so $s_1 > 1$. Set paths $\alpha_t = [0,t]$ and $\beta_t = [t,1]$. By Lemma 2.1.2(iii) we have

$$\begin{aligned}
\mathrm{L}_{\mathbf{s}}(1) - \mathrm{L}_{\mathbf{s}}(t) &= \sum_{k_1>\cdots>k_d\geq 1} \frac{1-t^{k_1}}{k_1^{s_1}\cdots k_d^{s_d}} \\
&\leq \sum_{k_1>\cdots>k_d\geq 1} \frac{1-t^{k_1}}{k_1^2 k_1\cdots k_d} \\
&= \mathrm{L}_{(2,\{1\}^{d-1})}(1) - \mathrm{L}_{(2,\{1\}^{d-1})}(t) \\
&= \sum_{j=1}^{d} \int_t^1 \frac{\mathrm{d}t}{t}\left(\frac{\mathrm{d}t}{1-t}\right)^{j-1} \int_0^t \left(\frac{\mathrm{d}t}{1-t}\right)^{d-j}
\end{aligned}$$

by applying Lemma 2.1.2(ii) to $\mathrm{L}_{(2,\{1\}^{d-1})}(1) = \int_0^1 \frac{\mathrm{d}t}{t}\left(\frac{\mathrm{d}t}{1-t}\right)^{d-1}$. Now

$$\int_0^t \left(\frac{\mathrm{d}t}{1-t}\right)^{d-j} = \log^{d-j}\left(\frac{1}{1-t}\right)$$

and by the change of variables $t_j \to 1-t_j$ and Lemma 2.1.2(iv)

$$-\int_t^1 \frac{\mathrm{d}t}{t}\left(\frac{\mathrm{d}t}{t-1}\right)^{j-1} = \int_0^{1-t}\left(\frac{\mathrm{d}t}{t}\right)^{j-1}\frac{\mathrm{d}t}{1-t} = \mathrm{L}_j(1-t) = O(1-t).$$

Thus

$$\mathrm{L}_{\mathbf{s}}(t) = \zeta(\mathbf{s}) + O\left((1-t)\log^d\left(\frac{1}{1-t}\right)\right).$$

Since $\zeta_{\sqcup\!\sqcup}(\mathbf{s};T) = \zeta(\mathbf{s})$ this proves the lemma when $\ell = 0$. The induction step can now be finished by using the shuffle relation, similar to the proof of Lemma 3.3.19. We leave out the details. This finishes the proof of the lemma. □

3.3.3 *Main Statement*

The following result provides the *regularized double shuffle relations* (regularized DBSFs).

Theorem 3.3.21. *For all* $\mathbf{w} \in \mathfrak{A}^1$, *one has*

$$(\zeta_{\sqcup\!\sqcup} - \rho \circ \zeta_*)(\mathbf{w}) = 0. \tag{3.22}$$

Proof. Let $J = \mathrm{dp}(\mathbf{s})$. Notice

$$\begin{aligned}
\mathrm{L}_{\mathbf{s}}(t) &= \sum_{m=1}^{\infty} \sum_{m>k_2>\cdots>k_d\geq 1} \frac{t^m}{m^{s_1}k_2^{s_2}\cdots k_d^{s_d}} \\
&= \sum_{m=1}^{\infty} (H_m(\mathbf{s}) - H_{m-1}(\mathbf{s}))t^m \\
&= (1-t)\sum_{m=1}^{\infty} H_m(\mathbf{s})t^m
\end{aligned}$$

since $H_0(\mathbf{s}) = 0$. Thus

$$\begin{aligned}
&\zeta_{\sqcup\!\sqcup}\left(\mathbf{s}; \log\frac{1}{1-t}\right) \xlongequal{\text{Lem. 3.3.20}} \mathrm{L}_{\mathbf{s}}(t) + O\left((1-t)\log^J\left(\frac{1}{1-t}\right)\right) \\
&\xlongequal{\text{Lem. 3.3.19}} (1-t)\sum_{m=1}^{\infty} \zeta_*(\mathbf{s}; \log m + \gamma)t^m + O\left((1-t)\sum_{m=1}^{\infty} \frac{\log^J m}{m} t^{m-1}\right) \\
&\xlongequal{\text{Lem. 2.3.9}} (1-t)\sum_{m=1}^{\infty} \zeta_*(\mathbf{s}; \log m + \gamma)t^m + O\left((1-t)\log^{J+1}\left(\frac{1}{1-t}\right)\right) \\
&\xlongequal{\text{Lem. 3.3.17}} tQ\left(\log\frac{1}{1-t}\right) + O\left((1-t)\log^{J+1}\left(\frac{1}{1-t}\right)\right) \\
&\xlongequal{t=1+(t-1)} Q\left(\log\frac{1}{1-t}\right) + O\left((1-t)\log^{J+1}\left(\frac{1}{1-t}\right)\right)
\end{aligned}$$

with $Q(T) = \rho(\zeta_*(\mathbf{s}; T))$ since $\deg_T \zeta_*(\mathbf{s}; T) \leq J$. Thus $\zeta_{\sqcup\!\sqcup}(\mathbf{s}; T) = \rho(\zeta_*(\mathbf{s}; T))$. The theorem is now proved since $\mathbf{s} \in \mathbb{N}^d$ is arbitrary. □

Remark 3.3.22. One can compute ρ explicitly as follows. First expand $\Lambda(u)$ into a power series:

$$\Lambda(u) = \exp\left(\sum_{n=2}^{\infty} \frac{(-1)^n}{n}\zeta(n)u^n\right) = \sum_{k=0}^{\infty} c_k u^k,$$

where $c_0 = 1, c_1 = 0, c_2 = \zeta(2)/2, c_3 = -\zeta(3)/3, c_4 = 9\zeta(4)/16, \cdots$. Then

$$\rho(T^n) = \sum_{k=0}^{n} k!c_k \binom{n}{k} T^{n-k}, \quad \forall n \geq 0.$$

Corollary 3.3.23. *For all* $\mathbf{w}_0 \in \mathfrak{A}^0$ *and* $\mathbf{w}_1 \in \mathfrak{A}^1$, *one has*

$$\zeta_{\sqcup\!\sqcup}(\mathbf{w}_1 \sqcup\!\sqcup \mathbf{w}_0 - \mathbf{w}_1 * \mathbf{w}_0) = 0. \tag{3.23}$$

Proof. We can get the desired result by multiplying $\zeta(\mathbf{w}_0)$ on both sides of $\zeta_{\sqcup\!\sqcup}(\mathbf{w}_1) = \rho(\zeta_*(\mathbf{w}_1))$ and using $\mathbb{R}$-linearity of ρ. □

Definition 3.3.24. The *regularized double shuffle relations* (regularized DBSFs) of the MZVs are those $\mathbb{Q}$-linear relations obtained by setting $T = 0$ in Eq. (3.22) for all words $\mathbf{w} \in \mathfrak{A}^1$.

Conjecture 3.3.25. *The non-finite regularized DBSFs of the MZVs are generated by the Hoffman relations* $\zeta(\mathtt{x}_1 \sqcup\!\sqcup \mathbf{w} - \mathtt{x}_1 * \mathbf{w})$ *for all admissible words* $\mathbf{w}$.

If $\mathbf{w} \in \mathfrak{A}^1 \setminus \mathfrak{A}^0$ then the leading letter of $\mathbf{w}$ must be $\mathtt{x}_1$. The conjecture says that to produce all the non-finite regularized shuffle relations we only need to regularize divergent MZVs with only one leading $\mathtt{x}_1$.

Example 3.3.26. The simplest regularized DBSF is obtained by considering the word $\mathtt{x}_1\mathtt{x}_0\mathtt{x}_1 \in \mathfrak{A}^1$. First, since $\zeta(2) = -\zeta_{\sqcup\!\sqcup}(\mathtt{x}_0\mathtt{x}_1)$ and $T = -\zeta_{\sqcup\!\sqcup}(\mathtt{x}_1)$

$$\zeta(2)T = \zeta_{\sqcup\!\sqcup}(\mathtt{x}_0\mathtt{x}_1 \sqcup\!\sqcup \mathtt{x}_1) = \zeta_{\sqcup\!\sqcup}(\mathtt{x}_1\mathtt{x}_0\mathtt{x}_1) + 2\zeta_{\sqcup\!\sqcup}(\mathtt{x}_0\mathtt{x}_1^2) = \zeta_{\sqcup\!\sqcup}(\mathtt{x}_1\mathtt{x}_0\mathtt{x}_1) + 2\zeta(2,1).$$

On the other hand,

$$\zeta(2)T = \zeta_*(\mathtt{x}_0\mathtt{x}_1)\zeta_*(\mathtt{x}_1) = \zeta_*(\mathtt{x}_0\mathtt{x}_1 * \mathtt{x}_1) = \zeta_*(\mathtt{x}_1\mathtt{x}_0\mathtt{x}_1) - \zeta_*(\mathtt{x}_0^2\mathtt{x}_1) + \zeta_*(\mathtt{x}_0\mathtt{x}_1^2).$$

Observe that $-\zeta_*(\mathtt{x}_0\mathtt{x}_0\mathtt{x}_1) = \zeta(3) = \zeta(2,1) = \zeta_*(\mathtt{x}_0\mathtt{x}_1\mathtt{x}_1)$ by (3.24). Applying Thm. 3.3.21 to $\mathbf{w} = \mathtt{x}_1\mathtt{x}_0\mathtt{x}_1$ we get

$$\zeta(2)T - 2\zeta(2,1) = \rho(\zeta(2)T - \zeta(3) - \zeta(2,1)) = \zeta(2)T - \zeta(3) - \zeta(2,1)$$

since $\rho(T) = T$ by Eq. (3.15). This leads to the famous identity first discovered by Euler:

$$\zeta(3) = \zeta(2,1). \tag{3.24}$$

3.4 Dimension Conjectures Over $\mathbb{Q}$

One of the guiding principles in the study of the MZVs is the following widely believed conjecture.

Conjecture 3.4.1. *There are no nontrivial* $\mathbb{Q}$*-linear relations among the MZVs of different weights.*

This means that every MZV identity involving different weights is a sum of a bunch of identities each of which has pure weight.

For the MZVs of the same weight, by Example 3.3.2 and Example 3.3.26, often there exist $\mathbb{Q}$-linear relations. Let MZ_w be the $\mathbb{Q}$-vector space spanned by all the MZVs of weight w. Zagier made the following famous conjecture.

Conjecture 3.4.2. *Define the sequence D_w $(w \geq 1)$ by*

$$D_0 = 1, D_1 = 0, D_2 = 1, D_w = D_{w-2} + D_{w-3} \ \forall w \geq 3. \tag{3.25}$$

Then $\dim \mathsf{MZ}_w = D_w$ *for all* $w \geq 1$.

Remark 3.4.3. The numbers D_w are called Padovan numbers. According to OEIS [518] the Padovan sequence A000931 is

$$1, 0, 0, 1, 0, 1, 1, 1, 2, 2, 3, 4, 5, 7, 9, 12, 16, 21, \dots.$$

Using the finite and regularized DBSFs one can show that each MZV of weight less than 8 is a polynomial of ordinary Riemann zeta values with $\mathbb{Q}$ coefficients. But already $\zeta(3,5)$ of weight 8 seems to fail this rule even if no proof has been found. In fact, this would be the case if the following conjecture could be verified.

Conjecture 3.4.4. *Every linear relation among the MZVs can be proved by the finite and regularized DBSFs.*

It is quite fruitful to reinterpret some of the above results in terms of a certain graded Hopf algebra (see Appendix A). Let $\mathcal{L}$ denote the free Lie algebra generated by one element e_{2n+1} in every odd degree $-2n-1$ for $n > 1$. Graded by the degree the underlying $\mathbb{Q}$-vector space is generated by,

$$e_3; \quad e_5; \quad e_7; \quad [e_3, e_5]; \quad e_9; \quad [e_3, e_7]; \quad e_{11}, [e_3, [e_5, e_3]]; \quad \cdots.$$

Let $\mathcal{UL}$ be its universal enveloping algebra, and let $\mathcal{UL}^\vee$ be its graded dual, which is a commutative Hopf algebra. One may think of $\mathcal{UL}^\vee$ as the set of all non-commutative words in letters f_{2n+1} in degree $2n+1$ dual to e_{2n+1}, equipped with the shuffle product $ш$. Finally, set

$$\mathcal{A} = \mathbb{Q}[f_2] \otimes \mathbb{Q}(\langle f_3, f_5, f_7, \dots \rangle, ш)$$

where f_2 is a new generator of degree 2 which commutes with all the others. The generators in each weight w with $w \leq 9$ are

$$f_2; \quad f_3; \quad f_2^2; \quad f_5, f_2 f_3; \quad f_2^3, f_3^2; \quad f_7, f_2 f_5, f_2^2 f_3;$$
$$f_2^4, f_2 f_3^2, f_3 ш f_5, f_3 f_5; \quad f_2^3 f_3, f_3^3, f_2 f_7, f_2^2 f_5, f_9$$

which correspond to the following MZVs:

$$\zeta(2);\ \zeta(3);\ \zeta(2)^2;\ \zeta(5),\zeta(2)\zeta(3);\ \zeta(2)^3,\zeta(3,3);\ \zeta(7),\zeta(2)\zeta(5),\zeta(2)^2\zeta(3);$$
$$\zeta(2)^4,\zeta(2)\zeta(3,3),\zeta(3)\zeta(5),\zeta(3,5);\ \zeta(2)^3\zeta(3),\zeta(3)^3,\zeta(2)\zeta(7),\zeta(2)^2\zeta(5),\zeta(9)$$

The next conjecture is a more precise version of Zagier's Conjecture 3.4.2 .

Conjecture 3.4.5. *The algebra spanned by the MZVs over* $\mathbb{Q}$ *is isomorphic to* $\mathscr{A}$.

Actually, we only need the elements f_2 and f_3 to generate all the others by the rule

$$f_{2n+1} \to f_3 f_2^{n-1} \qquad \forall n \geq 1.$$

Hence if we denote $\mathscr{F}(2,3)$ the free graded Lie algebra generated by two elements of degree -2 and -3, and $\mathscr{F}(2,3)^\vee$ its graded dual, then with the same notation we have a canonical isomorphism of graded vector spaces

$$\mathscr{A} \cong \mathscr{U}\mathscr{F}(2,3)^\vee.$$

Let d'_w be the dimension of the graded w piece of $\mathscr{A}$. Then it is clear that

$$d'_1 = 0, d'_2 = 1, d'_3 = 1, d'_w = d'_{w-2} + d'_{w-3}\ \forall w \geq 4.$$

One can even consider MZVs with both depth and weight fixed. The next conjecture provides the hypothetical dimensions of $\mathbb{Q}$-spaces generated by such values. Let $\mathsf{MZ}_{w,\leq d}$ be the $\mathbb{Q}$-vector space spanned by all the MZVs of weight w and depth at most d. Let

$$\mathsf{MZ}_{w,+,\leq d} := \sum_{\substack{w_1,w_2\geq 1,\ w_1+w_2=w \\ 1\leq d_1,d_2<d,\ d_1+d_2\leq d}} \mathsf{MZ}_{w_1,\leq d_1} \cdot \mathsf{MZ}_{w_2,\leq d_2}$$

and

$$\mathsf{MZ}_{w,d} := \mathsf{MZ}_{w,\leq d} \Big/ \Big(\mathsf{MZ}_{w,\leq d-1} + \mathsf{MZ}_{w,+,\leq d}\Big)$$

be the $\mathbb{Q}$-vector space generated by the MZVs of weight w and depth d that are not $\mathbb{Q}$-linear combination of lower depth MZVs and products of lower weight MZVs.

Conjecture 3.4.6. *We have* $\dim_{\mathbb{Q}} \mathsf{MZ}_{w,d} = D_{w,d}$ *where* $D_{2,1} = 1$ *and*

$$\prod_{w\geq 3}\prod_{d\geq 1}(1 - x^w y^d)^{D_{w,d}} = 1 - \frac{x^3 y}{1-x^2} + \frac{x^{12}y^2(1-y^2)}{(1-x^4)(1-x^6)}.$$

Remark 3.4.7. The conjectured term

$$\frac{x^3y}{1-x^2}=\sum_{n\geq 2} x^{2n-1}y$$

originates from the Riemann zeta values $\zeta(2n)$ for $n \geq 2$ since all these values can be generated by $\zeta(2)$. For the other term, we have the expansion

$$\frac{x^{12}}{(1-x^4)(1-x^6)}=\sum_{n\geq 12, 2|n} \mathrm{sp}(n)x^n$$

where $\mathrm{sp}(n) = \lfloor (n-4)/4 \rfloor - \lfloor (n-2)/6 \rfloor$ is the dimension of the space of weight n cusp forms on the full modular group. So MZVs are expected to be intimately related to the modular forms. See Sec. 5.9.

Similar to the situation for $\dim_{\mathbb{Q}} \mathsf{MZ}_w$, it is unlikely to prove $\dim_{\mathbb{Q}} \mathsf{MZ}_{w,d} \geq D_{w,d}$ in the near future. However, Conjecture 3.4.6 implies that $D_{w,d} = 0$ unless $w \geq 3d$ and $2 \mid (w-d)$. It also implies the following conjectured formulas for depth $d \leq 3$:

$$D_{w,1} = 1, \quad D_{w,2} = \left[\frac{w-2}{6}\right], \quad D_{w,3} = \left[\frac{(w-3)^2-1}{48}\right] \tag{3.26}$$

for all w having the same parity as the depth. Note that according to the Parity Principle (see Cor. 3.5.7) $D_{w,2} = 0$ if w is odd and $D_{w,3} = 0$ if w is even.

3.5 Linearized Double Shuffle and Parity Principle

In this section we will study the DBSFs modulo elements of lower depths and products. Recall that $\mathsf{MZ}_{w,d}$ is the $\mathbb{Q}$-vector space generated by the MZVs of weight w and depth d that cannot be expressed as a $\mathbb{Q}$-linear combination of lower depth MZVs, augmented by products of lower weight MZVs. Set

$$\mathcal{M} = \sum_{w,n\geq 1} \mathsf{MZ}_{w,n}.$$

Of course, this is a direct sum conjecturally. Write $V_n = \mathbb{Q}[x_1, \dots, x_n]$ and denote by $V_n(d)$ the homogeneous polynomials of degree d. For $f(x_1, \dots, x_n) \in V_n$, define the action of $M \in \mathrm{GL}_n(\mathbb{Z})$ on $f \in V_n$ by

$$(f|M)(x_1, \dots, x_n) = f((x_1, \dots, x_n) \cdot M^{-1}).$$

Note that $f|(M_1M_2) = (f|M_1)|M_2$ for all $M_1, M_2 \in \mathrm{GL}_n(\mathbb{Z})$. We can then extend this action to $\mathbb{Z}[\mathrm{GL}_n(\mathbb{Z})]$ by linearity.

Let $\mathfrak{S}_n$ be the symmetry group of order n regarded as a subgroup of $\mathrm{GL}_n(\mathbb{Z})$ so that the action of $\sigma \in \mathfrak{S}_n$ on $f \in V_n$ is given by

$$(f|\sigma)(x_1, \dots, x_n) = f(x_{\sigma^{-1}(1)}, \dots, x_{\sigma^{-1}(n)}).$$

Explicitly, the matrix M_σ for $\sigma \in \mathfrak{S}_n$ can be obtained by applying σ to the columns of the $n \times n$ identity matrix, namely, its (i,j)th entry is $\delta_{i,\sigma(i)}$. In what follows we abuse notation and just write σ for M_σ by identifying σ with its image in $\mathrm{GL}_n(\mathbb{Z})$.

For any $j \le n-1$, define the shuffle element

$$\mathrm{sh}_j := \mathrm{sh}_j^{(n)} = \sum_{\sigma \in \mathrm{Sh}_j^{(n)}} \sigma \tag{3.27}$$

where $\mathrm{Sh}_j^{(n)} = \{\sigma \in \mathfrak{S}_n | \sigma(1) < \cdots < \sigma(j), \sigma(j+1) < \cdots < \sigma(w)\}$. Further, for all $f \in V_n$ we set $f^\sharp = f|P$ where

$$P = \begin{pmatrix} 1 & & & \\ -1 & 1 & & \\ & \ddots & \ddots & \\ & & -1 & 1 \end{pmatrix}, \qquad P^{-1} = \begin{pmatrix} 1 & & & \\ 1 & 1 & & \\ \vdots & \vdots & \ddots & \\ 1 & 1 & \cdots & 1 \end{pmatrix}. \tag{3.28}$$

Clearly

$$f^\sharp(x_1, \dots, x_n) = f(x_1 + \cdots + x_n, x_2 + \cdots + x_n, \dots, x_{n-1} + x_n, x_n).$$

Definition 3.5.1. The *double shuffle space* DSh_n of V_n is defined by

$$\mathsf{DSh}_n = \left\{ f \in V_n \,\middle|\, f|\,\mathrm{sh}_j^{(n)} = f^\sharp|\,\mathrm{sh}_j^{(n)} = 0 \text{ for } j = 1, \dots, n-1 \right\}$$

for all $n \ge 2$. In particular $\mathsf{DSh}_1 = \mathbb{Q}[x_1]$. Denote by $\mathsf{DSh}_n(d)$ the degree d part of DSh_n.

Setting $S^\sharp = PSP^{-1}$ for all $S \in \mathbb{Z}[\mathrm{GL}_n(\mathbb{Z})]$, we see that $f^\sharp|\,\mathrm{sh}_j = 0$ is equivalent to $f|\,\mathrm{sh}_j^\sharp = 0$. Our goal in this section is to study DSh_n by considering the following two generating functions

$$F_{ш}^{(n)}(x_1, \dots, x_n) = \sum_{s_1, \dots, s_n \ge 1} \zeta_{ш}(s_1, \dots, s_n) x_1^{s_1-1} \dots x_n^{s_n-1},$$

$$F_*^{(n)}(x_1, \dots, x_n) = \sum_{s_1, \dots, s_n \ge 1} \zeta_*(s_1, \dots, s_n) x_1^{s_1-1} \dots x_n^{s_n-1}.$$

We regard these as elements in $\widetilde{\mathcal{M}}[x_1, \dots, x_n] = \widetilde{\mathcal{M}} \otimes_{\mathbb{Q}} V_n$ where $\widetilde{\mathcal{M}} = \mathcal{M} + \mathbb{Q}$ is a $\mathbb{Q}$-vector space regarded as a $\mathbb{Q}$-algebra (again, conjecturally $\widetilde{\mathcal{M}} = \mathcal{M} \oplus \mathbb{Q}$).

Lemma 3.5.2. *If* $1 \le j < n$ *then* $(F_{\sqcup\!\sqcup}^{(n)})^\sharp|\,\mathrm{sh}_j = 0$ *in* $\mathcal{M} \otimes_{\mathbb{Q}} V_n$.

Proof. For $\mathbf{s} = (s_1, \ldots, s_n) \in \mathbb{N}^n$ with $|\mathbf{s}| = w$ and $\varepsilon < 1$, we define

$$\zeta_\varepsilon(\mathbf{s}) = \int\cdots\int_{1-\varepsilon>t_1>\cdots>t_w>0} \omega(t_1)\cdots\omega(t_w)$$

where the integrand is the same as in the ordinary MZV defined by Eq. (3.1). Now applying the formula

$$\int\cdots\int_{a>x_1>\cdots>x_r>b} \frac{\mathrm{d}x_1}{x_1}\cdots\frac{\mathrm{d}x_r}{x_r} = \frac{1}{r!}\left(\log\frac{a}{b}\right)^r$$

to $(x_0, b, r) = (1-\varepsilon, u_1, s_1-1), (u_1, u_2, s_2-1), \ldots, (u_{n-1}, u_n, s_n-1)$, where $u_1 = t_{s_1}$, $u_2 = t_{s_1+s_2}, \ldots,$ $u_n = t_w$, we find

$$\zeta_\varepsilon(\mathbf{s}) = \frac{1}{(s_1-1)!\cdots(s_n-1)!}\int\cdots\int_{1-\varepsilon>u_1>\cdots>u_n>0}\left(\log\frac{1-\varepsilon}{u_1}\right)^{s_1-1}\frac{\mathrm{d}u_1}{1-u_1}$$
$$\left(\log\frac{u_1}{u_2}\right)^{s_2-1}\frac{\mathrm{d}u_2}{1-u_2}\cdots\left(\log\frac{u_{n-1}}{u_n}\right)^{s_n-1}\frac{\mathrm{d}u_n}{1-u_n}.$$

Easy computation using $\sum_{i\ge0} t^n/n! = e^t$ yields

$$F_{\sqcup\!\sqcup,\varepsilon}^{(n)}(x_1, \ldots, x_n) := \sum_{s_1,\ldots,s_n\ge1} \zeta_\varepsilon(s_1, \ldots, s_n)x_1^{s_1-1}\ldots x_n^{s_n-1}$$
$$= (1-\varepsilon)^{x_1}\int\cdots\int_{1-\varepsilon>u_1>\cdots>u_n>0}\frac{u_1^{-x_1+x_2}\mathrm{d}u_1}{1-u_1}\cdots\frac{u_{n-1}^{-x_{n-1}+x_n}\mathrm{d}u_{n-1}}{1-u_{n-1}}\frac{u_n^{-x_n}\mathrm{d}u_n}{1-u_n}.$$

Thus

$$(F_{\sqcup\!\sqcup,\varepsilon}^{(n)})^\sharp(x_1, \ldots, x_n) = F_{\sqcup\!\sqcup,\varepsilon}^{(n)}(x_1+\cdots+x_n, \ldots, x_{n-1}+x_n, x_n)$$
$$= (1-\varepsilon)^{x_1+\cdots+x_n}\int\cdots\int_{1-\varepsilon>u_1>\cdots>u_n>0}\frac{u_1^{-x_1}\mathrm{d}u_1}{1-u_1}\cdots\frac{u_{n-1}^{-x_{n-1}}\mathrm{d}u_{n-1}}{1-u_{n-1}}\frac{u_n^{-x_n}\mathrm{d}u_n}{1-u_n}.$$

Hence $(F_{\sqcup\!\sqcup,\varepsilon}^{(n)})^\sharp$ satisfies the shuffle relation

$$(F_{\sqcup\!\sqcup,\varepsilon}^{(j)})^\sharp(x_1, \ldots, x_j)(F_{\sqcup\!\sqcup,\varepsilon}^{(n-j)})^\sharp(x_{j+1}, \ldots, x_n) = \big((F_{\sqcup\!\sqcup,\varepsilon}^{(n)})^\sharp|\,\mathrm{sh}_j\big)(x_1, \ldots, x_n)$$

for any $\varepsilon < 1$ and $1 \le j < n$. The lemma now follows from the definition of $\zeta_{\sqcup\!\sqcup}(\mathbf{s}; T)$. □

Theorem 3.5.3. *For all $n \geq 1$, we have*

$$F_{\sqcup\!\sqcup}^{(n)}(x_1, \ldots, x_n) = F_*^{(n)}(x_1, \ldots, x_n) \tag{3.29}$$

in $\mathcal{M} \otimes_{\mathbb{Q}} V_n$. Moreover, both belong to the subspace $\sum_w \mathsf{MZ}_{w,\leq n} \otimes_{\mathbb{Q}} \mathsf{DSh}_n$ modulo $\sum_w \mathsf{MZ}_{w,+,\leq n}$.

Proof. First, we establish Eq. (3.29). If $n = 1$ we see by the definition that $\zeta_{\sqcup\!\sqcup}(s) = \zeta_*(s) = (1 - \delta_{1,s})\zeta(s)$ for all positive integer s, where $\delta_{1,s}$ is the Kronecker symbol. Thus the two generating functions $F_{\sqcup\!\sqcup}^{(1)}(x_1)$ and $F_*^{(1)}(x_1)$ are exactly the same. When $n \geq 2$ by Thm. 3.3.21, we see that each coefficient of $F_{\sqcup\!\sqcup}^{(n)}(x_1, \ldots, x_n) - F_*^{(n)}(x_1, \ldots, x_n)$ comes from the products of some MZV factors with the coefficients of $\rho(T^i)$. Now the coefficients of $\rho(T^i)$ are all $\mathbb{Q}$-linear combinations of at most i Riemann zeta values (see Eq. (3.15)) which can be written as $\mathbb{Q}$-linear combinations of the MZVs of depth at most i. Then MZV factors can be easily seen to have depths at most $n - i$ (for each T the weight of the corresponding MZVs must be decreased by 1).

To prove the second sentence in the theorem, we have

$$\begin{aligned}
0 &= F_*^{(j)}(x_1, \ldots, x_j) F_*^{(n-j)}(x_{j+1}, \ldots, x_n) \\
&= \sum_{s_1, \ldots, s_n \geq 1} \zeta_*(s_1, \ldots, s_j)\zeta_*(s_{j+1}, \ldots, s_n) x_1^{s_1-1} \ldots x_n^{s_n-1} \\
&= \sum_{s_1, \ldots, s_n \geq 1} \sum_{\sigma \in \mathrm{Sh}_j^{(n)}} \zeta_*(s_{\sigma^{-1}(1)}, \ldots, s_{\sigma^{-1}(n)}) x_1^{s_1-1} \ldots x_n^{s_n-1} \\
&= (F_*^{(n)} | \, \mathrm{sh}_j)(x_1, \ldots, x_n),
\end{aligned}$$

since the "stuffing" part of the stuffle product always produces the MZVs of lower depth. By Eq. (3.29) and Lemma 3.5.2 the theorem follows immediately. □

The following bound on the dimension of $\mathsf{MZ}_{w,d}$ (see Sec. 3.4) is an easy consequence of Thm. 3.5.3. The proof is left as an exercise (see Exercise 3.14).

Corollary 3.5.4. *For all $w > d > 0$, we have*

$$D_{w,d} \leq \dim_{\mathbb{Q}} \mathsf{DSh}_d(w - d).$$

Our next goal is to prove the Parity Principle in Cor. 3.5.7. The idea is to define a suitable space $\mathsf{ShC}_n(d)$ containing the double shuffle space $\mathsf{DSh}_n(d)$ and satisfying $\mathsf{ShC}_n(d) = \{0\}$ when d is odd and $n \geq 1$. In order to do so we need to define an action of $\mathfrak{S}_{n+1}$ on V_n (extending that of $\mathfrak{S}_n$) by

$$(F|\sigma)(x_1,\dots,x_n) := F(x_{\sigma^{-1}(1)},\dots,x_{\sigma^{-1}(n)}), \quad \text{for all } \sigma \in \mathfrak{S}_{n+1},$$

where $x_{n+1} = -x_1 - x_2 - \cdots - x_n$. Let

$$C_{n+1} = \begin{pmatrix} 1 & \cdots & n-1 & n & n+1 \\ 2 & \cdots & n & n+1 & 1 \end{pmatrix} \in \mathfrak{S}_{n+1}. \tag{3.30}$$

Then $(f|C_{n+1})(x_1,\dots,x_n) = f(x_{n+1},x_1,\dots,x_{n-1})$ so that the matrix form of C_{n+1} in $\mathrm{GL}_n(\mathbb{Z})$ is

$$C_{n+1} = \begin{pmatrix} 0 & \cdots & 0 & -1 \\ 1 & & & -1 \\ & \ddots & & \vdots \\ & & 1 & -1 \end{pmatrix}, \quad \text{and} \quad C_{n+1}^{-1} = \begin{pmatrix} -1 & 1 & & \\ -1 & 0 & \ddots & \\ \vdots & \vdots & \ddots & 1 \\ -1 & 0 & \cdots & 0 \end{pmatrix}.$$

Let $\nu(x_j) = -x_j$ for all $j \le n$ then its matrix form in $\mathrm{GL}_n(\mathbb{Z})$ is $-I_n$ where I_n is the $n \times n$ identity matrix. Clearly $\nu(x_{n+1}) = -x_{n+1}$.

Lemma 3.5.5. *For $0 \le j \le n$, define the involutions*

$$T_j = \begin{pmatrix} 1 & \cdots & j & j+1 & \cdots & n \\ 1 & \cdots & j & n & \cdots & j+1 \end{pmatrix} \in \mathfrak{S}_n. \tag{3.31}$$

Then

$$\sum_{j=1}^{n-1} (-1)^{j-1} \operatorname{sh}_j T_j = T_0 + (-1)^n.$$

Proof. For $1 \le j \le n$, we define

$$R_j = \sum_{\sigma \in \mathfrak{S}_n, \sigma(1) < \cdots < \sigma(j) > \sigma(j+1) > \cdots > \sigma(n)} \sigma.$$

Then $\operatorname{sh}_0 T_0 = R_1$, $\operatorname{sh}_n T_n = R_n$, and for $1 \le j \le n-1$, we have

$$\operatorname{sh}_j T_j = \sum_{\sigma \in \mathfrak{S}_n, \sigma(1) < \cdots < \sigma(j), \sigma(j+1) > \cdots > \sigma(n)} \sigma = R_j + R_{j+1},$$

since either $\sigma(j) > \sigma(j+1)$ or $\sigma(j) < \sigma(j+1)$. The lemma follows immediately. □

Theorem 3.5.6. *For $n, d \ge 1$, we define*

$$\mathsf{ShC}_n(d) := \Big\{ f \in V_n(d) \,\Big|\, f^\sharp | \operatorname{sh}_j = 0 \ (1 \le j < n), f^\sharp | C_{n+1} = \nu(f^\sharp) \Big\}. \tag{3.32}$$

Then

(i) $\mathsf{DSh}_n(d) \subset \mathsf{ShC}_n(d)$;
(ii) $\mathsf{ShC}_n(d) = 0$ *if d is odd.*

Proof. (i) By Lemma 3.5.5, for all $f \in \mathsf{DSh}_n(d)$ we have

$$f|T_0 = f|T_0^\sharp = (-1)^{n-1} f. \tag{3.33}$$

Define the involution

$$T' = \begin{pmatrix} 1 & 2 & \cdots & n-1 & n & n+1 \\ n-1 & n-2 & \cdots & 1 & n+1 & n \end{pmatrix} \in \mathfrak{S}_{n+1}. \tag{3.34}$$

Then $(f|T')(x_1,\dots,x_n) = f(x_{n-1},\dots,x_1,x_{n+1})$ so that the matrix form of T' in $\mathrm{GL}_n(\mathbb{Z})$ is

$$T' = T'^{-1} = \begin{pmatrix} & & 1 & -1 \\ & \iddots & & \vdots \\ 1 & & & -1 \\ 0 & \cdots & 0 & -1 \end{pmatrix}.$$

It is not hard to use the matrix forms to show that (see also Excercise 3.15)

$$\text{(A)}.\ T_0T' = C_{n+1}, \qquad \text{and} \qquad \text{(B)}.\ PT' = \nu T_0 P. \tag{3.35}$$

Thus

$$f^\sharp|C_{n+1} \overset{(3.35)(\mathrm{A})}{=\!=\!=} f^\sharp|T_0T' \overset{\mathrm{defn}}{=\!=} (f|P)|T_0T' = (f|PT_0)|T' \overset{\mathrm{defn}}{=\!=} (f|T_0^\sharp P)|T'$$
$$\overset{(3.33)}{=\!=} (-1)^{n-1}(f|P)|T' \overset{(3.35)(\mathrm{B})}{=\!=\!=} (-1)^{n-1}\nu(f|T_0P) \overset{(3.33)}{=\!=} \nu(f|P) = \nu(f^\sharp).$$

So $f \in \mathsf{ShC}_n(d)$ and therefore (i) is proved.

(ii) Let $\tau_{1,n+1}$ be the transposition $1 \leftrightarrow n+1$. It is straightforward to check (see Exercise 3.16) that in $\mathbb{Z}[\mathfrak{S}_{n+1}]$

$$1 + \mathrm{sh}_1^{(n)}\, C_{n+1} = C_{n+1}(1 + \mathrm{sh}_1^{(n)}\, \tau_{1,n+1}). \tag{3.36}$$

Now suppose $f \in \mathsf{ShC}_n(d)$ and let $F = f^\sharp$. Then $F|\,\mathrm{sh}_1 = 0$ and $F|C_{n+1} = \nu F$ by the definition. Thus by Eq. (3.36)

$$F = F|C_{n+1}(1 + \mathrm{sh}_1^{(n)}\, \tau_{1,n+1}) = \nu\big(F|(1 + \mathrm{sh}_1^{(n)}\, \tau_{1,n+1})\big) = \nu F = -F$$

if the degree d is odd. Hence $F = 0$ and thus $f = F|P^{-1} = 0$. This completes the proof of the theorem. ☐

Corollary 3.5.7. (Parity Principle) *If the weight w and depth n of $\zeta(\mathbf{s})$ have different parities then $\zeta(\mathbf{s})$ is a $\mathbb{Q}$-linear combination of lower depth MZVs and products of lower weight MZVs whose sum of depths is bounded by that of* $\mathbf{s}$.

Proof. If $|\mathbf{s}|$ and n have different parities then the degree of $x_1^{s_1-1}\dots x_n^{s_n-1}$ is odd. So the corollary follows from part (ii) of the above theorem. □

The next result is another kind of Parity Principle.

Corollary 3.5.8. *If the $w = |\mathbf{s}|$ and n have the same parity then both $\zeta_*(\mathbf{s}) + (-1)^w\zeta_*(\overleftarrow{\mathbf{s}})$ and $\zeta_{Ш}(\mathbf{s}) + (-1)^w\zeta_{Ш}(\overleftarrow{\mathbf{s}})$ are $\mathbb{Q}$-linear combinations of lower depth MZVs and products of lower weight MZVs.*

Proof. By Eq. (3.33) we see that

$$\sum_{\mathbf{s}\in\mathbb{N}^n} \zeta_*(\overleftarrow{\mathbf{s}})\mathbf{x}^{|\mathbf{s}|-\mathbf{1}} = \sum_{\mathbf{s}\in\mathbb{N}^n} \zeta_*(\mathbf{s})\mathbf{x}^{|\overleftarrow{\mathbf{s}}|-\mathbf{1}} = (F_*^{(n)}|T_0)(\mathbf{x}) = (-1)^{n-1}F_*^{(n)}(\mathbf{x}).$$

So we get

$$\zeta_*(\mathbf{s}) + (-1)^w\zeta_*(\overleftarrow{\mathbf{s}}) \equiv \zeta_*(\mathbf{s}) + (-1)^{w+n-1}\zeta_*(\mathbf{s}) \equiv 0$$

modulo $\mathsf{MZ}_{w,+,\leq n}$. Similar argument is valid for the Ш-version, too. This completes the proof of the corollary. □

3.6 Historical Notes

Even though some special cases of the theory of the multiple zeta values were discovered by Goldbach and Euler, their formal definition and study was initiated only recently by Hoffman [288] and Zagier [598] independently. Incidentally, in early 1980s, while developing a theory of resurgence in complex analysis, Ecalle studied some quite general mathematical objects called "moulds" (functions with variable number of variables) of which the MZVs are one of the examples [185, 186, p. 429]. He even mentioned their iterated integral representation [185, p. 431, Remark 4] without explicitly producing it. For an introduction to this theory, see Schneps' exposition [502].

In recent years, the study of the MZVs has opened up quite a lot fascinating connections to other branches of mathematics such as knot theory, Mahler measures and the theory of motives (see [83, 102, 104, 164, 368]). MZVs have also emerged unexpectedly in high energy physics such as Feyman diagrams, amplitude, mirror symmetry and quantum field theory (see Chap. 15 and Refs. [83, 90, 108, 291, 500]).

One of the most important algebraic approaches to the study of the MZVs is through the connection between the two kinds of products: shuffle and stuffle, namely, the regularized DBSFs. The name "stuffle" first appeared in [71, 72] although such algebras appeared implicitly already in Cartier's work [116] in early 1970s. It is also called quasi-shuffle in [290] using recursive definition, harmonic shuffle in [541], or sticky shuffle in [302].

It is equivalent to the algebra structure called mixable shuffle treated in [263] which generalizes the shuffle product in terms of permutations. In a series of papers [289–291], Hoffman first realized that one may equip the MZVs with two different algebra structures reflecting the two different ways to multiply the MZVs. This has become one of the cornerstones in various studies of the MZVs and their generalizations.

It turns out the MZV relations obtained by the shuffle product of the iterated integrals can often be proved using other methods. We have derived Euler's decomposition formula (*i.e.*, Thm. 3.1.1) by using the shuffle product. In [191], Eie extended this in several ways by varying the binomial coefficients with carefully chosen integrals. However, Euler originally discovered his formula with partial fractions (see the proof of Thm. 5.3.1). This is important because the integral representation is not always easily available in some generalizations of the MZVs, but the partial fraction technique can still be applied. For example, Thm. 3.1.1 were enhanced to a certain q-analogs (see Chap. 11) by the author [615] with the iterated Jackson q-integrals in a very complicated manner. But by using partial fractions, Bradley [79] found a much cleaner form with a shorter proof. Another q-analog of Euler's decomposition formula can be found in [119]. This formula is later generalized to the product of any two MZVs by the algebraic method of double shuffle product by Guo and Xie [262]. With different approaches, it is also extended to the cases involving the MZVs of the form $\zeta(m, \{1\}^n)$ in [193, 265]. Moreover, in [261] Guo and Xie showed that the class of fractions produced from MZVs has a natural shuffle algebra structure which explains why the partial fractions can be used as surrogates of the iterated integrals.

On the other hand, the stuffle product leads to so-called Delannoy numbers (see Exercise 3.1). K.-W. Chen [136] derived some interesting relations among these numbers by using the stuffle structure of the MVZs. From Euler's decomposition formula, he and his collaborators Chung and Eie [138] also produced several combinatorial identities extending that of Pascal.

Incidentally, both the shuffle and stuffle structures appear unexpectedly in a few other fields in recent years. For instance, in [155] Curry *et al.* used this to study the algebra of repeated integrals of semimartingales and proved that a finite family of independent Lévy processes that have finite moments generates a minimal family. In real world applications, efficient simulation of Lévy processes have been regarded as one of the key steps in finance, economics and insurance modeling.

Proposition 3.2.6 appears first in [71]. We will use it to establish some

MZV identity in Theorem 5.8.1. A similar result due to Yee [587] is given in Exercise 3.3.

Lemma-Definition 3.3.5 is a special case of Chen–Fox–Lyndon Theorem since it can be derived from [134], but the first explicit statement for the algebra of words representing the MZVs did not appear until the publication of [506], although Lemma 3.3.7 and a few other results in Sec. 3.3 can be found in [289, 482]. For Remark 3.3.13 on the classical Möbius Inversion Formula, see [404, (1.3.7)].

Theorem 3.3.21 giving the regularized DBSFs was stated by Ihara, Kaneko and Zagier in Ref. [308] parts of which had circulated as preprints long before its formal publication. Incidentally, the regularized DBSFs are called the extended DBSFs in [308]. In the same paper they also considered the derivation relations which we will discuss in Chap. 5. Z. Li [395] treated some variation of the shuffle regularization using the single-variable MPLs and the Drinfeld associator. In [338] Kaneko *et al.* conjectured that a certain concrete subset of the regularized DBSFs can generate all the regularized DBSFs which was verified by Combariza [145] with some numerical data. Note that the regularized DBSFs themselves are not explicit formulas of the MZVs, neither are the finite ones. In [262] Guo and Xie described an idea of how to write down these explicitly and carried this out for products of two double zeta values. Hoang Ngoc *et al.* further conjectured in [285] that all the $\mathbb{Q}$-linear relations among the MZVs can be produced by the finite DBSFs plus those of the form $\zeta(\mathbf{u} * \mathtt{x}_1 - \mathbf{u} ⧢ \mathtt{x}_1) = 0$ with admissible $\mathbf{u}$. Note also that by emulating Tate's idea Horozov [299] gave an interpretation of the MZV DBSFs using iterated integrals over the finite ideles and over the idele class group. As he remarked, this idea might be modified to handle the DBSFs of function field MZVs which are currently missing.

The folklore Conjecture 3.4.1 is a consequence of some deep conjectures in the theory of algebraic geometry and motives (see [246, Conjecture B]). Zagier further proposed Conjecture 3.4.4 while Conjecture 3.3.25 is a much stronger form of it. Clearly, Conjecture 3.4.5 also implies the next.

Conjecture 3.6.1. (Hoffman's Basis Conjecture) *The $\mathbb{Q}$-space generated by the MZVs has a basis consisting of the MZVs whose arguments are equal to either* 2 *or* 3 *(such MZVs are called the Hoffman elements).*

A deep result of Brown [102] (see also [162] for an exposition) implies that every MZV is a $\mathbb{Q}$-linear combination of the Hoffman elements.

Proposition 3.2.5 is due to Borwein, Bradley and Broadhurst (see [71, Prop. 1]). The idea to use relations between the words in the Hoffman algebra is one of the most powerful tools to investigate the $\mathbb{Q}$-linear relations among the MZVs.

Zagier's Dimension Conjecture 3.4.2 has been one of the driving forces in the study of the MZVs. Its motivic version, Conjecture 3.4.5, was stated in [250]. Since Conjecture 3.4.2 is an equality one may approach it by verifying two inequalities. Goncharov and Terasoma (see [250, 540]) independently proved that $\dim_{\mathbb{Q}} \mathsf{MZ}_w \leq D_w$. This inequality also follows from Brown's recent result mentioned above, which confirms only half of Hoffman's Basis Conjecture.

To prove the other half of Conjecture 3.4.2 one now needs to show $\dim_{\mathbb{Q}} \mathsf{MZ}_w \geq D_w$ which can be deduced from some notoriously difficult conjectures in transcendental number theory (more precisely, the Grothendieck's period conjecture, see [164, § 5.27] for details). For example, we know π is irrational by Lampert's work in [374] of 1768 (and transcendental by Lindemann [401] more than a century later). Almost another century had passed before Apéry [19] proved $\zeta(3)$ is irrational in 1978. An additional twenty plus years later, Ball and Rivoal [35, 484] showed that infinitely many zeta values $\zeta(2n+1)$ are irrational. More recently, Zudilin [642] and others made some remarkable refinements in this direction. The progress seems to be quite slow, and in fact, until today we don't even know whether $\zeta(5)$ is irrational or not.

In earlier 1990, Bailey *et al.* already noticed, on the basis of extensive numerical experiments, that some double zeta values such as $\zeta(6,2)$ cannot be expressed as a $\mathbb{Q}$-linear combination of products of the Riemann zeta values (assuming $\zeta(0) = -1/2$ is also a Riemann zeta value). To detect the fine structure of the MZVs, Broadhurst and his collaborators computed extensively with the aid of computers and observed a lot of subtle properties. In [83] he gave the precise Conjecture 3.4.6 for the number of irreducible MZVs of given weight and depth just by observing the patterns of the known dimensions. This formula has since been explained partly by a hidden relation between the MZVs and the modular forms (see [233] and Sec. 5.9). Note that each of the three conjectural formulas in Eq. (3.26) is known to give an upper bound for the corresponding true dimension (see [597] for $d = 2$ and [249] for $d = 3$ or [305] by a different method). The ideas from [305] and some homological computations involving finite subgroups of $\mathrm{GL}_4(\mathbb{Z})$ by Yasuda [586] may be useful to treat the $d = 4$ case. In the general case, Carr, Gangl and Schneps have found some convincing evidence in [115].

The $\mathbb{Q}$-algebra generated by the MZVs and the algebra $\mathscr{F}(2,3)$ are closely related to some stable derivation algebra coming from braid groups (see [159, 173, 215, 309–311, 552]). This provides the key links Brown used in [102] (see also [162]) to prove the Deligne-Ihara conjecture on the outer action of $\mathrm{Gal}(\overline{\mathbb{Q}}/\mathbb{Q})$ on the pro-fundamental group of $\mathbb{P}^1 \setminus \{0, 1, \infty\}$. For the study of the MZVs from the algebro-geometric point of view, one can consult [221, 223, 521–523].

Tsumura's original proof in [550] of the Parity Principle (Cor. 3.5.7) contains heavy analysis. An algebraic proof was later given by Ihara, Kaneko and Zagier (see [308, Thm. 7]) which we have modified and adopted in Sec. 3.5. Note that the case $d = 2$ was already known to Tornheim [547] and the case $d = 3$ was proved by Borwein and Girgensohn [69]. For lower depth cases it is possible to explicitly write down the reduction formulas, for example, see [118, 460, 461, 601]. Similar parity result for other kind of the multiple zeta functions should exist. For example, Okamoto [452] discovered such a result for the Witten zeta function $\zeta_2(\mathbf{s}; G_2)$ for Lie algebra G_2 in some extended sense.

There are a variety of ways to generalize multiple zeta functions and their special values, such as those of the Hurwitz-type [8, 75, 75, 170, 322, 345,351,418,509,634]. In particular, if the parameters in these Hurwitz-type special values are rational numbers with denominators equal to a divisor of a fixed positive integer N, then they become essentially so-called MZVs of level N. For pertinent research along this line, see [334, 438, 590, 591].

For general surveys of the results on the MZVs up to the beginning of twenty-first century, the interested reader may peruse Refs. [117, 563, 643] by Cartier, Waldschmidt and Zudilin, respectively.

Exercises

3.1. Let $\mathbf{u}$ and $\mathbf{v}$ be two words. Let $a = \mathrm{dp}(\mathbf{u})$ and $b = \mathrm{dp}(\mathbf{v})$. Show that $\mathrm{dp}(\mathbf{u} ш \mathbf{v}) = \binom{a+b}{a}$ and $\mathrm{dp}(\mathbf{u} * \mathbf{v}) = D(a, b)$ is the Delannoy number (see [146, p. 81]) that satisfies $D(a, 0) = 1$ for all $a \geq 0$ and

$$D(a, b) = D(a, b-1) + D(a-1, b) + D(a-1, b-1) \quad \forall a, b \geq 1.$$

3.2. Let T^j_{p+q} be defined as in Prop. 3.2.5. Show that the factor $\mathtt{x}_1^2$ appears exactly j times.

3.3. In the word $(\mathtt{x}_0^4\mathtt{x}_1^2)^n$, there are $2n-1$ ways in which one can exchange a letter $\mathtt{x}_0$ with an adjacent letter $\mathtt{x}_1$. Let $\Xi_n(k)$ denote the sum of the $\binom{2n-1}{k}$

words obtained from $(\mathrm{x}_0^4\mathrm{x}_1^2)^n$ by making k such transpositions. Show that

$$\sum_{r=-n}^{n}(-1)^r\left[(\mathrm{x}_0^2\mathrm{x}_1)^{n-r} ш (\mathrm{x}_0^2\mathrm{x}_1)^{n+r}\right]=3^n\sum_{k=0}^{2n-1}2^{2n-k}\Xi_n(k).$$

3.4. Prove Prop. 3.2.8 by induction.

3.5. Complete the proof of Lemma 3.3.7.

3.6. Complete the proof of Thm. 3.3.12.

3.7. Find the finite DBSF by computing $\zeta(2)\zeta(3)$ in two different ways.

3.8. Using Definition 3.3.3 how many Lyndon words of weight six can you find in $\mathbb{Q}\langle \mathrm{x}_0, \mathrm{x}_1\rangle$?

3.9. (1). Find the Lyndon factorization of $\mathrm{x}_1^2\mathrm{x}_0^2\mathrm{x}_1$ and

$$\mathrm{x}_1\mathrm{x}_0^3\mathrm{x}_1^3\mathrm{x}_0^5\mathrm{x}_1^2\mathrm{x}_0^2\mathrm{x}_1\mathrm{x}_0^5\mathrm{x}_1\mathrm{x}_0^3.$$

(2). Express each word in (1) as a $\mathbb{Q}$-linear combination of $*$-products of Lyndon words.

(3). Express each word in (1) as a $\mathbb{Q}$-linear combination of ш-products of Lyndon words.

3.10. Let f and g be two real functions defined on $\mathbb{N}$. If $g(n)=\sum_{d|n}f(d)$ for all $n\in\mathbb{N}$, prove that $f(n)=\sum_{d|n}f(d)\mu(n/d)$ for all $n\in\mathbb{N}$ where μ is the Möbius function defined by Eq. (3.12). This is called the Möbius inversion formula.

3.11. Find the regularized DBSF by computing $\zeta(3)T$ in two different ways using Thm. 3.3.21 and Remark 3.3.22.

3.12. By hand computation verify Zagier's Dimension Conjectuer 3.4.2 provides the upper bound for $\dim_{\mathbb{Q}}\mathsf{MZ}_w$ for $w\le 5$.

3.13. This exercise is for Maple users. First, download the software mzv16.m (or mzv12.txt if your computer has moderate RAM) from Petitot's website http://www.lifl.fr/ petitot/recherche/MZV/. Then

- Verify that Broadhurst Conjecture 3.4.6 provides the upper bound for $\dim_{\mathbb{Q}}\mathsf{MZ}_{w,d}$ for $d<w\le 12$.
- Verify the Parity Principle (Cor. 3.5.7) for all MZVs of weight 10.

3.14. Prove Cor. 3.5.4.

3.15. Recall that P, T' and T_0 are defined by Eqs. (3.28), (3.34) and (3.31), respectively. Show that for all $f \in V_n$

$$\begin{aligned}((f|P)T')(x_1,\dots,x_n) &= (f|T_0P)(x_1,\dots,x_n)\\ &= f(-x_n,-x_n-x_{n-1},\dots,-x_n-\cdots-x_1).\end{aligned}$$

3.16. Recall that $\tau_{1,n+1}$ is the transposition $1 \leftrightarrow n+1$, and $\mathrm{sh}_1^{(n)}$ and C_{n+1} are defined by Eq. (3.32) and Eq. (3.30), respectively. Show that by regarding $a_1 \dots a_{n+1}$ as a word

$$\begin{aligned}&(1+\mathrm{sh}_1^{(n)}\, C_{n+1})(a_1 \dots a_{n+1})\\ &\quad= C_{n+1}(1+\mathrm{sh}_1^{(n)}\, \tau_{1n+1})(a_1 \dots a_{n+1})\\ &\quad= a_1 \dots a_{n+1} + a_{n+1}a_1 \dots a_n + a_1[a_{n+1} ⧢ (a_2 \dots a_{n-1})]a_n\\ &\quad= a_{n+1} ⧢ (a_1 \dots a_n).\end{aligned}$$

3.17. For any $w, d \in \mathbb{N}$, show that

$$\zeta^\star(w,\{1\}^{d-1}) = \sum_{t=1}^{d} \sum_{\substack{c_1+\cdots+c_{t-1}=d\\ c_j\ge 0\,\forall j=1,\dots,t-1}} \zeta(c_{t-1}+w, c_{t-2}+1,\dots,c_1+1).$$

3.18. Suppose $f \in \mathbb{Q}[x_1,x_2]$ has homogeneous degree d. Show that $f \in \mathsf{DSh}_2(d)$ if and only

$$f(x_1,x_2)+f(x_2,x_1)=0 \quad \text{and} \quad f(x_1+x_2,x_1)+f(x_1+x_2,x_2)=0.$$

Chapter 4

Drinfeld Associator and Single-Valued MZVs

A crucial observation by Deligne is that one can attach tangential base point on $\mathbb{P}\setminus\{0,1,\infty\}$ to study the torsors of paths and the motivic fundamental group. Let 1_0 and $(-1)_1$ be the tangential base point at 0 (resp. at 1) pointing to the positive (resp. negative) direction. Then the de Rham version of the Betty path from 1_0 going straight to $(-1)_1$ gives rise to the Drinfeld associator whose coefficients are all expressible by the MZVs.

4.1 Drinfeld Associator

4.1.1 *KZ-Equations: A General Setup*

For future reference, we consider the general case by setting $\mathsf{X} = \{\mathtt{x}_1,\dots,\mathtt{x}_N\}$ to be an alphabet of $N+1$ letters and X^* the set of words on X including the empty word 1. Let $\mathbb{C}\langle\langle\mathsf{X}\rangle\rangle$ be the algebra of non-commutative formal power series in X^*. Fix an injection $j:\mathsf{X}\hookrightarrow\mathbb{C}$ and set $p_i=j(\mathtt{x}_i)$ for all $i=0,\dots,N$. Consider the differential equation

$$F'(z)=\sum_{i=0}^{N}\frac{\mathtt{x}_i}{z-p_i}F(z). \tag{4.1}$$

For each i, fix a half line $\ell(p_i)$ starting from p_i such that $\bigcap_{i=0}^{N}\ell(p_i)=\emptyset$. Let $U=\mathbb{C}\setminus\bigcup_{i=0}^{N}\ell(p_i)$ and fix a branch of logarithm $\log(z-p_0)$ on $\mathbb{C}\setminus\ell(p_0)$.

The following general theorem guarantees the existence of the Drinfeld associator.

Theorem 4.1.1. *The differential equation* (4.1) *has a unique solution $L(z)$ on U such that*

$$L(z)=f_0(z)\exp\big(\mathtt{x}_0\log(z-p_0)\big),$$

where $f_0(z)$ is a holomorphic function on $\mathbb{C}\setminus\bigcup_{i=1}^{N}\ell(p_i)$ with $f_0(p_0)=1$. Moreover, every holomorphic solution of (4.1) *on $\mathbb{C}\setminus\bigcup_{i=0}^{N}\{p_i\}$ is equal to $L(z)C$ for some $C\in\mathbb{C}\langle\langle\mathsf{X}\rangle\rangle$ depending only on p_i's but not on z.*

Proof. Set $L(z) = \sum_{\mathbf{w}\in\mathsf{X}^*} L_{\mathbf{w}}(z)\mathbf{w}$. Note that $L(z)$ satisfying (4.1) is equivalent to

$$\frac{\partial}{\partial z} L_{\mathrm{x}_i\mathbf{w}}(z) = \frac{L_{\mathbf{w}}(z)}{z-p_i}$$

for all $i = 0, \dots, N$ and all $\mathbf{w} \in \mathsf{X}^*$. We now recursively define functions $L_{\mathbf{w}}(z)$ satisfying (4.1) for all non-empty words $\mathbf{w} \in \mathsf{X}^*$ not ending in the letter x_0. For such words, the limiting condition is just $\lim_{z\to p_0} L_{\mathbf{w}}(z) = 0$. If we write $\mathbf{w} = \mathrm{x}_0^{s_1-1}\mathrm{x}_{i_1}\dots\mathrm{x}_0^{s_d-1}\mathrm{x}_{i_d}$, where $1 \le i_1, \dots, i_d \le N$, then in a neighborhood of p_0 the following function satisfies (4.1):

$$L_{\mathbf{w}}(z) = \sum_{k_1>\cdots>k_d} \frac{(-1)^d}{k_1^{s_1}\cdots k_d^{s_d}} \prod_{m=1}^{d} \left(\frac{z-p_0}{p_{i_m}-p_0}\right)^{k_m-k_{m+1}} \qquad (k_{d+1}=0),$$

which converges absolutely for $|z-p_0| < \min\{|p_{i_1}-p_0|, \dots, |p_{i_d}-p_0|\}$. The functions $L_{\mathbf{w}}(z)$ can be extended analytically to the whole of U by the recursive formula

$$L_{\mathrm{x}_i\mathbf{w}}(z) = \int_0^z \frac{L_{\mathbf{w}}(t)}{t-p_i}\mathrm{d}t, \quad \forall i = 0, \dots, N.$$

Now we write $f_0(z) = \sum_{\mathbf{w}\in\mathsf{X}^*} \Lambda_{\mathbf{w}}(z)\mathbf{w}$ where

$$\Lambda_{\mathbf{w}}(z) = \begin{cases} 1, \text{ if } \mathbf{w} = 1; \\ 0, \text{ if } \mathbf{w} = \mathrm{x}_0^n,\ n \in \mathbb{N}; \end{cases} \tag{4.2}$$

and if $\mathbf{w} = \mathrm{x}_0^{s_1}\mathrm{x}_{i_1}\dots\mathrm{x}_0^{s_d}\mathrm{x}_{i_d}\mathrm{x}_0^n$, where $s_1, \dots, s_d \ge 0$ and $1 \le i_1, \dots, i_d \le N$, then

$$\Lambda_{\mathbf{w}}(z) = (-1)^n \sum_{s_j\le k_j, |\mathbf{k}|=|\mathbf{s}|+n} \prod_{j=1}^{d} \binom{k_j}{s_j} L_{\mathrm{x}_0^{k_1}\mathrm{x}_{i_1}\dots\mathrm{x}_0^{k_d}\mathrm{x}_{i_d}}(z). \tag{4.3}$$

Then it is not hard to see that $L(z) = f_0(z)\exp\big(\mathrm{x}_0\log(z-p_0)\big)$ satisfies equation (4.1). Note that $f_0(z)$ is holomorphic in a neighborhood of p_0 since its coefficients are just linear combinations of the functions $L_{\mathbf{w}}(z)$.

Finally, the uniqueness can be proved in a straightforward manner. This completes the proof of the theorem. □

Definition 4.1.2. Let x_0 and x_1 be two non-commuting symbols. Let G_0 and G_1 be the analytic solutions of the *Knizhnik–Zamolodchikov (KZ) differential equation*

$$G'(z) = \left(\frac{\mathrm{x}_0}{z} + \frac{\mathrm{x}_1}{z-1}\right) G(z), \tag{4.4}$$

that are defined for $0 < z < 1$, such that

$$G_0(z) \sim z^{x_0} \text{ near } z = 0 \quad \text{and} \quad G_1(z) \sim (1-z)^{x_1} \text{ near } z = 1.$$

This means both $G_0(z)$ and $G_1(z)$ are formal series in x_0 and x_1 with coefficients given by analytic functions of z in $(0, 1)$. Moreover, $G_0(z)e^{-\mathrm{x}_0 \log z}$ (resp. $G_1(z)e^{-\mathrm{x}_1 \log(1-z)}$) has an analytic continuation into a neighborhood of $z = 0$ (reps. $z = 1$) and becomes 1 at this point. The *Drinfeld associator* is define by

$$\Phi(\mathrm{x}_0, \mathrm{x}_1) := \sum_{\mathbf{w} \in \mathsf{X}^*} \Phi[\mathbf{w}]\mathbf{w} := G_1^{-1} G_0. \tag{4.5}$$

Here and in what follows $f[\mathbf{w}]$ always means the coefficient of $\mathbf{w}$ in f.

Remark 4.1.3. The existence of $G_0(z)$ and $G_1(z)$ follows from Thm. 4.1.1 by taking $p_0 = 0$ and 1, respectively.

Note that $\Phi(\mathrm{x}_0, \mathrm{x}_1)$ should only depend on x_0 and x_1. In order to understand this we may consider the analogous but much easier differential equation $f'(z) = f(z)$ with two solutions $f_0(z)$ and $f_1(z)$ determined by the condition that $f_0(0) = 1$ and $f_1(1) = 1$. Then $f_0(z) = e^z$ and $f_1(z) = e^{z-1}$. So $f_1(z)^{-1} f_0(z) = e$ which is a constant.

Theorem 4.1.4. *The Drinfeld associator satisfies the following duality relation:*

$$\Phi(\mathrm{x}_0, \mathrm{x}_1)^{-1} = \Phi(\mathrm{x}_1, \mathrm{x}_0). \tag{4.6}$$

Proof. First observe that by $z \leftrightarrow 1 - z$ we can rewrite Eq. (4.4) as

$$H'(z) = \left(\frac{\mathrm{x}_0}{z-1} + \frac{\mathrm{x}_1}{z} \right) H(z), \quad H(z) = G(1-z).$$

Thus $\Phi^{-1} = G_0^{-1} G_1 = H_1^{-1} H_0$ which implies that Φ^{-1} can be obtained from Φ by the substitution $\mathrm{x}_0 \leftrightarrow \mathrm{x}_1$. □

Remark 4.1.5. Here is another way to understand $\Phi(\mathrm{x}_0, \mathrm{x}_1)$. Since $1/z$ and $1/(1-z)$ are both analytic in $(0, 1)$ for any $0 < u < 1$ there is a unique solution to Eq. (4.4) with $G(u) = 1$. Let $Z_u^v(\mathrm{x}_0, \mathrm{x}_1) = G(v)$ for $v \in (0, 1)$. Since $Z_u^u(\mathrm{x}_0, \mathrm{x}_1) = 1$ we can write

$$Z_u^v(\mathrm{x}_0, \mathrm{x}_1) = \sum_{\mathbf{w} \in \mathsf{X}^*} f_{\mathbf{w}}(u, v)\mathbf{w},$$

where $\mathfrak{A}$ is defined in Definition 3.2.1 and $f_{\mathbf{w}}(u, v)$ is an analytic function in u and v. Define the iterated integral $I_{\mathbf{w}}(u, v) = \int_u^v \mathbf{w}(\mathrm{d}t/t, \mathrm{d}t/(t-1))$

for any word $\mathbf{w} = \mathbf{w}(\mathrm{x}_0, \mathrm{x}_1)$. By the induction on the weight of the word we see easily that $f_{\mathbf{w}}(u, v) = I_{\mathbf{w}}(u, v)$. If $\mathbf{w} = \mathrm{x}_0^{s_1-1}\mathrm{x}_1 \cdots \mathrm{x}_0^{s_d-1}\mathrm{x}_1 \in \mathfrak{A}^0$ (*i.e.*, $s_1 \geq 2$, see Definition 3.2.2) then we have $\lim_{\varepsilon\to 0^+} I_{\mathbf{w}}(\varepsilon, 1-\varepsilon) = (-1)^d\zeta(s_1, \dots, s_d)$ by Eq. (3.1). But if $\mathbf{w}$ is not admissible we have to regularize $I_{\mathbf{w}}(\varepsilon, 1-\varepsilon)$. It is possible to compute this explicitly (see [385, Appendix]) but this would take us too far afield. Another high powered machinery to do this is to use some Hopf algebra fact whose proof will be omitted.

4.1.2 *Some Hopf Algebras and Their Group-Like Elements*

Let $\mathsf{X} = \{\mathrm{x}_0, \mathrm{x}_1\}$ and $\mathscr{R} := \mathbb{C}\langle\langle \mathsf{X} \rangle\rangle$ be the Hopf algebra as defined in Example A.12 with concatenation product and with coproduct $\Delta_ш$ such that $\Delta_ш(\mathrm{x}_i) = 1 \otimes \mathrm{x}_i + \mathrm{x}_i \otimes 1$ for $i = 0, 1$. The counit $\epsilon_ш$ is defined by $\epsilon_ш(\mathbf{w}) = 0$ for all nonempty word $\mathbf{w}$ and $\epsilon_ш(1) = 1$. Note that $\mathscr{R}$ can be obtained from $\mathfrak{A}$ by an extension of scalars to $\mathbb{C}$ and then by completion with respect to the weight grading.

We now consider $\mathscr{R}_ш := \mathscr{R}/\mathscr{R}\mathrm{x}_0$ which can be identified with the subalgebra of $\mathscr{R}$ generated by words ending with x_1. It can be thought as the completion of $\mathfrak{A}^1_ш \otimes \mathbb{C}$ with respect to the weight grading. By the above we see that the Drinfeld associator is naturally associated with the shuffle regularization of the MZVs. Similarly we can treat the stuffle version using a $*$-version of Φ.

Let $\mathscr{R}_* := \mathbb{C}\langle\langle \mathrm{z}_n,\ n \in \mathbb{N} \rangle\rangle$ be the Hopf algebra with $\Delta_*(\mathrm{z}_n) = \sum_{k=0}^n \mathrm{z}_k \otimes \mathrm{z}_{n-k}$ for all $n \in \mathbb{N}$, where $\mathrm{z}_0 = 1$. Clearly, this is the completion of $\mathfrak{A}^1_* \otimes \mathbb{C}$ with respect to the weight grading. Define the projection $\pi : \mathscr{R} \longrightarrow \mathscr{R}_ш$ by forgetting the terms corresponding to the words ending with x_0. Note that if we set $\mathrm{z}_n = \mathrm{x}_0^{n-1}\mathrm{x}_0$ for all $n \in \mathbb{N}$ then as concatenation algebras $\mathscr{R}_ш = \mathscr{R}_*$. Define

$$\Phi_ш = \pi(\Phi).$$

Theorem 4.1.6. *Let $f_* = 1 + \sum_{\mathbf{w}\in \mathsf{X}^*\mathrm{x}_1} f_*[\mathbf{w}]\mathbf{w} \in \mathscr{R}_*$. Then f_* satisfies stuffle relation if and only if $\Delta_*(f_*) = f_* \otimes f_*$,* i.e., *$f_*$ is group-like for Δ_*.*

Proof. Throughout this proof, by abuse of notation, for any composition $\mathbf{s} = (s_1, \dots, s_d) \in \mathbb{N}^d$ we also identify it with $(-1)^d\mathrm{x}_0^{s_1-1}\mathrm{x}_1 \dots \mathrm{x}_0^{s_d-1}\mathrm{x}_1 \in \mathfrak{A}^1_*$ when we write $\mathbf{s} \in \mathfrak{A}^1_*$. We also write

$$(r_1, \dots, r_\ell) \Rightarrow (s_1, \dots, s_d)$$

if $\ell \geq d$ and $(r_1, \ldots, r_\ell)$ can be obtained from $(s_1, \ldots, s_d)$ by inserting some 0's. By the definition, $\Delta_*(f_*) = f_* \otimes f_*$ means for all $\mathbf{s}, \mathbf{t} \in \mathfrak{A}^1_*$

$$f_*[\mathbf{s}]f_*[\mathbf{t}] = \sum_{\substack{(r'_1,\ldots,r'_\ell)\Rightarrow \mathbf{s}, (r''_1,\ldots,r''_\ell)\Rightarrow \mathbf{t} \\ \max\{\mathrm{dp}(\mathbf{s}),\mathrm{dp}(\mathbf{t})\}\leq \ell \leq \mathrm{dp}(\mathbf{s})+\mathrm{dp}(\mathbf{t}) \\ r'_j+r''_j>0\ \forall j=1,\ldots,\ell}} f_*[r'_1 + r''_1, \ldots, r'_\ell + r''_\ell]. \tag{4.7}$$

Extending f_* linearly to $\mathfrak{A}^1_*$ we need to show Eq. (4.7) is equivalent to

$$f_*[\mathbf{s}]f_*[\mathbf{t}] = f_*[\mathbf{s} * \mathbf{t}] \text{ for all } \mathbf{s}, \mathbf{t} \in \mathfrak{A}^1_*. \tag{4.8}$$

Assuming Eq. (4.7) we now establish Eq. (4.8) by induction on $|\mathbf{s}| + |\mathbf{t}|$. If either $\mathbf{s}$ or $\mathbf{t}$ is 1 then this is obvious. In general, we have

$$f_*[m, \mathbf{s}]f_*[n, \mathbf{t}] = \sum_{\substack{(r'_1,\ldots,r'_\ell)\Rightarrow (m,\mathbf{s}), (r''_1,\ldots,r''_\ell)\Rightarrow (n,\mathbf{t}) \\ \max\{\mathrm{dp}(\mathbf{s}),\mathrm{dp}(\mathbf{t})\}+1\leq \ell \leq \mathrm{dp}(\mathbf{s})+\mathrm{dp}(\mathbf{t})+2 \\ r'_j+r''_j>0\ \forall j=1,\ldots,\ell}} f_*[r'_1 + r''_1, \ldots, r'_\ell + r''_\ell].$$

On the right-hand side we have three cases to consider: $(r'_1, r''_1) = (m, 0)$, $(r'_1, r''_1) = (0, n)$, and $(r'_1, r''_1) = (m, n)$. By the induction assumption we then get

$$\begin{aligned} f_*[m, \mathbf{s}]f_*[n, \mathbf{t}] &= f_*\big[m, \mathbf{s} * (n, \mathbf{t})\big] + f_*\big[n, (m, \mathbf{s}) * \mathbf{t}\big] + f_*\big[m + n, \mathbf{s} * \mathbf{t}\big] \\ &= f_*\big[(m, \mathbf{s}) * (n, \mathbf{t})\big]. \end{aligned}$$

This completes the proof that $\Delta_*(f_*) = f_* \otimes f_*$ implies Eq. (4.8). On the other hand, it is not too hard to check that Eq. (4.8) implies Eq. (4.7) (see Exercise 4.3) which means $\Delta_*(f_*) = f_* \otimes f_*$. This finishes the proof of the theorem. □

Since the $*$-regularized MZVs $\zeta_*(s_1, \ldots, s_d; T)$ satisfy stuffle relations the next result follows immediately.

Corollary 4.1.7. *Let*

$$\Phi_* = 1 + \sum_{d=1}^{\infty} \sum_{s_1,\ldots,s_d \in \mathbb{N}} (-1)^d \zeta_*(s_1, \ldots, s_d) \mathbf{z}_{s_1} \ldots \mathbf{z}_{s_d}.$$

Then $\Delta_*(\Phi_*) = \Phi_* \otimes \Phi_*$.

Theorem 4.1.8. *Let* $f = 1 + \sum_{\mathbf{w} \in \mathsf{x}_0 \mathsf{X}^* \mathsf{x}_1} f[\mathbf{w}]\mathbf{w} \in (\mathscr{R}_{\mathrm{cv}}, \Delta_{\sqcup\!\sqcup})$. *Then* $f[\mathbf{w}]$ *satisfies the shuffle relation if and only if* $\Delta_{\sqcup\!\sqcup}(f) = f \otimes f$, i.e., f *is group-like. Similar result holds for* $(\mathscr{R}_{\sqcup\!\sqcup}, \Delta_{\sqcup\!\sqcup})$, *too.*

Proof. The computation is straightforward so we leave it as an exercise (see Exercise 4.4). □

We leave the proof of next two technical lemmas to the interested reader (see Exercise 4.5).

Lemma 4.1.9. *Let $\mathcal{R}_*^T = \mathcal{R}_* \otimes_{\mathbb{C}} \mathbb{C}[T]$. Any group-like element $f_{\rm cv}$ of $(\mathcal{R}_{\rm cv}, \Delta_*)$ is the image of a group-like element of $(\mathcal{R}_*^T, \Delta_*)$ under the specialization $T = 0$. Moreover, two such elements f_1 and f_2 are related by*

$$f_2 = \exp((\lambda_2 - \lambda_1)\mathbf{z}_1)f_1,$$

where λ_1 and λ_2 are the coefficients of $\mathbf{z}_1$ in f_1 and f_2 respectively.

Similarly one can prove,

Lemma 4.1.10. *Let $\mathcal{R}_{ш}^T = \mathcal{R}_{ш} \otimes_{\mathbb{C}} \mathbb{C}[T]$. Any group-like element $f_{\rm cv}$ of $(\mathcal{R}_{\rm cv}, \Delta_{ш})$ is the image of a group-like element of $(\mathcal{R}_{ш}^T, \Delta_{ш})$ under the specialization $T = 0$. Moreover, two such elements f_1 and f_2 are related by*

$$f_2 = \exp((\lambda_2 - \lambda_1)\mathbf{x}_1)f_1,$$

where λ_1 and λ_2 are the coefficients of $\mathbf{x}_1$ in f_1 and f_2 respectively.

Theorem 4.1.11. *The Drinfeld associator $\Phi(\mathbf{x}_0, \mathbf{x}_1)$ is the only element in the Hopf algebra $(\mathcal{R}, \Delta_{ш}, \epsilon_{ш})$ such that*

(i) $\Delta_{ш}(\Phi) = \Phi \otimes \Phi$ (i.e. Φ *is group-like, see Definition* A.6)
(ii) $\Phi[\mathbf{x}_0] = \Phi[\mathbf{x}_1] = 0$
(iii) $\Phi[\mathbf{w}] = (-1)^d \zeta(s_1, \dots, s_d) \quad \forall \mathbf{w} = \mathbf{x}_0^{s_1-1}\mathbf{x}_1 \cdots \mathbf{x}_0^{s_d-1}\mathbf{x}_1 \in \mathfrak{A}^0$.

Proof. This follows from Lemma 4.1.10 easily. □

Remark 4.1.12. In Chap. 13 we will present a more general statement in Thm. 13.4.1.

It is easy to see that the shuffle relations between the MZVs are direct consequences of the fact that Φ is group-like, namely, $\Delta_{ш}(\Phi) = \Phi \otimes \Phi$. See Thm. 4.1.8 and Exercise 4.4.

Using Thm. 4.1.11 we can prove the following result which provides the recursive relations one may use to compute all the coefficients of Φ.

Proposition 4.1.13. *We have*

(i) *For* $\mathbf{w} = \mathbf{x}_0^{s_1-1}\mathbf{x}_1 \cdots \mathbf{x}_0^{s_d-1}\mathbf{x}_1 \in \mathfrak{A}^0$, *we have* $\Phi[\mathbf{w}] = (-1)^d \zeta(s_1, \dots, s_d)$.

(ii) *For* $\mathbf{w} = \mathrm{x}_1^m \mathbf{v} \mathrm{x}_0^n \notin \mathfrak{A}^1$ *with* $m+n>0$, *we may set* $\mathbf{w} = \mathrm{x}_1^m \mathbf{v} \mathrm{x}_0^n = \mathrm{x}_1^m l_1 \cdots l_p \mathrm{x}_0^n$ *where* $l_j \in \{\mathrm{x}_0, \mathrm{x}_1\}$ *and if* $p>0$ *then* $l_1 \neq \mathrm{x}_1, l_p \neq \mathrm{x}_0$. *Further set* $l_1 \cdots l_p \mathrm{x}_0^s = l_1 \cdots l_q$. *Then*

$$\Phi[\mathbf{w}] = \begin{cases} 0, & \text{if } mn = p = 0; \\ -\dfrac{1}{m} \displaystyle\sum_{i=1}^{q} \Phi[\mathrm{x}_1^{m-1} l_1 \cdots l_i \mathrm{x}_1 l_{i+1} \dots l_q], & \text{if } m > 0; \\ -\dfrac{1}{n} \displaystyle\sum_{i=1}^{p} \Phi[l_1 \cdots l_{i-1} \mathrm{x}_0 l_i \cdots l_p \mathrm{x}_0^{n-1}], & \text{if } m = 0, n > 0. \end{cases} \tag{4.9}$$

Proof. We only need to prove (ii). Since Φ is group-like one has

$$\Delta_{ш} \Phi = \Phi \otimes \Phi. \tag{4.10}$$

First we assume $mn = p = 0$. We may further assume $m = 0$ ($n = 0$ is completely similar). We now show that $\Phi[\mathrm{x}_0^n] = 0$ for all $n \in \mathbb{N}$ by induction on n. Suppose $\Phi[\mathrm{x}_0^{n-1}] = 0$ for some $n \geq 2$. By Thm. 4.1.11 we have $\Phi[\mathrm{x}_0] = 0$ and by Eq. (4.10)

$$\begin{aligned} &\Delta_{ш}(1) + \Phi[\mathrm{x}_0^2](\Delta_{ш}(\mathrm{x}_0))^2 + \cdots \\ &\quad = 1 \otimes 1 + \Phi[\mathrm{x}_0^2](\mathrm{x}_0 \otimes 1 + 1 \otimes \mathrm{x}_0)^2 + \cdots \\ &\quad = (1 + \Phi[\mathrm{x}_0]\mathrm{x}_0 + \cdots) \otimes (1 + \Phi[\mathrm{x}_0]\mathrm{x}_0 + \cdots). \end{aligned}$$

We now compare the coefficient of $\mathrm{x}_0^{n-1} \otimes \mathrm{x}_0$ on the two sides. The only way to produce this on the left-hand side is by $\Phi[\mathrm{x}_0^n]\Delta_{ш}(\mathrm{x}_0^n)$ which in fact produces n copies of it. On the right-hand side the only way is to use $\Phi[\mathrm{x}_0^{n-1}]\mathrm{x}_0^{n-1} \otimes \Phi[\mathrm{x}_0]\mathrm{x}_0$. Hence $n\Phi[\mathrm{x}_0^n] = \Phi[\mathrm{x}_0^{n-1}]\Phi[\mathrm{x}_0] = 0$ and therefore $\Phi[\mathrm{x}_0^n] = 0$.

By similar argument if $m > 0$ then one can compare the coefficient of $\mathrm{x}_1 \otimes (\mathrm{x}_1^{m-1} l_1 \cdots l_q)$ of the two sides of Eq. (4.10) (which is 0) and find the relation (4.9). Finally, if $m = 0$ and $n > 0$ then one may similarly consider the coefficient of $(l_1 \cdots l_p \mathrm{x}_0^{n-1}) \otimes \mathrm{x}_0$ in Eq. (4.10). This finishes the proof of the proposition. □

Theorem 4.1.14. *If* $\mathbf{w} = \{0\}^{s_1-1}1 \cdots \{0\}^{s_d-1}1\{0\}^{n-1}$, *then*

$$\Phi[\mathbf{w}] = (-1)^n \sum_{k_1+\cdots+k_d=n} \prod_{j=1}^{d} \binom{k_j + s_j - 1}{k_j} \zeta(s_1 + k_1, \dots, s_d + k_d).$$

Proof. This can be proved easily by induction on n using Prop. 4.1.13 so it is left to the interested reader. □

Theorem 4.1.15. *We have* $\Phi_{ш} = \Lambda(\mathtt{x}_1)\Phi_*$ *where* Λ *is defined by Eq.* (3.17)*:*

$$\Lambda(u) = \exp\left(\sum_{n=2}^{\infty} \frac{(-1)^n\zeta(n)}{n} u^n\right).$$

Proof. Let $\mathscr{R}_{\text{cv}} := \mathscr{R}_*/\mathtt{z}_1\mathscr{R}_* \cong \mathbb{C}\langle\langle \mathtt{z}_n, n \geq 2\rangle\rangle$ which is also equal to $\mathscr{R}_*/\mathtt{x}_1\mathscr{R}_*$. Let $\pi_{\text{cv}} : \mathscr{R}_* \to \mathscr{R}_{\text{cv}}$ be the projection map by forgetting all the terms corresponding to the words beginning with $\mathtt{x}_1$. The generating series of $\zeta(s_1, \ldots, s_d)$ is defined as follows:

$$\begin{aligned}\mathbf{L}_{\text{cv}} &:= \sum_{s_1,\cdots,s_d\in\mathbb{N}, s_1\geq 2} (-1)^d\zeta(s_1,\ldots,s_d)\mathtt{z}_{s_1}\cdots\mathtt{z}_{s_d} \in (\mathscr{R}_{\text{cv}}, \Delta_*),\\ &:= \sum_{s_1,\cdots,s_d\in\mathbb{N}, s_1\geq 2} (-1)^d\zeta(s_1,\ldots,s_d)\mathtt{x}_0^{s_1-1}\mathtt{x}_1\cdots\mathtt{x}_0^{s_d-1}\mathtt{x}_1 \in (\mathscr{R}_{\text{cv}}, \Delta_{ш}).\end{aligned}$$

Since MZVs satisfy the stuffle relations we see that $\mathbf{L}_{\text{cv}}$ is a group-like elements in $(\mathscr{R}_{\text{cv}}, \Delta_*)$ by a similar argument as in the proof of Thm. 4.1.6. On the other hand, since MZVs also satisfy the shuffle relations $\mathbf{L}_{\text{cv}}$ is also a group-like elements in $(\mathscr{R}_{\text{cv}}, \Delta_{ш})$ by Thm. 4.1.8.

Let $\mathbf{L} := \exp(T\mathtt{z}_1)\cdot\Phi_* \in \mathscr{R}_*^T$ and $\mathbf{I} := \exp(T\mathtt{x}_1)\cdot\Phi_{ш} \in \mathscr{R}_{ш}^T$. Clearly $\mathbf{L}$ is group-like for Δ_* and $\mathbf{I}$ is group-like for $\Delta_{ш}$. Moreover, it is straightforward to check the coefficients of $\mathtt{z}_1 = \mathtt{x}_1$ in $\mathbf{L}$ and $\mathbf{I}$ are both equal to T. By the uniqueness guaranteed by Lemma 4.1.9 and Lemma 4.1.10

$$\mathbf{L} = 1 + \sum_{\mathbf{w}\in\mathsf{X}^*\mathtt{x}_1} \zeta_*(\mathbf{w};T)\mathbf{w}, \quad \mathbf{I} = 1 + \sum_{\mathbf{w}\in\mathsf{X}^*\mathtt{x}_1} \zeta_{ш}(\mathbf{w};T)\mathbf{w}$$

since both of the right-hand sides are group-like (with respect to Δ_* by Thm. 4.1.6 and $\Delta_{ш}$ by Thm. 4.1.8, respectively). By Thm. 3.3.21 we see that

$$\rho(\mathbf{L}) = \mathbf{I} = \exp(T\mathtt{x}_1)\cdot\Phi_{ш}.$$

But by the definition of ρ in Lemma 3.3.17

$$\rho(\mathbf{L}) = \rho(\exp(T\mathtt{z}_1)\cdot\Phi_*) = \Lambda(\mathtt{x}_1)\exp(T\mathtt{x}_1)\Phi_* = \exp(T\mathtt{x}_1)\cdot\Lambda(\mathtt{x}_1)\Phi_*.$$

The theorem follows immediately. □

4.1.3 *Grothendieck–Teichmüller Group*

To have a better understanding of the Drinfeld associator Φ and also for future reference we need the following definitions.

Definition 4.1.16. Let $\mathbb{K}$ be a field of characteristic 0. For $n \geq 2$, denote by $\mathfrak{t}_n$ the completed $\mathbb{K}$-Lie algebra with generators t^{ij} $(i \neq j,\ 1 \leq i, j \leq n)$ subject to the following *(infinitesimal) pure braids relations*[1]:

$$t^{ij} = t^{ji}, \ [t^{ij}, t^{ik} + t^{jk}] = 0 \text{ and } [t^{ij}, t^{kl}] = 0 \quad \text{for all distinct } i,\ j,\ k,\ l. \tag{4.11}$$

Let $\mathfrak{t}_n^0$ be the Lie subalgebra of $\mathfrak{t}_n$ with the generators t^{ij} and the same relations as $\mathfrak{t}_n$ plus one more relation $\sum_{1\leq i<j\leq n} t^{ij} = 0$.

Example 4.1.17. Observe that $\mathfrak{t}_2$ is the 1-dimensional abelian Lie algebra generated by t^{12}. Further, $\mathfrak{t}_3^0$ is the free Lie algebra $\mathbb{K}\langle \mathtt{x}_0, \mathtt{x}_1 \rangle$ of rank 2 with generators $\mathtt{x}_0 := t^{12}$ and $\mathtt{x}_1 := t^{23}$. See Exercise 4.6.

Definition 4.1.18. The *Grothendieck–Teichmüller Lie algebra* $\mathfrak{grt}_1(\mathbb{K})$ is defined to be the set of $\varphi = \varphi(\mathtt{x}_0, \mathtt{x}_1) \in \mathfrak{t}_3^0$ satisfying the *duality and hexagon relations* in $\mathfrak{t}_3^0$

$$\varphi(\mathtt{x}_0, \mathtt{x}_1) + \varphi(\mathtt{x}_1, \mathtt{x}_0) = 0, \quad \varphi(\mathtt{x}_0, \mathtt{x}_1) + \varphi(\mathtt{x}_1, \mathtt{x}_\infty) + \varphi(\mathtt{x}_\infty, \mathtt{x}_0) = 0 \tag{4.12}$$

with $\mathtt{x}_0 + \mathtt{x}_1 + \mathtt{x}_\infty = 0$, and the *pentagon relation* in $\mathfrak{t}_4^0$

$$\begin{aligned}\varphi(t^{12}, t^{23} + t^{24}) &+ \varphi(t^{13} + t^{23}, t^{34}) \\ &= \varphi(t^{23}, t^{34}) + \varphi(t^{12} + t^{13}, t^{24} + t^{34}) + \varphi(t^{12}, t^{23}).\end{aligned} \tag{4.13}$$

See Exercise 4.7 for the definition of the corresponding Lie bracket.

Proposition 4.1.19. *If* $\varphi \in \mathfrak{grt}_1(\mathbb{K})$ *then it satisfies the* special derivation condition *in* $\mathfrak{t}_3^0$

$$[\mathtt{x}_1, \varphi(\mathtt{x}_0, \mathtt{x}_1)] + [\mathtt{x}_\infty, \varphi(\mathtt{x}_0, \mathtt{x}_\infty)] = 0 \tag{4.14}$$

for $\mathtt{x}_0 + \mathtt{x}_1 + \mathtt{x}_\infty = 0$.

Proof. Let $s(\mathtt{x}_1, \mathtt{x}_\infty) = [\mathtt{x}_1, \varphi(\mathtt{x}_0, \mathtt{x}_1)] + [\mathtt{x}_\infty, \varphi(\mathtt{x}_0, \mathtt{x}_\infty)]$. Then $s(\mathtt{x}_1, \mathtt{x}_\infty) = s(\mathtt{x}_\infty, \mathtt{x}_1)$. Now let Y_1, Y_2 and Y_3 be any generators of $\mathfrak{t}_4$ and set

$$u(Y_1, Y_2, Y_3) := s(Y_1, Y_2) - s(Y_1, Y_2 + Y_3) + s(Y_1 + Y_2, Y_3) - s(Y_2, Y_3).$$

By Eq. (4.12) we have

$$u(t^{14}, t^{24}, t^{34}) = [t^{14} + t^{24} + t^{34}, m^{1234}] - [t^{14} + t^{24}, m^{1243}] + [t^{14}, m^{1423}], \tag{4.15}$$

where $m^{1234} = \{\text{left-hand side of (4.13)}\} - \{\text{right-hand side of (4.13)}\} = 0$ (see Exercise 4.13). Hence $u(Y_1, Y_2, Y_3) = 0$ and by Thm. B.13 we see that $s = 0$ since s is a Lie polynomial by Thm. B.1. This finishes the proof of the proposition. □

[1] Due to its relation to the braid groups this Lie algebra is called the *Lie algebra of pure braids*. But we will not go into the details of these relations in this book.

Definition 4.1.20. The *Grothendieck–Teichmüller group* $\mathrm{GRT}_1(\mathbb{K})$ is defined to be the set of series $g \in \exp \mathfrak{t}_3^0$ satisfying the *duality and hexagon relations* in $\exp \mathfrak{t}_3^0$

$$g(\mathrm{x}_0, \mathrm{x}_1)g(\mathrm{x}_1, \mathrm{x}_0) = 1, \qquad g(\mathrm{x}_\infty, \mathrm{x}_0)g(\mathrm{x}_1, \mathrm{x}_\infty)g(\mathrm{x}_0, \mathrm{x}_1) = 1 \tag{4.16}$$

with $\mathrm{x}_0 + \mathrm{x}_1 + \mathrm{x}_\infty = 0$, and the *pentagon equation* in $\exp \mathfrak{t}_4^0$

$$g(t^{12}, t^{23} + t^{24})g(t^{13} + t^{23}, t^{34}) = g(t^{23}, t^{34})g(t^{12} + t^{13}, t^{24} + t^{34})g(t^{12}, t^{23}). \tag{4.17}$$

See Exercise 4.8 for the definition of the corresponding group multiplication. Note that the associated Lie algebra is $\mathfrak{grt}_1(\mathbb{K})$.

Definition 4.1.21. The *associator set* $\mathrm{M}_\lambda(\mathbb{K})$ is defined to be the set of series $g \in \exp \mathfrak{t}_3^0$ satisfying the pentagon relation (4.17) and the following duality and λ-hexagon relations

$$g(\mathrm{x}_0, \mathrm{x}_1)g(\mathrm{x}_1, \mathrm{x}_0) = 1, \quad e^{\lambda \mathrm{x}_0/2}g(\mathrm{x}_\infty, \mathrm{x}_0)e^{\lambda \mathrm{x}_\infty/2}g(\mathrm{x}_1, \mathrm{x}_\infty)e^{\lambda \mathrm{x}_1/2}g(\mathrm{x}_0, \mathrm{x}_1) = 1 \tag{4.18}$$

with $\mathrm{x}_0 + \mathrm{x}_1 + \mathrm{x}_\infty = 0$ and $\lambda \in \mathbb{K}^\times$.

It can be proved that (see Exercise 4.9) $\mathrm{M}_\lambda(\mathbb{K})$ is a left $\mathrm{GRT}_1(\mathbb{K})$-torsor for any fixed $\lambda \in \mathbb{K}^\times$. Moreover, it is clear that $\mathrm{GRT}_1(\mathbb{K}) = \mathrm{M}_0(\mathbb{K})$. This is different from the original definition of $\mathrm{GRT}_1(\mathbb{K})$ where *special action condition* is also required. But by the next proposition due to Drinfeld in [173] we see this condition is not needed.

Proposition 4.1.22. *If* $g(\mathrm{x}_0, \mathrm{x}_1) \in \mathrm{GRT}_1(\mathbb{K})$ *then it satisfies the* special action condition *in* $\exp \mathfrak{t}_3^0$

$$\mathrm{x}_0 + g(\mathrm{x}_0, \mathrm{x}_1)^{-1}\mathrm{x}_1 g(\mathrm{x}_0, \mathrm{x}_1) + g(\mathrm{x}_0, \mathrm{x}_\infty))^{-1}\mathrm{x}_\infty g(\mathrm{x}_0, \mathrm{x}_\infty) = 0, \tag{4.19}$$

for $\mathrm{x}_0 + \mathrm{x}_1 + \mathrm{x}_\infty = 0$.

Proof. Let $g(\mathrm{x}_0, \mathrm{x}_1) \in \mathrm{GRT}_1(\mathbb{K})$, then $\log(g(\mathrm{x}_0, \mathrm{x}_1)) \in \mathfrak{t}_3^0$. For each positive integer n, let φ_n be the degree $\le n$ part of $\log(g(\mathrm{x}_0, \mathrm{x}_1))$. Then $\varphi_n \in \mathfrak{grt}_1(\mathbb{K})$ so it satisfies the special derivation condition Eq. (4.14) by Prop. 4.1.19. Since $\exp(-\varphi_n)g(\mathrm{x}_0, \mathrm{x}_1) \equiv 1$ (see Exercise 4.9 for the multiplication definition) modulo degree $n+1$ we see that Eq. (4.19) is true modulo degree $n+1$. This concludes the proof of the proposition since n is arbitrary. □

The next result of Furusho shows that the pentagon relation is in some sense more fundamental than the other relations. We refer the interested reader to his original proof in [219].

Theorem 4.1.23. (i) *Let* $\varphi \in \mathfrak{t}_3^0$ *with* $\varphi[\mathrm{x}_1] = \varphi[\mathrm{x}_0\mathrm{x}_1] = 0$ *(*i.e.*, both coefficients of* x_1 *and* $\mathrm{x}_0\mathrm{x}_1$ *vanish). If* φ *satisfies the pentagon equation* (4.13)*, then it also satisfies the duality and hexagon equations* (4.12)*.*

(ii) *Let* $g \in \exp \mathfrak{t}_3^0$ *with* $g[\mathrm{x}_1] = g[\mathrm{x}_0\mathrm{x}_1] = 0$*. If* g *satisfies the pentagon equation* (4.17)*, then it also satisfies the duality and hexagon equations* (4.16)*.*

(iii) *Let* $g \in \exp \mathfrak{t}_3^0$ *with* $g[\mathrm{x}_1] = 0$ *and* $g[\mathrm{x}_0\mathrm{x}_1] \in \mathbb{K}^\times$*. If* g *satisfies the pentagon equation* (4.17)*, then the duality and hexagon equations* (4.18) *hold for* $g(\mathrm{x}_0/\alpha, \mathrm{x}_1/\alpha)$ *with* $\alpha = \pm\sqrt{24g[\mathrm{x}_0\mathrm{x}_1]} \in \bar{\mathbb{K}}$*.*

4.1.4 *Pentagon and Hexagon Relations*

We shall see that the Drinfeld associator Φ satisfies both the pentagon and hexagon relations. It is easier to study these relations in the context of quasi-Hopf (for pentagon) and quasi-triangular (for hexagon) algebras. We will not define these formally although we will use some of the key ideas related to these structures in this subsection.

Let $\mathbb{K}$ be a field containing $\mathbb{Q}$. Let $h > 0$ be a formal real variable (which is germain to the Planck constant in theoretical physics) and $\hbar = h/2\pi i$. Let $\mathfrak{g}$ be a Lie algebra (see Appendix B) over $\mathbb{K}[\![h]\!]$ which as a $\mathbb{K}[\![h]\!]$-module is isomorphic to $\mathscr{V}[\![h]\!]$ for some vector space $\mathscr{V}$ over $\mathbb{K}$. Let $t \in \mathfrak{g} \otimes \mathfrak{g}$ be a $\mathfrak{g}$-invariant symmetric tensor, *i.e.*, $t = \tau(t)$ with $\tau(x \otimes y) = y \otimes x$. For any positive integer $n \geq 3$, we denote by t^{ij} the image of t under the (i, j)-th embedding $\mathfrak{g} \otimes \mathfrak{g} \to \mathfrak{g}^{\otimes n}$. For example, if $n = 3$ then $t^{12} = t \otimes 1, t^{23} = 1 \otimes t \in \mathfrak{g}^{\otimes 3}$.

Lemma 4.1.24. *The* t^{ij} *satisfy all the pure braids relations in Eq.* (4.11)*.*

Proof. This is straightforward by the symmetry of t. □

Definition 4.1.25. Let $\hbar > 0$ be a real number. Let $\Phi_h = \Phi_h(t^{12}, t^{23}) := G_1^{-1}G_0$, where G_0 and G_1 are the analytic solutions of the KZ differential equation

$$G'(z) = \hbar \left(\frac{t^{12}}{z} + \frac{t^{23}}{z-1} \right) G(z), \tag{4.20}$$

that are defined for $0 < z < 1$, such that

$$G_0(z) \sim z^{\hbar t^{12}} \text{ near } z = 0 \quad \text{and} \quad G_1(z) \sim (1-z)^{\hbar t^{23}} \text{ near } z = 1.$$

Here, $f(z) \sim 1$ near $z = 0$ means that there is an expansion of $f(z) = 1 + f_1(z)h + f_2(z)h^2 + \cdots$ such that $f_j(z)$ are all analytic near $z = 0$.

Remark 4.1.26. By the definition Φ_h is an invertible element of $(U\mathfrak{g})^{\otimes 3}$. From Definition 4.1.2 we see that if $h = 2\pi i$ (or, equivalently, if $\hbar = 1$) then $\Phi_{2\pi i}(t^{12}, t^{23}) = \Phi(t^{12}, t^{23})$.

The KZ system (4.20) has can be generalized to

$$\frac{\partial W}{\partial z_i} = \hbar \sum_{j \neq i} \frac{t^{ij}}{z_i - z_j} W, \quad i = 1, \ldots, n, \tag{4.21}$$

where $W(z_1, \ldots, z_n) \in (U\mathfrak{g})^{\otimes n}$. Since $\partial W/\partial z_1 + \cdots + \partial W/\partial z_n = 0$ the function W depends only on the difference $z_i - z_j$. Another restriction $\sum_{i=1}^n z_i \partial W/\partial z_i = \hbar \sum_{1=i<j\leq n} t^{ij} W$ further reduces (4.21) to a system of $n-2$ variables. In particular, if $n = 3$ the solution of (4.20) has the form $(z_3 - z_1)^{\hbar(t^{12}+t^{13}+t^{23})} G((z_2 - z_1)/(z_3 - z_1))$ where G satisfies (4.20). Hence

$$W_1 = W_2 \cdot \Phi_h \tag{4.22}$$

where W_1 and W_2 are the solutions of (4.21) for $n = 3$ in the region $z_1 < z_2 < z_3$ of $\mathbb{R}^3$ with the asymptotic behaviors

$$\begin{aligned} W_1 &\sim (z_2 - z_1)^{\hbar t^{12}} (z_3 - z_1)^{\hbar(t^{13}+t^{23})} \ \text{ for } \ z_2 - z_1 \ll z_3 - z_1, \\ W_2 &\sim (z_3 - z_2)^{\hbar t^{23}} (z_3 - z_1)^{\hbar(t^{13}+t^{23})} \ \text{ for } \ z_3 - z_2 \ll z_3 - z_1. \end{aligned} \tag{4.23}$$

Lemma 4.1.27. *For $n = 4$, there exist unique solutions $W_1, \ldots, W_5$ of the system (4.21) with the following given asymptotic behaviors in the given zones:*

$$\begin{aligned}
W_1 &\sim (z_2 - z_1)^{\hbar t^{12}} (z_3 - z_1)^{\hbar(t^{13}+t^{23})} (z_4 - z_1)^{\hbar(t^{14}+t^{24}+t^{34})} \\
&\quad \textit{for } \ z_2 - z_1 \ll z_3 - z_1 \ll z_4 - z_1, \\
W_2 &\sim (z_3 - z_2)^{\hbar t^{23}} (z_3 - z_1)^{\hbar(t^{12}+t^{13})} (z_4 - z_1)^{\hbar(t^{14}+t^{24}+t^{34})} \\
&\quad \textit{for } \ z_3 - z_2 \ll z_3 - z_1 \ll z_4 - z_1, \\
W_3 &\sim (z_3 - z_2)^{\hbar t^{23}} (z_4 - z_2)^{\hbar(t^{24}+t^{34})} (z_4 - z_1)^{\hbar(t^{12}+t^{13}+t^{14})} \\
&\quad \textit{for } \ z_3 - z_2 \ll z_4 - z_2 \ll z_4 - z_1, \\
W_4 &\sim (z_4 - z_3)^{\hbar t^{34}} (z_4 - z_2)^{\hbar(t^{23}+t^{24})} (z_4 - z_1)^{\hbar(t^{12}+t^{13}+t^{14})} \\
&\quad \textit{for } \ z_4 - z_3 \ll z_4 - z_2 \ll z_4 - z_1, \\
W_5 &\sim (z_2 - z_1)^{\hbar t^{12}} (z_4 - z_3)^{\hbar t^{34}} (z_4 - z_1)^{\hbar(t^{13}+t^{14}+t^{23}+t^{24})} \\
&\quad \textit{for } \ z_2 - z_1 \ll z_4 - z_1, z_4 - z_3 \ll z_4 - z_1.
\end{aligned}$$

Proof. For W_1 we set $u = (z_2 - z_1)/(z_3 - z_1)$ and $v = (z_3 - z_1)/(z_4 - z_1)$. Then we have

$$uv = \frac{z_2 - z_1}{z_4 - z_1}, \quad 1 - uv = \frac{z_4 - z_2}{z_4 - z_1} \quad 1 - u = \frac{z_3 - z_2}{z_3 - z_1}, \quad 1 - v = \frac{z_4 - z_3}{z_4 - z_1}.$$

Define $W = g(u,v)(z_4 - z_1)^{\hbar T}$ where $T = \sum_{1\le i<j\le n} t^{ij}$. Then

$$\frac{\partial g}{\partial z_2} = \frac{1}{z_3 - z_1}\frac{\partial g}{\partial u}, \qquad \frac{\partial g}{\partial z_3} = \frac{z_1 - z_2}{(z_3-z_1)^2}\frac{\partial g}{\partial u} + \frac{1}{z_4 - z_1}\frac{\partial g}{\partial v}$$
$$\Longrightarrow \frac{\partial g}{\partial u} = (z_3 - z_1)\frac{\partial g}{\partial z_2}, \qquad \frac{\partial g}{\partial v} = u(z_4 - z_1)\frac{\partial g}{\partial z_2} + (z_4 - z_1)\frac{\partial g}{\partial z_3}.$$

Hence g satisfies

$$\frac{\partial g}{\partial u} = \hbar\left(\frac{t^{12}}{u} - \frac{vt^{24}}{1-uv} - \frac{t^{23}}{1-u}\right)g = \hbar\left(\frac{A}{u} + R(u,v)\right)g, \tag{4.24}$$

$$\begin{aligned}\frac{\partial g}{\partial v} &= \hbar\left(\frac{t^{12}}{v} - \frac{ut^{23}}{v(1-u)} - \frac{ut^{24}}{1-uv} + \frac{t^{13}}{v} + \frac{t^{23}}{v(1-u)} - \frac{t^{34}}{1-v}\right)g(u,v)\\ &= \hbar\left(\frac{B}{v} - \frac{ut^{24}}{1-uv} - \frac{t^{34}}{1-v}\right)g = \hbar\left(\frac{B}{v} + S(u,v)\right)g,\end{aligned} \tag{4.25}$$

where $A = t^{12}$ and $B = t^{12} + t^{13} + t^{23}$. Let $g(u,v) = f(u,v)u^A v^B$ then we need to show the following system has a solution:

$$f^{-1}\nabla_u f = hAu^{-1}, \quad f^{-1}\nabla_v f = hBv^{-1},$$

where the linear differential operators $\nabla_u = h(Au^{-1} + R(u,v)) - \partial/\partial u$ and $\nabla_v = h(Bv^{-1} + S(u,v)) - \partial/\partial u$. Note that $[A,B] = 0$ by Lemma 4.1.24 so the above system has a solution if we note that

$$R[u^i v^{j+1}] = S[u^{i+1}v^j] = -\delta_{i,j}t^{24} \quad \forall i,j \ge 0, \tag{4.26}$$

where $\delta_{i,j}$ is the Kronecker symbol. To derive the uniqueness, we only need to find the unique power series expansion $f(u,v) = \sum_{i,j\ge 0} c_{i,j}u^i v^j$ so that

$$g(u,v) = f(u,v)u^A v^B = f(u,v)\left(\frac{z_2 - z_1}{z_4 - z_1}\right)^{\hbar t^{12}}\left(\frac{z_3 - z_1}{z_4 - z_1}\right)^{\hbar(t^{13}+t^{23})}$$

satisfies Eqs. (4.24) and (4.25). Starting from $c_{0,0} = 1$ all the coefficients of f can be determined uniquely by iterations. Thus W_1 exists and is unique. See Exercise 4.10 for details.

The proofs of the other solutions are similar so we leave them to the interested reader. □

Lemma 4.1.28. *Let $W_1, \dots, W_5$ be the analytic continuation of the functions in Lemma 4.1.27 to the whole region $z_1 < z_2 < z_3 < z_4$ in $\mathbb{R}^4$. Then*

$$W_1 = W_2(\Phi_h \otimes 1), \quad W_2 = W_3(\mathrm{id} \otimes \Delta \otimes \mathrm{id})(\Phi_h), \quad W_3 = W_4(1 \otimes \Phi_h),$$
$$W_1 = W_5(\Delta \otimes \mathrm{id} \otimes \mathrm{id})(\Phi_h), \quad \textit{and} \quad W_5 = W_4(\mathrm{id} \otimes \mathrm{id} \otimes \Delta)(\Phi_h).$$

Proof. Define two analytic functions on $z_1 < z_2 < z_3 < z_4$:

$$V_1 = W_1(z_4 - z_1)^{-\hbar(t^{14}+t^{24}+t^{34})},$$
$$V_2 = W_2(z_4 - z_1)^{-\hbar(t^{14}+t^{24}+t^{34})}(\Phi_h \otimes 1).$$

In fact, since the factor $z_4 - z_1$ does not appear in the asymptotics anymore, the two functions are still well-defined for $z_4 < z_1$. We now prove $V_1 = V_2$. First, both V_1 and V_2 satisfy the equations

$$\frac{\partial V}{\partial z_2} = \hbar \sum_{j \neq i} \frac{t^{ij}}{z_i - z_j} V, \quad i \neq 2, 3, \tag{4.27}$$

$$\frac{\partial V}{\partial z_1} = \hbar \sum_{j \neq 1} \frac{t^{1j}}{z_1 - z_j} V - \hbar V \cdot \frac{t^{14} + t^{24} + t^{34}}{z_1 - z_4}, \tag{4.28}$$

$$\frac{\partial V}{\partial z_4} = \hbar \sum_{j \neq 4} \frac{t^{4j}}{z_4 - z_j} V - \hbar V \cdot \frac{t^{14} + t^{24} + t^{34}}{z_4 - z_1}. \tag{4.29}$$

Now as $z_4 \to \infty$ we see that the differential equations in Eqs. (4.27) and (4.28) are exactly the same as those of Eq. (4.21) when $n = 3$. Moreover,

$$V_1 \sim (z_2 - z_1)^{\hbar t^{12}} (z_3 - z_1)^{\hbar(t^{13}+t^{23})},$$
$$V_2 \cdot \Phi_h^{-1} \sim (z_3 - z_2)^{\hbar t^{23}} (z_3 - z_1)^{\hbar(t^{12}+t^{13})},$$

which are exactly the same as those of Eq. (4.23). By Eq. (4.22) we see that $V_1 = V_2$ as $z_4 \to \infty$. Together with Eq. (4.29) this implies $V_1 = V_2$.

We leave the proofs of the other cases to the interested reader since they are almost identical to the above (see Exercise 4.11). □

Lemma 4.1.28 quickly yields the pentagon relations for Φ_h.

Corollary 4.1.29. *We have*

$$(\mathrm{id} \otimes \mathrm{id} \otimes \Delta)(\Phi_h)(\Delta \otimes \mathrm{id} \otimes \mathrm{id})(\Phi_h) = (1 \otimes \Phi_h)(\mathrm{id} \otimes \Delta \otimes \mathrm{id})(\Phi_h)(\Phi_h \otimes 1).$$

Or, equivalently,

$$\Phi_h(t^{12}, t^{23} + t^{24})\Phi_h(t^{13} + t^{23}, t^{34})$$
$$= \Phi_h(t^{23}, t^{34})\Phi_h(t^{12} + t^{13}, t^{24} + t^{34})\Phi_h(t^{12}, t^{23}).$$

We now turn to the hexagon relations.

Lemma 4.1.30. *For $n = 3$, define the unique solutions $W_1, \dots, W_6$ of the system (4.21) with the following given asymptotic behaviors in the given zones in $\{(z_1, z_2, z_3) \in \mathbb{C}^3 : \mathrm{Im}(z_1) \le \mathrm{Im}(z_2) \le \mathrm{Im}(z_3), z_1 \neq z_2 \neq z_3 \neq z_1\}$:*

$$W_1 \sim (z_2 - z_1)^{\hbar t^{12}} (z_3 - z_1)^{\hbar(t^{13}+t^{23})} \quad \textit{for } |z_2 - z_1| \ll |z_3 - z_1|,$$

$$\begin{aligned}
W_2 &\sim (z_3 - z_2)^{\hbar t^{23}}(z_3 - z_1)^{\hbar(t^{12}+t^{23})} \quad \textit{for} \quad |z_3 - z_2| \ll |z_3 - z_1|, \\
W_3 &\sim (z_2 - z_3)^{\hbar t^{23}}(z_2 - z_1)^{\hbar(t^{12}+t^{13})} \quad \textit{for} \quad |z_3 - z_2| \ll |z_2 - z_1|, \\
W_4 &\sim (z_3 - z_1)^{\hbar t^{13}}(z_2 - z_1)^{\hbar(t^{12}+t^{23})} \quad \textit{for} \quad |z_1 - z_3| \ll |z_2 - z_1|, \\
W_5 &\sim (z_1 - z_3)^{\hbar t^{13}}(z_2 - z_3)^{\hbar(t^{12}+t^{23})} \quad \textit{for} \quad |z_1 - z_3| \ll |z_2 - z_3|, \\
W_6 &\sim (z_2 - z_1)^{\hbar t^{12}}(z_2 - z_3)^{\hbar(t^{13}+t^{23})} \quad \textit{for} \quad |z_2 - z_1| \ll |z_3 - z_2|.
\end{aligned}$$

Then we have

$$W_1 = W_2\Phi_h, \qquad W_4 = W_3\Phi_h^{132}, \qquad W_5 = W_6\Phi_h^{312}, \tag{4.30}$$

$$W_2 = W_3 e^{ht^{23}/2}, \quad W_4 = W_5 e^{ht^{13}/2}, \quad W_1 = W_6 e^{h(t^{13}+t^{23})/2}. \tag{4.31}$$

Proof. By Eqs. (4.23) and (4.22) we see that $W_1 = W_2\Phi_h$ when $z_1 < z_2 < z_3$. Similarly, by transposition of indices $2 \leftrightarrow 3$ we see that $W_4 = W_3\Phi_h^{132}$ when $z_1 < z_3 < z_2$, and by a further transposition of indices $1 \leftrightarrow 3$ we see that $W_5 = W_6\Phi_h^{312}$ when $z_3 < z_1 < z_2$.

Now define $V_2 = W_2(z_3 - z_2)^{-\hbar t^{23}}$ and $V_3 = W_3(z_3 - z_2)^{-\hbar t^{23}} e^{ht^{23}/2} = W_3(z_2 - z_3)^{-\hbar t^{23}}$. We now show that $V_2 = V_3$ as follows. First, we see that both functions satisfy

$$\begin{aligned}
\frac{\partial V}{\partial z_1} &= \hbar\left(\frac{t^{12}}{z_1 - z_2}\frac{t^{13}}{z_1 - z_3}\right) + V, \\
\frac{\partial V}{\partial z_2} &= \hbar\sum_{j\neq 2}\frac{t^{2j}}{z_2 - z_j}V - \hbar V\cdot\frac{t^{23}}{z_2 - z_3} = \frac{t^{12}V}{z_2 - z_1} + \frac{[t^{23}, V]}{z_2 - z_3}, \\
\frac{\partial V}{\partial z_3} &= \hbar\sum_{j\neq 3}\frac{t^{3j}}{z_3 - z_j}V - \hbar V\cdot\frac{t^{23}}{z_3 - z_2} = \frac{t^{13}V}{z_3 - z_1} + \frac{[t^{23}, V]}{z_3 - z_2}.
\end{aligned}$$

Thus $V_2 - V_3$ is an absolute constant. Second, both functions have the same asymptotics as $z_2 \to z_3$. Hence $V_2 = V_3$.

The other two identities are left as an exercise for the interested reader (see Exercise 4.12). □

Lemma 4.1.30 immediately yields the h-hexagon relations for Φ_h given by Eq. (4.18).

Corollary 4.1.31. *We have*

$$e^{ht^{13}/2}\Phi_h(t^{23}, t^{13})e^{ht^{23}/2}\Phi_h(t^{12}, t^{23})e^{-h(t^{13}+t^{23})/2}\Phi_h(t^{13}, t^{12}) = 1.$$

Proof. In $\exp(\mathfrak{t}_3^0)$ we have $t^{12} = -t^{13} - t^{23}$. □

The following theorem due to Drinfeld shows in particular that the associator Φ satisfies the pentagon relation (4.13). Together with Thm. 4.1.14 one may derive a lot of relations among the MZVs.

Theorem 4.1.32. *The Drinfeld associator* $\Phi(a,b) \in \mathrm{M}_{2\pi i}(\mathbb{C})$. *Moreover, it satisfies the special action condition*

$$\mathrm{x}_0 + \Phi(\mathrm{x}_0,\mathrm{x}_1)^{-1}\mathrm{x}_1\Phi(\mathrm{x}_0,\mathrm{x}_1) + \Phi(\mathrm{x}_0,\mathrm{x}_\infty)^{-1}\mathrm{x}_\infty\Phi(\mathrm{x}_0,\mathrm{x}_\infty) = 0 \qquad (4.32)$$

if $\mathrm{x}_0 + \mathrm{x}_1 + \mathrm{x}_\infty = 0$.

Proof. By Remark 4.1.26 this follows from Cor. 4.1.29 and 4.1.31. □

In particular, the Drinfeld associator Φ does not lie in the Grothendieck–Teichmüller group $\mathrm{GRT}_1(\mathbb{C})$ but instead it is an element in the associator set $\mathrm{M}_{2\pi i}(\mathbb{C})$ by its duality relation Eq. (4.6) and Thm. 4.1.32.

4.2 Single-Variable Multiple Polylogs

Recall that in the study of the MZVs in the last chapter, the single-variable MPLs defined by Eq. (3.20) plays an important role. By the definition, it is clear that $\mathrm{L}_{\mathbf{s}}(z) = \mathrm{Li}_{\mathbf{s}}(z,1,\dots,1)$.

Proposition 4.2.1. *We have*

$$\frac{\mathrm{d}}{\mathrm{d}z}\mathrm{L}_{\mathbf{s}}(z) = \begin{cases} \mathrm{L}_{s_1-1,s_2,\dots,s_d}(z)/z, & \text{if } s_1 > 1; \\ \mathrm{L}_{s_2,s_3,\dots,s_d}(z)/(1-z), & \text{if } s_1 = 1. \end{cases}$$

Proof. The case $s_1 > 1$ is obvious. An easy application of geometric series summation formula yields the case $s_1 = 1$ immediately. □

Let $\omega_0(t) = \mathrm{d}t/t$ and $\omega_1(t) = \mathrm{d}t/(t-1)$. For any path $\gamma : [0,1] \to \mathbb{C}\setminus\{0,1\}$ from $y = \gamma(0)$ to $z = \gamma(1)$ and $a_1,\dots,a_n \in \{0,1\}$, we define

$$I_\gamma(y;a_1\dots a_n;z) = \int_{0<t_1<\dots<t_n<1} \gamma^*\omega_{a_1}(t_1)\wedge\dots\wedge\gamma^*\omega_{a_n}(t_n)$$

where $\gamma^*\omega$ is the pullback of ω by γ (see Sec. 2.1). Put $I_\gamma(y;\emptyset;z) = 1$. Let ϕ be one of the six Möbius transformations

$$\phi(z) = z, 1-z, 1-z^{-1}, (1-z^{-1})^{-1}, (1-z)^{-1}, z^{-1}. \qquad (4.33)$$

Then we can extend ϕ linearly to words and to iterated integrals:

$$\begin{aligned} 1-z &: \omega_0 \mapsto \omega_1, & \omega_1 &\mapsto \omega_0, \\ 1-z^{-1} &: \omega_0 \mapsto -\omega_0+\omega_1, & \omega_1 &\mapsto -\omega_0, \\ (1-z^{-1})^{-1} &: \omega_0 \mapsto \omega_0-\omega_1, & \omega_1 &\mapsto -\omega_1, \\ (1-z)^{-1} &: \omega_0 \mapsto -\omega_1, & \omega_1 &\mapsto \omega_0-\omega_1, \\ z^{-1} &: \omega_0 \mapsto -\omega_0, & \omega_1 &\mapsto -\omega_0+\omega_1. \end{aligned}$$

Lemma 4.2.2. *For any path γ from y to z, all the iterated integrals $\{I_\gamma(y; \mathbf{w}; z) : \mathbf{w} \in \{0,1\}^*\}$ form a shuffle algebra. Moreover, for every ϕ in* (4.33), *we have*

$$I_{\phi(\gamma)}(\phi(y); \mathbf{w}; \phi(z)) = I_\gamma(y; \phi^* \mathbf{w}; z) \tag{4.34}$$

and for any word $\mathbf{w} \in \{0,1\}^*$ *and* $a \in \{0,1\}$

$$\partial_z I(y; a\mathbf{w}; z) = \frac{1}{z-a} I(y; \mathbf{w}; z), \quad \partial_y I(y; a\mathbf{w}; z) = -\frac{1}{y-a} I(y; \mathbf{w}; z).$$

Proof. These follow easily from the properties of the iterated integrals. □

In the rest of this chapter let $\mathsf{X} = \{\mathtt{x}_0, \mathtt{x}_1\}$ and let X^* be the set of words over X. Denote by 1 the empty word and set $I_\gamma(y; 1; z) = 1$. Write $I_\gamma(y; \mathtt{x}_{a_1} \dots \mathtt{x}_{a_n}; z) = I_\gamma(y; a_1 \dots a_n; z)$ for any $\mathtt{x}_{a_1} \dots \mathtt{x}_{a_n} \in \mathsf{X}^*$. We now define the generating series

$$S_{\mathsf{X}}(\gamma) = \sum_{\mathbf{w} \in \mathsf{X}^*} I_\gamma(y; \mathbf{w}; z)\, \mathbf{w} \in \mathbb{C}\langle\langle \mathsf{X} \rangle\rangle. \tag{4.35}$$

By the properties of the iterated integrals one can prove

Lemma 4.2.3. *If $\gamma(0) = y$ and $\gamma(1) = z$ then*

$$\partial_z S_{\mathsf{X}}(\gamma) = \left(\frac{\mathtt{x}_0}{z} + \frac{\mathtt{x}_1}{z-1} \right) S_\gamma, \quad \partial_y S_{\mathsf{X}}(\gamma) = -S_\gamma \left(\frac{\mathtt{x}_0}{y} + \frac{\mathtt{x}_1}{y-1} \right),$$

Write $S_{\mathsf{X}}(y, z)$ when the homotopy class of γ is clear from the context.

Proposition 4.2.4. *The following limits are well defined*

$$F_0(z) := \lim_{\varepsilon \to 0^+} S_{\mathsf{X}}(\varepsilon, z) e^{\mathtt{x}_0 \log \varepsilon}, \qquad F_0(y)^{-1} := \lim_{\varepsilon \to 0^+} e^{-\mathtt{x}_0 \log \varepsilon} S_{\mathsf{X}}(y, \varepsilon),$$
$$F_1(z) := \lim_{\varepsilon \to 0} S_{\mathsf{X}}(1-\varepsilon, z) e^{\mathtt{x}_1 \log \varepsilon}, \quad F_1(y)^{-1} := \lim_{\varepsilon \to 0^+} e^{-\mathtt{x}_1 \log \varepsilon} S_{\mathsf{X}}(y, 1-\varepsilon).$$

Proof. This follows immediately from Thm 4.1.1. □

For simplicity, henceforth we will simply write $S_{\mathsf{X}}(0, z) = S_{\mathsf{X}}(z, 0)^{-1} = F_0(z)$ and $S_{\mathsf{X}}(1, z) = S_{\mathsf{X}}(z, 1)^{-1} = F_1(z)$. In particular we have

$$S_{\mathsf{X}}(0, 1) = \Phi(\mathtt{x}_0, \mathtt{x}_1).$$

Remark 4.2.5. One can also establish the proposition directly from the regularization procedure where all powers of $\log \varepsilon$ are assigned to be 0 in the limit $\varepsilon \to 0$. This is also equivalent to the regularization by tangential base points used by Deligne [159].

For simplicity, in what follows we will sometimes identify the words in X^* with the words in $\{0,1\}^*$ by identifying $\mathtt{x}_a$ with $a \in \{0,1\}$. It should be clear which type we mean from the context.

Definition 4.2.6. The generating series of the one-varable MPLs (in word form) is given by

$$\mathrm{L}_{\mathsf{X}}(z) := \sum_{\mathbf{w}\in\mathsf{X}^*} \mathrm{L}_{\mathbf{w}}(z)\mathbf{w}, \tag{4.36}$$

where

$$\mathrm{L}_{\{0\}^{s_1-1}1\cdots\{0\}^{s_d-1}1}(z) = (-1)^d\,\mathrm{L}_{s_1,\dots,s_d}(z) = \sum_{k_1>\cdots>k_d\geq 1} \frac{(-1)^d z^{k_1}}{k_1^{s_1}\cdots k_r^{s_d}}. \tag{4.37}$$

By Thm. 4.1.1, $\mathrm{L}_{\mathsf{X}}(z)$ is the unique solution of the KZ equation

$$\partial_z\,\mathrm{L}_{\mathsf{X}}(z) = \left(\frac{\mathtt{x}_0}{z} + \frac{\mathtt{x}_1}{z-1}\right)\mathrm{L}_{\mathsf{X}}(z) \tag{4.38}$$

satisfying the asymptotic condition

$$\mathrm{L}_{\mathsf{X}}(z) = h_0(z)e^{\mathtt{x}_0\log z} \quad \text{if } z \text{ is near } 0, \tag{4.39}$$

where $h_0(z)$ is a holomorphic function at 0 taking values in $\mathbb{C}\langle\langle\mathsf{X}\rangle\rangle$ with $h_0(0) = 1$. Hence

$$\mathrm{L}_{\mathsf{X}}(z) = G_0(z) \tag{4.40}$$

which appears in the Definition 4.1.2 for the Drinfeld associator. Moreover, by Lemma 4.2.3

$$\lim_{z\to 0^+} F_0(z)e^{-\mathtt{x}_0\log z} = \lim_{z=\varepsilon\to 0^+} S_{\mathsf{X}}(z,z) = 1.$$

Since $F_0(z)$ is also a solution of (4.38) we get $\mathrm{L}_{\mathsf{X}}(z) = G_0(z) = F_0(z)$.

Lemma 4.2.7. *The series* $\mathrm{L}_{\mathsf{X}}(z)$ *is group-like in* $(\mathbb{C}(z)\langle\langle\mathsf{X}\rangle\rangle, \Delta_{\sqcup\!\sqcup})$, *where* $\Delta_{\sqcup\!\sqcup}(\mathtt{x}_a) = 1\otimes\mathtt{x}_a + \mathtt{x}_a\otimes 1$.

Proof. By Eq.(4.35) we see that the series S_{X} is group-like (cf. Thm. 4.1.8). Furthermore, $e^{\mathtt{x}_0\log\varepsilon}$ is obviously group-like. By Lemma 4.2.3, we have

$$\begin{aligned}\Delta_{\sqcup\!\sqcup}\big(F_0(z)\big) &= \lim_{\varepsilon\to 0^+} \Delta_{\sqcup\!\sqcup}\big(S_{\mathsf{X}}(\varepsilon,z)\big)\Delta_{\sqcup\!\sqcup}\big(e^{\mathtt{x}_0\log\varepsilon}\big)\\ &= \lim_{\varepsilon\to 0^+} \big(S_{\mathsf{X}}(\varepsilon,z)\otimes S_{\mathsf{X}}(\varepsilon,z)\big)\big(e^{\mathtt{x}_0\log\varepsilon}\otimes e^{\mathtt{x}_0\log\varepsilon}\big) = F_0(z)\otimes F_0(z).\end{aligned}$$

Thus, the series $\mathrm{L}_{\mathsf{X}}(z) = F_0(z)$ is group-like, as desired. □

Corollary 4.2.8. *For all* $\mathbf{u}, \mathbf{v} \in \mathsf{X}^*$, *we have*

$$\mathrm{L}_{\mathbf{u} ш \mathbf{v}}(z) = \mathrm{L}_{\mathbf{u}}(z)\, \mathrm{L}_{\mathbf{v}}(z). \tag{4.41}$$

Proof. This is clear from Lemma 4.2.7 (see Exercise 4.4). □

Lemma 4.2.9. *There exists a function* $h_1(z)$ *taking values in the series* $\mathbb{C}\langle\langle \mathsf{X} \rangle\rangle$, *which is holomorphic at* $z = 1$ *with* $h_1(1) = 1$, *such that*

$$\mathrm{L}_{\mathsf{X}}(z) = h_1(z) e^{\mathrm{x}_1 \log(1-z)} \Phi(\mathrm{x}_0, \mathrm{x}_1). \tag{4.42}$$

Proof. By Definition 4.1.2 of the Drinfeld associator, we know $\Phi(\mathrm{x}_0, \mathrm{x}_1) = G_1(z)^{-1} G_0(z)$ where $G_1(z) = h_1(z) e^{\mathrm{x}_1 \log(1-z)}$ for some $h_1(z)$ satisfying the condition for $h_1(z)$ stated in the lemma. Hence the lemma follows from Eq. (4.40) immediately. □

For $a = 0, 1$ let σ_a be the loop centered at a oriented positively. Let $M(\sigma_a)$ denote the analytic continuation around σ_a. Then $M(\sigma_a)$ acts on the series $\mathrm{L}_{\mathsf{X}}(z)$ and its complex conjugate $\mathrm{L}_{\mathsf{X}}(\bar{z})$, and commutes with multiplication, ∂_z, and $\partial_{\bar{z}}$.

Lemma 4.2.10. *The monodromy operators* $M(\sigma_0)$, $M(\sigma_1)$ *act as follows:*

$$\begin{aligned} M(\sigma_0)\, \mathrm{L}_{\mathsf{X}}(z) &= \mathrm{L}_{\mathsf{X}}(z) e^{2\pi i \mathrm{x}_0}, \\ M(\sigma_1)\, \mathrm{L}_{\mathsf{X}}(z) &= \mathrm{L}_{\mathsf{X}}(z) \Phi(\mathrm{x}_0, \mathrm{x}_1)^{-1} e^{2\pi i \mathrm{x}_1} \Phi(\mathrm{x}_0, \mathrm{x}_1). \end{aligned} \tag{4.43}$$

Proof. These follow from (4.39), (4.42) and the equations $M(\sigma_0) \log z = \log z + 2\pi i$, $M(\sigma_1) \log(1-z) = \log(1-z) + 2\pi i$. □

Remark 4.2.11. It is unfortunate that $\mathrm{L}_1(z)$ could have two different meanings: the word form $\mathrm{L}_{\mathrm{x}_1}(z) = -\mathrm{Li}_1(z)$ or the single-variable MPL $\mathrm{L}_s(z)$ when $s = 1$, *i.e.*, $\mathrm{Li}_1(z)$. So in this book we will always treat the subscript 1 in $\mathrm{L}_1(z)$ as a letter in $\{0, 1\}$, meaning $\mathrm{L}_1(z) = -\mathrm{Li}_1(z) = \log(1-z)$.

Lemma 4.2.12. *If* $\mathbf{w} = \{0\}^{s_1 - 1} 1 \cdots \{0\}^{s_d - 1} 1 \{0\}^n$ *then*

$$\mathrm{L}_{\mathbf{w}}(z) = \sum_{\substack{\ell \ge 0, k_i \ge s_i\ (i \ge 1) \\ |\mathbf{k}| + \ell = |\mathbf{s}| + n}} (-1)^{\ell + n + d} \prod_{i=1}^{d} \binom{k_i - 1}{s_i - 1} \frac{\log^\ell z}{\ell!} \mathrm{L}_{k_1, \dots, k_r}(z). \tag{4.44}$$

Proof. This directly follows from Eq. (4.37) and Thm. 4.1.1 in which $f_0(z)$ is explicitly given by Eqs. (4.2) and (4.3). □

Example 4.2.13. By the lemma we get (also see Exercise 4.21) in the word form

$$\mathrm{L}_0(z) = \log z, \qquad \mathrm{L}_1(z) = -\,\mathrm{Li}_1(z) = \log(1-z) \quad \text{(see Remark 4.2.11)},$$
$$\mathrm{L}_{01}(z) = -\,\mathrm{Li}_2(z), \quad \mathrm{L}_{10}(z) = \mathrm{Li}_2(z) + \log z \log(1-z).$$

4.3 Single-Valued Single-Variable Multiple Polylogs (SVMPLs)

In this section we construct the single-valued version of the single-variable MPLs defined by Eq. (3.20). This will be crucial in the study of several classes of the Feynman integrals in Chap. 15.

We begin with a technical lemma. Let $\mathsf{X} = \{\mathrm{x}_0, \mathrm{x}_1\}$ be the same as in the preceding section. Let $\mathsf{MZ}_\mathbb{Z}$ be the ring over $\mathbb{Z}$ generated by the MZVs and let $\mathsf{MZ}_\mathbb{Z}\langle\langle \mathsf{X}\rangle\rangle$ be the Hopf algebra with coefficients in $\mathsf{MZ}_\mathbb{Z}$ so that both x_0 and x_1 are primitive elements. Further, we define $\leftarrow: \mathsf{X}^* \to \mathsf{X}^*$ to be the anti-homomorphism such that $\overleftarrow{\mathbf{w}}$ is the order reversal of $\mathbf{w}$. Note that $\overleftarrow{\mathrm{x}_a} = \mathrm{x}_a$ for $a = 0, 1$.

Lemma 4.3.1. *There exists* $\mathrm{y}_1 \in \mathsf{MZ}_\mathbb{Z}\langle\langle \mathsf{X}\rangle\rangle$ *such that*

$$\overleftarrow{\Phi}(\mathrm{x}_0, \mathrm{y}_1)\mathrm{y}_1\overleftarrow{\Phi}(\mathrm{x}_0, \mathrm{y}_1)^{-1} = \Phi(\mathrm{x}_0, \mathrm{x}_1)^{-1}\mathrm{x}_1\Phi(\mathrm{x}_0, \mathrm{x}_1) \tag{4.45}$$

Moreover, $\mathrm{y}_1 \equiv \mathrm{x}_1$ *modulo words with at least two* x_1*'s (depth two) or modulo words of weight* ≥ 4.

Proof. Let $\Phi(\mathrm{x}_0, \mathrm{x}_1)$ be the Drinfeld associator. Define

$$H_\mathsf{X}(t) := \Phi(\mathrm{x}_0, \mathrm{x}_1)^{-1}e^{t\mathrm{x}_1}\Phi(\mathrm{x}_0, \mathrm{x}_1) \in \mathsf{MZ}_\mathbb{Z}[t]\langle\langle \mathsf{X}\rangle\rangle$$

because $\Phi(\mathrm{x}_0, \mathrm{x}_1)^{-1} = \Phi(\mathrm{x}_1, \mathrm{x}_0)$. Moreover, by Thm. B.7(i) and the famous Baker–Campbell–Hausdorff formula $H_\mathsf{X}(t) = e^{h(t)}$ is a Lie exponential. Since $h(0) = 0$ we have $H'_\mathsf{X}(0) = h'(0) \in \mathsf{MZ}_\mathbb{Z}\langle\langle \mathsf{X}\rangle\rangle$. This implies that

$$F(\mathrm{x}_0, \mathrm{x}_1) := \Phi(\mathrm{x}_0, \mathrm{x}_1)^{-1}\mathrm{x}_1\Phi(\mathrm{x}_0, \mathrm{x}_1) - \mathrm{x}_1 \in \mathsf{MZ}_\mathbb{Z}\langle\langle \mathsf{X}\rangle\rangle. \tag{4.46}$$

By straightforward computation (see Exercise 4.19) we have $F \equiv 0$ modulo words of depth two and $F \equiv \zeta(2)[[\mathrm{x}_0, \mathrm{x}_1], \mathrm{x}_1]$ modulo words of weight four. Now we can solve Eq. (4.45) recursively by setting

$$\mathrm{y}_1^{(0)} = \mathrm{x}_1, \quad \mathrm{y}_1^{(k+1)} = \mathrm{x}_1 + F(\mathrm{x}_0, \mathrm{x}_1) + F(-\mathrm{x}_0, -\mathrm{y}_1^{(k)}).$$

Because F has lowest word weight three, the recursion converges by weight filtration. Let $\mathrm{y}_1 = \lim_{k\to\infty} \mathrm{y}_1^{(k)}$ then

$$\mathrm{y}_1 = \mathrm{x}_1 + F(\mathrm{x}_0, \mathrm{x}_1) + F(-\mathrm{x}_0, -\mathrm{y}_1) \tag{4.47}$$

which yields Eq. (4.45). Here we need the fact that $\overleftarrow{\Phi}(-\mathrm{x}_0, -\mathrm{x}_1) = \Phi(\mathrm{x}_0, \mathrm{x}_1)^{-1}$ by Thm. B.7(ii). By induction every $\mathrm{y}_1^{(k)}$ is in $\mathsf{MZ}_\mathbb{Z}\langle\langle \mathsf{X}\rangle\rangle$ so must be y_1. This completes the proof of the lemma. □

Let ∂_z and $\partial_{\bar{z}}$ be the holomorphic and antiholomorphic differentiation and set

$$\mathsf{O} := \mathbb{Q}\left[z, \frac{1}{z}, \frac{1}{z-1}\right], \quad \overline{\mathsf{O}} := \mathbb{Q}\left[\bar{z}, \frac{1}{\bar{z}}, \frac{1}{\bar{z}-1}\right].$$

Theorem 4.3.2. *For each* $\mathbf{w} \in \mathsf{X}^*$, *there is a single-valued function* $\mathscr{L}_{\mathbf{w}}(z)$ *on* $\mathbb{C}\setminus\{0,1\}$, *called the single-valued single-variable multiple polylog (SVMPL for abbreviation), such that*

(i) $\mathscr{L}_{\mathbf{w}}(z)$ *is a linear combination of* $\mathrm{L}_{\mathbf{u}}(z)\,\mathrm{L}_{\mathbf{v}}(\bar{z})$, *where* $\mathbf{u}, \mathbf{v} \in \mathsf{X}^*$;

(ii) $\mathscr{L}_1(z) = 1$, $\mathscr{L}_{\{0\}^n}(z) = \dfrac{\log^n(z\bar{z})}{n!}$ *for all* $n \geq 1$, *and* $\lim_{z\to 0} \mathscr{L}_{\mathbf{w}}(z) = 0$ *for all* $\mathbf{w} \neq \{0\}^*$;

(iii) *they satisfy the recursive relations*

$$\partial_z \mathscr{L}_{a\mathbf{w}}(z) = \frac{\mathscr{L}_{\mathbf{w}}(z)}{z-a}, \qquad \text{for } a \in \{0,1\}; \tag{4.48}$$

(iv) *they satisfy the shuffle relations, and are linearly independent over* $\mathsf{O}\overline{\mathsf{O}}$;

(v) *every linear combination of the functions* $\mathrm{L}_{\mathbf{u}}(z)\,\mathrm{L}_{\mathbf{v}}(\bar{z})$, *where* $\mathbf{u}, \mathbf{v} \in \mathsf{X}^*$, *which is single-valued, can be written as a unique linear combination of the functions* $\mathscr{L}_{\mathbf{w}}(z)$.

Proof. Let $\mathsf{Y} = \{\mathrm{y}_0, \mathrm{y}_1\}$ and Y^* the set of words on Y. Denote by $\overleftarrow{\mathbf{w}}$ the reversal of $\mathbf{w} \in \mathsf{Y}^*$. By Lemma 4.3.1 we see that

$$\mathrm{y}_0 = \mathrm{x}_0, \quad \overleftarrow{\Phi}(\mathrm{y}_0, \mathrm{y}_1)e^{2\pi i \mathrm{y}_1}\overleftarrow{\Phi}(\mathrm{y}_0, \mathrm{y}_1)^{-1} = \Phi(\mathrm{x}_0, \mathrm{x}_1)^{-1}e^{2\pi i \mathrm{x}_1}\Phi(\mathrm{x}_0, \mathrm{x}_1) \tag{4.49}$$

has a solution $\mathrm{y}_1 = \mathrm{x}_1 + \epsilon$ where $\epsilon \neq 0$ is a series of words with length ≥ 4. Define $\psi : \mathsf{Y}^* \to \mathsf{X}^*$, $\mathrm{y}_0 \mapsto \mathrm{x}_0$ and $\mathrm{y}_1 \mapsto \mathrm{x}_1 + \epsilon$. Set

$$\overleftarrow{\mathrm{L}}_{\mathsf{Y}}(\bar{z}) = \sum_{\mathbf{w}\in\mathsf{Y}^*} \mathrm{L}_{\psi(\mathbf{w})}(\bar{z})\overleftarrow{\mathbf{w}} = 1 + \mathrm{L}_{\mathrm{x}_0}(\bar{z})\mathrm{x}_0 + \mathrm{L}_{\mathrm{x}_1}(\bar{z})(\mathrm{x}_1 + \epsilon) + \cdots.$$

We now prove that

$$\mathscr{L}_{\mathsf{X}}(z) = \sum_{\mathbf{w}\in\mathsf{X}^*} \mathscr{L}_{\mathbf{w}}(z) := \mathrm{L}_{\mathsf{X}}(z)\overleftarrow{\mathrm{L}}_{\mathsf{Y}}(\bar{z}) \tag{4.50}$$

is single-valued. In fact, by monodromy computation

$$\begin{aligned}
M(\sigma_0)\,\mathscr{L}_{\mathsf{X}}(z) &= \mathrm{L}_{\mathsf{X}}(z)e^{2\pi i \mathrm{x}_0}e^{-2\pi i \mathrm{x}_0}\overleftarrow{\mathrm{L}}_{\mathsf{Y}}(\bar{z}) = \mathscr{L}_{\mathsf{X}}(z), \\
M(\sigma_1)\,\mathscr{L}_{\mathsf{X}}(z) &= \mathrm{L}_{\mathsf{X}}(z)\Phi(\mathrm{x}_0, \mathrm{x}_1)^{-1}e^{2\pi i \mathrm{x}_1}\Phi(\mathrm{x}_0, \mathrm{x}_1) \\
&\qquad \times \overleftarrow{\Phi}(\mathrm{y}_0, \mathrm{y}_1)e^{-2\pi i \mathrm{y}_1}\overleftarrow{\Phi}(\mathrm{y}_0, \mathrm{y}_1)^{-1}\overleftarrow{\mathrm{L}}_{\mathsf{Y}}(\bar{z}) = \mathscr{L}_{\mathsf{X}}(z)
\end{aligned}$$

by Eq. (4.49) and the fact that the word reversal map $\leftarrow$ is an anti-homomorphism.

(i) is clear from the definition (4.50) of $\mathscr{L}_{\mathbf{w}}(z)$.

(ii) By setting $\mathtt{x}_1 = 0$ (*i.e.*, killing all words involving $\mathtt{x}_1$), we have

$$\begin{aligned}\mathscr{L}_{\mathtt{x}_0,0}(z) = \mathrm{L}_{\mathsf{X}}(0)\overleftarrow{\mathrm{L}}_{\mathsf{Y}}(0) &= \mathrm{L}_{\mathtt{x}_0,0}(z)\overleftarrow{\mathrm{L}}_{\mathtt{x}_0,0}(\bar{z})\\ &= \mathrm{L}_{\mathtt{x}_0,0}(z)\,\mathrm{L}_{\mathtt{x}_0,0}(\bar{z}) = e^{\mathtt{x}_0\log z}e^{\mathtt{x}_0\log\bar{z}} = e^{\mathtt{x}_0\log z\bar{z}}.\end{aligned}$$

(iii) is equivalent to showing that

$$\partial_z\,\mathscr{L}_{\mathsf{X}}(z) = \left(\frac{\mathtt{x}_0}{z} + \frac{\mathtt{x}_1}{z-1}\right)\mathscr{L}_{\mathsf{X}}(z)$$

which follows from Eq. (4.38) immediately since $\partial_z\overleftarrow{\mathrm{L}}_{\mathsf{Y}}(\bar{z}) = 0$.

(iv) By Lemma 4.2.7, both $\mathrm{L}_{\mathsf{X}}(z)$ and $\overleftarrow{\mathrm{L}}_{\mathsf{Y}}(\bar{z})$ are group-like, so is their product $\mathscr{L}_{\mathsf{X}}(z)$. Hence the functions $\mathscr{L}_{\mathbf{w}}(z)$ satisfy the shuffle relations. The second half of (iv) and (v) are not needed in the rest of the book so we refer the interested reader to Brown's original manuscripts [97, 98] or a similar work of Hoang Ngoc [282]. □

Example 4.3.3. At weight one and two we obtain the following SVMPLs by using Example 4.2.13:

$$\begin{aligned}
\mathscr{L}_0(z) &= \mathrm{L}_0(z) + \mathrm{L}_0(\bar{z}) = \log z + \log\bar{z} = 2\log|z|,\\
\mathscr{L}_1(z) &= \mathrm{L}_1(z) + \mathrm{L}_1(\bar{z}) = \log(1-z) + \log(1-\bar{z}) = 2\log|1-z|,\\
\mathscr{L}_{00}(z) &= \frac{1}{2}\mathscr{L}_0(z)^2 = \frac{1}{2}\log^2(z\bar{z}) = 2\log^2|z|,\\
\mathscr{L}_{01}(z) &= \mathrm{L}_{01}(z) + \mathrm{L}_0(z)\,\mathrm{L}_1(\bar{z}) + \mathrm{L}_{10}(\bar{z}) = 2\log|z|\log|1-z| - 2i\,\mathscr{L}i_2(z),\\
\mathscr{L}_{10}(z) &= \mathrm{L}_{10}(z) + \mathrm{L}_1(z)\,\mathrm{L}_0(\bar{z}) + \mathrm{L}_{01}(\bar{z}) = 2\log|z|\log|1-z| + 2i\,\mathscr{L}i_2(z),\\
\mathscr{L}_{11}(z) &= \frac{1}{2}\mathscr{L}_1(z)^2 = \frac{1}{2}\log(1-z)(1-\bar{z}) = 2\log^2|1-z|,
\end{aligned}$$

where $\mathscr{L}i_2(z)$ is the Bloch–Wigner dilog (see Eq. (2.45) and also Exercise 4.22). So there is only one algebraically independent SVMPL of weight two, which can also be seen by the shuffle relation $\mathscr{L}_{01} + \mathscr{L}_{10} = \mathscr{L}_0\mathscr{L}_1$.

In higher weights, due to the appearance of $\mathtt{y}_1$, the expression of $\mathscr{L}_{\mathbf{w}}$ often involves the MZVs. Let the total weight of a product of the MZVs with holomorphic and antiholomorphic MPLs be the sum of the individual weights.

Proposition 4.3.4. *$\mathscr{L}_{\mathbf{w}}(z)$ is homogeneous of total weight $|\mathbf{w}|$.*

Proof. Assign weight -1 to $\mathtt{x}_0$ and $\mathtt{x}_1$. By its construction in the proof of Lemma 4.3.1 $\mathtt{y}_1$ has weight -1, the same as $\mathtt{x}_1$. Then $\mathrm{L}_{\mathsf{X}}(z)$ and $\overleftarrow{\mathrm{L}}_{\mathsf{Y}}(\bar{z})$ both have weight 0, so does $\mathscr{L}_{\mathsf{X}}(z)$, which yields the proposition. □

The shuffle-algebra over $\mathbb{C}$ spanned by SVMPLs graded by the total weight is denoted by

$$\mathsf{Ls} = \bigoplus_{n\geq 0} \mathsf{Ls}_n := \mathbb{C}\langle \mathscr{L}_{\mathbf{w}}(s) : \mathbf{w} \in \mathsf{X}^*\rangle.$$

For example, $\mathsf{Ls}_1 = \mathbb{C}\langle \log|z|, \log|1-z|\rangle$ and

$$\mathsf{Ls}_2 = \mathbb{C}\langle \mathscr{L}i_2(z), \log^2|z|, \log^2|1-z|, \log|z|\log|1-z|\rangle.$$

It turns out, however, the most natural coefficient ring for SVMPLs is much smaller than $\mathbb{C}$. We will consider this in the next section.

4.4 Single-Valued MZVs and Integration of SVMPLs

The SVMPLs are not defined at 0, 1 and ∞. We now extend SVMPLs to these points by applying regularizations.

Definition 4.4.1. For a single-valued function on $\mathbb{C}\setminus\{0,1\}$ written in the form $f(z) = \sum_{j=0}^n f_j(z)\log^j|z-a|$ ($a = 0$ or 1), where $f_j(z)$ are all single-valued continuous functions which are finite at a, we define the regularized value of $f(z)$ at a to be

$$\mathrm{reg}_{z=a} f(z) = f_0(a).$$

At ∞, if there is $M > 0$ such that $f(z) = \sum_{j=0}^n f_j(z)\log^j|z|$ where $f_j(z)$ are all single-valued function continuous for all $|z| > M$ and have finite limits when $z \to \infty$ then we define the regularized value of $f(z)$ at ∞ to be

$$\mathrm{reg}_{z=\infty} f(z) = \lim_{z\to\infty} f_0(z).$$

Lemma 4.4.2. *For any single-valued functions f and g defined near $a \in \mathbb{C}\cup\infty$, we have*

$$\mathrm{reg}_{z=a}\big(f(z)g(z)\big) = \mathrm{reg}_{z=a} f(z)\cdot \mathrm{reg}_{z=a} g(z),$$

provided all the regularized values are defined.

Proof. This is clear from the definition. □

By the definition we see that

$$\mathscr{L}_{\mathbf{w}}(0) := \mathrm{reg}_{z=0}\,\mathscr{L}_{\mathbf{w}}(z) = 0 \quad \forall \mathbf{w} \neq 1. \tag{4.51}$$

Definition 4.4.3. We set $\mathscr{L}_{\mathbf{w}}(1) := \operatorname{reg}_{z=1} \mathscr{L}_{\mathbf{w}}(z)$ and call it a *single-valued MZV* (SVMZV).

Lemma 4.4.4. *Let* $\mathsf{MZ}^{\rm sv}_{\mathbb{Z}}$ *be the set of all* $\mathbb{Z}$*-linear combinations of SVMZVs. Then* $\mathsf{MZ}^{\rm sv}_{\mathbb{Z}}$ *is a sub-ring of* $\mathsf{MZ}_{\mathbb{Z}}$*, namely,* $\mathscr{L}_{\mathbf{u}}(1)\,\mathscr{L}_{\mathbf{v}}(1) \in \mathsf{MZ}^{\rm sv}_{\mathbb{Z}}$ *for all* $\mathbf{u}, \mathbf{v} \in \mathsf{X}^*$.

Proof. Without loss of generality, let $\mathbf{u}, \mathbf{v} \in \mathsf{X}^*$ be two non-empty words. Then by Lemma 4.4.2

$$\begin{aligned}\mathscr{L}_{\mathbf{u}}(1)\,\mathscr{L}_{\mathbf{v}}(1) &= \operatorname{reg}_{z=1} \mathscr{L}_{\mathbf{u}}(z) \cdot \operatorname{reg}_{z=1} \mathscr{L}_{\mathbf{v}}(z)\\ &= \operatorname{reg}_{z=1} \big(\mathscr{L}_{\mathbf{u}}(z)\,\mathscr{L}_{\mathbf{v}}(z)\big) = \operatorname{reg}_{z=1} \mathscr{L}_{\mathbf{u}\sqcup\mathbf{v}}(z) = \mathscr{L}_{\mathbf{u}\sqcup\mathbf{v}}(1)\end{aligned}$$

where we have extended $\mathscr{L}$ over $\mathbb{Z}\langle X\rangle$ as a $\mathbb{Z}$-linearly map. So the set $\mathsf{MZ}^{\rm sv}_{\mathbb{Z}}$ is closed under multiplication, as desired. $\square$

Example 4.4.5. An easy computation by Maple shows that up to weight ten $\mathsf{MZ}^{\rm sv}_{\mathbb{Q}} := \mathsf{MZ}^{\rm sv}_{\mathbb{Z}} \otimes \mathbb{Q}$ is generated by $\zeta(3)$, $\zeta(5)$, $\zeta(7)$, $\zeta(9)$. At weight 11 we have two generators: $\zeta(11)$ and

$$g_{533} = \zeta(5,3,3) - \frac{4}{7}\zeta(5)\zeta(2)^3 + \frac{6}{5}\zeta(7)\zeta(2)^2 + 45\zeta(9)\zeta(2).$$

More precisely, up to weight 11 the ring $\mathsf{MZ}^{\rm sv}_{\mathbb{Z}}$ over $\mathbb{Z}$ is spanned by $2\zeta(3)$, $\zeta(5)$, $2\zeta(3)^2$, $\frac{1}{8}\zeta(7)$, $\zeta(3)\zeta(5)$, $\frac{1}{72}\zeta(9)$, $\frac{4}{3}\zeta(3)^3$, $\frac{1}{8}\zeta(3)\zeta(7)$, $\frac{1}{2}\zeta(5)^2$, $\frac{1}{384}\zeta(11)$, $\frac{1}{5}(g_{533} + \zeta(11)) + \frac{3}{2}\zeta(3)^2\zeta(5)$, and $2\zeta(3)^2\zeta(5)$.

Let $\mathsf{Ls}^{\rm sv}$ be the shuffle-algebra over $\mathsf{MZ}^{\rm sv}_{\mathbb{Z}}$ spanned by SVMPLs. Recall that $\mathsf{O} = \mathbb{Q}\left[z, \frac{1}{z}, \frac{1}{z-1}\right]$ and $\overline{\mathsf{O}} = \mathbb{Q}\left[\bar z, \frac{1}{\bar z}, \frac{1}{\bar z-1}\right]$. Define two bi-differential algebras

$$\mathsf{A} := \mathsf{O} \cdot \overline{\mathsf{O}} \cdot \mathsf{Ls}, \quad \mathsf{A}^{\rm sv} := \mathsf{O} \cdot \overline{\mathsf{O}} \cdot \mathsf{Ls}^{\rm sv}. \tag{4.52}$$

Let $\mathsf{MZ}^{\rm sv}_{\mathbb{Z},n}(\mathbb{Z})$, $\mathsf{Ls}^{\rm sv}_n$ and $\mathsf{A}^{\rm sv}_n$ denote the subspaces of total weight n. Due to the difficulty in transcendental number theory we do not know whether these subspaces for different n are $\mathbb{Q}$-linearly independent.

Remark 4.4.6. The only difference between A (resp. Ls) and $\mathsf{A}^{\rm sv}$ (resp. $\mathsf{Ls}^{\rm sv}$) is the coefficients for the former are in $\mathbb{C}$ while those for the latter are in $\mathsf{MZ}^{\rm sv}_{\mathbb{Z}}$.

By the definition, $\mathsf{A}^{\rm sv}$ is closed under ∂_z by Thm. 4.3.2(iii). For $\partial_{\bar z}$, by definition (4.50) we see that

$$\partial_{\bar z}\,\mathscr{L}_{\mathsf{X}}(z) = \mathscr{L}_{\mathsf{X}}\left(\frac{\mathrm{x}_0}{\bar z} + \frac{\mathrm{y}_1}{\bar z - 1}\right)$$

which yields

$$\partial_{\bar{z}}\mathscr{L}_{\mathbf{w}a}(z) = \frac{\delta_{a,0}}{\bar{z}}\mathscr{L}_{\mathbf{w}}(z) + \frac{1}{\bar{z}-1}\sum_{\mathbf{w}a=\mathbf{u}\mathbf{v}}\mathscr{L}_{\mathbf{u}}(z)\mathrm{y}_1[\mathbf{v}]$$

$$= \frac{\mathscr{L}_{\mathbf{w}}(z)}{\bar{z}-a} + \frac{1}{\bar{z}-1}\sum_{\mathbf{w}a=\mathbf{u}\mathbf{v}}\mathscr{L}_{\mathbf{u}}(z)(\mathrm{y}_1[\mathbf{v}] - \delta_{\mathrm{x}_1,\mathbf{v}}), \quad (4.53)$$

where $\mathrm{y}_1[\mathbf{v}]$ is the coefficient of the word $\mathbf{v}$ in the series y_1.

Definition 4.4.7. For $P \in \mathsf{A}$, let $\int_0 P(z)\mathrm{d}z$ be the unique function $F \in \mathsf{A}$, if it exists, such that $\partial_z F(z) = P(z)$ with $\int_0$ denoting the regularized value $\mathrm{reg}_{z=0} F(z) = 0$. Similarly, $\int_0 P(z)\mathrm{d}\bar{z}$ is the unique function $G(z) \in \mathsf{A}$, if it exists, such that $\partial_{\bar{z}} G(z) = P(z)$. For $b \in \{1, \infty\}$, we define $\int_b P(z)\mathrm{d}z = F(z) - F(b)$ and $\int_b P(z)\mathrm{d}\bar{z} = G(z) - G(b)$, where $F(b)$ and $G(b)$ are regularized values.

Lemma 4.4.8. *We have a well-defined integral operator on* A,

$$\int_0 \mathrm{d}z : \mathsf{A} \longrightarrow \mathsf{A}, \quad F(z) \longmapsto \int_0 F(z)\mathrm{d}z.$$

Moreover, $\int_0 \mathrm{d}z$ *maps* A^{sv} *into* A^{sv}.

Proof. For the first sentence, we want to show that the images of the operator lie in A. First, note that for $a \in \{0,1\}$ and $n \neq -1$

$$\int_0 \frac{\mathscr{L}_1(z)}{z-a}\mathrm{d}z = \mathscr{L}_1(z-a), \quad \int_0 \mathscr{L}_1(z)(z-a)^n\mathrm{d}z = \frac{(z-a)^{n+1}}{n+1},$$

both of which are in A^{sv}. For general $\mathbf{w}$, by Thm. 4.3.2(iii), we have

$$\int_0 \frac{\mathscr{L}_{\mathbf{w}}(z)}{z-a}\mathrm{d}z = \mathscr{L}_{a\mathbf{w}}(z) \in \mathsf{A}^{\mathrm{sv}}.$$

For any $n \in \mathbb{Z}$ $(n \neq -1)$ and $\mathbf{w} = a\mathbf{w}'$ $(a = 0, 1)$, we have

$$\int_0 \mathscr{L}_{\mathbf{w}}(z)(z-a)^n\mathrm{d}z = \mathscr{L}_{\mathbf{w}}(z)\frac{(z-a)^{n+1}}{n+1} - \int_0 \mathscr{L}_{\mathbf{w}'}(z)\frac{(z-a)^{n+1}}{n+1}\mathrm{d}z \in \mathsf{A}$$

by induction on $|\mathbf{w}|$. This shows that $\int_0 \mathrm{d}z$ is a well-defined operator on A. The claim for the stability of A^{sv} under $\int_0 \mathrm{d}z$ follows from the same argument. □

Proposition 4.4.9. *Let* $\mathfrak{S}_3$ *be the group of Möbius transformations permuting* $\{0, 1, \infty\}$. *Then* $\mathsf{Ls}_n^{\mathrm{sv}}$ *is stable under* $\mathfrak{S}_3$. *The integral operator* $\int_b \mathrm{d}z/(z-a)$ $(a \in \{0,1\}, b \in \{0,1,\infty\})$ *satisfies*

$$\int_b \mathrm{d}z/(z-a) : \mathsf{Ls}_n^{\mathrm{sv}} \hookrightarrow \mathsf{Ls}_{n+1}^{\mathrm{sv}}. \quad (4.54)$$

Furthermore,

$$\partial_z \mathsf{Ls}_n^{\mathrm{sv}} = \frac{\mathsf{Ls}_{n-1}^{\mathrm{sv}}}{z} + \frac{\mathsf{Ls}_{n-1}^{\mathrm{sv}}}{z-1}. \tag{4.55}$$

Proof. First, Eq. (4.55) follows easily from Thm. 4.3.2(iii). The stability of $\mathsf{Ls}^{\mathrm{sv}}$ under $z \to \frac{z}{z-1}$ and under $\int_0 \mathrm{d}z/z$, $\int_0 \mathrm{d}z/(z-1)$ follows from straightforward computations (see Exercises 4.27 and Lemma 4.4.8). If $|\mathbf{w}| = n$ and $b = 1$ or ∞ we have

$$\int_b \frac{\mathscr{L}_{\mathbf{w}}(z)}{z-a} \mathrm{d}z = \int_0 \frac{\mathscr{L}_{\mathbf{w}}(z)}{z-a} \mathrm{d}z + \mathscr{L}_{a\mathbf{w}}(b) \in \mathsf{Ls}_{n+1}^{\mathrm{sv}},$$

which implies Eq. (4.54).

Since the group $\mathfrak{S}_3$ is generated by $z \to \frac{z}{z-1}$ and $z \to 1-z$ we now only need to prove the stability of $\mathsf{Ls}^{\mathrm{sv}}$ under $z \to 1-z$. We proceed by induction on the total weight. In weight zero this is trivial. Assume $P \in \mathsf{Ls}_n^{\mathrm{sv}}$. Then by Eq. (4.55),

$$P(z) = \int_1 \left(\frac{Q_0(z)}{z} + \frac{Q_1(z)}{z-1} \right) \mathrm{d}z + P(1)$$

where $Q_0, Q_1 \in \mathsf{Ls}_{n-1}^{\mathrm{sv}}$. Since $Q_0(1-z), Q_1(1-z) \in \mathsf{Ls}_{n-1}^{\mathrm{sv}}$ by the induction we finally get

$$P(1-z) = \int_0 \left(\frac{Q_0(1-z)}{z-1} + \frac{Q_1(1-z)}{z} \right) \mathrm{d}z + P(1) \in \mathsf{Ls}_n^{\mathrm{sv}}$$

by Eq. (4.54). □

Proposition 4.4.10. *The series* $\mathrm{y}_1 \in \mathsf{MZ}_{\mathbb{Z}}^{\mathrm{sv}}\langle\langle \mathsf{X} \rangle\rangle$.

Proof. We show that $\mathrm{y}_1[\mathbf{w}] \in \mathsf{MZ}_{\mathbb{Z}}^{\mathrm{sv}}\langle\langle \mathsf{X} \rangle\rangle$ by the induction on the weight of $\mathbf{w} \in \mathsf{X}^*$. Suppose $\mathbf{w}$ has weight n. By (4.53), for any word $\mathbf{u}$, $\lim_{z\to 0} \bar{z}\partial_{\bar{z}} \mathscr{L}_{\mathbf{u}a}(z) = \delta_{a,0} \mathscr{L}_{\mathbf{u}}(0) \in \mathsf{MZ}_{\mathbb{Z}}^{\mathrm{sv}}$. By the stability of $\mathsf{Ls}^{\mathrm{sv}}$ under $z \to 1-z$ we obtain

$$\lim_{z\to 1} (\bar{z}-1)\partial_{\bar{z}} \mathscr{L}_{\mathbf{w}}(z) = \lim_{z\to 0} \bar{z}\partial_{\bar{z}} \mathscr{L}_{\mathbf{w}}(1-z) \in \mathsf{MZ}_{\mathbb{Z}}^{\mathrm{sv}}.$$

On the other hand, from (4.53) we obtain

$$\lim_{z\to 1} (\bar{z}-1)\partial_{\bar{z}} \mathscr{L}_{\mathbf{w}}(z) = \sum_{\mathbf{w}=\mathbf{u}\mathbf{v}} \mathscr{L}_{\mathbf{u}}(1)\mathrm{y}_1[\mathbf{v}] = \mathrm{y}_1[\mathbf{w}] + \sum_{\substack{\mathbf{w}=\mathbf{u}\mathbf{v} \\ |\mathbf{u}|<|\mathbf{w}|}} \mathscr{L}_{\mathbf{u}}(1)\mathrm{y}_1[\mathbf{v}].$$

Hence $\mathrm{y}_1 \in \mathsf{MZ}_{\mathbb{Z}}^{\mathrm{sv}}\langle\langle \mathsf{X} \rangle\rangle$ by induction. □

Lemma 4.4.11. *We have a well-defined integral operator on* A,

$$\int_0 \mathrm{d}\bar{z} : \mathsf{A} \longrightarrow \mathsf{A}, \quad F(z) \longmapsto \int_0 F(z)\mathrm{d}\bar{z}.$$

Moreover, $\int_0 \mathrm{d}\bar{z}$ *maps* A^{sv} *into* A^{sv}.

Proof. As in the proof of Lemma 4.4.8 we first show that the images of $\int_0 \mathrm{d}\bar{z}$ lie in A. Note that for $a \in \{0,1\}$ and $n \neq -1$

$$\int_0 \frac{\mathscr{L}_1(z)}{\bar{z}-a}\mathrm{d}\bar{z} = \mathscr{L}_1(z-a), \quad \int_0 \mathscr{L}_1(z)(\bar{z}-a)^n \mathrm{d}\bar{z} = \frac{(\bar{z}-a)^{n+1}}{n+1}.$$

both of which are in A^{sv}. For general $\mathbf{w}$, note that all the terms on the right-hand side of Eq. (4.53) involving $\mathbf{u}$ vanish unless $|\mathbf{u}| \leq |\mathbf{w}a| - 4$ according to the property that $\mathtt{y}_1 \equiv \mathtt{x}_1$ modulo words of weight at least 4. Thus $\int_0 \mathscr{L}_{\mathbf{w}}(z)(\bar{z}-a)^{-1}\mathrm{d}\bar{z} \in \mathsf{A}^{\mathrm{sv}}$ by Eq. (4.53) and Prop. 4.4.10. If $-1 \neq n \in \mathbb{Z}$ then by induction on $|\mathbf{w}|$ we can show that $\int_0 \mathscr{L}_{\mathbf{w}}(z)(\bar{z}-a)^n \mathrm{d}\bar{z} \in \mathsf{A}$ for any $\mathbf{w}$ by integration by parts using Eq. (4.53).

The claim for the stability of A^{sv} under $\int_0 \mathrm{d}z$ follows from the same argument by using Prop. 4.4.10. □

Proposition 4.4.12. *The integral operator* $\int_b \mathrm{d}\bar{z}/(\bar{z}-a)$ $(a \in \{0,1\}, b \in \{0,1,\infty\})$ *satisfies*

$$\int_b \mathrm{d}\bar{z}/(\bar{z}-a) : \mathsf{Ls}_n^{\mathrm{sv}} \hookrightarrow \mathsf{Ls}_{n+1}^{\mathrm{sv}}. \tag{4.56}$$

Furthermore,

$$\partial_{\bar{z}} \mathsf{Ls}_n^{\mathrm{sv}} = \frac{\mathsf{Ls}_{n-1}^{\mathrm{sv}}}{\bar{z}} + \frac{\mathsf{Ls}_{n-1}^{\mathrm{sv}}}{\bar{z}-1}. \tag{4.57}$$

Proof. Eqation (4.57) follows readily from Eq. (4.53) by using Prop. 4.4.10. Eqation (4.53) also yields the inductively defined antiholomorphic integral

$$\int_0 \frac{\mathscr{L}_{\mathbf{w}}(z)}{\bar{z}-a}\mathrm{d}\bar{z} = \mathscr{L}_{\mathbf{w}a}(z) - \sum_{\mathbf{w}a=\mathbf{u}\mathbf{v}} (\mathtt{y}_1[\mathbf{u}] - \mathtt{x}_1[\mathbf{u}]) \int_0 \frac{\mathscr{L}_{\mathbf{v}}(z)}{\bar{z}-1}\mathrm{d}\bar{z}, \tag{4.58}$$

where $a \in \{0,1\}$. This implies Eq. (4.56) by induction on the weight. We leave the details to the interested reader. □

Proposition 4.4.13. *For all* $n \in \mathbb{N}$, *we have* $\mathsf{Ls}_n^{\mathrm{sv}}(0) = \mathsf{Ls}_n^{\mathrm{sv}}(1) = \mathsf{Ls}_n^{\mathrm{sv}}(\infty) = \mathsf{MZ}_{\mathbb{Z},n}^{\mathrm{sv}} \subset \mathsf{MZ}_{\mathbb{Z},n}$.

Proof. By (4.51) and the definition $\mathsf{Ls}^{\rm sv}(0) = \mathscr{L}_1(0)\mathsf{MZ}_{\mathbb{Z}}^{\rm sv} = \mathsf{MZ}_{\mathbb{Z}}^{\rm sv}$ (see Remark 4.4.6). The equality $\mathsf{Ls}^{\rm sv}(1) = \mathsf{MZ}_{\mathbb{Z}}^{\rm sv}$ follows from Lemma 4.4.4. Finally, $\mathsf{Ls}^{\rm sv}(\infty) = \mathsf{MZ}_{\mathbb{Z}}^{\rm sv}$ since $\mathsf{Ls}^{\rm sv}(\infty) = \mathsf{Ls}^{\rm sv}(1)$ (see Exercise 4.27).

We also need to show that $\mathsf{MZ}_{\mathbb{Z},n}^{\rm sv} \subset \mathsf{MZ}_{\mathbb{Z},n}$, namely, all SVMZVs are indeed MZVs. Indeed, $\mathrm{y}_1 \in \mathsf{MZ}_{\mathbb{Z}}\langle\langle \mathsf{X}\rangle\rangle$ by Lemma 4.3.1, and by the definition, $\sum_{\mathbf{w}\in \mathsf{X}^*} \mathscr{L}_{\mathbf{w}}(z)\mathbf{w} = \mathrm{L}_{\mathrm{x}_0,\mathrm{x}_1}(z)\overleftarrow{\mathrm{L}}_{\mathrm{x}_0,\mathrm{y}_1}(\bar{z})$. This readily implies $\mathscr{L}_{\mathbf{w}}(0) \subset \mathsf{MZ}_{\mathbb{Z}}$ since $\mathscr{L}_{\mathbf{w}}(0)$ is a $\mathbb{Z}$-linear combination of products of the cofficients of L_X and $\overleftarrow{\mathrm{L}}_Y$. □

4.5 Generating Series of Hoffman Elements

According to Hoffman's Basis Conjecture 3.6.1 partly proved by Brown [102] all the MZVs can be expressed as a $\mathbb{Q}$-linear combination of the MZVs with arguments equal to 2 or 3. If exactly k of 3's appear in such an MZV then it is called a *k-Hoffman element.* It is clear that the word representation of a convergent 0-Hoffman element is an alternating word containing neither x_0^2 nor x_1^2. These reduce to even powers of π according to Cor. 5.6.4.

The word of a 1-Hoffman element contains exactly one subword x_0^2 but no subword x_1^2. The exact evaluation of such MZVs is given by Thm. 5.7.4.

Definition 4.5.1. Let $\mathsf{X} = \{\mathrm{x}_0, \mathrm{x}_1\}$. For $a = 0$ or 1, denote by H^a the set of words $\mathbf{w} \in \mathsf{X}^*$ containing the subword x_0^2 exactly a times but no subword x_1^2. Let H be the complement to the subspace of $\mathbb{C}\langle\langle \mathsf{X}\rangle\rangle$ generated by the set $\mathsf{H}^0 \cup \mathsf{H}^1$ and $I_H \subset \mathbb{C}\langle\langle \mathsf{X}\rangle\rangle$ denote the ideal generated by $\mathsf{X}^* \setminus (\mathsf{H}^0 \cup \mathsf{H}^1)$.

Remark 4.5.2. To show any identity in $\mathbb{C}\langle\langle \mathsf{X}\rangle\rangle$ it suffices to prove it under both the projection to H and modulo I_H, since H is stable under the deconcatenation coproduct and the coalgebra H is dual to $\mathbb{C}\langle\langle \mathsf{X}\rangle\rangle/I_H$.

Definition 4.5.3. Let $\mathsf{S}_{1,0}^1(m,n) = 4\binom{2m+2n}{2m}\zeta(2m+2n+1)$ for all $m \geq 1, n \geq 0$. Define a formal power series $\mathsf{S}_{\mathsf{X}} \in \mathbb{R}\langle\langle \mathsf{X}\rangle\rangle$ by

$$\mathsf{S}_{\mathsf{X}} = 1 + \mathsf{S}_{0,0}^0 + \mathsf{S}_{0,1}^1 + \mathsf{S}_{1,0}^1 + \mathsf{S}_{0,0}^1,$$

where $\mathsf{S}_{0,1}^1 = \overleftarrow{\mathsf{S}_{1,0}^1}$, and

$$\begin{aligned} \mathsf{S}_{0,0}^0 &= 4\sum_{n\geq 1} \zeta(2n+1)\,(\mathrm{x}_0\mathrm{x}_1)^n\mathrm{x}_0, \\ \mathsf{S}_{1,0}^1 &= \sum_{m\geq 1, n\geq 0} \mathsf{S}_{1,0}^1(m,n)(\mathrm{x}_1\mathrm{x}_0)^m\mathrm{x}_0(\mathrm{x}_1\mathrm{x}_0)^n \qquad (4.59) \\ \mathsf{S}_{0,0}^1 &= \frac{1}{2}\big(\mathsf{S}_{0,0}^0\,\mathsf{S}_{0,0}^0 + \mathsf{S}_{0,0}^0\,\mathsf{S}_{1,0}^1 + \mathsf{S}_{0,1}^1\,\mathsf{S}_{0,0}^0\big). \end{aligned}$$

In order to compute the Feynman amplitudes of so-called zig-zag graphs in Chap. 15 we also need to study another family of series in $\mathbb{R}\langle\langle \mathsf{X}\rangle\rangle$.

Definition 4.5.4. Let $\hat{\mathsf{S}}^1_{1,0}(m,n) = 4(1-4^{-m-n})\binom{2m+2n}{2m+1}\zeta(2m+2n+1)$ for all $m \geq 0, n \geq 1$. Define a formal power series $\hat{\mathsf{S}}_\mathsf{X} \in \mathbb{R}\langle\langle \mathsf{X}\rangle\rangle$ by

$$\hat{\mathsf{S}}_\mathsf{X} = 1 + \hat{\mathsf{S}}^0_{0,0} + \hat{\mathsf{S}}^1_{0,1} + \hat{\mathsf{S}}^1_{1,0} + \hat{\mathsf{S}}^1_{0,0}\,,$$

where $\hat{\mathsf{S}}^0_{0,0} = \mathsf{S}^0_{0,0}$, $\hat{\mathsf{S}}^1_{0,1} = \overleftarrow{\hat{\mathsf{S}}^1_{1,0}}$, and

$$\hat{\mathsf{S}}^1_{1,0} = \sum_{m\geq 0, n\geq 1} \hat{\mathsf{S}}^1_{1,0}(m,n)(\mathtt{x}_1\mathtt{x}_0)^m\mathtt{x}_1(\mathtt{x}_1\mathtt{x}_0)^n, \tag{4.60}$$

$$\hat{\mathsf{S}}^1_{0,0} = \frac{1}{2}\big(\hat{\mathsf{S}}^0_{0,0}\,\hat{\mathsf{S}}^1_{1,0} + \hat{\mathsf{S}}^1_{0,1}\,\hat{\mathsf{S}}^0_{0,0}\big).$$

Let $\hat{\mathsf{H}}^a$ (resp. $\hat{H}$, resp. $I_{\hat{H}}$) denote the set of elements obtained from H^a (resp. H, resp. I_H) by interchanging $\mathtt{x}_0$ and $\mathtt{x}_1$.

For convenience, if $T = \sum_{\mathbf{w}\in\mathsf{X}^*} T_\mathbf{w}\mathbf{w} \in \mathbb{C}\langle\langle \mathsf{X}\rangle\rangle$ then for $a, b = 0$ or 1, we set

$$T_{a,b} := \delta_{a,b}T_{\mathtt{x}_a}\mathtt{x}_a + \sum_{\mathbf{w}\in\mathtt{x}_a\mathsf{X}^*\mathtt{x}_b} T_\mathbf{w}\mathbf{w},$$

and for $c = 0$ or 1, we set $T^c := \sum_{\mathbf{w}\in\mathsf{H}^c} T_\mathbf{w}\mathbf{w}$. So there are eight possible subseries $T^c_{a,b}$ for $a, b, c = 0$ or 1. For example, we clearly have $\mathsf{S}^0_{1,1} = \mathsf{S}^1_{1,1} = \mathsf{S}^0_{1,0} = \mathsf{S}^0_{0,1} = 0$.

Let $V = \Phi^{-1}\mathtt{x}_1\Phi \in \mathbb{C}\langle\langle \mathsf{X}\rangle\rangle$ where $\Phi = \Phi(\mathtt{x}_0, \mathtt{x}_1)$ is the Drinfeld associator defined by Eq. (4.5).

Lemma 4.5.5. *We have*

(i) $[\mathsf{S}_\mathsf{X}, \mathtt{x}_0] \equiv 0 \pmod{I_H}$ *and* $[\hat{\mathsf{S}}_\mathsf{X}, \mathtt{x}_0] \equiv 0 \pmod{I_{\hat{H}}}$;

(ii) *for* $V_- := V(-x_0, -x_1)$, $\mathsf{S}_\mathsf{X}V_- + V\mathsf{S}_\mathsf{X} \equiv 0 \pmod{I_H}$ *and* $\hat{\mathsf{S}}_\mathsf{X}V_- + V\hat{\mathsf{S}}_\mathsf{X} \equiv 0 \pmod{I_{\hat{H}}}$;

(iii) *both* S_X *and and* $\hat{\mathsf{S}}_\mathsf{X}$ *are real and they satisfy* $\mathsf{S}_\mathsf{X} \equiv \overleftarrow{\mathsf{S}_\mathsf{X}} \pmod{I_H}$ *and* $\hat{\mathsf{S}}_\mathsf{X} \equiv \overleftarrow{\hat{\mathsf{S}}_\mathsf{X}} \pmod{I_{\hat{H}}}$.

Proof. (i) To prove $[\mathsf{S}_\mathsf{X}, \mathtt{x}_0] \equiv 0 \pmod{I_H}$, by Definition 4.5.3 we only need to show

$$\mathsf{S}^0_{0,0}\,\mathtt{x}_0 + \mathsf{S}^1_{0,1}\,\mathtt{x}_0 = \mathtt{x}_0\mathsf{S}^0_{0,0} + \mathtt{x}_0\mathsf{S}^1_{1,0}\,. \tag{4.61}$$

This can be established easily by comparing the coefficients of the words of the form $(\mathtt{x}_0\mathtt{x}_1)^a\mathtt{x}_0^2(\mathtt{x}_1\mathtt{x}_0)^b$. Similarly we can show that $[\hat{\mathsf{S}}_\mathsf{X}, \mathtt{x}_0] \equiv 0 \pmod{I_{\hat{H}}}$. We leave the details to the interested reader.

(ii) For S_X, it suffices to prove $(V_-\mathsf{S}_\mathsf{X}+\mathsf{S}_\mathsf{X}V)_{a,b} \equiv 0 \pmod{I_H}$ for $a, b = 0$ or 1. Breaking these further into the odd and even weight parts we get seven nontrivial equations all of which follow from Exercise 4.29. Similar conclusions can be drawn for $\hat{\mathsf{S}}_\mathsf{X}$ by using Exercise 4.30.

(iii) is obvious by the definitions. □

Define the generating series

$$F(z) = \sum_{\mathbf{w}\in\mathsf{X}^*} F_\mathbf{w}(z)\mathbf{w} = \mathrm{L}_\mathsf{X}(z)\mathsf{S}_\mathsf{X}\overleftarrow{\mathrm{L}_\mathsf{X}}(\bar{z}). \tag{4.62}$$

Theorem 4.5.6. *The functions $F_\mathbf{w}(z)$ are single-valued. Moreover, for all $\mathbf{w} \in H$*

$$\partial_z\partial_{\bar{z}}F_{\mathrm{x}_a\mathbf{w}\mathrm{x}_b}(z) = \frac{F_\mathbf{w}(z)}{(z-a)(\bar{z}-b)} \tag{4.63}$$

for all $a, b \in \{0, 1\}$, and they satisfy $F_\mathbf{w}(z) = F_{\overleftarrow{\mathbf{w}}}(\bar{z})$.

Proof. After projecting to the subspace H we clearly have $F(z) = \mathrm{L}_\mathsf{X}(z)\overleftarrow{\mathrm{L}_\mathsf{X}}(\bar{z})$ which has no monodromy at 0 and 1 by Lemma 4.2.10. By Remark 4.5.2 it suffices to prove

$$\begin{aligned} M(\sigma_0)F(z) &\equiv F(z) \pmod{I_H}, \\ M(\sigma_1)F(z) &\equiv F(z) \pmod{I_H}. \end{aligned} \tag{4.64}$$

For the monodromy at 0, by Lemma 4.5.5(i)

$$e^{2\pi i\mathrm{x}_0}\mathsf{S}_\mathsf{X} \equiv \mathsf{S}_\mathsf{X}e^{2\pi i\mathrm{x}_0} \pmod{I_H}.$$

Thus

$$\begin{aligned} M(\sigma_0)F(z) &= \mathrm{L}_\mathsf{X}(z)e^{2\pi i\mathrm{x}_0}\mathsf{S}_\mathsf{X}e^{-2\pi i\mathrm{x}_0}\overleftarrow{\mathrm{L}_\mathsf{X}}(\bar{z}) \\ &\equiv \mathrm{L}_\mathsf{X}(z)\mathsf{S}_\mathsf{X}\overleftarrow{\mathrm{L}_\mathsf{X}}(\bar{z}) \equiv F(z) \pmod{I_H}. \end{aligned} \tag{4.65}$$

For the monodromy at 1, first we note that $W := \Phi^{-1}e^{2\pi i\mathrm{x}_1}\Phi \equiv 1+2\pi iV \pmod{I_H}$ since $\mathrm{x}_1^2 \in I_H$. Moreover, from $\overleftarrow{\Phi}(-\mathrm{x}_0, -\mathrm{x}_1) = \Phi(\mathrm{x}_0, \mathrm{x}_1)^{-1}$ by Thm. B.7(ii)

$$\overleftarrow{V}(\mathrm{x}_0, \mathrm{x}_1) = \Phi(-\mathrm{x}_0, -\mathrm{x}_1)^{-1}\mathrm{x}_1\Phi(-\mathrm{x}_0, -\mathrm{x}_1) = -V_-. \tag{4.66}$$

Since $V^2 \equiv \Phi^{-2}\mathrm{x}_1^2\Phi^2 \equiv 0 \pmod{I_H}$, by multiplying Lemma 4.5.5(ii) by V from the left and using Eq. (4.66) we get $V\mathsf{S}_\mathsf{X}\overleftarrow{V} \equiv 0 \pmod{I_H}$. Therefore

$$W\mathsf{S}_\mathsf{X}\overleftarrow{\overline{W}} - \mathsf{S}_\mathsf{X} \equiv 2\pi i(-\mathsf{S}_\mathsf{X}\overleftarrow{V} + V\mathsf{S}_\mathsf{X}) + (2\pi i)^2V\mathsf{S}_\mathsf{X}\overleftarrow{V} \equiv 0 \pmod{I_H}$$

by Lemma 4.5.5(ii) and Eq. (4.66) again. By Lemma 4.2.10 we obtain

$$M(\sigma_1)F(z) = \mathsf{L}_\mathsf{X}(z)W\mathsf{S}_\mathsf{X}\overleftarrow{\overleftarrow{W}}\overleftarrow{\mathsf{L}_\mathsf{X}}(\bar{z}) \equiv \mathsf{L}_\mathsf{X}(z)\mathsf{S}_\mathsf{X}\overleftarrow{\mathsf{L}_\mathsf{X}}(\bar{z}) \equiv F(z) \pmod{I_H}.$$

From this and (4.65) we see that $F_\mathbf{w}(z)$ are single-valued for all $\mathbf{w} \in \mathsf{X}^*$.

Equation (4.63) follows easily from Prop. 4.2.1 and the last sentence of the theorem follows easily from Lemma 4.5.5(iii). □

Define the generating series

$$\hat{F}(z) = \sum_{\mathbf{w}\in\mathsf{X}^*} \hat{F}_\mathbf{w}(z)\mathbf{w} = \mathsf{L}_\mathsf{X}(z)\hat{\mathsf{S}}_\mathsf{X}\overleftarrow{\mathsf{L}_\mathsf{X}}(\bar{z}). \tag{4.67}$$

Parallel to Thm. 4.5.6 we have

Theorem 4.5.7. *The functions $\hat{F}_\mathbf{w}(z)$ are single-valued. Moreover, for all $\mathbf{w} \in \hat{H}$ and for all $a, b \in \{0, 1\}$, we have*

$$\partial_z\partial_{\bar{z}}\hat{F}_{\mathbf{x}_a\mathbf{w}\mathbf{x}_b}(z) = \frac{\hat{F}_\mathbf{w}(z)}{(z-a)(\bar{z}-b)}. \tag{4.68}$$

They satisfy $\hat{F}_\mathbf{w}(z) = \hat{F}_{\overleftarrow{\mathbf{w}}}(\bar{z})$.

Proof. We only need to show the triviality of the monodromy at 1. The proofs of the triviality of the monodromy at 0 and all the other statements in the theorem are exactly the same as those of Thm. 4.5.6.

For the monodromy at 1, we have from Lemma 4.5.5(ii)

$$V\hat{\mathsf{S}}_\mathsf{X} + \hat{\mathsf{S}}_\mathsf{X}V_- \equiv 0 \pmod{I_{\hat{H}}}. \tag{4.69}$$

Multiplying this on the left (resp. right) by V^2 (resp. V_-^2), we get

$$V^2\hat{\mathsf{S}}_\mathsf{X}V_- \equiv V\hat{\mathsf{S}}_\mathsf{X}V_-^2 \equiv 0 \pmod{I_{\hat{H}}}. \tag{4.70}$$

Now let $W = \Phi^{-1}e^{2\pi i\mathbf{x}_1}\Phi$ as before. Then $W \equiv 1 + 2\pi iV + \frac{1}{2}(2\pi i)^2V^2$ and $\overleftarrow{\overleftarrow{W}} \equiv 1 + 2\pi iV_- + \frac{1}{2}(2\pi i)^2V_-^2 \pmod{I_{\hat{H}}}$ by Eq. (4.66). Thus

$$W\hat{\mathsf{S}}_\mathsf{X}\overleftarrow{\overleftarrow{W}} - \hat{\mathsf{S}}_\mathsf{X} \equiv 2\pi i(V\hat{\mathsf{S}}_\mathsf{X} + \hat{\mathsf{S}}_\mathsf{X}V_-) - 4i\pi^3(V\hat{\mathsf{S}}_\mathsf{X}V_-^2 + V^2\hat{\mathsf{S}}_\mathsf{X}V_-)$$
$$-2\pi^2\big(V(V\hat{\mathsf{S}}_\mathsf{X}+\hat{\mathsf{S}}_\mathsf{X}V_-)+(V\hat{\mathsf{S}}_\mathsf{X}+\hat{\mathsf{S}}_\mathsf{X}V_-)V_-\big)+4\pi^4V^2\hat{\mathsf{S}}_\mathsf{X}V_-^2\big) \equiv 0 \pmod{I_{\hat{H}}},$$

by Eqs. (4.69) and (4.70). Hence

$$M(\sigma_1)\hat{F}(z) = \mathsf{L}_\mathsf{X}(z)W\hat{\mathsf{S}}_\mathsf{X}\overleftarrow{\overleftarrow{W}}\overleftarrow{\mathsf{L}_\mathsf{X}}(\bar{z}) \equiv \mathsf{L}_\mathsf{X}(z)\hat{\mathsf{S}}_\mathsf{X}\overleftarrow{\mathsf{L}_\mathsf{X}}(\bar{z}) \equiv \hat{F}(z) \pmod{I_{\hat{H}}}.$$

This completes the proof of the theorem. □

4.6 Historical Notes

By studying the twisting (a certain gauge transformation) of quasi-Hopf algebras Drinfeld (see [172, 173]) was able to give a beautiful explanation of Kohno's Monodromy Theorem in [350]. More importantly, he defined two important pro-algebraic groups, the Grothendieck–Teichmüller group GT and its graded version GRT. Both groups act freely and transitively on the associator sets $\{\mathrm{M}_\lambda(\mathbb{K}) : \lambda \in \mathbb{K}^\times\}$, though from opposite side (see a detailed survery of this subject by Furusho [217]). The group GRT_1 defined in the book is just the unipotent part of GRT. One of Drinfeld's main results is that $M_1(\mathbb{Q}) \neq \emptyset$. Therefore, there a formal Lie series P with rational coefficients such that his associator $\Phi(t^{12}, t^{23}) = \exp P(t^{12}, t^{23})$ satisfies both the pentagon relation (4.13) and the hexagon relation (4.12). Moreover, he showed that the special derivation condition (4.14) for $\varphi \in \mathfrak{t}_3^0$ (resp. (4.19) for $g \in \exp \mathfrak{t}_3^0$) follows from the duality and hexagon equations (4.12) and the pentagon equation (4.13) in [173, Prop. 5.7] (resp. (4.16) and (4.17) in [173, Prop. 5.9]). Our proof of the special derivation condition in Props. 4.1.19 and 4.1.22 follows Drinfeld's original one closely.

In [219] Furusho established that the hexagon relation is a consequence of the pentagon relation in the MZV setting. Later, in [220] he further proved that the pentagon relation implies all the regularized DBSFs for MZVs. Some numerical computation up to weight 19 in [205] showed that in some sense the opposite inclusion also holds. The computation of all the coefficients of Φ including its regularized terms, which generalizes Thm. 4.1.14, was stated first by Le and Murakami [385] but revised later by Furusho as [215, Prop. 3.2.3] using Le–Murakami's method. The p-adic version of Lemma 4.2.12 can be found as [216, Thm. 3.15]. Our Lie algebra approach is influenced by [164, 477]. For example, Lemma 4.1.9 is proved by Racinet as Cor. 2.4.4 of [477] by using Poincaré–Birkhoff–Witt Theorem.

Using higher order of roots of unity Enriquez [199] extended Drinfeld associator to the cyclotomic setting by defining the notion of a quasi-reflection algebra. These associators satisfy some relations called the octagon and the mixed pentagon relations. We will see in Chap. 13 that these generalizations provide relations among special values of MPLs at roots of unity. In [93] Brochier computed explicitly the monodromy representations of the cyclotomic KZ differential equation system and obtained the corresponding cyclotomic Kohno–Drinfeld Theorem. Enriquez and Furusho [202] further considered various implications among these and some more relations, one of which is the special action condition we have referred to in Eq. (4.32).

An explicit formula of the Drinfeld associator can be found in [385] and its logarithm in [203]. A generalization of this was defined by Enriquez in [201] in which he constructed the genus one analogue of the theory of associators and the Grothendieck–Teichmüller group.

As one of the most beautiful applications of the theory of the MZVs and MPLs, the SVMPLs and the SVMZVs have appeared in a few computations in mathematical physics in the past few years, such as Brown and Schnetz's solution of the zig-zag conjecture [108] (see Chap. 15), the expression of the coefficients of the closed super-string tree-level amplitude [500], the study of wrapping functions1 in $N = 4$ super Yang–Mills [389] and some other related works [128, 171]. The fact that the generating series $\mathsf{L}_{\mathsf{X}}(z)$ is group-like can be derived from [482, Cor. 3.5]. Essentially the same serie was considered by Furusho [218, (2.20)] (also see Thm. 2.27 of loc. cit.). Our proof of Thm. 4.1.1 follows from that of [98, Thm. 2.1] and [254, Prop. 1.2]. For the proof of the zig-zag conjecture, the two generating series in Sec. 4.5 will play crucial roles. In both Thm. 4.5.6 and Thm. 4.5.7 the single-valuedness is the most techinical part. Note that there are two possible ways to write down the KZ-equations because of the non-commutativity $G' = \Omega G$ or $G' = G\Omega$ ($\Omega = \frac{\mathsf{x}_0}{z} + \frac{\mathsf{x}_1}{z-1}$). We choose the same convention in this book as that of [97], but opposite to that of [108]. This is why the coefficients of both S and $\hat{\mathsf{S}}$ series here have opposite signs to those in Brown and Schnetz's original paper [108].

Exercises

4.1. Show that $\Phi(\mathsf{x}_0, \mathsf{x}_1)$ depends only on x_0 and x_1 but not on z.

4.2. Use the method of undetermined coefficients to find the first few terms of $G_0(x)$ and $G_1(x)$ satisfying Eq. (4.4). Then compute the first few term of $G_1^{-1}G_0$ and show that the Drinfeld associator has the following expansion:

$$\begin{aligned}\Phi(\mathsf{x}_0, \mathsf{x}_1) = 1 &- \zeta(2)[\mathsf{x}_0, \mathsf{x}_1] - \zeta(3)\Big([\mathsf{x}_0, [\mathsf{x}_0, \mathsf{x}_1]] - [[\mathsf{x}_0, \mathsf{x}_1], \mathsf{x}_1]\Big)\\ &+ \zeta(4)[\mathsf{x}_0, [\mathsf{x}_0, [\mathsf{x}_0, \mathsf{x}_1]]] + \zeta(3,1)[\mathsf{x}_0, [[\mathsf{x}_0, \mathsf{x}_1], \mathsf{x}_1]]\\ &- \zeta(2,1,1)[[[\mathsf{x}_0, \mathsf{x}_1], \mathsf{x}_1], \mathsf{x}_1] + \frac{1}{2}\zeta(2)^2[\mathsf{x}_0, \mathsf{x}_1]^2 + \cdots .\end{aligned}$$

4.3. Show that if $f_*[\ ] : \mathfrak{A}_*^1 \to \mathbb{R}$ satisfies the stuflle relations Eq. (4.7) then it satisfies Eq. (4.7) which implies that $f_* = 1 + \sum_{\mathsf{w} \in \mathsf{X}^*\mathsf{x}_1} f_*[\mathbf{w}]\mathbf{w}$ is group-like for Δ_*.

4.4. Let $f = 1 + \sum_{\mathbf{w}\in \mathsf{x}_0\mathsf{X}^*\mathsf{x}_1} f[\mathbf{w}]\mathbf{w} \in (\mathscr{R}_{\rm cv}, \Delta_{ш})$. Prove that f satisfies shuffle relation if and only if $\Delta_{ш}(f) = f \otimes f$ (*i.e.*, f is group-like for $\Delta_{ш}$) by expanding both $\Delta_{ш}(f)$ and $f \otimes f$ and then compare the corresponding terms. Prove the above result is still valid if we consider $(\mathscr{R}_{ш}, \Delta_{ш})$. This proves Thm. 4.1.8.

4.5. Prove that Lemma 4.1.9 and Lemma 4.1.10 both follow from the Cor. B.11 of the Poincaré–Birkhoff–Witt Theorem B.10.

4.6. Prove that $z_n = \sum_{1\le i<j\le n} t^{ij}$ is central in $\mathfrak{t}_n$ and that $\mathfrak{t}_n \cong \mathfrak{t}_n^0 \oplus \mathbb{K}\cdot z_n$.

4.7. Show that the Grothendieck–Teichmüller Lie algebra $\mathfrak{grt}_1(\mathbb{K})$ indeed forms a Lie algebra with the Lie bracket given by

$$\langle \varphi_1, \varphi_2\rangle = [\varphi_1, \varphi_2] + D_{\varphi_2}(\varphi_1) - D_{\varphi_1}(\varphi_2), \tag{4.71}$$

where D_φ is the derivation of $\mathfrak{t}_3^0$ defined by

$$D_\varphi(\mathsf{x}_0) = [\varphi, \mathsf{x}_0] \quad \text{and } D_\varphi(\mathsf{x}_1) = 0.$$

The Lie algebra structure can be realized as an embedding

$$\mathfrak{grt}_1(\mathbb{K}) \hookrightarrow \mathrm{Der}(\mathfrak{t}_3^0)$$

by sending $\varphi \mapsto D_\varphi$.

4.8. Show that the Grothendieck–Teichmüller group $\mathrm{GRT}_1(\mathbb{K})$ indeed forms a group by the multiplication

$$g_2 \circ g_1 = g_1(g_2^{-1}\mathsf{x}_0 g_2, \mathsf{x}_1)\cdot g_2(\mathsf{x}_0,\mathsf{x}_1) = g_2(\mathsf{x}_0,\mathsf{x}_1)\cdot g_1(\mathsf{x}_0, g_2^{-1}\mathsf{x}_1 g_2). \tag{4.72}$$

The group structure can be realized as an embedding (with anti-homomorphism)

$$\mathrm{GRT}_1(\mathbb{K}) \hookrightarrow \mathrm{Aut}\exp\mathfrak{t}_3^0$$

by sending g to the automorphism A_g defined by

$$\mathsf{x}_0 \mapsto \mathsf{x}_0 \quad \text{and } \mathsf{x}_1 \mapsto g^{-1}\mathsf{x}_1 g.$$

4.9. Recall that for a group G, a left G-torsor is a non-empty set X on which G acts freely and transitively on the left, meaning that for any $x, y \in X$ there exists a unique $g \in G$ such that $g * x = y$ where $*$ denotes the left action of G on X. For any fixed $\lambda \in \mathbb{K}^\times$, show that the associator set $\mathrm{M}_\lambda(\mathbb{K})$ is a left $\mathrm{GRT}_1(\mathbb{K})$-torsor by (4.72) defined in Exercise 4.8 with $g_1 \in \mathrm{M}_\lambda(\mathbb{K})$ and $g_2 \in \mathrm{GRT}_1(\mathbb{K})$.

4.10. Complete the proof of Lemma 4.1.27. For the uniqueness of W_1, let $\mathbf{w}$ be any word neither beginning nor ending with A or B. First show that the coefficient $c^{i,j}_{k,l,m,n}$ of $A^kB^l\mathbf{w}A^mB^nu^iv^j$ for $f(u,v)$ satisfies

$$ic^{i,j}_{k,l,m,n} = c^{i,j}_{k-1,l,m,n} - c^{i,j}_{k,l,m-1,n} + (Rf)[A^kB^l\mathbf{w}A^mB^nu^{i-1}v^j] \quad (4.73)$$

and

$$jc^{i,j}_{k,l,m,n} = c^{i,j}_{k,l-1,m,n} - c^{i,j}_{k,l,m,n-1} + (Sf)[A^kB^l\mathbf{w}A^mB^nu^iv^{j-1}] \quad (4.74)$$

for all $i+j \geq 0$ and $k+l+m+n \geq 0$. Here if a word contains either A^{-1} or B^{-1} the corresponding term is vacuous. Then use double induction starting with $k+l+m+n=0$ to show that $c^{i,j}_{0,0,0,0}$ can be determined uniquely, with the inner induction loop on $i+j$. Then treat the general $k+l+m+n>0$ case by induction. In fact, if $i=j$ then Eq. (4.73) coincides with Eq. (4.74) by Eq. (4.26). This determines recursively the coefficient $c^{i,i}_{k,l,m,n}$. Furthermore, for $i,j>0$ the constraint $(1/i)\times$Eq. (4.73)$=(1/j)\times$Eq. (4.74) determines recursively the coefficient $c^{i-1,j}_{k,l,m,n}$ (resp. $c^{i,j-1}_{k,l,m,n}$) if $i<j$ (resp. $i>j$).

4.11. Complete the proof of Lemma 4.1.28.

4.12. Complete the proof of Eq. (4.31) by mimicking the proof of the first identity $W_2 = W_3e^{ht^{23}/2}$.

4.13. Prove Eq. (4.15) by using Eq. (4.12).

4.14. Prove that $f_0(z)$ defined by (4.2) and (4.3) indeed provides a solution in Thm. 4.1.1.

4.15. Prove the uniqueness of the solution in Thm. 4.1.1.

4.16. Show that by the change of variables $z-p_0 \to 1/(z-p_0)$ one obtains a solution $L^\infty(z)$ to Eq. (4.1) on U which corresponds to the point at infinity. Fix any branch of $\log z$ on U. Show that $L^\infty(z) \sim \exp(\mathrm{x}_\infty \log z)$ where $\mathrm{x}_\infty = \sum_{i=0}^N \mathrm{x}_i$.

4.17. Show that for any $\mathbf{w} \in \mathsf{X}^*$ and $a \in \{0,1\}$

$$\partial_{\bar z}\mathscr{L}_{\mathbf{w}a}(z) = \frac{\mathscr{L}_{\mathbf{w}}(z)}{\bar z - a} + \frac{1}{\bar z - 1}\sum_{\mathbf{w}a=\mathbf{uv}} \mathscr{L}_{\mathbf{u}}(z)(\mathrm{y}_1[\mathbf{v}] - \delta_{\mathrm{x}_1,\mathbf{v}}).$$

4.18. For any $\mathbf{w} \in \{0,1\}^*$ (or $\mathbf{w} \in \mathsf{X}^*$), show that

$$\mathrm{L}_{a\mathbf{w}}(z)(=\mathrm{L}_{\mathrm{x}_a\mathbf{w}}(z)) = \int \frac{\mathrm{L}_{\mathbf{w}}(t)}{t-a}\mathrm{d}t, \quad a \in \{0,1\},$$

with the initial value $\mathrm{L}_{\mathbf{w}}(0) = 0$ unless $\mathbf{w} = \{0\}^n (= \mathrm{x}_0^n)$ when $\mathrm{L}_{\mathbf{w}}(0) = \frac{\log^n z}{n!}$.

4.19. Show that $F(\mathrm{x}_0, \mathrm{x}_1)$ defined by Eq. (4.46) satisfies $F \equiv 0$ modulo words of depth two. Moreover, it has the following explicit expansion

$$F(\mathrm{x}_0, \mathrm{x}_1) = \mathrm{x}_1 + \zeta(2)[[\mathrm{x}_0, \mathrm{x}_1], \mathrm{x}_1] + \zeta(3)[\mathrm{x}_1, \mathbf{u}_3] + \cdots$$

where $\mathbf{u}_3$ is the coefficient of $\zeta(3)$ in $\Phi(\mathrm{x}_0, \mathrm{x}_1)$ given by Exercise 4.2. Then show that by Eq. (4.47)

$$\mathrm{y}_1 = \mathrm{x}_1 + 2\zeta(3)[\mathrm{x}_1, \mathbf{u}_3] + \cdots$$

4.20. Show that if $\mathbf{z}_1$ satisfies Eq. (4.47) then it satisfies Eq. (4.45).

4.21. Show that

$$\mathrm{L}_{\{0\}^m 1\{0\}^n}(z) = \sum_{k=0}^{n} (-1)^{k+1} \binom{m+k}{m} \frac{\log^{n-k} z}{(n-k)!} \mathrm{Li}_{m+k+1}(z).$$

4.22. Use Exercise 4.21 to show that

$$\begin{aligned}\mathscr{L}_{\{0\}^m 1\{0\}^n}(z) = &\sum_{k=0}^{n} (-1)^{k+1} \binom{m+k}{m} \frac{\log^{n-k}(z\bar{z})}{(n-k)!} \mathrm{Li}_{m+k+1}(z) \\ &+ \sum_{k=0}^{m} (-1)^{k+1} \binom{n+k}{n} \frac{\log^{m-k}(z\bar{z})}{(m-k)!} \mathrm{Li}_{n+k+1}(\bar{z}).\end{aligned}$$

4.23. Show that the regularizations in Definition 4.4.1 are well defined (*i.e.*, unique).

4.24. Show that if $f(z) \in \mathsf{Ls}$ then $f(z) = o(|z|^{-\epsilon})$ as $z \to 0$, i.e.,

$$\lim_{z \to 0} |z|^{\epsilon} f(z) = 0.$$

4.25. Show that if $f(z) \in \mathsf{Ls}$ then $f(z) = o(|z|^{\epsilon})$ as $z \to \infty$, i.e.,

$$\lim_{z \to \infty} |z|^{-\epsilon} f(z) = 0.$$

4.26. Let $P \in \mathsf{Ls}$ and $a, b \in \{0, 1\}$. Show that

$$\begin{aligned}&\left(\int_0 \frac{dz}{z-a} \int_0 \frac{d\bar{z}}{\bar{z}-b} - \int_0 \frac{d\bar{z}}{\bar{z}-b} \int_0 \frac{dz}{z-a} \right) P(z) \\ &= \left(\int_0 \frac{d\bar{z}}{\bar{z}-b} P(z) \bigg|_{z=a} - \int_0 \frac{dz}{z-a} P(z) \bigg|_{z=b} \right) \cdot \mathscr{L}_1(z).\end{aligned}$$

4.27. Prove the stability of $\mathsf{Ls}^{\mathrm{sv}}$ under the transformations $z \to \frac{z}{z-1}$ by using the following identities:

$$\begin{aligned}\mathscr{L}_{\mathrm{x}_0,\mathrm{x}_1}(1-z) &= \mathrm{L}_{\mathrm{x}_1,\mathrm{x}_0}(z) \Phi_{\mathrm{x}_0,\mathrm{x}_1} \overleftarrow{\Phi}_{\mathrm{x}_0,\mathrm{y}_1} \overleftarrow{\mathrm{L}}_{\mathrm{y}_1,\mathrm{x}_0}(\bar{z}) \\ &= \mathscr{L}_{\mathrm{x}_1,\mathrm{x}_0}(z) \mathscr{L}_{\mathrm{x}_0,\mathrm{x}_1}(1) = \mathscr{L}_{\mathrm{x}_0,\mathrm{x}_1}(1) \mathscr{L}_{\mathrm{y}_1,\overleftarrow{\mathrm{y}_1}(\mathrm{y}_1,\mathrm{x}_0)}(z),\end{aligned}$$

$$\begin{aligned}
\mathscr{L}_{\mathrm{x}_0,\mathrm{x}_1}\left(\frac{z-1}{z}\right) &= \mathrm{L}_{-\mathrm{x}_0-\mathrm{x}_1,\mathrm{x}_0}(z)\Phi_{\mathrm{x}_0,-\mathrm{x}_0-\mathrm{x}_1}\overleftarrow{\Phi}_{\mathrm{x}_0,-\mathrm{x}_0-\mathrm{y}_1}\overleftarrow{\mathrm{L}}_{-\mathrm{x}_0-\mathrm{y}_1,\mathrm{x}_0}(\bar{z}) \\
&= \mathscr{L}_{-\mathrm{x}_0-\mathrm{x}_1,\mathrm{x}_0}(z)\,\mathscr{L}_{\mathrm{x}_0,-\mathrm{x}_0-\mathrm{x}_1}(1) \\
&= \mathscr{L}_{\mathrm{x}_0,-\mathrm{x}_0-\mathrm{x}_1}(1)\,\mathscr{L}_{-\mathrm{x}_0-\mathrm{y}_1,\overleftarrow{\mathrm{y}_1}(-\mathrm{x}_0-\mathrm{y}_1,\mathrm{x}_0)}\,(z), \\
\mathscr{L}_{\mathrm{x}_0,\mathrm{x}_1}\left(\frac{z}{z-1}\right) &= \mathrm{L}_{\mathrm{x}_0,-\mathrm{x}_0-\mathrm{x}_1}(z)\overleftarrow{\mathrm{L}}_{\mathrm{x}_0,-\mathrm{x}_0-\mathrm{y}_1}(\bar{z}) = \mathscr{L}_{\mathrm{x}_0,-\mathrm{x}_0-\mathrm{x}_1}(z), \\
\mathscr{L}_{\mathrm{x}_0,\mathrm{x}_1}\left(\frac{1}{1-z}\right) &= \mathrm{L}_{\mathrm{x}_1,-\mathrm{x}_0-\mathrm{x}_1}(z)\Phi_{\mathrm{x}_0,\mathrm{x}_1}\overleftarrow{\Phi}_{\mathrm{x}_0,\mathrm{y}_1}\overleftarrow{\mathrm{L}}_{\mathrm{y}_1,-\mathrm{x}_0-\mathrm{y}_1}(\bar{z}) \\
&= \mathscr{L}_{\mathrm{x}_1,-\mathrm{x}_0-\mathrm{x}_1}(z)\,\mathscr{L}_{\mathrm{x}_0,\mathrm{x}_1}(1) \\
&= \mathscr{L}_{\mathrm{x}_0,\mathrm{x}_1}(1)\,\mathscr{L}_{\mathrm{y}_1,\overleftarrow{\mathrm{y}_1}(\mathrm{y}_1,-\mathrm{x}_0-\mathrm{y}_1)}\,(z), \\
\mathscr{L}_{\mathrm{x}_0,\mathrm{x}_1}\left(\frac{1}{z}\right) &= \mathrm{L}_{-\mathrm{x}_0-\mathrm{x}_1,\mathrm{x}_1}(z)\Phi_{\mathrm{x}_0,-\mathrm{x}_0-\mathrm{x}_1}\overleftarrow{\Phi}_{\mathrm{x}_0,-\mathrm{x}_0-\mathrm{y}_1}\overleftarrow{\mathrm{L}}_{-\mathrm{x}_0-\mathrm{y}_1,\mathrm{y}_1}(\bar{z}) \\
&= \mathscr{L}_{-\mathrm{x}_0-\mathrm{x}_1,\mathrm{x}_1}(z)\,\mathscr{L}_{\mathrm{x}_0,-\mathrm{x}_0-\mathrm{x}_1}(1) \\
&= \mathscr{L}_{\mathrm{x}_0,-\mathrm{x}_0-\mathrm{x}_1}(1)\,\mathscr{L}_{-\mathrm{x}_0-\mathrm{y}_1,\overleftarrow{\mathrm{y}_1}(-\mathrm{x}_0-\mathrm{y}_1,\mathrm{y}_1)}\,(z).
\end{aligned}$$

4.28. Show that modulo I_H the series $V = \Phi^{-1}\mathrm{x}_1\Phi$ satisfies

$$V \equiv (1 - \Phi^0_{0,0} + \Phi^0_{0,1} + \Phi^1_{0,0} - \Phi^1_{0,1})\mathrm{x}_1(1 + \Phi^0_{0,0} + \Phi^0_{1,0} + \Phi^1_{0,0} + \Phi^1_{1,0}).$$

4.29. Show that modulo I_H

$$\begin{aligned}
2V^0_{0,1} + \mathrm{S}^0_{0,0}V^0_{1,1} \equiv 2V^0_{1,0} - V^0_{1,1}\mathrm{S}^0_{0,0} \equiv V^0_{0,1}\mathrm{S}^0_{0,0} + \mathrm{S}^0_{0,0}V^0_{1,0} &\equiv 0\,, \\
2V^1_{1,1} - V^0_{1,1}\mathrm{S}^1_{0,1} + \mathrm{S}^1_{1,0}V^0_{1,1} &\equiv 0, \\
V^0_{0,1}\mathrm{S}^1_{0,1} + \mathrm{S}^1_{0,1}V^0_{0,1} + \mathrm{S}^0_{0,0}(V^0_{0,1} + V^1_{1,1}) + \mathrm{S}^1_{0,0}V^0_{1,1} &\equiv 0, \\
V^0_{1,0}\mathrm{S}^1_{1,0} + \mathrm{S}^1_{1,0}V^0_{1,0} + (V^0_{1,0} + V^1_{1,1})\mathrm{S}^0_{0,0} - V^0_{1,1}\mathrm{S}^1_{0,0} &\equiv 0, \\
\mathbf{W} + (\mathrm{S}^0_{0,0} + \mathrm{S}^1_{0,1})V^0_{0,0} - V^0_{0,0}(\mathrm{S}^0_{0,0} + \mathrm{S}^1_{1,0}) &\equiv 0,
\end{aligned}$$

where $\mathbf{W} = 2V^1_{0,0} + V^0_{0,1}\mathrm{S}^1_{0,0} - V^1_{0,1}\mathrm{S}^0_{0,0} + \mathrm{S}^0_{0,0}V^1_{1,0} + \mathrm{S}^1_{0,0}V^0_{1,0}$.

4.30. Decompose V as $V = \sum_{a,b,c\in\{0,1\}}\hat{V}^a_{b,c}$ in $\hat{H}$ where $\hat{V}^a$ is the sum of the subset of words of V that begins with x_b, ends with x_c, and contains the subword x_1^2 exactly a times. Show that modulo $I_{\hat{H}}$

$$\begin{aligned}
2\hat{V}^0_{0,1} + \hat{\mathrm{S}}^0_{0,0}\hat{V}^0_{1,1} \equiv 2\hat{V}^1_{1,1} - \hat{V}^0_{1,1}\hat{\mathrm{S}}^1_{0,1} + \hat{\mathrm{S}}^1_{1,0}\hat{V}^0_{1,1} &\equiv 0, \\
\hat{V}^0_{0,1}\hat{\mathrm{S}}^1_{0,1} + \hat{\mathrm{S}}^1_{0,1}\hat{V}^0_{0,1} + \hat{\mathrm{S}}^0_{0,0}\hat{V}^1_{1,1} + \hat{\mathrm{S}}^1_{0,0}\hat{V}^0_{1,1} &\equiv 0, \\
\mathbf{W} - \hat{V}^0_{0,0}\hat{\mathrm{S}}^1_{1,0} + \hat{\mathrm{S}}^1_{0,1}\hat{V}^0_{0,0} &\equiv 0,
\end{aligned}$$

where $\mathbf{W} = 2\hat{V}^1_{0,0} + \hat{V}^0_{0,1}\hat{\mathrm{S}}^1_{0,0} + \hat{\mathrm{S}}^1_{0,0}\hat{V}^0_{1,0} - \hat{V}^1_{0,1}\hat{\mathrm{S}}^0_{0,0} + \hat{\mathrm{S}}^0_{0,0}\hat{V}^1_{1,0}$.

Chapter 5

Multiple Zeta Value Identities

There are plenty of known relations among the MZVs most of which have been proved rigorously in recent years with a variety of techniques. In Sec. 3.3 we have seen that the (regularized) DBSFs give rise to many linear relations over $\mathbb{Q}$ and conjecturally these imply all the other linear relations. In this chapter we first collect a few special families of these relations and then present some conjectural ones at the end. We will often resort to the properties of Bernoulli numbers thanks to Euler's famous evaluation in Eq. (1.2).

5.1 Duality Relations

Let k be a positive integer and $\mathbf{s} = (s_1, \dots, s_d)$ of weight $w = |\mathbf{s}|$. Write

$$\mathbf{s} = (a_1 + 1, \{1\}^{b_1 - 1}, \dots, a_\ell + 1, \{1\}^{b_\ell - 1})$$

for $a_1, b_1, \dots, a_\ell, b_\ell \geq 1$. Then its dual is defined by

$$\mathbf{s}^* = (b_\ell + 1, \{1\}^{a_\ell - 1}, \dots, b_1 + 1, \{1\}^{a_1 - 1}). \tag{5.1}$$

An important relation between $\mathbf{s}$ and its dual is the following:

$$\mathrm{dp}(\mathbf{s}) + \mathrm{dp}(\mathbf{s}^*) = |\mathbf{s}|. \tag{5.2}$$

Example 5.1.1. One can check easily that

$$(n + 1, \{1\}^{k-1})^* = (k + 1, \{1\}^{n-1}).$$

Recall from Definition 3.2.1 that $\mathfrak{A}$ is the $\mathbb{Q}$-algebra of words generated over $\mathtt{x}_0$ and $\mathtt{x}_1$. In this chapter, to save space, we will write $\mathtt{x}_0 = \mathtt{a}$ and $\mathtt{x}_1 = \mathtt{b}$. The following result provides so-called *duality relation* of the MZVs.

Theorem 5.1.2. *For every admissible composition* $\mathbf{s}$*, we have*

$$\zeta(\mathbf{s}) = \zeta(\mathbf{s}^*). \tag{5.3}$$

Using admissible words (see Definition 3.2.2), this can be expressed as

$$\zeta(\mathbf{w}) = \zeta(\tau\mathbf{w}) \quad \forall \mathbf{w} \in \mathfrak{A}^0, \tag{5.4}$$

where τ *is the anti-automorphism of* $\mathfrak{A}^0$ *by exchanging* a *and* b.

Proof. By Eq. (5.1), given an admissible word $\mathbf{w} = \mathsf{a}^{s_1-1}\mathsf{b}\cdots\mathsf{a}^{s_d-1}\mathsf{b}$ (see Definition 3.2.1) and $\mathbf{s}^* = \mathbf{t} = (t_1,\dots,t_\ell)$ one checks that $\mathsf{a}^{t_1-1}\mathsf{b}\cdots\mathsf{a}^{t_\ell-1}\mathsf{b} = \mathsf{a}\mathsf{b}^{s_d-1}\cdots\mathsf{a}\mathsf{b}^{s_1-1} = \tau(\mathbf{w})$ (see Exercise 5.2). Set the 1-forms $\mathsf{a} = dt/t$ and $\mathsf{b} = dt/(1-t)$ as before. By Chen's formula on the reversal of the iterated integrals Lemma 2.1.2(iii) followed by the substitution $t \to 1-t$ one gets (notice $d(1-t) = -dt$):

$$\begin{aligned}\zeta(\mathbf{s}) = (-1)^{|\mathbf{s}|}\int_0^1 \mathbf{w} &= \int_1^0 \mathsf{b}\mathsf{a}^{s_d-1}\cdots\mathsf{b}\mathsf{a}^{s_1-1}\\ &= (-1)^{|\mathbf{s}|}\int_0^1 \mathsf{a}\mathsf{b}^{s_d-1}\mathsf{b}\cdots\mathsf{a}\mathsf{b}^{s_1-1} = \zeta(\mathbf{s}^*),\end{aligned}$$

as desired. □

5.2 Cyclic Sum Relation

First we define a cyclic equivalence classes of multiple indices on the set

$$I(w,d) = \{(k_1,\dots,k_d) | k_1 + \cdots + k_d = w,\ k_1,\dots,k_d \geq 1\}.$$

Definition 5.2.1. We say two elements of $I(w,d)$ are *cyclically equivalent* if they are cyclic permutations of each other, *i.e.*, for the permutation $\sigma = (1,\dots,d)$ and $j = 1,\dots,d$, we define $(k_1,\dots,k_d) \equiv (\sigma^j(k_1),\dots,\sigma^j(k_d))$ whose equivalence class is denoted by $[(k_1,\dots,k_d)]$. Let $\mathscr{C}(w,d)$ be the set of cyclic equivalence classes of $I(w,d)$.

The main idea of the proof of the cyclic sum relation is to use partial fractions. Already applied by Euler this technique is ubiquitous in a lot of proofs of the MZV identities. We first need a lemma. Let $\mathbb{N}_0 = \mathbb{N} \cup \{0\}$ be the set of nonnegative integers. For $k_1,\dots,k_d \in \mathbb{N}$ and $k_{d+1} \in \mathbb{N}_0$, let

$$A(k_1,\dots,k_d) = \sum_{n_1>n_2>\cdots>n_d>n_{d+1}\geq 0} \frac{1}{(n_1-n_{d+1})n_1^{k_1}\dots n_d^{k_d}}$$

and

$$B(k_1,\dots,k_d,k_{d+1}) = \sum_{n_1>n_2>\cdots>n_d>n_{d+1}>0} \frac{1}{(n_1-n_{d+1})n_1^{k_1}\dots n_d^{k_d}n_{d+1}^{k_{d+1}}}.$$

Lemma 5.2.2. *The infinite sum* $A(k_1,\dots,k_d)$ *converges when one of* $k_1,\dots,k_d$ *is greater than* 1, *while* $B(k_1,\dots,k_d,k_{d+1})$ *converges when one of* $k_1,\dots,k_d,k_{d+1}+1$ *is greater than* 1.

Proof. By the definition

$$B(k_1,\dots,k_d,0)=A(k_1,\dots,k_d)-\zeta(k_1+1,k_2,\dots,k_d). \tag{5.5}$$

Applying the partial fraction identity

$$\frac{1}{n_1(n_1-n_{d+1})}=\frac{1}{n_{d+1}}\left(\frac{1}{n_1-n_{d+1}}-\frac{1}{n_1}\right) \tag{5.6}$$

to $B(k_1,\dots,k_d,k_{d+1})$ we get

$$B(k_1,\dots,k_d,k_{d+1})=B(k_1-1,k_2,\dots,k_d,k_{d+1}+1)-\zeta(k_1,\dots,k_d,k_{d+1}+1). \tag{5.7}$$

Applying Eq. (5.6) to $B(1,k_2,\dots,k_d,k_{d+1})$ we get

$$\begin{aligned}
&\sum_{n_1>n_2>\cdots>n_d>n_{d+1}>0}\frac{1}{n_2^{k_2}\cdots n_d^{k_d}n_{d+1}^{k_{d+1}+1}}\left(\frac{1}{n_1-n_{d+1}}-\frac{1}{n_1}\right)\\
&=\sum_{n_2>\cdots>n_d>n_{d+1}>0}\frac{1}{n_2^{k_2}\cdots n_d^{k_d}n_{d+1}^{k_{d+1}+1}}\sum_{n_1=n_2+1}^{\infty}\left(\frac{1}{n_1-n_{d+1}}-\frac{1}{n_1}\right)\\
&=\sum_{n_2>\cdots>n_d>n_{d+1}>0}\frac{1}{n_2^{k_2}\cdots n_d^{k_d}n_{d+1}^{k_{d+1}+1}}\sum_{j=0}^{n_{d+1}-1}\frac{1}{n_2-j}\\
&=\sum_{n_2>\cdots>n_d>n_{d+1}>j\geq 0}\frac{1}{(n_2-j)n_2^{k_2}\cdots n_d^{k_d}n_{d+1}^{k_{d+1}+1}}
\end{aligned}$$

and so

$$B(1,k_2,\dots,k_d,k_{d+1})=A(k_2,\dots,k_d,k_{d+1}+1). \tag{5.8}$$

By Eq. (5.5)

$$B(k_1,\dots,k_d,k_{d+1})\leq B(k_1,\dots,k_d,0)\leq A(k_1,\dots,k_d).$$

If $k_1\geq 2$ then $B(k_1,\dots,k_{d+1})$ converges whenever $A(k_1,\dots,k_d)$ does; otherwise, if $k_1=1$ then Eq. (5.8) yields $B(k_1,\dots,k_{d+1})=A(k_2,\dots,k_d,k_{d+1}+1)$. So the statement about the B's follows from the one about the A's.

Turning to the first assertion of lemma we clearly only need to consider the case $k_1+\cdots+k_d=d+1$ when $A(k_1,\dots,k_d)$ are among the largest. Now

$$A(2,1,\dots,1)=\sum_{n_1>n_2>\cdots>n_{d+1}\geq 0}\frac{1}{n_1^2(n_1-n_{d+1})n_2\cdots n_d}$$

$$\le \sum_{\substack{n_1>n_2>\cdots>n_d>0\\ n_1\ge j>0}} \frac{1}{n_1^2 j n_2 \ldots n_d}$$

$$= \zeta(3,\{1\}^{d-1}) + d\zeta(2,\{1\}^d) + \sum_{i=1}^{d-1} \zeta(2,\{1\}^{i-1},2,\{1\}^{d-i-1})$$

by inserting j into all possible places inside $(0, n_1+1)$. So $A(2,1,\ldots,1)$ converges and by Eqs. (5.5) and (5.8) we have

$$\begin{aligned} A(1,2,1,\ldots,1) &= B(1,2,1,\ldots,1,0) + \zeta(2,2,1,\ldots,1)\\ &= A(2,1,\ldots,1,1) + \zeta(2,2,1,\ldots,1). \end{aligned}$$

We can repeat this idea and show that all the sums $A(1,\ldots,1,2,1,\ldots,1)$ converge. This completes the proof of the lemma. □

The next result provides the *cyclic sum relations* of the MZVs.

Theorem 5.2.3. *For any* $\mathbf{k} = (k_1,\ldots,k_d) \in \mathbb{N}^d$ *with some* $k_i \ge 2$,

$$\sum_{j=1}^{d} \zeta(k_j+1, k_{j+1},\ldots,k_d,k_1,\ldots,k_{j-1})$$
$$= \sum_{\{j:k_j\ge 2\}} \sum_{q=0}^{k_j-2} \zeta(k_j-q, k_{j+1},\ldots,k_d,k_1,\ldots,k_{j-1},q+1). \quad (5.9)$$

Equivalently,

$$\sum_{(p_1,\ldots,p_d)\in[\mathbf{k}]} \zeta(p_1+1,p_2,\ldots,p_d) = \sum_{(q_1,\ldots,q_{|\mathbf{k}|-d})\in[\mathbf{k}^*]} \zeta(q_1+1,q_2,\ldots,q_{|\mathbf{k}|-d}). \quad (5.10)$$

Proof. We will prove Eq. (5.9) and leave the interested reader to check that Eq. (5.9) and Eq. (5.10) are equivalent (see Exercise 5.4). Let us first show that

$$A(k_1,\ldots,k_d) - A(k_2,\ldots,k_d,k_1)$$
$$= \zeta(k_1+1,k_2,\ldots,k_d) - \sum_{j=0}^{k_1-2} \zeta(k_1-j,k_2,\ldots,k_d,j+1) \quad (5.11)$$

where the sum on the right is understood as 0 if $k_1 = 1$. We have

$$A(k_1,\ldots,k_d) - \zeta(k_1+1,k_2,\ldots,k_d)$$
$$= B(k_1,\ldots,k_d,0) \qquad \text{(by Eq. (5.5))}$$

$$
\begin{aligned}
&=B(k_1-1,k_2,\dots,k_d,1)-\zeta(k_1,\dots,k_d,1)=\cdots && \text{(by Eq. (5.7))}\\
&=B(1,k_2,\dots,k_d,k_1-1)-\sum_{j=0}^{k_1-2}\zeta(k_1-j,k_2,\dots,k_d,j+1) && \text{(by Eq. (5.7))}\\
&=A(k_2,\dots,k_d,k_1)-\sum_{j=0}^{k_1-2}\zeta(k_1-j,k_2,\dots,k_d,j+1) && \text{(by Eq. (5.8)),}
\end{aligned}
$$

which yields Eq. (5.11). Equation (5.9) now follows from adding Eq. (5.11) over all cyclic permutations of the sequence $(k_1,\dots,k_d)$. This completes the proof of the theorem. □

An interesting corollary of the cyclic sum relations is so-called *sum relations*:

Corollary 5.2.4. *For any positive integers w and d, we have*

$$\sum_{k_1+\cdots+k_d=w}\zeta(k_1+1,k_2,\dots,k_d)=\zeta(w+1). \tag{5.12}$$

Proof. Indeed, if $d=1$ then Eq. (5.12) is trivial. Suppose it is true when the depth on the left-hand side is d. Applying cyclic permutation of the indices $(k_1,\dots,k_d)$ to the left-hand side of Eq. (5.12), we get by Thm. 5.2.3

$$
\begin{aligned}
d\zeta(w+1)&=\sum_{|\mathbf{k}|=w}\sum_{j=1}^{d}\zeta(k_j+1,k_{j+1},\dots,k_d,k_1,\dots,k_{j-1})\\
&=\sum_{|\mathbf{k}|=w}\sum_{\{j|k_j\ge 2\}}\sum_{q=0}^{k_j-2}\zeta(k_j-q,k_{j+1},\dots,k_d,k_1,\dots,k_{j-1},q+1)\\
&=d\sum_{|\mathbf{k}|=w}\sum_{q=1}^{k_1-1}\zeta(k_1-q+1,k_2,\dots,k_d,q)\\
&=d\sum_{|\mathbf{k}|+q=w}\zeta(k_1+1,k_2,\dots,k_d,q).
\end{aligned}
$$

This proves the sum relation Eq. (5.12) by the induction. □

With the notation of the Hoffman algebras as in Sec. 3.2, we define, for $k\ge 1$,

$$\mathfrak{A}^1(k):=\langle \mathbf{w}:\ \text{word } \mathbf{w}\in\{\mathtt{a},\mathtt{b}\}^*\mathtt{ab}, |\mathbf{w}|=k\rangle_{\mathbb{Q}}.$$

So an element in $\mathfrak{A}^1(k)$ is a homogeneous polynomial of degree k with each monomial given by a word ending with b but not a b-power. Define a $\mathbb{Q}$-linear map $\rho_0 : \bigcup_{k\geq 1} \mathfrak{A}^1(k) \to \mathfrak{A}^0$ by setting

$$\rho_0\big(\mathtt{z}_{k_1}\cdots\mathtt{z}_{k_\ell}\big)=\sum_{j=1}^{\ell}\sum_{i=1}^{k_j-1}\mathtt{z}_{k_j-i+1}\mathtt{z}_{k_{j+1}}\cdots\mathtt{z}_{k_{j+\ell-1}}\mathtt{z}_i-\sum_{j=1}^{\ell}\mathtt{z}_{k_j+1}\mathtt{z}_{k_{j+1}}\cdots\mathtt{z}_{k_{j+\ell-1}},$$

where $\mathtt{z}_m = \mathtt{z}_n$ if $m \equiv n \bmod \ell$. Define recursively

$$\rho_{n+1}(\mathbf{w}) = \rho_n\big((\mathtt{a}+\mathtt{b})\mathbf{w}\big) \quad \forall n \geq 0.$$

Recall we have the map $\zeta : \mathfrak{A}^0 \to \mathbb{R}$ defined by $\zeta(\mathbf{w}) = \int_0^1 \mathbf{w}$ (see Eq. (3.1)).

Theorem 5.2.5. *If $n \in \mathbb{N}_0$ and $k \in \mathbb{N}$, then*

$$\{0\} = \rho_{k-1}(\mathfrak{A}^1_1) \subset \rho_{k-2}(\mathfrak{A}^1_2) \subset \cdots \subset \rho_0(\mathfrak{A}^1_k) \subset \ker\zeta \cap \mathfrak{A}^0_{k+1},$$

where $\mathfrak{A}^0_{k+1}$ is the homogeneous weight $k+1$ piece of $\mathfrak{A}^0$. Moreover, if $k > n$ then

$$\dim \rho_n(\mathfrak{A}^1_{k-n}) = \frac{1}{k}\sum_{m|k}\varphi\left(\frac{k}{m}\right)(2^m - L_{m,n}) - 2,$$

where φ is Euler's totient function and

$$L_{m,n} = \begin{cases} 2^m - 1, & \text{if } m = 1, \dots, n; \\ L_{m-1,n} + \cdots + L_{m-n,n}, & \text{if } m \geq n+1. \end{cases}$$

We omit the proof of this theorem here. See [494, Cor. 12 and Thm. 15].

5.3 Euler's Decomposition Formula Revisited

The following result is just Euler's decomposition formula, presented already in Thm.3.1.1 whose proof uses shuffle product of the iterated integrals. But here we present another proof using partial fractions. We can also regard it as a kind of weighted sum relations of the double zeta values.

Theorem 5.3.1. *For all positive integers $2 \leq s \leq w-2$, we have*

$$\sum_{k=2}^{w-1}\left\{\binom{k-1}{s-1}+\binom{k-1}{w-s-1}\right\}\zeta(k, w-k) = \zeta(s)\zeta(w-s).$$

Proof. We start with an easy but crucial observation: for any positive integers m and n

$$\frac{1}{mn} = \frac{m+n}{(m+n)mn} = \frac{1}{(m+n)m} + \frac{1}{(m+n)n}. \tag{5.13}$$

Repeating this procedure one quickly finds that for any positive integers s and t

$$\frac{1}{m^s n^t} = \sum_{j,k>0,\, j+k=s+t} \binom{k-1}{s-1} \frac{1}{(m+n)^k n^j} + \binom{k-1}{t-1} \frac{1}{(m+n)^k m^j}. \tag{5.14}$$

This follows from the induction on w since

$$\begin{aligned}\frac{1}{m^{s+1} n^t} &= \frac{1}{m} \sum_{j,k>0,\, j+k=s+t} \binom{k-1}{s-1} \frac{1}{(m+n)^k n^j} + \binom{k-1}{t-1} \frac{1}{(m+n)^k m^j} \\ &= \sum_{j,k>0,\, j+k=s+t} \binom{k-1}{s-1} \frac{1}{(m+n)^k n^j m} + \binom{k-1}{t-1} \frac{1}{(m+n)^k m^{j+1}}.\end{aligned}$$

Thus we can finish the induction by the easily checked identity

$$\frac{1}{mn^j} = \frac{1}{(m+n)^j m} + \sum_{i=1}^{j} \frac{1}{(m+n)^{j-i+1} n^i}$$

using induction on j and Eq. (5.13). Now the theorem follows from taking the sum $\sum_{m,n=1}^{\infty}$ on both sides of Eq. (5.14). □

5.4 Weighted Sum Relations

The next result is a kind of weighted form of the sum relations of the MZVs considered in Cor. 5.2.4. We shall prove it with some generating series of the MZVs.

Theorem 5.4.1. *For all integer $w \geq 3$,*

$$\sum_{k=2}^{w-1} 2^k \zeta(k, w-k) = (w+1)\zeta(w). \tag{5.15}$$

Proof. By the stuffle relation and Thm. 5.3.1 the product $\zeta(s)\zeta(r)$ can be expressed in two different ways:

$$\zeta(r,s) + \zeta(s,r) + \zeta(r+s) = \sum_{k=2}^{r+s-1} \left\{ \binom{k-1}{r-1} + \binom{k-1}{s-1} \right\} \zeta(k, r+s-k). \tag{5.16}$$

For any positive integer d, we define the generating function of the MZVs by

$$G_d(x_1, \ldots, x_d) = \sum_{s_1,\ldots,s_d \in \mathbb{N}, s_1>1} x_1^{s_1-1} \cdots x_d^{s_d-1} \zeta(s_1, \ldots, s_d). \tag{5.17}$$

Then by multiplying $x^{a-1}y^{b-1}$ on Eq. (5.16) and then summing over a and b we get

$$G_2(x+y,x)+G_2(x+y,y)=G_2(x,y)+G_2(y,x)+\frac{G_1(x)-G_1(y)}{x-y}. \quad (5.18)$$

Replacing (x,y) by (tx,ty) and considering the coefficients of t^{w-2} we get

$$\sum_{k=2}^{w-1}(x+y)^{k-1}(x^{w-k-1}+y^{w-k-1})\zeta(k,w-k)$$
$$=\sum_{k=2}^{w-1}(x^{k-1}y^{w-k-1}+y^{k-1}x^{w-k-1})\zeta(k,w-k)+\zeta(w)\sum_{i=0}^{w-2}x^iy^{w-2-i}.$$

Taking $x=y=1$ and using the sum relation Cor. 5.2.4 we arrive at Eq. (5.15). □

There are some other families of weighted sum relations implicitly related to the above. The results in the next theorem are called *restricted sum formulas.* But they can be regarded as weighted sum formulas with weight factors given by either 0 or 1.

Theorem 5.4.2. *For all $n\ge 2$, we have*

$$\sum_{k=1}^{n-1}\zeta(2k,2n-2k)=\frac{3}{4}\zeta(2n), \quad (5.19)$$

$$\sum_{k=1}^{n-1}\zeta(2k+1,2n-2k-1)=\frac{1}{4}\zeta(2n). \quad (5.20)$$

Proof. Clearly Eq. (5.20) follows from Eq. (5.19) and the sum relation Eq. (5.12) which says in our case

$$\sum_{k=2}^{2n-1}\zeta(k,2n-k)=\zeta(2n).$$

Now we prove Eq. (5.19) using the following well-known formula for Bernoulli numbers (see for e.g., [143, p. 119]):

$$(2n+1)B_{2n}=-\sum_{k=1}^{n-1}\binom{2n}{2k}B_{2k}B_{2n-2k}, \qquad \forall n\ge 2. \quad (5.21)$$

By Euler's famous evaluation Eq. (1.2) we get

$$\sum_{k=1}^{n-1}\zeta(2k)\zeta(2n-2k)=\frac{(2\pi i)^{2n}}{4(2n)!}\sum_{k=1}^{n-1}\binom{2n}{2k}B_{2k}B_{2n-2k}$$
$$=\frac{(2\pi i)^2n}{4(2n)!}(2n+1)B_{2n}=\frac{2n+1}{2}\zeta(2n).$$

Now adding the following stuffle relations

$$\zeta(2k)\zeta(2n-2k) = \zeta(2k, 2n-2k) + \zeta(2n-2k, 2k) + \zeta(2n) \tag{5.22}$$

for $k = 1$ up to $n-1$ we get

$$2\sum_{k=1}^{n-1} \zeta(2k, 2n-2k) + (n-1)\zeta(2n) = \frac{2n+1}{2}\zeta(2n),$$

which yields Eq. (5.19) immediately. This finishes the proof of the theorem. □

Theorem 5.4.3. *For all positive integer $n \geq 2$, we have*

$$\sum_{k=1}^{n-1}(4^k + 4^{n-k})\zeta(2k, 2n-2k) = \left(n + \frac{4}{3} + \frac{4^n}{6}\right)\zeta(2n). \tag{5.23}$$

For all positive integer $n \geq 2$, we have

$$\sum_{k=2}^{n-2}(2k-1)(2n-2k-1)\zeta(2k, 2n-2k) = \frac{3}{4}(n-3)\zeta(2n). \tag{5.24}$$

Proof. We have the key identity of Bernoulli polynomials (see [143, p. 119]):

$$\sum_{i=0}^{n}\binom{n}{i}B_i(x)B_{n-i}(y) = n(x+y-1)B_{n-1}(x+y) - (n-1)B_n(x+y) \tag{5.25}$$

which can be proved by observing that (setting $z = x + y$)

$$\frac{te^{xt}}{e^t-1}\frac{te^{yt}}{e^t-1} = \frac{t^2e^{zt}}{(e^t-1)^2} = \frac{(z-1)t^2e^{zt}}{e^t-1} - t^2\frac{\mathrm{d}}{\mathrm{d}t}\left(\frac{e^{zt}}{e^t-1}\right).$$

Taking $x = 1/2$ and $y = 0$ and using the fact that $B_n(1/2) = -(1 - 2^{1-n})B_n$, we get

$$\sum_{k=1}^{n-1}\binom{2n}{2k}(2^{1-2k}-1)B_{2k}B_{2n-2k} = 2n(1-2^{1-2n})B_{2n} - B_{2n}.$$

By Eq. (5.21) this reduces to

$$\sum_{k=1}^{n-1}\binom{2n}{2k}2^{1-2k}B_{2k}B_{2n-2k} = -(2+4^{1-n}n)B_{2n}.$$

Rewriting this using zeta values according to Eq. (1.2), we get

$$\sum_{k=1}^{n-1}4^{1-k}\zeta(2k)\zeta(2n-2k) = (2+4^{1-n}n)\zeta(2n).$$

By the stuffle relation of Eq. (5.22), this quickly leads to Eq. (5.23).

Let $f(t) = 1/(e^t - 1) = \sum_{j=0}^{\infty} B_j t^{j-1}/j!$. Then

$$f'(t) = \frac{-e^t}{(e^t-1)^2} = \sum_{j=0}^{\infty} (j-1)B_j \frac{t^{j-2}}{j!}$$

has only even powers of t. We have

$$\begin{aligned} f''(t) &= -f'(t) + \frac{2e^t}{(e^t-1)^3} = \sum_{j=0}^{\infty} (j-1)(j-2)B_j \frac{t^{j-3}}{j!}, \\ f'''(t) &= f'(t) - \frac{6e^{2t}}{(e^t-1)^4} = \sum_{j=0}^{\infty} (j-1)(j-2)(j-3)B_j \frac{t^{j-4}}{j!}. \end{aligned}$$

Therefore

$$t^4\big(f'(t)\big)^2 = \frac{t^4 e^{2t}}{(e^t-1)^4} = \frac{t^4}{6}\big(f'(t) - f'''(t)\big).$$

Thus

$$\left(\sum_{j=0}^{\infty} (j-1)B_j \frac{t^j}{j!}\right)^2 = \frac{1}{6}\sum_{j=0}^{\infty}\left\{(j-1)B_j\frac{t^{j+2}}{j!} - (j-1)(j-2)(j-3)B_j\frac{t^j}{j!}\right\}.$$

Comparing the coefficients of t^k for even $k \ge 4$, we get

$$\sum_{i+j=k} (i-1)(j-1)B_iB_j\binom{k}{i} = \frac{1}{6}\Big((k-3)(k-1)kB_{k-2} - (k-1)(k-2)(k-3)B_k\Big).$$

Magically, the B_{k-2}-terms cancel each other on the two sides since $B_2 = 1/6$. So we have

$$6\sum_{i,j\ge 4, i+j=k} (i-1)(j-1)B_iB_j\binom{k}{i} = -(k-1)(k^2-5k-6)B_k. \quad (5.26)$$

Setting $k = 2n$, multiplying both sides by $(2\pi i)^{2n}/4$ and using Eq. (1.2), we get

$$6\sum_{k=2}^{n-2} (2k-1)(2n-2k-1)\zeta(2k)\zeta(2n-2k) = (n-3)(4n^2-1)\zeta(2n). \quad (5.27)$$

By the stuffle relation Eq. (5.22) and the fact

$$\sum_{k=2}^{n-2} (2k-1)(2n-2k-1) = \frac{(n-3)(2n^2-5)}{3},$$

we finally arrive at Eq. (5.24) and conclude the proof of the theorem. □

Theorem 5.4.4. *For all positive integer r and $n \geq 2r$, we have*

$$\sum_{k=r+1}^{n-r-1} \frac{(2n-2r-2)!}{(2k-r-1)!(2n-2k-r-1)!} \frac{B_{2k}}{2k} \frac{B_{2n-2k}}{2n-2k}$$
$$= -\left(\frac{2(-1)^r r!(2n-2r-2)!}{(2n-r-1)!} + \frac{(r!)^2}{(2r+1)!} \right) \frac{B_{2n}}{2n}$$
$$- \sum_{k=\lceil r/2 \rceil+1}^{r} 2\left\{ (-1)^r \binom{r}{2k-r-1} + \binom{2n-2r-2}{2k-r-1} \right\} \frac{B_{2k}}{2k} \frac{B_{2n-2k}}{2n-2k}. \quad (5.28)$$

Proof. Define a zeta function

$$Z_r(s) = \sum_{\alpha,\beta=1}^{\infty} \alpha^r \beta^r (\alpha+\beta)^{-s} = \sum_{n=1}^{\infty} \sum_{\beta=1}^{n-1} \beta^r (n-\beta)^r n^{-s}$$

by substitution $n = \alpha + \beta$. Let $B_j(x)$ be the jth Bernoulli polynomial. Then we have

$$\begin{aligned} Z_r(s) &= \sum_{j=0}^{r} \binom{r}{j} \sum_{n=1}^{\infty} (-1)^j n^{r-j-s} \sum_{\beta=1}^{n-1} \beta^{r+j} \\ &= \sum_{j=0}^{r} \binom{r}{j} \sum_{n=1}^{\infty} (-1)^j n^{r-j-s} \frac{B_{r+j+1}(n) - B_{r+j+1}}{r+j+1} \\ &= \sum_{j=0}^{r} \binom{r}{j} \sum_{n=1}^{\infty} (-1)^j n^{r-j-s} \sum_{k=0}^{r+j} \binom{r+j+1}{k} \frac{B_k}{r+j+1} n^{r+j+1-k} \\ &= \sum_{j=0}^{r} \binom{r}{j} \sum_{k=0}^{r+j} \binom{r+j+1}{k} (-1)^j \frac{B_k}{r+j+1} \zeta(s+k-2r-1). \end{aligned}$$

Now change the order of summations and break it into three parts: (i) $\sum_{k=0}^{0} \sum_{j=0}^{r}$, (ii) $\sum_{k=1}^{r} \sum_{j=0}^{r}$, and (iii) $\sum_{j=1}^{r} \sum_{k=r+1}^{r+j} = \sum_{k=r+1}^{2r} \sum_{j=k-r}^{r}$.

(i) Define

$$f(x) = \sum_{j=0}^{r} \binom{r}{j} (-1)^j \frac{x^{r+j+1}}{r+j+1}.$$

Then

$$f'(x) = \sum_{j=0}^{r} \binom{r}{j} (-1)^j x^{r+j} = x^r (1-x)^r.$$

Thus using beta function $B(a,b) = \int_0^1 x^{a-1}(1-x)^{b-1}\, dx = \Gamma(\mathtt{a})\Gamma(\mathtt{b})/\Gamma(\mathtt{a}+\mathtt{b})$ we have

$$\sum_{j=0}^{r} \binom{r}{j} (-1)^j \frac{x^{r+j+1}}{r+j+1} = \int_0^1 x^r (1-x)^r\, dx = B(r+1, r+1) = \frac{(r!)^2}{(2r+1)!}.$$

(ii) We want to show this sum is 0. Since $\binom{r+j+1}{k}/(r+j+1)$ is a polynomial of j of degree $k-1$ for all $k \ge 1$ it suffices to prove that for all nonnegative integer $d < r$ we have $g(1)=0$ where

$$g(x)=\sum_{j=0}^{r} j^d \binom{r}{j}(-1)^j x^j = \left(x\frac{\mathrm{d}}{\mathrm{d}x}\right)^d \sum_{j=0}^{r}\binom{r}{j}(-1)^j x^j = \left(x\frac{\mathrm{d}}{\mathrm{d}x}\right)^d (1-x)^r.$$

Since $r > d$ by Leibnitz rule it is now obvious that $g(1)=0$.

(iii) For each fixed integer $k \in [r+1, 2r]$, we claim that

$$\sum_{j=k-r}^{r} (-1)^j \binom{r}{j}\binom{r+j}{k-1} = (-1)^r(1+(-1)^k)\binom{r}{k-r-1}. \tag{5.29}$$

To show this we set

$$h(x) := \sum_{k=r+1}^{2r}\sum_{j=k-r}^{r} (-1)^j \binom{r}{j}\binom{r+j}{k-1} x^k = \sum_{j=1}^{r}\sum_{k=r+1}^{r+j} (-1)^j \binom{r}{j}\binom{r+j}{k-1} x^k.$$

For a polynomial $f(x)$, we denote by $\mathrm{Trunc}_d f(x)$ the truncation of $f(x)$ by throwing away all the terms of degree no greater than d. Then we see that

$$\begin{aligned} h(x) &= \mathrm{Trunc}_r \sum_{j=1}^{r}(-1)^j\binom{r}{j}\sum_{k=1}^{r+j}\binom{r+j}{k-1}x^k \\ &= \mathrm{Trunc}_r \sum_{j=0}^{r}(-1)^j\binom{r}{j}\sum_{k=0}^{r+j-1}\binom{r+j}{k}x^{k+1} \\ &= \mathrm{Trunc}_r \sum_{j=0}^{r}(-1)^j\binom{r}{j}\left(x(x+1)^{r+j} - x^{r+j+1}\right) \\ &= \mathrm{Trunc}_r \left\{x(x+1)^r(-x)^r - x^{r+1}(1-x)^r\right\} \\ &= (-1)^r x^{r+1}\left\{(x+1)^r - (x-1)^r)\right\}. \end{aligned}$$

Hence Eq. (5.29) follows immediately. Putting (i), (ii) and (iii) together we get

$$\begin{aligned} Z_r(s) &= \frac{(r!)^2}{(2r+1)!}\zeta(s-2r-1) \\ &\quad + (-1)^r \sum_{k=r+1}^{2r} (1+(-1)^k)\binom{r}{k-r-1}\frac{B_k}{k}\zeta(s+k-2r-1). \end{aligned} \tag{5.30}$$

On the other hand Eie [190] showed that functions like $Z_r(s)$ have analytic continuation over the whole complex plane and further they are defined at negative integers (see [189, 190]). In fact the expressions in (i) to (iii) shows clearly that this can be done using the Riemann zeta functions. More importantly, Eie showed that these special values at negative integers can also be computed using some integrals over clearly specified simplices. In our situation his theory implies that

$$Z(2r+2-2n)=J^2(x^r y^r (x+y)^{2n-2r-2})+2J^1\left(\int_0^{-x} x^r y^r (x+y)^{2n-2r-2}\,\mathrm{d}y\right)$$

by symmetry between x and y, where for positive integers a_j $(j=1,\dots,m)$

$$J^m(x_1^{a_1}\dots x_m^{a_m})=\prod_{j=1}^m (-1)^{a_j}\frac{B_{a_j+1}}{a_j+1}. \tag{5.31}$$

For the integral, we may use substitution $y=-xt$ to get

$$\begin{aligned}\int_0^{-x} x^r y^r (x+y)^{2n-2r-2} &=(-1)^{r+1}x^{2n-1}\int_0^1 t^r(1-t)^{2n-2r-2}\,\mathrm{d}t\\ &=(-1)^{r+1}x^{2n-1}\frac{r!(2n-2r-2)!}{(2n-r-1)!}.\end{aligned}$$

Applying operator J, we get

$$\begin{aligned}Z(2r+2-2n)=&\sum_{a+b=2n-2r-2}\frac{(2n-2r-2)!}{a!b!}\frac{B_{a+r+1}}{a+r+1}\frac{B_{b+r+1}}{b+r+1}\\ &+\frac{2(-1)^r r!(2n-2r-2)!}{(2n-r-1)!}\frac{B_{2n}}{2n}.\end{aligned}$$

Comparing with evaluation of Eq. (5.30) at $s=2r+2-2n$ we can now finish the proof of the theorem by noticing finally that in Eq. (5.28) if r is odd and $k=(r+1)/2$ then the term happens to be zero. Thus we can improve the lower limit of k from $\lceil (r+1)/2\rceil$ to $\lceil r/2\rceil+1$. □

Remark 5.4.5. In [7] Agoh and Dilcher gave a formula for $(B_k+B_m)^n$ (with the classical umbral calculus notation) which leads to very similar results when $k=m$.

Example 5.4.6. Taking $r=3$ in Thm. 5.4.4, we get, for all $n\ge 2r$

$$\begin{aligned}&\sum_{k=4}^{n-4}\frac{(2n-8)!}{(2k-4)!(2n-2k-4)!}\frac{B_{2k}}{2k}\frac{B_{2n-2k}}{2n-2k}\\ &=-\frac{(n-6)(2n+1)(2n^2-11n+35)}{140(n-2)(n-3)(2n-5)(2n-7)}\frac{B_{2n}}{2n}-\frac{2n-11}{6}B_6B_{2n-6}.\end{aligned}$$

Theorem 5.4.7. *For all positive integer r and $n \geq 2r$, we have*

$$\sum_{k=r+1}^{n-r-1} \zeta(2k)\zeta(2n-2k) \prod_{\alpha+1}^{r} \Big\{(2k-\alpha)(2n-2k-\alpha)\Big\}$$
$$= \left((-1)^r r! + \frac{(r!)^2}{2(2r+1)!} \prod_{\beta=r+1}^{2r+1} (2n-\beta) \right) \zeta(2n) \prod_{\alpha=1}^{r} (2n-\alpha)$$
$$- \sum_{k=\lceil r/2 \rceil+1}^{r} \prod_{\alpha=1}^{r} (2k-\alpha) \prod_{\beta=2k+1}^{2r+1} (2n-\beta)$$
$$\times \left\{ \frac{2(-1)^r r!}{(2r-2k+1)!} + \frac{2(2n-2r-2)!}{(2n-2k-r-1)!} \right\} \zeta(2k)\zeta(2n-2k). \qquad (5.32)$$

Proof. This follows from Thm. 5.4.4 and Eq. (1.2) easily. □

Example 5.4.8. When $r = 1$ we recover the formula Eq. (5.27). When $r = 2$ we find for all $n \geq 4$

$$\sum_{j,k\geq 3, j+k=n} (2j-1)(2j-2)(2k-1)(2k-2)\zeta(2j)\zeta(2k)$$
$$= \frac{1}{15}(n-1)(4n^2-1)(2n^2-13n+30)\zeta(2n)$$
$$- 24(n-2)(2n-5)\zeta(4)\zeta(2n-4).$$

When $r = 3$ for all $n \geq 6$

$$\sum_{j,k\geq 4, j+k=n} (2j-1)(2j-2)(2j-3)(2k-1)(2k-2)(2k-3)\zeta(2j)\zeta(2k)$$
$$= \frac{1}{35}(n-6)(2n-3)(n-1)(4n^2-1)(2n^2-11n+35)\zeta(2n)$$
$$- 240(2n-11)(n-3)(2n-7)\zeta(6)\zeta(2n-6).$$

When $r = 4$ for all $n \geq 8$

$$\sum_{j,k\geq 5, j+k=n} \zeta(2j)\zeta(2k) \prod_{\alpha=1}^{4} \Big\{(2j-\alpha)(2k-\alpha)\Big\}$$
$$= \frac{8}{315}\binom{n-1}{2}(2n-3)(4n^2-1)(4n^4-72n^3+521n^2-1923n+3780)\zeta(2n)$$
$$- 960(n-3)(n-4)(2n-7)(2n-9)\zeta(6)\zeta(2n-6)$$
$$- 6720(n-4)(2n-9)(2n^2-25n+81)\zeta(8)\zeta(2n-8).$$

Corollary 5.4.9. *For all $n \geq 4$, we have*

$$\sum_{k=3}^{n-3} \zeta(2k, 2n-2k) \prod_{\alpha=1}^{2} \Big\{(2n-2k-\alpha)(2k-\alpha)\Big\}$$
$$= \frac{1}{2}(57n^2 - 279n + 366)\zeta(2n) - 12(n-2)(2n-5)\zeta(4)\zeta(2n-4). \quad (5.33)$$

For all $n \geq 6$,

$$\sum_{k=4}^{n-4} \zeta(2k, 2n-2k) \prod_{\alpha=1}^{3} \Big\{(2n-2k-\alpha)(2k-\alpha)\Big\}$$
$$= \frac{1}{2}(1005n^3 - 12222n^2 + 48243n - 62946)\zeta(2n)$$
$$- 120(2n-11)(n-3)(2n-7)\zeta(6)\zeta(2n-6). \quad (5.34)$$

For all $n \geq 8$,

$$\sum_{k=5}^{n-5} \zeta(2k, 2n-2k) \prod_{\alpha=1}^{4} \Big\{(2n-2k-\alpha)(2k-\alpha)\Big\}$$
$$= \frac{1}{2}(31116n^4 - 631800n^3 + 4846020n^2 - 16543800n + 21168864)\zeta(2n)$$
$$- 480(n-3)(n-4)(2n-7)(2n-9)\zeta(6)\zeta(2n-6) \quad (5.35)$$
$$- 3360(n-4)(2n-9)(2n^2 - 25n + 81)\zeta(8)\zeta(2n-8).$$

Corollary 5.4.10. *For all $n \geq 2$, we have*

$$\sum_{k=1}^{n-1} k(n-k)\zeta(2k, 2n-2k) = \frac{n}{16}\zeta(2n) + \frac{2n-3}{4}\zeta(2)\zeta(2n-2). \quad (5.36)$$

Proof. Let S be the left-hand side of Eq. (5.36). By expanding $(2k-1)(2n-2k-1)$ in Eq. (5.24) we see that

$$\sum_{k=1}^{n-1} (4k(n-k) - 2n + 1)\zeta(2k, 2n-2k)$$
$$= \frac{3}{4}(n-3)\zeta(2n) + (2n-3)(\zeta(2, 2n-2) + \zeta(2n-2, 2))$$
$$= \frac{3-5n}{4}\zeta(2n) + (2n-3)\zeta(2)\zeta(2n-2).$$

By Eq. (5.19) we have

$$4S = (2n-1)\sum_{k=1}^{n-1} \zeta(2k, 2n-2k) + \frac{3-5n}{4}\zeta(2n) + (2n-3)\zeta(2)\zeta(2n-2)$$
$$= \frac{n}{4}\zeta(2n) + (2n-3)\zeta(2)\zeta(2n-2)$$

which yields the corollary at once. □

Similarly we can replace the factor $k(n-k)$ in Cor. 5.4.10 by $k^r(n-k)^r$ for any positive integer r. For example, we have

Corollary 5.4.11. *We have for all* $n \geq 2$

$$\sum_{j+k=n,j,k>0} k^2 j^2 \zeta(2k,2j) = \frac{n}{64}\zeta(2n) + \frac{(2n-3)(6n-5)}{16}\zeta(2)\zeta(2n-2) - \frac{3(2n-5)}{4}\zeta(4)\zeta(2n-4),$$

$$\sum_{j+k=n,j,k>0} k^3 j^3 \zeta(2k,2j) = +\frac{45}{8}(2n-7)\zeta(6)\zeta(2n-6) - \frac{1}{256}n(2n^2-3)\zeta(2n) + \frac{1}{64}(2n-3)(28n^2-48n+21)\zeta(2)\zeta(2n-2) + \frac{3}{16}(2n-5)(2n^2-25n+35)\zeta(4)\zeta(2n-4),$$

$$\sum_{j+k=n,j,k>0} k^4 j^4 \zeta(2k,2j) = -\frac{315}{4}(2n-9))\zeta(8)\zeta(2n-8) - \frac{n(16n^2-17)}{1024}\zeta(2n) + \frac{1}{256}(2n-3)(6n-5)(20n^2-36n+17)\zeta(2)\zeta(2n-2) + \frac{3}{64}(2n-5)(40n^3-420n^2+1050n-777)\zeta(4)\zeta(2n-4) - \frac{15}{16}(2n-7)(8n^2-98n+189)\zeta(6)\zeta(2n-6).$$

Proof. Suppose $j+k=n$. Then we easily see that

$$\prod_{\alpha=1}^{2}\Big\{(2j-\alpha)(2k-\alpha)\Big\} = 16j^2k^2 + 4(5-6n)jk + 4(2n-1)(n-1).$$

The corollary follows from Eq. (5.19), Eq. (5.33), and Eq. (5.36) immediately. □

Equation (5.21) can be generalized as follows.

Theorem 5.4.12. *Let* $N \in \mathbb{N}$ *and define the* unsigned Stirling numbers of the first kind *by*

$$\sum_{k=1}^{N} c(N,k)x^k = x(x+1)\cdots(x+N-1). \tag{5.37}$$

Then for any positive integer $m \geq N$ we have

$$\sum_{\substack{j_1+\cdots+j_N=m\\ j_1,\dots,j_N\geq 0}} \binom{m}{j_1,\cdots,j_N} B_{j_1}\cdots B_{j_N} = (-1)^{N-1}N\binom{m}{N}\sum_{k=1}^{N} c(N,k)\frac{B_{m+k-N}}{m+k-N}.$$

Proof. For any function $\sum_{n\geq 0} a_n t^n$ of t, we define its associated zeta function by $\sum_{n\geq 0} a_n/n^s$. Then the associated zeta function of $F(t) := (1-t)^{-N}$ is

$$Z_F(s) = \sum_{n_1=0}^{\infty}\cdots\sum_{n_N=0}^{\infty}(n_1+\cdots+n_N)^{-s}.$$

By Eie and Lai's result [192, Prop. 5] we see that

$$Z_F(-m) = (-1)^N \sum_{\substack{j_1+\cdots+j_N=m+N\\ j_1,\dots,j_N\geq 0}} \binom{m}{j_1,\cdots,j_N} B_{j_1}\cdots B_{j_N}. \tag{5.38}$$

On the other hand, by the definition

$$F(t) = \frac{1}{(1-t)^N} = \sum_{n=0}^{\infty}\binom{-N}{n}(-t)^n = \sum_{n=0}^{\infty}\frac{t^n}{(N-1)!}\sum_{k=1}^{N}c(N,k)n^{k-1}.$$

The associated zeta function is

$$Z_F(s) = \sum_{n=0}^{\infty}\frac{1}{(N-1)!}\sum_{k=1}^{N}c(N,k)n^{k-1-s}.$$

Thus

$$Z_F(-m) = \sum_{k=1}^{N}\frac{c(N,k)}{(N-1)!}\frac{-B_{m+k}}{m+k}. \tag{5.39}$$

Changing the index m to $m-N$ we see that the theorem follows from Eqs. (5.38) and (5.39) immediately. $\square$

For example, putting $b_{a_1,\dots,a_k} = \prod_{j=1}^{k}\left(B_{2a_j}/(2a_j)!\right)$, we have

$$\sum_{\substack{a,b,c\geq 1,\\ a+b+c=2n}} (2n)!b_{a,b,c} = \binom{2n+2}{2}B_{2n} + n\left(n-\frac{1}{2}\right)B_{2n-2}, \tag{5.40}$$

$$\sum_{\substack{a,b,c,d\geq 1,\\ a+b+c+d=2n}} (2n)!b_{a,b,c,d} = -\binom{2n+3}{3}B_{2n} + \frac{4}{3}n^2(2n-1)B_{2n-2}. \tag{5.41}$$

These clearly yield identities between the Riemann zeta values at positive even integers. It also plays a role in the next general restricted sum relations.

Theorem 5.4.13. *Set $\zeta(0) = -1/2$. Then for all positive integer $d \le n$, we have*

$$\sum_{\substack{j_1,\dots,j_d>0\\ j_1+\dots+j_d=n}} \zeta(2j_1,\dots,2j_d)$$
$$= -\sum_{j=0}^{\lfloor\frac{d-1}{2}\rfloor} \frac{1}{2^{2d-3}(2j+1)B_{2j}}\binom{2d-2j-1}{d}\zeta(2j)\zeta(2n-2j).$$

Proof. See [295] for details. □

5.5 Derivation Relations

Let $\mathsf{a} = \mathsf{x}_0$, $\mathsf{b} = \mathsf{x}_1$, and $\mathfrak{A} = \mathbb{Q}\langle \mathsf{a}, \mathsf{b}\rangle$ be the noncommutative polynomial algebra as in Definition 3.2.1, regarded as a graded $\mathbb{Q}$-algebra with a and b both of degree one. For any word $\mathbf{w}$ of $\mathfrak{A}$, denote by $|\mathbf{w}|$ its total degree and by $\mathrm{dp}(\mathbf{w})$ the *depth* of $\mathbf{w}$ which is the number of occurrences of b in $\mathbf{w}$. Some authors use length here, but it has an established meaning in the combinatorics of words (which is called the *weight* in this book). We call $|\mathbf{w}| - \mathrm{dp}(\mathbf{w})$ the codepth of $\mathbf{w}$, which gives the number of occurrences of a in $\mathbf{w}$.

As in Definition 3.2.2 we let $\mathfrak{A}^1 = \mathbb{Q}1 \oplus \mathfrak{A}b$ and $\mathfrak{A}^0 = \mathbb{Q}1 \oplus \mathsf{a}\mathfrak{A}\mathsf{b}$. Denote by $\widehat{\mathfrak{A}}$, $\widehat{\mathfrak{A}}^1$ and $\widehat{\mathfrak{A}}^0$ the completions of the corresponding objects with respect to the weight. Let τ be the anti-automorphism of $\widehat{\mathfrak{A}}$ exchanging a and b. Applied to words, τ preserves weight and exchanges depth and codepth: note that $\widehat{\mathfrak{A}}^0$ (but not $\widehat{\mathfrak{A}}^1$) is closed under τ. It is easy to check that $\tau(\mathsf{z}_\mathbf{s}) = \mathsf{z}_{\mathbf{s}^*}$ where $\mathsf{z}_\mathbf{s} = \mathsf{z}_{s_1}\cdots\mathsf{z}_{s_d}$ for $\mathbf{s} = (s_1,\dots,s_d)$.

As usual, by a derivation of $\widehat{\mathfrak{A}}$ we mean a map $\delta : \widehat{\mathfrak{A}} \to \widehat{\mathfrak{A}}$ (of graded $\mathbb{Q}$-vector spaces) such that $\delta(\mathbf{uv}) = \delta(\mathbf{u})\mathbf{v} + \mathbf{u}\delta(\mathbf{v})$ for all $\mathbf{u}, \mathbf{v} \in \widehat{\mathfrak{A}}$. It is straightforward to see that the commutator of two derivations is a derivation, so the set of derivations of $\widehat{\mathfrak{A}}$ is a Lie algebra graded by the degree, denoted by $\mathrm{Der}(\widehat{\mathfrak{A}})$.

Lemma 5.5.1. *If δ is a derivation, then $\bar{\delta} := \tau\delta\tau$ is also a derivation (of the same degree).*

Proof. Let $\mathbf{s}, \mathbf{t} \in \mathbb{N}^d$, and $\mathbf{u} = \mathsf{z}_\mathbf{s}\mathsf{a}^i, \mathbf{v} = \mathsf{z}_\mathbf{t}\mathsf{a}^j \in \mathfrak{A}$. Then

$$\begin{aligned}\tau\delta\tau(\mathbf{uv}) &= \tau\delta\Big(\tau\big(\mathsf{z}_\mathbf{s}\mathsf{a}^i\mathsf{z}_\mathbf{t}\mathsf{a}^j\big)\Big)\\ &= \tau\delta\Big(\tau\big(\mathsf{a}^j\big)\tau\big(\mathsf{z}_\mathbf{t}\big)\tau\big(\mathsf{a}^i\big)\tau\big(\mathsf{z}_\mathbf{s}\big)\Big)\end{aligned}$$

$$\begin{aligned}
&= \tau\delta\Big(\mathtt{b}^j\mathtt{z}_{\mathtt{t}^*}\mathtt{b}^i\mathtt{z}_{\mathtt{s}^*}\Big) \\
&= \tau\Big(\delta\big(\mathtt{b}^j\mathtt{z}_{\mathtt{t}^*}\big)\mathtt{b}^i\mathtt{z}_{\mathtt{s}^*}\Big) + \tau\Big(\mathtt{b}^j\mathtt{z}_{\mathtt{t}^*}\delta\big(\mathtt{b}^i\mathtt{z}_{\mathtt{s}^*}\big)\Big) \\
&= \mathtt{z}_{\mathtt{s}}\mathtt{a}^i\tau\Big(\delta\tau\big(\mathtt{z}_{\mathtt{t}}\mathtt{a}^j\big)\Big) + \tau\Big(\delta\tau\big(\mathtt{z}_{\mathtt{s}}\mathtt{a}^i\big)\Big)\mathtt{z}_{\mathtt{t}}\mathtt{a}^j \\
&= \mathbf{u}\tau\delta\tau(\mathbf{v}) + \tau\delta\tau(\mathbf{u})\mathbf{v},
\end{aligned}$$

as desired. □

Lemma 5.5.2. *There is an isomorphism*

$$\exp : \mathrm{Der}(\widehat{\mathfrak{A}}) \longrightarrow \mathrm{Aut}(\widehat{\mathfrak{A}})$$
$$\delta \longmapsto \sum_{n=0}^{\infty} \frac{\delta^n}{n!},$$

whose inverse is given by

$$\log : \mathrm{Aut}(\widehat{\mathfrak{A}}) \longrightarrow \mathrm{Der}(\widehat{\mathfrak{A}})$$
$$\sigma \longmapsto -\sum_{n=1}^{\infty} \frac{(1-\sigma)^n}{n}.$$

Proof. The proof is straightforward. □

To state the next theorem we now define $\mathfrak{d}$ to be the derivation on $\widehat{\mathfrak{A}}$ such that

$$\mathfrak{d}(\mathbf{w}) = \mathtt{b} \sqcup\!\sqcup\, \mathbf{w} - \mathtt{b}\mathbf{w} \quad \forall \mathbf{w} \in \widehat{\mathfrak{A}}. \tag{5.42}$$

It can be verified that (see Exercise 5.12) for all $n \geq 1$ and $\mathbf{w} \in \widehat{\mathfrak{A}}$

$$\mathfrak{d}(\mathtt{a}) = \mathtt{ab},\ \mathfrak{d}(\mathtt{b}) = \mathtt{b}^2,\ \text{and}\ \frac{1}{n!}\mathfrak{d}^n(\mathbf{w}) = \mathtt{b}^n \sqcup\!\sqcup\, \mathbf{w} - \mathtt{b}(\mathtt{b}^{n-1} \sqcup\!\sqcup\, \mathbf{w}). \tag{5.43}$$

Let $\mathfrak{z}$ be the (possibly infinite) $\mathbb{Q}$-linear span of $\mathtt{z}_k$ for $k \geq 1$. For all $z \in \mathfrak{z}$, we define the derivation δ_z on $\widehat{\mathfrak{A}}$ (see Exercise 5.13) such that

$$\delta_z(\mathtt{a}) = 0 \quad \text{and} \quad \delta_z(\mathtt{b}) = (\mathtt{a}+\mathtt{b})z, \tag{5.44}$$

and a map $\chi_z : \widehat{\mathfrak{A}}^1 \to \widehat{\mathfrak{A}}^1$ by

$$\chi_z(\mathbf{w}) = (1-z)\Big((1-z)^{-1} * \mathbf{w}\Big) \quad \forall \mathbf{w} \in \widehat{\mathfrak{A}}^1. \tag{5.45}$$

Lemma 5.5.3. (i) *The vector space $\mathfrak{z}$ becomes a commutative and associative algebra with respect to the multiplication $\circ$ defined by*

$$z \circ z' = z * z' - zz' - z'z, \quad \forall z, z' \in \mathfrak{z}.$$

(ii) *The* $\mathbb{Q}$*-linear map*

$$\gamma : X\mathbb{Q}[\![X]\!] \longrightarrow \mathfrak{z}, \quad \gamma(X^k) = \mathsf{z}_k \quad \forall k \geq 1, \tag{5.46}$$

is an algebra isomorphism for $\circ$*. Moreover, for every* $f \in X\mathbb{Q}[\![X]\!]$ *we have*

$$\delta_{\gamma(f)}(\mathsf{a}) = 0, \quad \delta_{\gamma(f)}(\mathsf{b}) = (\mathsf{a}+\mathsf{b})\frac{f(\mathsf{a})}{\mathsf{a}}\mathsf{b}. \tag{5.47}$$

Proof. (i) By the definition of $*$ we see that

$$\mathsf{z}_k \circ \mathsf{z}_l = \mathsf{z}_k * \mathsf{z}_l - \mathsf{z}_k\mathsf{z}_l - \mathsf{z}_l\mathsf{z}_k = \mathsf{z}_{k+l}.$$

Thus (i) follows immediately.

(ii) The first sentence is straightforward (see Exercise 5.14). To prove Eq. (5.47) we only need to assume $f(X) = X^k$ for some $k \geq 1$. Then $\gamma(X^k) = \mathsf{z}_k$ and by the definition in Eq. (5.44)

$$\delta_{\mathsf{z}_k}(\mathsf{a}) = 0, \quad \delta_{\mathsf{z}_k}(\mathsf{b}) = (\mathsf{a}+\mathsf{b})\mathsf{z}_k = (\mathsf{a}+\mathsf{b})\mathsf{a}^{k-1}\mathsf{b} = (\mathsf{a}+\mathsf{b})\frac{f(\mathsf{a})}{\mathsf{a}}\mathsf{b}.$$

This completes the proof of the lemma. □

Proposition 5.5.4. *Let* $z \in \mathfrak{z}$*. Then we have*

(i) $\chi_z \in \mathrm{Aut}(\widehat{\mathfrak{A}}^1)$ *and*

$$\chi_z(\mathbf{w}) = \exp(\delta_t)(\mathbf{w}), \ t = \log_\circ(1+z) = -\sum_{n=1}^{\infty}(-1)^n\frac{z^{\circ n}}{n} \in \mathfrak{z}. \tag{5.48}$$

(ii) χ_z *can be extended to* $\mathrm{Aut}(\widehat{\mathfrak{A}})$ *with* $\chi_z(\mathsf{a}) = \mathsf{a}$ *and* $\chi_z(\mathsf{a}+\mathsf{b}) = (\mathsf{a}+\mathsf{b})(1-z)^{-1}$*. Further,* χ_z *induces an automorphism of* $\widehat{\mathfrak{A}}^0$*.*

Proof. (i) For any $t \in \mathfrak{z}$, we have (see Exercise 5.10)

$$\exp_*(t) = (2 - \exp_\circ(t))^{-1}. \tag{5.49}$$

Taking $t = \log_\circ(1+z)$ we thus get $\exp_*(t) = (1-z)^{-1}$. Also, for $t \in \mathfrak{z}$ and $\mathbf{w} \in \widehat{\mathfrak{A}}^1$ we have (see Exercise 5.11)

$$\exp(\delta_t)(\mathbf{w}) = (\exp_*(t))^{-1}(\exp_*(t) * \mathbf{w}) = (1-z)((1-z)^{-1} * \mathbf{w}) = \chi_z(\mathbf{w}). \tag{5.50}$$

(ii) By the definition $\delta_t(\mathsf{a}) = 0$ and $\delta_t(\mathsf{b}) = (\mathsf{a}+\mathsf{b})t$. So

$$\delta_t\big((\mathsf{a}+\mathsf{b})\mathbf{w}\big) = (\mathsf{a}+\mathsf{b})\big(t\mathbf{w} + \delta_t(\mathbf{w})\big).$$

By induction it is straightforward to show that $\delta_t^n(\mathsf{b}) = (\mathsf{a}+\mathsf{b})t^{*n}$ for all $n \in \mathbb{N}$. Hence by Eq. (5.48) $\chi_z(\mathsf{a}) = \exp(\delta_t)(\mathsf{a}) = \mathsf{a}$ and $\chi_z(\mathsf{a}+\mathsf{b}) = \exp(\delta_t)(\mathsf{a}+\mathsf{b}) = (\mathsf{a}+\mathsf{b})\exp_*(t) = (\mathsf{a}+\mathsf{b})(1-z)^{-1}$ by Eq. (5.49). To show

χ_z induces an automorphism of $\widehat{\mathfrak{A}}^0$, it suffices to show that $\chi_z(\mathbf{w}) \in \widehat{\mathfrak{A}}^0$ for all $\mathbf{w} \in \widehat{\mathfrak{A}}^0$ since clearly $\exp(-\delta_t)$ is the inverse. Now suppose $\mathbf{w} = \mathtt{a}\mathbf{u}\mathtt{b}$. Then $\chi_z(\mathtt{a}\mathbf{u}\mathtt{b}) = \mathtt{a}\chi_z(\mathbf{u})\chi_z(\mathtt{b})$. But $\chi_z(\mathtt{b}) = \mathtt{b} + \sum_{n=1}^{\infty}(\mathtt{a}+\mathtt{b})z^n$ so that each word ends with $\mathtt{b}$. Hence $\chi_z(\mathbf{w}) \in \widehat{\mathfrak{A}}^0$.

This completes the proof of the proposition. □

For $n \geq 1$ let ∂_n be the derivation with

$$\partial_n(\mathtt{a}) = \mathtt{a}(\mathtt{a}+\mathtt{b})^{n-1}\mathtt{b}, \quad \text{and} \quad \partial_n(\mathtt{b}) = -\mathtt{a}(\mathtt{a}+\mathtt{b})^{n-1}\mathtt{b}. \tag{5.51}$$

Theorem 5.5.5. (Drivation Relations). *For all $n \in \mathbb{N}$ and words $\mathbf{w} \in \mathfrak{A}^0$, $\zeta(\partial_n(\mathbf{w})) = 0$.*

Proof. By Thm. 3.3.21 for all positive integer n one has

$$\zeta_{ш}(\mathtt{b}^n) = \rho \circ \zeta_*(\mathtt{b}^n).$$

Multiplying $\zeta(\mathbf{w})$ on both sides and using the fact that both $\zeta_{ш}$ and ζ_* are homomorphisms extending ζ, we get

$$\zeta_{ш}(\mathtt{b}^n ш \mathbf{w}) = \rho \circ \zeta_*(\mathtt{b}^n * \mathbf{w}) = \zeta_{ш}(\mathtt{b}^n * \mathbf{w}),$$

where the second equality follows from Thm. 3.3.21 again. Recall that we have the isomorphism Eq. (3.14): $\mathfrak{A}^1_{ш} \cong \mathfrak{A}^0_{ш}[T]$. Let's denote by $\mathrm{reg}_{ш}$ the composition of the regularization map $\mathfrak{A}^1 \cong \mathfrak{A}^0_{ш}[T] \to \mathfrak{A}^0$ followed by the specialization $T = 0$. We see that

$$\mathrm{reg}_{ш}(\mathtt{b}^n ш \mathbf{w}) = \frac{1}{n!}\mathrm{reg}_{ш}(\mathtt{b}^{ш n} ш \mathbf{w}) = \frac{1}{n!}\mathrm{reg}_{ш}(\mathtt{b})^n\mathbf{w} = 0.$$

Hence $\zeta \circ \mathrm{reg}_{ш}(\mathtt{b}^n * \mathbf{w}) = 0$ for all positive integer n. So

$$\zeta\Big(\sum_{n=1}^{\infty} \mathrm{reg}_{ш}(\mathtt{b}^n * \mathbf{w})u^n\Big) = 0, \tag{5.52}$$

where u is a formal variable. Define $\Delta_u \in \mathrm{Aut}(\widehat{\mathfrak{A}})$ by

$$\Delta_u = \exp(-\mathfrak{d}u) \circ \chi_{\mathtt{b}u}. \tag{5.53}$$

Now multiplying Eq. (5.43) by u^n and summing over n yields (after replacing $\mathbf{w}$ by $\mathbf{v}$ there)

$$\exp(\mathfrak{d}u)(\mathbf{v}) = (1-\mathtt{b}u)\Big((1-\mathtt{b}u)^{-1} ш \mathbf{v}\Big), \quad \forall \mathbf{v} \in \widehat{\mathfrak{A}}. \tag{5.54}$$

Substituting $\Delta_u(\mathbf{w})$ for $\mathbf{v}$ we see that

$$(1-\mathtt{b}u)\Big((1-\mathtt{b}u)^{-1} ш \Delta_u(\mathbf{w})\Big) = \chi_{\mathtt{b}u}(\mathbf{w}) = (1-\mathtt{b}u)\Big((1-\mathtt{b}u)^{-1} * \mathbf{w}\Big).$$

Dividing by $1-\mathtt{b}u$ and then applying $\mathrm{reg}_{\sqcup\!\sqcup}$ we get

$$\mathrm{reg}_{\sqcup\!\sqcup}\big(\Delta_u(\mathbf{w})\big)=\sum_{n=0}^{\infty}\mathrm{reg}_{\sqcup\!\sqcup}(\mathtt{b}^n*\mathbf{w})u^n.$$

Here we use the fact that $\mathrm{reg}_{\sqcup\!\sqcup}\big((1-\mathtt{b}u)^{-1})\big)=1$. Since $\mathbf{w}\in\widehat{\mathfrak{A}}^0$ and both $\exp(\mathfrak{d}u)$ and $\chi_{\mathtt{b}u}$ induce automorphisms on $\widehat{\mathfrak{A}}^0$ (see Excercise 5.12 and Prop. 5.5.4(ii)) we see that $\Delta_u(\mathbf{w})\in\widehat{\mathfrak{A}}^0$. Hence $\mathrm{reg}_{\sqcup\!\sqcup}\big(\Delta_u(\mathbf{w})\big)=\Delta_u(\mathbf{w})$ and Eq. (5.52) gives $\zeta(\Delta_u(\mathbf{w}))=\zeta(\mathbf{w})$ for all $\mathbf{w}\in\widehat{\mathfrak{A}}^0$. The theorem now follows from the following lemma. □

Lemma 5.5.6. *As an automorphism of* $\widehat{\mathfrak{A}}$

$$\Delta_u=\exp\left(\sum_{n=1}^{\infty}(-1)^n\frac{\partial_n}{n}u^n\right). \tag{5.55}$$

Proof. We only need to check their images on the generators $\mathtt{a}$ and $\mathtt{b}$ are the same. First, by Prop. 5.5.4(ii)

$$\begin{aligned}\Delta_u(\mathtt{a})&=\exp(-\mathfrak{d}u)\circ\chi_{\mathtt{b}u}(\mathtt{a})=\exp(-\mathfrak{d}u)(\mathtt{a})\\&=\sum_{n=0}^{n}(-1)^n\frac{\mathfrak{d}^n(\mathtt{a})}{n!}u^n=\sum_{n=0}^{n}(-1)^n\mathtt{a}\mathtt{b}^nu^n=\mathtt{a}(1+\mathtt{b}u)^{-1}.\end{aligned}\tag{5.56}$$

Now by Eq. (5.43) we have $\exp(\mathfrak{d}u)(\mathtt{a}+\mathtt{b})=(\mathtt{a}+\mathtt{b})(1-\mathtt{b}u)^{-1}$ (see Exercise 5.13) so that

$$\begin{aligned}\Delta_u(\mathtt{a}+\mathtt{b})&=\exp(-\mathfrak{d}u)\circ\chi_{\mathtt{b}u}(\mathtt{a}+\mathtt{b})\\&=\exp(-\mathfrak{d}u)\big((\mathtt{a}+\mathtt{b})(1-\mathtt{b}u)^{-1}\big)=\mathtt{a}+\mathtt{b}.\end{aligned}\tag{5.57}$$

Comparing to Prop. 5.5.4(ii) we see that

$$\Delta_u=\varepsilon\chi_{\mathtt{b}u}\varepsilon \tag{5.58}$$

where ε is the involution on $\widehat{\mathfrak{A}}$ such that $\varepsilon(\mathtt{a})=\mathtt{a}+\mathtt{b},\varepsilon(\mathtt{b})=-\mathtt{b}$. Therefore, Prop. 5.5.4(i) yields

$$\Delta_u=\varepsilon\exp\big(\delta_{\log_\circ(1+\mathtt{b}u)}\big)\varepsilon.$$

Hence

$$\begin{aligned}\Delta_u(\mathtt{a})=&\varepsilon\exp\big(\delta_{\log_\circ(1+\mathtt{b}u)}\big)(\mathtt{a}+\mathtt{b})\\=&\varepsilon\exp\big(\delta_{\gamma(\log(1+uX))}\big)(\mathtt{a}+\mathtt{b}) && \text{(by Lemma 5.5.3(ii))}\\=&\varepsilon\exp\left((\mathtt{a}+\mathtt{b})\frac{\log(1+\mathtt{a}u)}{\mathtt{a}}\mathtt{b}\right) && \text{(by Eq. (5.47))}\end{aligned}$$

$$= \exp\left(-a\frac{\log(1+(\mathsf{a}+\mathsf{b})u)}{\mathsf{a}+\mathsf{b}}\mathsf{b}\right)$$

$$= \exp\left(\sum_{n=1}^{\infty}(-1)^n\frac{\mathsf{a}(\mathsf{a}+\mathsf{b})^{n-1}\mathsf{b}}{n}u^n\right)$$

$$= \exp\left(\sum_{n=1}^{\infty}(-1)^n\frac{\partial_n}{n}u^n\right)(\mathsf{a}) \qquad \text{(by Eq. (5.51))},$$

and

$$\Delta_u(\mathsf{a}+\mathsf{b}) = \varepsilon\exp\left(\delta_{\log_\circ(1+\mathsf{b}u)}\right)(\mathsf{a})$$

$$= \varepsilon(\mathsf{a}) = \mathsf{a}+\mathsf{b} = \exp\left(\sum_{n=1}^{\infty}(-1)^n\frac{\partial_n}{n}u^n\right)(\mathsf{a}+\mathsf{b}),$$

since $\partial_n(\mathsf{a}+\mathsf{b}) = 0$ for all $n \geq 1$. The lemma is now proved. □

Corollary 5.5.7. *For all $m \geq 0$ and $\mathbf{w} = \mathsf{a}\mathbf{w}_0 \in \mathfrak{A}^0$, the regularization map satisfies*

$$\mathrm{reg}_{ш}(\mathsf{b}^m\mathbf{w}) = \frac{(-1)^m}{m!}\mathfrak{d}^m(\mathbf{w}) = (-1)^m\mathsf{a}(\mathsf{b}^m ш \mathbf{w}_0). \tag{5.59}$$

Proof. First note that $(1-\mathsf{b}u)^{-1} = \exp_{ш}(\mathsf{b}u)$. Taking $\mathbf{v} = \exp(-\mathfrak{d}u)(\mathbf{w})$ in Eq. (5.54) and multiplying $(1-\mathsf{b}u)^{-1}$ from the left, we have

$$(1-\mathsf{b}u)^{-1}\mathbf{w} = (1-\mathsf{b}u)^{-1} ш \exp(-\mathfrak{d}u)(\mathbf{w}).$$

Applying $\mathrm{reg}_{ш}$, we get

$$\mathrm{reg}_{ш}\left((1-\mathsf{b}u)^{-1}\mathbf{w}\right) = \mathrm{reg}_{ш}\left((1-\mathsf{b}u)^{-1}\right)\mathrm{reg}_{ш}\left(\exp(-\mathfrak{d}u)(\mathbf{w})\right)$$

$$= \mathrm{reg}_{ш}\left(\exp(-\mathfrak{d}u)(\mathbf{w})\right).$$

By comparing the coefficients of u^m on both sides, we see easily that the first equality in Eq. (5.59) holds.

Now observing that $1/m!\mathfrak{d}^m(\mathsf{a}) = \mathsf{a}\mathsf{b}^m$ and $1/m!\mathfrak{d}^m(\mathsf{b}) = \mathsf{b}^{m+1}$ (see Exercise 5.12), we get

$$\exp(\mathfrak{d}u)(\mathsf{a}) = \mathsf{a}(1-\mathsf{b}u)^{-1}, \quad \exp(\mathfrak{d}u)(\mathsf{b}) = \mathsf{b}(1-\mathsf{b}u)^{-1}. \tag{5.60}$$

Taking $\mathbf{v} = \mathsf{a}\mathbf{w}_0$ in Eq. (5.54) and using the first identity of Eq. (5.60) (with $u \to -u$), we get

$$\exp(-\mathfrak{d}u)(\mathbf{w}) = \exp(-\mathfrak{d}u)(x)\exp(-\mathfrak{d}u)(\mathbf{w}_0) = \mathsf{a}\Big((1+\mathsf{b}u)^{-1} ш \mathbf{w}_0\Big).$$

The second equality in Eq. (5.59) follows by a comparison of the coefficients of u^m on both sides. We now have completed the proof of the corollary. □

Set $D_n(\mathtt{a}) = 0$ and $D_n(\mathtt{b}) = \mathtt{a}^n\mathtt{b}$ for all $n \geq 1$. For each $n \geq 1$, define the linear maps σ_n and $\overline{\sigma_n} : \widehat{\mathfrak{A}} \to \widehat{\mathfrak{A}}$ as the homogeneous components of degree n of the homomorphisms:

$$\sigma = \exp\left(\sum_{n\geq 1} \frac{D_n}{n}\right) = \sum_{n=1}^{\infty} \sigma_n \quad \text{and} \quad \overline{\sigma} = \tau\sigma\tau = \exp\left(\sum_{n\geq 1} \frac{\overline{D_n}}{n}\right) = \sum_{n=1}^{\infty} \overline{\sigma_n},$$

respectively.

Lemma 5.5.8. *Set* $\partial = \sum_{n=1}^{\infty} \frac{\partial_n}{n}$. *Then*

$$\exp(\partial) = \overline{\sigma}\sigma^{-1}. \tag{5.61}$$

Proof. Let's check the two automorphisms' images on the generators $\mathtt{a}$ and $\mathtt{a}+\mathtt{b}$ are the same. First observe that $\exp(\partial) = \Delta_{-1}$. Thus by Eqs. (5.56) and (5.57)

$$\exp(\partial)(\mathtt{a}) = \Delta_{-1}(\mathtt{a}) = \mathtt{a}(1-\mathtt{b})^{-1} \quad \text{and} \quad \exp(\partial)(\mathtt{a}+\mathtt{b}) = \mathtt{a}+\mathtt{b}. \tag{5.62}$$

Now we set

$$D = \sum_{n\geq 1} \frac{D_n}{n}.$$

Then $D(\mathtt{a}) = \sum_{n\geq 1} D_n(\mathtt{a})/n = 0$ and

$$D(\mathtt{b}) = \sum_{n\geq 1} \frac{D_n(\mathtt{b})}{n} = \sum_{n\geq 1} \frac{\mathtt{a}^n\mathtt{b}}{n} = (-\log(1-\mathtt{a}))\mathtt{b}.$$

Hence $D^n(\mathtt{a}) = 0$ and $D^n(\mathtt{b}) = (-\log(1-\mathtt{a}))^n\mathtt{b}$ and therefore

$$\sigma(\mathtt{a}) = \exp(D)(\mathtt{a}) = \sum_{n\geq 0} \frac{D^n(\mathtt{a})}{n!} = \mathtt{a}, \tag{5.63}$$

so $\sigma^{-1}(\mathtt{a}) = \mathtt{a}$. Further,

$$\begin{aligned} \sigma(\mathtt{b}) = \exp(D)(\mathtt{b}) &= \sum_{n\geq 0} \frac{D^n(\mathtt{a})}{n!} \\ &= \sum_{n\geq 0} \frac{(-\log(1-\mathtt{a}))^n}{n!}\mathtt{b} = (1-\mathtt{a})^{-1}\mathtt{b} = \exp(\partial)(\mathtt{a}). \end{aligned} \tag{5.64}$$

Similarly $\sigma^{-1}(\mathtt{b}) = (1-\mathtt{a})\mathtt{b}$. Thus

$$\overline{\sigma}\sigma^{-1}(\mathtt{a}) = \overline{\sigma}(\mathtt{a}) = \tau(\sigma(\mathtt{b})) = \mathtt{a}(1-\mathtt{b})^{-1}$$

and

$$\begin{aligned} \overline{\sigma}\sigma^{-1}(\mathtt{b}) &= \overline{\sigma}((1-\mathtt{a})\mathtt{b}) = \tau\sigma(\mathtt{a}(1-\mathtt{b})) \\ &= \tau\big(\mathtt{a}(1-(1-\mathtt{a})^{-1}\mathtt{b})\big) = (1-\mathtt{a}(1-\mathtt{b})^{-1})\mathtt{b}. \end{aligned}$$

Adding the above two equations, we get $\overline{\sigma}\sigma^{-1}(\mathtt{a}+\mathtt{b}) = \mathtt{a}+\mathtt{b}$. By comparing with Eq. (5.62) this proves the lemma. □

We now prove the next theorem using derivations.

Theorem 5.5.9. (Ohno's relations). *Suppose* $\mathbf{s} = (s_1, \dots, s_d)$ *with* $s_1 > 1$ *and its dual* $\mathbf{s}^*$ *has depth* d'. *Then for any integer* $n \geq 0$

$$\sum_{\mathbf{e}\in(\mathbb{N}_0)^d, |\mathbf{e}|=n} \zeta(\mathbf{s}+\mathbf{e}) = \sum_{\mathbf{e}'\in(\mathbb{N}_0)^{d'}, |\mathbf{e}'|=n} \zeta(\mathbf{s}^*+\mathbf{e}'). \tag{5.65}$$

Proof. By Eqs. (5.63) and (5.64) we see that

$$\begin{aligned}\sigma(\mathsf{z}_{s_1}\cdots\mathsf{z}_{s_d}) &= \mathsf{a}^{s_1-1}(1-\mathsf{a})^{-1}\mathsf{b}\mathsf{a}^{s_2-1}(1-\mathsf{a})^{-1}\mathsf{b}\cdots\mathsf{a}^{s_d-1}(1-\mathsf{a})^{-1}\mathsf{b}\\ &= \sum_{n=0}^{\infty}\sum_{\substack{e_1+\cdots+e_d=n\\ e_1,\dots,e_d\in\mathbb{N}_0}} \mathsf{z}_{s_1+e_1}\cdots\mathsf{z}_{s_d+e_d}.\end{aligned}$$

Thus

$$\sigma_n(\mathsf{z}_{s_1}\cdots\mathsf{z}_{s_d}) = \sum_{\substack{e_1+\cdots+e_d=n\\ e_1,\dots,e_d\in\mathbb{N}_0}} \mathsf{z}_{s_1+e_1}\cdots\mathsf{z}_{s_d+e_d}.$$

Hence, in terms of derivation, Ohno's relation says that

$$\zeta\big((\sigma_n - \sigma_n\tau)\mathbf{w}\big) = 0 \quad \forall \mathbf{w} \in \mathfrak{A}^0, n \geq 1.$$

By the duality of Eq. (5.4) we only need to show that

$$\zeta\big((\sigma_n - \overline{\sigma_n})\mathbf{w}\big) = 0 \quad \forall \mathbf{w} \in \mathfrak{A}^0, n \geq 1. \tag{5.66}$$

But by Lemma 5.5.8 we get $\sigma - \overline{\sigma} = (1 - \exp(\partial))\sigma$. Hence Eq. (5.66) follows from Thm. 5.5.5 componentwise. The theorem is now proved. □

5.6 Fixed Weight, Height and Depth Relations

Another important technique in the study of the MZVs is the use of generating functions, which frequently leads to the generalized hypergeometric functions (see Appendix C for a brief overview of these functions and [517] for a comprehensive treatment). For example, we will use Lemma C.1 to prove the main theorem of this section.

First we need to introduce some new notation.

Definition 5.6.1. Let $\mathbf{s} = (s_1, \dots, s_d) \in \mathbb{N}^d$. We define the *height* of $\mathbf{s}$ to be the number of components of $\mathbf{s}$ which are greater than 1, denoted by $\mathrm{ht}(\mathbf{s})$. For nonnegative integers w, d and h, set

$$\begin{aligned}I(w,d,h) &= \big\{\mathbf{s} = (s_1,\dots,s_d) \in \mathbb{N}^d : |\mathbf{s}| = w, \mathrm{ht}(\mathbf{s}) = h\big\},\\ I_0(w,d,h) &= \big\{\mathbf{s} = (s_1,\dots,s_d) \in \mathbb{N}^d : |\mathbf{s}| = w, \mathrm{ht}(\mathbf{s}) = h, s_1 \geq 2\big\}.\end{aligned}$$

Theorem 5.6.2. *For any nonnegative integers w, d and h, let*

$$G_0(w,d,h) = \sum_{\mathbf{s}\in I_0(w,d,h)} \zeta(\mathbf{s}).$$

Then

$$\sum_{w,d,h\geq 1} G_0(w,d,h)x^{w-d-h}y^{d-h}z^{h-1} = \frac{1-\exp\left(\sum_{n=2}^{\infty}\frac{\zeta(n)}{n}S_n(x,y,z)\right)}{xy-z},$$

where

$$S_n(x,y,z) = x^n + y^n - \alpha^n - \beta^n, \quad \alpha,\beta = \frac{x+y\pm\sqrt{(x+y)^2-4z}}{2}.$$

Proof. The basic idea is to treat $G_0(w,d,h)$ as a special point in a one parameter family $G_0(w,d,h;t)$ and then deal with the generating function of the whole family. The advantage is that we can now take derivatives and solve the resulting differential equations.

Recall that we have defined the one variable multiple polylog $\mathrm{L}_{\mathbf{s}}(t)$ by Eq. (3.20). Further, we have set $\mathrm{L}_{\mathbf{s}}(t) = 0$ if $\mathbf{s}$ is empty. Now we define

$$G(w,d,h;t) = \sum_{\mathbf{s}\in I(w,d,h)} \mathrm{L}_{\mathbf{s}}(t), \quad G_0(w,d,h;t) = \sum_{\mathbf{s}\in I_0(w,d,h)} \mathrm{L}_{\mathbf{s}}(t).$$

By Prop. 4.2.1 we get

$$t\frac{\mathrm{d}}{\mathrm{d}t}G_0(w,d,h;t) = G(w-1,d,h-1;t) - G_0(w-1,d,h-1;t) + G_0(w-1,d,h;t), \tag{5.67}$$

$$\frac{\mathrm{d}}{\mathrm{d}t}\Big(G(w,d,h;t) - G_0(w,d,h;t)\Big) = \frac{1}{1-t}G_0(w-1,d-1,h;t). \tag{5.68}$$

Consider the generating functions

$$\Psi(x,y,z;t) = 1 + \sum_{w,d,h\geq 0} G(w,d,h;t)x^{w-d-h}y^{d-h}z^{h},$$

$$\Psi_0(x,y,z;t) = \sum_{w,d,h\geq 0} G_0(w,d,h;t)x^{w-d-h}y^{d-h}z^{h-1}.$$

Then Eqs. (5.67) and (5.68) yield

$$t\frac{\mathrm{d}\Psi}{\mathrm{d}t} = \frac{\Psi-1-z\Psi_0}{y} + x\Psi_0, \quad (1-t)\frac{\mathrm{d}}{\mathrm{d}t}(\Psi - z\Psi_0) = y\Psi.$$

Thus by elimination of Ψ we get

$$t(1-t)\frac{\mathrm{d}^2\Psi_0}{\mathrm{d}t^2} + \Big(1-x-(1-x+y)t\Big)\frac{\mathrm{d}\Psi_0}{\mathrm{d}t} + (xy-z)\Psi_0 = 1.$$

Clearly $\Psi_0(0) = 0$. To compute $\Psi_0'(0)$, first note that $\lim_{t\to 0} \mathrm{L}_{\mathbf{k}}(t)/t = 0$ unless $\mathrm{dp}(\mathbf{k}) = \mathrm{ht}(\mathbf{k}) = 1$ in which case $\lim_{t\to 0} \mathrm{L}_{\mathbf{k}}(t)/t = 1$. Thus $\Psi_0'(0) = \sum_{w=2}^{\infty} x^{w-2} = 1/(1-x)$. Setting $\alpha + \beta = x + y$ and $\alpha\beta = z$, we see that

$$\Psi_0(x, y, z; t) = \frac{1}{xy - z}\Big(1 - {}_2F_1(\alpha - x, \beta - x; 1 - x; t)\Big),$$

since they are both solutions of Eq. (C.3) satisfying the same initial conditions $y(0) = 0, y'(0) = 1/(1-x)$. Thus by Gauss summation theorem Eq. (C.2)

$$\Psi_0(x, y, z; 1) = \frac{1}{xy - z}\left(1 - \frac{\Gamma(1-x)\Gamma(1-y)}{\Gamma(1-\alpha)\Gamma(1-\beta)}\right).$$

Finally, the theorem follows from the expansion (see Eq. (3.17))

$$\Gamma(1-x) = \exp\left(\gamma x + \sum_{n=2}^{\infty} \frac{\zeta(n)}{n} x^n\right).$$

This concludes the proof of the theorem. □

Corollary 5.6.3. *For any nonnegative integers w, d and h, the quantity*

$$G_0(w, d, h) = \sum_{\mathbf{s} \in I_0(w,d,h)} \zeta(\mathbf{s})$$

can be expressed as a polynomial of $\zeta(2), \zeta(3), \ldots$ with rational coefficients.

Proof. This follows from the theorem immediately. □

Corollary 5.6.4. *For any positive integer m,*

$$\zeta(\{2\}^m) = \frac{\pi^{2m}}{(2m+1)!}.$$

Proof. In the generating function $\Psi_0(x, y, z; 1)$ if we set $x = y = 0$ then the only nonzero terms are given by $d = h$ and $w = 2d$ meaning that

$$\Psi_0(x, y, z; 1) = \sum_{s=1}^{\infty} \zeta(\{2\}^s) z^{s-1}.$$

By the theorem we get

$$\begin{aligned}
\sum_{m=1}^{\infty} \zeta(\{2\}^m) z^{m-1} &= \frac{1}{z}\left(\exp\left(-\sum_{n=1}^{\infty} \frac{\zeta(2n)}{n}(-z)^n\right) - 1\right) \\
&= \frac{1}{z}\left(\prod_{k=1}^{\infty} \exp\left(-\sum_{n=1}^{\infty} \frac{(-z/k^2)^n}{n}\right) - 1\right)
\end{aligned}$$

$$
\begin{aligned}
&= \frac{1}{z}\left(\prod_{k=1}^{\infty}\left(1+\frac{z}{k^2}\right)-1\right)\\
&= \frac{1}{z}\left(\frac{\sinh \pi\sqrt{z}}{\pi\sqrt{z}}-1\right)=\sum_{m=1}^{\infty}\frac{\pi^{2m}}{(2m+1)!}z^{m-1}.
\end{aligned}
$$

The corollary follows from this immediately. □

5.7 Zagier's 2-3-2 Formula of MZVs

In this section we will present a **2-3-2** formula of the MZVs, *i.e.*, a formula which involves the MZVs of the form

$$L(a,b) := \zeta(\{2\}^a, 3, \{2\}^b), \qquad a,b \in \mathbb{N}_0. \tag{5.69}$$

Zagier first proved this formula which has an important application in Brown's proof of the Deligne-Ihara Conjecture and part of Hoffman's Basis Conjecture (see Conjection 3.6.1) in [102]. The key step is to express the quantities in Thm. 5.7.4 using the hypergeometric functions. The formula in the theorem then follows from some transformation relations of the hypergeometric functions.

Lemma 5.7.1. *We have the generating function*

$$
\begin{aligned}
F_L(x,y) :=& \sum_{a,b=0}^{\infty}(-1)^{a+b+1}L(a,b)x^{2a+1}y^{2b+2}\\
=& \frac{\sin \pi x}{\pi}\frac{\mathrm{d}}{\mathrm{d}z}\bigg|_{z=0} {}_3F_2\left(\begin{matrix} y,-y,z\\ 1+x,1-x\end{matrix}\right),
\end{aligned}
$$

where the hypergeometric functions ${}_3F_2\left(\begin{smallmatrix}\alpha_1,\alpha_2,\alpha_3\\ \beta_1,\beta_2\end{smallmatrix}\right) = {}_3F_2\left(\begin{smallmatrix}\alpha_1,\alpha_2,\alpha_3\\ \beta_1,\beta_2\end{smallmatrix};1\right)$ *is defined in Appendix* C.

Proof. Using the definition of the MZVs and the Pochhammer symbol Eq. (1.13), we have

$$
\begin{aligned}
F_L(x,y) =& -xy^2\sum_{n=1}^{\infty}\prod_{l>n}\left(1-\frac{x^2}{l^2}\right)\frac{1}{n^3}\prod_{n>k>0}\left(1-\frac{y^2}{k^2}\right)\\
=& \frac{\sin \pi x}{\pi}\sum_{n=1}^{\infty}\frac{1}{n}\frac{(-y)_n(y)_n}{(1-x)_n(1+x)_n}
\end{aligned}
$$

which can be easily transformed into the formula in the lemma. □

For any nonnegative integers a and b, we define

$$R(a,b) := 2\sum_{n=1}^{a+b+1}(-1)^n\left[\binom{2n}{2b+2}-\left(1-\frac{1}{4^n}\right)\binom{2n}{2a+1}\right]$$
$$\times\,\zeta(2n+1)\zeta\big(\{2\}^{a+b+1-n}\big).$$

Lemma 5.7.2. *The generating function*

$$F_R(x,y) := \sum_{a,b=0}^{\infty}(-1)^{a+b+1}R(a,b)x^{2a+1}y^{2b+2}$$

satisfies the relation

$$\frac{\pi\cdot F_R(x,y)}{\sin\pi x} = \psi(1+x)+\psi(1-x)-\frac{1}{2}\sum_{\kappa,\lambda=\pm1}\psi(1+\kappa x+\lambda y)$$
$$-\frac{\sin\pi y}{2\sin\pi x}\sum_{\kappa,\lambda=\pm1}\kappa\lambda\left(\psi\Big(1+\frac{\kappa x+\lambda y}{2}\Big)-\psi(1+\kappa x+\lambda y)\right). \quad (5.70)$$

Proof. By the definition and Cor. 5.6.4, we have

$$F_R(x,y) = 2\sum_{k,n=1}^{\infty}\frac{(-1)^k\pi^{2k}}{(2k+1)!}\zeta(2n+1)\Bigg[\sum_{b=0}^{n-1}\binom{2n}{2b+2}x^{2k+2n-2b-1}y^{2b+2}$$
$$-\left(1-\frac{1}{4^n}\right)\sum_{a=0}^{n-1}\binom{2n}{2a+1}y^{2k+2n-2a}x^{2a+1}\Bigg]$$
$$=\frac{\sin\pi x}{\pi}\sum_{n=1}^{\infty}\zeta(2n+1)\big[(x+y)^{2n}+(x-y)^{2n}-2x^{2n}\big]$$
$$-\frac{\sin\pi y}{\pi}\sum_{n=1}^{\infty}\left(1-\frac{1}{4^n}\right)\zeta(2n+1)\big[(x+y)^{2n}-(x-y)^{2n}\big]$$
$$=\frac{\sin\pi x}{\pi}\big[A(x+y)+A(x-y)-2A(x)\big]-\frac{\sin\pi y}{\pi}\big[B(x+y)-B(x-y)\big],$$

where, for all $|z|<1$

$$A(z)=\sum_{n=1}^{\infty}\zeta(2n+1)z^{2n},\quad B(z)=\sum_{n=1}^{\infty}\left(1-\frac{1}{4^n}\right)\zeta(2n+1)z^{2n}=A(z)-A(\tfrac{z}{2}).$$

Moreover, using the expansion Eq. (1.5) of the digamma function $\psi(x)$, we can see easily that

$$A(z)=\psi(1)-\frac{1}{2}\psi(1+z)-\frac{1}{2}\psi(1-z).$$

This finishes the proof of the lemma. □

We also need a result concerning the digamma function $\psi(x)$ (see p. 2).

Lemma 5.7.3. *When $z \to 0$ we have*

$$\frac{\Gamma(x)}{\Gamma(x-z)} = 1 + \psi(x)z + o(z).$$

Proof. By L'Hôpital's Rule, we have

$$\lim_{z\to 0} \frac{\Gamma(x)-\Gamma(x-z)}{\Gamma(x-z)z} = \lim_{z\to 0} \frac{\Gamma'(x-z)}{\Gamma'(x-z)z+\Gamma(x-z)} = \frac{\Gamma'(x)}{\Gamma(x)} = \psi(x).$$

The lemma follows immediately. □

We are now ready to prove Zagier's **2-3-2** formula of the MZVs.

Theorem 5.7.4. *For any $a, b \in \mathbb{N}_0$, we have $L(a,b) = R(a,b)$. Namely,*

$$\zeta(\{2\}^a, 3, \{2\}^b) = 2 \sum_{n=1}^{a+b+1} (-1)^n \left[\binom{2n}{2b+2} - \left(1 - \frac{1}{4^n}\right) \binom{2n}{2a+1} \right] \times \zeta(2n+1)\zeta\big(\{2\}^{a+b+1-n}\big). \tag{5.71}$$

Proof. First we observe that

$$(y)_n(-y)_n = \frac{1}{2}(y)_n(1-y)_n + \frac{1}{2}(1+y)_n(-y)_n.$$

Thus

$${}_3F_2\left(\begin{matrix} y, -y, z \\ 1+x, 1-x \end{matrix}\right) = \frac{1}{2}\,{}_3F_2\left(\begin{matrix} y, 1-y, z \\ 1+x, 1-x \end{matrix}\right) + \frac{1}{2}\,{}_3F_2\left(\begin{matrix} 1+y, -y, z \\ 1+x, 1-x \end{matrix}\right). \tag{5.72}$$

Because of the symmetry $y \leftrightarrow -y$ on the right-hand side of Eq. (5.72), we only need to consider the first ${}_3F_2$-term. Applying the transformation formula Eq. (C.6) with $\alpha_1 = y, \alpha_2 = z, \alpha_3 = 1-y, \beta_1 = 1+x$ and $\beta_2 = 1-x$, we get

$$\begin{aligned} {}_3F_2\left(\begin{matrix} y, 1-y, z \\ 1+x, 1-x \end{matrix}\right) &= \frac{\Gamma(1+x)\Gamma(1+x-y-z)}{\Gamma(1+x-y)\Gamma(1+x-z)}\,{}_3F_2\left(\begin{matrix} y, y-x, z \\ y-x+z, 1-x \end{matrix}\right) \\ &\quad + \frac{\Gamma(1+x)\Gamma(1-x)\Gamma(y-x+z-1)\Gamma(1-z)}{\Gamma(y)\Gamma(z)\Gamma(y-x)\Gamma(2-y-z)} \\ &\quad \times {}_3F_2\left(\begin{matrix} 1+x-y, 1+x-z, 1-z \\ 2+x-y-z, 2-y-z \end{matrix}\right). \end{aligned} \tag{5.73}$$

Applying the transformation Eq. (C.7) to the first ${}_3F_2$-term on the right-hand side of Eq. (5.73) with $\alpha_1 = y, \alpha_2 = y-x, \alpha_3 = z, \beta_1 = y-x+z$ and $\beta_2 = 1-x$, we get

$${}_3F_2\left(\begin{matrix} y, 1-y, z \\ 1+x, 1-x \end{matrix}\right) = \frac{\Gamma(1+x)}{\Gamma(1+x-z)} \frac{\Gamma(1-x)}{\Gamma(1-x-z)} \frac{\Gamma(1-x-y)}{\Gamma(1-x-y+z)}$$

$$
\begin{aligned}
&\times \frac{\Gamma(1+x-y-z)}{\Gamma(1+x-y)}\, {}_3F_2\begin{pmatrix} -x+z, z, z \\ y-x+z, 1-x-y+z \end{pmatrix} \\
&+ \frac{x\Gamma(x)\Gamma(1-x)}{\Gamma(y)\Gamma(2-y-z)} \frac{\Gamma(y-x-1+z)}{\Gamma(y-x)} \\
&\times \frac{\Gamma(1-z)}{\Gamma(z)}\, {}_3F_2\begin{pmatrix} 1+x-y, 1+x-z, 1-z \\ 2+x-y-z, 2-y-z \end{pmatrix} \\
&= [1+\psi(1+x)z][1+\psi(1-x)z][1-\psi(1-x-y)z][1-\psi(1+x-y)z] \\
&+ \frac{x}{1-y} \frac{\sin \pi y}{\sin \pi x} \frac{z}{y-x-1}\, {}_3F_2\begin{pmatrix} 1+x-y, 1+x, 1 \\ 2+x-y, 2-y \end{pmatrix} + o(z)
\end{aligned}
$$

as $z \to 0$ by Lemma 5.7.3. Here, we have also used the reflection formula

$$\Gamma(t)\Gamma(1-t) = \frac{\pi}{\sin \pi t}$$

for the Γ-product coefficient of the second ${}_3F_2$-term.

Now taking $\alpha = 1+x-y$ and $\beta = 1+x$ in the summation formula of Prop. C.3, we get

$$
\begin{aligned}
&\frac{\mathrm{d}}{\mathrm{d}z}\bigg|_{z=0} {}_3F_2\begin{pmatrix} y, 1-y, z \\ 1+x, 1-x \end{pmatrix} = \psi(1+x)+\psi(1-x)-\psi(1+x-y)-\psi(1-x-y) \\
&- \frac{\sin \pi y}{\sin \pi x} \cdot \left[\psi(1+x-y) - \psi(1-x-y) - \psi\Big(1+\frac{x-y}{2}\Big) + \psi\Big(1-\frac{x+y}{2}\Big)\right].
\end{aligned} \tag{5.74}
$$

Combining Eqs. (5.72) and (5.74), we see that $F_L(x,y) = F_R(x,y)$ which yields the theorem immediately. □

If $k, n_1, \ldots, n_d \in \mathbb{N}$, we have defined the following shuffle-regularized MZVs (see Prop. 3.2.4):

$$\zeta_k(n_1, \ldots, n_d) := (-1)^d \zeta_{ш}(\mathsf{a}^{n_1-1}\mathsf{b} \cdots \mathsf{a}^{n_d-1}\mathsf{b}\mathsf{a}^k).$$

Corollary 5.7.5. *For any $n \in \mathbb{N}$, we have*

$$\zeta_1(\{2\}^n) = 2\sum_{i=1}^{n} (-1)^i \zeta(2i+1)\zeta(\{2\}^{n-i}).$$

Proof. By Thm. 4.1.14 we have

$$
\begin{aligned}
&\zeta_1(\{2\}^d) = 2\sum_{a=0}^{d-1} \zeta(\{2\}^a, 3, \{2\}^{d-a-1}) \\
&= 4\sum_{i=1}^{d} (-1)^i \zeta(2i+1)\zeta(\{2\}^{d-i}) \sum_{a=0}^{d-1} \left[\binom{2i}{2d-2a} - \Big(1-\frac{1}{4^i}\Big)\binom{2i}{2a+1}\right]
\end{aligned}
$$

by Thm. 5.7.4. So the corollary follows from the well-known combinatorial identities:

$$\sum_{b=0}^{i-1} \binom{2i}{2b+2} = 2^{2i-1} - 1 \quad \text{and} \quad \sum_{a=0}^{i-1} \binom{2i}{2a+1} = 2^{2i-1},$$

which can be proved, for e.g., by using generating functions or the WZ-method [472]. □

5.8 An Exotic Shuffle Relation

Using the shuffle relations of the MZVs we can prove the following identity. Usually, the shuffle product is defined on the algebra of words. But in the statement of the theorem (but not in the proof) we treat the numbers 1,2 and 3 as letters and then extend the zeta function ζ linearly over $\mathbb{Z}$.

Theorem 5.8.1. *For any nonnegative integers m and n,*

$$\zeta(\{2\}^m ш \{3,1\}^n) = \binom{2n+m}{m} \frac{\pi^{4n+2m}}{(2n+1)\cdot(4n+2m+1)!}. \tag{5.75}$$

Proof. The case $n = 0$ is just Cor. 5.6.4: for every positive integer r

$$\zeta(\{2\}^r) = \frac{\pi^{2r}}{(2r+1)!}. \tag{5.76}$$

Applying the evaluation map ζ defined in Prop. 3.2.4 to the two equations in Prop. 3.2.6 and using Eq. (5.76) we see that

$$\sum_{k=1-n}^{n} \frac{(-1)^k k^{2m-1} \pi^{4n-2}}{(2n+2k-1)!(2n-2k+1)!} = 4^n \sum_{j=1}^{m} x_{2n-1,n-j} \cdot a_{2m-1,j}, \tag{5.77}$$

$$\sum_{k=-n}^{n} \frac{(-1)^k k^{2m} \pi^{4n}}{(2n+2k+1)!(2n-2k+1)!} = 4^n \sum_{j=1}^{m} x_{2n,n-j} \cdot a_{2m,j}, \tag{5.78}$$

where $x_{2n-1,n-j} = -\int_0^1 T_{2n-1}^{n-j}$ and $x_{2n,n-j} = \int_0^1 T_{2n}^{n-j}$. Observe that $\zeta(\{2\}^r) = (-1)^r \zeta((\mathtt{ab})^r)$ and $\zeta(\{3,1\}^r) = \zeta((\mathtt{a}^2\mathtt{b}^2)^r)$, so it is not too hard to see that

$$x_{2n-1,n-j} = \zeta(\{2\}^{2j-1} ш \{3,1\}^{n-j}), \quad x_{2n,n-j} = \zeta(\{2\}^{2j} ш \{3,1\}^{n-j}).$$

Note that for arbitrary fixed n, $x_{2n-1,n-j}$ (resp. $x_{2n,n-j}$) are recursively defined by Eq. (5.77) (resp. Eq. (5.78)) when we take $m = 1, \ldots, n-1$.

Thus Thm. 5.8.1 is quickly reduced to the following combinatorial identities: for all $m, n \geq 1$, we have

$$\sum_{k=1-n}^{n} \frac{(-1)^k k^{2m-1}}{2n-2k+1} \binom{4n-1}{2n+2k-1} = \sum_{j=1}^{m} \sum_{k=1-j}^{j} \frac{4^{n-j}(-1)^k k^{2m-1}}{2n-2j+1} \binom{2n-1}{2j-1} \binom{2j-1}{j-k}, \tag{5.79}$$

$$\sum_{k=-n}^{n} \frac{(-1)^k k^{2m}}{2n-2k+1} \binom{4n+1}{2n+2k+1} = \sum_{j=1}^{m} \sum_{k=-j}^{j} \frac{4^{n-j}(-1)^k k^{2m}}{2n-2j+1} \binom{2n}{2j} \binom{2j}{j-k}. \tag{5.80}$$

Now observe that we may replace the outer sum of the right-hand side in both Eq. (5.79) and Eq. (5.80) by $\sum_{j=1}^{\infty}$ by the following computation: if $j > m$ then

$$\sum_{k=1-j}^{j} (-1)^k k^{2m-1} \binom{2j-1}{j-k} = \left(x \frac{\mathrm{d}}{\mathrm{d}x}\right)^{2m-1} \sum_{k=1-j}^{j} (-1)^k x^k \binom{2j-1}{j-k} \bigg|_{x=1}$$
$$= (-1)^{1-j} \left(x \frac{\mathrm{d}}{\mathrm{d}x}\right)^{2m-1} x^{1-j}(1-x)^{2j-1} \bigg|_{x=1} = 0.$$

Similarly, if $j > m$ then

$$\sum_{k=-j}^{j} (-1)^k k^{2m} \binom{2j}{j-k} = 0.$$

We therefore only need to prove that for all $m, n \geq 1$

$$\sum_{k=1-n}^{n} \frac{(-1)^k k^{m}}{2n-2k+1} \binom{4n-1}{2n+2k-1} = \sum_{j=1}^{\infty} \sum_{k=1-j}^{j} \frac{4^{n-j}(-1)^k k^{m}}{2n-2j+1} \binom{2n-1}{2j-1} \binom{2j-1}{j-k}, \tag{5.81}$$

$$\sum_{k=-n}^{n} \frac{(-1)^k k^{m}}{2n-2k+1} \binom{4n+1}{2n+2k+1} = \sum_{j=1}^{\infty} \sum_{k=-j}^{j} \frac{4^{n-j}(-1)^k k^{m}}{2n-2j+1} \binom{2n}{2j} \binom{2j}{j-k}. \tag{5.82}$$

In fact, we only need to consider the case m is odd in Eq. (5.81) and m is even in Eq. (5.82).

It is a well-known fact that $x\mathbb{Q}[x]$ has two bases over $\mathbb{Q}$: $\{x^m : m \geq 1\}$ and $\{\binom{x}{m} : m \geq 1\}$. So Thm. 5.8.1 follows immediately from the following Prop. 5.8.2. Note that we will exchange the indices j and k on the right-hand side. $\square$

Proposition 5.8.2. *For all $m, n \geq 1$, we have*

$$\sum_{k=1-n}^{n} \frac{(-1)^k}{2n-2k+1}\binom{k}{m}\binom{4n-1}{2n+2k-1}$$
$$= \sum_{k=1}^{\infty}\sum_{j=1-k}^{k} \frac{4^{n-k}(-1)^j}{2n-2k+1}\binom{j}{m}\binom{2n-1}{2k-1}\binom{2k-1}{k-j}, \tag{5.83}$$
$$\sum_{k=-n}^{n} \frac{(-1)^k}{2n-2k+1}\binom{k}{m}\binom{4n+1}{2n+2k+1}$$
$$= \sum_{k=1}^{\infty}\sum_{j=-k}^{k} \frac{4^{n-k}(-1)^j}{2n-2k+1}\binom{j}{m}\binom{2n}{2k}\binom{2k}{k-j}. \tag{5.84}$$

Proof. We first break the inner sum on the right-hand side of Eq. (5.83) as follows:

$$\sum_{j=1-k}^{k} (-1)^j\binom{j}{m}\binom{2k-1}{k-j}$$
$$= \sum_{j=1}^{k}(-1)^j\binom{j}{m}\binom{2k-1}{k-j} + \sum_{j=1}^{k-1}(-1)^{m-j}\binom{m+j-1}{m}\binom{2k-1}{j+k}.$$

Re-indexing and using the Chu–Vandermonde identity (see [472, p. 182]) we change the above to

$$\sum_{j=0}^{k-m}(-1)^m\binom{-m-1}{j}\binom{2k-1}{k-m-j} + \sum_{j=0}^{k-2}(-1)^{m-1}\binom{-m-1}{j}\binom{2k-1}{k-2-j}$$
$$= (-1)^m\binom{2k-m-2}{k-m} + (-1)^{m-1}\binom{2k-m-2}{k-2}.$$

Note that the sum is 0 when $k \geq m$, however, when $k < m$ only the first term in the sum is always 0. By denoting the left-hand (resp. right-hand) side of Eq. (5.83) by $L(m,n)$ (resp. $R(m,n)$) we get

$$R(m,n) = \sum_{k=1}^{n} \frac{(-1)^m 4^{n-k}}{(2n-2k+1)}\binom{2n-1}{2k-1}\left(\binom{2k-m-2}{k-m} - \binom{2k-m-2}{k-2}\right). \tag{5.85}$$

Set

$$\begin{aligned}
F_0(n,k) :=& \frac{(-1)^k}{2n-2k+1}\binom{k}{m}\binom{4n-1}{2n+2k-1},\\
F_1(n,k) :=& \frac{(-1)^{m+1}4^{n-k}}{2n-2k+1}\binom{2n-1}{2k-1}\binom{2k-m-2}{k-2},\\
F_2(n,k) :=& \frac{(-1)^m 4^{n-k}}{2n-2k+1}\binom{2n-1}{2k-1}\binom{2k-m-2}{k-m}.
\end{aligned}$$

Then clearly $\lim_{k\to\pm\infty} F_i(n,k) = 0$ for each n and

$$L(m,n) = \sum_{k\in\mathbb{Z}} F_0(n,k), \quad R(m,n) = \sum_{k\in\mathbb{Z}} \Big(F_1(n,k) + F_2(n,k)\Big).$$

Now by the WZ-method (see [472]) we can find functions $G_i(n,k)$ such that all of the three functions F_i above satisfy the following:

$$\begin{aligned}
&c_3F_i(n+3,k) - c_2F_i(n+2,k) + c_1F_i(n+1,k) - c_0F_i(n,k)\\
&\quad = G_i(n,k+1) - G_i(n,k), \quad \forall k \in \mathbb{Z},
\end{aligned} \tag{5.86}$$

where $\lim_{k\to\pm\infty} G_i(n,k) = 0$ for each fixed n and

$$\begin{aligned}
c_0 =& 32n(4n+1)(2n+1)(4n+3)(n+1)[144n^4 - 168n^3m\\
&+ 40n^2m^2 + 6nm^3 + 1128n^3 - 876n^2m + 97nm^2 + 10m^3\\
&+ 3188n^2 - 1457nm + 51m^2 + 3868n - 775m + 1704],\\
c_3 =& (2n+3)(n+3)(n+1)(2n-2m+7)(n-m+3)[144n^4\\
&- 168n^3m + 40n^2m^2 + 6nm^3 + 552n^3 - 372n^2m + 17nm^2\\
&+ 4m^3 + 668n^2 - 209nm - 6m^2 + 300n - 26m + 40].
\end{aligned}$$

Similarly we can find explicit expression of c_1 and c_2 (see Exercise 5.15). Now summing up Eq. (5.86) for $-\infty < k < \infty$ we find that both $L(m,n)$ and $R(m,n)$ satisfy the same recurrence relation

$$c_3A(n+3) = c_2A(n+2) - c_1A(n+1) + c_0A(n).$$

Furthermore it is easy to check by using Eq. (5.83) that for all $m \ge 1$

$$L(m,1) = R(m,1) = -\delta_{m,1}, \quad L(m,2) = R(m,2) = -5\delta_{m,1} - (-1)^m,$$

where $\delta_{m,1}$ is the Kronecker symbol. When $n = 3$ direct computation using Eq. (5.85) shows that

$$L(m,3) = R(m,3) = \begin{cases} -16 & \text{if } m = 1,\\ 0 & \text{if } m = 2,\\ 40/3 & \text{if } m = 3,\\ (-1)^m(m - 52/3) & \text{if } m \ge 4. \end{cases}$$

This implies that Eq. (5.83) is true.

The proof of Eq. (5.84) can be carried out by exactly the same method as above so we leave the details to the interested reader. This completes the proof of the proposition. □

5.9 Period Polynomial Relations of Double Zeta Values

In Remark 3.4.7 concerning Conjecture 3.4.6 about the dimensions of MZV space we mentioned that MZVs are expected to be tied up with the modular forms. In this section we will explore this important relation in some details.

We will always assume k to be an even positive integer throughout this section, except when we point out otherwise.

5.9.1 *Double Zeta Space*

Recall that the double zeta values satisfy the following stuffle and shuffle relations (see Eq. (5.16)):

$$\zeta(j)\zeta(w-j) = \zeta(j, w-j) + \zeta(w-j, j) + \zeta(w), \tag{5.87}$$

$$\zeta(j)\zeta(w-j) = \sum_{r=2}^{w-1} \left[\binom{r-1}{j-1} + \binom{r-1}{w-j-1} \right] \zeta(r, w-r). \tag{5.88}$$

for $2 \le j \le w/2$. Equating the right-hand side of these two relations we get

$$\zeta(j, w-j) + \zeta(w-j, j) + \zeta(w) = \sum_{r=2}^{w-1} \left[\binom{r-1}{j-1} + \binom{r-1}{w-j-1} \right] \zeta(r, w-r). \tag{5.89}$$

We define the *formal double zeta space* D_w to be the $\mathbb{Q}$-vector space generated by the formal symbols $Z_{r,s}$, $P_{r,s}$ and Z_w subject to the above two sets of relations, with $\zeta(r,s)$, $\zeta(r)\zeta(s)$ and $\zeta(w)$ replaced by $Z_{r,s}$, $P_{r,s}$ and Z_w, respectively, by formally allowing r and s to be 1. Clearly we can also write

$$\mathsf{D}_w = \frac{\mathbb{Q}\langle Z_{r,w-r}, Z_w : 1 \le r < w \rangle}{\mathbb{Q}\langle \text{relations (5.89)} \rangle},$$

with w generators and $\lfloor w/2 \rfloor$ relations, which shows that

$$\dim \mathsf{D}_w \ge \frac{w}{2}. \tag{5.90}$$

We will show in fact the equality holds in Thm. 5.9.11.

Theorem 5.9.1. *Let w be a positive even integer. Then*

$$\sum_{1<r<w,\ r\ odd} Z_{r,w-r} = \frac{1}{4}Z_w, \qquad \sum_{1<r<w,\ r\ even} Z_{r,w-r} = \frac{3}{4}Z_w. \tag{5.91}$$

Remark 5.9.2. This is the formal version of Thm. 5.4.2.

Proof. Consider the generating functions

$$\mathfrak{Z}_w(X,Y) = \sum_{r+s=w} Z_{r,s}X^{r-1}Y^{s-1}.$$

By (5.89) we see that

$$\mathfrak{Z}_w(X,Y)+\mathfrak{Z}_w(Y,X)+Z_w\frac{X^{w-1}-Y^{w-1}}{X-Y} = \mathfrak{Z}_w(X+Y,Y)+\mathfrak{Z}_w(X+Y,X).$$

Setting $(X,Y)=(1,0)$ and then $(X,Y)=(1,-1)$, we get, respectively,

$$\sum_{r=2}^{w-1} Z_{r,w-r} = Z_w, \tag{5.92}$$

$$2\sum_{r=2}^{w-1}(-1)^{r-1}Z_{r,w-r} + Z_w = 0. \tag{5.93}$$

By adding (resp. subtracting) twice of (5.92) to (resp. from) (5.93) we obtain (5.91). □

5.9.2 *Period Polynomials*

Let w be an even positive integer and V_w be the $\mathbb{Q}$-vector space of homogeneous polynomials of degree $w-2$ in $\mathbb{Q}[X,Y]$.

To study period polynomials, we need to use the following special elements in $\Gamma := \mathrm{PGL}_2(\mathbb{Z})$ (notation only valid in Sec. 5.9):

$$U := \begin{pmatrix} 1 & -1 \\ 1 & 0 \end{pmatrix}, \quad S := \begin{pmatrix} 0 & -1 \\ 1 & 0 \end{pmatrix}, \quad T := US = \begin{pmatrix} 1 & 1 \\ 0 & 1 \end{pmatrix},$$

$$\varepsilon := \begin{pmatrix} 0 & 1 \\ 1 & 0 \end{pmatrix}, \quad \delta := \begin{pmatrix} -1 & 0 \\ 0 & 1 \end{pmatrix}, \quad T' := U^2S = \begin{pmatrix} 1 & 0 \\ 1 & 1 \end{pmatrix}.$$

Let $I = \pm\left(\begin{smallmatrix} 1 & 0 \\ 0 & 1 \end{smallmatrix}\right)$. The above matrices satisfy the following relations:

$$\begin{aligned} S^2 = U^3 = I, \quad \varepsilon T\varepsilon = T', \quad \delta T\delta = T^{-1}, \\ S = \varepsilon\delta, \qquad \varepsilon U\varepsilon = U^2, \quad STS = {T'}^{-1}. \end{aligned} \tag{5.94}$$

For any $F(X,Y)\in V_w$ and $\gamma = \left(\begin{smallmatrix} a & b \\ c & d \end{smallmatrix}\right)\in\Gamma$ we define

$$F|_\gamma(X,Y) = F(a+bY, cX+dY).$$

Denote by $V_w^{\pm}$ the $(\pm)$-eigenspaces of ε, V_w^{ev} and V_w^{od} the $(\pm)$-eigenspaces of δ, respectively.

Proposition 5.9.3. *We have a surjective map:*

$$\pi^{\text{od}} := \frac{1}{2}(1-\delta) : V_w^+|_{T-I} \longrightarrow V_w^{\text{od}}.$$

Proof. We prove that the following diagram is commutative:

$$\begin{array}{ccccc} V_w^{T\delta} & \xrightarrow{T-I} & V_w^{\text{od}} & \longrightarrow & 0 \\ {\scriptstyle 1+\varepsilon-\varepsilon U}\downarrow & & \uparrow{\scriptstyle \pi^{\text{od}}} & & \\ V_w^+ & \xrightarrow{T-I} & V_w^+|_{T-I} & & \end{array}$$

and the top row is exact, namely, the top horizontal map is surjective.

(1). The two maps from $V_w^{T\delta}$ are well defined. Let $g \in V_w^{T\delta}$.

(i) Horizontal: Since $g|_{T\delta} = g$ and $T\delta T = \delta$ we have $g|_\delta = g|_{T\delta T} = g|_T$ and

$$(g|_{T-I})|_\delta = g|_{T\delta-\delta} = g|_{I-T\delta T} = g|_{I-T} = -g|_{T-I}. \tag{5.95}$$

Thus $g|_{T-I} \in V_w^{\text{od}}$.
(ii) Vertical: By Eq. (5.94) we have $\varepsilon U \varepsilon = U^2 = TSU$. Thus

$$(g|_{1+\varepsilon-\varepsilon U})|_\varepsilon = g|_{1+\varepsilon-TSU} = g|_{1+\varepsilon-\delta SU} = g|_{1+\varepsilon-\varepsilon U}.$$

Thus $g|_{1+\varepsilon-\varepsilon U} \in V_w^+$.

(2). The top horizontal map $T-I$ is surjective. Indeed, if $f \in V_w^{\text{od}}$ then there exists $g \in V_w$ such that $f = g|_{T-I}$ (see Exercise 5.21). But then

$$(g|_{T\delta-I})|_{I-T} = g|_{T\delta-I-T\delta T+T} = g|_{T\delta-I-\delta+T} = f|_{\delta+I} = 0.$$

Hence $g|_{T\delta-I}(X,Y) = g(Y-X,Y) - g(X,Y)$ must be a constant. Taking $(X,Y) = (1,2)$ we see that $g|_{T\delta-I} = 0$ which implies $g \in V_w^{T\delta}$.

(3). The square commutes. Direct computation shows that

$$(\varepsilon - \varepsilon U)(T-I)(I-\delta) = (I - T\delta)(\varepsilon - \varepsilon U)T(I-\delta).$$

Therefore,

$$(g|_{1+\varepsilon-\varepsilon U} - g)|_{T-I}|_{I-\delta} = g|_{(I-T\delta)(\varepsilon-\varepsilon U)T(I-\delta)} = 0.$$

But by Eq. (5.95) $(g|_{T-I})|_{I-\delta} = 2g|_{T-I}$. Hence

$$\pi^{\text{od}}\big((g|_{1+\varepsilon-\varepsilon U})|_{T-I}\big) = g|_{T-I}.$$

Combining (2) and (3) we can now prove the proposition easily. □

We now define the subspace $W_w \subset V_w$ by

$$W_w = \{P(X,Y) \in V_w : P|_{I+S} = P|_{I+U+U^2} = 0\}.$$

This is equivalent to saying that W_w consists of all the polynomials $P(X,Y)$ in V_w satisfying

$$P(X,Y) + P(-Y,X) = P(X,Y) + P(X-Y,X) + P(Y,Y-X) = 0.$$

Since $\varepsilon S \varepsilon = S$ and $\varepsilon U \varepsilon = U^2$ we see that W_w is invariant under ε, implying that $W^{\pm} = W \cap V_w^{\pm}$ are the $(\pm)$-eigenspaces of ε in W, $W^{\text{ev}} = W \cap V_w^{\text{ev}}$ and $W_w^{\text{od}} = W \cap V_w^{\text{od}}$ the $(\pm)$-eigenspaces of δ, respectively. As every polynomial in W is a sum of its symmetric part $P^+(X,Y) := (P(X,Y)+P(Y,X))/2$ and its anti-symmetric part $P^-(X,Y) := (P(X,Y)-P(Y,X))/2$, we see that $W_w = W_w^+ \oplus W_w^-$ where

$$W_w^{\pm} = \{P \in W_w : P|_{\varepsilon} = \pm P\}.$$

Noticing that $P|_{\delta} = P|_{S\varepsilon} = -P|_{\varepsilon}$ for all $P \in W_w$ we can also write $W_w^+ = W_w^{\text{od}}$ and $W_w^- = W_w^{\text{ev}}$ where

$$W_w^{\text{ev}} = \{P \in W_w : P|_{\delta} = P\} \quad \text{and} \quad W_w^{\text{od}} = \{P \in W_w : P|_{\delta} = -P\}.$$

Consequently, we have

Lemma 5.9.4. *If* $p(X,Y) = \sum \binom{w-2}{r-1} p_{r,s} X^{r-1} Y^{s-1} \in W_w^-$ *then* $p_{r,s} = 0$ *for all* r, s *even.*

Proposition 5.9.5. *Suppose* $p(X,Y) \in W_w^+$. *Then* $F(x) := p(x,1)$ *has the following properties.*

(i) $F(x)$ *is a palindrome polynomial of degree* $\le w-3$.
(ii) $F(x) + x^{w-2}F(1-1/x) + F(1-x) = 0$.
(iii) $F(0) = F(1) = F'(1) = 0$ *so* $F(x+1)$ *has neither constant nor linear terms.*

Proof. (i) This follows from the symmetry of $p(X,Y)$. Since the degree of F is bounded by $w-2$ and is odd, it must be bounded by $w-3$.

(ii) This follows from $p|_{I+U+U^2} = p(X,Y) + p(X-Y,X) + p(-Y,X-Y) = p(X,Y) + p(X-Y,X) + p(X-Y,-Y) = 0$ since $p|_{\varepsilon} = p$.

(iii) Clearly $F(0) = 0$ since $F(x)$ has only odd degree terms. Setting $x = 1$ in (ii) we get $F(1) = 0$. By (i) we have $F(x) = x^d F(1/x)$ where $d = \deg F$. Differentiating both sides yields $F'(x) = dx^{d-1}F(1/x) - x^{d-2}F'(1/x)$. Taking $x = 1$ we see that $F'(1) = 0$. □

Proposition 5.9.6. *Suppose $p(X,Y) \in W_w^-$. Let $F(x) := p(x,1)$ as above. Let $p_1(X,Y)$ and $p_{1,1}(X,Y)$ be the first and the second partial derivatives with respect to X, respectively. Then*

(i) $F(x) = -x^{w-2}F(1/x)$.
(ii) $F(x) + x^{w-2}F(1-1/x) - F(1-x) = 0$.
(iii) $F(1) = F'(0) = 0$.
(iv) $p_1(1,1) - p_1(1,0) = 0$.
(v) $p_{1,1}(1,1) - p_{1,1}(1,0) = 0$.

Proof. (i) We have $p(X,Y) = -p(Y,X)$ since $p(X,Y) \in W_w^-$. Since p has homogeneous degree $w-2$ (i) follows immediately.

(ii) This follows from $p|_{I+U+U^2} = p(X,Y) + p(X-Y,X) + p(Y,Y-X) = p(X,Y) + p(X-Y,X) - p(X-Y,Y) = 0$ since $p|_\varepsilon = -p$ and $p(X,Y) = p(-X,Y)$ by Lemma 5.9.4.

(iii) Clearly $F'(0) = 0$ since $F'(x)$ has only odd degree terms. Setting $x = 1$ in (i) we get $F(1) = 0$.

(iv) Let $p_2(X,Y)$ be the partial derivative of $p(X,Y)$ with respect to Y. Differentiating $p(Y,X) = -p(X,Y)$ and $p|_{I+U+U^2} = 0$ with respect to X we get $p_2(Y,X) = -p_1(X,Y)$ and $p_1(X,Y) + p_1(X-Y,X) + p_2(X-Y,X) - p_1(X-Y,Y) = 0$. Thus

$$p_1(X,Y) + p_1(X-Y,X) - p_1(X,X-Y) - p_1(X-Y,Y) = 0. \tag{5.96}$$

We can prove (iv) by setting $X = Y = 1$.

(v) Differentiating Eq. (5.96) with respect to X we get

$$\begin{aligned} &p_{1,1}(X,Y) + p_{1,1}(X-Y,X) + p_{1,2}(X-Y,X) \\ &\quad - p_{1,1}(X,X-Y) - p_{1,2}(X,X-Y) - p_{1,1}(Y-X,Y) = 0. \end{aligned}$$

Taking $X = Y = 1$ we get

$$p_{1,1}(1,1) + p_{1,2}(0,1) - p_{1,1}(1,0) - p_{1,2}(1,0) = 0.$$

Note that all the powers of X and Y of $p_{1,2}(X,Y)$ are odd by Lemma 5.9.4. So $p_{1,2}(0,1) = p_{1,2}(1,0) = 0$ which yields (v). □

5.9.3 *Modular Forms and Cusp Forms on* $\mathbf{SL}_2(\mathbb{Z})$

Recall that a modular form on $\mathrm{PSL}_2(\mathbb{Z})$ of weight w is a holomorphic function on the upper half-plane $\mathbb{H}$ satisfying the following two conditions:

(i) $f\left(\dfrac{az+b}{cz+d}\right) = (cz+d)^w f(z)$ for all $\left(\begin{smallmatrix} a & b \\ c & d \end{smallmatrix}\right) \in \mathrm{PSL}_2(\mathbb{Z})$;

(ii) The Fourier expansion of f at $i\infty$ has the form $\sum_{n=0}^{i\infty} a_n e^{2\pi i n z}$.

A modular form f is a *cusp form* if it further satisfies $a_0 = 0$. We denote by S_w (resp. M_w) the $\mathbb{Q}$-vector space generated by all the cusp (resp. modular) forms of weight w. For example, for every even integer $w \geq 4$, we have the Eisenstein series defined by

$$G_w(z) := \frac{1}{2} \sum_{(m,n)\in\mathbb{Z}^2\setminus\{(0,0)\}} \frac{1}{(m+nz)^w}.$$

This can be shown to be a modular form of weight w (see [375]). One can also show that for each w all the modular forms of weight w form a finite dimension vector space over $\mathbb{Q}$ and

$$\dim S_w = \dim M_w - 1, \quad \dim M_w = \begin{cases} \left[\frac{w}{12}\right] + 1, & \text{if } 12 \nmid w-2; \\ \left[\frac{w}{12}\right], & \text{if } 12 \mid w-2. \end{cases}$$

Definition 5.9.7. The period polynomial associated with a cusp form $f \in S_w$ is defined by

$$P_f(X,Y) := \int_0^{i\infty} (X - Y\tau)^{w-2} f(\tau)\, \mathrm{d}\tau. \tag{5.97}$$

Recall that V_w is the $\mathbb{Q}$-vector space of homogeneous polynomials of degree $w-2$ in $\mathbb{Q}[X,Y]$. By integration by parts one can show that $P_f(X,Y) \in V_w$ (see [375, Chap. VI]). The following result and Thm. 5.9.18 to be proved later provide the reason why W_w is called the *space of period polynomials*.

Theorem 5.9.8. *For any cusp form $f \in S_w$, its associated period polynomial $P_f(X,Y) \in W_w$.*

Proof. For any $\gamma = \left(\begin{smallmatrix} a & b \\ c & d \end{smallmatrix}\right) \in \mathrm{PSL}_2(\mathbb{Z})$, we have

$$P_f|_\gamma(X,Y) = \int_{\gamma^{-1}(0)}^{\gamma^{-1}(i\infty)} (X - Y\tau)^{w-2} f(\tau)\, \mathrm{d}\tau. \tag{5.98}$$

(See Exercise 5.19). Since $U^{-1} = \left(\begin{smallmatrix} 0 & -1 \\ 1 & -1 \end{smallmatrix}\right)$ and $S^{-1} = \left(\begin{smallmatrix} 0 & 1 \\ -1 & 0 \end{smallmatrix}\right)$, we get

$$P_f|_{I+S} = \left(\int_0^{i\infty} + \int_{i\infty}^0\right) (X - Y\tau)^{w-2} f(\tau)\, \mathrm{d}\tau = 0$$

and

$$P_f|_{I+U+U^2} = \left(\int_0^{i\infty} + \int_1^0 + \int_{i\infty}^1\right) (X - Y\tau)^{w-2} f(\tau)\, \mathrm{d}\tau = 0.$$

Thus $P_f(X,Y) \in W_w$ by the definition. $\square$

In order to study double zeta values using period polynomial we still need two auxiliary results.

Proposition 5.9.9. *Let w be any positive integer,* ***either even or odd****. Let $\lambda, a_{r,s} \in \mathbb{Q}$ for all $r,s \in \mathbb{N}$ and $r+s=w$. Then the following two statements are equivalent:*

(i) *The relation*

$$\sum_{r+s=w} a_{r,s} Z_{r,s} = \lambda Z_w \tag{5.99}$$

holds in D_w.

(ii) *The generating function*

$$A(X,Y) = \sum_{r+s=w} \binom{w-2}{r-1} a_{r,s} X^{r-1} Y^{s-1} \in V_w \tag{5.100}$$

can be written as $H|_{T'-I}(X,Y) = H(X,X+Y) - H(X,Y)$ for some $H \in V_w^+$, and

$$\lambda = \frac{w-1}{2} \int_0^1 H(t,1-t)\,\mathrm{d}t. \tag{5.101}$$

Proof. Since V_w^+ is spanned by symmetric polynomials of the form $X^{i-1}Y^{j-1} + X^{j-1}Y^{i-1}$ $(i+j=w)$ we may assume without loss of generality $H(X,Y) = X^{i-1}Y^{j-1} + X^{j-1}Y^{i-1}$. Then

$$\lambda = (w-1) \int_0^1 t^{i-1}(1-t)^{j-1}\mathrm{d}t = \frac{(i-1)!(j-1)!}{(w-2)!}$$

by beta integral. An easy binomial expansion shows that the $a_{r,s}$'s in (5.100) reduce Eq. (5.99) to Eq. (5.89). □

We now define a $\mathbb{Q}$-bilinear pairing $\langle -, - \rangle : V_w \times V_w \to \mathbb{Q}$ such that

$$\langle X^{r-1}Y^{s-1}, X^{m-1}Y^{n-1} \rangle := \frac{(-1)^r}{\binom{w-2}{s-1}} \delta_{(r,s),(n,m)}, \tag{5.102}$$

for all $r+s=m+n=w$. Watch out for the change of order in (n,m). For example,

$$\langle X^3Y, X^3Y \rangle = 0, \quad \langle X^3Y, XY^3 \rangle = \frac{1}{4}.$$

Proposition 5.9.10. *Let $a_{r,s}$ and λ be rational numbers with $a_{r,s} = a_{s,r}$. Then the following three statements are equivalent:*

(i) *The relation*

$$\sum_{r+s=w} a_{r,s} P_{r,s} = \lambda' Z_w$$

holds in D_w *for some* λ'.

(ii) *The generating function*

$$A(X,Y) = \sum_{r+s=w} \binom{w-2}{r-1} a_{r,s} X^{r-1} Y^{s-1} \in V_w$$

can be written as $H|_{I-S}(X,Y) = H(X,Y) - H(-Y,X)$ *for some* $H \in V_w^{U,+} := V_w^U \cap V_w^+$.

(iii) *The generating function* $A(X,Y) \in V_w^{\text{od}}$ *and* $A \perp W_w^+$ *with respect to the scalar product defined by Eq.* (5.102).

If these statements hold, then $a_{r,s} = 0$ *for all odd* r *and* s.

Proof. The equivalence of (i) and (ii) follows almost immediately from Prop. 5.9.9 except for the extra condition $a_{r,s} = a_{s,r}$. Then (ii) $\Rightarrow$ (i) is obvious. For (i) $\Rightarrow$ (ii), we note that if $H \in V_w^+$ then $H|_{(T'-I)(\varepsilon-I)} = H|_{T-T\varepsilon} = H|_{(I-U)T}$. Thus $H|_{T'-I}$ is symmetric $\Leftrightarrow$ $H = H|_U$ $\Leftrightarrow$ $H|_{T'-I} = H|_{S-I}$ since $T' = U^2 S = U^{-1} S$.

We now establish the equivalence of (ii) and (iii). We have

$$\begin{aligned}
&v \text{ is orthogonal to } V_w^{U,+}|_{I-S} = V_w|_{(I+U+U^2)(I-S)}\\
&\Leftrightarrow v|_{(I-S)(I+\varepsilon)(I+U+U^2)} = 0\\
&\Leftrightarrow v|_{(I-S)(I+\varepsilon)} \in \ker(I+U+U^2) \cap \ker(I+S) \cap \ker(I-\varepsilon) = W_w^+\\
&\Leftrightarrow v^{\text{od},+} \in W_w^+ \Leftrightarrow v \in W_w^+ + V_w^- + V_w^{\text{ev}}
\end{aligned}$$

by writing $v = v^{\text{ev},+} + v^{\text{ev},-} + v^{\text{od},+} + v^{\text{od},-}$. Therefore,

$$A \in V_w^{U,+}|_{I-S} \Leftrightarrow A \perp W_w^+ + V_w^- + V_w^{\text{ev}} \Leftrightarrow A \perp W_w^+ \text{ and } A \in V_w^{\text{od}}.$$

The last sentence of the proposition follows from $A \in V_w^{\text{od}}$ because this means that every even X-power term vanishes. □

5.9.4 *Even Weight Double Zeta Value Relations*

We are now ready to apply the results from the preceding subsection to the study of the double zeta values of even weight. The first implication concerns the formal double zeta space D_w.

Theorem 5.9.11. *Let* $w \geq 4$ *be an even integer. Then the set*

$$\left\{ Z_{r,w-r} : 0 < r < w,\ r \text{ is odd} \right\}$$

form a basis of D_w.

Proof. We already know $\dim \mathsf{D}_w \geq w/2$ and Z_w is a linear combination of $Z_{\text{od,od}}$'s by Thm. 5.9.1. So it suffices to prove that if r is even then $Z_{r,w-r}$ can be expressed as a $\mathbb{Q}$-linear combination of $Z_{\text{od,od}}$'s. In other words, for the choice $a_{r,w-r} = 1$ and $a_{j,w-j} = 0$ for all other even j, there exist suitable choices of $a_{\text{od,od}}$'s and λ such that Eq. (5.99) holds. But by Prop. 5.9.9 this means we need to show that the odd polynomial $X^{r-1}Y^{w-r-1}$ is the odd part of $H|_{T'-I}$ for some $H \in V_w^+$ or, equivalently, $X^{w-r-1}Y^{r-1}$ is the odd part of $H|_{T-I}$. Indeed, in general $F = 2\pi^{\text{od}}(H|_{T-I}) = H|_{(T-I)(I-\delta)}$ for $H \in V_w^+$ if and only if

$$F|\varepsilon = H|_{(T-I)(\varepsilon-\delta\varepsilon)} = H|_{(T\varepsilon-\varepsilon)(I-\delta)} = H|_{(T'-I)(I-\delta)} = 2\pi^{\text{od}}(H|_{T'-I})$$

since $\delta\varepsilon = \varepsilon\delta$ in $\mathrm{PGL}_2(\mathbb{Z})$ and $H|_\varepsilon = H$. Finally, the existence of such an $H \in V_w^+$ is guaranteed by Prop. 5.9.3. This completes the proof of the theorem. □

Remark 5.9.12. We caution the reader that the theorem is true only in the context of formal double zeta space. There will be non-trivial $\mathbb{Q}$-linear relations among double zeta values $\zeta(\text{od}, \text{od})$'s. In fact, this will be the content of Cor. 5.9.19 below.

Let $\mathcal{P}_w^{\text{ev}}$ be the space spanned by $P_{\text{ev,ev}}$ over $\mathbb{Q}$. We define a map $\phi : W_w^- \to \mathcal{P}_w^{\text{ev}}$ as follows. For each $p(X,Y) \in W_w^-$, we write

$$p(X,Y) = \sum_{r+s=w} \binom{w-2}{r-1} p_{r,s} X^{r-1}Y^{s-1},$$

$$p|_T(X,Y) = p(X+Y,Y) = \sum_{r+s=w} \binom{w-2}{r-1} q_{r,s} X^{r-1}Y^{s-1}.$$

Then we define

$$\phi\big(p(X,Y)\big) = \sum_{r+s=w,r,s \text{ even}} q_{r,s} Z_{r,s} \in \mathcal{P}_w^{\text{ev}}. \tag{5.103}$$

Lemma 5.9.13. *For all $r, s \in \mathbb{N}$, we have $p_{r,s} = q_{r,s} - q_{s,r}$. Moreover, $q_{r,s} = q_{s,r}$ for all r, s even so the map ϕ is well defined.*

Proof. Let $q = p|_T$. Since $T + \varepsilon T \varepsilon = I + (I + U + U^2)S - (I + S)S$ and $p|_\varepsilon = -p$ we have

$$q|_{I-\varepsilon} = p|_{T-T\varepsilon} = p|_{T+\varepsilon T\varepsilon} = p.$$

Thus $p_{r,s} = q_{r,s} - q_{s,r}$. Then $p_{r,s} = 0$ for all r, s even by Lemma 5.9.4. Now by the second equation of Eq. (5.91) we see that

$$Z_w = \frac{2}{w+1} \sum_{1<r<w,\ r \text{ even}} P_{r,w-r} \in \mathcal{P}_w^{\text{ev}}.$$

So for all r, s we have $Z_{r,s} + Z_{s,r} = P_{r,s} - Z_w \in \mathscr{P}_w^{\mathrm{ev}}$ and therefore the map ϕ is well-define. □

Theorem 5.9.14. *The map $\phi : W_w^- \to \mathscr{P}_w^{\mathrm{ev}}$ is an isomorphism. Moreover,*

$$\phi\big(p(X,Y)\big) = \sum_{\substack{r+s=w\\ r,s\ even}} q_{r,s} Z_{r,s} \equiv 3 \sum_{\substack{r+s=w\\ r,s\ odd}} q_{r,s} Z_{r,s} \pmod{Z_w}. \tag{5.104}$$

Conversely, an element $\sum_{r+s=w, r,s\ odd} c_{r,s} Z_{r,s} \in \mathsf{D}_w$ belongs to $\mathscr{P}_w^{\mathrm{ev}}$ only if $c_{r,s} = q_{r,s}$ arising in this way.

Proof. The injectivity of ϕ follows from Eq. (5.104) since the coefficients on the right-hand side of Eq. (5.104) is anti-symmetric. The surjectivity, which is equivalence to the last sentence in the statement of the theorem, follows from Cor. 5.9.22 since injectivity and $\dim W_w^- = \dim \mathscr{P}_w^{\mathrm{ev}}$ implies bijectivity.

We now prove Eq. (5.104). We decompose q as $q = q^{\mathrm{ev},+} + q^{\mathrm{ev},-} + q^{\mathrm{od},+} + q^{\mathrm{od},-}$. Then $q^{\mathrm{ev},-} = p/2$ and $q^{\mathrm{od},-} = 0$ by Lemma 5.9.13. Set $[a,b,c] = aq^{\mathrm{ev},+} + bq^{\mathrm{ev},-} + cq^{\mathrm{od},+}$. Then we have

$$\begin{aligned} [0,2,0]|_{T'} &= p|_{T'} = p|_{\varepsilon T \varepsilon} = -p|_{T\varepsilon} = -q|_{\varepsilon} = [-1,1,-1], \\ [1,-1,-1]|_{T'} &= q|_{ST'} = p|_{TST'} = p|_S = -p = [0,-2,0]. \end{aligned}$$

So

$$[2,0,-2]|_{T'-I} = [-1,-3,-1] - [2,0,-2] = [-3,-3,1].$$

Hence setting $H = 2(q^{\mathrm{ev},+} - q^{\mathrm{od},+}) \in V_w^+$, we get

$$q^{\mathrm{od}} - 3q^{\mathrm{ev}} = H|_{T'-I}.$$

So Eq. (5.104) follows from Prop. 5.9.9. This concludes the proof of the theorem. □

The proof above provides us with an explicit way to write any $\zeta(\mathrm{ev},\mathrm{ev})$ or $\zeta(\mathrm{ev})\zeta(\mathrm{ev})$ as a $\mathbb{Q}$-linear combination of $\zeta(\mathrm{od},\mathrm{od})$'s. In particular, every anti-symmetric polynomial in W_w^- leads to some nontrivial relation among the double zeta values.

Example 5.9.15. The polynomials $X^2Y^2(X^2 - Y^2)^3$ in W_{12}^- and $X^2Y^2(X^2 - Y^2)^3(2X^4 - X^2Y^2 + 2Y^4)$ in W_{16}^- yield

$$\begin{aligned} &28P_{8,4} + \frac{95}{3}P_{6,6} - \frac{167}{3}Z_{12} = 28Z_{9,3} + 150Z_{7,5} + 168Z_{5,7}, \\ &66P_{4,12} + 185P_{6,10} + \frac{364}{3}P_{8,8} - \frac{1081}{3}Z_{16} \end{aligned}$$

$$= 66Z_{13,3} + 375Z_{11,5} + 686Z_{9,7} + 675Z_{7,9} + 396Z_{5,11},$$

see Exercise 5.24. These imply the identities:

$$\tfrac{5197}{691}\zeta(12) = 28\zeta(9,3)+150\zeta(7,5)+168\zeta(5,7),$$
$$\tfrac{78967}{3617}\zeta(16) = 66\zeta(13,3)+375\zeta(11,5)+686\zeta(9,7)+675\zeta(7,9)+396\zeta(5,11).$$

They are by no means a couple of accidents that 691 and 3617 are exactly the numerators of B_{12} and B_{16}, respectively.

In order to prove the surjectivity of ϕ in Thm. 5.9.14 we need to study not only the space W_w but also its 1-dimensional extension $\widehat{W}_w$, called the *space of extended period polynomials*, which is defined by

$$\widehat{W}_w = \ker(I+S, \widehat{V}_w) \cap \ker(I+U+U^2, \widehat{V}_w)$$

where we have extended the maps S, U, *etc.* to

$$S, U, etc. : \widehat{V}_w := \bigoplus_{\substack{r+s=w \\ r,s\geq 0}} \mathbb{C}X^{r-1}Y^{s-1} \longrightarrow \mathbb{C}(X,Y).$$

We define $\widehat{W}_w^{\pm}$ to be the $(\pm)$-eigenspace of ε.

Lemma 5.9.16. *Let $f \in M_w$ with $f(i\infty) = a_0 \neq 0$. Let $\tau_0 \in \mathbb{H}$. Then*

$$\widehat{P}_f(X,Y) = \int_{\tau_0}^{i\infty} (X-Y\tau)^{w-2}(f(\tau)-a_0)\mathrm{d}\tau$$
$$+ \int_0^{\tau_0} (X-Y\tau)^{w-2}\Big(f(\tau)-\frac{a_0}{\tau^w}\Big)\mathrm{d}\tau + \frac{a_0}{w-1}\Big(\frac{1}{Y}-\frac{\tau^{1-w}}{X}\Big)(X-Y\tau)^{w-1}$$

is independent of τ_0 and lies in $\widehat{W}_w$.

Proof. It is straightforward to check that $\widehat{P}_f(X,Y)$ is independent of τ_0 by showing the derivative with respect to τ_0 vanishes. The first integral converges because of the fast decay of e^{-y} as $y \to \infty$. The second integral also converges because $f(-1/it) = (it)^w f(it)$ as $f \in M_w$. This finishes the proof of the lemma. ☐

One may wonder how the expression in the lemma is found in the first place. For this, we need to use the L-function $L(s,f) = \sum_{n\geq 1} a_n/n^s$ associated with f. For $\mathrm{Re}(s) \gg 0$, it is easy to see that

$$L^*(s,f) := \int_0^{\infty} (f(it)-a_0)t^{s-1}\mathrm{d}t = (2\pi)\Gamma(s)L(s,f)$$

which can be rewritten as (for any $t_0 > 0$)

$$L^*(s,f) = \int_{t_0}^{\infty}\Big(f(it)-a_0\Big)t^{s-1}\mathrm{d}t + \int_0^{t_0}\Big(f(it)-\frac{a_0}{(it)^w}\Big)t^{s-1}\mathrm{d}t \\ - a_0\Big(\frac{t_0^s}{s}+\frac{(-1)^{w/2}t_0^{w-s}}{w-s}\Big). \quad (5.105)$$

From Eq. (5.105) we see immediately that $L^*(s,f)$ can be analytically continued as a meromorphic function on $\mathbb{C}$ satisfying the functional equation $L^*(s,f) = (-1)^{w/2}L^*(w-s,f)$. It has simple poles at $s=0$ with residue $-a_0$ and at $s=w$ with residue $(-1)^{w/2}a_0$.

Now if $f \in S_w$ we have

$$\begin{aligned} P_f(X,Y) &= \int_0^{i\infty}(X-Y\tau)^{w-2}f(\tau)\mathrm{d}\tau \\ &= \int_0^{\infty}\sum_{n=0}^{w-2} i^{1-n}\binom{w-2}{n}X^{w-n-2}Y^n f(it)t^n\mathrm{d}t \\ &= \sum_{n=0}^{w-2} i^{1-n}\binom{w-2}{n}X^{w-n-2}Y^n L^*(n+1,f). \end{aligned}$$

For $f \in M_w$, by adding two terms in the above sum when $n \to (-1)^+$ and $n \to (w-1)^-$, we would have added

$$\begin{aligned} &\lim_{n\to c}\frac{\Gamma(w-1)}{\Gamma(n+1)\Gamma(w-n-1)}(-i)^n X^{w-n-2}Y^n L^*(n+1,f) \\ &\quad = \begin{cases} -a_0/(w-1)\cdot X^{w-1}/Y, & \text{if } c=(-1)^+; \\ -a_0/(w-1)\cdot Y^{w-1}/X, & \text{if } c=(w-1)^-. \end{cases} \end{aligned}$$

Thus the expression

$$\frac{a_0}{w-1}\Big(\frac{X^{w-1}}{Y}+\frac{X^{w-1}}{Y}\Big)+\sum_{n=0}^{w-2} i^{1-n}\binom{w-2}{n}X^{w-n-2}Y^n L^*(n+1,f) \quad (5.106)$$

is well defined in $\widehat{V}_w$. Now we can easily obtain the expression in the lemma by applying Eqs. (5.105) and (5.106).

Lemma 5.9.17. *The element*

$$\widehat{\mathscr{E}}_w = \sum_{r+s=w,r,s\geq 0}\binom{w}{r}B_r B_s X^{r-1}Y^{s-1} \in \widehat{W}_w^+.$$

Proof. The function is clearly symmetric. We leave the interested reader to check it is annihilated by both $I+S$ and $I+U+U^2$. □

The Eichler–Shimura–Manin theory completed by Kohnen–Zagier (see [349, 375] or [599]) provides the following fundamental result whose proof is far beyond the scope of this book.

Theorem 5.9.18. *Let w be an even positive integer. Let $P_f^{\pm}$ be the projection of P_f into $W_w^{\pm}$. Then we have*

(i) *The map $f \mapsto P_f^+$ gives an isomorphism $S_w \cong W_w^+$.*
(ii) *The map $f \mapsto P_f^-$ gives an isomorphism of S_w onto a codimension 1 subspace of W_w^-.*
(iii) *The pairing (5.102) induces a pairing $\widehat{W}_w^+ \times W_w^- \to \mathbb{C}, (P_f^+, P_g^-) \mapsto \langle P_f^+|_{T-T^{-1}}, P_g^-\rangle$ such that the codimension 1 subspace of W_w^- in (ii) is precisely the space orthogonal to $\widehat{\mathscr{E}}_w$.*

Corollary 5.9.19. *The values $\zeta(\text{od}, \text{od})$ of weight w satisfy at least $\dim S_w$ linearly independent relations. Therefore,*

$$\dim \mathsf{DZ}_w \le \frac{w}{2} - \dim S_w, \tag{5.107}$$

where DZ_w is the $\mathbb{Q}$-vector space generated by the double zeta values of weight w.

Proof. By Thm. 5.9.14 the values $\zeta(\text{od}, \text{od})$ of weight w and $\zeta(w)$ satisfy at least $\dim W_w^-$ linearly independent relations. Noticing that $\zeta(\text{ev})\zeta(w - \text{ev})$ is always a rational multiple of $\zeta(w)$, after eliminating all such terms we arrive at the first sentence of the corollary in view of Thm. 5.9.18(ii). Hence, Eq. (5.107) follows from Thm. 5.9.11 immediately. □

Corollary 5.9.20. *We have $1 + \dim W_w^+ = \dim W_w^- = \dim M_w$.*

Proof. It is well known that $M_w = S_w \oplus \mathbb{Q} \cdot G_w$ where G_w is the Eisenstein series of weight w, see [166, Chap. 4]. □

Theorem 5.9.21. *If $\{Z_{r,s}, P_{r,s}, Z_w\}$ satisfy Eqs. (5.87)-(5.89), then*

$$\sum_{\substack{r+s=w\\ r,s\ \text{even}}} P_{r,s} X^{r-1} Y^{s-1} - \frac{Z_w}{2}\left(\frac{X^{w-1}}{Y} + \frac{Y^{w-1}}{X}\right) \in \widehat{W}_w^+.$$

Moreover, every element of $\widehat{W}_w^+$ arises in this way.

Proof. Proposition 5.9.10(i) is equivalent to the statement that $\sum a_{r,s} p_{r,s} = 0$ for any realization $\omega : \mathsf{D}_w \to \mathbb{C}$ with $\omega(P_{r,s}) = p_{r,s}$ and $\omega(Z_w) = 0$, while (iii) means that all $a_{\text{od},\text{od}} = 0$ and $\sum a_{r,s} p_{r,s} = 0$ for all $\sum p_{r,s} X^{r-1} Y^{s-1} \in W_w^+$. The equivalence of (i) and (iii) implies that every collection of numbers

$\{p_{r,s} = p_{s,r} : r+s = w, r, s \text{ odd}\}$ can be extended to a realization $\omega : \mathsf{D}_w \to \mathbb{C}$ with $\omega(Z_w) = 0$, while every collection of numbers $\{p_{r,s} = p_{s,r} : r+s = w, r, s \text{ even}\}$ can be extended to a realization $\omega : \mathsf{D}_w \to \mathbb{C}$ with $\omega(Z_w) = 0$ if and only $\sum \omega(P_{r,s}) X^{r-1} Y^{s-1} \in W_w^+$. Thus the case of the theorem when $\omega(Z_w) = 0$ is proved.

In general, we need to show that given any extended symmetric function

$$\widehat{P} = \sum_{r+s=w;r,s\geq 0 \text{ and even}} p_{r,s} X^{r-1} Y^{s-1} \in \widehat{W}_w^+,$$

there is a realization $\omega : \mathsf{D}_w \to \mathbb{C}$ with

$$\omega(P_{r,s}) = \begin{cases} p_{s,r}, & \text{if } r, s \text{ even;} \\ 0, & \text{if } r, s \text{ odd,} \end{cases} \qquad \omega(Z_w) = -2p_{w,0}.$$

To see this, it suffices to check it for any ONE extended element $\widehat{P} \in \widehat{W}_w^+ \backslash W_w^+$ because the codimension of W_w^+ in $\widehat{W}_w^+$ is 1. But by taking $\widehat{\mathscr{E}}_w$ we see that $p_{r,s} = \binom{w}{r} B_r B_s = 4w! \beta_r \beta_s$, where $\beta_r = \zeta(r)/(2\pi i)^r = -B_r/(2 \cdot r!)$ for all even $r \geq 0$. This is exactly $4w!/(2\pi i)^w$ times the original double zeta realization (denoted by ω_0) of D_w since

$$\frac{4w!}{(2\pi i)^w} \omega_0(P_{r,s}) = \frac{4w! \zeta(r) \zeta(s)}{(2\pi i)^w} = 4w! \beta_r \beta_s,$$

$$\frac{4w!}{(2\pi i)^w} \omega_0(Z_w) = \frac{4w! \zeta(w)}{(2\pi i)^w} = 4w! \beta_w = -2(4w! \beta_w \beta_0).$$

This completes the proof of the theorem. □

Corollary 5.9.22. *We have* $\dim \mathscr{P}_w^{\text{ev}} = \dim \widehat{W}_w^+ = \dim W_w^-$.

Proof. This follows from Thm. 5.9.21 and Cor. 5.9.20. □

5.9.5 *Odd Weight Double Zeta Value Relations*

First of all, we want to remind the reader that by the Parity Principle of Sec. 3.5, all the double zeta values of odd weight can be expressed as a $\mathbb{Q}$-linear combination of products of the Riemann zeta values. The following theorem provides the explicit relations.

Theorem 5.9.23. *Let* $m \geq 2$, $n \geq 1$ *and the weight* $w = m + n = 2k + 1$ *be an odd number. Then*

$$\zeta(m, n) = (-1)^n \sum_{j=0}^{k-1} \left[\binom{w - 2j - 1}{m - 1} + \binom{w - 2j - 1}{n - 1} - \delta_{m,2j} + (-1)^n \delta_{j,0} \right] \times \zeta(2j) \zeta(w - 2j),$$

where $\zeta(0) = -1/2$.

Proof. Recall that in the formal double zeta space we have

$$P_{j,w-j} = Z_{j,w-j} + Z_{w-j,j} + Z_w, \tag{5.108}$$

$$P_{j,w-j} = \sum_{r=1}^{w-1} \left[\binom{r-1}{j-1} + \binom{r-1}{w-j-1} \right] Z_{r,w-r}, \tag{5.109}$$

for all $j = 1, \dots, w-1$. Define the generating function $(j + j' = w)$

$$\mathfrak{P}(X,Y) = \sum_{j=1}^{w-1} P_{j,j'} X^{j'-1} Y^{j-1}, \quad \mathfrak{D}(X,Y) = \sum_{j=1}^{w-1} Z_{j,j'} X^{j'-1} Y^{j-1},$$

and $\mathfrak{Z}(X,Y) = Z_w(X^{w-1} - Y^{w-1})/(X-Y)$. Then we have

$$\mathfrak{P}(X,Y) = \mathfrak{D}(X,Y) + \mathfrak{D}(Y,X) + \mathfrak{Z}(X,Y) = \mathfrak{D}(X,X+Y) + \mathfrak{D}(Y,X+Y).$$

Repeatedly using this we get

$$\mathfrak{D}(X,Y) = \mathfrak{D}(-X,-Y) + \mathfrak{Q}(X,Y) + \mathfrak{Q}(X-Y,-Y) + \mathfrak{Q}(X-Y,X),$$

where $\mathfrak{Q}(X,Y) = \mathfrak{P}(X,Y) - \mathfrak{P}(X,-Y) - \mathfrak{Z}(X,Y)$. Since the weight w is odd we get $\mathfrak{D}(-X,-Y) = -\mathfrak{D}(X,Y)$. Easy computation shows that $(j + j' = 2j + (2j)' = w$ and $U = X - Y)$

$$\mathfrak{Q}(X,Y) = 2\sum_{j=1}^{k} P_{2j,(2j)'} X^{(2j)'-1} Y^{2j-1} - Z_w \sum_{j=1}^{w-1} X^{j'-1} Y^{j-1},$$

$$\begin{aligned}
\mathfrak{Q}(U,-Y) &= 2\sum_{j=1}^{k} P_{2j,(2j)'} U^{(2j)'-1} (-Y)^{2j-1} - Z_w \sum_{j=1}^{w-1} U^{j'-1} (-Y)^{j-1} \\
&= 2\sum_{j=1}^{k} P_{2j,(2j)'} \sum_{n=1}^{(2j)'} (-1)^n \binom{(2j)'-1}{n-1} X^{n-1} Y^{w-n-1} \\
&\quad - Z_w \sum_{j=1}^{w-1} \sum_{n=1}^{j'} (-1)^n \binom{j'-1}{n-1} X^{n-1} Y^{w-n-1} \\
&= 2\sum_{n=1}^{w-1} \sum_{j=1}^{k} P_{2j,(2j)'} (-1)^n \binom{(2j)'-1}{n-1} X^{n-1} Y^{m-1} \\
&\quad - Z_w \sum_{n=1}^{w-1} (-1)^n \sum_{j=1}^{w-1} \binom{j'-1}{n-1} X^{n-1} Y^{m-1}
\end{aligned}$$

$$\mathfrak{Q}(U,X) = 2\sum_{j=1}^{k} P_{2j,(2j)'} U^{(2j)'-1} X^{2j-1} - Z_w \sum_{j=1}^{w-1} U^{j'-1} X^{j-1}$$

$$
\begin{aligned}
&= 2\sum_{j=1}^{k} P_{2j,(2j)'} \sum_{n=2j}^{w-1} (-1)^n \binom{(2j)'-1}{n-2j} X^{n-1}Y^{w-n-1} \\
&\quad - Z_w \sum_{j=1}^{w-1}\sum_{n=j}^{w-1} (-1)^n \binom{j'-1}{n-j} X^{n-1}Y^{w-n-1} \\
&= 2\sum_{n=1}^{w-1}\sum_{j=1}^{k} P_{2j,(2j)'}(-1)^n \binom{(2j)'-1}{m-1} X^{n-1}Y^{m-1} \\
&\quad - Z_w \sum_{n=1}^{w-1} (-1)^n \sum_{j=1}^{w-1} \binom{j'-1}{m-1} X^{n-1}Y^{m-1}.
\end{aligned}
$$

The theorem now follows from the identity

$$
\sum_{j=1}^{w-1} \binom{j'-1}{n-1} = \sum_{j=n}^{w-2} \binom{j-1}{n-1} = \binom{w-1}{n} = \binom{w-1}{m-1}
$$

and a similar identity with $n \leftrightarrow m$. □

The next result is an analog of Thm. 5.9.14 in the odd weight case.

Theorem 5.9.24. *Let $w \geq 12$ be an even integer. For $p \in W_w^+$, we define $q_{r,s}$ $(r+s=w+1, r,s\geq 1)$ by*

$$
p(X+Y,Y) = \sum_{r+s=w+1} \binom{w-1}{r-1} q_{r,s} X^{r-1}Y^{s-2}.
$$

Then

$$
\sum_{\substack{r+s=w+1\\ 4\leq r\leq w-2\ \text{even}}} (q_{r,s}-q_{s,r})Z_{r,s} \equiv 0 \pmod{Z_{w+1}}. \tag{5.110}
$$

Proof. Let $g(X,Y) := p|_T = p(X+Y,Y)$. Since $\varepsilon S = S\varepsilon$ in $\mathrm{PGL}_2(\mathbb{Z})$ we have $\varepsilon T\varepsilon S = \varepsilon TS\varepsilon = \varepsilon U\varepsilon = U^2$. Thus

$$
\begin{aligned}
\varepsilon T\varepsilon + TST' + \varepsilon T\varepsilon ST' &= (I+U+U^2)T' \\
\varepsilon T\varepsilon ST' - T &= U^2T' - T = US - T = 0.
\end{aligned}
$$

Hence

$$
\begin{aligned}
g|_{\varepsilon+ST'+\varepsilon ST'} &= p|_{T\varepsilon+TST'+T\varepsilon ST'} = p|_{\varepsilon T\varepsilon+TST'+\varepsilon T\varepsilon ST'} = 0, \\
g|_{\varepsilon ST'-I} &= p|_{T\varepsilon ST'-T} = p|_{\varepsilon T\varepsilon ST'-T} = 0.
\end{aligned} \tag{5.111}
$$

We now define

$$
f(X,Y) := g\cdot Y - g|_\varepsilon \cdot X = \sum_{r+s=w+1} \binom{w-1}{r-1} (q_{r,s}-q_{s,r}) X^{r-1}Y^{s-1}.
$$

Then

$$\begin{aligned} f|_{ST'} - f &= (g\cdot Y - g|_\varepsilon \cdot X)|_{ST'} - (g\cdot Y - g|_\varepsilon\cdot X)\\ &= g|_{ST'}\cdot X - g|_{\varepsilon ST'}\cdot(-(X+Y)) - g\cdot Y + g|_\varepsilon \cdot X\\ &= g|_{\varepsilon + ST' + \varepsilon ST'}\cdot X + g|_{\varepsilon ST' - I}\cdot Y = 0\end{aligned}$$

by Eq. (5.111). So we get $f = f|_{ST'}$. Moreover,

$$\begin{aligned}(f|_S)|_\varepsilon - f|_S &= (g\cdot Y - g|_\varepsilon\cdot X)|_{S\varepsilon} - (g\cdot Y - g|_\varepsilon \cdot X)|_S\\ &= g|_{\varepsilon S\varepsilon - S}\cdot X + g|_{S\varepsilon - \varepsilon S}\cdot Y = 0.\end{aligned}$$

Thus $f|_S$ is a symmetric homogeneous polynomial of degree $w-1$ which is odd. Therefore,

$$f|_{I-S} = \sum_{r+s=w+1, r \text{ even}, 4\le r\le w-2} 2\binom{w-1}{r-1}(q_{r,s}-q_{s,r})X^{r-1}Y^{s-1}.$$

Note that g does not have terms with X-degree equal to 1 or $w-1$ by Prop. 5.9.5 so neither does f. Finally, since

$$f|_{I-S} = f|_{ST'-S} = (f|_S)|_{T'-I},$$

the theorem follows from Prop. 5.9.9 immediately. □

Example 5.9.25. The space W_{12}^+ is 1-dimensional, spanned by the odd period polynomial $p(x) = x(x^2-4)(4x^2-1)(x^2-1)^2$. We have $p(x+1) = x^2(x^2-1)(2x+3)(2x+1)(x+3)(x+2)^2$, so the numbers $q_{r,s}$ of the theorem can be found by the table

r	10	9	8	7	6	5	4	3
s	3	4	5	6	7	8	9	10
$330q_{r,s}$	24	72	119	115	15	-161	-288	-216

There is only one relation in the theorem which becomes (after dividing by 10)

$$24Z_{10,3} + 28Z_{8,5} - 10Z_{6,7} - 36Z_{4,9} \equiv 0 \pmod{Z_{13}}.$$

In the double zeta value realization, this can be written as

$$24\zeta(10,3) + 28\zeta(8,5) - 10\zeta(6,7) - 36\zeta(4,9) = -3\zeta(13).$$

Theorem 5.9.26. *Let $w \ge 12$ be an even integer. With each $p \in W_w^-$ we associate the coefficients $q_{r,s}$ $(r+s=w-1)$ which are defined by*

$$\frac{\partial}{\partial X}p(X+Y,Y) = \sum_{r+s=w-1}\binom{w-3}{r-1}q_{r,s}X^{r-1}Y^{s-1}.$$

Then

$$\sum_{\substack{r+s=w-1\\ 4\le r\le w-4 \text{ even}}} (q_{r,s}-q_{s,r})Z_{r,s} \equiv 0 \pmod{Z_{w-1}}. \tag{5.112}$$

Proof. Let $g = \frac{\partial}{\partial X}p|_T = \frac{\partial}{\partial X}p(X+Y,Y)$. We now define

$$f(X,Y) := g|_{I-\varepsilon} = \sum_{r+s=w-1} \binom{w-3}{r-1}(q_{r,s} - q_{s,r})X^{r-1}Y^{s-1}.$$

Easy computation shows that $T(I-\varepsilon) = \left(\begin{smallmatrix} 0 & 0 \\ 1 & -1 \end{smallmatrix}\right)$. Thus

$$f|_{ST'-I} = g|_{(I-\varepsilon)(ST'-I)} = \frac{\partial}{\partial X}p|_{T(I-\varepsilon)(ST'-I)} = \frac{\partial}{\partial X}p(0,-3X) = 0$$

by Prop. 5.9.6(iii). So we get $f = f|_{ST'}$. Moreover, $(I-\varepsilon)S(\varepsilon - I) = \left(\begin{smallmatrix} 0 & 0 \\ 0 & 0 \end{smallmatrix}\right)$ so we get

$$(f|_S)|_\varepsilon - f|_S = g|_{(I-\varepsilon)S(\varepsilon-I)} = \frac{\partial}{\partial X}p(0,0) = 0$$

by Prop. 5.9.6(iii). Thus $f|_S$ is a symmetric homogeneous polynomial of degree $w-3$. Furthermore, easy computation shows that the coefficient of $X^{w-3}Y$ in f is $p_1(1,1) - p_1(1,0) = 0$ by Prop. 5.9.6(iv). Similarly, the coefficient of XY^{w-3} in f is $p_{1,1}(1,1) - p_{1,1}(1,0) = 0$ by Prop. 5.9.6(v). Hence

$$f|_{I-S} = \sum_{r+s=w-1, r \text{ even}, 4\le r\le w-4} 2\binom{w-3}{r-1}(q_{r,s} - q_{s,r})X^{r-1}Y^{s-1}.$$

Finally, since

$$f|_{I-S} = f|_{ST'-S} = (f|_S)|_{T'-I}$$

the theorem follows from Prop. 5.9.9 immediately. □

Example 5.9.27. The space W_{12}^- is 2-dimensional, spanned by the even period polynomial $p(x) = x^{10} - 1$ and $x^2(x^2-1)^3$. For the latter, we have $p'(x+1) = 8x^7 + 56x^6 + 150x^5 + 190x^4 + 112x^3 + 24x^2$, so the $q_{r,s}$ of the theorem are given (after multiplication by 63) by the table

r	8	7	6	5	4	3
s	3	4	5	6	7	8
$63q_{r,s}$	14	42	75	95	84	42

The relations in the theorem, divided by -1, become

$$28Z_{8,3} + 20Z_{6,5} - 42Z_{4,7} \equiv 0 \pmod{Z_{11}}.$$

In the double zeta value realization, this can be written as

$$28\zeta(8,3) + 20\zeta(6,5) - 42\zeta(4,7) = -3\zeta(11). \tag{5.113}$$

5.10 Conjectural Multiple Zeta Value Relations

In this section, we collect a few families of MZVs identities which are all numerically verified but not proved rigorously yet.

Hoffman posted on the internet the following conjecture and mentioned that J. Vermaseren had used the MZV data mine presented in [62] to check it through weight 22.

Conjecture 5.10.1. *For every nonnegative integer n,*

$$\zeta(2,1,\{2\}^n,3) = \zeta(\{2\}^{n+3}) + 2\zeta(\{2\}^n,3,3).$$

By tedious computation using an idea of Charlton [126] one can show that $\zeta(2,1,\{2\}^n,3) - 2\zeta(\{2\}^n,3,3)$ is a rational multiple of π^{2n+6} for all $n \in \mathbb{N}_0$.

During mid 1990s Borwein *et al.* started the development of EZ-Face (an abbreviation for Euler Zetas interFace, see Sec. 14.1), an on-line calculator for alternating MZVs called the Euler sums. They detected quite a few relations and the following is one such family (see [71]). First put

$$Z(m_0,\dots,m_{2n}) := \zeta(\{2\}^{m_0},3,\{2\}^{m_1},1,\{2\}^{m_2},\dots,3,\{2\}^{m_{2n-1}},1,\{2\}^{m_{2n}}), \tag{5.114}$$

with $\{2\}^{m_j}$ inserted after the jth element of the string $\{3,1\}^n$. For example, $Z(1,3,2) = \zeta(2,3,2,2,2,1,2,2)$.

Conjecture 5.10.2. *For any sequence $S = (m_0,\dots,m_{2n})$ of $2n+1$ nonnegative integers, we have*

$$\sum_{j=0}^{2n} Z(\mathscr{C}^j S) = \frac{\pi^{4n+2M}}{(4n+2M+1)!}, \tag{5.115}$$

where $M := \sum_{i=0}^{2n} m_i$ and $\mathscr{C}$ is the cyclic permutation operator, that is,

$$\mathscr{C}^j(m_0,\dots,m_{2n}) = (m_{2n-j+1},\dots,m_{2n},m_0,\dots,m_{2n-j}).$$

Note that by Eq. (5.75) the right-hand side of Eq. (5.115) is equal to $\zeta\left(\{2\}^{2n+M}\right)$. Borwein *et al.* proved Eq. (5.115) in [71] for the cases $M = 0$ (which is Eq. (5.75) when $m = 0$) and $M = 1$. For $n = 0$, Eq. (5.115) is exactly Eq. (5.75) when $n = 0$. If all m_i's are equal, Eq. (5.115) reduces to [70, Conjecture (18)].

There are a few conjectured identities between the MZVs with components consisting of only 1, 2 or 3's. Here is a typical one given near the end of [71]. One can verify it with the aid of Petitot's table in [471] up to weight 16 and by EZ-face numerically for larger weight.

Conjecture 5.10.3. *For any $a_1, a_2, a_3, b_1, b_2 \in \mathbb{N}_0$, we have*

$$\begin{aligned} &Z(a_1, b_1, a_2, b_2, a_3) + Z(a_2, b_1, a_3, b_2, a_1) + Z(a_3, b_1, a_1, b_2, a_2) \\ &\quad = Z(a_1, b_2, a_2, b_1, a_3) + Z(a_2, b_2, a_3, b_1, a_1) + Z(a_3, b_2, a_1, b_1, a_2). \end{aligned}$$

The following conjecture, which is corroborated by numerous examples in depth 2 and 3 (see [266]), provides a general form of the weighted sum relations.

Conjecture 5.10.4. *Let $F(x_1, \dots, x_m) \in \mathbb{Q}[x_1, \dots, x_m]$ be a symmetric polynomial of total degree r. Suppose $d = \deg_{x_1} F(x_1, \dots, x_m)$. Then for every positive integer $n \geq m$ we have*

$$\sum_{\substack{k_1+\cdots+k_m=n \\ k_1,\dots,k_m\geq 1}} F(k_1, \dots, k_m)\zeta(2k_1)\dots\zeta(2k_m) = \sum_{k=0}^{K} e_{F,k}(n)\zeta(2k)\zeta(2n-2k),$$

$$\sum_{\substack{k_1+\cdots+k_m=n \\ k_1,\dots,k_m\geq 1}} F(k_1, \dots, k_m)\zeta(2k_1, \dots, 2k_m) = \sum_{k=0}^{K} c_{F,k}(n)\zeta(2k)\zeta(2n-2k),$$

where $K = \max\{\lfloor (r+m-2)/2 \rfloor, \lfloor (m-1)/2 \rfloor\}$, $e_{F,k}(x), c_{F,k}(x) \in \mathbb{Q}[x]$ depend only on k and F, $\deg(e_{F,k}(x)) \leq r-1$ and $\deg(c_{F,k}(x)) \leq d$.

5.11 Historical Notes

Hoffman proved Thm. 5.1.2 for **s** having some special forms such as those in Example 5.1.1. The general case was first settled by the iterated integral approach discovered by Drinfeld, Kontsevich and Zagier (see also the Appendix of [385]). Ohno [443] generalized this to Thm. 5.5.9 by basically the same approach. Our proof here is adopted from Ref. [308] by Ihara *et al.* Another approach is presented in [457]. K.-W. Chen *et al.* [139] obtained a generalization of the duality relations by using some parametrized form of MZVs.

Using generating functions, Igarashi [304] extended Ohno's relation to a parametric form (see Exercise 5.6). This parametric form should have something to do with the one in [139] just mentioned above. Kawashima [342] showed another class of relations among the MZVs containing Ohno's relations. He further conjectured that his relations should contain all cyclic sum relations given by Thm. 5.2.3 which, in turn, was conjectured by Hoffman and confirmed by Ohno [298]. These relations were studied in details by Tanaka in [532]. Ohno's proof with partial fractions was reproduced here. For some other discussions of Ohno's relations, see [458]. In [343],

Kawashima *et al.* extended the cyclic sum relations to multiple L-values (essentially the colored MZVs which we will study in Chap. 13). Roughly at the same time, Igarashi [303] improved these relations to parametrized multiple series defined by

$$\sum_{k_1>\cdots>k_d\geq 0} \frac{(\alpha)_{k_d}}{k_d!} \frac{k_1!}{(\alpha)_{k_1}} \frac{1}{(k_1+\alpha)^{s_1}\cdots(k_d+\alpha)^{s_d}},$$

where $(\alpha)_k$ is the Pochhammer symbol defined by Eq. (1.13). These series clearly reduce to the MZVs when $\alpha = 1$.

The sum relations in Cor. 5.2.4 were conjectured by Moen [288] and by Schmidt and Markett [415] independently. The depth two case was known for quite a long time, see [441, III § 19 (1)]. Hoffman proved the general case under the assumption of the duality relation. The general case of Eq. (5.12) was established independently by A. Granville [256] and D. Zagier using generating functions. Ohno considered similar relations in [444] and provided an application to the Arakawa–Kaneko zeta functions defined in [21]. Okuda and Ueno [459] derived the sum relations in two different ways by using two formulas of the single-variable MPL $\mathrm{L}_{\mathbf{s}}(z)$.

Eie *et al.* [194] improved the sum relations in Eq. (5.12) in the following way:

$$\sum_{k_1+\cdots+k_d=w} \zeta(k_1+1,k_2,\ldots,k_d,\{1\}^e) = \sum_{l_1+\cdots+l_{e+1}=d+e} \zeta(l_1+(w-d)+1,l_2,\ldots,l_{e+1}). \tag{5.116}$$

This reduces to Eq. (5.12) if one sets $e = 0$. Tanaka [533] showed that this follows from the derivation relations in Thm. 5.5.5. In [494] S. Saito *et al.* gave a detailed analysis of the cyclic sum relations by stratifying them using the Hoffman algebras and then proved Thm. 5.2.5.

During their study of the multiple zeta star values (see Chap. 10) Ohno and Zudilin [448] discovered the weighted sum relations in Eq. (5.15) using the DBSFs. Our proof follows the idea in [308]. Equation (5.15) was subsequently generalized by Gue and Xie [260] as follows.

Theorem 5.11.1. *For positive integers $d \geq 2$ and $w \geq d+1$, we have*

$$\sum_{\substack{s_i\geq 1, s_1\geq 2\\ s_1+\cdots+s_d=w}} \left[2^{s_1-1} + (2^{s_1-1}-1)\Big(\Big(\sum_{i=2}^{d-1} 2^{S_i-s_1-(i-1)}\Big) + 2^{S_{d-1}-s_1-(d-2)}\Big)\right] \times \zeta(s_1,\ldots,s_d) = w\zeta(w),$$

where $S_i = s_1+\cdots+s_i$ for $i = 1,\ldots,d-1$.

Using some key ideas from [233], in a series of papers [406, 407, 410, 411] Machide further studied more general weighted sum formulas in depth two to four in which he used some parameters whose specializations would give the preceding formulas in these depths. See Exercises 5.16 and 5.17.

A number of MZV identities (and a few allied infinite sums) have been shown directly by the generating function method. For more research along this direction, one may consult Refs. [15, 16, 233, 295, 592, 592, 630, 634, 639].

Theorem 5.4.2 was discovered by Gangl *et al.* [233] during their study of the double zeta values and the modular forms. In fact, it is contained implicitly already in Euler's work [210]. Our proof here is due to Nakamura [434], which is much shorter than the original proof by Gangl *et al.* However, using their approach Gangl *et al.* further showed Thm. 5.9.11 and Cor. 5.9.19. Moreover, they conjectured that the relations in Cor. 5.9.19 are the only ones between $\zeta(\mathrm{od},\mathrm{od})$ values, which is consistent with Broadhurst's Conjecture 3.4.6. They also produced a more complicated identity.

Theorem 5.11.2. *For all odd integers $m, n \geq 2$, let $w = m + n$. Then we have*

$$2\sum_{i=0}^{n-1}\binom{-m}{i}B_i\zeta(n-i, m+i) = \sum_{j+k=w}(-1)^{k-1}\lambda_{m,n}(j,k)\zeta(j)\zeta(k)$$

where B_i is the ith Bernoulli number and

$$\lambda_{m,n}(j,k) = \sum_{i=0}^{n-1}\binom{m+i-1}{i}\binom{j-1}{n-i-1}B_i.$$

In a more precise way, Machide [408] discovered explicitly a set of generators yielding the upper bound of Eq. (5.107).

Theorem 5.4.3 is due to Nakamura [434]. He also obtained a formula for $\zeta(\{2m\}^n)$ in [433] where one can also find an expression of the Bernoulli numbers with just one summation. The identity Eq. (5.26) between Bernoulli numbers have been obtained by a few diverse methods. Rademacher [478, p. 121] derived it as a consequence of an identity among Eisenstein series of different weights. Shimura essentially did the same in his book [513, (11.10)]. Eie [189] proved it using the zeta function associated with some polynomials. His method can be generalized to obtain infinitely many families of Bernoulli number identities similar to Eq. (5.26) which lead to infinitely many different weighted sum formulas.

By manipulating higher derivatives of the function $f(t) = 1/(e^t - 1)$ Dilcher [167] generalized Eq. (5.21) in Thm. 5.4.12 involving arbitrary N

factors in the product, which was also discovered in [516] for $N = 3$ and $N = 4$, by Sankarayanan [496] for $N = 5$ and by Zhang [604] for $N \le 7$. Theorem 5.4.12 was also settled by Eie [189] in some special cases and by K.-W. Chen [135, 189] in general using completely different method. Our proof here follows that in [135].

With the aid of Eqs. (5.40) and (5.41), Shen and Cai [511] came up with similar statements as Thm. 5.4.2 in depth 3 and 4 cases. As a kind of corollary Shen and Cai [511] also found the following identity whose generalization to larger depth is unknown so far.

Theorem 5.11.3. *For any positive integer $n > 2$, we have*

$$\sum_{\substack{a+b+c=2n \\ a\ge 2, b,c\ge 1}} \big[(-1)^a + (-1)^b + (-1)^c\big]\zeta(a,b,c) = \frac{3}{2}\zeta(2n) - \zeta(2)\zeta(2n-2).$$

The case $r = 1$ of Example 5.4.6 gives [189, Prop. 1] (note that there is a typo there where $(2n-2)!$ on the left numerator should be $(2n-4)!$). Taking $r = 2$ we recover [190, Prop. 4.2.2]. Corollary 5.4.10 is a result used by Shen and Cai [511, Lemma 5] which they derived by some complicated method using Bernoulli numbers. More formulas in the same spirit and some involving alternating sums can be found in [565]. Theorem 5.4.13 was proved by Hoffman [295] using some recurrent system of partial differential equations. In [589] Yuan and the author found an analogous result when all the arguments are multiples of four.

Theorem 5.5.5 was due to K. Ihara and M. Kaneko [308] which generalizes the case $n = 1$ proved by Hoffman [288]. In fact, they showed that the claim in Thm. 5.5.5 is equivalent to the validity of the DBSFs. Our proof only goes in one direction and therefore is much shorter. Kaneko [327] defined a generalization, denoted by $\partial_n^{(c)}$, of the derivation ∂_n so that $\partial_n^{(0)} = \partial_n$ and he conjectured that $\zeta(\partial_n^{(c)}(\mathbf{w})) = 0$ for all $c \in \mathbb{Q}$ and $n \ge 1$. This conjecture was confirmed later by Tanaka [531].

The special case of Thm. 5.6.2 with $z = 0$ (*i.e.*, height 1 case) can be found in [18, 70, 173]. The general case was settled by Ohno and Zagier in [447] and was enhanced to i-height case by Z. Li [394], where i-height of a composition is the number of components greater than i. So the classical height is just 1-height. As a cognate result of Cor. 5.6.4 Kajikawa [324] showed that the relation

$$G_0(w,d,h) = \sum_{\mathbf{s}\in I_0(w,d,h)} \zeta(\mathbf{s}) = \sum_{\mathbf{s}\in I_0(w,d,h)} \zeta(\mathbf{s}^*)$$

can be derived from the finite and regularized DBSFs. But in general we don't know how to prove the duality relations using only the finite and regularized DBSFs. In [399], Li also gave an algebraic proof of the Ohno–Zagier relation using the regularized DBSFs.

Zagier's evaluation Thm. 5.7.4 from [601] was achieved originally by using a few deep results in complex analysis. Here, we choose to present Z.-H. Li's shorter hypergeometric function proof contained in [398]. Li also found another algebraic proof using the Hoffman algebras in [400]. Theorem 5.7.4 plays an important role in the work of F. Brown [102] on the mixed Tate motives over $\mathbb{Z}$. As a corollary, Brown proved that every MZV can be expressed as a $\mathbb{Q}$-linear combination of the MZVs whose arguments are either 2 or 3. Theorem 5.7.4 is also used in F. Brown and O. Schnetz's proof of the Zig-Zag Conjecture in [108]. Terasoma [542] further showed that Thm. 5.7.4 can be augmented to the coefficients of any associator, in particular, to the motivic multiple zeta values. This provides another proof of Brown's result.

Theorem 5.8.1 was a special case of Conjecture 5.10.2 and was proved first by Bowman and Bradley [77, Cor. 5.1]. Two simplified proofs can be found in [430] and [616]. The proof here is adopted from [616]. The case when $n = 0$, *i.e.*, Cor. 5.6.4, was known to Euler and re-proved by Hoffman [288]. The case $m = 0$ was conjectured by Zagier [598] and verified first by Broadhurst [71]. Charlton [126] improved Eq. (5.75) to show that the sum of inserting all possible permutations of some fixed blocks of 2's into $\zeta(3, 1, \ldots, 3, 1)$ produces a rational multiple of a power of π. The novelty of this proof is the use of the motivic MZV theory developed by Deligne, Goncharov and Brown [102]. However, it does not provide the explicit multiple of the π-power specified in Conjecture 5.10.2.

One of the most fascinating aspects of the MZVs is the intimate relations between the double zeta values, modular forms and the period polynomials. For the even weight case, we have adopted the framework of [233] in Sec. 5.9. Schneps gave a different approach to these results in [501]. Baumard and Schneps [40] further connected the identities in Cor. 5.9.19 to those in Ihara's double shuffle Lie algebra and showed how they can be deduced from each other by duality. For the odd case, Thm. 5.9.23 was proved by Cartier [118] and Zagier [601] independently with essentially the same algebraic approach which we have adopted in the book. Theorems 5.9.24 and 5.9.26 are both due to Ma (see [405]). Pertinent to this line of research, Tasaka [536] studied the structure of the MZVs at odd integer arguments and their relation to the period polynomials.

The MZVs are known to be linked to the multi-poly-Bernoulli numbers which were first defined by Hamahata and Masubuchi [272] (based on Kaneko's poly-Bernoulli numbers in [326, 328, 329]), extended further by Bayad and Hamahata [42], and modified by Imatomi *et al.* [312] using the generating function

$$\frac{\mathrm{Li}_{s_1,s_2,\ldots,s_d}(1-e^{-t})}{e^t-1} = \sum_{n=0}^{\infty} C_n(s_1,\ldots,s_d)\frac{t^n}{n!}. \tag{5.117}$$

This generating function is intimately related to the Arakawa–Kaneko zeta function defined in [21] whose special values can be expressed using modified Bell polynomials (see [137, 151, 335]). A kind of "twin sibling" of this function was studied in [335] by Kaneko and Tsumura.

The input from the knot theory stimulated quite a lot of research on the MZVs at the initial stage. For example, the knot invariant of the 3-braid torus knot 8_{19} at 8 crossings (left of Fig. 5.1, according to Rolfsen's well-known table [485]) leads to the conjecturally irreducible MZV $\zeta(5,3)$ of depth 2 occurring first at weight 8, while knot 10_{124} (right of Fig. 5.1) involves the double zeta $\zeta(7,3)$ (see [90]). Likewise the first (conjecturally)

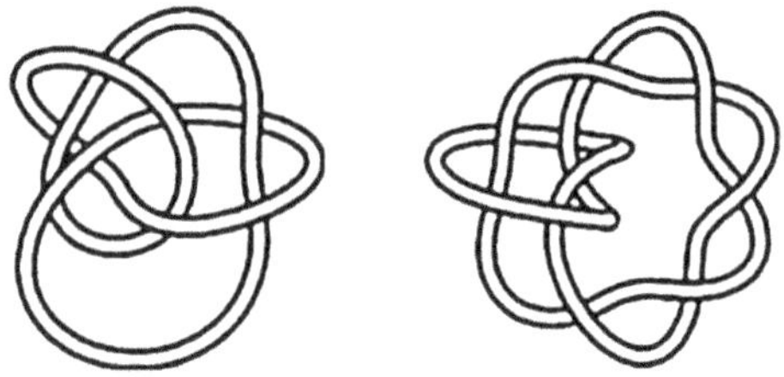

Figure 5.1: Knots 8_{19} and 10_{124} are related to irreducible depth 2 MZVs.

irreducible MZVs of depth 3 occurs at weight 11 (for e.g., $\zeta(3,3,5)$) by [69], matching the appearance of the first positive 4-braid at 11 crossings, obtained from skeining link diagrams that encode momentum flow in 7-loop counterterms [89]. Later, Broadhurst and his collaborators also found the first weight 12 MZV reduction that requires the alternating Euler sums in [92]. A more systematic study of the relation between the knots and the MZVs can be found in [90].

In [384] Le and Murakami dealt with Kontsevich integrals of Homfly polynomials using representations of the chord diagram algebra. In the process they found a few interesting relations of the MZVs. For example,

$$(1-2^{2-2n})\zeta(2n)=\sum_{k=2}^{2n-1}(-1)^k\zeta\big(k,\{1\}^{2n-k}\big)\quad \forall n\geq 2.$$

They also discovered the following

$$\sum_{\mathbf{s}:|\mathbf{s}|=2n,\ \mathrm{ht}(\mathbf{s})=k}(-1)^{\mathrm{dp}(\mathbf{s})}\zeta(\mathbf{s})=(-1)^n\zeta(\{2\}^n)\sum_{j=0}^{n-k}\binom{2n+1}{2j}(2-2^{2j})B_{2j}$$

by examining the Kontsevich integral of the unknot. This can also be explained by setting $y=-x$ in Thm. 5.6.2. More identities produced from knot theory can be found in [222, 306, 529].

Exercises

5.1. Find the dual $\mathbf{s}^*$ if $\mathbf{s}$ is given by
(i) $(4,2,1)$ (ii) $(3,\{2\}^n)$

5.2. Suppose $\mathbf{s}=(s_1,\dots,s_d)$ and $\mathbf{w}=\mathsf{a}^{s_1-1}\mathsf{b}\cdots\mathsf{a}^{s_d-1}\mathsf{b}\in\mathfrak{A}^0$. If $\mathbf{s}^*=(t_1,\dots,t_\ell)$ then prove that $\mathsf{a}^{t_1-1}\mathsf{b}\cdots\mathsf{a}^{t_\ell-1}\mathsf{b}=\mathsf{a}\mathsf{b}^{s_d-1}\cdots\mathsf{a}\mathsf{b}^{s_1-1}$.

5.3. Show that for every prime $p\geq 5$

$$\sum_{\substack{i+j+k=p\\ i,j,k>0}}\frac{1}{ijk}\equiv -2B_{p-3}\pmod p,$$

where B_{p-3} is the Bernoulli number.

5.4. Show that Eq. (5.9) and Eq. (5.10) are equivalent.

5.5. Prove the following shuffle-regularized sum formula: for any fixed $w,r\in\mathbb{N}$, we have

$$\sum_{\mathbf{s}\in\mathbb{N}^d,|\mathbf{s}|=w+d}\zeta_{\sqcup\!\sqcup}(\mathbf{s})=(-1)^{d-1}\zeta\big(w+1,\{1\}^{d-1}\big),$$

where $\zeta_{\sqcup\!\sqcup}$ is defined by Remark 3.3.18.

5.6. Let $\alpha\in\mathbb{C}$ such that $\mathrm{Re}(\alpha)>0$. Define

$$Z(\mathbf{s};\alpha)=\sum_{k_1>\cdots>k_d>0}\frac{(\alpha)_{k_1}}{k_1!}\frac{k_d!}{(\alpha)_{k_d+1}}\frac{1}{(k_1+\alpha)^{s_1}\cdots(k_d+\alpha)^{s_d}}.$$

where $(\alpha)_k$ is the Pochhammer symbol defined by Eq. (1.13). Then for any integer $k\geq 0$ admissible $\mathbf{s}=(s_1,\dots,s_d)\in\mathbb{N}^d$, show that

$$\sum_{\substack{\varepsilon_1+\cdots+\varepsilon_d=k\\ \varepsilon_1,\dots,\varepsilon_d\in\mathbb{N}_0}}Z(s_1+\varepsilon_1,\dots,s_d+\varepsilon_d;\alpha)=\sum_{\substack{\varepsilon'_1+\cdots+\varepsilon'_\ell=k\\ \varepsilon'_1,\dots,\varepsilon'_d\in\mathbb{N}_0}}Z(s'_1+\varepsilon'_1,\dots,s'_\ell+\varepsilon'_\ell;\alpha).$$

5.7. One may use the following idea to give an alternative elementary proof of Thm. 5.4.1. First, rewrite the double sum as

$$\sum_{k=2}^{w-1}\sum_{m,n=1}^{\infty}\frac{2^k}{n^{w-k}(n+m)^k}.$$

Then break the sum into two cases $m=n$ and $m\neq n$. Show that the first case yields $(w-2)\zeta(w)$ and the second yields $3\zeta(w)$ by summing over k first and then using telescoping series.

5.8. Let $w\geq 2$ be a positive integer. Using the generating function of the double zeta values to show the following restricted sum formulas:

$$\left(\sum_{k\equiv 3\,(6)}-\sum_{k\equiv 4\,(6)}-\sum_{k\equiv 5\,(6)}\right)\zeta(k,l)=\frac{1}{3}\sum_{k\text{ odd}}\zeta(k,l)\quad\text{if } w\equiv 0\ (3),$$

$$\left(\sum_{k\equiv 3\,(6)}+\sum_{k\equiv 4\,(6)}-\sum_{k\equiv 5\,(6)}\right)\zeta(k,l)=\frac{1}{3}\sum_{k\text{ even}}\zeta(k,l)\quad\text{if } w\equiv 1\ (3),$$

$$\frac{1}{6}\zeta(w)-\sum_{k\equiv 4\,(6)}\zeta(k,l)=\frac{1}{3}\sum_{k\text{ odd}}\zeta(k,l)\quad\text{if } w\equiv 2\ (3),$$

where in the sums we always assume $k\geq 2, l\geq 1$ and $k+l=w$.

5.9. Show that the commutator of two derivations is a derivation.

5.10. For any $t\in\mathfrak{z}$ define a power series $f(u)\in\mathfrak{z}[\![u]\!]$ by

$$f(u)=\exp_\circ(tu)-1=\sum_{n\geq 1}\frac{t^{\circ n}}{n!}u^n.$$

(i) Show that

$$\frac{\mathrm{d}}{\mathrm{d}u}\left(\frac{1}{1-f(u)}\right)=t*\left(\frac{1}{1-f(u)}\right)$$

by using the fact that $f'(u)=t\circ(1+f(u))$.

(ii) Show that $F(u)=(1-f(u))^{-1}$ satisfies $F'(u)=z*F(u)$ and $F(0)=1$. Consequently, one has

$$\exp_*(tu)=(2-\exp_\circ(tu))^{-1}.$$

5.11. Let $t\in\mathfrak{z}$ and $\mathbf{w}\in\widehat{\mathfrak{A}}^1$.

(i) Show that for any $z\in\mathfrak{z}$

$$\delta_t(z\mathbf{w})=(t\circ z)\mathbf{w}+z(t*\mathbf{w}).$$

(ii) Show that for any positive integer n

$$\delta_t^n(z\mathbf{w}) = \sum_{\alpha+\beta=n,\alpha,\beta\geq 0} \binom{n}{\alpha}(t^{\circ\alpha}\circ z)(t^{*\beta} * \mathbf{w}).$$

(iii) Show that

$$\exp(\delta_t)(\mathbf{w}) = (\exp_*(t))^{-1}(\exp_*(t) * \mathbf{w}).$$

5.12. Show that $\mathfrak{d}: \mathfrak{A} \to \mathfrak{A}$ defined by $\mathfrak{d}(\mathbf{w}) = \mathtt{b} \sqcup\!\sqcup \mathbf{w} - \mathtt{b}\mathbf{w}$ is a derivation on $\mathfrak{A}$ such that $\mathfrak{d}^m(\mathtt{a})/m! = \mathtt{a}\mathtt{b}^m$ and $\mathfrak{d}^m(\mathtt{b})/m! = \mathtt{b}^{m+1}$ for all $m \geq 1$. Further, show that $\mathfrak{d}$ induces a derivation on $\widehat{\mathfrak{A}}^0$ and then verify Eq. (5.43) by induction. In particular,

$$\exp(\mathfrak{d}u)(\mathtt{a}) = \mathtt{a}(1-\mathtt{b}u)^{-1}, \ \exp(\mathfrak{d}u)(\mathtt{b}) = \mathtt{b}(1-\mathtt{b}u)^{-1},$$

and

$$\exp(\mathfrak{d}u)(\mathtt{a}+\mathtt{b}) = (\mathtt{a}+\mathtt{b})(1-\mathtt{b}u)^{-1}.$$

5.13. Show that $\delta_z : \widehat{\mathfrak{A}}^1 \to \widehat{\mathfrak{A}}^1$ defined by $\delta(\mathbf{w}) = z * \mathbf{w} - z\mathbf{w}$ is a derivation on $\widehat{\mathfrak{A}}^1$ for every $z \in \mathfrak{z}$, and these derivations all commute. Furthermore, δ_z extends to a derivation on all of $\widehat{\mathfrak{A}}$, with

$$\delta_z(\mathtt{a}) = 0 \quad \text{and} \quad \delta_z(\mathtt{b}) = (\mathtt{a}+\mathtt{b})z.$$

5.14. Show that γ defined by Eq. (5.46) provides an algebra isomorphism

$$\gamma : X\mathbb{Q}[\![X]\!] \xrightarrow{\simeq} (\mathfrak{z}, \circ).$$

5.15. Use your favorite computer algebra system to find c_1 and c_2 in the proof of Prop. 5.8.2. One can download the free computer programs for the WZ-method from the internet after searching "WZ-method".

5.16. Use appropriate specializations of the generating function for double and triple zeta values to show that for all $w \geq 1$

$$\sum_{a\geq 2,b\geq 1,a+b<w} \Big(3^{a-1}2^b - 2^{a+b-1} - 2^{a-1}\Big)\zeta(a,b,w-a-b)$$
$$= \frac{(w-4)(w+1)}{6}\zeta(w), \tag{5.118}$$
$$\sum_{a\geq 2,b\geq 1,a+b<w} \Big(2^{a+b-1} + 2^{a-1} - 2^b\Big)\zeta(a,b,w-a-b) = w\zeta(w), \tag{5.119}$$
$$\sum_{a\geq 2,b\geq 1,a+b<w} 2^b\Big(3^{a-1} - 1\Big)\zeta(a,b,w-a-b) = \frac{(w-1)(w+4)}{6}\zeta(w).$$

Note that Eq. (5.119) is the depth $d = 3$ case of Thm. 5.11.1 by Guo and Xie.

5.17. Use the generating function for depth 4 MZVs to show that

$$\sum_{s_1+\cdots+s_4=w, s_1\geq 2, s_j,\geq 1} (3^{s_2}2^{s_1-1}-3^{s_2}-1)2^{s_1+s_3}\zeta(s_1,\dots,s_4) = \left(\frac{(w+7)(w+2)(w-3)}{12}+2\right)\zeta(w).$$

5.18. Let $k, r \geq 0$ be two even integers. Show that

$$\sum_{j=0}^{k}\sum_{\ell=0}^{r}\sum_{\substack{|\boldsymbol{\alpha}|=j+r-\ell+1\\ |\boldsymbol{\beta}|=k-j+\ell+1}} \zeta(\alpha_0,\alpha_1,\dots,\alpha_j+\beta_j,\beta_{j+1},\dots,\beta_k,2) = \sum_{|\boldsymbol{\alpha}|=k+r-2} \zeta(\alpha_0,\alpha_1,\dots,\alpha_k,2).$$

5.19. Prove Eq. (5.98) using the identity $(aX+bY)-(cX+dY)\tau = (a-c\tau)(X-\gamma^{-1}(\tau)Y)$ and the modularity of f.

5.20. Prove that in Part (ii) of Prop. 5.9.10 we can take

$$\lambda' = \left\langle H, \frac{X^{w-1}-Y^{w-1}}{X-Y}\right\rangle.$$

5.21. Show that if $f(x)$ is an odd polynomial in $\mathbb{Q}[x]$ of degree $2d-1$ then there is $g(x)\in\mathbb{Q}[x]$ of degree $\leq 2d$ such that $f(x)=g(x+1)-g(x)$.

5.22. By direct computation show that

$$\begin{aligned}(\varepsilon-\varepsilon U)(T-I)(I-\delta) &= (I-T\delta)(\varepsilon-\varepsilon U)T(I-\delta),\\ T+\varepsilon T\varepsilon &= I+(I+U+U^2)S-(I+S)S.\end{aligned}$$

5.23. Instead of using the generating function of the double zeta values, prove Thm. 5.9.1 by two suitable choices of H in Prop. 5.9.9.

5.24. Show that the Eq. (5.104) in Thm. 5.9.14 can be strengthened as

$$\sum_{\substack{r+s=w\\ r,s \text{ even}}} q_{r,s}Z_{r,s} + \sum_{r+s=w}(-1)^r q_{r,s}Z_w = 3\sum_{\substack{r+s=w\\ r,s\text{ odd}}} q_{r,s}Z_{r,s}.$$

5.25. Complete the proof of Lemma 5.9.17 by using the method of generating functions and the fact that

$$\left(\frac{X^{w-1}}{Y}+\frac{X^{w-1}}{Y}\right)\bigg|_{I+U+U^2} = \frac{X^{w-1}-(X-Y)^{w-1}}{Y} + \frac{Y^{w-1}+(X-Y)^{w-1}}{X} + \frac{X^{w-1}-Y^{w-1}}{X-Y}.$$

Chapter 6

Symmetrized Multiple Zeta Values (SMZVs)

In the preceding two chapters, we have considered many interesting properties of the MZVs. However, one unpleasant feature causes a lot of trouble, namely, admissible MZVs must start with an argument at least two. One effective way to overcome this difficulty is to define some *symmetrized* versions of the MZVs, which will be carried out in this chapter.

In Sec. 8.5.4 we will see a mysterious relation between the MZVs and so-called finite version of the MZVs. This relation will be established by using the symmetrized MZVs.

6.1 Definition

There are in fact two versions of the symmetrized MZVs. Recall that we have studied the regularized MZVs $\zeta(\mathbf{s};T)$ in Sec. 3.3. In general, these are polynomials in T.

Definition 6.1.1. For any $\mathbf{s}=(s_1,\dots,s_d)\in\mathbb{N}^d$, we define

$$\zeta^{\mathcal{S}}_{ш}(\mathbf{s})=\sum_{i=0}^{d}(-1)^{s_1+\cdots+s_i}\zeta_{ш}(s_i,\dots,s_1;T)\zeta_{ш}(s_{i+1},\dots,s_d;T), \tag{6.1}$$

$$\zeta^{\mathcal{S}}_{*}(\mathbf{s})=\sum_{i=0}^{d}(-1)^{s_1+\cdots+s_i}\zeta_{*}(s_i,\dots,s_1;T)\zeta_{*}(s_{i+1},\dots,s_d;T). \tag{6.2}$$

These are called the $ш$*-symmetrized MZVs* and $*$*-symmetrized MZVs*, respectively. Both are called symmetrized MZVs (SMZVs) if we don't need to distinguish the two versions.

Remark 6.1.2. Although perhaps looking strange at the first glance, the $*$-symmetrized MZVs have in fact a natural series definition (see Prop. 6.1.4).

First, we show that neither $\zeta^{\mathcal{S}}_{ш}(\mathbf{s})$ nor $\zeta^{\mathcal{S}}_{*}(\mathbf{s})$ depends on the variable T, namely, they are actually real numbers.

Proposition 6.1.3. *For all $s_1, \ldots, s_d \in \mathbb{N}$, both $\zeta^{\mathcal{S}}_{ш}(s_1, \ldots, s_d)$ and $\zeta^{\mathcal{S}}_{*}(s_1, \ldots, s_d)$ are finite real numbers. In particular,*

$$\zeta^{\mathcal{S}}_{ш}(\mathbf{s}) = \sum_{i=0}^{d} (-1)^{s_1+\cdots+s_i} \zeta_{ш}(s_i, \ldots, s_1)\zeta_{ш}(s_{i+1}, \ldots, s_d),$$

$$\zeta^{\mathcal{S}}_{*}(\mathbf{s}) = \sum_{i=0}^{d} (-1)^{s_1+\cdots+s_i} \zeta_{*}(s_i, \ldots, s_1)\zeta_{*}(s_{i+1}, \ldots, s_d).$$

Proof. We start with the shuffle version $\zeta^{\mathcal{S}}_{ш}(s_1, \ldots, s_d)$. If all $s_i \geq 2$ then clearly it is finite. If all $s_i = 1$ then

$$\zeta^{\mathcal{S}}_{ш}(\mathbf{s}) = \sum_{i=0}^{d} (-1)^{d-i} \frac{T^i}{i!} \frac{T^{d-i}}{(d-i)!} = 0.$$

Otherwise, we may assume for some i and $l \geq 1$ we have $s_i \geq 2$, $s_{i+1} = \cdots = s_{i+l} = 1$, and $s_{i+l+1} \geq 2$. We only need to show that

$$\sum_{j=0}^{l} (-1)^j \zeta_{ш}(s_{i+j}, \ldots, s_1)\zeta_{ш}(s_{i+j+1}, \ldots, s_d)$$

is finite. Consider the words in $\mathfrak{A}^1_{ш}$ (see Definition 3.2.2) corresponding to the above values. We may assume $\mathrm{x}_0\mathbf{u} = \mathrm{x}_0^{s_i-1}\mathrm{x}_1 \cdots \mathrm{x}_0^{s_1-1}\mathrm{x}_1$ and $\mathrm{x}_0\mathbf{v} = \mathrm{x}_0^{s_{i+l+1}-1}\mathrm{x}_1 \cdots \mathrm{x}_0^{s_d-1}\mathrm{x}_1$. So we only need to show that

$$\sum_{j=0}^{l} (-1)^j (\mathrm{x}_1^j \mathrm{x}_0 \mathbf{u}) ш (\mathrm{x}_1^{l-j} \mathrm{x}_0 \mathbf{v}) \in \mathfrak{A}^0. \tag{6.3}$$

Indeed,

$$\begin{aligned}
&\sum_{j=0}^{l} (-1)^j (\mathrm{x}_1^j \mathrm{x}_0 \mathbf{u}) ш (\mathrm{x}_1^{l-j} \mathrm{x}_0 \mathbf{v}) \\
&\quad = \mathrm{x}_0(\mathbf{u} ш \mathrm{x}_1^l \mathrm{x}_0 \mathbf{v}) + \sum_{j=1}^{l} (-1)^j \mathrm{x}_1(\mathrm{x}_1^{j-1}\mathrm{x}_0\mathbf{u} ш \mathrm{x}_1^{l-j}\mathrm{x}_0\mathbf{v}) \\
&\qquad + (-1)^l \mathrm{x}_0(\mathrm{x}_1^l \mathrm{x}_0 \mathbf{u} ш \mathbf{v}) + \sum_{j=0}^{l-1} (-1)^j \mathrm{x}_1(\mathrm{x}_1^j \mathrm{x}_0 \mathbf{u} ш \mathrm{x}_1^{l-j-1}\mathrm{x}_0\mathbf{v}) \\
&\quad = \mathrm{x}_0(\mathbf{u} ш \mathrm{x}_1^l \mathrm{x}_0 \mathbf{v}) + (-1)^l \mathrm{x}_0(\mathrm{x}_1^l \mathrm{x}_0 \mathbf{u} ш \mathbf{v}) \in \mathfrak{A}^0,
\end{aligned}$$

where we have used the substitution $j \to j+1$ in the first sigma summation which is canceled by the second sigma summation.

For the stuffle version $\zeta^{\mathcal{S}}_{*}(s_1, \ldots, s_d)$, we can use the same idea and show it's finite by noticing that the "stuffing" part of the stuffle product produces admissible words in $\mathfrak{A}^0$. We leave the details to the interested reader (see Exercise 6.2). □

Another useful way to define the $*$-symmetrized MZVs is to use some infinite series.

Proposition 6.1.4. *For any* $\mathbf{s} = (s_1, \dots, s_d) \in \mathbb{N}^d$, *the limit*

$$\lim_{M\to\infty} \sum_{\substack{\frac{1}{k_1} > \cdots > \frac{1}{k_d} \\ 0 \neq |k_j| \in \mathbb{Z}_{\leq M} \ \forall j=1,\dots,d}} \frac{1}{k_1^{s_1} \cdots k_d^{s_d}}$$

is equal to $(-1)^{|\mathbf{s}|} \zeta_*^{\mathcal{S}}(\mathbf{s})$.

Proof. We see that for all for $\frac{1}{k_1} > \cdots > \frac{1}{k_d}$ there is some i ($i = 0, \dots, d$) such that $\frac{1}{k_1} > \cdots \frac{1}{k_i} > 0 > \frac{1}{k_{i+1}} > \cdots > \frac{1}{k_d}$. Hence

$$\begin{aligned}
(-1)^{|\mathbf{s}|} & \sum_{\substack{\frac{1}{k_1} > \cdots > \frac{1}{k_d} \\ 0 \neq |k_j| \in \mathbb{Z}_{\leq M} \ \forall j=1,\dots,d}} \frac{1}{k_1^{s_1} \cdots k_d^{s_d}} \\
&= \sum_{i=0}^{d} \sum_{\substack{\frac{1}{k_1} > \cdots > \frac{1}{k_i} > 0 > \frac{1}{k_{i+1}} > \cdots > \frac{1}{k_d} \\ 0 \neq |k_j| \in \mathbb{Z}_{\leq M} \ \forall j=1,\dots,d}} \frac{1}{k_1^{s_1} \cdots k_d^{s_d}} \\
&= \sum_{i=0}^{d} (-1)^{s_1+\cdots+s_i} H_M(s_i, \dots, s_1) H_M(s_{i+1}, \dots, s_d)
\end{aligned}$$

where H_M is defined by (3.18). Clearly H_M satisfies the stuffle relation so that

$$\begin{aligned}
\sum_{i=0}^{d} & (-1)^{s_1+\cdots+s_i} H_M(s_i, \dots, s_1) H_M(s_{i+1}, \dots, s_d) \\
&= \sum_{i=0}^{d} (-1)^{s_1+\cdots+s_i} H_M\big((s_i, \dots, s_1) * (s_{i+1}, \dots, s_d)\big) \\
&= \sum_{i=0}^{d} (-1)^{s_1+\cdots+s_i} \zeta_*\big((s_i, \dots, s_1) * (s_{i+1}, \dots, s_d)\big) + O(T^{-1} \log^A(T))
\end{aligned}$$

by Lemma 3.3.19. Here $T = \log M + \gamma$, γ is Euler's constant, and A is some contant. We also have used the same proof as in Prop. 6.1.3 to reduce the expression in the stuffle by using only admissible compositions for which Lemma 3.3.19 applies. Since ζ_* satisfies the stuffle relation, we finally obtain the expression

$$\sum_{i=0}^{d} (-1)^{s_1+\cdots+s_i} \zeta_*(s_i, \dots, s_1; T) \zeta_*(s_{i+1}, \dots, s_d; T) + O(T^{-1} \log^A(T)).$$

By Prop. 6.1.3 we see that the sum above is also equal to $\zeta^{\mathcal{S}}_*(\mathbf{s})$. □

This proposition clearly implies the following corollary.

Corollary 6.1.5. *The $*$-symmetrized MZVs satisfy the stuffle relations.*

6.2 Identities Involving SMZVs

In this subsection, we will look at some relations among the SMZVs that are analogs of those for ordinary MZVs. First, we want to present a kind of sum formula, which requires a preliminary combinatorial lemma. For any $w, d, i \in \mathbb{N}$ with $1 \le i \le d \le w-1$, we set

$$J_{w,d,i} = \{(s_1, \dots, s_d) \in \mathbb{N}^n \mid s_1 + \cdots + s_d = w,\ s_i \ge 2\}.$$

Lemma 6.2.1. *Let f be a multiple integer variable function satisfying the stuffle relations, namely, $f(\mathbf{t}_1)f(\mathbf{t}_2) = f(\mathbf{t}_1 * \mathbf{t}_2)$ for any $\mathbf{t}_j \in \mathbb{N}^{d_j}$ $(j = 1, 2)$. Set*

$$S_{w,d,i} := \sum_{(s_1,\dots,s_d)\in J_{w,d,i}} f(s_1, \dots, s_d).$$

If $f(s) = 0$ for all $s \in \mathbb{N}$ then for any $w, d, i \in \mathbb{N}$ with $2 \le i+1 \le d \le w-1$, we have

$$(d-i)S_{w,d,i} + iS_{w,d,i+1} + (w-d)S_{w,d-1,i} = 0.$$

Proof. By stuffle relation

$$\begin{aligned} 0 &= f(l)f(s_1, \dots, s_{d-1}) \\ &= \sum_{j=1}^{d} f(s_1, \dots, s_{j-1}, l, s_j, \dots, s_{d-1}) + \sum_{j=1}^{d-1} f(s_1, \dots, s_{j-1}, s_j + l, s_{j+1}, \dots, s_{d-1}). \end{aligned}$$

Summing this over all $(s_1, \dots, s_{d-1}, l) \in J_{w,d,i}$ gives the desired recurrence relations. Indeed, the map

$$(s_1, \dots, s_{d-1}, l) \mapsto (s_1, \dots, s_{j-1}, l, s_j, \dots, s_{d-1})$$

defined on $J_{w,d,i}$ is a bijection to $J_{w,d,i+1}$ for $1 \le j \le i$ and is surjection onto $J_{w,d,i}$ for $i < j \le d$. Moreover, under the map

$$\begin{aligned} J_{w,d,i} \times \{1, \dots, d-1\} &\longrightarrow J_{w,d-1,i} \\ \big((s_1, \dots, s_{d-1}, l), j\big) &\longmapsto (s_1, \dots, s_{j-1}, s_j + l, s_{j+1}, \dots, s_{d-1}), \end{aligned}$$

the pre-images of each $(s'_1, \dots, s'_{d-1}) \in J_{w,d-1,i}$ form a set of cardinality

$$\sum_{\substack{1\le j\le d-1 \\ j\ne i}} (s'_j - 1) + (s'_i - 2) = w - d.$$

This completes the proof of the lemma. □

The identity in the next theorem is often called a sum formula. In the proof we will use Thm. 6.3.2 from next section, which one can check does not depend on the result of this section.

Theorem 6.2.2. *For any* $\mathbf{s} = (s_1, \ldots, s_d) \in \mathbb{N}^d$ *and* $w, i \in \mathbb{N}$ *with* $1 \le i \le d \le w-1$, *we have, modulo* $\zeta(2)$,

$$\sum_{|\mathbf{s}|=w,\ s_i \ge 2} \zeta_*^{\mathcal{S}}(\mathbf{s}) \equiv (-1)^{i-1}\left(\binom{w-1}{i-1} + (-1)^d \binom{w-1}{d-i}\right)\zeta(w). \tag{6.4}$$

Proof. We prove the theorem by backward induction on the depth d. When $d = w-1$ the right-hand side of Eq. (6.4) is 0 when w is even and is

$$(-1)^{i-1}\left(\binom{w-1}{i-1} + \binom{w-1}{i}\right)\zeta(w) = (-1)^{i-1}\binom{w}{i}\zeta(w)$$

when w is odd. Now consider the left-hand side of Eq. (6.4). Since $\zeta_*^{\mathcal{S}}$ satisfies the stuffle relation by Cor. 6.1.5 we may apply Lemma 6.2.1 with $f = \zeta_*^{\mathcal{S}}$. By Cor. 5.5.7 we see that

$$\mathrm{reg}_{ш}(\mathtt{x}_1^m \mathtt{x}_0 \mathtt{x}_1^j) = (-1)^m \mathtt{x}_0(\mathtt{x}_1^m ш \mathtt{x}_1^j) = (-1)^m \binom{j+m}{m} \mathtt{x}_0 \mathtt{x}_1^{m+j}.$$

Therefore

$$\begin{aligned}\zeta_{ш}(\{1\}^{i-1}, 2, \{1\}^{w-i-1}) &= (-1)^{i-1}\binom{w-1}{i-1}\zeta(2, \{1\}^{w-2}) \\ &= (-1)^{i-1}\binom{w-1}{i-1}\zeta(w)\end{aligned}$$

by the duality relation (5.3). Similarly

$$\zeta_{ш}(\{1\}^{w-i-1}, 2, \{1\}^{i-1}) = (-1)^{w-i-1}\binom{w-1}{i}\zeta(w).$$

Thus by the definition and Thm. 6.3.2

$$\begin{aligned}S_{w,w-1,i} &= \zeta_*^{\mathcal{S}}(\{1\}^{i-1}, 2, \{1\}^{w-i-1} \equiv \zeta_{ш}^{\mathcal{S}}(\{1\}^{i-1}, 2, \{1\}^{w-i-1}) \\ &\equiv \zeta_{ш}(\{1\}^{i-1}, 2, \{1\}^{w-i-1}) + (-1)^w \zeta_{ш}(\{1\}^{w-i-1}, 2, \{1\}^{i-1}) \\ &\equiv (-1)^{i-1}\binom{w}{i}\zeta(w) \qquad \pmod{\zeta(2)}.\end{aligned}$$

So the left-hand side of Eq. (6.4) agrees with the right-hand side.

Now we assume that Eq. (6.4) holds for some $d \le w-1$. By Lemma 6.2.1

$$\begin{aligned}&(d-w)S_{w,d-1,i} = (d-i)S_{w,d,i} + iS_{w,d,i+1} \\ &= (d-i)(-1)^{i-1}\left(\binom{w-1}{i-1} + (-1)^d\binom{w-1}{d-i}\right)\zeta(w)\end{aligned}$$

$$+ i(-1)^i\left(\binom{w-1}{i} + (-1)^d\binom{w-1}{d-i-1}\right)\zeta(w)$$
$$= (-1)^{i-1}\left((d-i)\binom{w-1}{i-1} + (w-d+i)(-1)^d\binom{w-1}{d-i-1}\right)\zeta(w)$$
$$+ (-1)^i\left((w-i)\binom{w-1}{i-1} + i(-1)^d\binom{w-1}{d-i-1}\right)\zeta(w)$$
$$= (d-w)(-1)^{i-1}\left(\binom{w-1}{i-1} + (-1)^{d-1}\binom{w-1}{d-i-1}\right)\zeta(w).$$

Thus Eq. (6.4) holds for $d-1$ and the theorem is proved. □

Recall that the multiple zeta star values are defined by

$$\zeta^\star(\mathbf{s}) = \sum_{k_1\ge\cdots\ge k_d>0} k_1^{-s_1}\cdots k_d^{-s_d}.$$

Lemma 6.2.3. *For positive integers k and n with $k \ge 2$, we have the duality in height 1:*

$$\zeta_{ш}(\{1\}^{d-1}, k) = (-1)^{d-1}\zeta^\star(k, \{1\}^{d-1}).$$

Proof. Set $(\dots, i, \{1\}^{-1}, j, \dots) = (\dots, i+j-1, \dots)$. By Cor. 5.5.7, we have

$$\zeta_{ш}(\{1\}^{d-1}, k) = (-1)^{d+k}\zeta_{ш}(\mathrm{x}_0(\mathrm{x}_1^{d-1} ш \mathrm{x}_0^{k-2}\mathrm{x}_1)) \quad (\text{word depth}=k-1)$$
$$= (-1)^{d+k}\sum_{\substack{e_1+\cdots+e_{k-1}=d-1\\ e_i\ge 0\,(i=1,2,\dots,k-1)}} (e_1+1)\zeta(\mathrm{x}_0\mathrm{x}_1^{e_{k-1}}\cdots\mathrm{x}_0\mathrm{x}_1^{e_1+1})$$
$$= (-1)^{d+k}\sum_{\substack{e_1+\cdots+e_{k-1}=d-1\\ e_i\ge 0\,(i=1,2,\dots,k-1)}} (e_1+1)\zeta(\mathrm{x}_0^{e_1+1}\mathrm{x}_1\cdots\mathrm{x}_0^{e_{k-1}}\mathrm{x}_1) \quad (\text{by Eq. (5.3)})$$
$$= (-1)^{d-1}\sum_{\substack{e_1+\cdots+e_{k-1}=d-1\\ e_i\ge 0\,(i=1,2,\dots,k-1)}} (e_1+1)\zeta(2+e_1, 1+e_2, \dots, 1+e_{k-1})$$
$$= (-1)^{d-1}\sum_{t=1}^{d}\sum_{\substack{e_1+\cdots+e_{k-1}=d-t\\ e_i\ge 0\,(i=1,2,\dots,k-1)}} \zeta(t+1+e_1, 1+e_2, \dots, 1+e_{k-1})$$
$$= (-1)^{d-1}\sum_{t=1}^{d}\sum_{\substack{e'_1+\cdots+e'_t=d-t\\ e'_j\ge 0\,(j=1,2,\dots,t)}} \zeta(k+e'_1, 1+e'_2, \dots, 1+e'_t).$$

by Thm. 5.5.9 of Ohno's Relations. It is not hard to see that the last sum can be expressed as (see Eq. (1.10) and Exercise 3.17)

$$(-1)^{d-1}\sum\zeta(k\circ\underbrace{1\circ\cdots\circ 1}_{(d-1)\text{-times}}) = (-1)^{d-1}\zeta^\star(k, \{1\}^{d-1}),$$

where $\circ$ is either "+" or ",". The lemma follows immediately. $\square$

The SMZVs have the following height 1 duality relations.

Theorem 6.2.4. *For any $k, n \in \mathbb{N}$, we have*

$$\zeta^{\mathcal{S}}_{\sqcup\!\sqcup}(k, \{1\}^{d-1}) - \zeta^{\mathcal{S}}_{\sqcup\!\sqcup}(d, \{1\}^{k-1})$$

is equal to $\zeta(2)$ times a polynomial of the Riemann zeta values with rational coefficients.

Proof. The claim is trivial if $k = 1$ or $d = 1$ so we may assume $k, d \geq 2$. By Lemma 6.2.3, we have

$$\begin{aligned}&\zeta^{\mathcal{S}}_{\sqcup\!\sqcup}(k, \{1\}^{d-1}) - \zeta^{\mathcal{S}}_{\sqcup\!\sqcup}(d, \{1\}^{k-1})\\ &\quad = \zeta(k, \{1\}^{d-1}) + (-1)^k \zeta^\star(k, \{1\}^{d-1}) - \zeta(d, \{1\}^{k-1}) - (-1)^d \zeta^\star(d, \{1\}^{k-1}).\end{aligned}$$

Taking $z = 0$ in Thm. 5.6.2, we get

$$\begin{aligned}1 - &\sum_{k,d\geq 1} \zeta(k+1, \{1\}^{d-1}) x^k y^d\\ &= \exp\left(\sum_{d\geq 2} \zeta(d) \frac{x^d + y^d - (x+y)^d}{d}\right) = \frac{\Gamma(1-x)\Gamma(1-y)}{\Gamma(1-x-y)}. \qquad (6.5)\end{aligned}$$

Hence by Eq. (1.5) we have

$$\begin{aligned}&\sum_{k,d\geq 2} \Big(\zeta(k, \{1\}^{d-1}) - \zeta(d, \{1\}^{k-1})\Big) x^{k-1} y^{d-1}\\ &\qquad = \left(\frac{1}{y} - \frac{1}{x}\right)\left(1 - \frac{\Gamma(1-x)\Gamma(1-y)}{\Gamma(1-x-y)}\right) + \psi(1-x) - \psi(1-y).\end{aligned}$$

On the other hand, by Cor. 10.7.6

$$\begin{aligned}&\sum_{k,d\geq 2} \Big((-1)^k \zeta^\star(k, \{1\}^{d-1}) - (-1)^d \zeta^\star(d, \{1\}^{k-1})\Big) x^{k-1} y^{d-1}\\ &\qquad = -\psi(x) + \psi(y) - \pi(\cot(\pi x) - \cot(\pi y)) \frac{\Gamma(1-x)\Gamma(1-y)}{\Gamma(1-x-y)}.\end{aligned}$$

Combining with the well-known reflection formula (1.6) of the digamma function, we have

$$\begin{aligned}&\sum_{k,d\geq 2} \Big(\zeta^{\mathcal{S}}_{\sqcup\!\sqcup}(k, \{1\}^{d-1}) - \zeta^{\mathcal{S}}_{\sqcup\!\sqcup}(n, \{1\}^{k-1})\Big) x^{k-1} y^{d-1}\\ &= \left(\frac{1}{y} - \frac{1}{x}\right)\left(1 - \frac{\Gamma(1-x)\Gamma(1-y)}{\Gamma(1-x-y)}\right) + \psi(1-x) - \psi(1-y)\end{aligned}$$

$$
\begin{aligned}
&- \psi(x) + \psi(y) - \pi(\cot(\pi x) - \cot(\pi y))\frac{\Gamma(1-x)\Gamma(1-y)}{\Gamma(1-x-y)} \\
&= \left(1 - \frac{\Gamma(1-x)\Gamma(1-y)}{\Gamma(1-x-y)}\right)(\psi(1-x) - \psi(1+x) - \psi(1-y) + \psi(1+y)) \\
&= -2\left(1 - \frac{\Gamma(1-x)\Gamma(1-y)}{\Gamma(1-x-y)}\right)\sum_{l\geq 1}\zeta(2l)(x^{2l-1} - y^{2l-1}),
\end{aligned}
$$

by Eq. (1.5). Since all the coefficients of $\Gamma(1-x)\Gamma(1-y)/\Gamma(1-x-y)$ can be expressed as polynomials of the Riemann zeta values with rational coefficients according to Eq. (6.5), Thm. 6.2.4 follows immediately. □

6.3 Drinfeld Associator and SMZVs

To investigate other useful properties of the SMZVs it is more enlightening to explore their connections to the Drinfeld associator discussed in Sec. 4.1. Recall that $\Phi_{ш} = \pi(\Phi(\mathtt{a},\mathtt{b}))$ where $\Phi(\mathtt{a},\mathtt{b})$ is the Drinfeld associator and $\pi : \mathscr{R} \to \mathscr{R}/\mathscr{R}\mathtt{a}$. Here, for simplicity, we rewrite $\mathtt{x}_0 = \mathtt{a}$ and $\mathtt{x}_1 = \mathtt{b}$. Then it follows from Thm. 4.1.11 that

$$\Phi_{ш}[\mathbf{w}] = (-1)^d\zeta_{ш}(\mathbf{w};0) \quad \text{for all } \mathbf{w} \in \mathfrak{A}^1_{ш}.$$

As before we put $\zeta_{ш}(s_1,\dots,s_d) := (-1)^d\zeta_{ш}(\mathtt{a}^{s_1-1}\mathtt{b}\cdots\mathtt{a}^{s_d-1}\mathtt{b};0)$ and $\zeta_*(s_1,\dots,s_d) := (-1)^d\zeta_*(\mathtt{a}^{s_1-1}\mathtt{b}\cdots\mathtt{a}^{s_d-1}\mathtt{b};0)$ as the constant terms of the regularized MZVs.

Theorem 6.3.1. *For every word* $\mathbf{w} = \mathtt{a}^{s_1-1}\mathtt{b}\cdots\mathtt{a}^{s_d-1}\mathtt{b} \in \mathfrak{A}^1_{ш}$, *the SMZV*

$$\zeta^{\mathcal{S}}_{ш}(s_1,\dots,s_d) = (-1)^d(\Phi^{-1}\mathtt{b}\Phi)[\mathtt{b}\mathbf{w}].$$

Proof. By the duality relation in Eq. (4.6) for Φ, for all $i = 0,\dots,d$

$$
\begin{aligned}
\Phi^{-1}[\mathtt{b}\mathtt{a}^{s_1-1}\cdots\mathtt{b}\mathtt{a}^{s_i-1}] &= \Phi[\mathtt{a}\mathtt{b}^{s_1-1}\cdots\mathtt{a}\mathtt{b}^{s_i-1}] \\
&= (-1)^{s_1+\cdots+s_i-i}\zeta_{ш}(\mathtt{a}\mathtt{b}^{s_1-1}\cdots\mathtt{a}\mathtt{b}^{s_i-1};0) \\
&= (-1)^{s_1+\cdots+s_i-i}\zeta_{ш}(\mathtt{a}^{s_i-1}\mathtt{b}\cdots\mathtt{a}^{s_1-1}\mathtt{b};0)
\end{aligned}
$$

by the duality relation of MZVs (see Thm. 5.1.2). Therefore

$$
\begin{aligned}
&(\Phi^{-1}\mathtt{b}\Phi)[\mathtt{b}\mathtt{a}^{s_1-1}\mathtt{b}\cdots\mathtt{a}^{s_d-1}\mathtt{b}] \\
&\quad = \sum_{i=0}^{d}\Phi^{-1}[\mathtt{b}\mathtt{a}^{s_1-1}\cdots\mathtt{b}\mathtt{a}^{s_i-1}]\Phi[\mathtt{a}^{s_{i+1}-1}\mathtt{b}\cdots\mathtt{a}^{s_d-1}\mathtt{b}] \\
&\quad = \sum_{i=0}^{d}(-1)^{s_1+\cdots+s_i+d}\zeta_{ш}(s_i,\dots,s_1)\zeta_{ш}(s_{i+1},\dots,s_d) = (-1)^d\zeta^{\mathcal{S}}_{ш}(\mathbf{s})
\end{aligned}
$$

by Prop. 6.1.3, as desired. □

We can also use this Hopf algebra set-up to give another proof of Cor. 6.1.5. For $\sharp = ш$ or $*$, we set

$$\Psi_\sharp(\mathsf{a},\mathsf{b}) = 1 + \sum_{\mathbf{w}\in \mathsf{X}^*\mathsf{b}} \zeta_\sharp(\mathbf{w};0)\mathbf{w}.$$

Define $\mathrm{inv}(1) = 1$ and

$$\mathrm{inv}(\mathsf{a}^{s_1-1}\mathsf{b}\dots\mathsf{a}^{s_r-1}\mathsf{b}) = (-1)^{s_1+\dots+s_r}\mathsf{a}^{s_r-1}\mathsf{b}\dots\mathsf{a}^{s_1-1}\mathsf{b} \tag{6.6}$$

for all $s_1,\dots,s_r \in \mathbb{N}$. Since Ψ_* is group-like for Δ_*, one can check (see Exercise 6.3) that Ψ_*^{inv} is also group-like, namely, $\Delta_*(\Psi_*^{\mathrm{inv}}) = \Psi_*^{\mathrm{inv}} \otimes \Psi_*^{\mathrm{inv}}$. Hence

$$\Delta_*(\Psi_*^{\mathrm{inv}}\Psi_*) = \Delta_*(\Psi_*^{\mathrm{inv}})\Delta_*(\Psi_*) = \Psi_*^{\mathrm{inv}}\Psi_* \otimes \Psi_*^{\mathrm{inv}}\Psi_*.$$

But clearly

$$\Psi_*^{\mathrm{inv}}\Psi_* = 1 + \sum_{\mathbf{w}\in \mathsf{X}^*\mathsf{b}} \zeta_*^{\mathcal{S}}(\mathbf{w})\mathbf{w}. \tag{6.7}$$

So Cor. 6.1.5 follows from Thm. 4.1.6 directly.

Similar to Eq. (6.7) we also have

$$\Psi_ш^{\mathrm{inv}}\Psi_ш = 1 + \sum_{\mathbf{w}\in \mathsf{X}^*\mathsf{b}} \zeta_ш^{\mathcal{S}}(\mathbf{w})\mathbf{w}. \tag{6.8}$$

Let MZ_w, $\mathsf{MZ}^{\mathcal{S}}_{*,w}$ and $\mathsf{MZ}^{\mathcal{S}}_{ш,w}$ be the $\mathbb{Q}$-vector space generated by all weight w MZVs, $*$-symmetrized MZVs and ш-symmetrized MZVs, respectively.

Theorem 6.3.2. *For all* $\mathbf{s} = (s_1,\dots,s_d) \in \mathbb{N}^d$, *the SMZVs*

$$\zeta_ш^{\mathcal{S}}(\mathbf{s}) \equiv \zeta_*^{\mathcal{S}}(\mathbf{s}) \pmod{\zeta(2)},$$

where $\zeta(2)$ *means either* $\zeta(2)\mathsf{MZ}_{|\mathbf{s}|-2}$, *or* $\zeta(2)\mathsf{MZ}^{\mathcal{S}}_{*,|\mathbf{s}|-2}$, *or* $\zeta(2)\mathsf{MZ}^{\mathcal{S}}_{ш,|\mathbf{s}|-2}$. *Consequently, in any statement modulo products of the MZVs we may write* $\zeta^{\mathcal{S}}(\mathbf{s})$ *for either one of the two symmetrized versions.*

Proof. According to Thm. 4.1.15 we have $\Psi_ш(\mathsf{a},\mathsf{b}) = \Lambda\Psi_*(\mathsf{a},\mathsf{b})$ where

$$\Lambda = \exp\Big(\sum_{n=2}^\infty (-1)^n \frac{\zeta(n)}{n}\mathsf{b}^n\Big) \text{ and } \Lambda^{\mathrm{inv}} = \exp\Big(\sum_{n=2}^\infty \frac{\zeta(n)}{n}\mathsf{b}^n\Big).$$

Since inv is an anti-automorphism

$$\Psi_ш^{\mathrm{inv}}\Psi_ш = \Psi_*^{\mathrm{inv}}\Lambda^{\mathrm{inv}}\Lambda\Psi_* = \Psi_*^{\mathrm{inv}}\exp\Big(\sum_{n=2}^\infty \frac{\zeta(2n)}{n}\mathsf{b}^{2n}\Big)\Psi_* \equiv \Psi_*^{\mathrm{inv}}\Psi_*$$

modulo $\zeta(2)\mathsf{MZ}^{\mathcal{S}}_{*,|\mathbf{s}|-2}$. Similarly, we have

$$\Psi_*^{\mathrm{inv}}\Psi_* = \Psi_ш^{\mathrm{inv}}(\Lambda^{\mathrm{inv}})^{-1}\Lambda^{-1}\Psi_ш = \Psi_ш^{\mathrm{inv}}\exp\Big(-\sum_{n=2}^\infty \frac{\zeta(2n)}{n}\mathsf{b}^{2n}\Big)\Psi_ш \equiv \Psi_ш^{\mathrm{inv}}\Psi_ш$$

modulo $\zeta(2)\mathsf{MZ}^{\mathcal{S}}_{ш,|\mathbf{s}|-2}$ (or $\zeta(2)\mathsf{MZ}_{|\mathbf{s}|-2}$ by Prop. 6.1.3). Together with Eqs. (6.7) and (6.8) this completes the proof of the theorem. □

Remark 6.3.3. More detailed analysis reveals that $\zeta(2)\mathsf{MZ}_{|\mathbf{s}|-2}$ can be replaced by $\zeta(2)\mathsf{MZ}_{|\mathbf{s}|-2,\le \mathrm{dp}(\mathbf{s})-1}$ where $\mathsf{MZ}_{w,\le n}$ is the subspace of MZ_w generated by the MZVs of depth at most n. Similar claims are true for the $*$- and $ш$-versions of the MZVs, too.

The next theorem shows that the SMZVs satisfy not only the stuffle relations but also some kind of the shuffle relations. Define $\tau(1)=1$ and

$$\tau(\mathsf{a}^{s_1-1}\mathsf{b}\dots\mathsf{a}^{s_d-1}\mathsf{b}) = (-1)^{s_1+\dots+s_r}\mathsf{a}^{s_d-1}\mathsf{b}\dots\mathsf{a}^{s_1-1}\mathsf{b}. \tag{6.9}$$

Note that τ and inv are the same as maps. But we should think inv as an anti-automorphism on $(\mathfrak{A}^1,*,\Delta_*)$ while τ is an anti-automorphism on $(\mathfrak{A}^1,ш)$.

Theorem 6.3.4. *For all words* $\mathbf{u},\mathbf{w},\mathbf{w}'\in\mathfrak{A}^1$, *we have*

(i) $\zeta^{\mathcal{S}}_{ш}(\mathbf{u}\,ш\,\mathbf{v}) = \zeta^{\mathcal{S}}_{ш}(\tau(\mathbf{u})\mathbf{v})$,
(ii) $\zeta^{\mathcal{S}}_{ш}((\mathbf{wu})\,ш\,\mathbf{v}) = \zeta^{\mathcal{S}}_{ш}(\mathbf{u}\,ш\,\tau(\mathbf{w})\mathbf{v})$,
(iii) $\zeta^{\mathcal{S}}_{ш}((\mathsf{a}^{s-1}\mathsf{b}\mathbf{u})\,ш\,\mathbf{v}) = (-1)^s\zeta^{\mathcal{S}}_{ш}(\mathbf{u}\,ш\,(\mathsf{a}^{s-1}\mathsf{b}\mathbf{v}))$.

Proof. Taking $\mathbf{u}=\emptyset$ and then $\mathbf{w}=\mathbf{u}$ we see that (ii) implies (i). Decomposing $\mathbf{w}$ into strings of the type $\mathsf{a}^{s-1}\mathsf{b}$ we see that (iii) implies (ii). To prove (iii), we observe that

$$\mathsf{b}\Big((\mathsf{a}^{s-1}\mathsf{b}\mathbf{u})\,ш\,\mathbf{v} - (-1)^s\mathbf{u}\,ш\,(\mathsf{a}^{s-1}\mathsf{b}\mathbf{v})\Big) = \sum_{i=0}^{s-1}(-1)^i(\mathsf{a}^{s-1-i}\mathsf{b}\mathbf{u})\,ш\,(\mathsf{a}^i\mathsf{b}\mathbf{v}). \tag{6.10}$$

By Thm. 6.3.1 it suffices to show that the image of the above under $E=\Phi^{-1}\mathsf{b}\Phi$ vanishes. Now

$$\Delta_{ш}(E) = (\Phi^{-1}\otimes\Phi^{-1})(\mathsf{b}\otimes 1 + 1\otimes\mathsf{b})(\Phi\otimes\Phi) = E\otimes 1 + 1\otimes E.$$

Thus E is a primitive element for $\Delta_{ш}$ so that we can regard it as a Lie element, namely, it acts on shuffle products like a "derivation". By the definition, the counit $\epsilon_{ш}$ satisfies $\epsilon_{ш}[\mathbf{w}]=0$ for all $\mathbf{w}\neq 1$. Hence for any nonempty words $\mathbf{u},\mathbf{v}\in\mathfrak{A}^1$,

$$E[\mathbf{u}\,ш\,\mathbf{v}] = E[\mathbf{u}]\epsilon_{ш}[\mathbf{v}] + \epsilon_{ш}[\mathbf{u}]E[\mathbf{v}] = 0.$$

This completes the proof by Thm. 6.3.1 since none of the factors in the shuffle products on the right-hand side of Eq. (6.10) is empty word as the letter b appears in every factor. □

Besides the stuffle and the shuffle relations, MZVs satisfy quite a few other well-known relations such as the duality relations and the sum relations presented in Chap. 5. For the SMZVs, there is in fact another kind of duality which we call *φ-dual.* On the level of words it corresponds to the involution

$$\mathsf{a}^\varphi = \mathsf{a}+\mathsf{b}, \quad \mathsf{b}^\varphi = -\mathsf{b}. \tag{6.11}$$

Theorem 6.3.5. *For all $\mathbf{w} \in \mathfrak{A}^1_{ш}$, we have*

$$\zeta^{\mathcal{S}}(\mathbf{w}) \equiv \zeta^{\mathcal{S}}(\mathbf{w}^\varphi) \pmod{\zeta(2)},$$

where $\zeta^{\mathcal{S}}(\mathbf{w})$ means either $\zeta^{\mathcal{S}}_{ш}(\mathbf{w})$ or $\zeta^{\mathcal{S}}_{*}(\mathbf{w})$.

Proof. Throughout this proof we write $\Phi_{\mathsf{a},\mathsf{b}} = \Phi(\mathsf{a},\mathsf{b})$. Set $\mathsf{c} = -(\mathsf{a}+\mathsf{b})$. For any $\mathbf{t} \in \mathbb{N}^d$, we first decipher the coefficient

$$\Phi_{\mathsf{a},\mathsf{c}}[\mathsf{a}^{t_1-1}\mathsf{b}\cdots\mathsf{b}\mathsf{a}^{t_d-1}\mathsf{b}].$$

Observe that the expansion of $\mathsf{a}^{r_1-1}(\mathsf{a}+\mathsf{b})\cdots\mathsf{a}^{r_\ell-1}(\mathsf{a}+\mathsf{b})$ contains $\mathsf{a}^{t_1-1}\mathsf{b}\cdots\mathsf{a}^{t_d-1}\mathsf{b}$ if and only if it has the form

$$\alpha_{1,1}\cdots\alpha_{1,t_1-1}(\mathsf{a}+\mathsf{b})\cdots\alpha_{d,1}\cdots\alpha_{d,t_d-1}(\mathsf{a}+\mathsf{b})$$

where $\alpha_{\mu,\nu} = \mathsf{a}$ or $\alpha_{\mu,\nu} = \mathsf{a}+\mathsf{b}$. Since the coefficient of $\mathsf{a}^{r_1-1}(\mathsf{a}+\mathsf{b})\cdots\mathsf{a}^{r_\ell-1}(\mathsf{a}+\mathsf{b})$ in $\Phi_{\mathsf{a},\mathsf{c}}$ is $(-1)^\ell\Phi_{\mathsf{a},\mathsf{b}}[\mathsf{a}^{r_1-1}\mathsf{b}\cdots\mathsf{a}^{r_\ell-1}\mathsf{b}]$ we get

$$\begin{aligned}\Phi_{\mathsf{a},\mathsf{c}}&[\mathsf{a}^{t_1-1}\mathsf{b}\cdots\mathsf{a}^{t_d-1}\mathsf{b}]\\ &= \sum_{\substack{\beta_{\mu,\nu}=\mathsf{a},\mathsf{b}\\ 1\le\mu\le d,1\le\nu<t_\mu}} (-1)^{d+\boldsymbol{\beta}(\mathsf{b})\}}\Phi_{\mathsf{a},\mathsf{b}}[\beta_{1,1}\cdots\beta_{1,t_1-1}\mathsf{b}\cdots\beta_{\ell,1}\cdots\beta_{d,t_d-1}\mathsf{b}],\end{aligned}$$

where $\boldsymbol{\beta}(x) = \sharp\{(\mu,\nu) : \beta_{\mu,\nu} = x\}$ for any letter x and $\sharp S$ is the cardinality of S. Putting $\mathsf{a}^\natural = \mathsf{a}-\mathsf{b}, \mathsf{b}^\natural = -\mathsf{b}$, we can rewrite this more succinctly as

$$\Phi_{\mathsf{a},\mathsf{c}}[\mathsf{a}^{t_1-1}\mathsf{b}\cdots\mathsf{a}^{t_d-1}\mathsf{b}] = \Phi_{\mathsf{a},\mathsf{b}}[(\mathsf{a}^{t_1-1}\mathsf{b}\cdots\mathsf{a}^{t_d-1}\mathsf{b})^\natural]. \tag{6.12}$$

For any $\mathbf{t} = (t_1,\dots,t_d) \in \mathbb{N}^d$ with $t_d \ge 2$, we turn to the coefficient

$$\Phi^{-1}_{\mathsf{a},\mathsf{c}}[\mathsf{b}\mathsf{a}^{t_1-1}\mathsf{b}\cdots\mathsf{b}\mathsf{a}^{t_d-1}].$$

Observe that the expansion of $(\mathsf{a}+\mathsf{b})^{r_1-1}\mathsf{a}\cdots(\mathsf{a}+\mathsf{b})^{r_\ell-1}\mathsf{a}$ contains $\mathsf{b}\mathsf{a}^{t_1-1}\cdots\mathsf{b}\mathsf{a}^{t_d-1}$ if and only if it has the form

$$(\mathsf{a}+\mathsf{b})\alpha_{1,1}\cdots\alpha_{1,t_1-1}\cdots(\mathsf{a}+\mathsf{b})\alpha_{d,1}\cdots\alpha_{d,t_d-1}$$

where $\alpha_{\mu,\nu} = \mathsf{a}$ or $\alpha_{\mu,\nu} = \mathsf{a}+\mathsf{b}$. Since the coefficient of $(\mathsf{a}+\mathsf{b})^{r_1-1}\mathsf{a}\cdots(\mathsf{a}+\mathsf{b})^{r_\ell-1}\mathsf{a}$ in $\Phi^{-1}_{\mathsf{a},\mathsf{c}}$ is $(-1)^{|\mathbf{r}|-\ell}\Phi_{\mathsf{a},\mathsf{b}}[\mathsf{a}^{r_1-1}\mathsf{b}\cdots\mathsf{a}^{r_\ell-1}\mathsf{b}]$, by the duality (essentially by Lemma 2.1.2(iii)) we get

$$\Phi^{-1}_{\mathsf{a},\mathsf{c}}[\mathsf{b}\mathsf{a}^{t_1-1}\mathsf{b}\cdots\mathsf{a}^{t_d-1}]$$

$$= \sum_{\substack{\beta_{\mu,\nu}=\mathsf{a},\mathsf{b}\\ 1\le\mu\le d, 1\le\nu<t_\mu}} (-1)^{d+\boldsymbol{\beta}(\mathsf{a})}\Phi_{\mathsf{a},\mathsf{b}}[\mathsf{a}\beta_{1,1}\cdots\beta_{1,t_1-1}\cdots\mathsf{a}\beta_{d,1}\cdots\beta_{d,t_d-1}]$$

$$= \sum_{\substack{\beta_{\mu,\nu}=\mathsf{a},\mathsf{b}\\ 1\le\mu\le d, 1\le\nu<t_\mu}} (-1)^{d+\boldsymbol{\beta}(\mathsf{b})+|\mathbf{t}|}\Phi_{\mathsf{a},\mathsf{b}}[\beta_{d,t_d-1}\cdots\beta_{d,1}\mathsf{b}\cdots\beta_{1,t_1-1}\cdots\beta_{1,1}\mathsf{b}],$$

where we have used the fact that for any admissible word $\mathbf{w} = l_1 \dots l_d$, $\Phi_{\mathsf{a},\mathsf{b}}[\mathbf{w}] = (-1)^{\sharp\{i:l_i=\mathsf{b}\}}\int_0^1 \mathbf{w}$. Hence

$$\Phi_{\mathsf{a},\mathsf{c}}^{-1}[\mathsf{b}\mathsf{a}^{t_1-1}\mathsf{b}\cdots\mathsf{a}^{t_d-1}] = (-1)^{|\mathbf{t}|}\Phi_{\mathsf{a},\mathsf{b}}[(\mathsf{a}^{t_d-1}\mathsf{b}\cdots\mathsf{a}^{t_1-1}\mathsf{b})^{\natural}]. \tag{6.13}$$

Combining Eqs. (6.12) and (6.13) we see that for any $\mathbf{s}\in\mathbb{N}^d$,

$$\begin{aligned}
&\Phi_{\mathsf{a},\mathsf{c}}^{-1}\mathsf{b}\Phi_{\mathsf{a},\mathsf{c}}[\mathsf{b}\mathsf{a}^{s_1-1}\mathsf{b}\cdots\mathsf{a}^{s_d-1}\mathsf{b}]\\
&= \sum_{\substack{0\le i\le d\\ s_i\ge 2}} \Phi_{\mathsf{a},\mathsf{c}}^{-1}[\mathsf{b}\mathsf{a}^{s_1-1}\mathsf{b}\cdots\mathsf{a}^{s_i-1}]\Phi_{\mathsf{a},\mathsf{c}}[\mathsf{a}^{s_{i+1}-1}\mathsf{b}\cdots\mathsf{a}^{s_d-1}\mathsf{b}]\\
&= \sum_{\substack{0\le i\le d\\ s_i\ge 2}} (-1)^{s_1+\cdots+s_i}\Phi_{\mathsf{a},\mathsf{b}}[(\mathsf{a}^{s_i-1}\mathsf{b}\cdots\mathsf{a}^{s_1-1}\mathsf{b})^{\natural}]\Phi_{\mathsf{a},\mathsf{b}}[(\mathsf{a}^{s_{i+1}-1}\mathsf{b}\cdots\mathsf{a}^{s_d-1}\mathsf{b})^{\natural}]\\
&= \sum_{\substack{0\le i\le d\\ s_i\ge 2}} (-1)^{d+s_1+\cdots+s_i}\zeta_{ш}((\mathsf{a}^{s_i-1}\mathsf{b}\cdots\mathsf{a}^{s_1-1}\mathsf{b})^{\varphi})\zeta_{ш}((\mathsf{a}^{s_{i+1}-1}\mathsf{b}\cdots\mathsf{a}^{s_d-1}\mathsf{b})^{\varphi})\\
&\equiv (-1)^d\zeta_{ш}^{\mathcal{S}}(\mathbf{w}^{\varphi})
\end{aligned} \tag{6.14}$$

modulo $\zeta(2)$ by Prop. 6.1.3, where $\mathsf{a}^{\varphi} = \mathsf{a}+\mathsf{b}, \mathsf{b}^{\varphi} = -\mathsf{b}$.

Mimicking the proof of Eq. (6.14) and using an argument similar to that of Eq. (6.3), one can show that (see Exercise 6.4)

$$\Phi_{\mathsf{a},\mathsf{c}}^{-1}\mathsf{a}\Phi_{\mathsf{a},\mathsf{c}}[\mathsf{b}\mathbf{w}] = \mathsf{a} \quad \forall \mathbf{w}\in\mathfrak{A}_{ш}^1. \tag{6.15}$$

So the theorem follows from Thm. 6.3.1 and Eq. (4.32). □

6.4* Space Generated by SMZVs

In this section we present a theorem of Yasuda which implies that no information is lost if we consider the SMZVs instead of the MZVs (see Thm. 6.4.19). In Chap. 8 this result will play a key role in a conjecture relating the MZVs to the finite version of the MZVs. We break the proof of this theorem into a series of lemmas and propositions.

6.4.1 *Natural SMZVs*

We will be interested in the behavior of the depth reduction of the MZVs so we let $\mathrm{Surj}(n,m)$ be the set of surjective maps from $\{1,\dots,n\}$ to $\{1,\dots,m\}$ for all positive integers $n \geq m$. For any $\phi \in \mathrm{Surj}(n,m)$ and $\mathbf{s} = (s_1,\dots,s_n) \in \mathbb{N}^n$, we set

$$\phi_*\mathbf{s} = \left(\sum_{i\in\phi^{-1}(1)} s_i, \sum_{i\in\phi^{-1}(2)} s_i, \dots, \sum_{i\in\phi^{-1}(m)} s_i \right).$$

Then we define

$$\zeta^{\natural,\mathcal{S}}(\mathbf{s}) := \sum_{1\leq m\leq n} \sum_{\phi\in\mathrm{Surj}(n,m)} \frac{1}{\prod_{j=1}^{m}(\sharp\phi^{-1}(j))!} \zeta_*^{\mathcal{S}}(\phi_*\mathbf{s}).$$

Remark 6.4.1. If we use $\zeta_{\sqcup\!\sqcup}^{\mathcal{S}}(\mathbf{s})$ instead of $\zeta_*^{\mathcal{S}}(\mathbf{s})$ to define $\zeta^{\natural,\mathcal{S}}(\mathbf{s})$ then the difference lies in $\zeta(2)\mathsf{MZ}_{|\mathbf{s}|-2,\leq \mathrm{dp}(\mathbf{s})-1}$ by Remark 6.3.3.

Example 6.4.2. We have

$$\begin{aligned}
\zeta^{\natural,\mathcal{S}}(s_1,s_2) &= \zeta_*^{\mathcal{S}}(s_1,s_2) + \frac{1}{2}\zeta_*^{\mathcal{S}}(s_1+s_2),\\
\zeta^{\natural,\mathcal{S}}(s_1,s_2,s_3) &= \zeta_*^{\mathcal{S}}(s_1,s_2,s_3) + \frac{1}{2}\zeta_*^{\mathcal{S}}(s_1+s_2,s_3)\\
&\quad + \frac{1}{2}\zeta_*^{\mathcal{S}}(s_1,s_2+s_3) + \frac{1}{6}\zeta_*^{\mathcal{S}}(s_1+s_2+s_3).
\end{aligned}$$

Recall that MZ_w (resp. $\mathsf{MZ}_{w,\leq d}$) is the $\mathbb{Q}$-vector space spanned by all the MZVs of weight w (resp. weight w and depth at most d). Now we further set

$$\begin{aligned}
\mathsf{MZ}_{w,+} &= \sum_{w',w''\geq 1,\ w'+w''=w} \mathsf{MZ}_{w'}\cdot\mathsf{MZ}_{w''} \subseteq \mathsf{MZ}_w,\\
\mathsf{MZ}_{w,+,\leq d} &= \sum_{\substack{w',w''\geq 1,\ w'+w''=w,\\ 1\leq k,l\leq d-1,\ k+l\leq d}} \mathsf{MZ}_{w',\leq k}\cdot\mathsf{MZ}_{w'',\leq l} \subseteq \mathsf{MZ}_{w,\leq d}.
\end{aligned}$$

For each $d \in \mathbb{N}$, define the $\mathbb{Q}$-vector space of formal generating series

$$M_d := \left\{ \sum_{\mathbf{s}=(s_1,\dots,s_d)\in\mathbb{N}^d} a_{\mathbf{s}}\mathbf{x}^{\mathbf{s}-\mathbf{1}} \;\middle|\; a_{\mathbf{s}} \in \mathsf{MZ}_{|\mathbf{s}|}\ \forall \mathbf{s}\in\mathbb{N}^d \right\}$$

where $\mathbf{x}^{\mathbf{s}-\mathbf{1}} = x_1^{s_1-1}\cdots x_d^{s_d-1}$, and its subspace

$$M_{d,+} := \left\{ \sum_{\mathbf{s}=(s_1,\dots,s_d)\in\mathbb{N}^d} a_{\mathbf{s}}\mathbf{x}^{\mathbf{s}-\mathbf{1}} \;\middle|\; a_{\mathbf{s}} \in \mathsf{MZ}_{|\mathbf{s}|,+}\cup\mathsf{MZ}_{|\mathbf{s}|,\leq 1}\ \forall \mathbf{s}\in\mathbb{N}^d \right\}.$$

One can similarly define subspaces $M_{d,\le m}$ and $M_{d,+,\le m}$. Similar as in Chap. 3 we define the generating functions

$$F_{\sqcup\!\sqcup}(x_1,\dots,x_n) = F_{\sqcup\!\sqcup}^{(n)}(x_1,\dots,x_n) := \sum_{\mathbf{s}\in\mathbb{N}^n} \zeta_{\sqcup\!\sqcup}(\mathbf{s})\mathbf{x}^{\mathbf{s}-\mathbf{1}},$$

$$F_*(x_1,\dots,x_n) = F_*^{(n)}(x_1,\dots,x_n) := \sum_{\mathbf{s}\in\mathbb{N}^n} \zeta_*(\mathbf{s})\mathbf{x}^{\mathbf{s}-\mathbf{1}},$$

$$F_\natural(x_1,\dots,x_n) = F_\natural^{(n)}(x_1,\dots,x_n) := \sum_{\mathbf{s}\in\mathbb{N}^n} \zeta^\natural(\mathbf{s})\mathbf{x}^{\mathbf{s}-\mathbf{1}},$$

where

$$\zeta^\natural(\mathbf{s}) = \sum_{1\le m\le n}\ \sum_{\phi\in\mathrm{Surj}(n,m)} \frac{1}{\prod_{j=1}^m (\sharp\phi^{-1}(j))!}\zeta_*(\phi_*\mathbf{s}).$$

By Thm. 3.3.21 and Remark 3.3.22 we could as well use $\zeta_{\sqcup\!\sqcup}(\phi_*\mathbf{s})$ to define $\zeta^\natural(\mathbf{s})$ when we consider identities modulo $\mathsf{MZ}_{|\mathbf{s}|,+}$. By convention, $F_{\sqcup\!\sqcup}^{(0)} = F_*^{(0)} = F_\natural^{(0)} = 1$.

6.4.2 *An Embedding* $\iota_n : \mathfrak{S}_{n+1} \hookrightarrow \mathbf{GL}_n(\mathbb{Z})$

In the proof of the Parity Principle of the MZVs, we defined an action of $\mathfrak{S}_{n+1}$ on V_n extending that of $\mathfrak{S}_n$. We now need to use a new construction. Recall that $-\nu_n$ is the $n\times n$ identity matrix and

$$P = P_n = \begin{pmatrix} 1 & & & \\ -1 & 1 & & \\ & \ddots & \ddots & \\ & & -1 & 1 \end{pmatrix}, \qquad P^{-1} = \begin{pmatrix} 1 & & & \\ 1 & 1 & & \\ \vdots & \vdots & \ddots & \\ 1 & 1 & \cdots & 1 \end{pmatrix}.$$

As before, by abuse of notation we let $T_0 = T_{n,0}$ denote the matrix corresponding to the reversal $T_{n,0} = \left(\begin{smallmatrix} 1 & 2 & \cdots & n \\ n & n-1 & \cdots & 1 \end{smallmatrix}\right) \in \mathfrak{S}_n$.

Lemma 6.4.3. *There exists a unique injective group homomorphism* $\iota_n : \mathfrak{S}_{n+1} \hookrightarrow \mathrm{GL}_n(\mathbb{Z})$ *which satisfies the following properties:*

- *If* $\sigma\in\mathfrak{S}_n$, *then* $\iota_n(\sigma) = M_\sigma$ *so that*

$$(x_{\sigma^{-1}(1)}\ \ x_{\sigma^{-1}(2)}\ \ \cdots\ \ x_{\sigma^{-1}(n)}) = (x_1\ \ x_2\ \ \cdots\ \ x_n)M_\sigma^{-1}.$$

- *The image under* ι_n *of the permutation*

$$T' = \begin{pmatrix} 1 & 2 & \cdots & n-1 & n & n+1 \\ n-1 & n-2 & \cdots & 1 & n+1 & n \end{pmatrix}$$

is equal to $\nu T_0 T_0^\sharp T_0$.

Proof. Let $\mathbb{Z}^n$ denote the abelian group of n-dimensional row vectors with integer components. By the right multiplication the group $\mathrm{GL}_n(\mathbb{Z})$ acts on $\mathbb{Z}^n$ from the right. This yields the embedding $\mathfrak{S}_n \hookrightarrow \mathrm{GL}_n(\mathbb{Z})$, $\sigma \to M_\sigma^{-1}$.

But the embedding $\mathbb{Z}^n \hookrightarrow \mathbb{Z}^{(n+1)}$, $(a_1, \dots, a_n) \to (a_1, \dots, a_n, -(a_1 + \cdots + a_n))$ has a stable image under the action of $\mathfrak{S}_{n+1}$, leading to a homomorphism $\iota_n : \mathfrak{S}_{n+1} \hookrightarrow \mathrm{GL}_n(\mathbb{Z})$. It is not hard to show that the image $\iota_n(T')$ is given by

$$\iota_n(T') = \gamma_2 := \begin{pmatrix} & & 1 & 0 \\ & \iddots & & \vdots \\ 1 & & & 0 \\ -1 & \cdots & -1 & -1 \end{pmatrix} = \nu T_0 T_0^\sharp T_0 \quad \text{with} \quad \gamma_2^{-1} = \gamma_2. \tag{6.16}$$

Finally, the map ι_n is unique since $\mathfrak{S}_{n+1}$ is generated by $\mathfrak{S}_n$ and T'. □

6.4.3 *Parity Results*

We have seen in Sec. 3.5 that if the depth and the weight of a MZV have different parities then it can be expressed as a $\mathbb{Q}$-linear combination of lower depth MZVs (of the same weight), augmented by the products of lower weight MZVs. For the SMZVs, we can say a little more. Even in the same parity case, they can be reduced.

Proposition 6.4.4. *For all* $\mathbf{s} = (s_1, \dots, s_n) \in \mathbb{N}^n$, *if* $w := |\mathbf{s}|$ *and* n *have the same parity, then* $\zeta^{\natural,\mathcal{S}}(\mathbf{s}) \in \mathsf{MZ}_{w,+,\le n}$.

Proof. First, Lemma 3.5.5 implies that $\zeta_*(\mathbf{s}) + (-1)^n \zeta_*(\overleftarrow{\mathbf{s}}) \equiv 0$ modulo $\mathsf{MZ}_{w,+,\le n}$. Thus $\zeta_*(\mathbf{s}) + (-1)^w \zeta_*(\overleftarrow{\mathbf{s}}) \equiv 0$. So for all $\phi \in \mathrm{Surj}(n, m)$, we have $\zeta_*(\phi_* \mathbf{s}) + (-1)^w \zeta_*(\phi_* \overleftarrow{\mathbf{s}}) \equiv 0$. Hence by the definition

$$\zeta^{\natural,\mathcal{S}}(\mathbf{s}) \equiv \zeta^\natural(\mathbf{s}) + (-1)^w \zeta^\natural(\overleftarrow{\mathbf{s}}) \equiv 0$$

modulo $\mathsf{MZ}_{w,+,\le n}$, as desired. □

Corollary 6.4.5. *For all* $\mathbf{s} = (s_1, \dots, s_n) \in \mathbb{N}^n$, *if* $w := |\mathbf{s}|$ *and* n *have the same parity, then*

$$\zeta_{\sqcup\!\sqcup}^{\mathcal{S}}(\mathbf{s}) \equiv \zeta_*^{\mathcal{S}}(\mathbf{s}) \equiv -\sum_{i=1}^{n-1} \zeta_*(s_1, \dots, s_{i-1}, s_i + s_{i+1}, s_{i+1}, \dots, s_n)$$

modulo $\mathsf{MZ}_{w,+,\le n} + \mathsf{MZ}_{w,\le n-2}$.

Proof. By the definition and Prop. 6.4.4

$$0 \equiv \zeta^{\natural,\mathcal{S}}(\mathbf{s}) \equiv \zeta_*^{\mathcal{S}}(\mathbf{s}) + \frac{1}{2}\sum_{i=1}^{n-1} \zeta_*^{\mathcal{S}}(s_1, \dots, s_{i-1}, s_i + s_{i+1}, s_{i+1}, \dots, s_n)$$

modulo $\mathsf{MZ}_{w,\le n-2}$. But for all $\mathbf{t}$ of depth $n-1$ we have

$$\zeta_*^{\mathcal{S}}(\mathbf{t}) \equiv \zeta_*(\mathbf{t}) + (-1)^w \zeta_*(\overleftarrow{\mathbf{t}}) \equiv \zeta_*(\mathbf{t}) + (-1)^{w+n}\zeta_*(\mathbf{t}) \equiv 2\zeta_*(\mathbf{t})$$

modulo $\mathsf{MZ}_{w,+,\le n}$. So the corollary follows. $\square$

The case of different parities is much more involved. In the rest of this subsection, we assume this is the case. By Thm. 3.5.3 and Eq. (3.33) we see that $F_*|T_0 \equiv F_*|T_0^\sharp \equiv (-1)^{n-1}F_* \pmod{\mathsf{MZ}_{w,+,\le n}}$. Similar congruences hold for $F_\natural$ and $F_ш$, too. The same remark can be applied throughout the rest of this subsection.

To save space, we put $M_n' = M_{n,+,\le n} + M_{n,\le n-2}$. By Eq. (3.35) (A) we have $T_0T' = C_{n+1}$ which yields $\iota(C_{n+1}) = \nu_n T_0^\sharp T_0$. Hence, by Eq. (3.36) we have

$$F_\natural \equiv F_\natural \big| \big[1 + \mathrm{sh}_1^{(n)}\, \nu_n T_0^\sharp T_0\big] = F_\natural \big| \big[\nu_n T_0^\sharp T_0 \big(1 + \mathrm{sh}_1^{(n)}\, \iota(\tau)\big)\big] \pmod{M_n'}$$

where τ is the transposition $(1, n+1)$. Further, since ν commutes with everything else we get

$$\begin{aligned} F_\natural \equiv{}& (-1)^{n-1}(F_\natural - F_*)\big|\big[T_0T_0^\sharp T_0\big(1+\mathrm{sh}_1^{(n)}\,\iota(\tau)\big)\nu\big] + F_*\big|\big[\big(1+\mathrm{sh}_1^{(n)}\,\iota(\tau)\big)\nu\big] \\ \equiv{}& (-1)^{n-1}(F_\natural - F_*)\big|\big[\tilde{T}\big(1+\mathrm{sh}_1^{(n)}\,\iota(\tau)\big)\nu\big] \\ &+ (F_* - F_\natural)\big|\big[\big(1+\mathrm{sh}_1^{(n)}\,\iota(\tau)\big)\nu\big] + F_\natural\big|\big[\big(1+\mathrm{sh}_1^{(n)}\,\iota(\tau)\big)\nu\big] \\ \equiv{}& F_\natural|\nu - (F_\natural - F_*)\big|\big[\big((-1)^n\tilde{T}+1\big)\big(1+\mathrm{sh}_1^{(n)}\,\iota(\tau)\big)\nu\big] \end{aligned} \tag{6.17}$$

modulo M_n', where $\tilde{T} = \tilde{T}_n := T_{n,0}T_{n,0}^\sharp T_{n,0}$. Now

$$F_\natural - F_* \equiv \frac{1}{2}\sum_{i=1}^{n-1} \frac{F_\natural(x_1, \dots, \widehat{x_{i+1}}, \dots, x_n) - F_\natural(x_1, \dots, \widehat{x}_i, \dots, x_n)}{x_i - x_{i+1}} \tag{6.18}$$

modulo M_n', where $F_\natural = F_\natural^{(n-1)}$ on the right-hand side. Note that

$$\left.\frac{F_\natural(x_1, \dots, x_i, \widehat{x_{i+1}}, \dots, x_n) - F_\natural(x_1, \dots, \widehat{x}_i, x_{i+1}, \dots, x_n)}{x_i - x_{i+1}}\right|\tilde{T}$$

$$= \begin{cases} \dfrac{F_\natural(y_1, \dots, \widehat{y_{i+1}}, \dots, y_n) - F_\natural(y_1, \dots, \widehat{y}_i, \dots, y_n)}{x_{n-i-1} - x_{n-i}} & \text{if } i \le n-2; \\ \dfrac{F_\natural(y_1, \dots, y_{n-1}) - F_\natural(y_1, \dots, y_{n-2}, x_n)}{-x_1} & \text{if } i = n-1, \end{cases}$$

where $y_i = x_n - x_{n-i}$ for $i = 1, \ldots, n$ and $x_0 = 0$. Since $F_\natural\big|T_{n-1,0} = F_\natural\big|T^\sharp_{n-1,0} \equiv (-1)^n F_\natural \pmod{M_{n-1,+,\le n-1}}$ we get

$$\begin{aligned} F_\natural(y_1, \ldots, \widehat{y_i}, \ldots, y_n) &= F_\natural\big|\tilde{T}_{n-1}(x_1, \ldots, \widehat{x_{n-i}}, \ldots, x_n) \\ &\equiv (-1)^n F_\natural(x_1, \ldots, \widehat{x_{n-i}}, \ldots, x_n) \end{aligned}$$

and similarly

$$\begin{aligned} F_\natural(y_1, \ldots, y_{n-1}) &= F_\natural\big|\tilde{T}_{n-1}(x_2 - x_1, \ldots, x_n - x_1) \\ &\equiv (-1)^n F_\natural(x_2 - x_1, \ldots, x_n - x_1) \end{aligned}$$

modulo M'_n. Hence

$$\left.\frac{F_\natural(x_1, \ldots, \widehat{x_{i+1}}, \ldots, x_n) - F_\natural(x_1, \ldots, \widehat{x_i}, \ldots, x_n)}{x_i - x_{i+1}}\right| (-1)^n\tilde{T}$$

$$\equiv \begin{cases} \dfrac{F_\natural(x_1, \ldots, \widehat{x_{n-i-1}}, \ldots, x_n) - F_\natural(x_1, \ldots, \widehat{x_{n-i}}, \ldots, x_n)}{x_{n-i-1} - x_{n-i}} & \text{if } i \le n-2; \\ \dfrac{F_\natural(x_2 - x_1, \ldots, x_n - x_1) - F_\natural(x_2, \ldots, x_n)}{-x_1} & \text{if } i = n-1, \end{cases}$$

modulo M'_n. Therefore, by cancelation Eq. (6.18) yields

$$\begin{aligned} 2(F^{(n)}_\natural - F^{(n)}_*)\big|((-1)^n\tilde{T}+1) \equiv{}& \frac{F_\natural(x_2, \ldots, x_n) - F_\natural(x_2 - x_1, \ldots, x_n - x_1)}{x_1} \\ &+ \frac{F_\natural(x_1, \ldots, x_{n-1}) - F_\natural(x_1, \ldots, \widehat{x_{n-1}}, x_n)}{x_{n-1} - x_n} \end{aligned} \tag{6.19}$$

modulo M'_n. Applying $\mathrm{sh}^{(n)}_1$ to both sides of Eq. (6.19), we obtain

$$\begin{aligned} 2(F_\natural - F_*)\big|((-1)^n\tilde{T}+1)\,\mathrm{sh}^{(n)}_1 \equiv{}& \frac{F_\natural(x_2, \ldots, x_n) - F_\natural(x_2 - x_1, \ldots, x_n - x_1)}{x_1} \\ &+ \frac{F_\natural(x_2, \ldots, \widehat{x_{n-1}}, x_n, x_1) - F_\natural(x_2, \ldots, x_{n-1}, x_1)}{x_{n-1} - x_n} \\ &+ 2 \cdot \frac{F_\natural(x_2, \ldots, x_{n-1}, x_1) - F_\natural(x_2, \ldots, x_n)}{x_1 - x_n} \end{aligned} \tag{6.20}$$

modulo M'_n. Define $\mathbf{1} = (1, \ldots, 1)$, $\mathbf{0} = (0, \ldots, 0)$, and

$$Q := Q_{n-1} = \nu_{n-1} T^\sharp_{n-1,0} T_{n-1,0} = \begin{pmatrix} -\mathbf{1} & -1 \\ I_{n-2} & \mathbf{0}^t \end{pmatrix} \in \mathrm{GL}_{n-1}(\mathbb{Z}),$$

where $\mathbf{0}^t$ is the transpose of $\mathbf{0}$. Then by applying

$$\iota(\tau) = \begin{pmatrix} -1 & -\mathbf{1} \\ \mathbf{0}^t & I_{n-1} \end{pmatrix} \in \mathrm{GL}_n(\mathbb{Z})$$

to Eq. (6.20) and using $F_\natural|Q^{-1} \equiv F_\natural|\nu$, we have

$$
\begin{aligned}
2(F_\natural - F_*)\big|((-1)^n\tilde{T}+1)\,\mathrm{sh}_1^{(n)}\,\iota(\tau)
&\equiv \frac{F_\natural(x_2,\dots,x_n) - F_\natural(x_2-x_1,\dots,x_n-x_1)}{x_1}\\
&\quad + \frac{F_\natural|Q^{-1}(x_1,\dots,\widehat{x_{n-1}},x_n) - F_\natural|Q^{-1}(x_1,\dots,x_{n-1})}{x_{n-1}-x_n}\\
&\quad + 2\cdot\frac{F_\natural|Q^{-1}(x_1,\dots,x_{n-1}) - F_\natural|Q^{-1}(x_1-x_n,\dots,x_{n-1}-x_n)}{-x_n}\\
&\equiv \frac{F_\natural(x_2,\dots,x_n) - F_\natural(x_2-x_1,\dots,x_n-x_1)}{x_1}\\
&\quad + \frac{F_\natural|\nu(x_1,\dots,\widehat{x_{n-1}},x_n) - F_\natural|\nu(x_1,\dots,x_{n-1})}{x_{n-1}-x_n}\\
&\quad + 2\cdot\frac{F_\natural|\nu(x_1,\dots,x_{n-1}) - F_\natural|\nu(x_1-x_n,\dots,x_{n-1}-x_n)}{-x_n}
\end{aligned}
$$

modulo M_n'. Together with Eq. (6.19) this yields

$$
\begin{aligned}
2(F_\natural - F_*)\big|((-1)^n\tilde{T}+1)(1+\mathrm{sh}_1^{(n)}\,\iota(\tau))
&\equiv 2\cdot\frac{F_\natural(x_2,\dots,x_n) - F_\natural(x_2-x_1,\dots,x_n-x_1)}{x_1}\\
&\quad - \frac{F_\natural\big|(1-\nu)(x_1,\dots,\widehat{x_{n-1}},x_n) - F_\natural\big|(1-\nu)(x_1,\dots,x_{n-1})}{x_{n-1}-x_n}\\
&\quad + 2\cdot\frac{F_\natural|\nu(x_1-x_n,\dots,x_{n-1}-x_n) - F_\natural|\nu(x_1,\dots,x_{n-1})}{x_n}
\end{aligned}
\tag{6.21}
$$

modulo M_n'. Hence by Eq. (6.17),

$$
F_\natural\big|(\nu-1) \equiv \frac{1}{2}\Big(\text{right-hand side of Eq.(6.21)}\Big) \tag{6.22}
$$

modulo M_n'. Let $F_\natural^{\mathrm{ev}}$ and $F_\natural^{\mathrm{od}}$ denote the even degree part and the odd degree part of $F_\natural$, respectively. Since $F_\natural^{\mathrm{od}}|\nu = -F_\natural^{\mathrm{od}}$ and $F_\natural^{\mathrm{ev}}|\nu = F_\natural^{\mathrm{ev}}$, by Eq. (6.22), we see that

$$
\begin{aligned}
2F_\natural^{\mathrm{od}}(x_1,\dots,x_n) &\equiv \frac{F_\natural^{\mathrm{ev}}(x_2-x_1,\dots,x_n-x_1) - F_\natural^{\mathrm{ev}}(x_2,\dots,x_n)}{x_1}\\
&\quad - \frac{F_\natural^{\mathrm{ev}}(x_1-x_n,\dots,x_{n-1}-x_n) - F_\natural^{\mathrm{ev}}(x_1,\dots,x_{n-1})}{x_n}
\end{aligned}
\tag{6.23}
$$

modulo M_n'.

Remark 6.4.6. If we define $\zeta^{\natural,\mathcal{S}}$ using $\zeta_{Ш}$ as mentioned in Remark 6.4.1 then all the arguements in this subsection are still valid modulo M_n'.

Proposition 6.4.7. *For all* $\mathbf{s} = (s_1, \ldots, s_n) \in \mathbb{N}^n$, *if* $w := |\mathbf{s}|$ *and* n *have different parities and* $|\mathbf{s}| \neq 2$, *then modulo* $\mathsf{MZ}_{w,+,\leq n} + \mathsf{MZ}_{w,\leq n-2}$ *we have:*

$$\begin{aligned}
&\zeta^{\natural,\mathcal{S}}(s_1, \ldots, s_n) \\
&\quad \equiv (-1)^{s_1} \sum_{\substack{\ell_2,\ldots,\ell_n \geq 0 \\ \ell_2+\cdots+\ell_n = s_n}} \prod_{j=2}^{n} \binom{s_j+\ell_j-1}{\ell_j} \zeta_*(s_2+\ell_2, \ldots, s_n+\ell_n) \\
&\qquad - (-1)^{s_n} \sum_{\substack{\ell_1,\ldots,\ell_{n-1} \geq 0 \\ \ell_1+\cdots+\ell_{n-1} = s_1}} \prod_{j=1}^{n-1} \binom{s_j+\ell_j-1}{\ell_j} \zeta_*(s_1+\ell_1, \ldots, s_{n-1}+\ell_{n-1}).
\end{aligned} \tag{6.24}$$

Proof. Comparing the coefficients of $x_1^{s_1-1} \cdots x_n^{s_n-1}$ on both sides of Eq. (6.23), we have

$$\begin{aligned}
&2\zeta_*^{\natural}(s_1, \ldots, s_n) \\
&\quad \equiv (-1)^{s_1} \sum_{\substack{\ell_2,\ldots,\ell_n \geq 0 \\ \ell_2+\cdots+\ell_n = s_1}} \prod_{j=2}^{n} \binom{s_j+\ell_j-1}{\ell_j} \zeta_*^{\natural}(s_2+\ell_2, \ldots, s_n+\ell_n) \\
&\qquad - (-1)^{s_n} \sum_{\substack{\ell_1,\ldots,\ell_{n-1} \geq 0 \\ \ell_1+\cdots+\ell_{n-1} = s_n}} \prod_{j=1}^{n-1} \binom{s_j+\ell_j-1}{\ell_j} \zeta_*^{\natural}(s_1+\ell_1, \ldots, s_{n-1}+\ell_{n-1})
\end{aligned}$$

modulo $\mathsf{MZ}_{w,+,\leq n} + \mathsf{MZ}_{w,\leq n-2}$. Since $\zeta_*^{\natural}(\mathbf{t}) \equiv \zeta_*(\mathbf{t})$ modulo $\mathsf{MZ}_{k,\leq n-2}$ for all $\mathbf{t} \in \mathbb{N}^{n-1}$ and

$$\zeta^{\natural,\mathcal{S}}(\mathbf{s}) \equiv \zeta_*^{\natural}(\mathbf{s}) + (-1)^w \zeta_*^{\natural}(\overleftarrow{\mathbf{s}}) \equiv \zeta_*^{\natural}(\mathbf{s}) + (-1)^{w+n-1} \zeta_*^{\natural}(\mathbf{s}) \equiv 2\zeta_*^{\natural}(\mathbf{s})$$

modulo $\mathsf{MZ}_{w,+,\leq n}$, the proposition follows immediately. □

Remark 6.4.8. In view of Remark 6.4.6 we may replace ζ_* by $\zeta_{\sqcup\!\sqcup}$ on the right-hand side of Eq. (6.24). This also follows from the fact that no index of the form $(1, \ldots, 1)$ appears there.

6.4.4 *Diagonal Translation Invariant Polynomials*

In order to prove the main result Thm. 6.4.19 of this section, we need to study a special class of polynomials.

Lemma 6.4.9. *Let* $f(x_1, \ldots, x_n) \in \mathbb{Q}[x_1, \ldots, x_n]$. *Then the following conditions are equivalent:*

(i) $f(x_1 + t, \ldots, x_n + t) = f(x_1, \ldots, x_n)$ *for all* $t \in \mathbb{R}$.

(ii) *There exists a polynomial g in $n-1$ variables such that*

$$f(x_1,\dots,x_n)=g(x_1-x_n,\dots,x_{n-1}-x_n).$$

(iii) $\displaystyle\sum_{i=1}^{n}\frac{\partial f}{\partial x_i}=0.$

Proof. (i)$\Longleftrightarrow$(iii) Let $h(t)=f(x_1+t,\dots,x_n+t)$. Then by the Chain Rule we have

$$\text{(i)} \Longleftrightarrow h(t)=h(0) \Longleftrightarrow h'(t)=0 \Longleftrightarrow \text{(iii)}.$$

(i)$\Longrightarrow$(ii) Take $t=-x_n$ and $g(x_1,\dots,x_{n-1})=f(x_1,\dots,x_{n-1},0)$.

(ii)$\Longrightarrow$(i) Trivial. □

Definition 6.4.10. A polynomial $f(x_1,\dots,x_n)\in\mathbb{Q}[x_1,\dots,x_n]$ is called *diagonal translation invariant* (DTI) if the equivalent conditions in Lemma 6.4.9 are satisfied.

Definition 6.4.11. A polynomial $f(x_1,\dots,x_n)\in\mathbb{Q}[x_1,\dots,x_n]$ is called *pseudo-symmetric* if $f|\sigma=\pm f$ for all permutations $\sigma\in\mathfrak{S}_n$.

Lemma 6.4.12. *Suppose $f,f_1,\dots,f_m\in\mathbb{Q}[x_1,\dots,x_n]$ are all DTI. Then*

(i) *$f|\sigma$ is DTI for every permutation $\sigma\in\mathfrak{S}_n$.*
(ii) *Every $\mathbb{Q}$-linear combination of $f_1,\dots,f_m$ is DTI.*

Proof. By assumption $h(x_1+t,\dots,x_n+t)=h(x_1,\dots,x_n)$ for all $t\in\mathbb{R}$ and $h=f,f_1,\dots,f_m$. So the lemma is obvious. □

Lemma 6.4.13. *Let $f(x_1,\dots,x_n)$ be a pseudo-symmetric polynomial in n variables. Suppose that $\frac{\partial}{\partial x_n}(x_nf)$ is DTI. Then f is constant.*

Proof. We may assume that f is homogeneous of degree d. Let σ be the transposition (j,n). Then by pseudo-symmetry and Lemma 6.4.12

$$\frac{\partial}{\partial x_j}(x_jf)=\pm\left(\frac{\partial}{\partial x_n}(x_nf)\right)\bigg|\,\sigma$$

is DTI for every positive integer $j\le n$. Hence, by Euler's Homogeneous Function Theorem and Lemma 6.4.12(ii)

$$f=\frac{1}{n+d}\sum_{i=1}^{n}\frac{\partial}{\partial x_i}(x_if)$$

is DTI. For any positive integer $j\le n$, by direct computation

$$\frac{\partial f}{\partial x_j}=\sum_i\frac{\partial}{\partial x_i}\frac{\partial}{\partial x_j}(x_jf)-\frac{\partial}{\partial x_j}\left(x_j\sum_i\frac{\partial f}{\partial x_i}\right)=0.$$

Thus f is constant, as desired. □

Lemma 6.4.14. *Let $f(x_1,\dots,x_n)$ be a homogeneous polynomial of degree $d\ge 2$. Suppose that*

$$\frac{1}{x_{n+1}}\big(f(x_1-x_{n+1},\dots,x_n-x_{n+1})-f(x_1,\dots,x_n)\big)$$

is a pseudo-symmetric polynomial in $n+1$ variables. Then f is DTI.

Proof. We set $h(x_1,\dots,x_{n+1}) = \frac{1}{x_{n+1}}\big(f(x_1 - x_{n+1},\dots,x_n - x_{n+1}) - f(x_1,\dots,x_n)\big)$. By substition $x_j \to x_j + t$ one checks easily that

$$\frac{\partial}{\partial x_{n+1}}(x_{n+1}h)=\frac{\partial}{\partial x_{n+1}}f(x_1-x_{n+1},\dots,x_n-x_{n+1})$$

is DTI. Further, h is a pseudo-symmetric by assumption, Lemma 6.4.13 then shows h is a constant of degree $d-1\ge 1$. So we must have $h=0$. Hence f is DTI by defnition (setting $x_{n+1}=t$). □

Recall that $\mathsf{DSh}_n(d)$ is the degree d part of the double shuffle space

$$\mathsf{DSh}_n=\Big\{f\in V_n\;\Big|\; f|\operatorname{sh}_j^{(n)}=f^\sharp|\operatorname{sh}_j^{(n)}=0 \text{ for } j=1,\dots,n-1\Big\}$$

defined by Definition 3.5.1.

Lemma 6.4.15. *Let $d\ge 2$ and $f\in\mathsf{DSh}_n(d)$. If*

$$\frac{1}{x_{n+1}}\big(f(x_1-x_{n+1},\dots,x_n-x_{n+1})-f(x_1,\dots,x_n)\big)$$

is invariant under the cyclic permutations then f is DTI.

Proof. Set

$$h(x_1,\dots,x_{n+1}):=\frac{1}{x_{n+1}}\big(f(x_1-x_{n+1},\dots,x_n-x_{n+1})-f(x_1,\dots,x_n)\big).$$

Step (1). Since $f\in\mathsf{DSh}_n(d)$ by the definition of h we have

$$\begin{aligned}
h|T_0 &= \frac{1}{x_1}\big(f(x_{n+1}-x_1,\dots,x_2-x_1)-f(x_{n+1},\dots,x_2)\big)\\
&\overset{(3.33)}{=\!=}\frac{(-1)^{n-1}}{x_1}\big(f(x_2-x_1,\dots,x_{n+1}-x_1)-f(x_2,\dots,x_{n+1})\big)\\
&=(-1)^{n-1}h(x_2,\dots,x_{n+1},x_1) && \text{(by defn. of } h)\\
&=(-1)^{n-1}h(x_1,x_2,\dots,x_{n+1}) && \text{(by cyclic invariance).}
\end{aligned}$$

Step (2). By assumption, h is invariant under the action of

$$\gamma_0:=\iota_{n+1}(C_{n+1})=\begin{pmatrix} & 1\\ I_n & \end{pmatrix}\in \mathrm{GL}_{n+1}(\mathbb{Z}).$$

Step (3). By setting

$$\gamma_1 := \begin{pmatrix} -1 & -\mathbf{1} \\ \mathbf{0}^t & I_n \end{pmatrix} \in \mathrm{GL}_{n+1}(\mathbb{Z})$$

we get

$$\begin{aligned} h|\gamma_1 &= h\big((x_1,\dots,x_{n+1})\gamma_1^{-1}\big) = h\big((x_1,\dots,x_{n+1})\gamma_1\big) \\ &= h(-x_1, x_2-x_1,\dots,x_{n+1}-x_1) \\ &= h(x_2-x_1,\dots,x_{n+1}-x_1,-x_1) && \text{(by cyclic invariance)} \\ &= \frac{f(x_2-x_1,\dots,x_{n+1}-x_1)-f(x_2,\dots,x_{n+1})}{x_1} && \text{(by defn. of } h) \\ &= h(x_2,\dots,x_{n+1},x_1) && \text{(by defn. of } h) \\ &= h(x_1,x_2,\dots,x_{n+1}) && \text{(by cyclic invariance).} \end{aligned}$$

Step (4). Recall that the image under ι_{n+1} of the permutation

$$T'_{n+1} = \begin{pmatrix} 1 & 2 & \cdots & n & n+1 & n+2 \\ n & n-1 & \cdots & 1 & n+2 & n+1 \end{pmatrix} \in \mathfrak{S}_{n+2}$$

is equal to $\gamma_2 := \nu T_0 T_0^\sharp T_0 = T_0\gamma_1\gamma_0$. Hence

$$h|\gamma_2 = h|T_0\gamma_1\gamma_0 \overset{Step(1)}{=\!=\!=} \pm h|\gamma_1\gamma_0 = \pm h \tag{6.25}$$

since h is invariant under both γ_0 and γ_1. So $h|\sigma = \pm h$ for all $\sigma \in \mathfrak{S}_{n+2}$ by the following fact (see Exercise 6.8):

$$\mathfrak{S}_{n+2} = \langle T_{n+1,0}, C_{n+1}, T'_{n+1}\rangle.$$

In particular, $h|\sigma = \pm h$ for all $\sigma \in \mathfrak{S}_{n+1}$ which means h is pseudo-symmetric. Finally, the proof is completed by applying Lemma 6.4.14. □

Proposition 6.4.16. *For any even integer $d \geq 0$ and $f \in \mathsf{DSh}_n(d)$, if*

$$\frac{1}{x_1}\Big(f(x_2-x_1,\dots,x_{n+1}-x_1) - f(x_2,\dots,x_{n+1})\Big) \\ = \frac{1}{x_{n+1}}\Big(f(x_1-x_{n+1},\dots,x_n-x_{n+1}) - f(x_1,\dots,x_n)\Big) \tag{6.26}$$

then $f = 0$.

Proof. We may assume $d \geq 2$ without loss of generality. Note that the condition Eq. (6.26) eaily implies that $\frac{1}{x_{n+1}}\big(f(x_1-x_{n+1},\dots,x_n-x_{n+1}) -$

$f(x_1, \ldots, x_n))$ is invariant under the cyclic permutations so Lemma 6.4.15 yields that f is DTI. By taking $t = -x_n$ we get

$$f(-x_n, x_1 - x_n, \ldots, x_{n-1} - x_n)) = f(0, x_1, \ldots, x_{n-1}). \tag{6.27}$$

On the other hands, the matrix $Q_n = \nu T_0^\sharp T_0$ acts on f by $(-1)^d = \text{id}$. Hence we have

$$\begin{aligned} f(x_1, \ldots, x_n) &= f|Q(x_1, \ldots, x_n) \\ &= f(-x_n, x_1 - x_n, \ldots, x_{n-1} - x_n) = f(0, x_1, \ldots, x_{n-1}) \end{aligned}$$

by Eq. (6.27). Repeatedly applying this we get

$$f(x_1, \ldots, x_n) = f(0, 0, x_1, \ldots, x_{n-2}) = \cdots = f(0, \ldots, 0).$$

But f has degree $d \geq 2$ so we must have $f = 0$, as desired. □

6.4.5 *SMZVs Generate Whole MZV Space*

For any two integers $w, n \geq 0$, we set

$$\mathsf{MZ}'_{w,n} = \mathsf{MZ}_{w,\leq n-1} + \mathsf{MZ}_{w,+,\leq n}.$$

Let $\mathsf{MZ}''_{w,n}$ denote the $\mathbb{Q}$-linear subspace of $\mathsf{MZ}_{w,\leq n}$ generated by the set

$$\left\{ \text{RHS of Eq. (6.24) for} (s_1, \ldots, s_{n+1}) \;\middle|\; \begin{array}{c} s_1, \ldots, s_{n+1} \geq 1, \\ s_1 + \cdots + s_{n+1} = w \end{array} \right\}.$$

Proposition 6.4.17. *For any two integers $w, n \geq 0$, we have*

$$\mathsf{MZ}_{w,\leq n} = \mathsf{MZ}'_{w,n} + \mathsf{MZ}''_{w,n}. \tag{6.28}$$

Proof. Not that if w and n have different parities, then the proposition follows from the Parity Principle (see Cor. 3.5.7).

We now assume $d = w - n$ is even. By a simple algebra argument we see that it suffices to show that the restriction map $\text{Hom}_{\mathbb{Q}}(\mathsf{MZ}_{w,\leq n}, \mathbb{Q}) \to \text{Hom}_{\mathbb{Q}}(\mathsf{MZ}'_{w,n} + \mathsf{MZ}''_{w,n}, \mathbb{Q})$ is injective. Suppose $\alpha : \mathsf{MZ}_{k,\leq n} \to \mathbb{Q}$ is a $\mathbb{Q}$-linear map vanishing on $\mathsf{MZ}'_{w,n} + \mathsf{MZ}''_{w,n}$. To prove $\alpha = 0$ it is enough to check that

$$f := \sum_{\substack{s_1, \ldots, s_n \geq 1 \\ s_1 + \cdots + s_n = w}} \alpha(\zeta_*(s_1, \ldots, s_n)) x_1^{s_1 - 1} \cdots x_n^{s_n - 1}$$

vanishes. Note that d is the degree of f. Since α vanishes on $\mathsf{MZ}_{w,+}$ we see that $f \in \mathsf{DSh}_n(d)$. Finally, α is 0 on $\mathsf{MZ}''_{w,n}$ so that Eq. (6.26) holds by Eq. (6.23) and the proof of Prop. 6.4.7 since now $n+1$ and w have different parities. Hence $\alpha = 0$ by Prop. 6.4.16. This completes the proof of the proposition. □

For any compositions of positive integers $\mathbf{s}$ and $\mathbf{t}$, let $\mathrm{St}(\mathbf{s},\mathbf{t})$ be the set of the stuffle products of $\mathbf{s}$ and $\mathbf{t}$ with multiplicity. For example,

$$\mathrm{St}((3),(3,2)) = \{(3,3,2),(3,3,2),(3,2,3),(6,2),(3,5)\}.$$

Lemma 6.4.18. *Let* $\mathbf{s} \in \mathbb{N}^m$ *and* $\mathbf{t} \in (\mathbb{Z}_{\geq 2})^n$. *Then we have*

$$\zeta_{ш}^{\mathcal{S}}(\mathbf{s})\zeta^{\mathcal{S}}(\mathbf{t}) = \sum_{\mathbf{r}\in \mathrm{St}(\mathbf{s},\mathbf{t})} \zeta_{ш}^{\mathcal{S}}(\mathbf{r}).$$

Proof. For $\mathbf{s} = (s_1, \dots, s_m)$, we set

$$I(\mathbf{s}) := \{((s_i, \dots, s_1), (s_{i+1}, \dots, s_m)) \mid 0 \leq i \leq m\}$$

and similarly for $\mathbf{t}$. Then by the definition

$$\zeta_{ш}^{\mathcal{S}}(\mathbf{s})\zeta^{\mathcal{S}}(\mathbf{t}) = \sum_{(\mathbf{s}_1,\mathbf{s}_2)\in I(\mathbf{s})} \sum_{(\mathbf{t}_1,\mathbf{t}_2)\in I(\mathbf{t})} (-1)^{|\mathbf{s}_1|+|\mathbf{t}_1|}\zeta_{ш}(\mathbf{s}_1)\zeta(\mathbf{t}_1)\cdot\zeta_{ш}(\mathbf{s}_2)\zeta(\mathbf{t}_2).$$

It follows from Cor 3.3.23 that

$$\zeta_{ш}(\mathbf{s}_j)\zeta(\mathbf{t}_j) = \sum_{\mathbf{r}_j\in \mathrm{St}(\mathbf{s}_j,\mathbf{t}_j)} \zeta_{ш}(\mathbf{r}_j)$$

for $j = 1, 2$. So the lemma follows easily from the identity

$$\coprod_{\mathbf{r}\in \mathrm{St}(\mathbf{s},\mathbf{t})} I(\mathbf{r}) = \coprod_{(\mathbf{s}_1,\mathbf{s}_2)\in I(\mathbf{s})} \coprod_{(\mathbf{t}_1,\mathbf{t}_2)\in I(\mathbf{t})} \mathrm{St}(\mathbf{s}_1,\mathbf{t}_1)\times \mathrm{St}(\mathbf{s}_2,\mathbf{t}_2)$$

where both sides are regarded as multisets of pairs of compositions. □

Recall that MZ_w, $\mathsf{MZ}^{\mathcal{S}}_{*,w}$ and $\mathsf{MZ}^{\mathcal{S}}_{ш,w}$ are the $\mathbb{Q}$-vector space generated by all weight w MZVs, $*$-SMZVs and ш-SMZVs, respectively.

Theorem 6.4.19. *For all positive interger* $w \geq 2$, *we have*

$$\mathsf{MZ}_w = \mathsf{MZ}^{\mathcal{S}}_{*,w} = \mathsf{MZ}^{\mathcal{S}}_{ш,w}.$$

Proof. First, since $\mathsf{MZ}^{\mathcal{S}}_{*,w} \subseteq \mathsf{MZ}_w$ by Prop. 6.1.3 we can prove $\mathsf{MZ}_w = \mathsf{MZ}^{\mathcal{S}}_{*,w}$ by showing the claim

$$\mathsf{MZ}_{w,\leq n} \subseteq \mathsf{MZ}^{\mathcal{S}}_{*,w} \quad \text{for all } n \geq 0. \tag{6.29}$$

We now proceed by induction on n. When $n = 0$ this is trivial for all w. Assume $n \geq 1$ and the inclusion holds for all w. If $w - n$ is even then w and $n+1$ have different parities and Eq. (6.29) follows from the Parity Principle and the induction. If $w - n$ is odd then by Prop. 6.4.17, we have

$$\begin{aligned}\mathsf{MZ}_{w,\leq n+1} &= \mathsf{MZ}'_{w,n+1} + \mathsf{MZ}''_{w,n+1} \\ &\subseteq \mathsf{MZ}_{w,\leq n} + \mathsf{MZ}_{w,+,\leq n+1} \qquad \text{(by Prop. 6.4.7)}\end{aligned}$$

$$\subseteq \mathsf{MZ}^{\mathcal{S}}_{*,w} + \mathsf{MZ}^{\mathcal{S}}_{*,w,+} \qquad \text{(by induction)}$$

$$\subseteq \mathsf{MZ}^{\mathcal{S}}_{*,w} + \sum_{w',w''\geq 1,\ w'+w''=w} \mathsf{MZ}^{\mathcal{S}}_{*,w'} \cdot \mathsf{MZ}^{\mathcal{S}}_{*,w''} \subseteq \mathsf{MZ}^{\mathcal{S}}_{*,w}$$

$$\text{(by Cor. 6.1.5)}.$$

Turning to the $⧢$-version, by Thm. 6.3.2 and Prop. 6.1.3, we see that

$$\mathsf{MZ}^{\mathcal{S}}_{*,w} \subseteq \mathsf{MZ}^{\mathcal{S}}_{⧢,w} + \zeta(2)\mathsf{MZ}^{\mathcal{S}}_{⧢,w-2} \subseteq \mathsf{MZ}_w = \mathsf{MZ}^{\mathcal{S}}_{*,w}$$

by what we have just proved. Thus, to prove $\mathsf{MZ}^{\mathcal{S}}_{*,w} = \mathsf{MZ}^{\mathcal{S}}_{⧢,w}$ it suffices to show $\zeta(2)\mathsf{MZ}^{\mathcal{S}}_{⧢,w-2} \subset \mathsf{MZ}^{\mathcal{S}}_{⧢,w}$ for all $w \geq 2$, which follows immediately from Lemma 6.4.18. The theorem is now proved. □

6.5 Historical Notes

Under a suggestion of Kontsevich the SMZVs were first introduced by M. Kaneko and D. Zagier in April of 2013 at Cambridge. Their goal was to establish a crucial bridge between the ordinary MZVs and their finite version to be explained in Chap. 8. Moreover, the SMZVs are nicer versions of the MZVs in the sense that they not only satisfy the stuffle relations and shuffle relations but also are defined for all positive integers. Furthermore, their shuffle relations are simpler to state than those for the MZVs.

Our presentation in Sec. 6.3 follows Jarossay's notes [317] closely. In addition, we have provided some results such as Thm. 6.2.2 and Thm. 6.2.4, due to Murahara [431], which are very similar to the ones for the finite MZVs in Chap. 8. The guiding principle here is a mysterious relation between the SMZVs and the finite MZVs to be discussed in Conjecture 8.5.15 of Chap. 8.

Theorem 6.4.19 was proved by Yasuda [585], however, Eq. (6.23) was also discovered independently by Brown.

Exercises

6.1. Show that

$$\zeta^{\mathcal{S}}_{⧢}(2,1) = 3\zeta(2,1)$$

by using the fact that $\mathtt{b}⧢\mathtt{ab} = \mathtt{bab} + 2\mathtt{abb}$. Similarly, show that $\zeta^{\mathcal{S}}_{⧢}(4,1) = 3\zeta(4,1) - \zeta(5) + \zeta(2)\zeta(3) = 5\zeta(5) - 2\zeta(2)\zeta(3)$ by using the DBSF $\zeta(4,1) = 2\zeta(5) - \zeta(2)\zeta(3)$.

6.2. Complete the proof of Prop. 6.1.3 by showing that $\zeta^{\mathcal{S}}_{*}(\mathbf{s})$ are finite real numbers for all positive compositions $\mathbf{s}$.

6.3. Let the map inv be defined by Eq. (6.6). Show that

$$\Delta_* \circ \mathrm{inv} = (\mathrm{inv} \otimes \mathrm{inv}) \circ \Delta_*.$$

6.4. Prove Eq. (6.15) by mimicking the proof of Eq. (6.14) and using an argument similar to the proof of Eq. (6.3).

6.5. Show that under the embedding $\mathbb{Z}^n \hookrightarrow \mathbb{Z}^{(n+1)}$, $(a_1, \ldots, a_n) \to (a_1, \ldots, a_n, -(a_1 + \cdots + a_n))$, the image $\iota_n(T')$ is given by the matrix γ_2 as defined by Eq. 6.16.

6.6. For any $n \in \mathbb{N}$, define

$$\angle_{n,\mathbb{R}} = \left\{ (t_1, \ldots, t_n) \in (\mathbb{R}^\times)^n \;\middle|\; \frac{1}{t_1} > \cdots > \frac{1}{t_n} \right\}.$$

For $\mathbf{x} = (x_1, \ldots, x_n) \in \mathbb{R}^n$, let $B(\mathbf{x}, \varepsilon)$ be the open ball in $\mathbb{R}^n$ of radius ε centered at $\mathbf{x}$. Set

$$v(\mathbf{x}) = \lim_{\varepsilon \to 0} \frac{\mathrm{vol}(B(\mathbf{x}, \varepsilon) \cap \angle_{n,\mathbb{R}})}{\mathrm{vol}(B(\mathbf{x}, \varepsilon))}.$$

Show that for any $\mathbf{s} \in \mathbb{N}^n$,

$$\zeta^{\natural,\mathcal{S}}(\mathbf{s}) = (-1)^{|\mathbf{s}|} \lim_{M \to \infty} \sum_{\substack{\frac{1}{k_1} > \cdots > \frac{1}{k_d} \\ 0 \neq |k_j| \in \mathbb{Z}_{\leq M} \ \forall j = 1, \ldots, d}} \frac{v((k_1, \ldots, k_d))}{k_1^{s_1} \cdots k_d^{s_d}}.$$

6.7. Prove Eq. (6.20) by applying

$$\mathrm{sh}_1^{(n)} = \sum_{i=1}^{n} \begin{pmatrix} 1 \; 2 \cdots & i & i+1 \cdots n \\ i \; 1 \cdots & i-1 & i+1 \cdots n \end{pmatrix},$$

to both sides of Eq. (6.19).

6.8. Show that the symmetry group $\mathfrak{S}_{n+1}$ is generated by the following three elements: $C_n, T_{n,0} \in \mathfrak{S}_n$ and T'_n defined by Eq. (3.34).

Chapter 7

Multiple Harmonic Sums (MHSs) and Alternating Version

A multiple harmonic sum (MHS) is a finite partial sum of the infinite series in Eq. (1.8) defining the MZV $\zeta(s_1, \dots, s_d)$ for $s_1, \dots, s_d \in \mathbb{N}$ (note that $s_1 = 1$ is allowed for partial sums). We have already encountered these numbers in previous chapters (see (3.18)). An alternating MHS (AMHS for abbreviation) is a finite partial sum of the infinite series defining a Euler sum, an alternating version of a MZV. All of these partial sums are rational numbers so it makes perfect sense to consider congruence properties modulo sufficiently large primes. This will be our first goal in this chapter. The second goal is to derive some identities using the binomial coefficients which will be useful to evaluate the MZVs and similar objects when the number of terms of the partial sums goes to infinity.

7.1 Definitions

We start by defining a sort of double cover of the set $\mathbb{N}_0$ of nonnegative integers.

Definition 7.1.1. Let $\mathbb{D}_0$ be the set of *signed numbers* $\mathbb{N}_0 \cup \overline{\mathbb{N}_0}$ where

$$\overline{\mathbb{N}_0} := \{\bar{k} : k \in \mathbb{N}_0\}.$$

Define the absolute value function $|\cdot|$ on $\mathbb{D}_0$ by $|k| = |\bar{k}| = k$ for all $k \in \mathbb{N}_0$ and the sign function by $\operatorname{sgn}(k) = 1$ and $\operatorname{sgn}(\bar{k}) = -1$ for all $k \in \mathbb{N}_0$. Numbers in $\overline{\mathbb{N}_0}$ are called *signed numbers* and numbers in $\mathbb{N}_0$ are called *unsigned numbers*. We make $\mathbb{D}$ a semi-group by defining a commutative and associative binary operation $\oplus$ (called *O-plus*) as follows: for all $a, b \in \mathbb{D}_0$

$$a \oplus b := \begin{cases} \overline{|a| + |b|}, & \text{if } \operatorname{sgn}(a) \neq \operatorname{sgn}(b); \\ |a| + |b|, & \text{if } \operatorname{sgn}(a) = \operatorname{sgn}(b). \end{cases} \tag{7.1}$$

Finally, we define $\mathbb{D} := \mathbb{D}_0 \setminus \{0, \bar{0}\}$.

Remark 7.1.2. We only use $\mathbb{D}$ in this chapter but we will need $\mathbb{D}_0$ to handle the q-analogs of the MHSs and the MH*Ss in Chap. 9.

Definition 7.1.3. For any $d \in \mathbb{N}$ and $\mathbf{s} = (s_1, s_2, \ldots, s_d) \in \mathbb{D}^d$, we define the *alternating multiple harmonic sums* by

$$H_n(s_1, s_2, \ldots, s_d) = \sum_{n \geq k_1 > k_2 > \ldots > k_d \geq 1} \prod_{j=1}^{d} \frac{\operatorname{sgn}(s_j)^{k_j}}{k_j^{|s_j|}}, \tag{7.2}$$

$$H_n^\star(s_1, s_2, \ldots, s_d) = \sum_{n \geq k_1 \geq k_2 \geq \ldots \geq k_d \geq 1} \prod_{j=1}^{d} \frac{\operatorname{sgn}(s_j)^{k_j}}{k_j^{|s_j|}}. \tag{7.3}$$

When $\mathbf{s} \in \mathbb{N}^d$ these are called the *multiple harmonic sums* (MHSs) and the *multiple harmonic star sums* (MH*Ss), respectively. Conventionally, we call $\operatorname{dp}(\mathbf{s}) := d$ the depth and $|\mathbf{s}| := \sum_{j=1}^{d} |s_j|$ the weight. For convenience, we set $H_n(\mathbf{s}) = 0$ if $n < \operatorname{dp}(\mathbf{s})$, $H_n(\emptyset) = H_n^\star(\emptyset) = 1$ for all $n \geq 0$.

For instance, the harmonic numbers and the alternating harmonic numbers have weight and depth both equal to 1:

$$H_n(1) = \sum_{k=1}^{n} \frac{1}{k}, \qquad H_n(\bar{1}) = \sum_{k=1}^{n} \frac{(-1)^k}{k}.$$

7.2 Stuffle Relations

Throughout this section we let n be a fixed positive integer and consider only the AMHSs of the form H_n or $H_n^\star$.

Recall the Hoffman algebra $\mathfrak{A}^1 = \mathbb{Q}1 \oplus \mathfrak{A}\mathsf{b}$ given in Definition 3.2.2 is a noncommutative polynomial algebra on generators $\mathsf{z}_k := \mathsf{a}^{k-1}\mathsf{b}$ for all $k \geq 1$. We further equip it with the stuffle product $*$ and regard it as a subalgebra of $\mathfrak{A}_*$ defined in Definition 3.2.7. To study both the MHSs and the AMHSs we need to generalize this algebra as sketched below. We omit the details because we will extend this framework even further to arbitrary level N in Chap. 13 by replacing ± 1 by Nth roots of unity (see Definition 13.3.2 there).

Definition 7.2.1. For $N \in \mathbb{N}$, we define the Hoffman–Racinet algebra of level N, denoted by $\mathfrak{A}_N$, to be the (weight) graded noncommutative polynomial $\mathbb{Q}$-algebra generated by $N+1$ letters $\mathsf{a}, \mathsf{b}_0, \ldots, \mathsf{b}_{N-1}$. The number of letters in a word $\mathbf{w}$ is called its *weight* and the number of $\mathsf{b}_0, \ldots, \mathsf{b}_{N-1}$, its *depth*. For every positive integer k, define

$$\mathsf{z}_k := \mathsf{a}^{k-1}\mathsf{b}_0 \quad \text{and, if } N = 2, \quad \mathsf{z}_{\bar{k}} := \mathsf{a}^{k-1}\mathsf{b}_1. \tag{7.4}$$

Let $\mathfrak{A}_N^1$ be the subalgebra of $\mathfrak{A}_N$ generated by words not ending with a.

Note that when $N = 2$, $\mathfrak{A}_2^1$ is the subalgebra of $\mathfrak{A}_2$ generated by words $\mathtt{z}_k$ with $k \in \mathbb{D}$.

Remark 7.2.2. For future reference, when we deal with the Euler sums we will need to consider the subalgebra $\mathfrak{A}_2^0$ of $\mathfrak{A}_2$ generated by words beginning with $\mathtt{a}$ or $\mathtt{b}_1$ and ending with $\mathtt{b}_0$ or $\mathtt{b}_1$. The words in $\mathfrak{A}_2^0$ are called *admissible words.*

To define the stuffle relations among AMHSs, a very simple example is already illuminating:

$$H_n(\bar{1})H_n(2) = H_n(\bar{1}, 2) + H_n(2, \bar{1}) + H_n(\bar{3}).$$

This follows immediately from the expansion

$$\sum_{n \ge j > 0} \cdot \sum_{n \ge k > 0} = \sum_{n \ge j > k > 0} + \sum_{n \ge k > j > 0} + \sum_{n \ge j = k > 0} .$$

Definition 7.2.3. For any $k \in \mathbb{N}$, we define the *toggling operator* $\tau_{\mathbf{z}_k}$ as follows: $\forall \mathbf{w} \in \mathfrak{A}_2^1$ we set $\tau_{\mathbf{z}_k}(\mathbf{w}) = \mathbf{w}$, and define $\tau_{\mathbf{z}_{\bar{k}}}(\mathbf{w})$ to be the word produced from $\mathbf{w}$ by exchanging $\mathbf{z}_l$ and $\mathbf{z}_{\bar{l}}$ for all l. We define a multiplication $*$ on $\mathfrak{A}_2^1$ by requiring that $*$ distribute over addition, that $1 * \mathbf{w} = \mathbf{w} * 1 = \mathbf{w}$ for each word $\mathbf{w} \in \mathfrak{A}_2^1$, and that, for all words $\mathbf{u}, \mathbf{v}$ and special words x and y equaling either $\mathbf{z}_k$ or $\mathbf{z}_{\bar{k}}$ for some k,

$$\begin{aligned} x\mathbf{u} * y\mathbf{v} = x\Big(\tau_x\big(\tau_x(\mathbf{u}) * y\mathbf{v}\big)\Big) + y\Big(\tau_y\big(x\mathbf{u} * \tau_y(\mathbf{v})\big)\Big) \\ + [x, y]\Big(\tau_{[x,y]}\big(\tau_x(\mathbf{u}) * \tau_y(\mathbf{v})\big)\Big) \end{aligned} \tag{7.5}$$

where

$$[\mathbf{z}_k, \mathbf{z}_l] = [\mathbf{z}_{\bar{k}}, \mathbf{z}_{\bar{l}}] = \mathbf{z}_{k+l} \quad \text{and} \quad [\mathbf{z}_{\bar{k}}, \mathbf{z}_l] = [\mathbf{z}_k, \mathbf{z}_{\bar{l}}] = \mathbf{z}_{\overline{k+l}}.$$

Using Eq. (7.5) we can further define a multiplication $*^\star$ on $\mathfrak{A}_2^1$: simply replace $*$ by $*^\star$ and substitute $-[x, y]$ for $+[x, y]$ in Eq. (7.5).

The proof of the following stuffle relations is similar to that of MZV using induction on depth and is thus omitted here.

Theorem 7.2.4. *For any word $\mathbf{w} = \mathbf{z}_{s_1} \dots \mathbf{z}_{s_d} \in \mathfrak{A}_2^1$, define $H_n(\mathbf{w}) := H_n(s_1, \dots, s_d)$ and $H_n^\star(\mathbf{w}) := H_n^\star(s_1, \dots, s_d)$, and then linearly extend both over $\mathfrak{A}_2^1$ (cf. Definition 3.2.2). Then for any words $\mathbf{u}, \mathbf{v} \in \mathfrak{A}_2^1$ we have*

$$H_n(\mathbf{u})H_n(\mathbf{v}) = H_n(\mathbf{u} * \mathbf{v}), \qquad H_n^\star(\mathbf{u})H_n^\star(\mathbf{v}) = H_n^\star(\mathbf{u} *^\star \mathbf{v}). \tag{7.6}$$

7.3 Quasi-Symmetric Functions with Signed Powers

To derive more relations between the MHSs, the MH*Ss and the AMHSs we turn to the theory of quasi-symmetric functions. To treat the AMHSs with signed indices in $\mathbb{D}$ we have to allow the powers in these quasi-symmetric functions to be signed numbers.

Definition 7.3.1. We denote by $\mathbb{Z}[x_1,\dots,x_n;\mathbb{D}]$ the set of polynomials in $x_1,\dots,x_n$ with signed powers, namely,

$$\mathbb{Z}[x_1,\dots,x_n;\mathbb{D}] = \left\{ \sum_{e_1=\overline{d_1}}^{d_1} \cdots \sum_{e_n=\overline{d_n}}^{d_n} c_{e_1,\dots,e_n} x_1^{e_1}\cdots x_n^{e_n} \,\middle|\, d_1,\dots,d_n \in \mathbb{N}_0, c_{e_1,\dots,e_n} \in \mathbb{Z} \right\}.$$

Here, $\overline{0}=0$ and $\sum_{e=\overline{d}}^{d}$ means e runs through the set $\{\overline{d},\dots,\overline{1},0,1,\dots,d\}$. Furthermore, $x_j^e x_j^{e'} = x_i^{e\oplus e'}$ for any $j \le n$ and $e,e' \in \mathbb{D}$. Also we set

$$\deg(x_1^{e_1}\cdots x_n^{e_n}) = |e_1| + \cdots + |e_n|.$$

For positive integer n, we define a graded algebra homomorphism

$$\begin{aligned} \psi_n: \quad \mathfrak{A}_2^1 \quad &\longrightarrow \quad \mathbb{Z}[x_1,\dots,x_n;\mathbb{D}] \\ \mathbf{z}_{s_1}\cdots\mathbf{z}_{s_d} &\longmapsto \sum_{n\ge k_1>k_2>\cdots>k_d\ge 1} x_{k_1}^{s_1}\cdots x_{k_d}^{s_d}, \quad \forall d \le n, \end{aligned}$$

where we have set $\psi_n(1)=1$, $\psi_n(\mathbf{w})=0$ if the depth of $\mathbf{w}$ is greater than n, and the degree of every x_i is equal to 1. It is easy to see we can arrange $(\psi_n)_{n\ge1}$ into a compatible system to obtain a homomorphism ψ from the Hoffman–Racinet algebra $\mathfrak{A}_2^1$ to the algebra QSym_2 of quasi-symmetric functions which are defined below.

Definition 7.3.2. Let $\mathbf{x}=(x_j)_{j\ge1}$. An element of finite degree $F(\mathbf{x}) \in \mathbb{Z}[\![\mathbf{x};\mathbb{D}]\!]$ is called a *quasi-symmetric function* if for any $i_1>i_2>\cdots>i_d$, $j_1>j_2>\cdots>j_d$ and any signed powers $e_1,\dots,e_d \in \mathbb{D}$ the coefficients of the monomials $x_{i_1}^{e_1}\cdots x_{i_d}^{e_d}$ and $x_{j_1}^{e_1}\cdots x_{j_d}^{e_d}$ are the same. The set of all such quasi-symmetric functions is denoted by QSym_2.

For example, $\sum_{i>j\ge1} x_i x_j^{\overline{2}} \in \mathsf{QSym}_2$ but is *not* a symmetric function since the monomial $x_2 x_1^{\overline{2}}$ appears in the sum but $x_1 x_2^{\overline{2}}$ does not. We may choose an integral basis for QSym_2 using

$$E_{\mathbf{s}} = E_{s_1,\dots,s_d} := \sum_{k_1\ge k_2\ge\cdots\ge k_d} x_{k_1}^{s_1}\cdots x_{k_d}^{s_d}, \quad \text{or}$$

$$M_{\mathbf{s}} = M_{s_1,\dots,s_d} := \sum_{k_1>k_2>\cdots>k_d} x_{k_1}^{s_1}\cdots x_{k_d}^{s_d} = \psi(\mathbf{z}_{s_1}\cdots\mathbf{z}_{s_d}), \tag{7.7}$$

(see Exercise 7.5). This yields the following theorem which can be compared to [292, Thm. 2.2].

Theorem 7.3.3. *The map ψ provides an isomorphism* $(\mathfrak{A}_2^1, *) \cong \mathsf{QSym}_2$.

We will see that the MHSs satisfy a different kind of duality from that of the MZVs. To derive it, we first define the power set of $\mathbf{s}$ to be the partial sum sequence: $P(\mathbf{s}) = (s_1, s_1+s_2, \dots, s_1+\dots+s_{d-1}, w)$ as an (increasingly ordered) d-subset (*i.e.*, subset of d elements) of $(1, 2, \dots, w)$. We see that P provides a one-to-one correspondence between the compositions of weight w and depth d and the d-subsets of $(1, 2, \dots, w)$. Let

$$\mathfrak{R}_w(a_1, \dots, a_\ell) = (w+1-a_\ell, \dots, w+1-a_1).$$

The v-dual of $\mathbf{s}$ is the composition of weight w corresponding to the complimentary subset of $\mathfrak{R}_w P(\mathbf{s})$ in $(1, 2, \cdots, w)$. Namely,

$$\mathbf{s}^\vee = P^{-1}\big((1,2,\cdots,w) \setminus \mathfrak{R}_w P(\mathbf{s})\big). \tag{7.8}$$

It's easy to see that $(\mathbf{s}^\vee)^\vee = \mathbf{s}$ so the v-dual is an involution. Another equivalent definition is the following (see Excercise 7.2).

Definition 7.3.4. For positive integers $r_1, \dots, r_\ell, t_1, \dots, t_\ell$, let

$$\mathbf{s} = (r_1, \{1\}^{t_1-1}, r_2+1, \{1\}^{t_2-1}, \dots, r_\ell+1, \{1\}^{t_\ell-1}).$$

We define the v-dual of $\mathbf{s}$ by

$$\mathbf{s}^\vee := (\{1\}^{r_1-1}, t_1+1, \{1\}^{r_2-1}, t_2+1, \dots, t_{\ell-1}+1, \{1\}^{r_\ell-1}, t_\ell). \tag{7.9}$$

From the definition we clearly have

$$\mathrm{dp}(\mathbf{s}) + \mathrm{dp}(\mathbf{s}^\vee) = |\mathbf{s}| + 1. \tag{7.10}$$

Remark 7.3.5. The v-dual can be easily explained using the conjugation on the ribbons (a kind of skew-Young diagrams). See Exercise 7.3.

Recall that the φ-dual is defined on $\mathfrak{A}_1^1$ by Eq. (6.11): $\mathrm{x}_0^\varphi = \mathrm{x}_0 + \mathrm{x}_1, \mathrm{x}_1^\varphi = -\mathrm{x}_1$. We transport it to QSym_1 by Eq. (7.7). Namely, we set

$$M_{\mathbf{s}}^\varphi := \psi(\mathbf{z}_{s_1}^\varphi \cdots \mathbf{z}_{s_d}^\varphi).$$

In the next theorem, the two kinds of dualities, v-dual and φ-dual, are related to each other in $\mathfrak{A}_1^1$.

Theorem 7.3.6. *Let* $\mathbf{s}$ *be a composition of positive integers. Then we have*

(i) $M_{\mathbf{s}}^\varphi = (-1)^{\mathrm{dp}(\mathbf{s})} \sum_{\mathbf{t} \succeq \mathbf{s}} M_{\mathbf{t}}$;
(ii) $E_{\mathbf{s}}^\varphi = -E_{\mathbf{s}^\vee}$.

Proof. (i) Suppose a word $\mathbf{w} \in \mathfrak{A}_1^1$ corresponds to the composition $\mathbf{s}$. Then a replacement of some particular $\mathtt{x}_0$ by $\mathtt{x}_1$ in $\mathbf{w}$ corresponds to a splitting of $\mathbf{s}$. Thus (i) follows.

(ii) Define $F_{\mathbf{s}} := \sum_{\mathbf{t}\succeq\mathbf{s}} M_{\mathbf{t}} = (-1)^{\mathrm{dp}(\mathbf{s})} M_{\mathbf{s}}^{\varphi}$ by (i). Then, by the definition

$$E_{\mathbf{s}}^{\varphi} = \sum_{\mathbf{t}\preceq\mathbf{s}} M_{\mathbf{t}}^{\varphi} = \sum_{\mathbf{t}\preceq\mathbf{s}} (-1)^{\mathrm{dp}(\mathbf{t})} F_{\mathbf{t}}. \tag{7.11}$$

Dualizing [235, Cor. 3.16] we get $S(F_{\mathbf{t}}) = (-1)^{|\mathbf{t}|} F_{\overleftarrow{\mathbf{t}}^\vee}$ where S is the antipode of QSym_2 as defined in Thm. A.15. Thus, applying S to Eq. (7.11) we get

$$S(E_{\mathbf{s}}^{\varphi}) = \sum_{\mathbf{t}\preceq\mathbf{s}} (-1)^{\mathrm{dp}(\mathbf{t})+|\mathbf{t}|} F_{\overleftarrow{\mathbf{t}}^\vee} = -\sum_{\mathbf{t}\preceq\mathbf{s}} (-1)^{\mathrm{dp}(\mathbf{t}^\vee)} F_{\overleftarrow{\mathbf{t}}^\vee} = -\sum_{\mathbf{r}\succeq\overleftarrow{\mathbf{s}}^\vee} (-1)^{\mathrm{dp}(\mathbf{r})} F_{\mathbf{r}}$$

because of Eq. (7.10). By Möbius inversion formula we see that $S(E_{\mathbf{s}}^{\varphi}) = -(-1)^{\mathrm{dp}(\overleftarrow{\mathbf{s}}^\vee)} M_{\overleftarrow{\mathbf{s}}^\vee}$. Applying S again, we finally get

$$E_{\mathbf{s}}^{\varphi} = -(-1)^{\mathrm{dp}(\overleftarrow{\mathbf{s}}^\vee)} S(M_{\overleftarrow{\mathbf{s}}^\vee}) = -E_{\mathbf{s}^\vee}$$

by Thm. A.15(i). □

Corollary 7.3.7. *For any composition* $\mathbf{s}$ *of positive integers,*

$$-E_{\mathbf{s}^\vee} = \sum_{\mathbf{r}\preceq\mathbf{s}} (-1)^{\mathrm{dp}(\mathbf{r})} \sum_{\mathbf{t}\succeq\mathbf{r}} M_{\mathbf{t}}. \tag{7.12}$$

Proof. By Thm. 7.3.6(ii) and the definition of $E_{\mathbf{s}}$ we get

$$-E_{\mathbf{s}^\vee} = \sum_{\mathbf{r}\preceq\mathbf{s}} M_{\mathbf{r}}^{\varphi}.$$

So Eq. (7.12) follows from Thm. 7.3.6(i). □

We now apply the above results concerning quasi-symmetric functions to AMHSs. In order to do so, for any $n \in \mathbb{N}$, we define the algebra homomorphism $\rho_n : (\mathfrak{A}_2^1, *) \to \mathbb{R}$ such that

$$\rho_n(M_{\mathbf{s}}) = H_n(\mathbf{s}). \tag{7.13}$$

Theorem 7.3.8. *For all* $\mathbf{s} = (s_1, \dots, s_d) \in \mathbb{D}^d$ *and positive integers* n,

$$H_n(\mathbf{s}) = \sum_{\mathbf{t}\preceq\mathbf{s}} (-1)^{\mathrm{dp}(\mathbf{s})-\mathrm{dp}(\mathbf{t})} H_n^\star(\mathbf{t}), \quad H_n^\star(\mathbf{s}) = \sum_{\mathbf{t}\preceq\mathbf{s}} H_n(\mathbf{t}), \tag{7.14}$$

$$H_n^\star(\overleftarrow{\mathbf{s}}) = (-1)^{\mathrm{dp}(\mathbf{s})} \sum_{\sqcup_{j=1}^{l} \mathbf{s}_j=\mathbf{s}} (-1)^r \prod_{j=1}^{r} H_n(\mathbf{s}_j), \tag{7.15}$$

$$H_n(\overleftarrow{\mathbf{s}}) = (-1)^{\mathrm{dp}(\mathbf{s})} \sum_{\sqcup_{j=1}^{l} \mathbf{s}_j = \mathbf{s}} (-1)^r \prod_{j=1}^{r} H_n^\star(\mathbf{s}_j). \tag{7.16}$$

Moreover, for all $\mathbf{s} = (s_1, \dots, s_d) \in \mathbb{N}^d$

$$H_n^\star(\mathbf{s}^\vee) = -\sum_{\mathbf{r} \preceq \mathbf{s}} (-1)^{\mathrm{dp}(\mathbf{r})} \sum_{\mathbf{t} \succeq \mathbf{r}} H_n(\mathbf{t}). \tag{7.17}$$

Proof. The four equations (7.14), (7.15), (7.16), and (7.17) follow from the definition of $E_{\mathbf{s}}$, Thm. A.15(ii) and (iii), and Cor. 7.3.7, respectively, after we apply ρ_n. □

Definition 7.3.9. We will call the relations in Eqs. (7.15) and (7.16) the *concatenation relations* between the MHSs and the MH*Ss.

7.4 Symmetric Sums

For a partition $\Pi = \{P_1, \dots, P_l\}$ of the set $\{1, 2, \dots, d\}$, let

$$c(\Pi) = \prod_{i=1}^{l} (\sharp P_i - 1)! \quad \text{and} \quad \tilde{c}(\Pi) = (-1)^{d-l} \prod_{i=1}^{l} (\sharp P_i - 1)!.$$

Let $\mathbf{s} = (s_1, \dots, s_d) \in \mathbb{D}^d$ be a composition of depth d. Set

$$H_n(\mathbf{s}; \Pi) = \prod_{i=1}^{l} H_n(\lambda_i), \quad H_n^\star(\mathbf{s}; \Pi) = \prod_{i=1}^{l} H_n^\star(\lambda_i),$$

where $\lambda_i = \oplus_{j \in P_i} s_j$. Further, let $\sigma \in \mathfrak{S}_d$ act on $\mathbf{s}$ by permuting the subscripts.

Theorem 7.4.1. *For any* $d, n \in \mathbb{N}$ *and any compositions* $\mathbf{s} \in \mathbb{D}^d$,

$$\sum_{\sigma \in \mathfrak{S}_d} H_n^\star\big(\sigma(\mathbf{s})\big) = \sum_{\Pi:\ \text{partitions of } \{1,2,\dots,d\}} \tilde{c}(\Pi) H_n(\mathbf{s}; \Pi), \tag{7.18}$$

$$\sum_{\sigma \in \mathfrak{S}_d} H_n\big(\sigma(\mathbf{s})\big) = \sum_{\Pi:\ \text{partitions of } \{1,2,\dots,d\}} c(\Pi) H_n(\mathbf{s}; \Pi). \tag{7.19}$$

Proof. We prove the theorem under the assumption that each of the components $s_1, \dots, s_d$ of $\mathbf{s}$ satisfies that $|s_j|$ is a positive real numbers greater than 1 (of course we should first extend $\mathbb{D}$ to a double cover of $\mathbb{R}$ in the obvious way). Then by taking limit we can prove the theorem easily. Further we may assume all $s_1, \dots, s_d$ are pairwise distinct since again we can take limit.

By the definition

$$\sum_{\sigma\in\mathfrak{S}_d} H_n(\sigma(\mathbf{s})) = \sum_{\sigma\in\mathfrak{S}_d}\sum_{k_1\geq\cdots\geq k_d} \frac{\operatorname{sgn}(s_1)^{k_{\sigma(1)}}\cdots\operatorname{sgn}(s_d)^{k_{\sigma(d)}}}{k_{\sigma(1)}^{|s_1|}\cdots k_{\sigma(d)}^{|s_d|}}.$$

Let $\mathfrak{S}_d$ be the symmetry group of order d. Let $\Xi(\mathbf{k})$ be the set of equivalence classes of $\{1,\dots,d\}$ with $i \overset{\mathbf{k}}{\sim} j$ if and only if $k_i = k_j$. Let $O(\mathbf{k}) = \{\sigma \in \mathfrak{S}_d | \sigma(i) \overset{\mathbf{k}}{\sim} i\ \forall i = 1,\dots,d\}$ be the isometry group of $\mathbf{k} = (k_1,\dots,k_d)$. Then

$$\frac{\operatorname{sgn}(s_1)^{k_{\sigma(1)}}\cdots\operatorname{sgn}(s_d)^{k_{\sigma(d)}}}{k_{\sigma(1)}^{|s_1|}\cdots k_{\sigma(d)}^{|s_d|}}$$

appears on the left-hand side of Eq. (7.18) exactly $\sharp O(\mathbf{k})$ times. On the right-hand side it appears exactly in those terms corresponding to partitions Π refining $\Xi(\mathbf{k})$, denoted by $\Pi \succeq \Xi(\mathbf{k})$, namely, $\Xi(\mathbf{k})$ can be obtained from Π by combining some of its equivalence classes (see Thm. A.15). Hence, to prove Eq. (7.18) it suffices to prove that for all $\mathbf{k} = (k_1,\dots,k_d)$

$$\sharp O(\mathbf{k}) = \sum_{\Pi\succeq\Xi(\mathbf{k})} c(\Pi).$$

Indeed, $c(\Pi)$ counts the permutations with the cycle type uniquely determined by Π while on the left-hand side every element of $O(\mathbf{k})$ determines a unique cycle type prescribed by some $\Pi \succeq \Xi(\mathbf{k})$.

We leave the proof of Eq. (7.19) as an exercise (see Exercise 7.6) for the interested reader. □

7.5 Binomial Identities

The binomial identities play very important roles in combinatorics and discrete mathematics in general. In this section we will consider some of these in the context of the MHSs and the MH*Ss.

We start with an old formula known to Euler already.

Proposition 7.5.1. *For all positive integers n, we have*

$$H_n(1) = -\sum_{k=1}^{n} \frac{(-1)^k}{k}\binom{n}{k}. \tag{7.20}$$

Proof. Let

$$f(x) = \sum_{k=1}^{n} \frac{x^k}{k}.$$

Then setting $y = 1 - x$

$$f'(x) = \sum_{k=1}^{n} x^{k-1} = \frac{1-x^n}{1-x} = \frac{1-(1-y)^n}{y} = -\sum_{k=1}^{n} (-1)^k y^{k-1} \binom{n}{k}.$$

Integrating from 1 to x we get

$$f(x) - f(1) = \sum_{k=1}^{n} (-1)^k \frac{(1-x)^k}{k} \binom{n}{k}.$$

So Eq. (7.20) follows by setting $x = 0$ as $f(1) = \sum_{k=1}^{n} \frac{1}{k} = H_n(1)$. □

We will see that more binomial identities come from Apéry's famous series

$$\zeta(2) = 3 \sum_{k=1}^{\infty} \frac{1}{k^2 \binom{2k}{k}}, \qquad \zeta(3) = \frac{5}{2} \sum_{k=1}^{\infty} \frac{(-1)^{k-1}}{k^3 \binom{2k}{k}}, \tag{7.21}$$

which were used in his irrationality proof of $\zeta(2)$ and $\zeta(3)$ (see [19] or [556]). These have been generalized by Leshchiner [387] to all the Riemann zeta values and Dirichlet beta values $\beta(n) = \sum_{k=0}^{\infty} (-1)^k/(2k+1)^n$. For example, he showed

$$\begin{aligned}\Big(1 - \frac{1}{2^{2m+1}}\Big)\zeta(2m+2) =& \frac{3}{2} \sum_{k=1}^{\infty} \frac{(-1)^m H_{k-1}(\{2\}^m)}{k^2 \binom{2k}{k}} \\ &+ 2 \sum_{j=1}^{m} \sum_{k=1}^{\infty} \frac{(-1)^{m-j} H_{k-1}(\{2\}^{m-j})}{k^{2j+2} \binom{2k}{k}}, \\ (-1)^{m-1} \cdot \zeta(2m+3) =& \frac{5}{2} \sum_{k=1}^{\infty} \frac{(-1)^k H_{k-1}(\{2\}^m)}{k^3 \binom{2k}{k}} \\ &+ 2 \sum_{j=1}^{m} \sum_{k=1}^{\infty} \frac{(-1)^{k-j} H_{k-1}(\{2\}^{m-j})}{k^{2j+3} \binom{2k}{k}}.\end{aligned}$$

In the rest of this section we state a few families of the binomial identities for the MH*Ss similar to Prop. 7.5.1. We will apply some of these results to a few congruences of the MHSs and the MH*Ss in Chap. 8 and then to some identities of the multiple zeta star values in Chap. 10.

Theorem 7.5.2. *Let $n, c \in \mathbb{N}$ and $a, b \in \mathbb{N}_0$. If $c \geq 3$ then*

$$\begin{aligned}H_n^\star(\{2\}^a, c, \{2\}^b) =& 2 \sum_{k=1}^{n} \frac{(-1)^{k-1} \binom{n}{k}}{k^{2a+2b+c} \binom{n+k}{k}} \\ &+ 4 \sum_{i,j,\mathbf{x}} 2^{\mathrm{dp}(\mathbf{x})} \sum_{k=1}^{n} \frac{H_{k-1}(\mathbf{x}, i+2b)(-1)^{k-1} \binom{n}{k}}{k^{2a+j} \binom{n+k}{k}},\end{aligned} \tag{7.22}$$

where in the second sum $i \geq 1$, $j \geq 2$ and $\mathbf{x} \in \mathbb{N}^r$ runs through all possible compositions of positive integers (or the empty set when $r = 0$) satisfying the restriction $i + j + |\mathbf{x}| = c$.

Remark 7.5.3. If we take $c = 3$ then we obtain so-called *Two-three formula*. Here "Two" is capitalized to denote a 2-string.

Theorem 7.5.4. *For all $n, c \in \mathbb{N}$ and $a, b \in \mathbb{N}_0$, we have*

$$\begin{aligned} H_n^\star(\{1\}^a, c, \{1\}^b) &= \sum_{k=1}^{n} \frac{(-1)^{k-1}\binom{n}{k}}{k^{a+b+c}} \\ &\quad + \sum_{\substack{i+j+|\mathbf{s}|=c \\ i\geq 1, j\geq 1, |\mathbf{s}|\geq 0}} \sum_{k=1}^{n} \frac{H_{k-1}(b+i, \mathbf{s})(-1)^{k-1}\binom{n}{k}}{k^{a+j}}. \end{aligned} \tag{7.23}$$

The following formulas are special cases of so-called *Two-one formula.*

Theorem 7.5.5. *Let $a \in \mathbb{N}_0$ and $b \in \mathbb{N}$. Then for any $n \in \mathbb{N}$, we have*

$$H_n^\star(\{2\}^a, 1) = 2\sum_{k=1}^{n} \frac{\binom{n}{k}}{k^{1+2a}\binom{n+k}{k}}, \tag{7.24}$$

$$H_n^\star(\{2\}^a, 1, \{2\}^b) = -2\sum_{k=1}^{n} \frac{(-1)^k\binom{n}{k}}{k^{1+2a+2b}\binom{n+k}{k}} - 4\sum_{k=1}^{n} \frac{H_{k-1}(\overline{2b})\binom{n}{k}}{k^{1+2a}\binom{n+k}{k}}. \tag{7.25}$$

The proofs of the above theorems are not given here since we will generalize them to the case with argument strings having arbitrary times of repetitions of certain patterns (for example, $(\{2\}^{a_1}, c_1, \dots, \{2\}^{a_r}, c_r)$ with $r \geq 1$) in Chap. 9, and moreover, generalize further to their q-analogs, which will be used in Chap. 10 where the multiple zeta star values are treated.

7.6 Historical Notes

Even before the appearance of renewed interest in MZVs in the early 1990s, physicists already started to study the MHSs and the AMHSs. In [165], Devoto and Duke listed a lot of concrete computations involving such sums. Later on, Blümlein and Kurth [58,61] and Vermaseren [560] discovered that they can be realized as the general basis for the Mellin transforms of all individual functions of the momentum fraction x emerging in the quantities of massless quantum electrodynamics (QED) and quantum chromodynamics (QCD), and as Mellin transforms and the inverse Mellin transforms for some functions that are encountered in Feynman diagram calculations, and some other similar functions with significant physical meanings. Incidentally, our binary operation $\oplus$ for the signed numbers is denoted by the ampersand & by Vermaseren in [560]. Later, it is also noticed that the MHS structure and its cyclotomic extensions are closely linked to high energy physics (see [3, 59, 60]) and the relations between these sums can be used to simplify certain results of higher-order calculations in QED and QCD.

The first systematic study of the MHSs from the number theoretical perspective was initiated by the author in the arXiv paper [609] in early 2003, and independently by Hoffman in the arXiv paper [292] in early 2004. In particular, the approach using algebras of words and quasi-symmetric functions to study MZVs was carried over to this new setting by Hoffman. Note that the star-version $H^\star$ is also denoted by S in [292] but it seems to be more appropriate to use $H^\star$ in this book due to its close connection with multiple zeta star values to be considered in Chap. 10.

Quasi-symmetric functions are generalizations of symmetric functions introduced by Gessel and Stanley in early 1980s to investigate the combinatorics of P-partitions and the counting of permutations with given descent sets, see [236, 237, 525]. They can be regarded as the dual algebra over the integers of the Leibniz-Hopf algebra (see [275]). In [289], Hoffman first treated MZVs from the point of view of two different product structures and later he extended these to the Hopf algebra on both the MZVs and the MHSs in [290, 293].

Our proof of Thm. 7.4.1 is adopted from that of Thms. 2.1 and 2.2 of [288]. Note that [288, Thm. 2.2] was stated as Prop. 9.4 in [368] whose proof is different from that contained in [288] but also works here in the alternating setting.

We saw a few different types of sum relations among the MZVs in Chap. 5. Theorem 7.4.1 is an analog of such formulas in the AMHS setting, which is a generalization of Thms. 2.1 and 2.2 of [288] concerning only the MHSs. Using binomial identities, Hessami Pilehroods and Tauraso [278] discovered many congruences for MHSs and MH*Ss appearing on the left-hand side of Eqs. (7.23)-(7.25) that will be discussed in Sec. 8.6.4.

Theorem A.15 generalizes [294, Thm. 6.2]. Known already to Euler [211], Prop. 7.5.1 was the starting point from where Wolstenholme (see [578]), in 1862, began to prove the famous result now bearing his name (see Thm. 8.2.1). Our proof of Thm. 7.3.6 follows that of [294, Thm. 6.3].

Finally, we note that the detailed proofs of Thms. 7.5.2 to 7.5.5 can be found in [278] which motivated a complete proof of the Two-one formula by the author in [628].

Exercises

7.1. By Definition 7.8 find the v-duals $\mathbf{s}^\vee$ if (i) $\mathbf{s} = (1,3,2)$ (ii) $\mathbf{s} = (4,2,1,3)$. Then verify them using Definition 7.3.4.

7.2. Show that Eq. (7.8) and Eq. (7.9) are equivalent.

7.3. In this exercise, we re-interprete the v-dual using the following operation on the ribbons, *i.e.*, skew-Young diagrams. For a composition $\mathbf{s} = (s_1, \dots, s_d)$, the ribbon $R_\mathbf{s}$ is defined to be the skew-Young diagram of d rows whose jth row starts below the last box of $(j-1)$st row and has exactly s_j boxes. Recall that the conjugate of a (skew-)Young diagram is the mirror image about the diagonal line going from the south-west corner to north-east. For example, the following two diagrams give the ribbon $R_{1,3,2}$ and its conjugate:

Show that in general the ribbon $R_{\mathbf{s}^\vee}$ is exactly the conjugate of the ribbon $R_{\overline{\mathbf{s}}}$. So $(2,3,1)^\vee = (1,2,1,2)$.

7.4. Prove Thm. 7.2.4.

7.5. Check that the set $\{M_{t_1,\dots,t_d} : d \in \mathbb{N}, (t_1,\dots,t_d) \in \mathbb{D}^d\}$ forms an integral basis for QSym_2.

7.6. Prove Eq. (7.19) by modifying the argument for the proof of Eq. (7.18).

Chapter 8

Finite Multiple Zeta Values and Finite Euler Sums

A general philosophy in studying congruences is to consider so-called van Hamme type congruences. One starts with an intriguing infinite series whose terms are given by rational numbers with some pattern and then, for suitable primes p, looks at the $(p-1)$-st partial sum modulo p, or even modulo higher powers of p which leads to super congruences.

A classical result in this spirit is a variant of Wolstenholme's Theorem dating back to the mid nineteenth century: for all prime $p \geq 5$ we have

$$\sum_{k=1}^{p-1} \frac{1}{k^2} \equiv 0 \pmod{p}, \qquad \sum_{k=1}^{p-1} \frac{1}{k} \equiv 0 \pmod{p^2}. \tag{8.1}$$

These can be proved rather easily as follows. First, the map $f(x) = x^{-2}$ is an automorphism of $(\mathbb{Z}/p\mathbb{Z})^\times$ so the first congruence is obvious. For the second, consider the equations in the ring $\mathbb{Z}_p$ of p-adic integers:

$$\begin{aligned} 2\sum_{k=1}^{p-1} \frac{1}{k} = \sum_{k=1}^{p-1} \left(\frac{1}{k} + \frac{1}{p-k}\right) &= \sum_{k=1}^{p-1} \left(\frac{1}{k} - \frac{1}{k} \cdot \left(1 - \frac{p}{k}\right)^{-1}\right) \\ &\equiv -p\sum_{k=1}^{p-1} \frac{1}{k^2} \equiv 0 \pmod{p^2} \end{aligned}$$

by the first congruence of (8.1).

These were improved to super congruences later (see Sec. 8.7). In the proof of these congruences one finds that Bernoulli numbers play the key roles by virtue of the Seki–Bernoulli formula[1]:

$$\sum_{j=1}^{n-1} j^d = \sum_{r=0}^{d} \binom{d+1}{r} \frac{B_r}{d+1} n^{d+1-r}, \quad \forall n, d \geq 1. \tag{8.2}$$

[1] See [22, p. 3]. This formula is also called Faulhaber's formula in the literature, see [6, § 23.1.4-7].

This chapter gives a brief introduction to the van Hamme type (super) congruences for the (alternating) multiple harmonic (star) sums. A general phenomenon is that in order to study the weight w MHSs (or MH*S) modulo p^{k+1} one must first study those objects of weight $w+k$ modulo p. A few important ideas have been used to investigate this structure by a number of mathematicians. Generally speaking, they all use some kind of combinations of the following four approaches:

(1) rearrange the sum,
(2) apply the stuffle, shuffle, reversal, concatenation, and duality relations,
(3) change to the sums of positive powers and then use the Bernoulli numbers/polynomials, and
(4) exploit the binomial identities.

Throughout this chapter we will follow the above approaches closely.

8.1 Definitions

Let $\mathscr{P}$ be the set of primes and let ℓ be a positive integer. Define

$$\mathscr{A}_\ell := \prod_{p\in\mathscr{P}} (\mathbb{Z}/p^\ell\mathbb{Z}) \Big/ \bigoplus_{p\in\mathscr{P}} (\mathbb{Z}/p^\ell\mathbb{Z}) \tag{8.3}$$

with componentwise addition and multiplication. Clearly the elements of $\mathscr{A}_\ell$ are represented by $(a_p)_{p\in\mathscr{P}}$ and two elements (a_p) and (b_p) represent the same element in $\mathscr{A}_\ell$ if $a_p = b_p$ for all but finitely many primes p. For every nonzero $r \in \mathbb{Q}$, by the fundamental Theorem of Arithmetic

$$r = \prod_{p\in\mathscr{P}} p^{\operatorname{ord}_p(r)}.$$

It is straightforward to see that $\mathscr{A}_\ell$ is a $\mathbb{Q}$-algebra after embedding $\mathbb{Q}$ in $\mathscr{A}_\ell$ "diagonally" using the following map:

$$\begin{aligned} \iota_\ell : \mathbb{Q} &\longrightarrow \quad \mathscr{A}_\ell \\ r &\longrightarrow \big(\iota_{p,\ell}(r)\big)_{p\in\mathscr{P}} \end{aligned} \tag{8.4}$$

where $\iota_\ell(0) = (0)_{p\in\mathscr{P}}$ and for all nonzero $r \in \mathbb{Q}$

$$\iota_{p,\ell}(r) := \begin{cases} 0, & \text{if } \operatorname{ord}_p(r) < 0; \\ r \pmod{p^\ell}, & \text{if } \operatorname{ord}_p(r) \ge 0. \end{cases}$$

Proposition 8.1.1. *The map ι_ℓ is a monomorphism of algebras. Namely, ι_ℓ gives an embedding of $\mathbb{Q}$ into $\mathscr{A}_\ell$ as an algebra.*

Proof. Suppose $r, s \in \mathbb{Q}$ with $\iota_\ell(r) = \iota_\ell(s)$. Note that $\mathrm{ord}_p(r) < 0$ or $\mathrm{ord}_p(s) < 0$ for only finitely many primes p. Thus $\iota_\ell(r) = \iota_\ell(s)$ implies that $r - s$ is divisible by infinitely many primes which immediately yields $r = s$. This shows ι_ℓ is injective. Since we clearly have $\iota_\ell(r)\iota_\ell(s) = \iota_\ell(rs)$ for all $r, s \in \mathbb{Q}$ the injection ι_ℓ is an algebra homomorphism. □

Remark 8.1.2. It is easy to see that the cardinality of $\mathscr{A}_\ell$ is $\aleph_1$, namely, the same as that of the real numbers. See Exercise 8.1. It is not too hard to construct an embedding of $\mathbb{R}$ into $\mathscr{A}_\ell$ as sets. But we don't know whether there is an embedding of $\mathbb{R}$ as an algebra.

For convenience, we abuse our notation by writing a_p for $(a_p)_{p\in\mathscr{P}}$ if no confusion arises. In particular, whenever p appears in an equation in $\mathscr{A}_\ell$ $(\ell \geq 2)$ it means the element $(p)_{p\in\mathscr{P}}$. Also, by writing $(a_p)_{p\geq k} \in \mathscr{A}_\ell$ we mean the element $(a_p)_{p\in\mathscr{P}}$ with $a_p = 0$ for all $p < k$.

Definition 8.1.3. We define a number $a \in \mathscr{A}_\ell$ to be *algebraic over* $\mathbb{Q}$ if there is a nontrivial polynomial $f(t) \in \mathbb{Q}[t]$ such that $f(a) = 0$. Non-algebraic numbers of $\mathscr{A}_\ell$ over $\mathbb{Q}$ are called *transcendental.* For finitely many numbers $a_1, \dots, a_n \in \mathscr{A}_\ell$, we say they are *algebraically independent over* $\mathbb{Q}$ if for any nontrivial polynomial $f(t_1, \dots, t_n) \in \mathbb{Q}[t_1, \dots, t_n]$ we have $f(a_1, \dots, a_n) \neq 0$. We call the elements in an infinite subset S of $\mathscr{A}_\ell$ *algebraically independent over* $\mathbb{Q}$ if any finitely many elements of S are.

Definition 8.1.4. For any nonnegative integer k, we define the $\mathscr{A}_1$-*Bernoulli numbers* $\beta_k := \Big(-\frac{B_{p-k}}{p-k} \equiv \frac{B_{p-k}}{k} \pmod p\Big)_{p>k}$. For all $k \geq 2$, define the kth ($\mathscr{A}_1$-)Fermat quotient

$$\mathfrak{q}_k := \left(\frac{k^{p-1}-1}{p} \pmod p\right)_{p>k}.$$

We have $\beta_{2k} = 0$ for all $k \geq 1$ since all Bernoulli-numbers $B_{2j+1} = 0$ when $j \geq 1$. As for the $\mathscr{A}_1$-Fermat quotient $\mathfrak{q}_2$, according to [348], we know that for all primes less than 1.25×10^{15} the super congruence $2^{p-1} \equiv 1 \pmod{p^2}$ is satisfied by only two primes 1093 and 3511 which are called Wieferich primes. It is also known that $\mathfrak{q}_k \neq 0$ $(k \geq 2)$ in $\mathscr{A}_1$ under *abc*-conjecture (see [514]). Moreover, note that we have $\mathfrak{q}_{k\ell} = \mathfrak{q}_k + \mathfrak{q}_\ell$ for all k, ℓ and $\mathfrak{q}_1 = 0$. So $\mathfrak{q}_k$ is an $\mathscr{A}_1$-analog of the logarithm value $\log k$.

By considering the mysterious relation between the Euler sums and their finite versions (see Sec. 8.6.3) we come to the following hypothesis.

Conjecture 8.1.5. *Put $\beta_1 = 1$ by abuse of notation. Suppose $n_1, \dots, n_r \in \mathbb{N}$ such that $\log n_1, \dots, \log n_r$ are $\mathbb{Q}$-linearly independent. Then the numbers in the set*

$$\bigcup_{k=0}^{\infty} \left\{ \beta_{2k+1}\mathfrak{q}_{n_1}, \dots, \beta_{2k+1}\mathfrak{q}_{n_r} \right\}$$

are algebraically independent.

One should compare this with the following

Conjecture 8.1.6. *Put $\zeta(1) = 1$ by abuse of notation. Suppose $n_1, \dots, n_r \in \mathbb{N}$ such that $\log n_1, \dots, \log n_r$ are $\mathbb{Q}$-linearly independent. Then the numbers in the set*

$$\bigcup_{k=0}^{\infty} \left\{ \zeta(2k+1)\log n_1, \dots, \zeta(2k+1)\log n_r \right\}$$

are algebraically independent.

By the same token, every object similarly defined for each prime can be put into $\mathscr{A}_1$, for examples,

- Wilson quotients $\frac{(p-1)!+1}{p}$
- p-adic zeta values $\zeta_p(k)$
- $F_{p-(\frac{5}{p})}/p$: the Fibonacci quotient (OEIS A092330)
- $U_{p-(\frac{2}{p})}(2,-1)/p$: the Pell quotient (OEIS A000129)
- $U_{p-(\frac{3}{p})}(4,1)/p$: a quotient related to the Lucas sequence 1, 4, 15, 56, 209, ... (OEIS A001353)
- $U_{p-(\frac{6}{p})}(10,1)/p$: a quotient related to the Lucas sequence 1, 10, 99, 980, 9701, ... (OEIS A004189)

Definition 8.1.7. For $\mathbf{s} = (s_1, \dots, s_d) \in \mathbb{D}^d$ (see Definition 7.1.1), we define

$$\zeta_{\mathscr{A}_\ell}(\mathbf{s}) := \sum_{p>k_1>\cdots>k_d\geq 1} \frac{\operatorname{sgn}(s_1)^{k_1}\cdots\operatorname{sgn}(s_d)^{k_d}}{k_1^{|s_1|}\cdots k_d^{|s_d|}} \in \mathscr{A}_\ell,$$

$$\zeta^\star_{\mathscr{A}_\ell}(\mathbf{s}) := \sum_{p>k_1\geq\cdots\geq k_n\geq 1} \frac{\operatorname{sgn}(s_1)^{k_1}\cdots\operatorname{sgn}(s_d)^{k_d}}{k_1^{|s_1|}\cdots k_d^{|s_d|}} \in \mathscr{A}_\ell.$$

These are called *finite Euler (star) sums*, abbreviated as FES. If all the s_j's are positive then they are called *finite multiple zeta (star) values*, abbreviated as FMZV. The number ℓ is called the *superbity*.

8.2 Reversal and Concatenation Relations

The famous Wolstenholme's Theorem stated in Eq. (8.1) can be rephrased in the following way.

Theorem 8.2.1. *For any prime $p > 3$, we have*

$$H_{p-1}(1) \equiv 0 \pmod{p^2} \qquad \text{and} \qquad H_{p-1}(2) \equiv 0 \pmod{p}.$$

In order to extend these results to the general MHSs, we start in this section with some simple properties of the MHSs, namely, the reversal and stuffle relations.

For any $\mathbf{s} = (s_1, \ldots, s_d) \in \mathbb{D}^d$, denote its *reversal* by $\overleftarrow{\mathbf{s}} = (s_d, \ldots, s_1)$ and set $\operatorname{sgn}(\mathbf{s}) = \prod_{j=1}^d \operatorname{sgn}(s_j)$. The following results are called *reversal relations.*

Theorem 8.2.2. *Let $\mathbf{s} = (s_1, \ldots, s_d) \in \mathbb{D}^d$. Then*

$$\zeta_{\mathscr{A}_1}(\overleftarrow{\mathbf{s}}) = (-1)^{|\mathbf{s}|} \operatorname{sgn}(\mathbf{s}) \zeta_{\mathscr{A}_1}(\mathbf{s}), \quad \zeta^\star_{\mathscr{A}_1}(\overleftarrow{\mathbf{s}}) = (-1)^{|\mathbf{s}|} \operatorname{sgn}(\mathbf{s}) \zeta^\star_{\mathscr{A}_1}(\mathbf{s}). \tag{8.5}$$

Proof. Let p be a prime at least 3. The theorem follows easily from the index substitution $k \to p-k$ and the following congruence which is valid for all $s \in \mathbb{D}$:

$$\begin{aligned} \frac{\operatorname{sgn}(s)^{p-k}}{(p-k)^{|s|}} &= (-1)^{|s|} \operatorname{sgn}(s) \frac{\operatorname{sgn}(s)^k}{k^{|s|}} \left(1 - \frac{p}{k}\right)^{-|s|} \\ &\equiv (-1)^{|s|} \operatorname{sgn}(s)^{k+1} \left(\frac{1}{k^{|s|}} + \frac{|s|p}{k^{|s|+1}} + \binom{|s|+1}{2} \frac{p^2}{k^{|s|+2}} \right) \pmod{p^3}. \end{aligned} \tag{8.6}$$

The rest is straightforward and is left to the interested reader. □

Sometimes this idea can be applied to obtain super congruences, namely, with moduli given by higher powers of p.

Theorem 8.2.3. *Let s be any positive integer. Then*

$$\zeta_{\mathscr{A}_1}(2s) = 0, \qquad \text{and} \qquad \zeta_{\mathscr{A}_2}(2s+1) = 0. \tag{8.7}$$

Proof. Suppose the multiplicative group $(\mathbb{Z}/p\mathbb{Z})^\times$ has a generator g_p. Then in $\mathscr{A}_1$

$$H_{p-1}(s) = \sum_{k=0}^{p-2} g_p^{sk} = \frac{1 - g_p^{s(p-1)}}{1 - g_p^n} = 0.$$

This argument fails if $s = p-1$ in which case $H_{p-1}(s) = -1$ in $\mathbb{Z}/p\mathbb{Z}$. But in $\mathscr{A}_1$ this is negligible since s is fixed and we can neglect any number of finitely many primes. Therefore $\zeta_{\mathscr{A}_1}(s) = 0$ for every positive integer s. Now if $s = 2m+1$ is odd then by taking reversal we see that in $\mathbb{Z}_p$

$$\begin{aligned} 2H_{p-1}(2m+1) &= \sum_{k=1}^{p-1} \left(\frac{1}{k^{2m+1}} + \frac{1}{(p-k)^{2m+1}} \right) \\ &\equiv \sum_{k=1}^{p-1} \frac{(2m+1)p}{k^{2m+2}} \equiv 0 \pmod{p^2}, \end{aligned}$$

because of the expansion in Eq. (8.6). This completes the proof of the theorem. □

We now lift Thm. 8.2.2 to superbity two by exactly the same idea.

Theorem 8.2.4. *Let* $\mathbf{s} = (s_1, \dots, s_d) \in \mathbb{D}^d$. *Let* $\mathbf{e}_i = (0, \dots, 0, 1, 0, \dots, 0)$ *where* 1 *appears at the* i*th component. Then*

$$(-1)^{|\mathbf{s}|} \operatorname{sgn}(\mathbf{s}) \zeta_{\mathscr{A}_2}(\overleftarrow{\mathbf{s}}) = \zeta_{\mathscr{A}_2}(\mathbf{s}) + p \sum_{i=1}^{d} |s_i| \zeta_{\mathscr{A}_2}(\mathbf{s} \oplus \mathbf{e}_i), \tag{8.8}$$

$$(-1)^{|\mathbf{s}|} \operatorname{sgn}(\mathbf{s}) \zeta^{\star}_{\mathscr{A}_2}(\overleftarrow{\mathbf{s}}) = \zeta^{\star}_{\mathscr{A}_2}(\mathbf{s}) + p \sum_{i=1}^{d} |s_i| \zeta^{\star}_{\mathscr{A}_2}(\mathbf{s} \oplus \mathbf{e}_i), \tag{8.9}$$

where the binary operation $\oplus$ *is carried out componentwise by Eq.* (7.1).

Proof. Take reversal and then apply Eq. (8.6). □

To find a result corresponding to Thm. 8.2.3 for signed numbers the Euler polynomials and the Euler numbers are indispensable tools. Recall that the Euler polynomials $E_n(x)$ are defined by the generating function

$$\frac{2e^{tx}}{e^t + 1} = \sum_{n=0}^{\infty} E_n(x) \frac{t^n}{n!}.$$

Lemma 8.2.5. *Let* $n \in \mathbb{N}_0$. *Then we have*

$$\sum_{i=1}^{d-1} (-1)^i i^n = \frac{1}{2}\Big((-1)^{d-1} E_n(d) + E_n(0) \Big) = \sum_{a=0}^{n} \binom{n}{a} F_{n,d,a} d^{n-a}, \tag{8.10}$$

where

$$F_{n,d,a} = \begin{cases} (-1)^{d-1} E_a(0)/2, & \text{if } a < n; \\ (1 - (-1)^d) E_n(0)/2, & \text{if } a = n > 0; \\ -(1 + (-1)^d)/2, & \text{if } a = n = 0. \end{cases}$$

Moreover, $E_0(0) = 1$ *and for all* $a \in \mathbb{N}$

$$E_a(0) = \frac{2^{a+1}}{a+1} \Big(B_{a+1}\Big(\frac{1}{2}\Big) - B_{a+1} \Big) = \frac{2}{a+1} (1 - 2^{a+1}) B_{a+1}. \tag{8.11}$$

Proof. Consider the generating function

$$\sum_{n=0}^{\infty}\left(\sum_{i=1}^{d-1}(-1)^i i^n\right)\frac{t^n}{n!}=\sum_{i=1}^{d-1}(-1)^i e^{ti}=\frac{(-e^t)^d-1}{-e^t-1}-1$$
$$=\frac{(-1)^{d-1}e^{dt}+1}{e^t+1}-1$$
$$=\frac{1}{2}\sum_{n=0}^{\infty}\Big((-1)^{d-1}E_n(d)+E_n(0)\Big)\frac{t^n}{n!}-1. \quad (8.12)$$

Now Eq. (8.10) follows from the notorious equation (see, for e.g., [6, § 23.1.7])

$$E_n(x)=\sum_{a=0}^{n}\binom{n}{a}E_a(0)x^{n-a} \quad (8.13)$$

for all $n>0$. For Eq. (8.11), see, for e.g., § 23.1.20 on p. 805 of loc. cit. □

Remark 8.2.6. The classical Euler numbers E_k is defined by

$$\frac{2}{e^t+e^{-t}}=\sum_{k=0}^{\infty}E_k\frac{t^k}{k!}.$$

They are related to $E_k(0)$ by the formula

$$E_m(0)=\sum_{k=0}^{m}\binom{m}{k}\frac{E_k}{2^k}\left(-\frac{1}{2}\right)^{m-k}.$$

See § 23.1.7 on p. 805 of loc. cit.

Theorem 8.2.7. *Let $s\in\mathbb{N}$. Then*

$$\zeta_{\mathscr{A}_1}(\bar{s})=\begin{cases}-2\mathfrak{q}_2, & \text{if } s=1;\\ -2(1-2^{1-s})\beta_s, & \text{if } s>2 \text{ is odd;}\end{cases} \quad (8.14)$$
$$\zeta_{\mathscr{A}_2}(\bar{s})=s(1-2^{-s})p\beta_{s+1} \qquad \text{if } s \text{ is even.} \quad (8.15)$$

Proof. Taking $d=p$ and $n=p(p-1)-s$ in Lemma 8.2.5 we see that

$$H_{p-1}(\bar{s})\equiv F_{p(p-1)-s,p,p(p-1)}+p(p(p-1)-s)F_{p(p-1)-s,p,p(p-1)-1-s}$$
$$\equiv E_{p(p-1)-s}(0)-\frac{1}{2}psE_{p(p-1)-1-s}(0) \pmod{p^2},$$

since all the coefficients in Eq. (8.10) are p-integral by Eq. (8.11). Note that B_m is not p-integral if and only if $p-1$ divides $m>0$. Then the theorem follows directly from Eq. (8.11) and the Kummer congruences

$$\frac{B_{p(p-1)-s}}{p(p-1)-s}\equiv\frac{B_{p-1-s}}{p-1-s},\quad \frac{B_{p(p-1)-s+1}}{p(p-1)-s+1}\equiv\frac{B_{p-s}}{p-s} \pmod{p}. \qquad □$$

Using the theory of quasi-symmetric functions one can get:

Theorem 8.2.8. *For all* $\mathbf{s} = (s_1, \dots, s_d) \in \mathbb{D}^d$ *and positive integer* ℓ,

$$\zeta^\star_{\mathscr{A}_\ell}(\mathbf{s}) = \sum_{\mathbf{t} \preceq \mathbf{s}} \zeta_{\mathscr{A}_\ell}(\mathbf{t}), \qquad \zeta_{\mathscr{A}_\ell}(\mathbf{s}) = \sum_{\mathbf{t} \preceq \mathbf{s}} (-1)^{\mathrm{dp}(\mathbf{s})-\mathrm{dp}(\mathbf{t})} \zeta^\star_{\mathscr{A}_\ell}(\mathbf{t}). \tag{8.16}$$

When $\ell = 1$ *we have*

$$(-1)^{\mathrm{dp}(\mathbf{s})} \zeta^\star_{\mathscr{A}_1}(\overleftarrow{\mathbf{s}}) = \sum_{\sqcup_{j=1}^r \mathbf{s}_j = \mathbf{s}} (-1)^r \prod_{j=1}^r \zeta_{\mathscr{A}_1}(\mathbf{s}_j), \tag{8.17}$$

$$(-1)^{\mathrm{dp}(\mathbf{s})} \zeta_{\mathscr{A}_1}(\overleftarrow{\mathbf{s}}) = \sum_{\sqcup_{j=1}^r \mathbf{s}_j = \mathbf{s}} (-1)^r \prod_{j=1}^r \zeta^\star_{\mathscr{A}_1}(\mathbf{s}_j). \tag{8.18}$$

Proof. These follow immediately from Thm. 7.3.8. □

Definition 8.2.9. We will call the relations in (8.17) and (8.18) the *concatenation relations* between FMZVs and FESs.

Example 8.2.10. By (8.16) and the concatenation relation (8.17) we have

$$\zeta_{\mathscr{A}_1}(1, \bar{2}) + \zeta_{\mathscr{A}_1}(\bar{3}) = \zeta^\star_{\mathscr{A}_1}(1, \bar{2}) = \zeta_{\mathscr{A}_1}(\bar{2})\zeta_{\mathscr{A}_1}(1) - \zeta_{\mathscr{A}_1}(\bar{2}, 1).$$

Thus we get $\zeta_{\mathscr{A}_1}(\bar{3}) = -2\zeta_{\mathscr{A}_1}(\bar{2}, 1)$ since $\zeta_{\mathscr{A}_1}(1) = 0$ and $\zeta_{\mathscr{A}_1}(1, \bar{2}) = \zeta_{\mathscr{A}_1}(\bar{2}, 1)$ by the reversal relation (8.5).

8.3 Stuffle Relations

We may obtain more congruences by combining the reversal relations with the stuffle relations in Thm. 7.2.4.

Definition 8.3.1. To find as many $\mathbb{Q}$-linear relations as possible in weight w we may choose all the known relations in weight $k < w$, multiply them by $\zeta_{\mathscr{A}_\ell}(\mathbf{s})$ for all compositions $\mathbf{s}$ of signed numbers of weight $w - k$, and then expand all the products using the stuffle relations (7.5). Such $\mathbb{Q}$-linear relations among the FESs of the same weight produced in this way are called *linear stuffle relations of the FESs.*

For example, if k and l have the same parity and same sign (or have different parities and different signs) then

$$0 = \zeta_{\mathscr{A}_1}(k)\zeta_{\mathscr{A}_1}(l) = \zeta_{\mathscr{A}_1}(k, l) + \zeta_{\mathscr{A}_1}(l, k) + \zeta_{\mathscr{A}_1}(k \oplus l) = 2\zeta_{\mathscr{A}_1}(k, l)$$

by Eqs. (8.7), (7.6) and (8.5). When $k, l > 0$ this is a special case of the more general Thm. 8.5.12.

An application of the stuffle relation yields:

Theorem 8.3.2. *Let s and t be two positive integers.*

(i) *If $s+t$ is odd then*

$$\zeta^{\star}_{\mathscr{A}_2}(s,t) = \zeta_{\mathscr{A}_2}(s,t). \tag{8.19}$$

(ii) *If t is even. Then*

$$\zeta^{\star}_{\mathscr{A}_2}(s,t,s) = \zeta_{\mathscr{A}_2}(s,t,s). \tag{8.20}$$

(ii) *If t is odd. Then*

$$\zeta^{\star}_{\mathscr{A}_3}(s,t,s) = \zeta_{\mathscr{A}_3}(s,t,s). \tag{8.21}$$

Proof. Part (i) follows easily from Thm. 8.2.3 since

$$\zeta^{\star}_{\mathscr{A}_2}(s,t) = \zeta_{\mathscr{A}_2}(s,t) + \zeta_{\mathscr{A}_2}(s+t).$$

For part (ii), by (8.16) and the stuffle relation we get

$$\begin{aligned}\zeta^{\star}_{\mathscr{A}_2}(s,t,s) &= \zeta_{\mathscr{A}_2}(s,t,s) + \zeta_{\mathscr{A}_2}(s+t,s) + \zeta_{\mathscr{A}_2}(s,t+s) + \zeta_{\mathscr{A}_2}(2s+t)\\ &= \zeta_{\mathscr{A}_2}(s,t,s) + \zeta_{\mathscr{A}_2}(s+t)\zeta_{\mathscr{A}_2}(s) = \zeta_{\mathscr{A}_2}(s,t,s)\end{aligned}$$

by Thm. 8.2.3, as desired. If t is odd then s and $s+t$ have different parity. Thus $\zeta_{\mathscr{A}_3}(s+t)\zeta_{\mathscr{A}_3}(s) = 0$ in $\mathscr{A}_3$. This proves part (iii). □

8.4 Shuffle Relations

It is not hard to see that the Euler sums can be expressed by iterated integrals, just like MZVs. Suppose $s_j \in \mathbb{D}$ and $\operatorname{sgn}(s_j) = \eta_j \in \{1,-1\}$ for all $j = 1,\dots,d$. Then

$$\zeta(s_1,\dots,s_n) = \int_0^1 \left(\frac{dt}{t}\right)^{|s_1|-1}\left(\frac{dt}{\xi_1 - t}\right)\cdots\left(\frac{dt}{t}\right)^{|s_d|-1}\left(\frac{dt}{\xi_d - t}\right), \tag{8.22}$$

where $\xi_i = \prod_{j=1}^{i} \eta_i$ for $i = 1,\dots,d$.

Recall from Definition 7.2.1 that $\mathfrak{A}_2^1$ is the algebra generated by $\mathbf{z}_s, s \in \mathbb{D}$. Write

$$\mathbf{z}_s = \mathbf{y}_{|s|,\operatorname{sgn}(s)}. \tag{8.23}$$

Then we define $\operatorname{tog}, \operatorname{tog}^{-1} : \mathfrak{A}_2^1 \to \mathfrak{A}_2^1$ by

$$\operatorname{tog}(\mathbf{y}_{s_1,\xi_1}\cdots\mathbf{y}_{s_d,\xi_d}) := \mathbf{y}_{s_1,\eta_1}\cdots\mathbf{y}_{s_d,\eta_d}, \tag{8.24}$$

where $\eta_i = \prod_{j=1}^{i} \xi_j$ for $i = 1,\dots,d$ and

$$\operatorname{tog}^{-1}(\mathbf{y}_{s_1,\eta_1}\cdots\mathbf{y}_{s_d,\eta_d}) := \mathbf{y}_{s_1,\xi_1}\cdots\mathbf{y}_{s_d,\xi_d}, \tag{8.25}$$

where $\xi_j = \eta_{j-1}^{-1}\eta_j$ for $j = 1,\dots,d$ if we set $\eta_0 = 1$.

For all $s_1, \dots, s_d \in \mathbb{D}$, we define the one-variable multiple polylog

$$\mathrm{L}_{s_1,\dots,s_d}(z) = \sum_{k_1>k_2>\dots>k_d\geq 1} z^{k_1} \prod_{j=1}^{d} \frac{\operatorname{sgn}(s_j)^{k_j}}{k_j^{|s_j|}}, \quad 0<|z|<1,$$

which generalizes Eq. (3.20) to the larger index set $\mathbb{D}$. It satisfies

$$\mathrm{L}_{s_1,\dots,s_d}(z) = \int_0^z \left(\frac{\mathrm{d}t}{t}\right)^{|s_1|-1} \left(\frac{\mathrm{d}t}{\xi_1 - t}\right) \cdots \left(\frac{\mathrm{d}t}{t}\right)^{|s_d|-1} \left(\frac{\mathrm{d}t}{\xi_d - t}\right), \quad (8.26)$$

where $\xi_i = \operatorname{sgn}(s_{i-1})\operatorname{sgn}(s_i)$, $i = 1, \dots, d$. Here we have set $\operatorname{sgn}(s_0) = 1$.

Recall from Eq. (8.23) that $\mathbf{z}_s = \mathrm{y}_{|s|,\operatorname{sgn}(s)}$ for all $s \in \mathbb{D}$ and $\mathbf{z}_{\mathbf{s}} = \mathbf{z}_{s_1} \dots \mathbf{z}_{s_d} \in \mathfrak{A}_2^1$ for all $\mathbf{s} = (s_1, \dots, s_d) \in \mathbb{D}^d$. Further, define the map $\mathrm{L}(-;t) : (\mathfrak{A}_2^1, *) \to \mathbb{R}[\![t]\!]$ by setting $\mathrm{L}(\mathbf{z}_{\mathbf{s}}; t) = \mathrm{L}_{\mathbf{s}}(t)$ for all $\mathbf{s} \in \mathbb{D}^d$ and the map $\mathrm{L}_{ш}(-;t) : (\mathfrak{A}_2^1, ш) \to \mathbb{R}[\![t]\!]$ by setting $\mathrm{L}_{ш}(\mathbf{w}; t) = \mathrm{L}(\operatorname{tog}(\mathbf{w}); t)$.

Proposition 8.4.1. *The map* $\mathrm{L}_{ш}(-;t) : (\mathfrak{A}_2^1, ш) \to \mathbb{R}[\![t]\!]$ *is an algebra homomorphism.*

Proof. This follows from (8.26) and the shuffle relation satisfied by the iterated integrals. □

Example 8.4.2. Suppose $\mathbf{s} = (\bar{2}, 1)$. Then $\mathbf{z}_{\mathbf{s}} = \mathrm{y}_{2,-1}\mathrm{y}_{1,1}$ and

$$\mathrm{L}(\mathrm{y}_{2,-1}\mathrm{y}_{1,1}; t) = \int_0^1 \left(\frac{\mathrm{d}x}{x}\right) \left(\frac{-\mathrm{d}x}{1+x}\right)^2 = \mathrm{L}_{\bar{2},1}(t).$$

Set $\zeta_{\mathscr{A}_1,ш} = \zeta_{\mathscr{A}_1,*} \circ \operatorname{tog} : \mathfrak{A}_2^1 \to \mathscr{A}_1$. Although $\zeta_{\mathscr{A}_1,ш}$ is not an algebra homomorphism from $(\mathfrak{A}_2^1, ш)$ we will see in the next theorem that it does provide a kind of shuffle relation. It is the FES analog of Thm. 6.3.4 for the SMZVs.

Recall that $\tau : \mathfrak{A}_1^1 \to \mathfrak{A}_1^1$ is defined by (6.9).

Theorem 8.4.3. *For all words* $\mathbf{w}, \mathbf{u} \in \mathfrak{A}_1^1$ *and* $\mathbf{v} \in \mathfrak{A}_2^1$, *we have*

(i) $\zeta_{\mathscr{A}_1,ш}(\mathbf{u} ш \mathbf{v}) = \zeta_{\mathscr{A}_1,ш}(\tau(\mathbf{u})\mathbf{v})$,
(ii) $\zeta_{\mathscr{A}_1,ш}((\mathbf{wu}) ш \mathbf{v}) = \zeta_{\mathscr{A}_1,ш}(\mathbf{u} ш \tau(\mathbf{w})\mathbf{v})$,
(iii) *For all* $s \in \mathbb{N}$, $\zeta_{\mathscr{A}_1,ш}((\mathrm{x}_0^{s-1}\mathrm{x}_1\mathbf{u}) ш \mathbf{v}) = (-1)^s \zeta_{\mathscr{A}_1,ш}(\mathbf{u} ш (\mathrm{x}_0^{s-1}\mathrm{x}_1\mathbf{v}))$.

Proof. Taking $\mathbf{u} = \emptyset$ and then setting $\mathbf{w} = \mathbf{u}$ we see that (ii) implies (i). Decomposing $\mathbf{w}$ into strings of the type $\mathrm{x}_0^{s-1}\mathrm{x}_1$ we see that (iii) implies (ii). So we only need to prove (iii).

For simplicity, let $\mathsf{a} = \mathrm{x}_0$ and $\mathsf{b} = \mathrm{x}_1$ in the rest of this chapter. Let $s_j \in \mathbb{D}$ with $\operatorname{sgn}(s_j) = \eta_j$ for $j = 1, \dots, d$. Let $\mathbf{u} = \mathbf{z}_{\mathbf{s}}$ and $\operatorname{tog}(\mathbf{u}) = \mathrm{y}_{t_1,\xi_1} \cdots \mathrm{y}_{t_d,\xi_d} \in \mathbb{D}^d$. Then clearly $\operatorname{tog}(\mathsf{b}\mathbf{u}) = \mathsf{b}\operatorname{tog}(\mathbf{u})$.

For any prime $p > 2$, the coefficient of t^p in $\mathrm{L}_{\sqcup\!\sqcup}(\mathbf{bu}; t)$ is given by

$$\mathrm{Coeff}_{t^p}\left[\mathrm{L}_{\sqcup\!\sqcup}(\mathbf{bu}; t)\right] = \frac{1}{p} \sum_{p>k_1>\cdots>k_d>0} \frac{\xi_1^{k_1}\cdots\xi_d^{k_d}}{k_1^{t_1}\cdots k_d^{t_d}} = \frac{1}{p} H_{p-1}(\mathrm{tog}(\mathbf{u})).$$

Observe that

$$\mathtt{b}\Big((\mathtt{a}^{s-1}\mathbf{bu}) \sqcup\!\sqcup \mathbf{v} - (-1)^s \mathbf{u} \sqcup\!\sqcup (\mathtt{a}^{s-1}\mathbf{bv})\Big) = \sum_{i=0}^{s-1} (-1)^i (\mathtt{a}^{s-1-i}\mathbf{bu}) \sqcup\!\sqcup (\mathtt{a}^i\mathbf{bv}).$$

By first applying $\mathrm{L}_{\sqcup\!\sqcup}(-; t)$ to the above and then extracting the coefficients of t^p from both sides we get

$$\begin{aligned}
&\frac{1}{p}\Big(H_{p-1} \circ \mathrm{tog}\left((\mathtt{a}^{s-1}\mathbf{bu}) \sqcup\!\sqcup \mathbf{v}\right) - (-1)^s H_{p-1} \circ \mathrm{tog}\left(\mathbf{u} \sqcup\!\sqcup (\mathtt{a}^{s-1}\mathbf{bv})\right)\Big) \\
&\quad = \sum_{i=0}^{s-1} (-1)^i \mathrm{Coeff}_{t^p}\left[\mathrm{L}_{\sqcup\!\sqcup}(\mathtt{a}^{s-1-i}\mathbf{bu}; t)\, \mathrm{L}_{\sqcup\!\sqcup}(\mathtt{a}^i\mathbf{bv}; t)\right] \\
&\quad = \sum_{i=0}^{s-1} (-1)^i \sum_{j=1}^{p-1} \mathrm{Coeff}_{t^j}\left[\mathrm{L}_{\sqcup\!\sqcup}(\mathtt{a}^{s-1-i}\mathbf{bu}; t)\right] \cdot \mathrm{Coeff}_{t^{p-j}}\left[\mathrm{L}_{\sqcup\!\sqcup}(\mathtt{a}^i\mathbf{bv}; t)\right]
\end{aligned}$$

by Prop. 8.4.1. Now the last sum is p-integral since $p - j < p$ and $j < p$ and therefore we get

$$H_{p-1} \circ \mathrm{tog}((\mathtt{a}^{s-1}\mathbf{bu}) \sqcup\!\sqcup \mathbf{v}) \equiv (-1)^s H_{p-1} \circ \mathrm{tog}(\mathbf{u} \sqcup\!\sqcup (\mathtt{a}^{s-1}\mathbf{bv})) \pmod p$$

which completes the proof of (iii). □

Definition 8.4.4. A relation produced by Thm. 8.4.3(i) is called a *linear shuffle relation* of the FESs. For each weight $w \geq 2$, by the *double shuffle relations* of the FESs of weight w we mean all the linear shuffle relations of weight w and all the linear stuffle relations of w defined in Definition 8.3.1.

Restricting to FMZVs, we obtain the *linear shuffle relations* and the *double shuffle relations* of the FMZVs.

8.5 Finite Multiple Zeta Values (FMZVs)

We now apply the results from the previous sections to the FMZVs.

8.5.1 *Homogeneous FMZVs*

The following homogeneous generalization of Thm. 8.2.3 will be very handy, even though we will extend it further.

Theorem 8.5.1. *Let s and d be two positive integers. Then*

$$\zeta_{\mathscr{A}_1}(\{s\}^d) = 0 \quad \text{if the weight } ds \text{ is even;} \tag{8.27}$$

$$\zeta_{\mathscr{A}_2}(\{s\}^d) = 0 \quad \text{if the weight } ds \text{ is odd.} \tag{8.28}$$

Proof. We may proceed by induction on d. The case $d = 1$ is just Thm. 8.2.3. Assume the theorem holds for $d \geq 2$. Then

$$0 = \zeta_{\mathscr{A}_2}(s)\zeta_{\mathscr{A}_2}(\{s\}^{d-1}) = d\zeta_{\mathscr{A}_2}(\{s\}^d) + \frac{1}{(d-2)!}\sum_{\sigma\in\mathfrak{S}_d}\zeta_{\mathscr{A}_2}(\sigma(2s,\{s\}^{d-2})).$$

So the theorem follows from Thm. 7.4.1 after counting the weight. □

Our goal in this section is to prove an improvement of Thm. 8.5.1. First, we need some preliminary results.

Lemma 8.5.2. *Let $p > 3$ be a prime and let r be a positive integer such that $r \leq p-3$. Then*

$$H_{p-1}(r) = \sum_{k=1}^{p-1}\frac{1}{k^r} \equiv \begin{cases} -\frac{r(r+1)}{2(r+2)}B_{p-r-2}p^2 \pmod{p^3} & \text{if } r \text{ is odd;} \\ \frac{r}{r+1}B_{p-r-1}p \pmod{p^2} & \text{if } r \text{ is even.} \end{cases}$$

The proof is standard so we leave it as Exercise 8.2.

Lemma 8.5.3. *Let $r, \alpha_1, \cdots, \alpha_n$ be positive integers, $r = \alpha_1 + \alpha_2 + \cdots + \alpha_n \leq p-3$, then*

$$\sum_{\substack{1\leq l_1,\cdots,l_n\leq p-1\\ l_i\neq l_j}}\frac{1}{l_1^{\alpha_1}\cdots l_n^{\alpha_n}} \equiv \begin{cases} \frac{(-1)^n n! r(r+1)}{2n(r+2)}B_{p-r-2}p^2 \pmod{p^3} & \text{if } r \text{ is odd;} \\ \frac{(-1)^{n-1} n! r}{n(r+1)}B_{p-r-1}p \pmod{p^2} & \text{if } r \text{ is even.} \end{cases}$$

Proof. If $n = 1$, the congruence is obvious by Lemma 8.5.2. We assume that the formula holds when the number of variables is less than n. Then by induction

$$\sum_{\substack{1\le l_1,\cdots,l_n<p\\ l_i\ne l_j}} \frac{1}{l_1^{\alpha_1}\cdots l_n^{\alpha_n}}$$

$$= \sum_{\substack{1\le l_1,\cdots,l_{n-1}<p\\ l_i\ne l_j}} \frac{1}{l_1^{\alpha_1}\cdots l_{n-1}^{\alpha_{n-1}}}\Big(\sum_{l_n=1}^{p-1}\frac{1}{l_n^{\alpha_n}}-\frac{1}{l_1^{\alpha_n}}-\cdots-\frac{1}{l_{n-1}^{\alpha_n}}\Big)$$

$$= \sum_{\substack{1\le l_1,\cdots,l_{n-1}<p\\ l_i\ne l_j}} \frac{1}{l_1^{\alpha_1}\cdots l_{n-1}^{\alpha_{n-1}}}\sum_{l_n=1}^{p-1}\frac{1}{l_n^{\alpha_n}} - \sum_{\substack{1\le l_1,\cdots,l_{n-1}<p\\ l_i\ne l_j}} \frac{1}{l_1^{\alpha_1+\alpha_n} l_2^{\alpha_2}\cdots l_{n-1}^{\alpha_{n-1}}}$$

$$-\cdots- \sum_{\substack{1\le l_1,\cdots,l_{n-1}<p\\ l_i\ne l_j}} \frac{1}{l_1^{\alpha_1} l_2^{\alpha_2}\cdots l_{n-1}^{\alpha_{n-1}+\alpha_n}}.$$

By Lemma 8.5.2 and the inductive assumption, we have

$$\sum_{\substack{1\le l_1,\ldots,l_{n-1}<p\\ l_i\ne l_j}} \frac{1}{l_1^{\alpha_1} l_2^{\alpha_2}\cdots l_{n-1}^{\alpha_{n-1}}}\sum_{l_n=1}^{p-1}\frac{1}{l_n^{\alpha_n}} \equiv \begin{cases} 0 & \pmod{p^3} \quad \text{if } 2\nmid r,\\ 0 & \pmod{p^2} \quad \text{if } 2|r.\end{cases}$$

If r is odd, then by the inductive assumption,

$$\begin{aligned}\sum_{\substack{1\le l_1,\ldots,l_n<p\\ l_i\ne l_j}} \frac{1}{l_1^{\alpha_1} l_2^{\alpha_2}\cdots l_n^{\alpha_n}} &\equiv -(n-1) \sum_{\substack{1\le l_1,\cdots,l_{n-1}<p\\ l_i\ne l_j}} \frac{1}{l_1^{\alpha_1+\alpha_n} l_2^{\alpha_2}\cdots l_{n-1}^{\alpha_{n-1}}}\\ &\equiv -(n-1)(-1)^{n-1}(n-2)!\frac{r(r+1)}{2(r+2)}B_{p-r-2}p^2\\ &\equiv (-1)^n(n-1)!\frac{r(r+1)}{2(r+2)}B_{p-r-2}p^2 \pmod{p^3}.\end{aligned}$$

Similarly, if r is even, then by the inductive assumption,

$$\sum_{\substack{1\le l_1,\cdots,l_n<p\\ l_i\ne l_j}} \frac{1}{l_1^{\alpha_1} l_2^{\alpha_2}\cdots l_n^{\alpha_n}} \equiv (-1)^{n-1}(n-1)!\frac{r}{r+1}B_{p-r-1}p \pmod{p^2}.$$

The lemma is now proved. □

Theorem 8.5.4. *Let $d,s\in\mathbb{N}$. If ds is even then*

$$\zeta_{\mathscr{A}_2}(\{s\}^d) = (-1)^{d-1}s\beta_{ds+1}p. \tag{8.29}$$

If ds is odd then

$$\zeta_{\mathscr{A}_3}(\{s\}^d) = (-1)^d\frac{s(ds+1)}{2}\beta_{ds+2}p^2. \tag{8.30}$$

Proof. We can permute $l_1,\ldots,l_d$ in the inequality $l_1>\cdots>l_d$ in $d!$ ways and then apply Lemma 8.5.3. This proves the theorem. □

8.5.2 Duality of FMZVs

The FMZVs satisfy a different kind of duality from that of the MZVs. Recall from Definition 7.3.4 that for all positive integers $r_1, \dots, r_\ell, t_1, \dots, t_\ell$ and

$$\mathbf{s} = (r_1, \{1\}^{t_1-1}, r_2+1, \{1\}^{t_2-1}, \dots, r_\ell+1, \{1\}^{t_\ell-1}),$$

we define the v-dual of $\mathbf{s}$ by

$$\mathbf{s}^\vee = (\{1\}^{r_1-1}, t_1+1, \{1\}^{r_2-1}, t_2+1, \dots, t_{\ell-1}+1, \{1\}^{r_\ell-1}, t_\ell).$$

Theorem 8.5.5. *If* $\mathbf{s} = (s_1, \dots, s_d) \in \mathbb{N}^d$ *then*

$$\zeta^\star_{\mathscr{A}_1}(\mathbf{s}) = -\zeta^\star_{\mathscr{A}_1}(\mathbf{s}^\vee). \tag{8.31}$$

Proof. These follow immediately from Eq. (7.17). □

Definition 8.5.6. For any sequence $\{f(n)\}_{n\ge0}$ with $f(0) = 0$, the summation operator Σ and the *conjugation* operator ∇ are defined by (using umbral calculus notation $f^k = f(k)$)

$$\Sigma f(n) = \sum_{i=0}^n f(i), \quad \nabla f(n) = (1-f)^n = \sum_{k=0}^n (-1)^k \binom{n}{k} f(k), \quad \forall n \ge 0.$$

For any function $g(n) = \sum_{k=0}^n f(k)$, we set $\Sigma^{-1} g = f$.

Lemma 8.5.7. *The two operators* Σ *and* ∇ *generate a dihedral group, i.e,* $\nabla^2 = \mathrm{id}$, $\Sigma\nabla = \nabla\Sigma^{-l}$, *and therefore* $(\Sigma\nabla)^2 = \mathrm{id}$.

Proof. Clearly $\nabla^2 f(n) = (1-(1-f))^n = f^n = f(n)$. Moreover by changing the orders of summation

$$\begin{aligned}
\Sigma\nabla\Sigma f(n) &= \sum_{k=0}^n \sum_{j=0}^k (-1)^j \binom{k}{j} \sum_{l=0}^j f(l) \\
&= \sum_{k=0}^n \sum_{l=0}^k \sum_{j=l}^k (-1)^j \binom{k}{j} f(l) \\
&= \sum_{k=0}^n \sum_{l=0}^k (-1)^l \binom{k-1}{l-1} f(l) \\
&= \sum_{l=0}^n \sum_{k=l}^n (-1)^l \binom{k-1}{l-1} f(l) \\
&= \sum_{l=0}^n (-1)^l \binom{n}{l} f(l) = \nabla f(n).
\end{aligned}$$

This easily implies $(\Sigma\nabla)^2 = \mathrm{id}$ and the lemma is proved. □

Lemma 8.5.8. *For any positive number n and composition $\mathbf{s}$ of positive integers, we have*

$$\Sigma\nabla H_n^\star(\mathbf{s}) = -H_n^\star(\mathbf{s}^\vee).$$

Proof. For $\mathbf{s}=(s_1,s_2,\dots,s_d)$, we see that

$$\begin{aligned}
\nabla H_n^\star(\mathbf{s}) &= \sum_{k=0}^{n}(-1)^k\binom{n}{k}\sum_{k_1=1}^{k}\frac{1}{k_1^{s_1}}\sum_{k_1\ge k_2\ge\cdots\ge k_d}\frac{1}{k_2^{s_2}\cdots k_d^{s_d}}\\
&= \sum_{k_1=1}^{n}\sum_{k=k_1}^{n}(-1)^k\binom{n}{k}\frac{1}{k_1^{s_1}}\sum_{k_1\ge k_2\ge\cdots\ge k_d}\frac{1}{k_2^{s_2}\cdots k_d^{s_d}}\\
&= \sum_{k_1=1}^{n}(-1)^{k_1}\binom{n-1}{k_1-1}\frac{1}{k_1^{s_1}}\sum_{k_1\ge k_2\ge\cdots\ge k_d}\frac{1}{k_2^{s_2}\cdots k_d^{s_d}}\\
&= \frac{1}{n}\nabla f(n),
\end{aligned}\tag{8.32}$$

where

$$f(n)=\frac{H_n^\star(s_2,\dots,s_d)}{n^{s_1-1}}=\begin{cases}H_n^\star(s_2,\dots,s_d), & \text{if } s_1=1;\\ \Sigma^{-1}H_n^\star(s_1-1,s_2,\dots,s_d), & \text{otherwise.}\end{cases}$$

We now prove the lemma by induction on the weight of $\mathbf{s}$. The weight one case is $\Sigma\nabla H_n^\star(1)=\nabla\Sigma^{-1}H_n^\star(1)=-H_n^\star(1)$, which is just Eq. (7.20). Assume the lemma holds if $|\mathbf{s}|<n$. Suppose $\mathbf{s}=(s_1,\dots,s_d)$ has weight n. We consider two cases: (i) $s_1=1$, and (ii) $s_1>1$. If $s_1=1$ then $\mathbf{s}^\vee=(j_1+1,j_2,\dots,j_r)$. Putting $(s_2,\dots,s_d)^\vee=\mathbf{t}=(j_1,\dots,j_r)$, using Eq. (8.32) and the induction hypothesis we get

$$\Sigma\nabla H_n^\star(\mathbf{s})=\Sigma\left(\frac{\nabla H_n^\star(\mathbf{t}^\vee)}{n}\right)=-\Sigma\left(\frac{\Sigma^{-1}H_n^\star(\mathbf{t})}{n}\right)=-H_n^\star(j_1+1,j_2,\dots,j_r).$$

Hence the lemma holds in this case. If $s_1>1$, $\mathbf{s}^\vee=(1,j_1,\dots,j_r)$. Putting $\mathbf{t}=(s_1-1,s_2,\dots,s_k)^\vee=(j_1,\dots,j_r)$, we have

$$\begin{aligned}
\Sigma\nabla H_n^\star(\mathbf{s}) =& \Sigma\left(\frac{1}{n}\nabla\Sigma^{-1}H_n^\star(\mathbf{t}^\vee)\right)=\Sigma\left(\frac{1}{n}\Sigma\nabla H_n^\star(\mathbf{t}^\vee)\right)\\
=& -\Sigma\left(\frac{1}{n}H_n^\star(\mathbf{t})\right)=-H_n^\star(1,j_1,\dots,j_r)=-H_n^\star(\mathbf{s}^\vee)
\end{aligned}$$

as desired. The lemma is now proved. □

Recall from Eq. (7.13) that $\rho_n:(\mathfrak{A}_2^1,*)\to\mathbb{R}$ is defined by $\rho_n(M_{\mathbf{s}})=H_n(\mathbf{s})$. Let $\rho=(\rho_n)_{n\ge1}$ be restricted to $\mathfrak{A}_1^1$.

Corollary 8.5.9. *The following diagram commutes:*

$$\begin{array}{ccc} \mathsf{QSym}_1 & \xrightarrow{\ \varphi\ } & \mathsf{QSym}_1 \\ \Big\downarrow \rho & & \Big\downarrow \rho \\ \mathbb{R}^{\mathbb{N}} & \xrightarrow{\ \Sigma\nabla\ } & \mathbb{R}^{\mathbb{N}} \end{array}$$

where the map φ *defined by Eq.* (6.11) *is an involution on* $\mathfrak{A}_1^1$ *such that* $\mathsf{a}^\varphi = \mathsf{a} + \mathsf{b}$ *and* $\mathsf{b}^\varphi = -\mathsf{b}$.

Proof. This follows immediately from Thm. 7.3.6(ii) and Lemma 8.5.8. □

Parallel to Thm. 6.3.5 for the SMZVs, the following result provides another evidence for Conjecture 8.5.15.

Theorem 8.5.10. *For all* $\mathbf{w} \in \mathfrak{A}_1^1$, *we have*

$$\zeta_{\mathscr{A}_1}(\mathbf{w}) = \zeta_{\mathscr{A}_1}(\mathbf{w}^\varphi).$$

Proof. This is clear from Cor. 8.5.9 by considering ρ_{p-1} for primes p. □

We now consider the duality property in superbity two.

Theorem 8.5.11. *Let* $\mathbf{s}$ *be any composition of positive integers of weight* w. *Then*

$$-\zeta^\star_{\mathscr{A}_2}(\mathbf{s}^\vee) = \zeta^\star_{\mathscr{A}_2}(\mathbf{s}) + p\left(\sum_{\mathbf{t}\preceq\mathbf{s}} \zeta_{\mathscr{A}_2}(1,\mathbf{t})\right). \tag{8.33}$$

Proof. By Lemma 8.5.8 we have

$$\begin{aligned} &-\zeta^\star_{\mathscr{A}_2}(\mathbf{s}^\vee) - \zeta^\star_{\mathscr{A}_2}(\mathbf{s}) \\ &= \Sigma\nabla H^\star_{p-1}(\mathbf{s}) - \zeta^\star_{\mathscr{A}_2}(\mathbf{s}) = \sum_{i=0}^{p-2} \binom{p}{i+1}(-1)^i H^\star_i(\mathbf{s}) \qquad (\text{in } \mathscr{A}_2) \\ &= p\left(\sum_{i=0}^{p-2} \frac{1}{i+1}\binom{p-1}{i}(-1)^i H^\star_i(\mathbf{s})\right) \\ &= p\left(\sum_{i=0}^{p-2} \frac{1}{i+1} H^\star_i(\mathbf{s})\right) = p\left(\sum_{i=0}^{p-2}\sum_{\mathbf{t}\preceq\mathbf{s}} \frac{H_i(\mathbf{t})}{i+1}\right) = p\left(\sum_{\mathbf{t}\preceq\mathbf{s}} \zeta_{\mathscr{A}_2}(1,\mathbf{t})\right), \end{aligned}$$

as desired. □

8.5.3 *FMZVs of Lower Depths*

In this section we use Fermat's Little Theorem and Bernoulli numbers to derive $\mathbb{Q}$-linear relations among FMZVs when the depth is two or three.

Theorem 8.5.12. *For all positive integers s and t,*

$$\zeta_{\mathscr{A}_1}(s,t) = (-1)^s \binom{s+t}{s} \beta_{s+t}. \tag{8.34}$$

Proof. The Seki–Bernoulli's formula Eq. (8.2) together with Fermat's Little Theorem yields

$$\zeta_{\mathscr{A}_1}(s,t) = \zeta^\star_{\mathscr{A}_1}(s,t) = \sum_{a=1}^{p-1}\sum_{b=1}^{a-1} \frac{1}{a^s b^t} = \sum_{a=1}^{p-1} a^{-s} \sum_{b=1}^{a-1} b^{p-1-t}$$

$$= \sum_{a=1}^{p-1} \frac{a^{-s}}{p-t} \sum_{k=0}^{p-1-t} \binom{p-t}{k} B_k a^{p-t-k} = \frac{1}{p-t} \sum_{k=0}^{p-1-t} \binom{p-t}{k} B_k \sum_{a=1}^{p-1} a^{p-s-t-k}.$$

In $\mathscr{A}_1$ we know $\sum_{a=1}^{p-1} a^{p-s-t-k} = 0$ unless $k = p-s-t$. Thus

$$\begin{aligned} \zeta_{\mathscr{A}_1}(s,t) = \zeta^\star_{\mathscr{A}_1}(s,t) &= (s+t)\frac{p-1}{p-t}\binom{p-t}{p-s-t}\beta_{s+t} \\ &= \frac{-(s+t)(p-t-1)!}{s!(p-s-t)!}\beta_{s+t}. \end{aligned}$$

Equation (8.34) follows from this easily since by Wilson's Theorem $\frac{1}{(p-m)!} \equiv (-1)^m (m-1)! \pmod p$ for any integer $m < p$. □

Theorem 8.5.13. *Let $(l,m,n) \in \mathbb{N}^3$. If $w = l+m+n$ is* odd *then*

$$-\zeta_{\mathscr{A}_1}(l,m,n) = \zeta^\star_{\mathscr{A}_1}(l,m,n) = \left[(-1)^l \binom{w}{l} - (-1)^n \binom{w}{n}\right] \frac{\beta_w}{2}.$$

Proof. From Eq. (8.16) and the concatenation relation Eq. (8.18) we have

$$\zeta_{\mathscr{A}_1}(l,m,n) = \zeta^\star_{\mathscr{A}_1}(l,m,n) - \zeta^\star_{\mathscr{A}_1}(l+m,n) - \zeta^\star_{\mathscr{A}_1}(l,m+n) + \zeta^\star_{\mathscr{A}_1}(l+m+n),$$

and

$$\begin{aligned} \zeta_{\mathscr{A}_1}(l,m,n) = {}& \zeta^\star_{\mathscr{A}_1}(n)\zeta^\star_{\mathscr{A}_1}(m)\zeta^\star_{\mathscr{A}_1}(l) - \zeta^\star_{\mathscr{A}_1}(n,m)\zeta^\star_{\mathscr{A}_1}(l) \\ & - \zeta^\star_{\mathscr{A}_1}(n)\zeta^\star_{\mathscr{A}_1}(m,l) + \zeta^\star_{\mathscr{A}_1}(n,m,l), \end{aligned} \tag{8.35}$$

respectively. Thus by Thm. 8.2.3 we have

$$\zeta^\star_{\mathscr{A}_1}(l,m,n) - \zeta^\star_{\mathscr{A}_1}(l+m,n) - \zeta^\star_{\mathscr{A}_1}(l,m+n) = \zeta^\star_{\mathscr{A}_1}(n,m,l).$$

Notice the weight is odd so the reversal relation Eq. (8.5) and Thm. 8.5.12 imply that

$$2\zeta^{\star}_{\mathscr{A}_1}(l,m,n)=\zeta^{\star}_{\mathscr{A}_1}(l+m,n)+\zeta^{\star}_{\mathscr{A}_1}(l,m+n)=\left[(-1)^l\binom{w}{l}-(-1)^n\binom{w}{n}\right]\beta_w.$$

This yields the second equation in the theorem. The first equation follows immediately from Eq. (8.35) by the reversal relation. This completes the proof of the theorem. □

8.5.4 *Structure of FMZVs*

We denote by $\mathsf{FMZ}_{w,\ell}$ the $\mathbb{Q}$-vector subspace of $\mathscr{A}_\ell$ generated by all weight w and superbity ℓ FMZVs. Further, we write $\mathsf{FMZ}_w=\mathsf{FMZ}_{w,1}$. From overwhelming numerical evidence, we can formulate the following conjecture.

Conjecture 8.5.14. *Let w be any positive integer.*

(i) *Set $d_0=1$ and $d_w=\dim\mathsf{FMZ}_w$ for all $w\ge 1$. Then $d_1=d_2=0$ and*

$$d_w=d_{w-2}+d_{w-3}\quad\forall w\ge 3.$$

(ii) *For all $w\ge 3$, FMZ_w has a basis*

$$\{\zeta_{\mathscr{A}_1}(1,2,a_1,\dots,a_r):a_1,\dots,a_r=2 \text{ or } 3, a_1+\cdots+a_r=w-3\}.$$

(iii) *All the $\mathbb{Q}$-linear relations among the FMZVs can be produced by the double shuffle relations.*

In fact, Kaneko and Zagier has proposed the following conjecture providing a relation between FMZVs and MZVs.

Conjecture 8.5.15. *There is a $\mathbb{Q}$-algebra isomorphism*

$$f_{\mathrm{KZ}}:\mathsf{FMZ}_{w,1}\longrightarrow \mathsf{MZ}_w/\zeta(2)\mathsf{MZ}_{w-2}$$
$$\zeta_{\mathscr{A}_1}(\mathbf{s})\longmapsto \zeta^{\mathcal{S}}_{\sqcup\!\sqcup}(\mathbf{s})$$

where $\zeta^{\mathcal{S}}_{\sqcup\!\sqcup}(\mathbf{s})$ is defined in Eq. (6.1).

It follows from this conjecture that $\dim\mathsf{FMZ}_{w,1}=\dim\mathsf{MZ}_w-\dim\mathsf{MZ}_{w-2}=\dim\mathsf{MZ}_{w-3}$ by Conjecture 3.4.2 of Zagier. However, even in weight five, it seems impossible to prove the map f_{KZ} is well defined and injective at the moment. Indeed, we have

$$\zeta_{\mathscr{A}_1}(4,1)=5\beta_5=(B_{p-5})_p.$$

But we know $\zeta^{\mathcal{S}}_{*}(4,1)=-\zeta(5)$ (see Exercise 6.1 of Chap. 6) is conjecturally nonzero on the right-hand side so that if f_{KZ} is well defined then

$\beta_5 \neq 0$ which would imply that $B_{p-5} \not\equiv 0 \pmod p$ for infinitely many primes p, a statement far from proved. On the other hand, even if we could prove $\beta_5 \neq 0$ by some other means, we still don't know whether $\zeta(2)\zeta(3)/\zeta(5) \in \mathbb{Q}$ is true or not, thus we still don't know whether $f_{\rm KZ}$ is injective. Note, however, Lemma 6.4.18 and Thm. 6.4.19 implies that the map in Conjecture 8.5.15 is surjective. Moreover, since $\zeta^{\mathcal{S}}(2) = 2\zeta(2)$ we see that $f_{\rm KZ}(0) = f_{\rm KZ}(\zeta_{\mathscr{A}_1}(2)\mathsf{FMZ}_{w-2,1}) = 2\zeta(2)\mathsf{MZ}_{w-2} = 0$ on the right.

A useful way to compare the structure of the FMZV and the MZV is that β_k in the finite setting should have weight k and correspond to the Riemann zeta value $\zeta(k)$ for odd k. It is conjectured that not every MZV can be expressed using Riemann zeta values, the first such instance should be $\zeta(6,2)$ (or $\zeta(2,6)$) in weight 8. On the FMZV side, it is likely that the first such value is also in weight 8. In Table E.2 except for the two values $\zeta_{\mathscr{A}_1}(1,3,1,3) = 0$ and $\zeta_{\mathscr{A}_1}(1,3,2,2) = (45/2)\beta_3\beta_5$. I expect none of the others can be expressed as a $\mathbb{Q}$ multiple of $\beta_3\beta_5$.

In higher weight cases, there should be linear independence among the FMZVs which are expressible as $\mathbb{Q}$-linear combinations of weight k products of Bernoulli numbers, as implied by Conjecture 8.1.5.

As a further support of Conjecture 8.5.14, we have:

Theorem 8.5.16. *For all* $3 \le w \le 13$,

$$\mathsf{FMZ}_w = \big\langle \zeta_{\mathscr{A}_1}(1,2,a_1,\dots,a_r) : a_1,\dots,a_r \in \{2,3\}, a_1+\cdots+a_r = w-3 \big\rangle.$$

Moreover, all the relations in these weights can be proved by the double shuffle relations of the FMZVs defined as in Definition 8.4.4.

Proof. This theorem can be proved with the help of Maple (see Appendix D.1). So we leave it to the interested reader. □

8.5.5 *FMZVs of Superbity 2*

We now move up to superbity 2, namely, we consider congruences modulo p^2. First, we improve on Thm. 8.5.12.

Theorem 8.5.17. *Suppose* $s,t \in \mathbb{N}$ *and* $w = s+t$ *is even. Then*

$$\zeta_{\mathscr{A}_2}(s,t) = p\left[(-1)^t s\binom{w+1}{t} - (-1)^t t\binom{w+1}{s} - w\right]\frac{\beta_{w+1}}{2},$$
$$\zeta^\star_{\mathscr{A}_2}(t,s) = p\left[(-1)^t s\binom{w+1}{t} - (-1)^t t\binom{w+1}{s} + w\right]\frac{\beta_{w+1}}{2}.$$

Proof. By the conditions on s and t we know from Thm. 8.2.3 and 8.5.4

$$\zeta_{\mathscr{A}_2}(s)\cdot\zeta_{\mathscr{A}_2}(t)=0, \qquad \zeta_{\mathscr{A}_2}(w)=pw\beta_{w+1}.$$

Therefore by the stuffle relation

$$\zeta_{\mathscr{A}_2}(s,t)+\zeta_{\mathscr{A}_2}(t,s)=\zeta_{\mathscr{A}_2}(s)\cdot\zeta_{\mathscr{A}_2}(t)-\zeta_{\mathscr{A}_2}(w)=-pw\beta_{w+1}. \tag{8.36}$$

Moreover, by superbity 2 reversal relation Eq. (8.8)

$$\begin{aligned}\zeta_{\mathscr{A}_2}(s,t) &= \zeta_{\mathscr{A}_2}(t,s)+ps\zeta_{\mathscr{A}_2}(t,s+1)+pt\zeta_{\mathscr{A}_2}(t+1,s)\\ &= \zeta_{\mathscr{A}_2}(t,s)+p\left[(-1)^{s+1}s\binom{w+1}{t}+(-1)^s t\binom{w+1}{t+1}\right]\beta_{w+1}.\end{aligned}$$

Combined with Eq. (8.36) this completes the proof of the equation for $\zeta_{\mathscr{A}_2}$. Then the $\zeta^\star_{\mathscr{A}_2}$ part follows from the identity $\zeta^\star_{\mathscr{A}_2}(s,t)=\zeta_{\mathscr{A}_2}(s,t)+\zeta_{\mathscr{A}_2}(w)$ and Thm. 8.5.4. □

So up to weight 4, we have the following table (see [619, Prop. 3.7] for detailed computation in weight 4). Note that $\zeta_{\mathscr{A}_2}(1)=\zeta_{\mathscr{A}_2}(3)=\zeta_{\mathscr{A}_2}(5)=\zeta_{\mathscr{A}_2}(\{1\}^3)=\zeta_{\mathscr{A}_2}(\{1\}^5)=0$. We have the next two theorems for higher weights.

Table 8.1: FMZV of superbity 2 in weight up to 4.

$\zeta_{\mathscr{A}_2}(2)$	$2p\beta_3$	$\zeta_{\mathscr{A}_2}(1,1)$	$-p\beta_3$	$\zeta_{\mathscr{A}_2}(1,2)$	$\zeta_{\mathscr{A}_2}(1,2)$
$\zeta_{\mathscr{A}_2}(2,1)$	$-\zeta_{\mathscr{A}_2}(1,2)$	$\zeta_{\mathscr{A}_2}(4)$	$4p\beta_5$	$\zeta_{\mathscr{A}_2}(1,3)$	$\frac{1}{2}p\beta_5$
$\zeta_{\mathscr{A}_2}(3,1)$	$-\frac{9}{2}p\beta_5$	$\zeta_{\mathscr{A}_2}(1,1,2)$	$3\beta_5$	$\zeta_{\mathscr{A}_2}(1,2,1)$	$-\frac{9}{2}p\beta_5$
$\zeta_{\mathscr{A}_2}(2,1,1)$	$\frac{11}{2}\beta_5$	$\zeta_{\mathscr{A}_2}(\{1\}^4)$	$-\beta_5$		

Theorem 8.5.18. *In weight 5 we have*

$$\zeta^\star_{\mathscr{A}_2}(1,3,1)=\zeta_{\mathscr{A}_2}(1,3,1)=0, \tag{8.37}$$

$$\zeta^\star_{\mathscr{A}_2}(2,1,2)=\zeta_{\mathscr{A}_2}(2,1,2)=-3p\beta_3^2, \tag{8.38}$$

$$\zeta^\star_{\mathscr{A}_2}(2,3)=\zeta_{\mathscr{A}_2}(2,3)=2\zeta^\star_{\mathscr{A}_2}(4,1)=2\zeta_{\mathscr{A}_2}(4,1), \tag{8.39}$$

$$2\zeta^\star_{\mathscr{A}_2}(2,2,1)-3\zeta_{\mathscr{A}_2}(4,1)=9p\beta_3^2, \tag{8.40}$$

and

$$2\zeta^\star_{\mathscr{A}_2}(1,1,3)=-2\zeta^\star_{\mathscr{A}_2}(3,1,1)=-2\zeta_{\mathscr{A}_2}(1,1,3)=2\zeta_{\mathscr{A}_2}(3,1,1)=\zeta_{\mathscr{A}_2}(4,1). \tag{8.41}$$

Proof. First, by applying Eq. (8.33) to $(1, 1, 3)$ we get

$$\begin{aligned}-\zeta^\star_{\mathscr{A}_2}(3,1,1) = \zeta^\star_{\mathscr{A}_2}(1,1,3) + p\Big(\zeta_{\mathscr{A}_1}(1,1,1,3) + \zeta_{\mathscr{A}_1}(1,2,3)\\ + \zeta_{\mathscr{A}_1}(1,1,4) + \zeta_{\mathscr{A}_1}(1,5)\Big) = \zeta^\star_{\mathscr{A}_2}(1,1,3),\end{aligned} \quad (8.42)$$

which yields the first "=" in Eq. (8.41) by looking up Table E.1 on page 540 and easy computation with Thm. 8.5.12 for weight six FMZV. Notice by Eq. (8.21) $\zeta^\star_{\mathscr{A}_2}(1,3,1) = \zeta_{\mathscr{A}_2}(1,3,1)$ and by reversal relations Eq. (8.8) in superbity 2

$$-\zeta_{\mathscr{A}_2}(1,1,3) - \zeta_{\mathscr{A}_2}(3,1,1) = p(\zeta_{\mathscr{A}_1}(2,1,3) + \zeta_{\mathscr{A}_1}(1,2,3) + 3\zeta_{\mathscr{A}_1}(1,1,4))$$

by Table E.1 in Appendix E.2. Thus Eq. (8.42) together with stuffle relations yields

$$\begin{aligned}0 = \zeta_{\mathscr{A}_2}(1,1)\zeta_{\mathscr{A}_2}(3) = \zeta_{\mathscr{A}_2}(1,3,1) + \zeta_{\mathscr{A}_2}(1,1,3)\\ + \zeta_{\mathscr{A}_2}(3,1,1) + \zeta_{\mathscr{A}_2}(2)\zeta_{\mathscr{A}_2}(3) - \zeta_{\mathscr{A}_2}(5) = \zeta_{\mathscr{A}_2}(1,3,1)\end{aligned}$$

which gives Eq. (8.37) using Eq. (8.21). Applying Eq. (8.33) to $\mathbf{s} = (2, 1, 2)$ we get

$$\begin{aligned}-\zeta^\star_{\mathscr{A}_2}(1,3,1) = p\Big(\zeta_{\mathscr{A}_1}(1,2,1,2) + \zeta_{\mathscr{A}_1}(1,3,2) + \zeta_{\mathscr{A}_1}(1,2,3) + \zeta_{\mathscr{A}_1}(1,5)\Big)\\ + \zeta^\star_{\mathscr{A}_2}(2,1,2) = \zeta^\star_{\mathscr{A}_2}(2,1,2) + 3p\beta_3^2\end{aligned}$$

using Table E.1. Thus Eq. (8.38) follows immediately from Eq. (8.37).

Turning to the proof of Eq. (8.39) we note that for any positive integer $k < p$ (setting $h_k = H_{k-1}$)

$$\begin{aligned}\frac{\binom{p-1}{k}}{\binom{p-1+k}{k}} &= \frac{(p-1)(p-2)\cdots(p-k)}{p(p+1)\cdots(p+k-1)} = \frac{(-1)^k k}{p}\prod_{j=1}^{k}\left(1-\frac{p}{j}\right)\prod_{j=1}^{k-1}\left(1+\frac{p}{j}\right)^{-1}\\ &\equiv \frac{(-1)^k k}{p}\Big(1 - pH_k(1) + p^2 H_k(1,1)\Big)\Big(1 - ph_k(1) + p^2 h_k(2) + p^2 h_k(1,1)\Big)\\ &\equiv \frac{(-1)^k k}{p}\left(1 - 2ph_k(1) - \frac{p}{k} + \frac{2p^2}{k}h_k(1) + 2p^2 h_k(2) + 4p^2 h_k(1,1)\right).\end{aligned} \quad (8.43)$$

Put $\tilde{a} = 2a + 1$ and $\tilde{b} = 2b + 1$. By Eq. (7.22) of Thm. 7.5.2, we obtain (with abuse of notation $H = H_{p-1}$)

$$\begin{aligned}&H^\star_{p-1}(\{2\}^a, 3, \{2\}^b)\\ &\equiv -\frac{2}{p}\sum_{k=1}^{p-1}\frac{1}{k^{\tilde{a}+\tilde{b}}}\left(1 - 2ph_k(1) - \frac{p}{k} + \frac{2p^2}{k}h_k(1) + 2p^2 h_k(2) + 4p^2 h_k(1,1)\right)\end{aligned}$$

$$
\begin{aligned}
&-\frac{4}{p}\sum_{k=1}^{p-1}\frac{h_k(\tilde b)}{k^{\tilde a}}\left(1-2ph_k(1)-\frac{p}{k}+\frac{2p^2}{k}h_k(1)+2p^2h_k(2)+4p^2h_k(1,1)\right)\\
=&-\frac{2}{p}H(\tilde a+\tilde b)-\frac{4}{p}H(\tilde a,\tilde b)+2H(\tilde a+\tilde b+1)\\
&+4H(\tilde a+\tilde b,1)+4H(\tilde a+1,\tilde b)+8H(\tilde a,1,\tilde b)+8H(\tilde a,\tilde b,1)+8H(\tilde a,\tilde b+1)\\
&-4p\Big[H(\tilde a+\tilde b+1,1)+H(\tilde a+\tilde b+1,2)+2H(\tilde a+\tilde b+1,1,1)]\\
&-8p\Big[H(\tilde a+1,1,\tilde b)+H(\tilde a+1,\tilde b,1)+H(\tilde a+1,2b+2)\\
&+H(\tilde a,2,\tilde b)+H(\tilde a,\tilde b,2)+H(\tilde a,\tilde b+2)+2H(\tilde a,1,\tilde b+1)\\
&+2H(\tilde a,\tilde b+1,1)+2H(\tilde a,1,1,\tilde b)+2H(\tilde a,1,\tilde b,1)+2H(\tilde a,\tilde b,1,1)\Big].
\end{aligned}
$$

Taking $a=1$ and $b=0$ we get (here $H=H_{p-1}$)

$$
\begin{aligned}
H^\star_{p-1}(2,3)=&-\frac{2}{p}H(4)-\frac{4}{p}H(3,1)+16H(3,1,1)+8H(3,2)+8H(4,1)\\
&-4p\Big[H(5,1)+3H(4,2)+6H(4,1,1)+6H(3,2,1)\\
&+6H(3,1,2)+12H(3,\{1\}^3)\Big].
\end{aligned}
\tag{8.44}
$$

By stuffle relation and using Eq. (8.37) we see that

$$
2\zeta_{\mathscr{A}_2}(3,1,1)+\zeta_{\mathscr{A}_2}(3,2)+\zeta_{\mathscr{A}_2}(4,1)=\zeta_{\mathscr{A}_2}(3,1)\zeta_{\mathscr{A}_2}(1)-\zeta_{\mathscr{A}_2}(1,3,1)=0.
$$

On the other hand

$$
\begin{aligned}
\frac{1}{p}\Big(H(4)+2H(3,1)\Big)&=\frac{1}{p}\Big(H(1)H(3)+H(3,1)-H(1,3)\Big)\\
&\equiv-\Big(3H(4,1)+H(3,2)+3pH(4,2)+6pH(5,1)\Big)
\end{aligned}
$$

by Eq. (8.8). Thus Eq. (8.44) is reduced to

$$
\begin{aligned}
&H^\star_{p-1}(2,3)\equiv 6H(4,1)+2H(3,2)+6pH(4,2)+12pH(5,1)\\
&-4p\Big[H(5,1)+3H(4,2)+6H(4,1,1)+6H(3,2,1)+6H(3,1,2)+12H(3,\{1\}^3)\Big]\\
&\qquad\equiv 6H(4,1)+2H(3)H(2)-2H(2,3)\equiv 6H(4,1)-2H(2,3)
\end{aligned}
$$

by using the reversal relation and then looking up Table E.1 for weight six FMZV on page 540. This together with Eq. (8.19) implies Eq. (8.39). Further, Eq. (8.39) shows that

$$
\begin{aligned}
\zeta^\star_{\mathscr{A}_2}(2,2,1)&=\zeta_{\mathscr{A}_2}(2,2,1)+\zeta_{\mathscr{A}_2}(2,3)+\zeta_{\mathscr{A}_2}(4,1)+\zeta_{\mathscr{A}_2}(5)\\
&=\zeta_{\mathscr{A}_2}(2,2,1)+3\zeta_{\mathscr{A}_2}(4,1).
\end{aligned}
$$

Combining this with Thm. 8.5.4 and Eq. (8.39) we have

$$\begin{aligned}-6\beta_3^2 &= \zeta_{\mathscr{A}_2}(2)\zeta_{\mathscr{A}_2}(2,1) = 2\zeta_{\mathscr{A}_2}(2,2,1)+\zeta_{\mathscr{A}_2}(4,1)+\zeta_{\mathscr{A}_2}(2,3)+\zeta_{\mathscr{A}_2}(2,1,2)\\ &= 2\zeta^\star_{\mathscr{A}_2}(2,2,1)-3\zeta_{\mathscr{A}_2}(4,1)+\zeta_{\mathscr{A}_2}(2,1,2) = 2\zeta^\star_{\mathscr{A}_2}(2,2,1)-3\zeta_{\mathscr{A}_2}(4,1)-3\beta_3^2\end{aligned}$$

by Eq. (8.38). This clearly implies Eq. (8.40).

Finally, the second "=" of Eq. (8.41) is an easy consequence of Eq. (8.35) since $\zeta_{\mathscr{A}_2}(3)=\zeta_{\mathscr{A}_2}(1)=0$. Therefore

$$\begin{aligned}\zeta^\star_{\mathscr{A}_2}(3,1,1) &= \zeta_{\mathscr{A}_2}(3,1,1)+\zeta_{\mathscr{A}_2}(3,2)+\zeta_{\mathscr{A}_2}(4,1)+\zeta_{\mathscr{A}_2}(5)\\ &= -\zeta^\star_{\mathscr{A}_2}(3,1,1)-\zeta_{\mathscr{A}_2}(4,1)\end{aligned}$$

by Eq. (8.39). So the last "=" of Eq. (8.41) is verified. This finishes the proof of the theorem. □

Similarly we can obtain the following results in weight six. We leave the detailed computation to the interested reader and provide essentially all the other values in Table E.5.

Theorem 8.5.19. *In weight* 6 *we have*

$$\begin{aligned}\zeta^\star_{\mathscr{A}_2}(1,1,4) = \zeta^\star_{\mathscr{A}_2}(4,1,1) &= \zeta_{\mathscr{A}_2}(1,1,4) = \zeta_{\mathscr{A}_2}(4,1,1),\\ \zeta^\star_{\mathscr{A}_2}(1,4,1) = \zeta_{\mathscr{A}_2}(1,4,1) &= -2\zeta_{\mathscr{A}_2}(4,1,1)+6p\beta_7,\\ \zeta^\star_{\mathscr{A}_2}(2,3,1) = \zeta_{\mathscr{A}_2}(1,3,2) &= 3\zeta_{\mathscr{A}_2}(4,1,1)-\frac{65}{4}p\beta_7,\\ \zeta^\star_{\mathscr{A}_2}(1,2,3) = \zeta_{\mathscr{A}_2}(3,2,1) &= -2\zeta_{\mathscr{A}_2}(4,1,1)+\frac{17}{4}p\beta_7,\\ \zeta^\star_{\mathscr{A}_2}(2,1,3) = \zeta_{\mathscr{A}_2}(3,1,2) &= -\zeta_{\mathscr{A}_2}(4,1,1)+11p\beta_7,\\ \zeta^\star_{\mathscr{A}_2}(1,3,2) = \zeta_{\mathscr{A}_2}(2,3,1) &= 3\zeta_{\mathscr{A}_2}(4,1,1)-\frac{9}{4}p\beta_7,\\ \zeta^\star_{\mathscr{A}_2}(3,2,1) = \zeta_{\mathscr{A}_2}(1,2,3) &= -2\zeta_{\mathscr{A}_2}(4,1,1)-\frac{11}{4}p\beta_7,\\ \zeta^\star_{\mathscr{A}_2}(3,1,2) = \zeta_{\mathscr{A}_2}(2,1,3) &= -\zeta_{\mathscr{A}_2}(4,1,1)+18p\beta_7.\end{aligned}$$

Corollary 8.5.20. *The FMZV space of weight* 6 *superbity* 2 *is given by*

$$\mathsf{FMZ}_{6,2} = \langle \zeta_{\mathscr{A}_2}(4,1,1), p\beta_7\rangle.$$

Proof. By Thm. 8.5.4 and Thm. 8.5.17 all depth 1 and 2 values are $\mathbb{Q}$ multiples of $p\beta_7$. All depth 3 values are presented in Thm. 8.5.19. By the duality of Eq. (8.33) and Table E.1 for weight 7 values (which are all $\mathbb{Q}$ multiples of β_7) we see that all values of larger depths are also linear combinations of $\zeta_{\mathscr{A}_2}(4,1,1)$ and $p\beta_7$. This completes the proof of the corollary. □

Using Maple one can compute FMZV of depth 2 up to weight 9 which are listed in the next theorem. For essentially complete tables of values of weight up to 6 (inclusive), see Tables 8.1, Tables E.4 and Tables E.5.

Theorem 8.5.21. *Setting* $\zeta_2 = \zeta_{\mathcal{A}_2}$ *we have*

$$\begin{aligned}
&\mathsf{FMZ}_{1,2} = \langle 0 \rangle, \quad \mathsf{FMZ}_{2,2} = \langle \zeta_2(2) \rangle, \quad \mathsf{FMZ}_{3,2} = \big\langle \zeta_2(1,2) \big\rangle,\\
&\mathsf{FMZ}_{4,2} = \langle \zeta_2(2,2) \rangle, \quad \mathsf{FMZ}_{5,2} = \big\langle \zeta_2(2,3), \zeta_2(1,2,2) \big\rangle,\\
&\mathsf{FMZ}_{6,2} = \big\langle \zeta_2(2,2,2), \zeta_2(1,2,3) \big\rangle,\\
&\mathsf{FMZ}_{7,2} = \left\langle \begin{matrix} \zeta_2(\{1\}^5,2), \zeta_2(\{1\}^3,4), \zeta_2(1,6), \\ \boldsymbol{\zeta}_2(2,3,1,1), \boldsymbol{\zeta}_2(3,1,1,2), \boldsymbol{\zeta}_2(3,3,1) \end{matrix} \right\rangle,\\
&\mathsf{FMZ}_{8,2} = \left\langle \begin{matrix} \zeta_2(\{1\}^6,2), \zeta_2(\{1\}^4,4), \zeta_2(1,1,6), \zeta_2(1,2,5), \\ \boldsymbol{\zeta}_2(5,1,1,1) \end{matrix} \right\rangle,\\
&\mathsf{FMZ}_{9,2} = \left\langle \begin{matrix} \zeta_2(\{1\}^7,2), \zeta_2(\{1\}^5,4), \zeta_2(\{1\}^3,6), \zeta_2(1,8), \zeta_2(1,2,6), \\ \boldsymbol{\zeta}_2(1,3,2,3), \boldsymbol{\zeta}_2(2,1,3,3), \boldsymbol{\zeta}_2(2,3,1,3), \boldsymbol{\zeta}_2(1,2,5,1) \end{matrix} \right\rangle.
\end{aligned}$$

We expect that the bold-faced FMZVs are not really needed for generating the corresponding spaces because of the conjectured relations in Exercise 8.14. In general, we have the following dimension conjecture parallel to Conjecture 3.4.2 for MZVs. See Table 8.2 for numerical support.

Conjecture 8.5.22. *Let* w *be any positive integer. Set* $d_{0,2} = 1$ *and* $d_{w,2} = \dim \mathsf{FMZ}_{w,2}$. *Then* $d_{1,2} = 0, d_{2,2} = 1$ *and for all* $w \geq 3$ *we have*

$$d_{w,2} = d_{w-2,2} + d_{w-3,2}.$$

8.5.6 *FMZVs of Superbities* ≤ 5

First, let us look at some numerical data given in Table 8.2.

Note that at every superbity the conjectured recurrence relation is always $f_w = f_{w-2} + f_{w-3}$ in general. The only exceptions occur at superbities at least four but only for very small weights where we have "too few" values. For example, at superbity four and weight two the dimension is supposed to be 2, but because of the trivial relation

$$\zeta_{\mathcal{A}_4}(2) + 2\zeta_{\mathcal{A}_4}(1,1) = \zeta_{\mathcal{A}_2}(1)^2 = 0 \in \mathcal{A}_4$$

the dimension is decreased to only 1. For another example, at superbity $\ell = 5$ or 6 we have another trivial relation

$$6\zeta_{\mathcal{A}_\ell}(1,1,1) + \zeta_{\mathcal{A}_\ell}(1,2) + \zeta_{\mathcal{A}_\ell}(2,1) + \zeta_{\mathcal{A}_\ell}(3) = \zeta_{\mathcal{A}_\ell}(1)^3 = 0.$$

Table 8.2: Numerically verified dimension upper bounds of MZ_w and $\mathsf{FMZ}_{w,\ell}$ for superbity $\ell \le 5$. Bold-faced numbers are all one less than they're supposed to be by the conjectured recurrence formula.

w	0	1	2	3	4	5	6	7	8	9	10	11	12
$\dim \mathsf{MZ}_w$	1	0	1	1	1	2	2	3	4	5	7	9	12
$\dim \mathsf{FMZ}_{w,1}$	1	0	0	1	0	1	1	1	2	2	3	4	5
$\dim \mathsf{FMZ}_{w,2}$	1	0	1	1	1	2	2	3	4	5	7	9	12
$\dim \mathsf{FMZ}_{w,3}$	1	1	1	2	2	3	4	5	7	9	12	16	21
$\dim \mathsf{FMZ}_{w,4}$	2	1	**1**	3	3	5	6	8	11	14	19	25	
$\dim \mathsf{FMZ}_{w,5}$	2	**1**	**2**	**3**	5	7	9	12	16	21			

8.5.7 *FMZVs of Arbitrary Depths*

The next result is a kind of sum formula for FMZVs which is an analog of Thm. 6.2.2 for the SMZVs. For $w, d, i \in \mathbb{Z}$ with $1 \le i \le d \le w-1$, set

$$I_{w,d,i} = \{\mathbf{s} = (s_1, \dots, s_d) \in \mathbb{N}^d : |\mathbf{s}| = w,\ s_i \ge 2\}.$$

Theorem 8.5.23. *Fix $w, d, i \in \mathbb{Z}$ with $1 \le i \le d \le w-1$ and write $\mathbf{s} = (s_1, \dots, s_d)$. Then we have*

$$\sum_{\mathbf{s} \in I_{w,d,i}} \zeta_{\mathscr{A}_1}(\mathbf{s}) = (-1)^{i-1} \left(\binom{w-1}{i-1} + (-1)^d \binom{w-1}{d-i} \right) \beta_w, \tag{8.45}$$

$$\sum_{\mathbf{s} \in I_{w,d,i}} \zeta^{\star}_{\mathscr{A}}(\mathbf{s}) = (-1)^{i-1} \left((-1)^d \binom{w-1}{i-1} + \binom{w-1}{d-i} \right) \beta_w. \tag{8.46}$$

Proof. To save space we let $\mathscr{A} = \mathscr{A}_1$ and

$$S^{\star}_{w,d,i} = \sum_{\mathbf{s} \in I_{w,d,i}} \zeta^{\star}_{\mathscr{A}}(\mathbf{s})$$

throughout this proof. First, by the stuffle relations we get

$$\begin{aligned} 0 = \zeta^{\star}_{\mathscr{A}}(s_1, \dots, s_{d-1}) \zeta^{\star}_{\mathscr{A}}(l) &= \sum_{j=1}^{d} \zeta^{\star}_{\mathscr{A}}(s_1, \dots, s_{j-1}, l, s_j, \dots, s_{d-1}) \\ &\quad - \sum_{j=1}^{d-1} \zeta^{\star}_{\mathscr{A}}(s_1, \dots, s_{j-1}, s_j + l, s_{j+1}, \dots, s_{d-1}). \end{aligned}$$

Adding the above equation when $(s_1, \dots, s_{d-1}, l)$ ranges over $I_{w,d,i}$ we get

$$(d-i) S^{\star}_{w,d,i} + i S^{\star}_{w,d,i+1} - (k-d) S^{\star}_{w,d-1,i} = 0. \tag{8.47}$$

We now prove Eq. (8.46) by backward induction on d. When $d = w-1$ by the v-duality Eq. (8.31), we have

$$S^\star_{w,w-1,i} = \zeta^\star_{\mathscr{A}}(\{1\}^{i-1}, 2, \{1\}^{w-i-1}) = -\zeta^\star_{\mathscr{A}}(i, w-i) = (-1)^{i-1}\binom{w}{i}\beta_w$$

by Eqs. (8.49) and (8.54). Now assume that Eq. (8.46) holds for d. Then by Eq. (8.47)

$$\begin{aligned}
&(w-d)S^\star_{w,d-1,i} = (d-i)S^\star_{w,d,i} + iS^\star_{w,d,i+1}\\
&= (d-i)(-1)^{i-1}\left((-1)^d\binom{w-1}{i-1} + \binom{w-1}{d-i}\right)\beta_w\\
&\quad + i(-1)^i\left((-1)^d\binom{w-1}{i} + \binom{w-1}{d-i-1}\right)\beta_w\\
&= (-1)^{i-1}\left((d-i)(-1)^d\binom{w-1}{i-1} + (w-d+i)\binom{w-1}{d-i-1}\right)\beta_w\\
&\quad + (-1)^i\left((w-i)(-1)^d\binom{w-1}{i-1} + i\binom{w-1}{d-i-1}\right)\beta_w\\
&= (w-d)(-1)^{i-1}\left((-1)^{d-1}\binom{w-1}{i-1} + \binom{w-1}{d-i-1}\right)\beta_w.
\end{aligned}$$

Therefore Eq. (8.46) holds for $d-1$. This completes the proof of Eq. (8.46). We leave the proof of Eq. (8.45) to the interested reader since the idea is basically the same. □

8.6 Finite Euler Sums (FESs)

We now turn to the study of the FESs. Most of the results for FMZVs generalize to these values although some, such as the duality relation, does not.

8.6.1 *Two Reduction Formulas*

We now prove two reduction formulas of $H(\mathbf{s})$ for arbitrary composition $\mathbf{s}$ of signed numbers, corresponding to the two cases where $\mathbf{s}$ ends with a positive or a negative signed number.

Theorem 8.6.1. *Let $a, d \in \mathbb{N}$, $\mathbf{s} = (s_1, \dots, s_d) \in \mathbb{D}^d$ and $\mathbf{s}' = (s_1, \dots, s_{d-1})$. For any prime $p \geq a+2$, write $H(-) = H_{p-1}(-)$. Then*

$$\begin{aligned}
H(\mathbf{s}, a) \equiv& -\frac{1}{a}H(\mathbf{s}', (a-1)\oplus s_d)\\
&+ \sum_{k=1}^{p-1-a}\binom{p-a}{k}\frac{B_k}{p-a}H(\mathbf{s}', (k+a-1)\oplus s_d) \pmod p. \quad (8.48)
\end{aligned}$$

Proof. By the definition and Fermat's Little Theorem

$$\begin{aligned}
H(\mathbf{s},a) &\equiv \sum_{p>j_1>\cdots>j_d} \frac{\operatorname{sgn}(s_1)^{j_1}\cdots\operatorname{sgn}(s_d)^{j_d}}{j_1^{|s_1|}\cdots j_d^{|s_d|}} \sum_{j=1}^{j_d-1} j^{p-1-a} \\
&\equiv \sum_{k=0}^{p-1-a} \binom{p-a}{k} \frac{B_k}{p-a} \sum_{p>j_1>\cdots>j_d\geq 1} \frac{\operatorname{sgn}(s_1)^{j_1}\cdots\operatorname{sgn}(s_d)^{j_d}}{j_1^{|s_1|}\cdots j_d^{|s_d|}} j_d^{p-a-k}
\end{aligned}$$

which is exactly the right-hand side of Eq. (8.48). □

Theorem 8.6.2. *Let $a,d \in \mathbb{N}$, $\mathbf{s} = (s_1,\ldots,s_d) \in \mathbb{D}^d$ and $\mathbf{s}' = (s_1,\ldots,s_{d-1})$. For any prime $p \geq a+2$, write $H(-) = H_{p-1}(-)$. Then*

$$\begin{aligned}
H(\mathbf{s},\overline{a}) \equiv &\frac{(1-2^{p-a})B_{p-a}}{p-a}\Big(H(\mathbf{s}) - H(\mathbf{s}',\overline{s_d})\Big) \\
&- \sum_{k=0}^{p-2-a} \binom{p-1-a}{k} \frac{(1-2^{k+1})B_{k+1}}{k+1} H\big(\mathbf{s}',(k+a)\oplus\overline{s_d}\big) \pmod p,
\end{aligned}$$

where $\overline{s_d} = t$ if $s_d = \overline{t} \in \overline{\mathbb{N}}$.

Proof. By the definition and Lemma 8.2.5

$$\begin{aligned}
H(\mathbf{s},a) &\equiv \sum_{p>j_1>\cdots>j_d} \frac{\operatorname{sgn}(s_1)^{j_1}\cdots\operatorname{sgn}(s_d)^{j_d}}{j_1^{|s_1|}\cdots j_d^{|s_d|}} \sum_{j=1}^{j_d-1} (-1)^j j^{p-1-a} \\
&\equiv \sum_{k=0}^{p-2-a} \binom{p-1-a}{k} \frac{E_k(0)}{2} \sum_{p>j_1>\cdots>j_d} \frac{\operatorname{sgn}(s_1)^{j_1}\cdots\operatorname{sgn}(s_{d-1})^{j_{d-1}}\operatorname{sgn}(s_d)^{j_d}}{(-1)^{j_d-1} j_1^{|s_1|}\cdots j_{d-1}^{|s_{d-1}|} j_d^{|s_d|+k+a}} \\
&\quad + \frac{E_{p-1-a}(0)}{2} \sum_{p>j_1>\cdots>j_d} \frac{\operatorname{sgn}(s_1)^{j_1}\cdots\operatorname{sgn}(s_d)^{j_d}}{j_1^{|s_1|}\cdots j_d^{|s_d|}} \Big(1-(-1)^{j_d}\Big) \\
&\equiv \frac{E_{a'}(0)}{2}\Big(H(\mathbf{s}) - H(\mathbf{s}',\overline{s_d})\Big) - \sum_{k=0}^{a'-1} \binom{a'}{k} \frac{E_k(0)}{2} H\Big(\mathbf{s}',(k+a)\oplus\overline{s_d}\Big),
\end{aligned}$$

where $a' = p-1-a$. The theorem follows from Eq. (8.11) easily. □

8.6.2 *FESs of Small Depths*

In this section we study FESs in depth 2 and 3. All but one case are given by very concise values involving Bernoulli numbers or Euler numbers (which are themselves related by the identity Eq. (8.11)).

Lemma 8.6.3. *Let $a, b \in \mathbb{N}$ and prime $p \geq w+3$. Let $w = a+b$. Then*

$$H_{p-1}(\bar{a}, \bar{b}) \equiv \begin{cases} (-1)^a \frac{2^{p-w}-1}{w}\binom{w}{a} B_{p-w} \pmod{p}, & \text{if } w \text{ is odd;} \\ \frac{(2^a-2)(2^b-2)}{2^{w-1}ab} B_{p-a}B_{p-b} \pmod{p}, & \text{if } w \text{ is even.} \end{cases}$$

Proof. Set $H = H_{p-1}$. By the change of variable $k_j \to p - k_j$ for odd indices k_j ($j = 1, 2$), we have

$$\begin{aligned} H(\bar{a}, \bar{b}) &+ H(a, b) \\ &\equiv 2 \sum_{p>k_1>k_2>0, 2|k_1, 2|k_2} \frac{1}{k_1^a k_2^b} + 2 \sum_{p>k_1>k_2>0, 2\nmid k_1, 2\nmid k_2} \frac{1}{k_1^a k_2^b} \pmod{p} \\ &\equiv \frac{2}{2^w}\Big(H_{\frac{p-1}{2}}(a, b) + (-1)^w H_{\frac{p-1}{2}}(b, a)\Big) \pmod{p}. \end{aligned}$$

Similarly,

$$H(a, b) \equiv H_{\frac{p-1}{2}}(a, b) + (-1)^w H_{\frac{p-1}{2}}(b, a) + (-1)^b H_{\frac{p-1}{2}}(a) H_{\frac{p-1}{2}}(b) \pmod{p}.$$

Thus

$$H_{\frac{p-1}{2}}(a, b) + (-1)^w H_{\frac{p-1}{2}}(b, a) \equiv H(a, b) - (-1)^b H_{\frac{p-1}{2}}(a) H_{\frac{p-1}{2}}(b) \pmod{p}.$$

By combining the above, we get

$$H(\bar{a}, \bar{b}) \equiv (2^{1-w} - 1) H(a, b) - \frac{(-1)^b}{2^{w-1}} H_{\frac{p-1}{2}}(a) H_{\frac{p-1}{2}}(b) \pmod{p}.$$

Observe that

$$H(a) + H(\bar{a}) \equiv \frac{2}{2^w} H_{\frac{p-1}{2}}(a) \pmod{p}.$$

The lemma follows from Thms. 8.2.3 and 8.2.7 immediately. □

Theorem 8.6.4. *Let $a, b \in \mathbb{N}$ and a prime $p \geq a+b+2$. If the weight $w = a+b$ is odd then we have*

$$\zeta^{\star}_{\mathscr{A}_1}(a, b) = \zeta_{\mathscr{A}_1}(a, b) = (-1)^a \binom{w}{a} \beta_w, \tag{8.49}$$

$$\zeta_{\mathscr{A}_1}(\bar{a}, \bar{b}) = (-1)^a (2^{p-w} - 1) \binom{w}{a} \beta_w, \tag{8.50}$$

$$\zeta^{\star}_{\mathscr{A}_1}(\bar{a}, \bar{b}) = (2^{1-w} - 1) \left(2 + (-1)^a \binom{w}{a}\right) \beta_w, \tag{8.51}$$

$$\zeta_{\mathscr{A}_1}(\bar{a}, b) = \zeta_{\mathscr{A}_1}(a, \bar{b}) = (1 - 2^{1-w}) \beta_w, \tag{8.52}$$

$$\zeta^{\star}_{\mathscr{A}_1}(\bar{a}, b) = \zeta^{\star}_{\mathscr{A}_1}(a, \bar{b}) = (2^{1-w} - 1) \beta_w. \tag{8.53}$$

If $a+b$ is even, then we have

$$\zeta^\star_{\mathscr{A}_1}(a,b)=\zeta_{\mathscr{A}_1}(a,b)=0, \tag{8.54}$$

$$\zeta^\star_{\mathscr{A}_1}(\bar{a},\bar{b})=\zeta_{\mathscr{A}_1}(\bar{a},\bar{b})=\frac{(2^a-2)(2^b-2)}{2^{w-1}}\beta_a\beta_b. \tag{8.55}$$

Moreover,

$$\zeta_{\mathscr{A}_2}(\overline{2a},\overline{2b})=\left(\frac{(a-b)(1-2^{-2w})}{2w+2}\binom{2w+2}{2a+1}-w\right)p\beta_{2w+1}. \tag{8.56}$$

Proof. Equations (8.49) and (8.54) follow from Thm. 8.5.12. Equations (8.50) and (8.55) follow from Lemma 8.6.3. To prove (8.52), by stuffle relation we see that for $(\alpha,\beta)=(\bar{a},b)$ or $(a,\bar{b})$, and odd w, we have

$$\begin{aligned}0=\zeta_{\mathscr{A}_1}(\alpha)\zeta_{\mathscr{A}_1}(\beta)&=\zeta_{\mathscr{A}_1}(\alpha,\beta)+\zeta_{\mathscr{A}_1}(\beta,\alpha)+\zeta_{\mathscr{A}_1}(\alpha\oplus\beta)\\&=\zeta_{\mathscr{A}_1}(\alpha,\beta)+\operatorname{sgn}(\alpha\beta)(-1)^w\zeta_{\mathscr{A}_1}(\alpha,\beta)+\zeta_{\mathscr{A}_1}(\alpha\oplus\beta)\\&=2\zeta_{\mathscr{A}_1}(\alpha,\beta)+\zeta_{\mathscr{A}_1}(\alpha\oplus\beta).\end{aligned}$$

Thus

$$\zeta_{\mathscr{A}_1}(a,b)=\zeta_{\mathscr{A}_1}(a,\bar{b})=-\frac{1}{2}\zeta_{\mathscr{A}_1}(\overline{w})=(1-2^{1-w})\beta_w$$

from Thm. 8.2.7. Finally, all $\zeta^\star_{\mathscr{A}_1}$ identities are easy consequence of those of $\zeta_{\mathscr{A}_1}$ by the relation $\zeta^\star_{\mathscr{A}_1}(\alpha,\beta)=\zeta_{\mathscr{A}_1}(\alpha,\beta)+\zeta_{\mathscr{A}_1}(\alpha\oplus\beta)$.

We now turn to Eq. (8.56). By superbity two reversal relation of Thm. 8.2.4

$$\zeta_{\mathscr{A}_2}(\overline{2a},\overline{2b})=\zeta_{\mathscr{A}_2}(\overline{2b},\overline{2a})+2bp\zeta_{\mathscr{A}_1}(\overline{2b+1},\overline{2a})+2ap\zeta_{\mathscr{A}_1}(\overline{2b},\overline{2a+1}).$$

So by Eq. (8.51)

$$\zeta_{\mathscr{A}_2}(\overline{2a},\overline{2b})=\zeta_{\mathscr{A}_2}(\overline{2b},\overline{2a})+\frac{2(a-b)(1-2^{-2w})}{2w+2}\binom{2w+2}{2a+1}p\beta_{2w+1}. \tag{8.57}$$

On the other hand, we have

$$\zeta_{\mathscr{A}_2}(\overline{2a},\overline{2b})+\zeta_{\mathscr{A}_2}(\overline{2b},\overline{2a})=\zeta_{\mathscr{A}_2}(\overline{2a})\zeta_{\mathscr{A}_2}(\overline{2b})-\zeta_{\mathscr{A}_2}(2w)$$

and therefore by Eqs. (8.15) and (8.29),

$$\zeta_{\mathscr{A}_2}(\overline{2a},\overline{2b})+\zeta_{\mathscr{A}_2}(\overline{2b},\overline{2a})=-\zeta_{\mathscr{A}_2}(2w)=-2pw\beta_{2w+1}. \tag{8.58}$$

Equation (8.56) now follows easily from Eqs. (8.57) and (8.58). □

Turning to FES of depth 3, we first observe that for any $\alpha, \beta, \gamma \in \mathbb{Z}$

$$\begin{aligned} H(\alpha,\beta,\gamma) =& H(\alpha)H(\beta)H(\gamma) - H(\gamma)H(\beta,\alpha) - H(\gamma)H(\beta\oplus\alpha) \\ & - H(\gamma,\beta)H(\alpha) - H(\gamma\oplus\beta)H(\alpha) + H(\gamma,\beta,\alpha) \\ & + H(\gamma\oplus\beta,\alpha) + H(\gamma,\beta\oplus\alpha) + H(\gamma\oplus\beta\oplus\alpha). \end{aligned}$$

This can be easily checked by the stuffle relations but the idea is hidden in the general framework set up by Hoffman [293]. Combining with the reversal relations one can obtain the following results without much difficulty. We leave its proof to the interested reader.

Theorem 8.6.5. *Let $a, b, c \in \mathbb{N}$. If the weight $w = a + b + c$ is even then*

$$2\zeta_{\mathscr{A}_1}(a,\bar{b},c) = \zeta_{\mathscr{A}_1}(\overline{b+c},a) + \zeta_{\mathscr{A}_1}(c,\overline{a+b}), \tag{8.59}$$

$$2\zeta_{\mathscr{A}_1}(\bar{a},b,c) = -\zeta_{\mathscr{A}_1}(\bar{a})\zeta_{\mathscr{A}_1}(c,b) + \zeta_{\mathscr{A}_1}(c,\overline{a+b}) + \zeta_{\mathscr{A}_1}(b+c,\bar{a}), \tag{8.60}$$

$$\begin{aligned} 2\zeta_{\mathscr{A}_1}(\bar{a},\bar{b},\bar{c}) =& -\zeta_{\mathscr{A}_1}(\bar{c})\zeta_{\mathscr{A}_1}(\bar{b},\bar{a}) - \zeta_{\mathscr{A}_1}(\bar{c},\bar{b})\zeta_{\mathscr{A}_1}(\bar{a}) \\ & + \zeta_{\mathscr{A}_1}(b+c,\bar{a}) + \zeta_{\mathscr{A}_1}(\bar{c},a+b). \end{aligned} \tag{8.61}$$

If the weight w is odd then

$$2\zeta_{\mathscr{A}_1}(\bar{a},\bar{b},c) = \zeta_{\mathscr{A}_1}(c,a+b) + \zeta_{\mathscr{A}_1}(\overline{b+c},\bar{a}) - \zeta_{\mathscr{A}_1}(\bar{a})\zeta_{\mathscr{A}_1}(c,\bar{b}), \tag{8.62}$$

$$\begin{aligned} 2\zeta_{\mathscr{A}_1}(\bar{a},b,\bar{c}) =& -\zeta_{\mathscr{A}_1}(\bar{c})\zeta_{\mathscr{A}_1}(b,\bar{a}) - \zeta_{\mathscr{A}_1}(\bar{c},b)\zeta_{\mathscr{A}_1}(\bar{a}) \\ & + \zeta_{\mathscr{A}_1}(\overline{a+b+c}) + \zeta_{\mathscr{A}_1}(\bar{c},\overline{a+b}). \end{aligned} \tag{8.63}$$

Because of the reversal relations when the weight is even there remains essentially only one more case to consider in depth 3. This is given by the next result.

Theorem 8.6.6. *Let a, b, c be positive integers such that $w := a + b + c$ is even. Then for any prime $p \geq w + 3$ we have*

$$\begin{aligned} H(c,\bar{b},\bar{a}) \equiv & -\sum_{k=2}^{p-w+1} \binom{p-a}{p-w-k+1}\binom{k+c-1}{c}\frac{(1-2^k)B_k B_{p-w-k+1}}{ak} \\ & - \sum_{k=p+1-b-c}^{p-c} \binom{p-a}{2p-w-k}\binom{k+c-1}{c}\frac{(1-2^k)B_k B_{2p-w-k}}{ak} \\ & - \frac{(1-2^{p-c})(1-2^{p-a-b})B_{p-a-b}B_{p-c}}{(a+b)c}. \end{aligned}$$

Proof. The proof is basically a repeated application of Thm. 8.6.1. But we spell out all the details below because there are some subtle details that we need to attend to.

By Eq. (8.2) and Fermat's Little Theorem we have, modulo p,

$$\begin{aligned}H(c,\bar{b},\bar{a}) &\equiv \sum_{i=1}^{p-1} \frac{(-1)^i}{i^c} \sum_{j=1}^{i-1} (-1)^j j^{p-1-b} \sum_{k=1}^{j-1} k^{p-1-a} \\ &\equiv \sum_{i=1}^{p-1} \frac{(-1)^i}{i^c} \sum_{k=0}^{p-1-a} \binom{p-a}{k} \frac{B_k}{p-a} \sum_{j=1}^{i-1} (-1)^j j^{\rho(k)+p-a-b-k},\end{aligned}$$

where $\rho(k)=0$ if $k<p-a-b$ and $\rho(k)=p-1$ if $k\ge p-a-b$ (to make sure all exponents are positive in the sum of the second line above). By Lemma 8.2.5

$$\begin{aligned}H(c,\bar{b},\bar{a}) &\equiv \sum_{k=0}^{p-1-a} \binom{p-a}{k} \frac{B_k}{p-a} \sum_{r=0}^{n} \binom{n}{r} \sum_{i=1}^{p-1} (-1)^i i^{n-r-c} F_{n,i,r} \\ &\equiv - \sum_{k=0}^{p-1-a} \binom{p-a}{k} \sum_{r=1}^{n-1} \binom{n}{r} \frac{(1-2^{r+1})B_k B_{r+1}}{(p-a)(r+1)} \sum_{i=1}^{p-1} i^{n-c-r} \\ &\quad + \sum_{k=0}^{p-1-a} \binom{p-a}{k} \frac{(1-2^{n+1})B_k B_{n+1}}{(p-a)(n+1)} \sum_{i=1}^{p-1} ((-1)^i - 1) i^{-c},\end{aligned}$$

where $n=\rho(k)+p-a-b-k$. Here we have used the fact that when $r=0$ we have $F_{n,i,r}=(-1)^{i-1}$ and thus the inner sum is $\sum_{i=1}^{p-1} i^{n-c}\equiv 0 \pmod p$ except when $p-1|n-c$, *i.e.*, except when $n=c$ and $k=p-w$. But then $B_k=0$ since w is even by assumption. Thus

$$\begin{aligned}H(c,\bar{b},\bar{a}) \equiv &\sum_{\substack{0\le k<p-a\\ c<n}} \binom{p-a}{k} \binom{n}{n-c} \frac{(1-2^{n-c+1})B_k B_{n-c+1}}{(p-a)(n-c+1)} \\ &+ H(\bar{c}) \sum_{k=0}^{p-1-a} \binom{p-a}{k} \frac{(1-2^{n+1})B_k B_{n+1}}{(p-a)(n+1)}.\end{aligned}$$

Now if c is even then $H(\bar{c})\equiv 0 \pmod p$. So we may assume c is odd in the last line above. Then $k+n+1$ is always odd so that $B_k B_{n+1}\neq 0$ if and only if $k=1$ and $n=p-a-b-1$. Hence

$$\begin{aligned}H(c,\bar{b},\bar{a}) \equiv &\sum_{k=0}^{p-w-1} \binom{p-a}{k} \binom{p-a-b-k}{p-w-k} \frac{(1-2^{p-w-k+1})B_k B_{p-w-k+1}}{(p-a)(p-w-k+1)} \\ &+ \sum_{k=p-a-b}^{p-1-a} \binom{p-a}{k} \binom{2p-1-a-b-k}{2p-1-w-k} \frac{(1-2^{2p-w-k})B_k B_{2p-w-k}}{(p-a)(2p-w-k)} \\ &- \frac{(1-2^{p-c})(1-2^{p-a-b})B_{p-a-b}B_{p-c}}{(a+b)c}.\end{aligned}$$

After substitutions $k \to p-w+1-k$ in the first sum and $k \to 2p-w-k$ in the second sum the theorem follows immediately from Thm. 8.2.7, □

Remark 8.6.7. The condition $p \geq w+3$ in Thm. 8.6.6 cannot be weakened since

$$H(1,\bar{2},\bar{3}) \equiv \text{RHS}+5 \not\equiv \text{RHS} \pmod 7.$$

8.6.3 *FESs of Superbity 1 and 2*

According to the Conjecture 14.2.3 the dimensions of the Euler sums (see Sec. 14.2) are given by the Fibonacci numbers. From what we have found so far we get the following table. Let ES be the $\mathbb{Q}$-space spanned by of the Euler sums.

Table 8.3: Conjectural dimensions of ES_w and $\mathsf{FES}_{w,\ell}$ for $\ell \leq 2$.

w	0	1	2	3	4	5	6	7
$\dim \mathsf{ES}_w$	1	1	2	3	5	8	13	21
$\dim \mathsf{FES}_{w,1}$	1	1	1	2	3	5	8	13
$\dim \mathsf{FES}_{w,2}$	1	1	2	4	7	12	20	33

In Table 8.3, we list some dimensions that we have verified numerically. We also provide $\dim \mathsf{ES}_w$ for comparison purpose, which have been proved by Deligne [161] (see also [240]) under the assumption of a variant Grothendieck period conjecture. Without assuming the conjecture he proved these numbers provide the upper bounds.

We conclude this section by a conjecture analogous to Conjecture 8.5.15. For $\mathbf{s}=(s_1,\dots,s_d)\in\mathbb{N}^d$, define the symmetrized version of the alternating Euler sums

$$\zeta_*^{\mathcal{S}}(\mathbf{s}) := \sum_{i=0}^{d} \left(\prod_{j=1}^{i} (-1)^{s_j} \operatorname{sgn}(s_j) \right) \zeta_*(s_i,\dots,s_1)\zeta_*(s_{i+1},\dots,s_d),$$

$$\zeta_{\sqcup\!\sqcup}^{\mathcal{S}}(\mathbf{s}) := \sum_{i=0}^{d} \left(\prod_{j=1}^{i} (-1)^{s_j} \operatorname{sgn}(s_j) \right) \zeta_{\sqcup\!\sqcup}(\mathbf{p}(s_i,\dots,s_1))\zeta_{\sqcup\!\sqcup}(\mathbf{p}(s_{i+1},\dots,s_d)),$$

where, similar to the MZV case,

$$\zeta_\sharp(s_1,\dots,s_d) := (-1)^d \zeta_\sharp(z_{|s_1|,\operatorname{sgn}(s_1)} \cdots z_{|s_d|,\operatorname{sgn}(s_d)};T)$$

($\sharp = *$ or $\sqcup\!\sqcup$) are regularized values given by Prop. 13.3.8, and

$$\mathbf{p}(s_1,\dots,s_r)=(t_1,\dots,t_r)\in\mathbb{D}^r,\quad |t_j|=|s_j|,\quad \operatorname{sgn}(t_j)=\prod_{k=1}^{j}\operatorname{sgn}(s_k).$$

The sign in the above is necessary since we have the reversal relation for the FES given by Thm. 8.2.4.

Proposition 8.6.8. *Let $d\in\mathbb{N}$ and $\mathbf{s}\in\mathbb{N}^d$. Then*

(i) $\zeta^{\mathcal{S}}_{\sqcup\!\sqcup}(\mathbf{s})$ *and* $\zeta^{\mathcal{S}}_{*}(\mathbf{s})$ *are all finite, and*
(ii) $\zeta^{\mathcal{S}}_{\sqcup\!\sqcup}(\mathbf{s})-\zeta^{\mathcal{S}}_{*}(\mathbf{s})\in\zeta(2)\mathsf{ES}_{w-2}$ *for all* $|\mathbf{s}|=w$.

Proof. Part (i) and (ii) can be proved by the same ideas used in the proof of Prop. 6.1.3 and Thm. 6.3.2, respectively. See Exercise 8.7. □

Conjecture 8.6.9. *There is an isomorphism*

$$\begin{aligned} f:\mathsf{FES}_{w,1} &\longrightarrow \frac{\mathsf{ES}_w}{\zeta(2)\mathsf{ES}_{w-2}},\\ \zeta_{\mathscr{A}_1}(\mathbf{s}) &\longmapsto \zeta^{\mathcal{S}}_{\sqcup\!\sqcup}(\mathbf{s}). \end{aligned}$$

This conjecture has been checked up to weight 5 assuming Conjecture 8.1.5 which concerns the linear independence of the $\mathscr{A}_1$-Bernoulli numbers and $\mathscr{A}_1$-Fermat quotients.

8.6.4 *Two Applications of MHS Binomial Identities*

In this section we derive some equations in $\mathscr{A}_1$ using the MHS binomial identities from Chap. 7. The first result is an application of the Two-three formula (see Remark 7.5.3).

Theorem 8.6.10. *Let a,b be nonnegative integers. Then*

$$\zeta^{\star}_{\mathscr{A}_1}(\{2\}^a,3,\{2\}^b)=\frac{2(b-a)}{a+1}\binom{2a+2b+3}{2a+1}\beta_{2a+2b+3},\tag{8.64}$$

$$\zeta_{\mathscr{A}_1}(\{2\}^a,3,\{2\}^b)=\frac{2(-1)^{a+b}(a-b)}{a+1}\binom{2a+2b+3}{2a+1}\beta_{2a+2b+3}.\tag{8.65}$$

Proof. By Thm. 7.5.2 we have

$$\begin{aligned}\zeta^{\star}_{\mathscr{A}_1}(\{2\}^a,3,\{2\}^b)=\Bigg(&2\sum_{k=1}^{p-1}\frac{(-1)^{k-1}\binom{p-1}{k}}{k^{2a+2b+3}\binom{p-1+k}{k}}\\ &-4\sum_{k=1}^{p-1}\frac{H_{k-1}(2b+1)(-1)^k\binom{p-1}{k}}{k^{2a+2}\binom{p-1+k}{k}}\Bigg)_p.\end{aligned}$$

Since

$$\frac{\binom{p-1}{k}}{\binom{p-1+k}{k}} = \frac{(p-1)(p-2)\cdots(p-k)}{p(p+1)\cdots(p+k-1)}$$
$$= \frac{(-1)^k k}{p}\prod_{j=1}^{k}\left(1-\frac{p}{j}\right)\prod_{j=1}^{k-1}\left(1+\frac{p}{j}\right)^{-1}$$
$$\equiv \frac{(-1)^k k}{p}(1-pH_k(1))(1-pH_{k-1}(1))$$
$$\equiv \frac{(-1)^k k}{p}\left(1-2pH_{k-1}(1)-\frac{p}{k}\right) \pmod{p}, \tag{8.66}$$

we obtain (by setting $\tilde{a}=2a+1, \tilde{b}=2b+1$)

$$\zeta^\star_{\mathscr{A}_1}(\{2\}^a,3,\{2\}^b)$$
$$= \left(-\frac{2}{p}\sum_{k=1}^{p-1}\left(\frac{1}{k^{\tilde{a}+\tilde{b}}}+\frac{2H_{k-1}(\tilde{b})}{k^{\tilde{a}}}\right)\left(1-2pH_{k-1}(1)-\frac{p}{k}\right)\right)_p$$
$$= -\frac{2}{p}\zeta_{\mathscr{A}_2}(2a+2b+2)-\frac{4}{p}\zeta_{\mathscr{A}_2}(\tilde{a},\tilde{b})+4\zeta_{\mathscr{A}_1}(\tilde{a}+\tilde{b},1)+4\zeta_{\mathscr{A}_1}(\tilde{a}+1,\tilde{b})$$
$$+8\zeta_{\mathscr{A}_1}(\tilde{a},1,\tilde{b})+8\zeta_{\mathscr{A}_1}(\tilde{a},\tilde{b},1)+8\zeta_{\mathscr{A}_1}(\tilde{a},\tilde{b}+1).$$

Here we have applied the stuffle relation

$$H_{k-1}(\tilde{b})H_{k-1}(1) = H_{k-1}(1,\tilde{b})+H_{k-1}(\tilde{b},1)+H_{k-1}(\tilde{b}+1).$$

Now Eq. (8.64) follows from Thms. 8.5.4, 8.5.13, 8.5.17 easily. To prove Eq. (8.65), we take $\mathbf{s} = (\{2\}^a,3,\{2\}^b)$ in the concatenation relation Eq. (8.18) then one of $\mathbf{s}_j = (\{2\}^\ell)$ for some positive ℓ and therefore $\zeta_{\mathscr{A}_1}(\mathbf{s}_j)=0$. Thus

$$(-1)^{a+b}\zeta^\star_{\mathscr{A}_1}(\{2\}^a,3,\{2\}^b) = \zeta_{\mathscr{A}_1}(\{2\}^b,3,\{2\}^a). \tag{8.67}$$

On the other hand, by reversal relation Eq. (8.5)

$$\zeta_{\mathscr{A}_1}(\{2\}^a,3,\{2\}^b) = -\zeta_{\mathscr{A}_1}(\{2\}^b,3,\{2\}^a). \tag{8.68}$$

Now from Eqs. (8.67) and (8.68) we get the required congruence and the proof is complete. □

The second result is an application of the Two-one formula.

Theorem 8.6.11. *Let $a,b\in\mathbb{N}_0$ and $w=2a+2b+1$. Then*

$$\zeta^\star_{\mathscr{A}_1}(\{2\}^a,1,\{2\}^b) = \frac{4(b-a)(1-2^{1-w})}{w+1}\binom{w+1}{2a+1}\beta_w, \tag{8.69}$$
$$\zeta_{\mathscr{A}_1}(\{2\}^a,1,\{2\}^b) = \frac{4(a-b)(1-2^{1-w})}{(-1)^{a+b}(w+1)}\binom{w+1}{2a+1}\beta_w. \tag{8.70}$$

Proof. Equation (8.70) follows easily from the Eq. (8.69) by using the concatenation relation Eq. (8.18). For the latter, we begin by considering the case $a \geq 1$. By Eq. (7.25) with $n = p-1$ and congruence Eq. (8.66) we have

$$\zeta^\star_{\mathscr{A}_1}(\{2\}^a, 1, \{2\}^b) = \left(-\frac{2}{p} \sum_{k=1}^{p-1} \frac{1}{k^{2a+2b}} \left(1 - 2pH_{k-1}(1) - \frac{p}{k}\right) \right)_p$$
$$- \frac{4}{p} \sum_{k=1}^{p-1} \frac{H_{k-1}(\overline{2b})(-1)^k}{k^{2a}} \left(1 - 2pH_{k-1}(1) - \frac{p}{k}\right).$$

Applying the stuffle relation

$$H_{k-1}(1)H_{k-1}(\overline{2b}) = H_{k-1}(1, \overline{2b}) + H_{k-1}(\overline{2b}, 1) + H_{k-1}(\overline{2b+1}),$$

we get

$$\zeta^\star_{\mathscr{A}_1}(\{2\}^a, 1, \{2\}^b) = -\frac{2}{p}\zeta_{\mathscr{A}_2}(2a+2b) - \frac{4}{p}\zeta_{\mathscr{A}_2}(\overline{2a}, \overline{2b}) + 4\zeta_{\mathscr{A}_1}(2a+2b, 1)$$
$$+ 4\zeta_{\mathscr{A}_1}(\overline{2a+1}, \overline{2b}) + 8\zeta_{\mathscr{A}_1}(\overline{2a}, 1, \overline{2b}) + 8\zeta_{\mathscr{A}_1}(\overline{2a}, \overline{2b}, 1) + 8\zeta_{\mathscr{A}_1}(\overline{2a}, \overline{2b+1}).$$

Now we can prove the case $a \geq 1$ by applying Eqs. (8.29), (8.34), (8.51), (8.56), (8.62), and (8.63).

The case $a = 0$ can be proved similarly (see Exercise 8.3). This finishes the proof of the theorem. □

Corollary 8.6.12. *Let p be a prime greater than 7. Then*

$$\zeta^\star_{\mathscr{A}_1}(2,2,2,1) = -\frac{189}{16}\beta_7, \quad \zeta^\star_{\mathscr{A}_1}(2,2,2,2,1) = -\frac{255}{16}\beta_9.$$

Proof. Both are consequences of Thm. 8.6.11 with $(a,b) = (3,0)$ and $(a,b) = (4,0)$, respectively. □

8.6.5 *FESs of Arbitrary Depths*

The signed version of Thm. 8.5.1, *i.e.*, the case for the homogeneous FESs, is considerably more involved, as can be seen already in Thm. 8.2.7 for depth 1. To state it, we denote by Odd(d) the subset of odd partitions $\Lambda = (\lambda_1, \dots, \lambda_r)$ of d (*i.e.*, $\sum_{j=1}^r \lambda_j = d$ and $\lambda_j > 0$ are odd for all j) and define

$$c(\Pi) = \prod_{j=1}^r (\lambda_j - 1)! \quad \text{and} \quad \tilde{c}(\Pi) = (-1)^{d-r} \prod_{j=1}^r (\lambda_j - 1)!.$$

Theorem 8.6.13. *Let $s,d\in\mathbb{N}$. For an odd partition $\Lambda=(\lambda_1,\dots,\lambda_r)\in \mathrm{Odd}(d)$, we put $\zeta_{\mathscr{A}_1}(\bar{s};\Lambda)=\prod_{i=1}^r\zeta_{\mathscr{A}_1}(\oplus_{j=1}^{\lambda_i}\bar{s})$ (notice $\oplus_{j=1}^{\lambda}\bar{s}$ equals $s\lambda$ for even λ and $\overline{s\lambda}$ for odd λ). Then*

$$\begin{aligned}\zeta_{\mathscr{A}_1}(\{\bar{s}\}^d)&=\frac{1}{d!}\sum_{\Lambda\in\mathrm{Odd}(d)}c(\Pi)\,\zeta_{\mathscr{A}_1}(\bar{s};\Lambda),\\ \zeta^\star_{\mathscr{A}_1}(\{\bar{s}\}^d)&=\frac{1}{d!}\sum_{\Lambda\in\mathrm{Odd}(d)}\tilde{c}(\Pi)\,\zeta_{\mathscr{A}_1}(\bar{s};\Lambda).\end{aligned}\tag{8.71}$$

In particular, if s is even then both sums in Eq. (8.71) are 0 in $\mathscr{A}_1$. If $s>1$ is odd then both sums in Eq. (8.71) are $\mathbb{Q}$-linear combinations of $\beta_{\lambda_1 s}\cdots\beta_{\lambda_r s}$ for odd partitions $\Lambda=(\lambda_1,\dots,\lambda_r)$. If $s=1$ then Eq. (8.71) is a $\mathbb{Q}$-linear combination of $\mathfrak{q}_2^\ell\beta_{\lambda_{\ell+1}}\cdots\beta_{\lambda_r}$ for odd partitions $\lambda=(\{1\}^\ell,\lambda_{\ell+1},\dots,\lambda_r)$ ($\lambda_j>1$ for all $j>\ell$), where $\mathfrak{q}_2^\ell=((2^{p-1}-1)/p)_{p\in\mathscr{P}}$ is an $\mathscr{A}_1$-Fermat quotient.

Proof. Apply Thm. 7.4.1 and then use Thm. 8.2.7. All are straightforward except in the case $s=1$ we need the famous Claussen–von Staudt Theorem which says $p\beta_1=-1$ in $\mathscr{A}_1$ (see [314, p. 233]). □

For example, by [538, Thm. 2.1] and [527, Thm. 5.2] we have

$$\zeta_{\mathscr{A}_3}(\bar{1})=-\zeta_{\mathscr{A}_3}(1)+H_{\frac{p-1}{2}}(1)=-2\mathfrak{q}_2+p\mathfrak{q}_2^2-\frac{2}{3}p^2\mathfrak{q}_2^3-\frac{3}{4}p^2\beta_3.\tag{8.72}$$

Observe that $\mathrm{Odd}(2)=\{(1,1)\},\mathrm{Odd}(3)=\{(1,1,1),(3)\},\mathrm{Odd}(4)=\{(1,1,1,1),(1,3)\}$. It is obvious that $c_{(1,\cdots,1)}=1$, $c_{(3)}=2$, $c_{(1,3)}=8$, $c_{(1,1,3)}=20$, and $c_{(1,1,1,3)}=c_{(3,3)}=40$, which implies that

$$\zeta_{\mathscr{A}_1}(\{\bar{1}\}^2)=2\mathfrak{q}_2^2,\quad \zeta_{\mathscr{A}_1}(\{\bar{1}\}^3)=-\frac{4}{3}\mathfrak{q}_2^3-\frac{1}{2}\beta_3,\quad \zeta_{\mathscr{A}_1}(\{\bar{1}\}^4)=\frac{2}{3}\mathfrak{q}_2^4+\mathfrak{q}_2\beta_3,$$
$$\zeta_{\mathscr{A}_1}(\{\bar{1}\}^5)=-\frac{4}{15}\mathfrak{q}_2^5-\mathfrak{q}_2^2\beta_3,\quad \zeta_{\mathscr{A}_1}(\{\bar{1}\}^6)=\frac{4}{45}\mathfrak{q}_2^6+\frac{2}{3}\mathfrak{q}_2^3\beta_3+\frac{1}{8}\beta_3^2.$$

The expression in the examples above are unique if we assume Conjecture 8.1.5.

8.7 Historical Notes

The following identity is one of the series related to π that Ramanujan listed in [480]:

$$\sum_{k=0}^{\infty}(4k+1)\binom{-1/2}{k}^3=\frac{2}{\pi}.$$

This was actually discovered earlier by Bauer in 1859. In his seminal paper [557] of 1997 van Hamme studied the $(p-1)$-st partial sums of over a dozen such series for odd primes p. Among the many conjectures he proposed one can find the following super congruence:

$$\sum_{k=0}^{(p-1)/2} (4k+1)\binom{-1/2}{k}^3 \equiv \left(\frac{-1}{p}\right) p \pmod{p^3}.$$

This was later settled by Mortenson [428] using the Gamma function and the hypergeometric functions. Since then this kind of congruences have been said of *van Hamme type*.

The Wolstenholme's Theorem was improved by Glaisher [239] to the super congruence

$$\sum_{k=1}^{p-1} \frac{1}{k} \equiv -\frac{1}{3}p^2 B_{p-3} \pmod{p^3} \text{ for all prime } p > 3, \tag{8.73}$$

and by Z.-W. Sun [527] as follows:

$$\sum_{k=1}^{p-1} \frac{1}{k} \equiv p^2 \left(\frac{2B_{p-3}}{p-3} - \frac{B_{2p-4}}{2p-4}\right) \pmod{p^4} \text{ for all prime } p > 5. \tag{8.74}$$

Ever since Wolstenholme published his work on the congruences Eq. (8.1) a plethora of generalizations are given along various approaches, the majority of which involve variations with composite moduli or changes of the summation range, see, for example, the work by Leudesdorf [388], Glaisher [238, 239], Vandiver [558], Bayat [43], Pan [465] and so on. For a more complete history, please refer to the survey article [426] by Mestrovic. Virtually none of these work, however, treated multiple variable cases. In an unpublished work R. Bruck seems to be the first who studied the congruence properties of the homogeneous multiple harmonic sums.

Generalizing Wolstenholme's Theorem to the more general MHSs (*i.e.*, the partial sums of the MZVs) was the initial motivation of [609, 619] by the author starting around 2002. For example, Thm. 8.5.12, 8.5.13, and 8.5.17 were proved in [619] as Thm. 3.1, 3.5, and 3.2, respectively. A little later, Hoffman independently studied these sums by his powerful algebraic setup using word algebras. In order to find a suitable theoretical framework in which this kind of results can be organized and further investigated, the author first defined in [625] an adèle-like structure essentially the same as Eq. (8.3) which was later defined, when the superbity is 1, by Kaneko and Zagier who called it poorman's adèle.

The first systematic investigation of congruences involving the AMHSs was carried out in the joint work [539]. We have recasted these sums as the finite Euler sums in this chapter due to their close relations to the classical Euler sums. Further generalization of these to some finite version of multiple polylogs can be found in [495].

Note that [578] contains one more statement which is equivalent to the two congruences in Wolstenholme's Thm. 8.2.1:

$$\binom{2p-1}{p-1} \equiv 1 \pmod{p^3}, \quad i.e., \quad \binom{2p}{p} \equiv 2 \pmod{p^3}.$$

Wolstenholme's Theorem was revisited by Gardiner [234] who also showed that the three statements are equivalent. In [508] Sekine extended some of the author's results of [609, 619] to the partial sums truncated at p^2 instead of p. A further study of congruence properties of MHS with varying summation ranges was carried out in [627] based on the earlier works of Boyd [78] and Eswarathasan and Levine [209].

Theorem 8.2.7 can also be obtained from [538, Thm. 2.1] combined with [527, Thm. 5.2]. Note that both terms in [538, Thm. 2.1] contribute nontrivially when s is even since the modulus is p^2. In this case, the theorem is also given by [123, (6.2)]. For the FMZVs, as an improvement of Thm. 8.5.1 Zhou and Cai [637] obtained the super congruences in Thm. 8.5.4.

Neither FMZV dualities given by Thm. 8.5.10 (φ-dual) and Thm. 8.5.11 (v-dual) is the same as that of the MZVs. These were discovered by Hoffman first in [292, 294]. Definition 8.5.6 comes from Hoffman's modification of an idea of Vermaseren described in [560, Appendix B]. Our proof of Lemma 8.5.8 here follows that of [292, Thm. 4.2] while Thm. 8.5.11 is [619, Thm. 2.11]. J. Rosen [486] extended this result to an asymptotic type formula with weights associated with the powers of p. This hints at a connection to the p-adic MZVs subsequently studied by Jarossay [317–319]. In particular, the shuffle relation of the SMZVs stated in Thm. 6.3.4 corresponds to the shuffle relation satisfied by the FMZVs. These are given precisely by Thm. 8.4.3, which also includes the case for FESs.

In the spring of 2013 after the author gave an Ober-Seminar talk at the Max Planck Institute for Mathematics at Bonn, Prof. D. Zagier mentioned that he had come to the dimension part of Conjecture 8.5.14 some time earlier [602]. He also informed the author of Conjecture 8.5.15 which was to appear in a joint paper with Kaneko [336]. Note that Thm. 6.4.19, proved by Yasuda [585], shows that the map in Conjecture 8.5.15 is surjective. In order to find the upper bounds of the FMZV and FES dimensions, it is essential to determine the correct analog of the DBSFs. This was carried out in [633] which is incorporated into Sections 8.4-8.6. A statement of the shuffle relations of the FMZVs similar to that of Thm. 8.4.3 first appeared in the Japanese survey paper [330] by Kaneko.

Theorem 8.5.23, an FMZV analog of the sum formula for the SMZVs (see Thm. 6.2.2) was proved by Saito and Wakabayashi in [492, 493], who also found the results in Exercise 8.10 which are the sum formulas for FMZVs in the same spirit as that of Conjecture 5.10.2 (for MZVs) or that of Thm. 10.9.3 (for MZ*Vs). The FMZV analog of the Ohno's relation of the MZVs (see Thm. 5.5.9) can be found in Exercise 8.11, which was conjectured by Kaneko and proved by Oyama [464]. In [312] Imatomi *et al.* expressed the FMZVs as the multi-poly-Bernoulli numbers (see Eq. (5.117) for the definition):

$$\zeta_{\mathscr{A}_1}(s_1, s_2, \dots, s_d) = \Big(- C_{p-2}(s_1 - 1, s_2, \dots, s_d) \pmod p \Big)_{p\in\mathscr{P}}.$$

Theorems 8.6.11 and 8.6.10 were proved by Hessami Pilehroods and Tauraso in [278], which led to the improvement of the upper bounds for the dimensions of $\mathsf{FMZ}_{7,1}$ and $\mathsf{FMZ}_{9,1}$ (see [278, Cor. 4.4]). This yields that $\dim \mathsf{FMZ}_{7,1} \le 1$ and $\dim \mathsf{FMZ}_{9,1} \le 2$ (see Table 8.2).

Exercises

8.1. For every positive integer ℓ, prove that the following procedure provides an embedding $\iota_\ell : \mathbb{R} \hookrightarrow \mathscr{A}_\ell$ as sets. If $r \in \mathbb{Q}$ then $\iota_\ell(r)$ is defined as in Eq. (8.4). Assume $r = n.d_1d_2\cdots$ is irrational where n is its integral part and $d_1, d_2, \dots$ are its decimal digits. Let $r_k = n.d_1d_2\dots d_k$ be the kth approximation of r. We define $\iota_\ell(r) = \big(\iota_{p_j,\ell}(r)\big)_{j\ge1}$ by the following.

$$\iota_{p_j,\ell}(r) = \iota_{p_j,\ell}(r_k) \text{ for all } j = k(k-1)+1, k(k-1)+2, \cdots, k(k+1).$$

Roughly speaking, this means that for each k we put the kth approximation r_k of r at $2k$ consecutive components.

8.2. Prove Lemma 8.5.2 using Eq. (8.2) and Fermat's little Theorem:

$$k^{p^2(p-1)} \equiv 1 \pmod{p^3}$$

for all k not divisible by p. You also need the fact that pB_{2m} is p-integral for all $m \in \mathbb{N}$ by von Staudt–Clausen Theorem [314, p. 233].

8.3. Prove the case $a = 0$ of Eq. (8.69) using Eq. (7.24), Thm. 8.5.1, and Thm. 8.6.4.

8.4. Prove Eq. (8.46) using the same idea as in the proof of Eq. (8.45).

8.5. Prove Eq. (8.46) by backward induction on d by proving the following key identity first:

$$S_{w,d,1} = \sum_{j=0}^{d-1}(-1)^j \binom{w-d+j-1}{j} S^{\star}_{w,d-j,1} = \left(1+(-1)^d\binom{w-1}{d-1}\right)\beta_w.$$

8.6. Compute

$$\mathrm{tog}(\mathtt{z}_{\bar{4}}\mathtt{z}_2\mathtt{z}_1\mathtt{z}_{\bar{3}}\mathtt{z}_2\mathtt{z}_{\bar{2}}\mathtt{z}_{\bar{1}}\mathtt{z}_7\mathtt{z}_{\bar{6}}) \quad \text{and} \quad \mathrm{tog}^{-1}(\mathtt{b}_1\mathtt{a}^2\mathtt{b}_1\mathtt{a}^3\mathtt{b}_0\mathtt{b}_1\mathtt{a}^4\mathtt{b}_0\mathtt{a}^4\mathtt{b}_1\mathtt{a}\mathtt{b}_0).$$

8.7. Prove Prop. 8.6.8 by mimicking the proof of Prop. 6.1.3 and Thm. 6.3.2. You will need regularized Euler sums in two ways from Chap. 13 given by Prop. 13.3.8 and apply their relation given by Thm. 13.3.31.

8.8. Let $a, b \in \mathbb{N}_0$, $\tilde{a} = 2a+1$ and $\tilde{b} = 2b+3$. Then show that

$$\begin{aligned}\zeta^{\star}_{\mathscr{A}_1}(\{2\}^a, 3, \{2\}^b, 3) =& 2\zeta_{\mathscr{A}_1}(\tilde{a}, 1, \tilde{b}+1) - 6\zeta_{\mathscr{A}_1}(\tilde{b}+1, \tilde{a}, 1)\\ &+ (8a-6)\zeta_{\mathscr{A}_1}(\tilde{a}+1, \tilde{b}, 1) + 16\zeta_{\mathscr{A}_1}(\tilde{a}+\tilde{b}, 1, 1)\\ &+ (8b+14)\zeta_{\mathscr{A}_1}(\tilde{a}, \tilde{b}+1, 1) - 16\zeta_{\mathscr{A}_1}(1, \tilde{a}, \tilde{b}, 1)\\ &- 2\zeta_{\mathscr{A}_1}(\tilde{a}+1, 1, \tilde{b}) - 2\zeta_{\mathscr{A}_1}(\tilde{b}, \tilde{a}+1, 1) + 4\zeta_{\mathscr{A}_1}(\tilde{a}, \tilde{b}, 2).\end{aligned}$$

8.9. Verify that $\dim \mathsf{ES}_2 \le 2$ and $\dim \mathsf{ES}_3 \le 3$ by using the linear DBSFs and reversal relations for FESs. Are DBSFs enough?

8.10. Suppose a and b are odd positive integers and c is an even positive integer. Show that for all nonnegative integers m and n with $(m, n) \neq (0, 0)$, we have

$$\sum_{\substack{\sum_{i=0}^{2m} n_i = n\\ n_0,\dots,n_{2m}\ge 0}} \zeta_{\mathscr{A}_1}(\{c\}^{n_0}, a, \{c\}^{n_1}, b, \{c\}^{n_2}, \dots, a, \{c\}^{n_{2m-1}}, b, \{c\}^{n_{2m}}) = 0,$$

$$\sum_{\substack{\sum_{i=0}^{2m} n_i = n\\ n_0,\dots,n_{2m}\ge 0}} \zeta^{\star}_{\mathscr{A}_1}(\{c\}^{n_0}, a, \{c\}^{n_1}, b, \{c\}^{n_2}, \dots, a, \{c\}^{n_{2m-1}}, b, \{c\}^{n_{2m}}) = 0.$$

8.11. Let $\mathbf{s} = (s_1, \dots, s_d) \in \mathbb{N}^d$ and suppose its v-dual $\mathbf{s}^{\vee}$ has depth d'. Then for any $n \in \mathbb{N}_0$, we have

$$\sum_{\mathbf{e}\in(\mathbb{N}_0)^d, |\mathbf{e}|=n} \zeta_{\mathscr{A}}(\mathbf{s}+\mathbf{e}) = \sum_{\mathbf{e}'\in(\mathbb{N}_0)^{d'}, |\mathbf{e}'|=n} \zeta_{\mathscr{A}}\big((\mathbf{s}^{\vee}+\mathbf{e}')^{\vee}\big).$$

You may want to read Sec. D.3 first before trying the following problems.

8.12. With your favorite computer algebra system, verify the following relations which give some evidence for Conjecture 8.6.9:

$$\begin{aligned}\zeta_{\mathscr{A}_1}(\{\bar{1}\}^3) &= 4\zeta_{\mathscr{A}_1}(\bar{2},1) + 4\zeta_{\mathscr{A}_1}(\bar{1},1,1),\\ \zeta_{\mathscr{A}_1}(\{\bar{1}\}^4) &= 4\zeta_{\mathscr{A}_1}(\bar{1},1,2) + 4\zeta_{\mathscr{A}_1}(\bar{1},2,1) + 8\zeta_{\mathscr{A}_1}(\bar{1},\{1\}^3),\end{aligned}$$

and

$$\begin{aligned}\zeta_*^{\mathcal{S}}(\{\bar{1}\}^3) &\equiv 4\zeta_*^{\mathcal{S}}(\bar{2},1) + 4\zeta_*^{\mathcal{S}}(\bar{1},1,1) \pmod{\zeta(2)},\\ \zeta_*^{\mathcal{S}}(\{\bar{1}\}^4) &\equiv 4\zeta_*^{\mathcal{S}}(\bar{1},1,2) + 4\zeta_*^{\mathcal{S}}(\bar{1},2,1) + 8\zeta_*^{\mathcal{S}}(\bar{1},\{1\}^3) \pmod{\zeta(2)}.\end{aligned}$$

Note that the verification can be done at two levels, numerically using congruences, or rigorously using the linear DBSFs and reversal relations for FESs (DBSFs are not enough!).

8.13. With your favorite computer algebra system, verify numerically that $\zeta_{\mathscr{A}_1}(\mathbf{s}) = 0$ if

(1) $\mathbf{s} = \big(\{\{1\}^m, 2, \{1\}^n, 2\}^q, \{1\}^m, 2, \{1\}^n\big)$ for $q, m, n \geq 0$, where either (i) q is odd, or (ii) q is even and $m+n$ is even.

(2) $\mathbf{s} = \big(\{2\}^m, \{3, \{2\}^m\}^n\big)$ for $m, n \geq 0$.

Can you prove these rigorously?

8.14. With your favorite computer algebra system, verify numerically the following identities in $\mathscr{A}_2$: (setting $\zeta_2 = \zeta_{\mathscr{A}_2}$)

$$\begin{aligned}
68\zeta_2(2,3,1,1) &= -480\zeta_2(\{1\}^5,2) + 716\zeta_2(\{1\}^3,4) - 843\zeta_2(1,6)\\
34\zeta_2(3,1,1,2) &= -585\zeta_2(\{1\}^5,2) + 998\zeta_2(\{1\}^3,4) - 1029\zeta_2(1,6)\\
68\zeta_2(3,3,1) &= 1730\zeta_2(\{1\}^5,2) - 2480\zeta_2(\{1\}^3,4) + 2319\zeta_2(1,6)\\
12\zeta_2(5,1,1,1) &= -73\zeta_2(\{1\}^6,2) + 12\zeta_2(1,1,6) - 3\zeta_2(1,2,5)\\
84\zeta_2(3,2,1,3) &= -995\zeta_2(1,8) + 952\zeta_2(\{1\}^3,6) - 1288\zeta_2(\{1\}^5,4)\\
&\quad - 437\zeta_2(\{1\}^7,2) - 624\zeta_2(1,2,6)\\
924\zeta_2(2,1,3,3) &= 2509\zeta_2(1,8) - 6356\zeta_2(\{1\}^3,6) + 5432\zeta_2(\{1\}^5,4)\\
&\quad - 180\zeta_2(\{1\}^7,2) + 801\zeta_2(1,2,6)\\
924\zeta_2(2,3,1,3) &= -9424\zeta_2(1,8) - 5824\zeta_2(\{1\}^3,6) + 11368\zeta_2(\{1\}^5,4)\\
&\quad - 3807\zeta_2(\{1\}^7,2) + 852\zeta_2(1,2,6)\\
36\zeta_2(1,2,5,1) &= 358\zeta_2(1,8) - 248\zeta_2(\{1\}^3,6) + 464\zeta_2(\{1\}^5,4)\\
&\quad + 5\zeta_2(\{1\}^7,2) + 147\zeta_2(1,2,6).
\end{aligned}$$

Can you prove any of these rigorously?

8.15. With your favorite computer algebra system, verify numerically the following identities in $\mathscr{A}_2$: (setting $\zeta_2 = \zeta_{\mathscr{A}_2}$)

$$\begin{aligned} 4\zeta_2(1,\bar{1}) &= -3\zeta_2(2) - 4\zeta_2(\bar{1},1), \\ 8\zeta_2(\bar{1},\bar{1}) &= 9\zeta_2(2) + 16\zeta_2(\bar{1},1), \\ 4\zeta_2(\bar{2}) &= 3\zeta_2(2), \\ 2\zeta_2(1,1) &= -\zeta_2(2). \end{aligned}$$

Can you prove all of these rigorously? These would imply that $\dim \mathsf{ES}_{2,2} \le 2$. Similarly, verify numerically

$$\begin{aligned} \zeta_2(1,2) &= 2\zeta_2(\bar{3}), \\ 2\zeta_2(1,\bar{1},1) &= \zeta_2(\bar{3}) - 2\zeta_2(\bar{1},2) - 4\zeta_2(\bar{1},1,1), \\ 4\zeta_2(\bar{1},\bar{1},1) &= 3\zeta_2(\bar{3}) + 6\zeta_2(\bar{1},2) + 12\zeta_2(\bar{1},1,1), \\ 2\zeta_2(\bar{2},1) &= -\zeta_2(\bar{3}), \\ \zeta_2(2,1) &= -2\zeta_2(\bar{3}), \\ 2\zeta_2(1,\bar{2}) &= -\zeta_2(\bar{3}), \\ \zeta_2(\bar{1},-2) &= \zeta_2(\bar{1},2) - \zeta_2(\bar{3}), \\ 16\zeta_2(1,\bar{1},\bar{1}) &= -17\zeta_2(\bar{3}) - 26\zeta_2(\bar{1},2) - 2\zeta_2(\bar{1},\bar{1},\bar{1}) - 40\zeta_2(\bar{1},1,1), \\ 4\zeta_2(\bar{2},\bar{1}) &= \zeta_2(\bar{3}) + 14\zeta_2(\bar{1},2) - 6\zeta_2(\bar{1},\bar{1},\bar{1}) + 24\zeta_2(\bar{1},1,1), \\ \zeta_2(1,1,\bar{1}) &= \zeta_2(\bar{3}) - 2\zeta_2(\bar{1},2) + \zeta_2(\bar{1},\bar{1},\bar{1}) - 3\zeta_2(\bar{1},1,1), \\ 16\zeta_2(\bar{1},1,\bar{1}) &= 17\zeta_2(\bar{3}) - 70\zeta_2(\bar{1},2) + 26\zeta_2(\bar{1},\bar{1},\bar{1}) - 104\zeta_2(\bar{1},1,1), \\ \zeta_2(2,\bar{1}) &= 5\zeta_2(\bar{1},2) - 2\zeta_2(\bar{3}) - 2\zeta_2(\bar{1},\bar{1},\bar{1}) + 8\zeta_2(\bar{1},1,1). \end{aligned}$$

These would imply that $\dim \mathsf{ES}_{3,2} \le 4$.

Chapter 9

q-Analogs of Multiple Harmonic (Star) Sums

We have seen from the preceding two chapters that the MHSs and MH*Ss not only have a lot of nice properties but also are closely intertwined with the MZVs and the MZ*Vs. In this chapter, we move on to study their q-analogs which again will have important applications in a number of results concerning the MZVs and the MZ*Vs.

9.1 Definitions and Notation

Throughout this chapter let m and n denote nonnegative integers and q a real number with $0 < q < 1$. For any real number a, we have the quantum factorial and quantum Pochhammer symbol defined by:

$$(a)_0 := (a;q)_0 := 1, \qquad (a)_n := (a;q)_n := \prod_{k=0}^{n-1}(1-aq^k), \quad n \geq 1.$$

As a convention throughout the chapter, we always use [ ] to represent q-analog objects. For example, the q-analog of a positive integer n is defined as

$$[n] = [n]_q := \sum_{k=0}^{n-1} q^k = \frac{1-q^n}{1-q},$$

and the Gaussian q-binomial coefficient

$$\begin{bmatrix} n \\ m \end{bmatrix} = \begin{bmatrix} n \\ m \end{bmatrix}_q := \begin{cases} \dfrac{(q)_n}{(q)_m (q)_{n-m}}, & \text{if } 0 \leq m \leq n, \\ 0, & \text{otherwise.} \end{cases}$$

Recall that the double cover $\mathbb{D}_0$ of $\mathbb{N}_0$ is defined in Definition 7.1.1. For $m \in \mathbb{N}_0$, $\mathbf{t} = (t_1, \ldots, t_m) \in \mathbb{Z}^m$ and $\mathbf{s} = (s_1, \ldots, s_m) \in \mathbb{D}_0^m$, we set $\mathbf{s} = \emptyset$ if

$m = 0$ and define the q-analogs of multiple harmonic (star) sums (q-MHS or q-MH*S) by

$$H_n^{\mathbf{t}}[\mathbf{s}] := \sum_{n\geq k_1>\cdots>k_m\geq 1} \prod_{j=1}^{m} \frac{\mathrm{sgn}(s_j)^{k_j} q^{k_j t_j}}{[k_j]^{|s_j|}},$$

$$H_n^{\star,\mathbf{t}}[\mathbf{s}] := \sum_{n\geq k_1\geq\cdots\geq k_m\geq 1} \prod_{j=1}^{m} \frac{\mathrm{sgn}(s_j)^{k_j} q^{k_j t_j}}{[k_j]^{|s_j|}},$$

with the convention that $H_n^{\mathbf{t}}[\mathbf{s}] = 0$ if $n < m$, and $H_n^{\star,\mathbf{t}}[\emptyset] = H_n^{\mathbf{t}}[\emptyset] = 1$ for all $n \geq 0$. Note that we allow s_j to be 0 or $\bar{0}$ in these sums. For the star-version, since we will consider only $H_n^{\star,\mathbf{1}}[\mathbf{s}]$ in this chapter, we will simply write $H_n^{\star}[\mathbf{s}] = H_n^{\star,\mathbf{1}}[\mathbf{s}]$, namely,

$$H_n^{\star}[\mathbf{s}] = \sum_{n\geq k_1\geq\cdots\geq k_m\geq 1} \prod_{j=1}^{m} \frac{\mathrm{sgn}(s_j)^{k_j} q^{k_j}}{[k_j]^{|s_j|}}.$$

Now we fix a symbol θ and for any $r \in \mathbb{Z} \cup \{\theta\}$ and $k \in \mathbb{N}$, define

$$Q(r,k) := \begin{cases} rk(k-1)/2, & \text{if } r > 0; \\ rk(k-1)/2 - k, & \text{if } r \leq 0; \\ 0, & \text{if } r = \theta. \end{cases}$$

Let $m \in \mathbb{N}_0$. For any $\mathbf{s} = (s_1, \ldots, s_m) \in \mathbb{D}_0^m$ and two *mollifiers* $\mathbf{t} = (t_1, \ldots, t_m) \in (\mathbb{N}_0)^m$ and $\mathbf{r} = (r_1, \ldots, r_m) \in (\mathbb{Z} \cup \{\theta\})^m$, we define the *mollified companion of* $H_n^{\mathbf{1}}[\mathbf{s}]$ by

$$\mathscr{H}_n[\mathbf{s};\mathbf{t};\mathbf{r}] := \sum_{n\geq k_1>\cdots>k_m\geq 1} \frac{\begin{bmatrix} n \\ k_1 \end{bmatrix}}{\begin{bmatrix} n+k_1 \\ k_1 \end{bmatrix}} \prod_{j=1}^{m} \frac{q^{t_j k_j + Q(r_j,k_j)}(1+q^{k_j})}{\mathrm{sgn}(s_j)^{k_j}[k_j]^{|s_j|}}, \tag{9.1}$$

where we set $\mathscr{H}_n[\mathbf{s};\mathbf{t};\mathbf{r}] = 1$ if $m = 0$. We call $(\mathbf{s};\mathbf{t};\mathbf{r})$ an *admissible triple of mollifiers* if the limit $\lim_{n\to\infty} \mathscr{H}_n[\mathbf{s};\mathbf{t};\mathbf{r}]$ exists.

Definition 9.1.1. Let $\boxplus$ be a non-commutative binary operation on $\mathbb{Z} \cup \{\theta\}$, called $\square$-*plus*, such that

- $\theta \boxplus a = a \boxplus \theta = a$ for all $a \in \mathbb{Z} \cup \{\theta\}$,
- $a \boxplus b = a + b$ for all $a, b \in \mathbb{Z}$ with $(a, b) \neq (1, -1), (-1, 1)$, and
- $1 \boxplus (-1) = \theta$ and $(-1) \boxplus 1 = 0$.

Lemma 9.1.2. *Let* $r \in \{\theta\} \cup \mathbb{Z} \setminus \{0\}$, $d \in \{0, -1\}$. *Then for any* $k \in \mathbb{Z}$

$$Q(r,k) + Q(1,k) = Q(r \boxplus 1, k), \tag{9.2}$$

$$k^2 + Q(d,k) = Q(2 \boxplus d, k). \tag{9.3}$$

Moreover, the projection

$$\begin{aligned}\pi : (\mathbb{Z} \cup \{\theta\}, \boxplus) &\longrightarrow (\mathbb{Z}, +)\\ a &\longmapsto a \quad \forall a \in \mathbb{Z},\\ \theta &\longmapsto 0,\end{aligned}$$

is a homomorphism of semi-groups and its restriction to $\mathbb{Z}^*$ *is injective.*

Proof. These follow from straightforward computation. □

For an admissible triple of mollifiers $(\mathbf{s}; \mathbf{t}; \mathbf{r})]$ of depth m, we define

$$\begin{aligned}&\Big\{s_1 \circ \cdots \circ s_m; t_1 \circ \cdots \circ t_m; r_1 \circ \cdots \circ r_m\Big\}\\ &\quad := \Big\{(\mathbf{p}; \widetilde{\mathbf{p}}; \widetilde{\widetilde{\mathbf{p}}}) \in \mathbb{D}^m \times \mathbb{Z}^m \times (\mathbb{Z} \cup \{\theta\})^m : (\mathbf{p}; \widetilde{\mathbf{p}}; \widetilde{\widetilde{\mathbf{p}}}) \preceq (\mathbf{s}; \mathbf{t}; \mathbf{r})\Big\}\end{aligned}$$

to be the set of triples of strings produced by replacing every $\circ$ in $\mathbf{s}$ by either comma "," or O-plus "$\oplus$" (see Eq. (7.1)), replacing every $\circ$ in $\mathbf{t}$ by either comma "," or the usual plus "+", and replacing every $\circ$ in $\mathbf{r}$ by either comma "," or □-plus "$\boxplus$". Moreover, the commas should be at the same positions for all the three compositions $\mathbf{s}$, $\mathbf{t}$ and $\mathbf{r}$. Finally, to save space, we define

$$a^{\circ k} = \underbrace{a \circ \ldots \circ a}_{k \text{ times}}, \quad \text{and} \quad \operatorname*{\mathbf{Cat}}_{j=1}^{m} \{a_j\} = (a_1, \ldots, a_m).$$

9.2 Some q-Combinatorial Identities

We start with a preliminary combinatorial lemma.

Lemma 9.2.1. *For integers* $n \geq 1$ *and* $l \geq 0$, *we have*

$$\sum_{k=l+1}^{n} (1+q^k) \frac{\begin{bmatrix} n \\ k \end{bmatrix}}{\begin{bmatrix} n+k \\ k \end{bmatrix}} (-1)^k q^{\binom{k}{2}} = \frac{[l]-[n]}{[n]} \frac{\begin{bmatrix} n \\ l \end{bmatrix}}{\begin{bmatrix} n+l \\ l \end{bmatrix}} (-1)^l q^{\binom{l}{2}}, \tag{9.4}$$

$$\sum_{k=l+1}^{n} (1+q^k)[k] \frac{\begin{bmatrix} n \\ k \end{bmatrix}}{\begin{bmatrix} n+k \\ k \end{bmatrix}} q^{k(k-1)} = ([n]-[l]) \frac{\begin{bmatrix} n \\ l \end{bmatrix}}{\begin{bmatrix} n+l \\ l \end{bmatrix}} q^{l^2}, \tag{9.5}$$

$$\sum_{k=1}^{n} \frac{1+q^k}{[k]} \frac{\begin{bmatrix} n \\ k \end{bmatrix}}{\begin{bmatrix} n+k \\ k \end{bmatrix}} q^{k^2} = \sum_{m=1}^{n} \frac{q^m}{[m]}. \tag{9.6}$$

Moreover, if $l \geq 1$ *then*

$$\sum_{k=l}^{n} \frac{q^k}{[k]^2} \frac{\begin{bmatrix} k \\ l \end{bmatrix}}{\begin{bmatrix} k+l \\ l \end{bmatrix}} = \frac{q^l}{[l]^2} \frac{\begin{bmatrix} n \\ l \end{bmatrix}}{\begin{bmatrix} n+l \\ l \end{bmatrix}}. \tag{9.7}$$

Proof. The general idea of the proof is to use WZ-method. First, for all positive integers k, l and n, we define

$$F_1(n,k) := (1+q^k)\,\frac{\begin{bmatrix} n \\ k \end{bmatrix}}{\begin{bmatrix} n+k \\ k \end{bmatrix}}\,(-1)^{k-1}q^{\binom{k}{2}}, \qquad G_1(n,k) := \frac{q^{n+k}-1}{q^k+1}\,F_1(n,k),$$

$$F_2(n,k) := \frac{\begin{bmatrix} n \\ k \end{bmatrix}}{\begin{bmatrix} n+k \\ k \end{bmatrix}}\,(1-q^{2k})q^{k(k-1)}, \qquad G_2(n,k) := \frac{1-q^{n+k}}{1-q^{2k}}\,F_2(n,k),$$

$$F_3(l,k) := \frac{\begin{bmatrix} l \\ k \end{bmatrix}}{\begin{bmatrix} l+k \\ k \end{bmatrix}}\,\frac{1+q^k}{1-q^k}\,q^{k^2}, \qquad G_3(l,k) := \frac{q^{l-k+1}}{1+q^k}\,\frac{1-q^k}{1-q^{l-k+1}}\,F_3(l,k),$$

$$F_4(l,k) := \frac{q^{k-l}}{[k]^2}\,\frac{\begin{bmatrix} l \\ k \end{bmatrix}}{\begin{bmatrix} l+k \\ k \end{bmatrix}}, \qquad G_4(l,k) := \frac{(1-q^{k-l})(1-q^{k+l})}{q^{k-l}}\,F_4(l,k).$$

Set the difference operator $\Delta_n f(n) := f(n+1) - f(n)$. Easy computation shows that

$$\begin{aligned} (1-q^n)F_1(n,k) &= \Delta_k G_1(n,k), & F_2(n,k) &= -\,\Delta_k G_2(n,k), \\ (1-q^l)^2 F_4(l,k) &= \Delta_k G_4(l,k), & \Delta_l F_3(l,k) &= -\,\Delta_k G_3(l,k). \end{aligned}$$

Therefore

$$\begin{aligned} (1-q^n)\sum_{k=l+1}^{n} F_1(n,k) &= \sum_{k=l+1}^{n} \Delta_k G_1(n,k) = -G_1(n,l+1) \\ &= \frac{1-q^{n+l+1}}{1+q^{l+1}}\,F_1(n,l+1) = (1-q^{n-l})\,\frac{\begin{bmatrix} n \\ l \end{bmatrix}}{\begin{bmatrix} n+l \\ l \end{bmatrix}}\,(-1)^l q^{l(l+1)/2}, \end{aligned}$$

$$\sum_{k=l+1}^{n} F_2(n,k) = G_2(n,l+1) = \frac{\begin{bmatrix} n \\ l \end{bmatrix}}{\begin{bmatrix} n+l \\ l \end{bmatrix}}\,(1-q^n)q^{l(l+1)},$$

$$\sum_{l=0}^{n-1} \Delta_k G_3(l,k) = -\sum_{l=0}^{n-1} \Delta_l F_3(l,k) = F_3(0,k) - F_3(n,k) = -F_3(n,k),$$

$$(1-q^l)^2 \sum_{k=l}^{n} F_4(l,k) = \sum_{k=l}^{n} \Delta_k G_4(l,k) = G_4(l,n+1) = (1-q)^2\frac{\begin{bmatrix} n \\ l \end{bmatrix}}{\begin{bmatrix} n+l \\ l \end{bmatrix}}.$$

So Eqs. (9.4), (9.5) and (9.7) are proved. Moreover,

$$\begin{aligned} \sum_{k=1}^{n} F_3(n,k) = -\sum_{m=0}^{n-1}\sum_{k=1}^{n} \Delta_k G_3(m,k) &= \sum_{m=0}^{n-1} \big(G_3(m,1) - G_3(m,n+1)\big) \\ &= \sum_{m=0}^{n-1} G_3(m,1) = \sum_{m=0}^{n-1} \frac{q^{m+1}}{1-q^{m+1}}, \end{aligned}$$

which implies Eq. (9.6). This completes the proof of the lemma. □

Lemma 9.2.2. *Let $a \in \mathbb{D}_0$, $b \in \mathbb{N}_0$, $c \in \mathbb{N}$, $r \in \{\theta\} \cup \mathbb{Z} \setminus \{0\}$, and $(\mathbf{x};\mathbf{y};\mathbf{z})$ an admissible triple of mollifiers. Then for any positive integer n,*

$$\frac{1}{[n]^c}\mathscr{H}_n[a,\mathbf{x};b,\mathbf{y};r \boxplus 1,\mathbf{z}] = \sum_{(\mathbf{p};\widetilde{\mathbf{p}};\widetilde{\widetilde{\mathbf{p}}})} \mathscr{H}_n[\mathbf{p},\mathbf{x};\widetilde{\mathbf{p}},\mathbf{y};\widetilde{\widetilde{\mathbf{p}}},\mathbf{z}].$$

where $(\mathbf{p};\widetilde{\mathbf{p}};\widetilde{\widetilde{\mathbf{p}}}) \preceq \big(\bar{0},\{1\}^{c-1},a\oplus\bar{1};\{0\}^c,b;\,1,\{\theta\}^{c-1},r\big)$ *in the sum.*

Proof. We prove the lemma by induction on c. Suppose all the mollifiers $\mathbf{x},\mathbf{y},\mathbf{z}$ have depth m. Set

$$A_{n,k} = (-1)^k(1+q^k)q^{k(k-1)/2}\frac{\begin{bmatrix} n \\ k\end{bmatrix}}{\begin{bmatrix} n+k \\ k\end{bmatrix}}. \tag{9.8}$$

Then by Eq. (9.4),

$$\frac{(1+q^l)}{[l]}\sum_{k=l+1}^{n} A_{n,k} = \left(\frac{1}{[n]}-\frac{1}{[l]}\right)A_{n,l} \tag{9.9}$$

which, together with Eq. (9.2) yields

$$\begin{aligned}
&\mathscr{H}_n[\bar{0},a\oplus\bar{1},\mathbf{x};0,b,\mathbf{y};1,r,\mathbf{z}]\\
&= \sum_{n\ge k>k_0>k_1>\cdots>k_m\ge 1} \frac{A_{n,k}q^{bk_0+Q(r,k_0)}(1+q^{k_0})}{(-\operatorname{sgn}(a))^{k_0}[k_0]^{|a|+1}}\prod_{j=1}^{m}\frac{q^{y_jk_j+Q(z_j,k_j)}(1+q^{k_j})}{\operatorname{sgn}(x_j)^{k_j}[k_j]^{|x_j|}}\\
&= \sum_{n\ge k_0>\cdots>k_m\ge 1} \frac{q^{bk_0+Q(r,k_0)}(1+q^{k_0})}{(-\operatorname{sgn}(a))^{k_0}[k_0]^{|a|+1}}\prod_{j=1}^{m}\frac{q^{y_jk_j+Q(z_j,k_j)}(1+q^{k_j})}{\operatorname{sgn}(x_j)^{k_j}[k_j]^{|x_j|}}\sum_{k=k_0+1}^{n}A_{n,k}\\
&= \frac{1}{[n]}\mathscr{H}_n[a,\mathbf{x};b,\mathbf{y};r\boxplus 1,\mathbf{z}] - \mathscr{H}_n[a\oplus 1,\mathbf{x};b,\mathbf{y};r\boxplus 1,\mathbf{z}].
\end{aligned}$$

This proves the lemma for $c=1$. Now suppose $c>1$. By the case $c=1$ we have just proved,

$$\begin{aligned}
\frac{1}{[n]^c}\mathscr{H}_n[a,\mathbf{x};b,\mathbf{y};r\boxplus 1,\mathbf{z}] &= \frac{1}{[n]^{c-1}}\left(\frac{1}{[n]}\mathscr{H}_n[a,\mathbf{x};b,\mathbf{y};r\boxplus 1,\mathbf{z}]\right)\\
&= \frac{1}{[n]^{c-1}}\mathscr{H}_n[a\oplus 1,\mathbf{x};b,\mathbf{y};r\boxplus 1,\mathbf{z}] + \frac{1}{[n]^{c-1}}\mathscr{H}_n[\bar{0},a\oplus\bar{1},\mathbf{x};0,b,\mathbf{y};1,r,\mathbf{z}].
\end{aligned}$$

For the first summand, we now apply induction assumption using case $c-1$ with a replaced by $a\oplus 1$. For the second summand, we apply case $c-1$ of Lemma 9.2.2 with $a=\bar{0}$, $b=0$, and $r=\theta$. Hence the above is equal to

$$\sum_{(\mathbf{p};\widetilde{\mathbf{p}};\widetilde{\widetilde{\mathbf{p}}})\preceq\left(\bar{0},\{1\}^{c-2},(a\oplus 1)\oplus\bar{1};\{0\}^{c-1},b;1,\{\theta\}^{c-2},r\right)} \mathscr{H}_n[\mathbf{p},\mathbf{x};\widetilde{\mathbf{p}},\mathbf{y};\widetilde{\widetilde{\mathbf{p}}},\mathbf{z}]$$

$$+\sum_{(\mathbf{p};\widetilde{\mathbf{p}};\widetilde{\widetilde{\mathbf{p}}})\preceq\left(\bar{0},\{1\}^{c-1};\{0\}^{c};1,\{\theta\}^{c-1}\right)} \mathscr{H}_n[\mathbf{p},a\oplus\bar{1},\mathbf{x};\widetilde{\mathbf{p}},b,\mathbf{y};\widetilde{\widetilde{\mathbf{p}}},r,\mathbf{z}]$$
$$=\sum_{(\mathbf{p};\widetilde{\mathbf{p}};\widetilde{\widetilde{\mathbf{p}}})\preceq\left(\bar{0},\{1\}^{c-1},a\oplus\bar{1};\{0\}^{c},b;1,\{\theta\}^{c-1},r\right)} \mathscr{H}_n[\mathbf{p},\mathbf{x};\widetilde{\mathbf{p}},\mathbf{y};\widetilde{\widetilde{\mathbf{p}}},\mathbf{z}],$$

since $(a\oplus 1)\oplus\bar{1}=1\oplus(a\oplus\bar{1})$ for all $a\in\mathbb{D}_0$. We have now completed the proof of the lemma. □

The next corollary is the degenerate case of the preceding lemma.

Corollary 9.2.3. *Let* $c\in\mathbb{N}_0$ *and* $\Pi(c)=(\bar{0},\{1\}^c;\{0\}^{c+1};1,\{\theta\}^c)$. *Then*

$$\frac{1}{[n]^c}=-\sum_{(\mathbf{p};\widetilde{\mathbf{p}};\widetilde{\widetilde{\mathbf{p}}})\preceq\Pi(c)} \mathscr{H}_n[\mathbf{p};\widetilde{\mathbf{p}};\widetilde{\widetilde{\mathbf{p}}}].$$

Proof. The case $c=0$ follows from by Eq. (9.4) by setting $l=0$. For $c\geq 1$, using the $c=0$ case we get

$$\frac{1}{[n]^c}=-\frac{1}{[n]^c}\mathscr{H}_n[\bar{0};0;1]=-\sum_{(\mathbf{p};\widetilde{\mathbf{p}};\widetilde{\widetilde{\mathbf{p}}})\preceq\Pi(c)} \mathscr{H}_n[\mathbf{p};\widetilde{\mathbf{p}};\widetilde{\widetilde{\mathbf{p}}}]$$

by taking $a=\bar{0}$, $b=0$, $r=\theta$ and $\mathbf{x}=\mathbf{y}=\mathbf{z}=\emptyset$ in Lemma 9.2.2. Hence the corollary is proved. □

9.3 q-MH*S Identities

In this section we prove some general rules which explain what to expect for the q-MH*S $H_n^\star[\mathbf{s}]$ when we add an argument component in front of the positive composition $\mathbf{s}$. This allows us to extend expansion formulas from the two base cases of depth one, namely Eqs. (9.10) and (9.11), to an arbitrary positive composition $\mathbf{s}$.

For simplicity, throughout this section we always set $\underline{\mathbf{p}}=(\mathbf{p};\widetilde{\mathbf{p}};\widetilde{\widetilde{\mathbf{p}}})$ to be a triple of mollifiers.

Theorem 9.3.1. *Let* $n,c\in\mathbb{N}$, $c\geq 2$ *and* $\Pi_1(c)=(\bar{2},\{1\}^{c-2};1,\{0\}^{c-2};1,\{\theta\}^{c-2})$. *Then*

$$H_n^\star[1]=\mathscr{H}_n[1;1;2]=\sum_{k=1}^{n}\frac{1+q^k}{[k]}\frac{\begin{bmatrix}n\\k\end{bmatrix}}{\begin{bmatrix}n+k\\k\end{bmatrix}}q^{k^2}, \tag{9.10}$$

$$H_n^\star[c]=-\sum_{\underline{\mathbf{p}}\preceq\Pi_1(c)}\mathscr{H}_n[\underline{\mathbf{p}}]. \tag{9.11}$$

Proof. Equation (9.10) is just Eq. (9.6). To prove Eq. (9.11) we proceed by induction on n. When $n=1$ we have $H_1^\star[c]=q$. On the other hand,

$$\sum_{\underline{\mathbf{p}}\preceq\Pi_1(c)} \mathscr{H}_1[\underline{\mathbf{p}}]=\mathscr{H}_1[\overline{c};1;1]=-q,$$

since the depth of every mollifier in $\underline{\mathbf{p}}$ must be 1 in order to have non-trivial contribution. Therefore the formula is true when $n=1$. Suppose the statement is true for $n-1$. Then by the definition

$$H_n^\star[c]=H_{n-1}^\star[c]+\frac{q^n}{[n]}.$$

Applying inductive hypothesis, we obtain

$$H_n^\star[c]=-\sum_{\underline{\mathbf{p}}\preceq\Pi_1(c)} \mathscr{H}_{n-1}[\underline{\mathbf{p}}]+\frac{q^n}{[n]^c}. \tag{9.12}$$

Set $\Pi'(c)=(0,\{1\}^{c-2};\{0\}^{c-1};\{\theta\}^{c-1})$. To save space, for any string $\boldsymbol{\lambda}=(\lambda_1,\ldots,\lambda_m)$ we write the substring $\boldsymbol{\lambda}_{\hat{1}}=(\lambda_2,\ldots,\lambda_m)$. Then the sum on the right-hand side of Eq. (9.12) becomes

$$\begin{aligned}\sum_{\underline{\mathbf{p}}\preceq\Pi_1(c)} \mathscr{H}_{n-1}[\underline{\mathbf{p}}]&=\sum_{\underline{\mathbf{p}}\preceq\Pi'(c)} \mathscr{H}_{n-1}[\overline{2+p_1},\mathbf{p}_{\hat{1}};1+\widetilde{p}_1,\widetilde{\mathbf{p}}_{\hat{1}};1\boxplus\widetilde{\widetilde{p}}_1,\widetilde{\widetilde{\mathbf{p}}}_{\hat{1}}]\\&=\sum_{\underline{\mathbf{p}}\preceq\Pi'(c)}\sum_{n>k_1>\ldots>k_m\ge1}\frac{q^{k_1(1+\widetilde{p}_1)}A_{n-1,k_1}}{[k_1]^{2+p_1}}\prod_{j=2}^m\frac{q^{\widetilde{p}_jk_j}(1+q^{k_j})}{[k_j]^{p_j}},\end{aligned}$$

where $A_{n,k}$ is defined by Eq. (9.8). Plugging this into Eq. (9.12) and using the formula

$$A_{n-1,k}=A_{n,k}\left(1-\frac{[k]^2}{[n]^2}q^{n-k}\right), \tag{9.13}$$

we obtain

$$\begin{aligned}H_n^\star[c]=&-\sum_{\underline{\mathbf{p}}\preceq\Pi'(c)}\sum_{n\ge k_1>\cdots>k_m\ge1}\frac{q^{k_1(\widetilde{p}_1+1)}A_{n,k_1}}{[k_1]^{2+p_1}}\prod_{j=2}^m\frac{q^{\widetilde{p}_jk_j}(1+q^{k_j})}{[k_j]^{p_j}}\\&+\frac{q^n}{[n]^2}\sum_{\underline{\mathbf{p}}\preceq\Pi'(c)}\sum_{n\ge k_1>\cdots>k_m\ge1}\frac{q^{k_1\widetilde{p}_1}A_{n,k_1}}{[k_1]^{p_1}}\prod_{j=2}^m\frac{q^{\widetilde{p}_jk_j}(1+q^{k_j})}{[k_j]^{p_j}}+\frac{q^n}{[n]^c},\end{aligned}$$

which implies

$$H_n^\star[c]=-\sum_{\underline{\mathbf{p}}\preceq\Pi_1(c)} \mathscr{H}_n[\underline{\mathbf{p}}]+\frac{q^n}{[n]^2}\sum_{\underline{\mathbf{p}}\preceq\Pi''(c)} \mathscr{H}_n[\underline{\mathbf{p}}]+\frac{q^n}{[n]^c},$$

where $\Pi''(c)=(\overline{0},\{1\}^{c-2};\{0\}^{c-1};1,\{\theta\}^{c-2})$. Hence the theorem follows from Cor. 9.2.3 immediately since $\Pi_1(c)=\Pi(c-2)$ there. □

We now turn to the general rules. For any string $\boldsymbol{\lambda} = (\lambda_1, \dots, \lambda_m)$, we set $\boldsymbol{\lambda}_{\hat{1}} = (\lambda_2, \dots, \lambda_m)$ as above.

Theorem 9.3.2. *Let $n \in \mathbb{N}$ and $\mathbf{s} = (s_1, \dots, s_d) \in \mathbb{N}^d$. Set $\delta(\mathbf{s}) = 1$ if $s_d = 1$ and $\delta(\mathbf{s}) = -1$ if $s_d > 1$. Suppose there is a triple of mollifiers $\underline{\boldsymbol{\lambda}} := (\boldsymbol{\lambda}; \widetilde{\boldsymbol{\lambda}}; \approx{\boldsymbol{\lambda}}) = (\lambda_1, \dots, \lambda_m; \widetilde{\lambda}_1, \dots, \widetilde{\lambda}_m; \approx{\lambda}_1, \dots, \approx{\lambda}_m)$ satisfying*

$$\approx{\lambda}_1 \boxplus \cdots \boxplus \approx{\lambda}_j \in \{1, 2\} \quad \forall j \geq 1, \tag{9.14}$$

such that there is an expansion of the form

$$H_n^\star[\mathbf{s}] = \delta(\mathbf{s}) \sum_{\underline{\mathbf{p}} \preceq \underline{\boldsymbol{\lambda}}} \mathscr{H}_n[\underline{\mathbf{p}}].$$

Then for any positive integer $c \in \mathbb{N}$, we have

$$H_n^\star[c, \mathbf{s}] = \delta(\mathbf{s}) \sum_{\underline{\mathbf{p}} \preceq \Pi_c(\underline{\boldsymbol{\lambda}})} \mathscr{H}_n[\underline{\mathbf{p}}], \tag{9.15}$$

where[1] $\Pi_1(\underline{\boldsymbol{\lambda}}) = (1, \boldsymbol{\lambda}; 1, \widetilde{\boldsymbol{\lambda}}; 2, \approx{\lambda}_1 \boxplus (-2), \approx{\boldsymbol{\lambda}}_{\hat{1}})$, $\Pi_2(\underline{\boldsymbol{\lambda}}) = (2 \oplus \lambda_1, \boldsymbol{\lambda}_{\hat{1}}; 1 + \widetilde{\lambda}_1, \widetilde{\boldsymbol{\lambda}}_{\hat{1}}; \approx{\boldsymbol{\lambda}})$, *and* $\Pi_c(\underline{\boldsymbol{\lambda}}) = (\bar{2}, \{1\}^{c-3}, \lambda_1 \oplus \bar{1}, \boldsymbol{\lambda}_{\hat{1}}; 1, \{0\}^{c-3}, \widetilde{\boldsymbol{\lambda}}; 1, \{\theta\}^{c-3}, \approx{\lambda}_1 \boxplus (-1), \approx{\boldsymbol{\lambda}}_{\hat{1}})$ *for $c \geq 3$. Moreover, in all the sets $\Pi_c(\underline{\boldsymbol{\lambda}})$ the third component still satisfy Eq.* (9.14).

Note that the condition in Eq. (9.14) essentially guarantees that all the triples of mollifiers considered in this section are admissible.

Proof. Write $\Pi_c = \Pi_c(\underline{\boldsymbol{\lambda}})$ throughout the proof. We prove the identity in the theorem by induction on n. When $n = 1$ the theorem is clear. Assume Eq. (9.15) is true for $n - 1$.

First, we prove Eq. (9.15) for $c = 1$. By the definition, we have

$$H_n^\star[1, \mathbf{s}] = H_{n-1}^\star[1, \mathbf{s}] + \frac{q^n}{[n]} H_n^\star[\mathbf{s}].$$

Applying induction assumption, we obtain

$$\delta(\mathbf{s}) H_n^\star[1, \mathbf{s}] = \sum_{\underline{\mathbf{p}} \preceq \Pi_1} \mathscr{H}_{n-1}[\underline{\mathbf{p}}] + \frac{q^n}{[n]} \sum_{\underline{\mathbf{p}} \preceq \underline{\boldsymbol{\lambda}}} \mathscr{H}_n[\underline{\mathbf{p}}]. \tag{9.16}$$

Setting $\Pi_0 = (\boldsymbol{\lambda}; \widetilde{\boldsymbol{\lambda}}; \approx{\lambda}_1 \boxplus (-2), \approx{\boldsymbol{\lambda}}_{\hat{1}})$. To save space, for $\underline{\mathbf{p}} \preceq \Pi_0$ and $\mathbf{k} = (k_1, \dots, k_m)$ we write

$$f_j(\underline{\mathbf{p}}, \mathbf{k}) = \frac{\operatorname{sgn}(p_j)^{k_j} q^{\widetilde{p}_j k_j + Q(\approx{p}_j, k_j)} (1 + q^{k_j})}{[k_j]^{|p_j|}}.$$

[1]We have abused the notation here by writing $(1, \boldsymbol{\lambda}; \dots)$ for the concatenation $(1, \lambda_1, \dots, \lambda_m; \dots)$. No confusion should arise.

Then

$$\sum_{\underline{\mathbf{p}}\preceq\Pi_1}\mathscr{H}_{n-1}[\underline{\mathbf{p}}]$$

$$=\sum_{\underline{\mathbf{p}}\preceq\Pi_0}\Big(\mathscr{H}_{n-1}[1,\mathbf{p};1,\widetilde{\mathbf{p}};2,\widetilde{\widetilde{\mathbf{p}}}]+\mathscr{H}_{n-1}[1\oplus p_1,\mathbf{p}_{\hat{1}};1+\widetilde{p}_1,\widetilde{\mathbf{p}}_{\hat{1}};2\boxplus\widetilde{\widetilde{p}}_1,\widetilde{\widetilde{\mathbf{p}}}_{\hat{1}}]\Big)$$

$$=\sum_{\underline{\mathbf{p}}\preceq\Pi_0}\left(\sum_{n>k_0>k_1>\cdots>k_m\geq 1}\frac{\begin{bmatrix}n-1\\k_0\end{bmatrix}}{\begin{bmatrix}n-1+k_0\\k_0\end{bmatrix}}\frac{(1+q^{k_0})q^{k_0^2}}{[k_0]}\prod_{j=1}^m f_j(\underline{\mathbf{p}},\mathbf{k})\right.$$

$$\left.+\sum_{n>k_1>\cdots>k_m\geq 1}\frac{\begin{bmatrix}n-1\\k_1\end{bmatrix}}{\begin{bmatrix}n-1+k_1\\k_1\end{bmatrix}}\frac{(1+q^{k_1})q^{(1+\widetilde{p}_1)k_1+Q(2\boxplus\widetilde{\widetilde{p}}_1,k_1)}}{\operatorname{sgn}(p_1)^{k_1}[k_1]^{1+|p_1|}}\prod_{j=2}^m f_j(\underline{\mathbf{p}},\mathbf{k})\right).$$

Plugging the above expression into Eq. (9.16) and using Eq. (9.13), we obtain

$$\delta(\mathbf{s})H_n^\star[1,\mathbf{s}]-\frac{q^n}{[n]}\sum_{\underline{\mathbf{p}}\preceq\boldsymbol{\lambda}}\mathscr{H}_n[\underline{\mathbf{p}}]$$

$$=\sum_{\underline{\mathbf{p}}\preceq\Pi_0}\left(\sum_{n\geq k_0>k_1>\cdots>k_m\geq 1}\frac{\begin{bmatrix}n\\k_0\end{bmatrix}}{\begin{bmatrix}n+k_0\\k_0\end{bmatrix}}\frac{(1+q^{k_0})q^{k_0^2}}{[k_0]}\prod_{j=1}^m f_j(\underline{\mathbf{p}},\mathbf{k})\right.$$

$$-\frac{q^n}{[n]^2}\sum_{n\geq k_0>k_1>\cdots>k_m\geq 1}\frac{\begin{bmatrix}n\\k_0\end{bmatrix}}{\begin{bmatrix}n+k_0\\k_0\end{bmatrix}}(1+q^{k_0})q^{k_0(k_0-1)}[k_0]\prod_{j=1}^m f_j(\underline{\mathbf{p}},\mathbf{k})$$

$$+\sum_{n\geq k_1>\cdots>k_m\geq 1}\frac{\begin{bmatrix}n\\k_1\end{bmatrix}}{\begin{bmatrix}n+k_1\\k_1\end{bmatrix}}\frac{(1+q^{k_1})q^{(1+\widetilde{p}_1)k_1+Q(2\boxplus\widetilde{\widetilde{p}}_1,k_1)}}{\operatorname{sgn}(p_1)^{k_1}[k_1]^{2a+1+|p_1|}}\prod_{j=2}^m f_j(\underline{\mathbf{p}},\mathbf{k})$$

$$\left.-\frac{q^n}{[n]^2}\sum_{n\geq k_1>\cdots>k_m\geq 1}\frac{\begin{bmatrix}n\\k_1\end{bmatrix}}{\begin{bmatrix}n+k_1\\k_1\end{bmatrix}}\frac{(1+q^{k_1})q^{\widetilde{p}_1k_1+Q(2\boxplus\widetilde{\widetilde{p}}_1,k_1)}}{\operatorname{sgn}(p_1)^{k_1}[k_1]^{|p_1|-1}}\prod_{j=2}^m f_j(\underline{\mathbf{p}},\mathbf{k})\right).$$

Noting that the first and third sums on the right-hand side add up to $\sum_{\underline{\mathbf{p}}\preceq\Pi_1}\mathscr{H}_n[\underline{\mathbf{p}}]$, we have

$$\delta(\mathbf{s})H_n^\star[1,\mathbf{s}]=\frac{q^n}{[n]}\sum_{\underline{\mathbf{p}}\preceq\boldsymbol{\lambda}}\mathscr{H}_n[\underline{\mathbf{p}}]+\sum_{\underline{\mathbf{p}}\preceq\Pi_1}\mathscr{H}_n[\underline{\mathbf{p}}]$$

$$-\frac{q^n}{[n]^2}\sum_{\underline{\mathbf{p}}\preceq\Pi_0}\sum_{n\geq k_1>\cdots>k_m\geq 1}\left(\prod_{j=1}^m f_j(\underline{\mathbf{p}},\mathbf{k})\sum_{k_0=k_1+1}^{n}\frac{\begin{bmatrix}n\\k_0\end{bmatrix}}{\begin{bmatrix}n+k_0\\k_0\end{bmatrix}}\frac{(1+q^{k_0})q^{2\binom{k_0}{2}}}{[k_0]^{-1}}\right.$$

$$\left.+\frac{\begin{bmatrix}n\\k_1\end{bmatrix}}{\begin{bmatrix}n+k_1\\k_1\end{bmatrix}}\frac{(1+q^{k_1})q^{\widetilde{p}_1k_1+Q(2\boxplus\widetilde{\widetilde{p}}_1,k_1)}}{\operatorname{sgn}(p_1)^{k_1}[k_1]^{|p_1|-1}}\prod_{j=2}^m f_j(\underline{\mathbf{p}},\mathbf{k})\right).$$

Summing the multiple sum in the above over k_0 by Eq. (9.5) and noticing that for $\underline{\mathbf{p}} \preceq \Pi_0$, by Eq. (9.14), the first component $\widetilde{\widetilde{p}}_1$ can take only values equal to -1 and 0, with the help of Eq. (9.3) we obtain

$$\delta(\mathbf{s})H_n^\star[1,\mathbf{s}] = \sum_{\underline{\mathbf{p}} \preceq \Pi_1} \mathscr{H}_n[\underline{\mathbf{p}}].$$

This proves Eq. (9.15) in the case $c = 1$ by induction.

When $c = 2$, by the definition we have

$$H_n^\star[2,\mathbf{s}] = H_{n-1}^\star[\mathbf{s}] + \frac{q^n}{[n]^2} H_n^\star[\mathbf{s}].$$

The induction assumption implies that

$$\delta(\mathbf{s})H_n^\star[2,\mathbf{s}] = \sum_{\underline{\mathbf{p}} \preceq \Pi_2} \mathscr{H}_{n-1}[\underline{\mathbf{p}}] + \frac{q^n}{[n]^2} \sum_{\underline{\mathbf{p}} \preceq \boldsymbol{\lambda}} \mathscr{H}_n[\underline{\mathbf{p}}]. \tag{9.17}$$

Expanding the first sum on the right-hand side, we get

$$\begin{aligned}
\sum_{\underline{\mathbf{p}} \preceq \Pi_2} \mathscr{H}_{n-1}[\underline{\mathbf{p}}] &= \sum_{\underline{\mathbf{p}} \preceq \boldsymbol{\lambda}} \mathscr{H}_{n-1}[2 \oplus p_1, \mathbf{p}_{\hat{1}}; 1 + \widetilde{p}_1, \widetilde{\mathbf{p}}_{\hat{1}}; \widetilde{\widetilde{\mathbf{p}}}] \\
&= \sum_{\underline{\mathbf{p}} \preceq \boldsymbol{\lambda}} \sum_{n > k_1 > \cdots > k_r \geq 1} \frac{\begin{bmatrix} n-1 \\ k_1 \end{bmatrix}}{\begin{bmatrix} n-1+k_1 \\ k_1 \end{bmatrix}} \frac{q^{k_1}}{[k_1]^2} \prod_{j=1}^r f_j(\underline{\mathbf{p}}, \mathbf{k}).
\end{aligned}$$

Hence

$$\begin{aligned}
\delta(\mathbf{s})H_n^\star[2,\mathbf{s}] &= \frac{q^n}{[n]^2} \sum_{\underline{\mathbf{p}} \preceq \boldsymbol{\lambda}} \mathscr{H}_n[\underline{\mathbf{p}}] \\
&\quad + \sum_{\underline{\mathbf{p}} \preceq \boldsymbol{\lambda}} \sum_{n \geq k_1 > \cdots > k_r \geq 1} \frac{\begin{bmatrix} n \\ k_1 \end{bmatrix}}{\begin{bmatrix} n+k_1 \\ k_1 \end{bmatrix}} \left(\frac{q^{k_1}}{[k_1]^2} - \frac{q^n}{[n]^2} \right) \prod_{j=1}^r f_j(\underline{\mathbf{p}}, \mathbf{k})
\end{aligned}$$

which yields Eq. (9.15) when $c = 2$ by the definition and straightforward cancelation.

Finally, we turn to the case when $c \geq 3$. By the definition we have

$$H_n^\star[c,\mathbf{s}] = H_{n-1}^\star[c,\mathbf{s}] + \frac{q^n}{[n]^c} H_n^\star[\mathbf{s}].$$

By the induction assumption, we see that

$$\delta(\mathbf{s})H_n^\star[c,\mathbf{s}] = \sum_{\underline{\mathbf{p}} \preceq \Pi_c} \mathscr{H}_{n-1}[\underline{\mathbf{p}}] + \frac{q^n}{[n]^c} \sum_{\underline{\mathbf{p}} \preceq \boldsymbol{\lambda}} \mathscr{H}_n[\underline{\mathbf{p}}]. \tag{9.18}$$

Set $\Pi' = (0, \{1\}^{c-3}, \lambda_1 \oplus \bar{1}, \boldsymbol{\lambda}_{\hat{1}}; \{0\}^{c-2}, \widetilde{\boldsymbol{\lambda}}; \{\theta\}^{c-2}, \widetilde{\widetilde{\lambda}}_1 \boxplus (-1), \widetilde{\widetilde{\boldsymbol{\lambda}}}_{\hat{1}})$. Then

$$\sum_{\underline{\mathbf{p}} \preceq \Pi_c} \mathscr{H}_{n-1}[\underline{\mathbf{p}}] = \sum_{\underline{\mathbf{p}} \preceq \Pi'} \mathscr{H}_{n-1}[\bar{2} \oplus p_1, \mathbf{p}_{\hat{1}}; 1 + \widetilde{p}_1, \widetilde{\mathbf{p}}_{\hat{1}}; 1 \boxplus \widetilde{\widetilde{p}}_1, \widetilde{\widetilde{\mathbf{p}}}_{\hat{1}}]$$

$$= \sum_{\underline{\mathbf{p}} \preceq \Pi'} \sum_{n > k_1 > \cdots > k_r \geq 1} \frac{\begin{bmatrix} n-1 \\ k_1 \end{bmatrix}}{\begin{bmatrix} n-1+k_1 \\ k_1 \end{bmatrix}} \frac{q^{k_1(1+\widetilde{p}_1) + Q(1 \boxplus \widetilde{\widetilde{p}}_1, k_1)}}{(-\operatorname{sgn}(p_1))^{k_1} [k_1]^{2+|p_1|}} \prod_{j=2}^{r} f_j(\underline{\mathbf{p}}, \mathbf{k}).$$

Plugging this into Eq. (9.18) and using Eq. (9.13), we obtain

$$\delta(\mathbf{s}) H_n^{\star}[c, \mathbf{s}] = \sum_{\underline{\mathbf{p}} \preceq \Pi'} \mathscr{H}_n[\bar{2} \oplus p_1, \mathbf{p}_{\hat{1}}; 1 + \widetilde{p}_1, \widetilde{\mathbf{p}}_{\hat{1}}; 1 \boxplus \widetilde{\widetilde{p}}_1, \widetilde{\widetilde{\mathbf{p}}}_{\hat{1}}]$$

$$+ \frac{q^n}{[n]^c} \sum_{\underline{\mathbf{p}} \preceq \underline{\boldsymbol{\lambda}}} \mathscr{H}_n[\underline{\mathbf{p}}] - \frac{q^n}{[n]^2} \sum_{\underline{\mathbf{p}} \preceq \Pi'} \mathscr{H}_n[\bar{p}_1, \mathbf{p}_{\hat{1}}; \widetilde{\mathbf{p}}; 1 \boxplus \widetilde{\widetilde{\mathbf{p}}}_{\hat{1}}],$$

which implies

$$\delta(\mathbf{s}) H_n^{\star}[c, \mathbf{s}] = \sum_{\underline{\mathbf{p}} \preceq \Pi_c} \mathscr{H}_n[\underline{\mathbf{p}}] - \frac{q^n}{[n]^2} \sum_{\underline{\mathbf{p}} \preceq \Pi''} \mathscr{H}_n[\underline{\mathbf{p}}] + \frac{q^n}{[n]^c} \sum_{\underline{\mathbf{p}} \preceq \underline{\boldsymbol{\lambda}}} \mathscr{H}_n[\underline{\mathbf{p}}], \tag{9.19}$$

where $\Pi'' = (\bar{0}, \{1\}^{c-3}, \lambda_1 \oplus \bar{1}, \boldsymbol{\lambda}_{\hat{1}}; \{0\}^{c-2}, \widetilde{\boldsymbol{\lambda}}; 1, \{\theta\}^{c-3}, \widetilde{\widetilde{\lambda}}_1 \boxplus (-1), \widetilde{\widetilde{\boldsymbol{\lambda}}}_{\hat{1}})$. Expanding the second sum from Eq. (9.19), we see that

$$\sum_{\underline{\mathbf{p}} \preceq \Pi''} \mathscr{H}_n[\underline{\mathbf{p}}] = \sum_{\underline{\mathbf{p}} \preceq \Pi'''} \sum_{\substack{(\mathbf{w}; \widetilde{\mathbf{w}}; \widetilde{\widetilde{\mathbf{w}}}) \preceq (\bar{0}, \{1\}^{c-3}, p_1 \oplus \bar{1}; \\ \{0\}^{c-2}, \widetilde{p}_1; 1, \{\theta\}^{c-3}, \widetilde{\widetilde{p}}_1)}} \mathscr{H}_n[\mathbf{w}, \mathbf{p}_{\hat{1}}; \widetilde{\mathbf{w}}, \widetilde{\mathbf{p}}_{\hat{1}}; \widetilde{\widetilde{\mathbf{w}}}, \widetilde{\widetilde{\mathbf{p}}}_{\hat{1}}],$$

where $\Pi''' = (\boldsymbol{\lambda}; \widetilde{\boldsymbol{\lambda}}; \widetilde{\widetilde{\lambda}}_1 \boxplus (-1), \widetilde{\widetilde{\boldsymbol{\lambda}}}_{\hat{1}})$. Applying Lemma 9.2.2 to the inner sum with $(a, b, r) = (p_1, \widetilde{p}_1, \widetilde{\widetilde{p}}_1)$, $(\mathbf{x}; \mathbf{y}; \mathbf{z}) = (\mathbf{p}_{\hat{1}}; \widetilde{\mathbf{p}}_{\hat{1}}; \widetilde{\widetilde{\mathbf{p}}}_{\hat{1}})$, and c replaced by $c - 2$, we finally arrive at

$$\sum_{\underline{\mathbf{p}} \preceq \Pi''} \mathscr{H}_n[\underline{\mathbf{p}}] = \sum_{\underline{\mathbf{p}} \preceq \Pi'''} \frac{1}{[n]^{c-2}} \mathscr{H}_n[p_1, \mathbf{p}_{\hat{1}}; \widetilde{\mathbf{p}}; \widetilde{\widetilde{p}}_1 \boxplus 1, \widetilde{\widetilde{\mathbf{p}}}_{\hat{1}}] = \frac{1}{[n]^{c-2}} \sum_{\underline{\mathbf{p}} \preceq \underline{\boldsymbol{\lambda}}} \mathscr{H}_n[\underline{\mathbf{p}}]. \tag{9.20}$$

To justify the last equality above we need to show that for the components of $\widetilde{\widetilde{\boldsymbol{\lambda}}}$ satisfying Eq. (9.14) we have $\widetilde{\widetilde{\lambda}}_1 \boxplus (-1) \boxplus 1 = \widetilde{\widetilde{\lambda}}_1$ and for any $j \geq 2$,

$$\widetilde{\widetilde{\lambda}}_1 \boxplus (-1) \boxplus \widetilde{\widetilde{\lambda}}_2 \boxplus \cdots \boxplus \widetilde{\widetilde{\lambda}}_j \boxplus 1 = \widetilde{\widetilde{\lambda}}_1 \boxplus \widetilde{\widetilde{\lambda}}_2 \boxplus \cdots \boxplus \widetilde{\widetilde{\lambda}}_j.$$

These can be proved by using the projection π of Lemma 9.1.2 and the fact that $\pi(\widetilde{\widetilde{\lambda}}_1 \boxplus (-1) \boxplus \cdots \boxplus \widetilde{\widetilde{\lambda}}_j \boxplus 1) = \pi(\widetilde{\widetilde{\lambda}}_1 \boxplus \cdots \boxplus \widetilde{\widetilde{\lambda}}_j) \in \{1, 2\}$ by Eq. (9.14). Now by Eq. (9.19) and Eq. (9.20), we see that Eq. (9.15) is true for all $c \geq 3$.

We have completed the proof of the theorem. □

We now state the main q-MH*S identity that will lead to a few applications later. First, we need some extra notation. For any two compositions $\mathbf{a} = (a_1, \dots, a_m)$ and $\mathbf{b} = (b_1, \dots, b_n)$, we define binary operations

$$\mathbf{a} \bigodot \mathbf{b} = (a_1, \dots, a_{m-1}, a_m \bigodot b_1, b_2, \dots, b_n)$$

where $\bigodot = \oplus, +, \boxplus$ for the three components of the triple mollifiers. For $\mathbf{s} \in \mathbb{N}^\ell$, we define a sequence of admissible triple mollifiers $\underline{\mathbf{m}}_i(\mathbf{s}) := (\mathbf{m}_i, \widetilde{\mathbf{m}}_i, \widetilde{\widetilde{\mathbf{m}}}_i)$ $(i = \ell, \dots, 2, 1)$ (backward) inductively as follows.

$$\mathbf{m}_\ell = \begin{cases} (1), & \text{if } s_\ell = 1; \\ (\overline{2}, \{1\}^{s_\ell - 2}), & \text{if } s_\ell \geq 2, \end{cases} \qquad \mathbf{m}_i = \begin{cases} (1, \mathbf{m}_{i+1}), & \text{if } s_i = 1; \\ (2) \oplus \mathbf{m}_{i+1}, & \text{if } s_i = 2; \\ (\overline{2}, \{1\}^{s_i - 3}, \overline{1}) \oplus \mathbf{m}_{i+1}, & \text{if } s_i \geq 3, \end{cases}$$

$$\widetilde{\mathbf{m}}_\ell = \begin{cases} (1), & \text{if } s_\ell = 1; \\ (1, \{0\}^{s_\ell - 2}), & \text{if } s_\ell \geq 2, \end{cases} \qquad \widetilde{\mathbf{m}}_i = \begin{cases} (1, \widetilde{\mathbf{m}}_{i+1}), & \text{if } s_i = 1; \\ (1) + \widetilde{\mathbf{m}}_{i+1}, & \text{if } s_i = 2; \\ (1, \{0\}^{s_i - 3}, \widetilde{\mathbf{m}}_{i+1}), & \text{if } s_i \geq 3, \end{cases}$$

$$\widetilde{\widetilde{\mathbf{m}}}_\ell = \begin{cases} (2), & \text{if } s_\ell = 1; \\ (1, \{\theta\}^{s_\ell - 2}), & \text{if } s_\ell \geq 2, \end{cases} \qquad \widetilde{\widetilde{\mathbf{m}}}_i = \begin{cases} (2, -2) \boxplus \widetilde{\widetilde{\mathbf{m}}}_{i+1}, & \text{if } s_i = 1; \\ \widetilde{\widetilde{\mathbf{m}}}_{i+1}, & \text{if } s_i = 2; \\ (1, \{0\}^{s_i - 3}, -1) \boxplus \widetilde{\widetilde{\mathbf{m}}}_{i+1}, & \text{if } s_i \geq 3, \end{cases}$$

for all $i < \ell$.

Theorem 9.3.3. *For all* $\mathbf{s} = (s_1, \dots, s_\ell) \in \mathbb{N}^\ell$, *we have*

$$H_n^\star[\mathbf{s}] = \delta(\mathbf{s}) \sum_{\underline{\mathbf{p}} \preceq \underline{\mathbf{m}}_1(\mathbf{s})} \mathscr{H}_n[\underline{\mathbf{p}}], \quad \textit{where } \delta(\mathbf{s}) = \begin{cases} 1, & \textit{if } s_\ell = 1; \\ -1, & \textit{if } s_\ell \geq 2. \end{cases}$$

Proof. The theorem follows from Thms. 9.3.1 and 9.3.2 easily. □

9.4 Some Applications

By repeatedly applying Thm. 9.3.3 we can now obtain a number of different types of identities.

The following two results give us the q-analog of so-called **2**-$\underline{1}$-**2** formulas (or simply the *Two-one formulas*). Here the underline means the string type $(\{2\}^a, 1)$ may be repeated an arbitrary number of times where a may vary in the repetitions.

Corollary 9.4.1. *Suppose* $\ell \in \mathbb{N}_0$ *and* $a_1, \dots, a_\ell \in \mathbb{N}_0$. *Then*

$$H_n^\star[\{2\}^{a_1}, 1, \dots, \{2\}^{a_\ell}, 1] = \sum_{\underline{\mathbf{p}} \preceq \Pi} \mathscr{H}_n[\underline{\mathbf{p}}] \tag{9.21}$$

where $\Pi = \Big(\operatorname*{\mathbf{Cat}}_{j=1}^{\ell} \{2a_j + 1\}; \operatorname*{\mathbf{Cat}}_{j=1}^{\ell} \{a_j + 1\}; 2, \{0\}^{\ell - 1} \Big)$.

Corollary 9.4.2. *Suppose $\ell \in \mathbb{N}_0$, $a_1, \dots, a_\ell \in \mathbb{N}_0$ and $a_{\ell+1} \in \mathbb{N}$. Then*

$$H_n^\star[\{2\}^{a_1}, 1, \dots, \{2\}^{a_\ell}, 1, \{2\}^{a_{\ell+1}}] = -\sum_{\underline{\mathbf{p}} \preceq \Pi} \mathscr{H}_n[\underline{\mathbf{p}}] \tag{9.22}$$

where $\Pi = \Big(\operatorname*{\mathbf{Cat}}_{j=1}^{\ell}\{2a_j+1\}, \overline{2a_{\ell+1}}; \operatorname*{\mathbf{Cat}}_{j=1}^{\ell}\{a_j+1\}, a_{\ell+1}; 2, \{0\}^{\ell-1}, -1 \Big)$.

We can also get the following q-analog of the **2-c-2** ($c \geq 3$) formula.

Corollary 9.4.3. *Suppose $\ell, a_j, c_j \in \mathbb{N}_0$ and $c_j \geq 3$ for all $1 \leq j \leq \ell+1$. Set $\mathbf{s} = (\{2\}^{a_1}, c_1, \dots, \{2\}^{a_\ell}, c_\ell, \{2\}^{a_{\ell+1}})$. Then*

$$H_n^\star[\mathbf{s}] = -\sum_{\underline{\mathbf{p}} \preceq \Pi(\mathbf{s})} \mathscr{H}_n[\underline{\mathbf{p}}], \tag{9.23}$$

where
$$\Pi(\mathbf{s}) = \Big(\overline{2a_1+2}, \{1\}^{c_1-3}, \operatorname*{\mathbf{Cat}}_{j=2}^{\ell}\big\{2a_j+3, \{1\}^{c_j-3}\big\}, 2a_{\ell+1}+1; \operatorname*{\mathbf{Cat}}_{j=1}^{\ell}\big\{a_j+1, \{0\}^{c_j-3}\big\}, a_{\ell+1};\ 1, \operatorname*{\mathbf{Cat}}_{j=1}^{\ell}\{\theta\}^{c_j-2} \Big).$$

Remark 9.4.4. If we take $c_j = 3$ for all j then we obtain the q-analog of so-called *Two-three formula.*

9.5 Congruences of q-MHSs

From last chapter we see that it is quite fruitful to study the van Hamme type congruence of the MHSs. In this section, abiding by the same principle we will consider some congruences of the q-MHSs of the same type. But this time we treat q-MHSs as rational functions in q.

Note that $[p] = [p]_q$ is an irreducible polynomial in $\mathbb{Z}[q]$ when p is a prime. So we have the q-analog of Wolstenhomle's Theorem Eq. (8.1) given below.

Theorem 9.5.1. *For any prime $p \geq 5$,*

$$\sum_{j=1}^{p-1} \frac{1}{[j]_q} \equiv \frac{p-1}{2}(1-q) + \frac{p^2-1}{24}(1-q)^2[p] \pmod{[p]_q^2},$$
$$\sum_{j=1}^{p-1} \frac{1}{[j]_q^2} \equiv -\frac{(p-1)(p-5)}{12}(1-q)^2 \pmod{[p]}.$$

Or, equivalently,

$$\sum_{j=1}^{p-1} \frac{1}{1-q^j} \equiv \frac{p-1}{2} + \frac{p^2-1}{24}(1-q)[p] \pmod{[p]_q^2}, \tag{9.24}$$

$$\sum_{j=1}^{p-1} \frac{1}{(1-q^j)^2} \equiv -\frac{(p-1)(p-5)}{12} \pmod{[p]}. \tag{9.25}$$

Actually we will prove some generalizations of the above congruences. As in the previous sections we will suppress the dependence on q and write $[j] = [j]_q$ for all $j \in \mathbb{N}$. For $\mathbf{s} = (s_1, \dots, s_d) \in \mathbb{N}^d$ and $\mathbf{t} = (t_1, \dots, t_d) \in \mathbb{Z}^d$, we set

$$H_n^{\mathbf{t}}[\mathbf{s}] := \sum_{n \geq k_1 > \cdots > k_d \geq 1} \frac{q^{k_1 t_1 + \cdots + k_d t_d}}{[k_1]^{s_1} \cdots [k_d]^{s_d}}, \qquad \mathfrak{h}_n^{\mathbf{t}}[\mathbf{s}] := \frac{H_n^{(\mathbf{t})}[\mathbf{s}]}{(1-q)^{|\mathbf{s}|}}. \tag{9.26}$$

Note that $H_n^{(s_1-1,\dots,s_d-1)}[\mathbf{s}]$ are the partial sums of the type I q-analogs of the multiple zeta functions to be defined in Chap. 11. For trivial $\mathbf{t}$, we set

$$H_n^{\mathbf{0}}[\mathbf{s}] := H_n^{(0,\dots,0)}[\mathbf{s}], \qquad \mathfrak{h}_n^{\mathbf{0}}[\mathbf{s}] = H_n^{\mathbf{0}}[\mathbf{s}]/(1-q)^{|\mathbf{s}|}.$$

To state the first main theorem of this section we define the *degenerate Bernoulli numbers* $\beta_n(\lambda)$ by the generating series

$$\frac{x}{(1+\lambda x)^{1/\lambda} - 1} = \sum_{n=0}^{\infty} \beta_n(\lambda) \frac{x^n}{n!}.$$

Clearly we have $\lim_{\lambda \to 0} \beta_n(\lambda) = B_n$.

Remark 9.5.2. The degenerate Bernoulli numbers $\beta_n(\lambda)$ are polynomials of λ of degree n. More precisely, Howard proves in [301, Thm. 3.1] that $\beta_0(\lambda) = 1$, $\beta_1(\lambda) = (\lambda - 1)/2$ and

$$\beta_n(\lambda) = \tilde{b}_n \lambda^n + n \sum_{j=1}^{[n/2]} \frac{B_{2j}}{2j} s(n-1, 2j-1) \lambda^{n-2j} \quad \forall n \geq 2, \tag{9.27}$$

where $s(n,j)$ are the Stirling numbers of the first kind (cf. Eq. (5.37)) defined by

$$x(x-1)(x-2)\cdots(x-n+1) = \sum_{j=0}^{n} s(n,j) x^j,$$

and $\tilde{b}_n$ are the Bernoulli numbers of the second kind defined by

$$\frac{x}{\log(1+x)} = \sum_{n=0}^{\infty} \tilde{b}_n \frac{x^n}{n!} = 1 + \frac{1}{2}\frac{x}{1!} - \frac{1}{6}\frac{x^2}{2!} + \frac{1}{4}\frac{x^3}{3!} - \frac{19}{24}\frac{x^4}{4!} + \cdots.$$

Theorem 9.5.3. *If $p > 3$ is a prime, then for all integers $n > 1$ we have*

$$\mathfrak{h}^{(1)}_{p-1}[n] = \sum_{j=1}^{p-1} \frac{q^j}{(1-q^j)^n} \equiv K_n(p) \pmod{[p]},$$

where

$$\begin{aligned} K_n(p) &= \frac{(-1)^{n+1}}{n!} p^n \beta_n\left(\frac{1}{p}\right) \\ &= (-1)^{n+1} \frac{\tilde{b}_n}{n!} - \frac{(-1)^n}{(n-1)!} \sum_{j=1}^{[n/2]} \frac{B_{2j}}{2j} s(n-1, 2j-1) p^{2j}. \end{aligned} \tag{9.28}$$

Proof. Set $f_n(q) = \mathfrak{h}^{(1)}_{p-1}[n]_q$. Let $\mu = \mu_p = \exp(2\pi\sqrt{-1}/p)$ be the primitive pth root of unity. Then as a polynomial in $\mathbb{Z}[q]$ we have

$$[p]_q = \prod_{k=1}^{p-1} (q - \mu^k).$$

Hence, by the Remainder Theorem we only need to show that $f_n(\mu^k) = K_n(p)$ for all $k = 1, \ldots, p-1$. But

$$f_n(\mu^k) = \sum_{j=1}^{p-1} \frac{\mu^{jk}}{(1-\mu^{jk})^n} = \sum_{j=1}^{p-1} \frac{\mu^j}{(1-\mu^j)^n} = f_n(\mu).$$

Now one can check that (see Exercise 9.6)

$$\begin{aligned} \frac{p}{1-z^p} - \frac{1}{1-z} &= \sum_{j=1}^{p-1} \frac{\mu^j}{\mu^j - z} = \sum_{j=1}^{p-1} \frac{\mu^j}{\mu^j - 1} \frac{1}{1 - \frac{z-1}{\mu^j - 1}} \\ &= \sum_{j=1}^{p-1} \mu^j \sum_{n=0}^{\infty} \frac{(z-1)^n}{(\mu^j - 1)^{n+1}} = \sum_{n=0}^{\infty} (-1)^{n+1} f_{n+1}(\mu)(z-1)^n. \end{aligned} \tag{9.29}$$

On the other hand, by definition

$$\begin{aligned} \sum_{n=0}^{\infty} (-1)^{n+1} K_{n+1}(p)(z-1)^n &= \frac{1}{1-z} \sum_{n\geq 0} \frac{\beta_{n+1}(1/p)}{(n+1)!} \big(p(z-1)\big)^{n+1} \\ &= \frac{p}{1-z^p} - \frac{1}{1-z}. \end{aligned}$$

We now finish the proof of the theorem by comparing the above two generating functions. Note that the second equation in Eq. (9.28) follows from Remark 9.5.2. □

Corollary 9.5.4. *If $p>3$ is a prime, then for every integer $n\ge 0$ we have*

$$\mathfrak{h}^{(0)}_{p-1}[n]\equiv p+\sum_{j=0}^{n}K_j(p)\pmod{[p]}.$$

Proof. If $n=0$ then $\mathfrak{h}^{(0)}_{p-1}[n]=p-1$ while $K_0(p)=-1$. Thus the corollary is true. Suppose $n\ge 1$. Then we have

$$\mathfrak{h}^{(0)}_{p-1}[n]=\sum_{j=1}^{p-1}\frac{1}{(1-q^j)^n}=\sum_{j=1}^{p-1}\frac{1-q^j+q^j}{(1-q^j)^n}\equiv \mathfrak{h}^{(0)}_{p-1}[n-1]+K_n(p)\pmod{[p]}.$$

Hence the corollary follows by an easy induction. □

Corollary 9.5.5. *If $p>3$ is a prime, then for every integer $n>t\ge 1$ we have*

$$\mathfrak{h}^{(t)}_{p-1}[n]=\sum_{j=1}^{p-1}\frac{q^{tj}}{(1-q^j)^n}\equiv\sum_{i=0}^{t-1}\binom{t-1}{i}(-1)^iK_{n-i}(p)\pmod{[p]}.\qquad(9.30)$$

Moreover,

$$\sum_{j=1}^{p-1}\frac{1}{(1-q^j)^n}\equiv\left(\frac{p-1}{2}+\sum_{j=2}^{n}K_j(p)\right)\pmod{[p]}.\qquad(9.31)$$

Proof. If $t>1$ it is clear that

$$q^{tj}=q^j\big(1-(1-q^j)\big)^{t-1}=q^j\sum_{i=0}^{t-1}\binom{t-1}{i}(-1)^i(1-q^j)^i.$$

So Eq. (9.30) follows from Thm. 9.5.3 while Eq. (9.31) follows from Cor. 9.5.4 and Eq. (9.27). □

Theorem 9.5.6. *Let s be a positive integer and $\mu_s=\exp(2\pi\sqrt{-1}/s)$ the primitive sth root of unity. Then*

$$\sum_{d=0}^{\infty}\mathfrak{h}^{\mathbf{0}}_{p-1}[\{s\}^d]x^d\equiv\frac{(-1)^s}{p^sx}\prod_{n=0}^{s-1}\Big(1-(1-\mu_s^n(-x)^{1/s})^p\Big)\pmod{[p]}.$$

Proof. Let $\mu=\exp(2\pi\sqrt{-1}/p)$ be the primitive pth root of unity and set

$$P_n=\sum_{j=1}^{p-1}\frac{1}{(1-\mu^j)^n}-p\delta_{0,n}\quad\forall n\ge 0,\qquad(9.32)$$

where $\delta_{0,n}$ is the Kronecker symbol. Similar to the argument in the proof of Thm. 9.5.3 we see that $\mathfrak{h}^{\mathbf{0}}(n) \equiv P_n \pmod{[p]}$ for all $n \geq 1$. By partial fraction expansion in Eq. (9.29):

$$\begin{aligned}
g(x) := \sum_{n=0}^{\infty} P_n x^n &= \sum_{j=1}^{p-1} \frac{1-\mu^j}{1-x-\mu^j} - p \\
&= \sum_{j=1}^{p-1} \frac{\mu^{-j}}{z\mu^{-j}-1} - \sum_{j=1}^{p-1} \frac{\mu^j}{z-\mu^j} - p \quad (z = 1-x) \\
&= -z^{-1}\left(\frac{p}{z^{-p}-1} - \frac{1}{z^{-1}-1}\right) - \left(\frac{p}{z^p-1} - \frac{1}{z-1}\right) - p \\
&= -\frac{px(x-1)^{p-1}}{1-(1-x)^p}.
\end{aligned}$$

Let $a_d = \mathfrak{h}^{\mathbf{0}}_{p-1}\big[\{s\}^d\big]$ for all $d \geq 0$. Let $w(x) = \sum_{d=0}^{\infty} a_d x^d$ be its generating function. By

$$\mathfrak{h}^{\mathbf{0}}_{p-1}\big[\{s\}^d\big] = \frac{1}{d}\sum_{k=0}^{d-1}(-1)^{d-k-1}\mathfrak{h}^{\mathbf{0}}_{p-1}\big[(d-k)s\big] \cdot \mathfrak{h}^{\mathbf{0}}_{p-1}\big[\{s\}^k\big] \tag{9.33}$$

(see Exercise 9.5), we get

$$w(x) = \sum_{d=0}^{\infty} a_d x^d \equiv 1 + \sum_{d=1}^{\infty} \frac{1}{d} \sum_{k=0}^{d-1} (-1)^{d-k-1} P_{(d-k)s} a_k x^d \pmod{[p]}.$$

Differentiating both sides and changing index $d \to d+1$, we get, modulo $[p]$,

$$w'(x) \equiv \sum_{d=0}^{\infty}\sum_{k=0}^{d}(-1)^{d-k}P_{(d-k+1)s}a_k x^d \equiv \sum_{k=0}^{\infty}\sum_{d=k}^{\infty}(-1)^{d-k}P_{(d-k+1)s}a_k x^d.$$

By $d \to d+k$ and then changing the order of summation, we get

$$\begin{aligned}
w'(x) \equiv& w(x)\sum_{d=0}^{\infty} P_{(d+1)s}(-x)^d \equiv \frac{w(x)}{-x}\left(\sum_{d=0}^{\infty} P_{ds}(-x)^d + 1\right) \\
\equiv& \frac{w(x)}{-sx}\left(s + \sum_{n=0}^{s-1}\sum_{d=0}^{\infty} P_d(\mu_s^n(-x)^{1/s})^d\right) \\
\equiv& \frac{w(x)}{-sx}\left(s + \sum_{n=0}^{s-1} g\big(\mu_s^n(-x)^{1/s}\big)\right) \\
\equiv& \frac{w(x)}{-sx}\left(s - \sum_{n=0}^{s-1} \frac{p\mu^n(-x)^{1/s}(\mu^n(-x)^{1/s}-1)^{p-1}}{1-(1-\mu_s^n(-x)^{1/s})^p}\right).
\end{aligned}$$

Thus

$$(\ln w(x))' = \left(-(\ln x)' + \sum_{n=0}^{s-1} \frac{(1-(1-\mu^n(-x)^{1/s})^p)'}{1-(1-\mu_s^n(-x)^{1/s})^p} \right).$$

Therefore by comparing the constant term we get

$$w(x) \equiv \frac{(-1)^s}{p^s x} \prod_{n=0}^{s-1} \Big(1-(1-\mu_s^n(-x)^{1/s})^p\Big) \pmod{[p]}$$

as desired. □

Corollary 9.5.7. *For every positive integer $d < p$, we have*

$$\mathfrak{h}_{p-1}^{\mathbf{0}}[\{1\}^d] \equiv \frac{1}{d+1}\binom{p-1}{d} \qquad \pmod{[p]_q}.$$

Proof. By the theorem we get

$$\begin{aligned}\sum_{d=0}^{\infty} \mathfrak{h}_{p-1}^{\mathbf{0}}[\{1\}^d]x^d \equiv \frac{(1+x)^p-1}{px} &\equiv \frac{1}{px}\sum_{d=0}^{\infty}\binom{p}{d+1}x^{d+1} \\ &\equiv \sum_{d=0}^{\infty}\frac{1}{d+1}\binom{p-1}{d}x^d \pmod{[p]_q}.\end{aligned}$$

The corollary follows immediately. □

Corollary 9.5.8. *For every positive integer $d < p$, we have*

$$\begin{aligned}\mathfrak{h}_{p-1}^{\mathbf{0}}[\{2\}^d] &\equiv \frac{(-1)^d}{p^2}\left[\sum_{\substack{j+k=d \\ 0\le j,k<p/2}} \binom{p}{2j+1}\binom{p}{2k+1} - \sum_{\substack{j+k=d+1 \\ 1\le j,k<p/2}} \binom{p}{2j}\binom{p}{2k} \right] \\ &\equiv (-1)^d \frac{2\cdot d!}{(2d+2)!}\binom{p-1}{d} \cdot F_{2,d}(p)\cdot(1-q)^{2d} \pmod{[p]_q}, \qquad (9.34)\end{aligned}$$

where $F_{2,d}(p)$ is a monic polynomial in p of degree d.

Proof. By Thm. 9.5.6 we have modulo $[p]_q$

$$\begin{aligned}\sum_{d=0}^{\infty} \mathfrak{h}_{p-1}^{\mathbf{0}}[\{2\}^d]x^d &\equiv \frac{1}{p^2 x}\Big(1-(1-\sqrt{-1}\sqrt{x})^p\Big)\Big(1-(1+\sqrt{-1}\sqrt{x})^p\Big) \\ &\equiv \frac{1}{p^2 x}\left| \sum_{j=1}^{(p-1)/2}\binom{p}{2j}(-1)^j x^j + \sqrt{-1}\sqrt{x}\sum_{j=0}^{(p-1)/2}\binom{p}{2j+1}(-1)^j x^j \right|^2,\end{aligned}$$

which yields Eq. (9.34) easily. In the first sum of Eq. (9.34) if $j+k=d+1$ and $1 \le j,k < p/2$ then we may assume $j > d/2$. Then $(d+1)!\binom{p}{d+1}$ is a factor of $(2j+1)!\binom{p}{2j+1}$ as a polynomial of p, so is $d!\binom{p-1}{d}$. Similarly we can see that $d!\binom{p-1}{d}$ is a factor of the second sum.

In order to determine the leading coefficient we set

$$C_1(x) = \sum_{j=0}^{d} \frac{(2d+2)!\,x^{2j+1}}{(2j+1)!(2l-2j+1)!} = \frac{(x+1)^{2d+2}-(x-1)^{2d+2}}{2},$$

$$C_2(x) = \sum_{j=0}^{d+1} \frac{(2d+2)!\,x^{2j}}{(2j)!(2l-2j+2)!} = \frac{(x+1)^{2d+2}+(x-1)^{2d+2}}{2}.$$

Hence

$$\sum_{\substack{j+k=d\\ 0\le j,k<p/2}} \frac{1}{(2j+1)!(2k+1)!} - \sum_{\substack{j+k=d+1\\ 1\le j,k<p/2}} \frac{1}{(2j)!(2k)!}$$
$$= \frac{C_1(1)-(C_2(1)-2)}{(2d+2)!} = \frac{2}{(2d+2)!}.$$

This finishes the proof of the corollary. □

Corollary 9.5.9. *Let d be a positive integer. Set $\delta_d = (1+(-1)^d)$ and $L = 3d+3$. Then for every prime $p \ge L$, modulo $[p]_q$ we have*

$$\mathfrak{h}^{\mathbf{0}}_{p-1}[\{3\}^d] \equiv \frac{1}{p^3}\left[3\delta_d\binom{p}{L} + (-1)^d\cdot 3\sum_{k\ge1}\binom{p}{L-k}\binom{L-k}{k} + \binom{p}{d+1}\right] \tag{9.35}$$

$$\equiv \begin{cases} \dfrac{-3\cdot d!}{(3d+1)!}\dbinom{p-1}{d}\cdot F_{3,d}(p), & \text{if } d \text{ is odd},\\[2ex] \dfrac{6\cdot d!}{(3d+3)!}\dbinom{p-1}{d}\cdot F_{3,d}(p), & \text{if } d \text{ is even}, \end{cases} \tag{9.36}$$

where $F_{3,d}(p)$ is a monic polynomial in p of degree $2d-1$ if d is odd and of degree $2d$ if d is even.

Proof. Let $\mu = \exp(2\pi i/3)$. Then $\mu^2+\mu+1=0$. By Thm. 9.5.6 we have

$$\sum_{d=0}^{\infty} \mathfrak{h}^{\mathbf{0}}_{p-1}[\{3\}^d]x^d \equiv \frac{-1}{p^3 x}\prod_{a=0}^{2}\Big(1-(1-\mu^a\sqrt[3]{-x})^p\Big). \tag{9.37}$$

We now use two ways to expand this. Set $y = \sqrt[3]{-x}$. First, the product on the right-hand side of Eq. (9.37) can be expressed as

$$\begin{aligned}
1 - \sum_{a=0}^{2}(1-\mu^a y)^p + \sum_{a=0}^{2}(1-\mu^a y)^p(1-\mu^{a+1}y)^p - \prod_{a=0}^{2}(1-\mu^a y)^p \\
= 1 - \sum_{j=0}^{p}\binom{p}{j}\sum_{a=0}^{2}\mu^{aj}y^j + \sum_{a=0}^{2}(1+\mu^a y+\mu^{a+1}y^2)^p - (1+x)^p \\
= 1 - 3\sum_{j=0}^{[p/3]}\binom{p}{3j}x^j + 3\sum_{\substack{j,k\ge 0, j+k<p\\ 2j+k\equiv 0(3)}}\frac{p!\,(-x)^{(j+2k)/3}}{j!k!(p-j-k)!} - (1+x)^p.
\end{aligned}$$

This leads to Eq. (9.35) for $d > 0$. Note that if d is odd then the degree of the polynomial is reduced to $3d-1$ with leading coefficient given by

$$(-1)^d \cdot 3\frac{1}{(L-1)!}\binom{L-1}{1} = \frac{-3}{(L-2)!} = \frac{-3}{(3d+1)!}$$

as we wanted.

To prove $d!\binom{p}{d}$ is a factor we use the expansion of Eq. (9.37):

$$\sum_{d=0}^{\infty}\frac{1}{p^3x}\sum_{j,k,n\ge 1}(-1)^{j+k+n}\binom{p}{j}\binom{p}{k}\binom{p}{n}x^{(j+k+n)/3}\mu^{k+2n}.$$

Thus

$$\mathfrak{h}^{\mathbf{0}}_{p-1}\left[\{3\}^d\right] \equiv \frac{1}{p^3}\sum_{\substack{1\le j,k,n\le p\\ j+k+n=3d+3}}(-1)^{j+k+n}\binom{p}{j}\binom{p}{k}\binom{p}{n}\mu^{k+2n} \pmod{[p]_q}.$$

Note that $j+k+n = 3d+3$ implies one of the indices, say j, is at least $d+1$. Then clearly $\binom{p}{j}$ contains $d!\binom{p}{d}$ as a factor, therefore so does $\mathfrak{h}^{\mathbf{0}}_{p-1}\left[\{3\}^d\right] \pmod{[p]_q}$. This completes the proof of the corollary. □

One can also consider a few non-homogeneous cases.

Theorem 9.5.10. *Let $m, n \in \mathbb{N}$. Then for every prime p, we have*

$$\mathfrak{h}^{(m,n)}_{p-1}[2m, 2n] \equiv \frac{1}{2}\{f(m;p)f(n;p) - f(m+n;p)\} \pmod{[p]},$$

where

$$f(N;p) = \sum_{i=0}^{N-1}\binom{N-1}{i}(-1)^i K_{2N-i}(p).$$

Proof. By the definition and substitution $i \to p-i$ and $j \to p-j$ we have

$$\begin{aligned}
\mathfrak{h}_{p-1}^{(m,n)}[2m,2n] &= \sum_{1\le j<i<p} \frac{q^{mi+nj}}{(1-q^i)^{2m}(1-q^j)^{2n}} \\
&= \sum_{1\le i<j<p} \frac{q^{pm+pn-mi-nj}}{(1-q^{p-i})^{2m}(1-q^{p-j})^{2n}} \\
&\equiv \sum_{1\le i<j<p} \frac{q^{mi+nj}}{(q^i-q^p)^{2m}(q^j-q^p)^{2n}} && \pmod{[p]} \\
&\equiv \sum_{1\le i<j<p} \frac{q^{mi+nj}}{(1-p^i)^{2m}(1-p^j)^{2n}} && \pmod{[p]} \\
&\equiv \mathfrak{h}_{p-1}^{(n,m)}[2n,2m] && \pmod{[p]}, \quad (9.38)
\end{aligned}$$

since $q^p \equiv 1 \pmod{[p]}$. By the shuffle relation we have

$$\mathfrak{h}_{p-1}^{(m)}[2m]\mathfrak{h}_{p-1}^{(n)}[2n] = \mathfrak{h}_{p-1}^{(m,n)}[2m,2n] + \mathfrak{h}_{p-1}^{(n,m)}[2n,2m] + \mathfrak{h}_{p-1}^{(m+n)}[2m+2n].$$

Together with Eq. (9.38) this yields

$$2\mathfrak{h}_{p-1}^{(m,n)}[2m,2n] \equiv \mathfrak{h}_{p-1}^{(m)}[2m]\mathfrak{h}_{p-1}^{(n)}[2n] - \mathfrak{h}_{p-1}^{(m+n)}[2m+2n] \pmod{[p]}.$$

Our theorem follows from Eq. (9.30) quickly. □

In the study of the q-MZVs the following value appears naturally (see Eq. (12.78)):

$$\varphi_q[n] = \sum_{k=1}^{\infty}(k-1)\frac{q^{(n-1)k}}{[k]^n} = \sum_{k=1}^{\infty} \frac{kq^{(n-1)k}}{[k]^n} - \zeta_q^I[n],$$

where $\zeta_q^I[n] = \sum_{k=1}^{\infty} \frac{q^{(n-1)k}}{[k]^n}$ is the q-Riemann zeta value.

Proposition 9.5.11. *For every prime p, we have*

$$\sum_{k=1}^{p-1} \frac{kq^k}{[k]^2} \equiv -\frac{p(p-1)(p+1)}{24}(1-q)^2 \pmod{[p]}. \quad (9.39)$$

Proof. To save space, all congruences are modulo $[p]$ throughout this proof.

We can check the congruence Eq. (9.39) for $p=2$ and $p=3$ easily by hand. Now we assume $p \ge 5$. By the definition we have

$$\mathfrak{h}_{p-1}^{\mathbf{0}}[1,2] = \sum_{p>i>j<1} \frac{1}{(1-q^i)(1-q^j)^2}.$$

With substitution $i \to p-i$ and $j \to p-j$ we get

$$\begin{aligned}
-\mathfrak{h}^{\mathbf{o}}_{p-1}[1,2] &= -\sum_{1\le j<i<p} \frac{q^{2i}\cdot q^j}{(q^i-q^p)^2(q^j-q^p)} \\
&\equiv -\sum_{1\le j<i<p} \frac{q^{2i}\cdot q^j}{(q^i-1)^2(q^j-1)} \\
&\equiv -\sum_{1\le j<i<p} \frac{(q^i-1)^2+2(q^i-1)+1}{(q^i-1)^2}\cdot\frac{1-q^j-1}{1-q^j} \\
&\equiv \mathfrak{h}^{\mathbf{o}}_{p-1}[2,1]-2\mathfrak{h}^{\mathbf{o}}_{p-1}[1,1]+\sum_{k=1}^{p-1}\frac{p-3+k}{1-q^k}-\sum_{k=1}^{p-1}\frac{k-1}{(1-q^k)^2}-\binom{p-1}{2} \\
&\equiv \mathfrak{h}^{\mathbf{o}}_{p-1}[2,1]-2\mathfrak{h}^{\mathbf{o}}_{p-1}[1,1]+(p-3)\mathfrak{h}^{\mathbf{o}}_{p-1}[1]+\mathfrak{h}^{\mathbf{o}}_{p-1}[2] \\
&\quad -\binom{p-1}{2}-\sum_{k=1}^{p-1}\frac{kq^k}{(1-q^k)^2}.
\end{aligned}$$

Note that we have the stuffle relations

$$\mathfrak{h}^{\mathbf{o}}_{p-1}[2,1]+\mathfrak{h}^{\mathbf{o}}_{p-1}[1,2]=\mathfrak{h}^{\mathbf{o}}_{p-1}[1]\mathfrak{h}^{\mathbf{o}}_{p-1}[2]-\mathfrak{h}^{\mathbf{o}}_{p-1}[3],$$
$$2\mathfrak{h}^{\mathbf{o}}_{p-1}[1,1]=\mathfrak{h}^{\mathbf{o}}_{p-1}[1]^2-\mathfrak{h}^{\mathbf{o}}_{p-1}[2].$$

Hence

$$\begin{aligned}
\sum_{k=1}^{p-1}\frac{kq^k}{(1-q^k)^2} &\equiv (\mathfrak{h}^{\mathbf{o}}_{p-1}[1]+2)\mathfrak{h}^{\mathbf{o}}_{p-1}[2] \\
&\quad -\mathfrak{h}^{\mathbf{o}}_{p-1}[3]-\mathfrak{h}^{\mathbf{o}}_{p-1}[1]^2+(p-3)\mathfrak{h}^{\mathbf{o}}_{p-1}[1]-\binom{p-1}{2}.
\end{aligned}$$

Taking $d=1$ in Eq. (9.35) of Cor. 9.5.9 we get

$$\mathfrak{h}^{\mathbf{o}}_{p-1}[3]\equiv-\frac{(p-1)(p-3)}{8}.$$

The proposition now follows from Eqs. (9.24) and (9.25) immediately. □

9.6 Historical Notes

The study of the q-series and q-analogs of number theoretical objects has been an active and fruitful area in mathematics for quite a long time. In recent years, q-calculus has gradually developed into an interdisciplinary subject, so much so that some would call it the q-disease [204]. It shows up in a plethora of mathematical branches such as the partitions, elliptic curves and quantum group theory. Many classical identities possess q-analogs, some of which can even have combinatorial explanations. Note that the notation of quantum factorial and quantum Pochhammer symbol used in this chapter are only semi-standard (for e.g., see [12, 204]).

The results of Lemma 9.2.1, due to Hessami Pilehroods (see [277]), are generalizations of the ordinary combinatorial identities contained in [278]. Lemma 9.2.2 is the q-analog of [628, Lemma 3.1]. Like in the study of classical combinatorics, the WZ method (more precisely, its q-analog, see [362]) often plays an important role in the proof of these combinatorial results in the q-world.

The q-analogs of Two-one formulas, Cors. 9.4.1 and 9.4.2 were first proved by Hessami Pilehroods in [277]. The main results of Sections 9.1-9.3 were established in our joint work [280].

Theorem 9.5.1 is due to Shi and Pan [465, 512] who strengthened an earlier result of Andrews [13]. In [168] Dilcher extended it again to higher weight in Thm. 9.5.3 by using the degenerate Bernoulli numbers $\beta_n(\lambda)$ which were first designed by Carlitz [114]. The author discovered Cor. 9.5.5 and the homogeneous case, Thm. 9.5.6, in [629]. Note that congruence (9.31) is a variant of [168, (5.11)].

Comparing to the results from Chap. 8, we see that a key link is missing here: the q-analog of Bernoulli numbers. Even though there are several candidates in the literature (see, for e.g., Eq. (11.23) or [113, 549]), they have not shown to be related to the q-MHSs via any congruence relations. It would be very interesting to work out a solution to this problem. One possible approach is to consider the p-adic objects in [226] and take suitable congruences.

A certain modified form (type I form in the terminology of Chap. 11) of Cors. 9.5.8 and 9.5.9 was discovered by Hessami Pilehroods and Tauraso in [279] (see Exercises 9.7 and 9.8). Moreover, Hessami Pilehroods and Tauraso obtained some "quasi-homogeneous" results for both H- and $H^\star$-versions (see Exercises 9.9 and 9.10). They also found a general duality result (see Exercises 9.11) which, in its reversal q-binomial form, appears in [80].

Exercises

9.1. For any integers $n \geq 1$ and $l \geq 0$, show that

$$2 \sum_{k=l+1}^{n} (-1)^k \frac{\binom{n}{k}}{\binom{n+k}{k}} = \frac{l-n}{n}(-1)^l \frac{\binom{n}{l}}{\binom{n+l}{l}},$$

$$2 \sum_{k=l+1}^{n} \frac{k\binom{n}{k}}{\binom{n+k}{k}} = (n-l)\frac{\binom{n}{l}}{\binom{n+l}{l}},$$

$$2\sum_{k=1}^{n} \frac{\binom{n}{k}}{k\binom{n+k}{k}} = \sum_{m=1}^{n} \frac{1}{m}.$$

Moreover, if $l \geq 1$ then

$$\sum_{k=l}^{n} \frac{\binom{k}{l}}{k^2\binom{k+l}{l}} = \frac{\binom{n}{l}}{l^2\binom{n+l}{l}}.$$

9.2. Prove Cor. 9.4.1 and Cor. 9.4.3.

9.3. Suppose $\ell \in \mathbb{N}_0$, $n \in \mathbb{N}$ and $a_0, a_j, b_j, c_j - 3 \in \mathbb{N}_0$ for all $j \geq 1$. Consider the following two possible types of compositions:

$$(\underline{\mathbf{2}\text{-}c\text{-}\mathbf{2}\text{-}1}) : \mathbf{s} = \Big(\mathop{\mathbf{Cat}}_{j=1}^{\ell} \big\{\{2\}^{b_j}, c_j, \{2\}^{a_j}, 1\big\}\Big), \quad \ell \in \mathbb{N},$$

$$\Pi[\mathbf{s}] = \Big(\mathop{\mathbf{Cat}}_{j=1}^{\ell} \big\{\overline{2b_j+2}, \{1\}^{c_j-3}, \overline{2a_j+2}\big\};$$

$$\mathop{\mathbf{Cat}}_{j=1}^{\ell} \big\{(b_j+1), \{0\}^{c_j-3}, a_j+1\big\};$$

$$1, \{\theta\}^{c_1-3}, 1, \mathop{\mathbf{Cat}}_{j=2}^{\ell} \big\{-1, \{\theta\}^{c_j-3}, 1\big\}\Big);$$

$$(\mathbf{2}\text{-}1\text{-}\underline{\mathbf{2}\text{-}c\text{-}\mathbf{2}\text{-}1}) : \mathbf{s} = \Big(\{2\}^{a_0}, 1, \mathop{\mathbf{Cat}}_{j=1}^{\ell} \big\{\{2\}^{b_j}, c_j, \{2\}^{a_j}, 1\big\}\Big), \quad \ell \in \mathbb{N}_0,$$

$$\Pi[\mathbf{s}] = \Big(2a_0+1, \mathop{\mathbf{Cat}}_{j=1}^{\ell} \big\{\overline{2b_j+2}, \{1\}^{c_j-3}, \overline{2a_j+2}\big\};$$

$$a_0+1, \mathop{\mathbf{Cat}}_{j=1}^{\ell} \big\{b_j+1, \{0\}^{c_j-3}, a_j+1\big\};$$

$$2, \mathop{\mathbf{Cat}}_{j=1}^{\ell} \big\{-1, \{\theta\}^{c_j-3}, 1\big\}\Big).$$

Then in each case show that

$$H_n^\star[\mathbf{s}] = \sum_{\underline{\mathbf{p}} \preceq \Pi[\mathbf{s}]} \mathscr{H}_n[\underline{\mathbf{p}}].$$

9.4. Suppose $\ell \in \mathbb{N}_0$, $n, a_{\ell+1} \in \mathbb{N}$, and $a_0, a_j, b_j, c_j - 3 \in \mathbb{N}_0$ for all $1 \le j \le \ell$. Consider the following two possible types of compositions:

$$(\underline{\mathbf{2}\text{-}c\text{-}\mathbf{2}\text{-}1}\text{-}\mathbf{2}) : \mathbf{s} = \Big(\mathop{\mathbf{Cat}}_{j=1}^{\ell}\big\{\{2\}^{b_j}, c_j, \{2\}^{a_j}, 1\big\}, \{2\}^{a_{\ell+1}}\Big), \quad \ell \in \mathbb{N},$$

$$\begin{aligned}\Pi[\mathbf{s}] = \Big(&\mathop{\mathbf{Cat}}_{j=1}^{\ell}\big\{\overline{2b_j+2}, \{1\}^{c_j-3}, \overline{2a_j+2}\big\}, \overline{2a_{\ell+1}};\\ &\mathop{\mathbf{Cat}}_{j=1}^{\ell}\big\{b_j+1, \{0\}^{c_j-3}, a_j+1\big\}, a_{\ell+1};\\ &1, \{\theta\}^{c_1-3}, 1, \mathop{\mathbf{Cat}}_{j=2}^{\ell}\big\{-1, \{\theta\}^{c_j-3}, 1\big\}, -1\Big);\end{aligned}$$

$$(\mathbf{2}\text{-}1\text{-}\underline{\mathbf{2}\text{-}c\text{-}\mathbf{2}\text{-}1}\text{-}\mathbf{2}) : \mathbf{s} = \Big(\{2\}^{a_0}, 1, \mathop{\mathbf{Cat}}_{j=1}^{\ell}\big\{\{2\}^{b_j}, c_j, \{2\}^{a_j}, 1\big\}, \{2\}^{a_{\ell+1}}\Big), \quad \ell \in \mathbb{N}_0,$$

$$\begin{aligned}\Pi[\mathbf{s}] = \Big(&2a_0+1, \mathop{\mathbf{Cat}}_{j=1}^{\ell}\big\{\overline{2b_j+2}, \{1\}^{c_j-3}, \overline{2a_j+2}\big\}, \overline{2a_{\ell+1}};\\ &a_0+1, \mathop{\mathbf{Cat}}_{j=1}^{\ell}\big\{(b_j+1), \{0\}^{c_j-3}, a_j+1\big\}, a_{\ell+1};\\ &2, \mathop{\mathbf{Cat}}_{j=1}^{\ell}\big\{-1, \{\theta\}^{c_j-3}, 1\big\}, -1\Big).\end{aligned}$$

Then in each case show that

$$H_n^\star[\mathbf{s}] = -\sum_{\underline{\mathbf{p}} \preceq \Pi[\mathbf{s}]} \mathscr{H}_n[\underline{\mathbf{p}}].$$

9.5. Prove Eq. (9.33) by induction on the depth d.

9.6. Verify the partial fraction expansion in Eq. (9.29).

In the rest of the exercises we set $\mathfrak{h}_{p-1}^I[\mathbf{s}] = \mathfrak{h}_{p-1}^{\mathbf{s}-\mathbf{1}}[\mathbf{s}]$ (see Eq. (9.26)) and similarly for $\mathfrak{h}_{p-1}^{\star,I}[\mathbf{s}]$. Also, all the congruences are modulo $[p]$.

9.7. For any positive integer n and for any prime $p > n$, show that

$$\mathfrak{h}_{p-1}^I\big[\{2\}^n\big] \equiv \frac{(-1)^n}{p(n+1)}\binom{p+n}{2n+1}.$$

Moreover, prove that

$$\mathfrak{h}_{p-1}^{\mathbf{0}}\big[\{2\}^n\big] \equiv \frac{1}{p(n+1)}\left((-1)^n\binom{p-1}{2n+1} + \binom{p-1}{n}\right).$$

9.8. For any positive integer n and for any prime $p > n$, prove that

$$\mathfrak{h}_{p-1}^I\big[\{3\}^n\big] \equiv \frac{1}{p^2(n+1)}\left(\binom{p+2n+1}{3n+2} + \binom{-p+2n+1}{3n+2}\right).$$

9.9. Suppose $a, b \in \mathbb{N}_0$ and $t = a + b + 1$. For any prime $p > 2t + 1$, show that

$$\mathfrak{h}^{I}_{p-1}\big[\{2\}^a, 3, \{2\}^b\big] + \mathfrak{h}^{\mathbf{0}}_{p-1}\big[\{2\}^b, 3, \{2\}^a\big] \equiv \frac{(-1)^{t-1}}{p(t+1)}\binom{p+t}{2t+1}.$$

9.10. Suppose $a, b \in \mathbb{N}_0$ and $w = a + b + 2$. For any prime $p \geq w + 1$, show that

$$\mathfrak{h}^{I}_{p-1}\big[\{1\}^a, 2, \{1\}^b\big] + \mathfrak{h}^{I}_{p-1}\big[\{1\}^b, 2, \{1\}^a\big] \equiv -\frac{1}{p}\binom{p+1}{w+1}.$$

9.11. For any prime p and composition $\mathbf{s}$, show that

$$H^{\star,I}_{p-1}\big[\overleftarrow{\mathbf{s}}\big] \equiv -(-1)^{|\mathbf{s}|} H^{\star,I}_{p-1}\big[\mathbf{s}^{\vee}\big]$$

where the v-dual $\mathbf{s}^{\vee}$ is defined by Eq. (7.8).

9.12. Let $e_0, e_1, \ldots, e_n \in \mathbb{N}_0$. Prove the following statements.

(i) For any prime $p > w + 1$, where $w = 2n + \sum_{i=0}^{n} e_i$,

$$\sum_{i=0}^{n} \mathfrak{h}^{I}_{p-1}\left[\{1\}^{e_i}, 2, \{1\}^{e_{i+1}}, 2, \ldots, 2, \{1\}^{e_{i+n}}\right] \equiv \frac{(-1)^n}{p}\binom{p+n}{w+1}. \quad (9.40)$$

(ii) For any prime $p > w + 1$, where $w = 3n + \sum_{i=0}^{n} 2e_i$,

$$\sum_{i=0}^{n} \mathfrak{h}^{I}_{p-1}\left[\{2\}^{e_i}, 3, \{2\}^{e_{i+1}}, 3, \ldots, 3, \{2\}^{e_{i+n}}\right] \equiv \frac{1}{p} f(p, w, n), \quad (9.41)$$

where f is independent of q.

In both congruences we assume $e_i = e_j$ if $i \equiv j \pmod{n+1}$.

Chapter 10

Multiple Zeta Star Values (MZ*Vs)

The multiple zeta star values are a kind of variation of the MZVs. For any fixed positive integer d and $\mathbf{s} = (s_1, \ldots, s_d) \in \mathbb{N}^d$, $s_1 \geq 2$, the *multiple zeta star function* is defined by

$$\zeta^\star(\mathbf{s}) = \sum_{k_1 \geq \cdots \geq k_d > 0} k_1^{-s_1} \cdots k_d^{-s_d} \tag{10.1}$$

where $s_1 + \cdots + s_j > j$ for all $j = 1, \ldots, d$. Again, the number $d = \mathrm{dp}(\mathbf{s})$ is called the *depth* (or *length*), and $|\mathbf{s}| = s_1 + \cdots + s_d$ the *weight* when $\mathbf{s} \in \mathbb{N}^d$, and in this case the values $\zeta^\star(\mathbf{s})$ are called the *multiple zeta star values* (MZ*Vs for abbreviation). In this chapter we present some of the most important results concerning the MZ*Vs.

10.1 An Integral Expression

Recall from Sec. 5.1 that the power set of $\mathbf{s} = (s_1, \ldots, s_d)$ of weight $w = |\mathbf{s}|$ is $P(\mathbf{s}) = (s_1, s_1 + s_2, \ldots, s_1 + \cdots + s_{d-1}, w)$.

Theorem 10.1.1. *Let* $\mathbf{s} = (s_1, \ldots, s_d) \in \mathbb{N}^d$ *and*

$$\Delta(\mathbf{s}) = \left\{ (t_1, \ldots, t_w) \in [0,1]^w \,\middle|\, \begin{matrix} t_j < t_{j+1}, & \text{if } j \notin P(\mathbf{s}); \\ t_j > t_{j+1}, & \text{if } j \in P(\mathbf{s}). \end{matrix} \right\}$$

Then for every $N \in \mathbb{N}$ *the MH*S* $H_N^\star(\mathbf{s})$ *(see Definition (7.3)) is given by the iterated integral*

$$H_N^\star(s_1, \ldots, s_d) = \int_{\Delta(\mathbf{s})} \omega_{\delta(w)} \cdots \omega_{\delta(2)}(t^{N-1}\mathrm{d}t)$$

where $\omega_0 = \frac{\mathrm{d}t}{t}$, $\omega_1 = \frac{\mathrm{d}t}{1-t}$ *and*

$$\delta(j) = \delta(j, \mathbf{s}) = \begin{cases} 0, & \text{if } j - 1 \notin P(\mathbf{s}); \\ 1, & \text{if } j - 1 \in P(\mathbf{s}). \end{cases}$$

Proof. For $j = 1, \dots, d$, we set $p_j = s_1 + \cdots + s_j$. We have

$$\Delta(\mathbf{s}) = \{0 < t_1 < \cdots < t_{p_1} > t_{p_1+1} < t_{p_1+2} < \cdots \\ < t_{p_{d-1}} > t_{p_{d-1}+1} < \cdots < t_w < 1\}.$$

Now we prove the theorem by induction on d. When $d = 1$ it is not hard to see the theorem is true. Now suppose $d \geq 2$ and the theorem is true for depth less than d. Then

$$\begin{aligned}
&\int_{t_{p_1+1}}^{1} \omega_{\delta(p_1+1)} \cdots \omega_{\delta(2)}(t^{N-1}\mathrm{d}t) \\
&\quad = \left(\int_t^1 \left(\int_0^{t_{p_1}} \left(\frac{\mathrm{d}t}{t}\right)^{p_1-2} (t^{N-1}\mathrm{d}t)\right) \frac{\mathrm{d}t_{p_1}}{t_{p_1}}\right) \frac{\mathrm{d}t_{p_1+1}}{1-t_{p_1+1}} \\
&\quad = \left(\int_{t_{p_1+1}}^1 \frac{t_{p_1}^{N-1}\mathrm{d}t_{p_1}}{N^{p_1-1}}\right) \frac{\mathrm{d}t_{p_1+1}}{1-t_{p_1+1}} \\
&\quad = \frac{1}{N^{p_1}} \frac{(1-t_{p_1+1}^N)dt_{p_1+1}}{1-t_{p_1+1}} \\
&\quad = \sum_{N \geq m} \frac{1}{N^{p_1}} t_{p_1+1}^{m-1} dt_{p_1+1}.
\end{aligned}$$

Setting $\mathbf{s}' = (s_2, \dots, s_d)$ we see that

$$\begin{aligned}
\int_{\Delta(\mathbf{s})} \omega_{\delta(w)} \cdots \omega_{\delta(2)}(t^{N-1}dt) &= \sum_{N \geq m} \frac{1}{N^{s_1}} \int_{\Delta(\mathbf{s}')} \omega_{\delta(w)} \cdots \omega_{\delta(p_1+2)}(t^{m-1}dt) \\
&= \sum_{N \geq m} \frac{1}{N^{s_1}} \int_{\Delta(\mathbf{s}')} \omega_{\delta'(w-s_1)} \cdots \omega_{\delta'(2)}(t^{m-1}dt)
\end{aligned}$$

where $\delta'(j) = \delta(j, \mathbf{s}')$. Now the theorem follows immediately from the induction assumption. □

10.2 Relation Between MZ*Vs and MZVs

For compositions $\mathbf{s}$ and $\mathbf{t}$, we say $\mathbf{s}$ refines $\mathbf{t}$ (denoted by $\mathbf{s} \succ \mathbf{t}$) if $\mathbf{t}$ can be obtained from $\mathbf{s}$ by combining some of its parts.

Proposition 10.2.1. *For all admissible* $\mathbf{s} \in \mathbb{N}^d$, *we have*

$$\zeta^\star(\mathbf{s}) = \sum_{\mathbf{s} \succeq \mathbf{t}} \zeta(\mathbf{t}), \tag{10.2}$$

$$\zeta(\mathbf{s}) = \sum_{\mathbf{s} \succeq \mathbf{t}} (-1)^{\mathrm{dp}(\mathbf{s})-\mathrm{dp}(\mathbf{t})} \zeta^\star(\mathbf{t}). \tag{10.3}$$

Proof. The first formula Eq. (10.2) is clear by the definition. The second formula Eq. (10.3) follows from the Inclusion-Exclusion Principle. □

Fix a positive integer d. For a partition $\Pi = \{P_1, \ldots, P_\ell\}$ of the set $\{1, 2, \cdots, d\}$, we denote by $\ell(\Pi) = \ell$ its number of parts and set

$$c(\Pi) := \prod_{k=1}^{\ell} (\sharp P_k - 1)!,$$

where $\sharp S$ denotes the cardinality of a set S. The next theorem provides a kind of sum formula for both the MZVs and the MZ*Vs.

Theorem 10.2.2. *Let* $\mathbf{s} = (s_1, \ldots, s_d) \in \mathbb{R}^d_{>1}$. *Then*

$$\sum_{\sigma \in \mathfrak{S}_d} \zeta^\star(s_{\sigma(1)}, \ldots, s_{\sigma(d)}) = \sum_{\Pi:\,\text{partition of } \{1,\ldots,d\}} c(\Pi)\zeta(\mathbf{i}, \Pi), \tag{10.4}$$

$$\sum_{\sigma \in \mathfrak{S}_d} \zeta(s_{\sigma(1)}, \ldots, s_{\sigma(d)}) = \sum_{\Pi:\,\text{partition of } \{1,\ldots,d\}} (-1)^{d-\ell(\Pi)} c(\Pi)\zeta(\mathbf{i}, \Pi), \tag{10.5}$$

where $\mathfrak{S}_d$ *is the symmetric group of* d *letters and* $\zeta(\mathbf{i}, \Pi) = \prod_{k=1}^{\ell} \zeta\Big(\sum_{j \in P_k} s_j\Big)$ *for the partition* $\Pi = \{P_1, \ldots, P_\ell\}$.

Proof. Note that all the components in $\mathbf{s}$ are real numbers so we may assume they are all distinct since otherwise we may take limits.

Let us prove Eq. (10.4) first. By the definition it is easy to see

$$\sum_{\sigma \in \mathfrak{S}_d} \zeta^\star(s_{\sigma(1)}, \ldots, s_{\sigma(d)}) = \sum_{\sigma \in \mathfrak{S}_d} \sum_{k_1 \geq \cdots \geq k_d > 0} \frac{1}{k_{\sigma(1)}^{s_1} \cdots k_{\sigma(d)}^{s_d}}.$$

For each d-tuple $\mathbf{k} = (k_1, \ldots, k_d)$, let $\mathfrak{S}_d(\mathbf{k})$ be its stabilizer and $\Pi(\mathbf{k})$ the associated partition of $\{1, \ldots, d\}$. By considering each term $1/k_{\sigma(1)}^{s_1} \cdots k_{\sigma(d)}^{s_d}$ we only need to show for every $\mathbf{k}$

$$\sharp\mathfrak{S}_d(\mathbf{k}) = \sum_{\Pi \succeq \Pi(\mathbf{k})} c(\Pi).$$

Now every element in $\mathfrak{S}_d$ can be written as a product of disjoint cycles (called its cycle type) uniquely, up to permutation of the cycles. So $\sharp\mathfrak{S}_d(\mathbf{k})$ counts the number of those elements whose cycle types are specified by a partition of $\{1, \ldots, d\}$ that refines $\Pi(\mathbf{k})$. But by the definition $c(\Pi)$ is exactly the number of cycle types that are specified by Π. This concludes the proof of Eq. (10.4).

Similarly, to prove Eq. (10.5) we need to show that for every $\mathbf{k}$

$$\sum_{\Pi\succeq\Pi(\mathbf{k})} (-1)^{d-\ell(\Pi)} c(\Pi) = \begin{cases} 0, & \text{if } \ell(\Pi) < d; \\ 1, & \text{if } \ell(\Pi) = d. \end{cases} \tag{10.6}$$

This is because on the left-hand side of Eq. (10.5) all of the components of $\mathbf{k}$ are distinct. A moment reflection leads to the conclusion that $d-\ell(\Pi)$ is even if and only if the permutations of cycle type Π is even. For example, $\ell(\Pi) = d$ if and only if the permutations are trivial. Hence the left-hand side of Eq. (10.6) is the signed sum of the number of even and odd permutations in the stabilizer $\mathfrak{S}_d(\Pi)$. If $\mathfrak{S}_d(\Pi)$ is not trivial then there must be at least one transposition which is an odd permutation. Hence the numbers of even and odd permutations are the same. After cancelation the only nontrivial contribution is from $\Pi = \{\{1\},\{2\},\dots,\{d\}\}$ producing 1. This finishes the proof of Eq. (10.5) and the theorem. □

10.3 Sum Relations

We have seen the sum relation of the MZVs in Cor. 5.2.4. We have a similar result for the MZ*Vs.

Theorem 10.3.1. *For every pair of positive integers $d \le w$, we have*

$$\sum_{k_1+\cdots+k_d=w} \zeta^\star(k_1,\dots,k_d) = \binom{w-1}{d-1}\zeta(w). \tag{10.7}$$

Proof. First, for each fixed $\mathbf{s}$ of depth ℓ and weight w we want to count how many copies of $\zeta(\mathbf{s})$ the left-hand side of Eq. (10.7) can produce. For fixed positive number d, this number turns out to be

$$\sharp\{\mathbf{k} = (k_1,\dots,k_d) \in \mathbb{N}^d : k_1 > 1, \mathbf{k} \succeq \mathbf{s}\} = \binom{w-d-1}{d-\ell}.$$

This can be verified as follows. In the sum $s_1+\cdots+s_\ell = w$ one can think of each s_j as a string of 1's of length s_j. Then each refinement $\mathbf{k} \succeq (s_1,\dots,s_\ell)$ corresponds to choosing from $(s_1-2)+(s_2-1)+\cdots+(s_\ell-1) = w-\ell-1$ positions between the 1's to insert $d-\ell$ '+' signs (notice $s_1, k_1 \ge 2$). Therefore

$$\sum_{k_1+\cdots+k_d=w} \zeta^\star(k_1,\dots,k_d) = \sum_{\ell=1}^{d} \binom{w-\ell-1}{d-\ell} \sum_{|\mathbf{s}|=w,\mathrm{dp}(\mathbf{s})=\ell} \zeta(\mathbf{s}). \tag{10.8}$$

Since we have proved the following sum relation for MSVs (see Eq. (5.12))

$$\sum_{|\mathbf{s}|=w,\mathrm{dp}(\mathbf{s})=\ell} \zeta(\mathbf{s}) = \zeta(w),$$

it follows from Eq. (10.8) and an easy combinatorial formula that

$$\sum_{k_1+\cdots+k_d=w} \zeta^\star(k_1,\ldots,k_d) = \sum_{\ell=1}^{d} \binom{w-\ell-1}{d-\ell}\zeta(w) = \binom{w-1}{d-1}\zeta(w)$$

as desired. This finishes the proof the theorem. □

10.4 Cyclic Sum Relations

The cyclic sum relation for the MZVs was considered in Sec. 5.2. Similarly, to prove the cyclic sum relation for the MZ⋆Vs we need some auxiliary series. Define

$$C(k_1,\ldots,k_d) := \sum_{n_1\geq n_2\geq\cdots\geq n_d\geq n_{d+1}\geq 1,\ n_1\neq n_{d+1}} \frac{1}{(n_1-n_{d+1})n_1^{k_1}\ldots n_d^{k_d}}$$

Lemma 10.4.1. *Suppose $d, k_1, \ldots, k_d \in \mathbb{N}$. If $k_i > 1$ for some i then we have*

$$\begin{aligned} &C(k_1,k_2,\ldots,k_d) - C(k_2,k_3,\ldots,k_d,k_1) \\ &\quad = k_1\zeta(|\mathbf{k}|+1) - \sum_{i=0}^{k_1-2} \zeta^\star(k_1-i,k_2,k_3,\ldots,k_d,i+1), \end{aligned}$$

where the sum on the right vanishes if $k_1 = 1$.

Proof. By modifying the proof of Lemma 5.2.2 one can show that (see Exercise 10.1): for any nonnegative integer $i \leq k_1 - 2$, we have

$$\begin{aligned} &\sum_{n_1\geq n_2\geq\cdots\geq n_d\geq n_{d+1}\geq 1,\ n_1\neq n_{d+1}} \frac{1}{(n_1-n_{d+1})n_1^{k_1-i}n_2^{k_2}\ldots n_d^{k_d}n_{d+1}^{i}} \\ &\quad = \sum_{n_1\geq n_2\geq\cdots\geq n_d\geq n_{d+1}\geq 1,\ n_1\neq n_{d+1}} \frac{1}{(n_1-n_{d+1})n_1^{k_1-i-1}n_2^{k_2}\ldots n_d^{k_d}n_{d+1}^{i+1}} \\ &\qquad - \zeta^\star(k_1-i,k_2,k_3,\ldots,k_d,i+1) + \zeta(|\mathbf{k}|+1). \end{aligned}$$

Adding up this equality for $i = 0, 1, \ldots, k_1 - 2$, we get

$$\begin{aligned} C(k_1,k_2,\ldots,k_d) = &\sum_{\substack{n_1\geq n_2\geq\cdots\geq n_d\geq n_{d+1}\geq 1\\ n_1\neq n_{d+1}}} \frac{1}{(n_1-n_{d+1})n_1n_2^{k_2}\ldots n_d^{k_d}n_{d+1}^{k_1-1}} \\ &- \sum_{i=0}^{k_1-2} \zeta^\star(k_1-i,k_2,k_3,\ldots,k_d,i+1) + (k_1-1)\zeta(|\mathbf{k}|+1). \end{aligned}$$

We can now rewrite the first sum on the right-hand side as

$$\begin{aligned}
&\sum_{\substack{n_1\geq n_2\geq\cdots\geq n_d\geq n_{d+1}\geq 1\\ n_1\neq n_{d+1}}} \frac{1}{n_2^{k_2}\dots n_d^{k_d}n_{d+1}^{k_1}}\left(\frac{1}{n_1-n_{d+1}}-\frac{1}{n_1}\right)\\
&=\sum_{\substack{n_1\geq n_2\geq\cdots\geq n_d\geq n_{d+1}\geq 1\\ n_2\neq n_{d+1}}} \frac{1}{n_2^{k_2}\dots n_d^{k_d}n_{d+1}^{k_1}}\sum_{n_1\geq n_2}\left(\frac{1}{n_1-n_{d+1}}-\frac{1}{n_1}\right)\\
&\quad+\sum_{n_2=\cdots=n_{d+1}\geq 1} \frac{1}{n_2^{k_2}\dots n_d^{k_d}n_{d+1}^{k_1}}\sum_{n_1>n_2}\left(\frac{1}{n_1-n_{d+1}}-\frac{1}{n_1}\right)\\
&=\sum_{\substack{n_1\geq n_2\geq\cdots\geq n_d\geq n_{d+1}\geq 1\\ n_2\neq n_{d+1}}} \frac{1}{n_2^{k_2}\dots n_d^{k_d}n_{d+1}^{k_1}}\sum_{n_{d+2}=1}^{n_{d+1}}\left(\frac{1}{n_2}-\frac{1}{n_{d+2}}\right)\\
&\quad+\sum_{n_2=\cdots=n_{d+1}\geq 1} \frac{1}{n_2^{k_2}\dots n_d^{k_d}n_{d+1}^{k_1}}\sum_{n_{d+2}=0}^{n_{d+1}}\left(\frac{1}{n_2}-\frac{1}{n_{d+2}}\right)\\
&=C(k_2,k_3,\dots,k_d,k_1)+\zeta(|\mathbf{k}|+1)
\end{aligned}$$

by splitting the term corresponding to $n_{d+2}=0$ away from the last sum as $\zeta(|\mathbf{k}|+1)$. The lemma now follows immediately. □

Keeping the same notation as that in Sec. 5.2 we have the following *cyclic sum relations* of the MZ*Vs.

Theorem 10.4.2. *For any index set $S\in\mathscr{C}(w,d)$ with $0<d<w$, we have*

$$\sum_{(k_1,k_2,\dots,k_d)\in S}\ \sum_{i=0}^{k_1-2}\zeta^\star(k_1-i,k_2,\dots,k_d,i+1)=\frac{w\sharp S}{d}\zeta(w+1) \tag{10.9}$$

where the inner sum on the left-hand side is treated as 0 whenever $k_1=1$.

Proof. We can prove the theorem by applying Lemma 10.4.1 to all cyclic permutations of $(k_1,\dots,k_d)\in S$ and adding them up. □

Example 10.4.3. Applying the cyclic sum formula Eq. (10.9) to $\big(\{2\}^n\big)$ and $\big(3,\{2\}^{n-2}\big)$ respectively, we get

$$\zeta^\star(\{2\}^n,1)=2\zeta(2n+1), \tag{10.10}$$

$$\zeta^\star(\{2\}^n)+\sum_{i+j=n-2}\zeta^\star(\{2\}^i,3,\{2\}^j,1)=(2n-1)\zeta(2n). \tag{10.11}$$

Notice Eq. (10.11) is obviously true even for $n=1$.

10.5 Weighted Sum Relations

Similar to the MZV case we have the following result.

Theorem 10.5.1. *For any integer $w \geq 3$, we have*

$$\sum_{i=2}^{w-1} 2^i \zeta^\star(i, w-i) = (2^w + w - 3)\zeta(w). \tag{10.12}$$

Proof. By Eq. (10.2) we have $\zeta^\star(i, w-i) = \zeta(i, w-i) + \zeta(w)$. It follows from the weighted sum relations of MZV Eq. (5.15) that

$$\sum_{i=2}^{w-1} 2^i \zeta^\star(i, w-i) = \left(w + 1 + \sum_{i=2}^{w-1} 2^i \right) \zeta(w) = \big(2^w + w - 3\big)\zeta(w).$$

Hence Eq. (10.12) is proved. □

However, a parallel result corresponding to Thm. 5.11.1 by Guo and Xie for arbitrary depth is not available. In fact we have the following equation.

Example 10.5.2. When the depth $d = 3$ and $w = 5$ or 6 we have

$$\sum_{\substack{a\geq 2, b\geq 1\\ a+b<5}} \left(2^{a+b-1} + 2^{a-1} - 2^{b-1}\right)\zeta^\star(a, b, 5-a-b) = 6\zeta(2)\zeta(3) + 25\zeta(5),$$

$$\sum_{\substack{a\geq 2, b\geq 1\\ a+b<6}} \left(2^{a+b-1} + 2^{a-1} - 2^{b-1}\right)\zeta^\star(a, b, 6-a-b) = 9\zeta(3)^2 + 86\zeta(6).$$

Conjecturally, $\zeta(2)\zeta(3)$ and $\zeta(5)$ are linearly independent over $\mathbb{Q}$, and the same holds for $\zeta(3)^2$ and $\zeta(6)$.

10.6 Fixed Weight, Height and Depth Relations

In this section we will consider the formula of the MZ*Vs analogous to that of the MZVs from Sec. 5.6. The basic idea is still to use generating functions. However, it turns out that these generating functions can be expressed succinctly by some hypergeometric functions. Naturally, the proof depends critically on the transformation properties of such functions so the reader is advised to take a brief look at Appendix C before reading this section.

Recall from Definition 5.6.1 that, for nonnegative integers w, d and h, we have set

$$\begin{aligned} I(w,d,h) &= \big\{\mathbf{s} = (s_1,\dots,s_d) \in \mathbb{N}^d : |\mathbf{s}| = w, \operatorname{ht}(\mathbf{s}) = h\big\}, \\ I_0(w,d,h) &= \big\{\mathbf{s} = (s_1,\dots,s_d) \in \mathbb{N}^d : |\mathbf{s}| = w, \operatorname{ht}(\mathbf{s}) = h, s_1 \geq 2\big\}. \end{aligned}$$

We first consider the sum

$$G_0^\star(w,d,h) = \sum_{\mathbf{s}\in I_0(w,d,h)} \zeta^\star(\mathbf{s}).$$

In fact, similar to the proof of Thm. 5.6.2, we will treat this as a special point in a one parameter family $X_0(w,d,h;t)$ and consider the whole family's generating function. So we define the following variation of one variable multiple polylog function

$$\mathrm{Le}_{\mathbf{s}}(t) = \mathrm{Le}_{s_1,\dots,s_d}(t) := \sum_{k_1\ge k_2\ge\cdots\ge k_d\ge 1} \frac{t^{k_1}}{k_1^{s_1}\cdots k_d^{s_d}},$$

which converges uniformly for all $|t|<1$. Moreover, it is easy to see that for all admissible $\mathbf{s}$ we have $\mathrm{Le}_{\mathbf{s}}(1) = \zeta^\star(\mathbf{s})$. Since $\mathrm{Le}_{\mathbf{s}}(t)$ makes sense for all positive compositions $\mathbf{s}$ we define

$$G^\star(w,d,h;t) = \sum_{\mathbf{s}\in I(w,d,h)} \mathrm{Le}_{\mathbf{s}}(t), \quad G_0^\star(w,d,h;t) = \sum_{\mathbf{s}\in I_0(w,d,h)} \mathrm{Le}_{\mathbf{s}}(t),$$

and their generating functions

$$\Psi^\star(t) = \Psi^\star(x,y,z;t) := \sum_{w\ge d+h, d\ge h\ge 0} G^\star(w,d,h;t)x^{w-d-h}y^{d-h}z^{2h},$$

$$\Psi_0^\star(t) = \Psi_0^\star(x,y,z;t) := \sum_{w\ge d+h, d\ge h\ge 1} G_0^\star(w,d,h;t)x^{w-d-h}y^{d-h}z^{2h-2}.$$

Lemma 10.6.1. *The formal power series $\Psi_0^\star(t)$ is a unique power series solution of*

$$t^2(1-t)\frac{\mathrm{d}^2 f}{\mathrm{d}t^2} + t[(1-t)(1-x)-y]\frac{\mathrm{d}f}{\mathrm{d}t} + (xy-z^2)f = t, \quad f(0)=0. \tag{10.13}$$

Proof. It is straightforward to check that (see Exercise 10.3)

$$\frac{\mathrm{d}}{\mathrm{d}t}\mathrm{Le}_1(t) = \frac{1}{1-t}, \quad \frac{\mathrm{d}}{\mathrm{d}t}\mathrm{Le}_{\mathbf{s}}(t) = \begin{cases} \dfrac{1}{t}\mathrm{Le}_{s_1-1,\dots,s_d}(t), & \text{if } s_1\ge 2; \\ \dfrac{1}{t(1-t)}\mathrm{Le}_{s_2,\dots,s_d}(t), & \text{if } s_1 = 1, d\ge 2. \end{cases} \tag{10.14}$$

Hence

$$\begin{aligned} \frac{\mathrm{d}}{\mathrm{d}t}G_0^\star(w,d,h;t) =& \frac{1}{t}\Big(G^\star(w-1,d,h-1;t) - G_0^\star(w-1,d,h-1;t) \\ &+ G_0^\star(w-1,d,h;t)\Big), \end{aligned}$$

$$\frac{\mathrm{d}}{\mathrm{d}t}\Big(G^\star(w,d,h;t) - G_0^\star(w,d,h;t)\Big) = \frac{1}{t(1-t)}G^\star(w-1,d-1,h;t).$$

Therefore the generating functions satisfy

$$\frac{\mathrm{d}\Psi_0^\star}{\mathrm{d}t} = \frac{1}{yt}\Big(\Psi^\star - 1 - z^2\Psi_0^\star\Big) + \frac{x}{t}\Psi_0^\star,$$

$$\frac{\mathrm{d}}{\mathrm{d}t}\Big(\Psi^\star - z^2\Psi_0^\star\Big) = \frac{y}{t(1-t)}\Big(\Psi^\star - 1\Big) + \frac{y}{1-t}.$$

Solving $\Psi^\star$ from the first equation and plugging it into the second yields the lemma. □

Theorem 10.6.2. *Let α and β be determined by $\alpha+\beta = x+y$ and $\alpha\beta = xy - z^2$. We have*

$$\begin{aligned}
&\Psi_0^\star(x,y,z;t)\\
&= \frac{\Gamma(\beta-\alpha)\Gamma(1-y)}{\Gamma(1-\alpha)\Gamma(1+x-\alpha)}\int_0^t v^{-\beta}(1-v)^{y-1}{}_2F_1(\alpha,\alpha-x;\alpha-\beta+1;v)\,dv\\
&\quad+\frac{\Gamma(\alpha-\beta)\Gamma(1-y)}{\Gamma(1-\beta)\Gamma(1+x-\beta)}\int_0^t v^{-\alpha}(1-v)^{y-1}{}_2F_1(\beta,\beta-x;\beta-\alpha+1;v)\,dv;\\
&\Psi_0^\star(x,y,z;1)\\
&= \frac{\Gamma(\beta-\alpha)\Gamma(1-\beta)\Gamma(y)\Gamma(1-y)}{\Gamma(1-\alpha)\Gamma(1+x-\alpha)\Gamma(1+\alpha-x)}\sum_{n=0}^\infty \frac{(\alpha)_n(1-\beta)_n}{n!(1+\alpha-\beta)_n}\frac{\alpha-x}{n+\alpha-x}\\
&\quad+\frac{\Gamma(\alpha-\beta)\Gamma(1-\alpha)\Gamma(y)\Gamma(1-y)}{\Gamma(1-\beta)\Gamma(1+x-\beta)\Gamma(1+\beta-x)}\sum_{n=0}^\infty \frac{(\beta)_n(1-\alpha)_n}{n!(1+\beta-\alpha)_n}\frac{\beta-x}{n+\beta-x}.
\end{aligned}$$

Proof. Since $\Psi_0^\star(x,y,z;t)$ satisfies the second order ordinary differential equation (10.13) we only need to solve it and then specialize at $t=1$. By Lemma C.1 one checks easily that the two solutions for the corresponding homogeneous differential equation are given by

$$\varphi_j = t^{\alpha_j}{}_2F_1(\alpha_j,\alpha_j-x;\alpha_j-\alpha_{1-j}+1;t),\quad j=0,1, \tag{10.15}$$

where α_0 and α_1 are the two roots of the quadratic equation

$$t^2-(x+y)t+xy-z^2=0.$$

By variations of parameters (see Exercise 10.4) one can find a solution of Eq. (10.13):

$$\Psi_0^\star(x,y,z;t) = u_0(t)\varphi_0(t) + u_1(t)\varphi_1(t), \tag{10.16}$$

where

$$u_j(t) = \frac{1}{\alpha_{1-j}-\alpha_j}\int_0^t v^{-\alpha_{1-j}}(1-v)^{y-1}{}_2F_1(\alpha_j,\alpha_j-x;\alpha_j-\alpha_{1-j}+1;v)\,dv$$

for $j = 0, 1$. Finally, taking $t = 1$ we get

$$\Psi_0^\star(x, y, z) = f(1) = u_0(1)\varphi_0(1) + u_1(1)\varphi_1(1)$$

where

$$\varphi_j(1) = \frac{\Gamma(\alpha_j - \alpha_{1-j} + 1)\Gamma(1 - y)}{\Gamma(1 - \alpha_{1-j})\Gamma(1 + x - \alpha_{1-j})} \quad \text{for } j = 0, 1,$$

by the Gauss summation formula Eq. (C.2). This proves the first equation in the theorem. For the second, we can expand the ${}_2F_1$ by the definition and then use the beta integral to evaluate. This completes the proof of the theorem. □

Corollary 10.6.3. *With the same notation as in Thm.* 10.6.2, *we have*

$$\Psi_0^\star(x, y, z) = \frac{1}{(1-\beta)(1-y)}\,{}_3F_2(1-\beta, 1-\beta+x, 1; 2-y, 2-\beta; 1).$$

Proof. Using linear transformations (see Exercise 10.5), we see that

$$\begin{aligned}
\Psi_0^\star(x, y, z) &= \frac{1}{1-y}\int_0^1 v^{-\beta}\,{}_2F_1(1-\beta, 1-\beta+x, 2-y; 1-v)\,dv \\
&= \frac{1}{1-y}\int_0^1 (1-v)^{-\beta}\,{}_2F_1(1-\beta, 1-\beta+x, 2-y; v)\,dv \qquad (10.17)\\
&= \frac{1}{1-y}\sum_{n=0}^{\infty}\int_0^1 (1-v)^{-\beta} v^n \frac{(1-\beta)_n(1-\beta+x)_n}{(2-y)_n n!}\,dv \\
&= \frac{1}{1-y}\sum_{n=0}^{\infty}\frac{(1-\beta)_n(1-\beta+x)_n\Gamma(1-\beta)}{\Gamma(2-\beta+n)(2-y)_n} \qquad \text{(beta integral)}\\
&= \frac{1}{(1-\beta)(1-y)}\sum_{n=0}^{\infty}\frac{(1-\beta)_n(1-\beta+x)_n}{(2-\beta)_n(2-y)_n} \qquad \text{(by Eq. (1.13))}\\
&= \frac{1}{(1-\beta)(1-y)}\,{}_3F_2(1-\beta, 1-\beta+x, 1; 2-y, 2-\beta; 1),
\end{aligned}$$

as desired. □

We have another useful power series expansion of one of the generating functions in Thm. 10.6.2.

Theorem 10.6.4. *We have*

$$\Psi_0^\star(x, y, z; t) = \sum_{n=1}^{\infty}\frac{t^n}{n(n-x)}\prod_{j=1}^{n}\frac{j(j-x)}{(j-x)(j-y)-x^2}.$$

Proof. Assume the power-series solution at $t = 0$ is given by

$$\Phi_0(x, y, z; t) = \sum_{n=0}^{\infty} a_n t^n.$$

Then by using the differential equation (10.13) we find

$$a_0 = 0, \ a_1 = \frac{1}{(1-x)(1-y)-z^2}, \ a_n = \frac{(n-1)(n-1-x)a_{n-1}}{(n-x)(n-y)-z^2} \ \forall n \geq 2.$$

The yields the theorem easily. □

Let $I_0(w, *, h) = \bigcup_{d\geq 1} I_0(w, d, h)$. The next result is the sum formula of the MZ⋆Vs of fixed weight and height.

Corollary 10.6.5. *Let h and w be integers such that $h \geq 1$ and $w \geq 2h$. Then*

$$\sum_{\mathbf{s}\in I_0(w,*,h)} \zeta^\star(\mathbf{s}) = 2\binom{w-1}{2h-1}(1-2^{1-w})\zeta(w).$$

Proof. By Thm. 10.6.4, we have

$$\Psi_0^\star(x, x, z; t) = \sum_{n=1}^{\infty} \frac{t^n}{n(n-x)} \prod_{j=1}^{n} \frac{j(j-x)}{(j-x)^2 - z^2} := \sum_{n=1}^{\infty} a_n t^n.$$

It is easy to see that

$$a_n = \frac{\Gamma(n)\Gamma(n-x)\Gamma(1-x-z)\Gamma(1-x+z)}{\Gamma(1-x)\Gamma(1-x-z+n)\Gamma(1-x+z+n)}.$$

By partial fraction expansion we may assume

$$a_n = \sum_{\ell=1}^{\infty} \left(\frac{A_{n,\ell}^+}{x+z-\ell} + \frac{A_{n,\ell}^-}{x-z-\ell} \right)$$

where

$$A_{n,\ell}^+ = \lim_{z\to\ell-x} (x+z-\ell)a_n = (-1)^\ell \binom{n-1}{\ell-1} \frac{\Gamma(n-x)}{\Gamma(1-x)(1+\ell-2x)_n}.$$

Similar computation leads to $A_{n,\ell}^- = A_{n,\ell}^+$. Hence specializing to $t = 1$ gives

$$\begin{aligned}
\Psi_0^\star(x, x, z; 1) &= \sum_{n=1}^{\infty} a_n = \sum_{\ell=1}^{\infty}\sum_{n=\ell}^{\infty} \left(\frac{A_{n,\ell}^+}{x+z-\ell} + \frac{A_{n,\ell}^-}{x-z-\ell} \right) \\
&= \sum_{\ell=1}^{\infty} (-1)^\ell \left(\frac{1}{x+z-\ell} + \frac{1}{x-z-\ell} \right) \sum_{n=0}^{\infty} \frac{\binom{n+\ell-1}{\ell-1}\Gamma(n+\ell-x)}{\Gamma(1-x)(1+\ell-2x)_{n+\ell}}.
\end{aligned}$$

Note that the sum over n can be rewritten as

$$\frac{(1-x)\cdots(\ell-1-x)}{(1+\ell-2x)\cdots(2\ell-2x)}{}_2F_1(\ell,\ell-x,1+2\ell-2x;1)$$
$$=\frac{(1-x)\cdots(\ell-1-x)}{(1+\ell-2x)\cdots(2\ell-2x)}\frac{\Gamma(1+2\ell-2x)\Gamma(1-x)}{\Gamma(1+\ell-2x)\Gamma(1+\ell-x)}=\frac{1}{\ell-x}$$

by Gauss summation formula Eq. (C.1). Therefore

$$\begin{aligned}\Psi_0^\star(x,x,z;1)&=\sum_{\ell=1}^\infty\frac{(-1)^\ell}{\ell-x}\left(\frac{1}{x+z-\ell}+\frac{1}{x-z-\ell}\right)\\&=\frac{1}{z}\sum_{\ell=1}^\infty(-1)^\ell\left(\frac{1}{x+z-\ell}-\frac{1}{x-z-\ell}\right).\end{aligned}$$

Expanding and taking the coefficient of $x^{w-2h}z^{2h-2}$ we get

$$2\binom{w-1}{2h-1}\sum_{\ell=1}^\infty\frac{(-1)^{\ell-1}}{\ell^w}=2\binom{w-1}{2h-1}(1-2^{1-w})\zeta(w).$$

This finishes the proof of the corollary. □

Let $I_0(w,d,*)=\bigcup_{h\ge1}I_0(w,d,h)$. The next result is the sum formula of the MZ*Vs of fixed weight and depth.

Corollary 10.6.6. *Let d and w be integers such that $d\ge1$ and $w\ge d+1$. Then*

$$\sum_{\mathbf{s}\in I_0(w,d,*)}\zeta^\star(\mathbf{s})=\binom{w-1}{d-1}\zeta(w).$$

Proof. Take $z^2=xy$, $\alpha=0$, $\beta=x+y$, by Eq. (10.17), we have

$$\begin{aligned}\Psi_0^\star(x,y,z)&=\frac{1}{1-y}\int_0^1(1-v)^{-x-y}{}_2F_1(1-x-y,1-y,2-y;v)\,dv\\&=\sum_{n=0}^\infty\frac{(1-x-y)_n}{n!(n+1-y)}\int_0^1(1-v)^{-x-y}v^n\,dv\\&=\sum_{n=0}^\infty\frac{1}{(n+1-y)(n+1-x-y)}\\&=\frac{1}{x}\sum_{n=1}^\infty\left(\frac{1}{n-x-y}-\frac{1}{n-y}\right)\\&=\frac{1}{x}\sum_{n=1}^\infty\sum_{w=1}^\infty\frac{1}{n^{w+1}}\left((x+y)^w-y^w\right)\end{aligned}$$

$$= \frac{1}{x} \sum_{w=1}^{\infty} \zeta(w+1) \sum_{d=0}^{w-1} \binom{w}{d} x^{w-d} y^d$$
$$= \sum_{w \geq d+1 \geq 2}^{\infty} \zeta(w+1) \binom{w-1}{d-1} x^{w-d-1} y^{d-1}.$$

Comparing the coefficient of $x^{w-d-1}y^{d-1}$ we complete the proof of the corollary. □

10.7 Duality Relations

Unlike the MZVs the duality for the MZ*Vs is not a simple equation, instead, the duality (see Thm. 10.7.6) says that a simple linear combination of a MZ*V and its dual can be expressed as a polynomial of the Riemann zeta values with rational coefficients.

The notation from the last section is still in force. Define a and b by the conditions $a + b = -x + y$ and $ab = -xy - z^2$. Set

$$F_1(x, y, a, b)$$
$$= \frac{(x+a)\Gamma(b-a)\Gamma(1+a-b)}{(x+b)\Gamma(1-x-a)\Gamma(1+x+a)} \left(\frac{x\Gamma(y)\Gamma(1-y)}{b\Gamma(a)\Gamma(1-a)} - \frac{y\Gamma(x)\Gamma(1-x)}{a\Gamma(b)\Gamma(1-b)} \right),$$
$$F_2(x, y, a, b)$$
$$= \frac{xy(a-b)\Gamma(b-a)}{(x+b)\Gamma(1-x-a)\Gamma(1+x+a)}$$
$$\times \left(\frac{\Gamma(y)\Gamma(1-y)\Gamma(-b)}{\Gamma(1-a)} - \frac{\Gamma(x)\Gamma(1-x)\Gamma(a)}{\Gamma(1+b)} \right) {}_3F_2 \left({a, -b, x+a \atop a-b, 1+x+a} \right),$$

where we put ${}_3F_2(\cdots) = {}_3F_2(\cdots ; 1)$ to save space. Further we define

$$A(x, y, a, b) = A(x, y, b, a)$$
$$= \frac{1}{2\pi} \left\{ \frac{\cos \pi x}{\sin \pi y} - \frac{\cos \pi y}{\sin \pi x} + \cos \pi(a-b)(\cot \pi x - \cot \pi y) \right\} \tag{10.18}$$
$$= \frac{1}{\Gamma(x+a)\Gamma(1-x-a)} \left(\frac{\Gamma(y)\Gamma(1-y)}{\Gamma(a)\Gamma(1-a)} + \frac{\Gamma(x)\Gamma(1-x)}{\Gamma(b)\Gamma(1-b)} \right), \tag{10.19}$$

which can be checked easily using the well-known reflection formula

$$\Gamma(\gamma)\Gamma(1-\gamma) = \frac{\pi}{\sin(\pi\gamma)}. \tag{10.20}$$

Lemma 10.7.1. *We have*

$$F_1(x, y, a, b) + F_1(x, y, b, a)$$
$$= \frac{x-y}{ab} + \frac{(a-b)xy}{ab(x+a)(x+b)} \Gamma(b-a)\Gamma(1+a-b)A(x, y, a, b).$$

Proof. By (10.20) we see that

$$\begin{aligned}F_1(x,y,a,b)+F_1(x,y,b,a) &= \Gamma(b-a)\Gamma(1+a-b)\\ &\quad\times\left\{\frac{\sin\pi(x+a)}{\pi(x+b)}\left(\frac{x\sin\pi a}{b\sin\pi y}-\frac{y\sin\pi b}{a\sin\pi x}\right)\right.\\ &\quad\left.-\frac{\sin\pi(x+b)}{\pi(x+a)}\left(\frac{x\sin\pi b}{a\sin\pi y}-\frac{y\sin\pi a}{b\sin\pi x}\right)\right\}.\end{aligned}$$

With $x+a=y-b$ the terms within the curly brackets can be reduced to

$$\begin{aligned}&\frac{1}{2\pi(x+b)}\left(\frac{x(\cos\pi x-\cos\pi y\cos\pi(a-b)+\sin\pi y\sin\pi(a-b))}{b\sin\pi y}\right.\\ &\left.-\frac{y(\cos\pi x\cos\pi(a-b)-\sin\pi x\sin\pi(a-b)-\cos\pi y)}{a\sin\pi x}\right)\\ &-\frac{1}{2\pi(x+a)}\left(\frac{x(\cos\pi x-\cos\pi y\cos\pi(b-a)+\sin\pi y\sin\pi(b-a))}{a\sin\pi y}\right.\\ &\left.-\frac{y(\cos\pi x\cos\pi(b-a)-\sin\pi x\sin\pi(b-a)-\cos\pi y)}{b\sin\pi x}\right).\end{aligned}\tag{10.21}$$

By using the identities

$$\begin{aligned}\frac{1}{b(x+b)}-\frac{1}{a(x+a)}&=\frac{y(a-b)}{ab(x+a)(x+b)},\\ \frac{1}{a(x+b)}-\frac{1}{b(x+a)}&=\frac{x(b-a)}{ab(x+a)(x+b)},\\ \frac{x}{b(x+b)}+\frac{y}{a(x+b)}+\frac{x}{a(x+a)}+\frac{y}{b(x+a)}&=\frac{2(y-x)}{ab},\end{aligned}$$

we see that the expression in Eq. (10.21) becomes

$$\frac{xy(a-b)}{ab(x+a)(x+b)}A(x,y,a,b)+\frac{(y-x)\sin\pi(a-b)}{ab\pi}$$

by Eq. (10.18). We now have completed the proof of the lemma. □

Lemma 10.7.2. *We have*

$$\begin{aligned}&F_2(x,y,a,b)+F_2(x,y,b,a)\\ &\quad=\frac{(b-a)xy}{ab(x+a)(x+b)}\Gamma(b-a)\Gamma(1+a-b)A(x,y,a,b)\\ &\qquad+A(x,y,a,b)\frac{\Gamma(a)\Gamma(1-a)\Gamma(b)\Gamma(1-b)\Gamma(x+a)\Gamma(x+b)}{\Gamma(x)\Gamma(y)}.\end{aligned}$$

Proof. Applying Prop. C.2 to $F_2(x,y,a,b)$, we get

$$\begin{aligned}
&F_2(x,y,a,b)\\
&\quad= \frac{(a-b)\Gamma(a-b)\Gamma(b-a)\Gamma(x+b)}{\Gamma(x)\Gamma(y)\Gamma(1-x-a)}\left(\frac{\Gamma(y)\Gamma(1-y)}{\Gamma(a)\Gamma(1-a)}-\frac{\Gamma(x)\Gamma(1-x)}{\Gamma(-b)\Gamma(1+b)}\right)\\
&\qquad\times\left(\psi(1+y)-\psi(a)-\psi(-b)-\gamma-\sum_{n=1}^{\infty}\frac{(a)_n(1+b)_n}{nn!(1+y)_n}\right)\\
&\quad= A(x,y,a,b)\frac{\Gamma(b-a)\Gamma(1+a-b)\Gamma(x+a)\Gamma(x+b)}{\Gamma(x)\Gamma(y)}\\
&\qquad\times\left(\psi(1+y)-\psi(a)-\psi(-b)-\gamma-\sum_{n=1}^{\infty}\frac{(a)_n(1+b)_n}{nn!(1+y)_n}\right)
\end{aligned}$$

using definition Eq. (10.19). By $A(x,y,a,b)=A(x,y,b,a)$ we see that

$$\begin{aligned}
&F_2(x,y,a,b)+F_2(x,y,b,a)\\
&= A(x,y,a,b)\frac{\Gamma(b-a)\Gamma(1+a-b)\Gamma(x+a)\Gamma(x+b)}{\Gamma(x)\Gamma(y)}\\
&\quad\times\left\{\sum_{n=1}^{\infty}\left(\frac{(1+a)_n(b)_n}{n(1+y)_n n!}-\frac{(a)_n(1+b)_n}{n(1+y)_n n!}\right)+\psi(b)-\psi(-b)+\psi(-a)-\psi(a)\right\}.
\end{aligned}$$

Further,

$$\begin{aligned}
&\sum_{n=1}^{\infty}\left(\frac{(1+a)_n(b)_n}{nn!(1+y)_n}-\frac{(a)_n(1+b)_n}{n(1+y)_n n!}\right)\\
&\qquad=\frac{b-a}{ab}\sum_{n=1}^{\infty}\frac{(a)_n(b)_n}{n!(1+y)_n}\\
&\qquad=\frac{b-a}{ab}\left(\frac{\Gamma(1+x)\Gamma(1+y)}{\Gamma(1+x+a)\Gamma(1+x+b)}-1\right)\\
&\qquad=\frac{(b-a)\Gamma(1+x)\Gamma(1+y)}{ab\Gamma(1+x+a)\Gamma(1+x+b)}-\frac{1}{a}+\frac{1}{b}
\end{aligned}$$

by Gaussian summation formula Eq. (C.2). So the lemma follows from the identities

$$\pi\cot\pi a-\pi\cot\pi b=\frac{\pi\sin\pi(b-a)}{\sin(\pi a)\sin(\pi b)}=\frac{\Gamma(a)\Gamma(1-a)\Gamma(b)\Gamma(1-b)}{\Gamma(b-a)\Gamma(1+a-b)},$$

and the functional equation of the digamma function $\psi(-x)-\psi(x)-1/x=\pi\cot\pi x$ which can be derived easily from Eq. (1.6). ☐

Theorem 10.7.3. *We have*

$$x\Psi_0^\star(-x,y,z) - y\Psi_0^\star(-y,x,z) = \frac{x-y}{ab} + A(x,y,a,b)\frac{\Gamma(a)\Gamma(1-a)\Gamma(b)\Gamma(1-b)\Gamma(x+a)\Gamma(x+b)}{\Gamma(x)\Gamma(y)}. \tag{10.22}$$

Proof. Replacing x by $-x$ in the second equation in Thm. 10.6.2 we see that

$$\begin{aligned}\Psi_0^\star(-x,y,z) &= \frac{\Gamma(b-a)\Gamma(1-b)\Gamma(y)\Gamma(1-y)}{\Gamma(1-a)\Gamma(1-x-a)\Gamma(1+a+x)}\sum_{n=0}^\infty \frac{(a)_n(1-b)_n}{n!(1+a-b)_n}\frac{a+x}{n+a+x}\\ &\quad+\frac{\Gamma(a-b)\Gamma(1-a)\Gamma(y)\Gamma(1-y)}{\Gamma(1-b)\Gamma(1-x-b)\Gamma(1+b+x)}\sum_{n=0}^\infty \frac{(b)_n(1-a)_n}{n!(1+b-a)_n}\frac{b+x}{n+b+x}.\end{aligned}$$

Thus

$$x\Psi_0^\star(-x,y,z) - y\Psi_0^\star(-y,x,z) = F(x,y,a,b)+F(x,y,b,a), \tag{10.23}$$

where

$$\begin{aligned}F(x,y,a,b) &= \frac{x\Gamma(y)\Gamma(1-y)\Gamma(1-b)\Gamma(b-a)}{\Gamma(1-a)\Gamma(1-x-a)\Gamma(1+x+a)}\sum_{n=0}^\infty \frac{(a)_n(1-b)_n}{n!(1+a-b)_n}\frac{x+a}{n+x+a}\\ &\quad-\frac{y\Gamma(x)\Gamma(1-x)\Gamma(1+a)\Gamma(b-a)}{\Gamma(1+b)\Gamma(1-x-a)\Gamma(1+x+a)}\sum_{n=0}^\infty \frac{(1+a)_n(-b)_n}{n!(1+a-b)_n}\frac{x+a}{n+x+a}.\end{aligned}$$

But

$$\sum_{n=0}^\infty \frac{(a)_n(1-b)_n}{n!(1+a-b)_n}\frac{x+a}{n+x+a} = \sum_{n=0}^\infty \frac{(a)_n(-b)_n}{n!(a-b)_n}\frac{(a-b)(x+a)(n-b)}{(-b)(n+a-b)(n+x+a)},$$

and

$$\frac{-(n-b)}{(n+a-b)(n+x+a)} = \frac{a}{x+b}\frac{1}{n+a-b} - \frac{y}{x+b}\frac{1}{n+x+a},$$

we see that

$$\begin{aligned}\sum_{n=0}^\infty &\frac{(a)_n(1-b)_n}{n!(1+a-b)_n}\frac{x+a}{n+x+a}\\ &= \frac{a(x+a)}{b(x+b)}\,{}_2F_1\left(\begin{matrix}a,-b\\1+a-b\end{matrix}\right) - \frac{y(a-b)}{b(x+b)}\,{}_3F_2\left(\begin{matrix}a,-b,x+a\\a-b,1+x+a\end{matrix}\right)\\ &= \frac{a(x+a)\Gamma(1+a-b)}{b(x+b)\Gamma(1+a)\Gamma(1-b)} - \frac{y(a-b)}{b(x+b)}\,{}_3F_2\left(\begin{matrix}a,-b,x+a\\a-b,1+x+a\end{matrix}\right)\end{aligned}$$

by the Gauss summation formula Eq. (C.2). Exchanging $(a,x) \leftrightarrow (-b,y)$ and noticing that $x+a=y-b$, we get

$$\sum_{n=0}^{\infty} \frac{(1+a)_n(-b)_n}{n!(1+a-b)_n} \frac{x+a}{n+x+a}$$
$$= \frac{b(x+a)\Gamma(1+a-b)}{a(x+b)\Gamma(1+a)\Gamma(1-b)} + \frac{x(a-b)}{a(x+b)} {}_3F_2\left(\begin{matrix} a, -b, x+a \\ a-b, 1+x+a \end{matrix}\right).$$

Therefore

$$F(x,y,a,b) = F_1(x,y,a,b) + F_2(x,y,a,b).$$

So the theorem follows from this together with Eq. (10.23), Lemma 10.7.2 and Lemma 10.7.1. □

The theorem has an immediate application to the sum formulas of the MZ*Vs.

Corollary 10.7.4. *For any w,d,h with $w,d \geq h$, the difference*

$$D_{w,d,h} := (-1)^w X_0^\star(w+d+1,d+1,h) - (-1)^d X_0^\star(w+d+1,w+1,h)$$

can be expressed as a polynomial of the Riemann zeta values with rational coefficients.

Proof. We have

$$x\Psi_0^\star(-x,y,z) - y\Psi_0^\star(-y,x,z) = \sum_{w,d \geq h \geq 1} (-1)^h D_{w,d,h} x^{w+1-h} y^{d+1-h} z^{2h-2}$$
$$+ \sum_{d \geq h \geq 1} (-1)^{d+h} X_0^\star(d+h,h,h)(x^{d+1-h} - y^{d+1-h}) z^{2h-2}.$$

So the corollary follows from Theorem 10.7.3 using the Gamma function expansion given by Eq. (3.17). □

Corollary 10.7.5. *Let $\psi(X)$ be the digamma function (see Eq. (1.5)). Then*

$$x\Psi_0^\star(-x,y,0) - x\Psi_0^\star(-x,0,0) - y\Psi_0^\star(-y,x,0) + y\Psi_0^\star(-y,0,0)$$
$$= -\psi(x) + \psi(y) - \pi\big(\cot(\pi x) - \cot(\pi y)\big)\frac{\Gamma(1-x)\Gamma(1-y)}{\Gamma(1-x-y)}.$$

Proof. Setting $\alpha = x$ and $\beta = y$ in Thm. 10.6.2 we find

$$\Psi_0^\star(x,y,0) = \frac{\Gamma(y-x)\Gamma(1-y)}{\Gamma(1-x)} \int_0^t v^{-y}(1-v)^{y-1} dv$$
$$+ \frac{1}{x-y} \int_0^t v^{-x}(1-v)^{y-1} {}_2F_1(y, y-x; y-x+1; v)\, dv.$$

Clearly we have

$$\begin{aligned} y\Psi_0^\star(-y,0,0)-x\Psi_0^\star(-x,0,0) &= \int_0^1 \frac{v^x}{1-v}-\frac{v^y}{1-v}\,dv \\ &= \psi(y+1)-\psi(x+1)=\psi(y)+\frac{1}{y}-\psi(x+1)-\frac{1}{x} \qquad (10.24) \end{aligned}$$

by Eq. (1.5) and Eq. (1.6). To compute $x\Psi_0^\star(-x,y,0)-y\Psi_0^\star(-y,x,0)$ we take $z=0,(a,b)=(-x,y)$ in Thm. 10.7.3. Thus we have

$$\begin{aligned} &x\Psi_0^\star(-x,y,0)-y\Psi_0^\star(-y,x,0) \\ &=\frac{1}{x}-\frac{1}{y}+\left(\frac{\Gamma(y)\Gamma(1-y)}{\Gamma(a)\Gamma(1-a)}+\frac{\Gamma(x)\Gamma(1-x)}{\Gamma(b)\Gamma(1-b)}\right)\frac{\Gamma(a)\Gamma(1-a)\Gamma(b)\Gamma(1-b)}{\Gamma(x)\Gamma(y)\Gamma(x+y)^{-1}} \\ &=\frac{1}{x}-\frac{1}{y}-\left(\frac{\sin\pi y}{\sin\pi x}-\frac{\sin\pi x}{\sin\pi y}\right)\frac{\pi\Gamma(1-x)\Gamma(1-y)}{\Gamma(1-x-y)\sin\pi(x+y)} \\ &=\frac{1}{x}-\frac{1}{y}-\big(\cot(\pi x)-\cot(\pi y)\big)\frac{\pi\Gamma(1-x)\Gamma(1-y)}{\Gamma(1-x-y)} \end{aligned}$$

by an easy manipulation of trigonometric identities. This completes the proof of the corollary. □

Corollary 10.7.6. *Let $\psi(X)$ be the digamma function. Then*

$$\begin{aligned} &\sum_{\ell,d\ge 2}\Big((-1)^\ell\zeta^\star(\ell,\{1\}^{d-1})-(-1)^d\zeta^\star(d,\{1\}^{\ell-1})\Big)x^{\ell-1}y^{d-1} \\ &\qquad=-\psi(x)+\psi(y)-\pi(\cot(\pi x)-\cot(\pi y))\frac{\Gamma(1-x)\Gamma(1-y)}{\Gamma(1-x-y)}. \end{aligned}$$

Consequently, $(-1)^\ell\zeta^\star(\ell,\{1\}^{d-1})-(-1)^d\zeta^\star(d,\{1\}^{\ell-1})$ can always be expressed as a $\mathbb{Q}$-linear combination of products of the Riemann zeta values.

Proof. The first part follows from Cor. 10.7.5 immediately. For the last sentence, it suffices to note that

$$\begin{aligned} \psi(x) &= -\frac{1}{x}-\sum_{\ell\ge 2}\zeta(\ell)x^{\ell-1}-\gamma \quad \text{(by Eq. (1.5) and Eq. (1.6))}, \\ \pi\cot(\pi x) &= \frac{1}{x}-\sum_{\ell\ge 1}\zeta(2\ell)x^{2\ell-1} \qquad \text{(see, for e.g., [6, § 4.3.70])}, \end{aligned}$$

and the expansion (6.5) of $\Gamma(1-x)\Gamma(1-y)/\Gamma(1-x-y)$. □

10.8 Some Applications of Binomial Identities of q-MZ*Vs

In Chap. 12, we will prove some binomial identities of the q-analog of the MZ*Vs by using results from Chap. 9. By taking the limit $q \to 1$ in Thm. 12.9.4 we will obtain the following result. Let $\mathbf{s} \in \mathbb{D}^d$ with $s_1 \neq 1$. Define

$$\zeta^\sharp(\mathbf{s}) = \sum_{\mathbf{p} \preceq \mathbf{s}} 2^{\mathrm{dp}(\mathbf{p})} \zeta(\mathbf{p}), \tag{10.25}$$

where $\mathbf{p} \preceq \mathbf{s}$ means $\mathbf{p}$ can be obtained from $\mathbf{s}$ by combining some of its parts using $\oplus$.

Theorem 10.8.1. *For all $\mathbf{s} = (s_1, \dots, s_\ell) \in \mathbb{N}^\ell$, we define a sequence of compositions $\mathbf{m}_i(\mathbf{s})$ with entries in $\mathbb{D}$ (see Definition7.1.1) for $i = \ell, \dots, 2, 1$ (backward) inductively as follows:*

$$\mathbf{m}_\ell(\mathbf{s}) = \begin{cases} (1), & \text{if } s_\ell = 1; \\ (\overline{2}, \{1\}^{s_\ell - 2}), & \text{if } s_\ell \geq 2, \end{cases}$$

$$\mathbf{m}_i(\mathbf{s}) = \begin{cases} (1, \mathbf{m}_{i+1}(\mathbf{s})), & \text{if } s_i = 1; \\ (2) \oplus \mathbf{m}_{i+1}(\mathbf{s}), & \text{if } s_i = 2; \\ (\overline{2}, \{1\}^{s_i - 3}, \overline{1}) \oplus \mathbf{m}_{i+1}(\mathbf{s}), & \text{if } s_i \geq 3, \end{cases}$$

for all $i < l$. If $s_1 > 1$ then we have

$$\zeta^\star(\mathbf{s}) = \delta(\mathbf{s}) \zeta^\sharp\big(\mathbf{m}_1(\mathbf{s})\big), \qquad \text{where } \delta(\mathbf{s}) = \begin{cases} 1, & \text{if } s_\ell = 1; \\ -1, & \text{if } s_\ell \geq 2. \end{cases}$$

Remark 10.8.2. This theorem implies that there are many nontrivial combinations of Euler sums that are in fact lie in the world of multiple zeta values, which is a kind of descent result.

The following result is called the *Two-one formula.*

Proposition 10.8.3. *For any positive integers $s_1, \dots, s_\ell$, define the composition $\mathbf{s}' = (2s_1 + 1, \dots, 2s_\ell + 1)$. Then we have*

$$\zeta^\star(\{2\}^{s_1}, 1, \dots, \{2\}^{s_\ell}, 1) = \zeta^\sharp(\mathbf{s}') \tag{10.26}$$

$$= \sum_{\mathbf{p} \preceq \mathbf{s}'} (-1)^{\ell - \mathrm{dp}(\mathbf{p})} 2^{\mathrm{dp}(\mathbf{p})} \zeta^\star(\mathbf{p}). \tag{10.27}$$

Proof. Equation Eq. (10.26) follows easily from Thm. 10.8.1. To prove Eq. (10.27), by Eq. (10.2), we have

$$\sum_{\mathbf{p} \preceq \mathbf{s}'} (-1)^{\ell - \mathrm{dp}(\mathbf{p})} 2^{\mathrm{dp}(\mathbf{p})} \zeta^\star(\mathbf{p}) = \sum_{\mathbf{p} \preceq \mathbf{s}'} (-1)^{\ell - \mathrm{dp}(\mathbf{p})} 2^{\mathrm{dp}(\mathbf{p})} \sum_{\mathbf{t} \preceq \mathbf{p}} \zeta(\mathbf{t}).$$

Now for each fixed $\mathbf{t} \preceq \mathbf{s}'$ and fixed number r between $d := \mathrm{dp}(\mathbf{t})$ and $\ell = \mathrm{dp}(\mathbf{s}')$ we need to count how many ways to choose $\mathbf{p}$ so that $\mathbf{t} \preceq \mathbf{p} \preceq \mathbf{s}'$ and $\mathrm{dp}(\mathbf{p}) = r$. Easy combinatorial consideration gives this number as $\binom{\ell-d}{\ell-r}$. Thus

$$\sum_{\mathbf{p}\preceq\mathbf{s}'}(-1)^{\ell-\mathrm{dp}(\mathbf{p})}2^{\mathrm{dp}(\mathbf{p})}\sum_{\mathbf{t}\preceq\mathbf{p}}\zeta(\mathbf{t}) = \sum_{\mathbf{t}\preceq\mathbf{s}'}(-1)^{\ell}\sum_{r=\mathrm{dp}(\mathbf{t})}^{\ell}(-2)^r\binom{\ell-\mathrm{dp}(\mathbf{t})}{\ell-r}\zeta(\mathbf{t})$$
$$= \sum_{\mathbf{t}\preceq\mathbf{s}'}2^{\mathrm{dp}(\mathbf{t})}\zeta(\mathbf{t}) = \zeta^{\sharp}(\mathbf{s}').$$

Hence Eq. (10.27) is proved and so is the proposition. □

Using exactly the same idea we can get the following **2**-<u>**1**</u>-**2** formula from Thm. 10.8.1.

Proposition 10.8.4. *Suppose $\ell, a_j \in \mathbb{N}_0$ for all $j \le \ell$ and $a_{\ell+1} \in \mathbb{N}$. Set $\mathbf{s} = (\{2\}^{a_1}, 1, \dots, \{2\}^{a_\ell}, 1, \{2\}^{a_{\ell+1}})$. Then we have*

$$\zeta^{\star}(\mathbf{s}) = -\zeta^{\sharp}\big(2a_1+1, \dots, 2a_\ell+1, \overline{2a_{\ell+1}}\big).$$

For example,

$$\zeta^{\star}(\{2\}^n) = -\zeta^{\sharp}(\overline{2n}) = -2\zeta(\overline{2n}) \quad \text{for all } n \ge 1. \tag{10.28}$$

The **2**-<u>c</u>-**2** ($c \ge 3$) formula can be proved similarly using Thm. 10.8.1 or, more directly, Thm. 12.9.8. When all $c = 3$ we get so-called *Two-three formula.*

Proposition 10.8.5. *Suppose $\ell, a_j, c_j \in \mathbb{N}_0$ and $c_j \ge 3$ for all $1 \le j \le \ell+1$. Set $\mathbf{s} = (\{2\}^{a_1}, c_1, \dots, \{2\}^{a_\ell}, c_\ell, \{2\}^{a_{\ell+1}})$. Then*

$$\zeta^{\star}(\mathbf{s}) = -\zeta^{\sharp}\Big(\overline{2a_1+2}, \{1\}^{c_1-3}, \operatorname*{\mathbf{Cat}}_{j=2}^{\ell}\{2a_j+3, \{1\}^{c_j-3}\}, 2a_{\ell+1}+1\Big).$$

For example, for all $m, n \ge 0$ and $n \ge 1$, we have the **2**-<u>**3**</u>-**2** formula

$$\zeta^{\star}(\{2\}^m, 3, \{2\}^n) = -4\zeta(\overline{2m+2}, 2n+1) - 2\zeta(\overline{2m+2n+3}). \tag{10.29}$$

Remark 10.8.6. The general **2**-<u>**3**</u>-**2** formula will be crucial in investigating the following conjecture which is the star version of Hoffman's Basis Conjecture 3.6.1.

Conjecture 10.8.7. *Let $w \ge 2$ be a positive integer. Then the set*

$$\Big\{\zeta^{\star}(\mathbf{s}) : \mathbf{s} \in \{2,3\}^d,\ d \in \mathbb{N},\ |\mathbf{s}| = w\Big\}$$

form a $\mathbb{Q}$-basis of MZ_w, the space generated by the MZVs of weight w.

We leave the proof of the next two propositions to the interested reader.

Proposition 10.8.8. *Let $\ell \in \mathbb{N}$ and $a_j, b_j, c_j - 3 \in \mathbb{N}_0$ for all $j \geq 1$.*

(**2**-*c*-**2**-1): *For* $\mathbf{s} = \Big(\operatorname*{Cat}_{j=1}^{\ell} \{\{2\}^{b_j}, c_j, \{2\}^{a_j}, 1\}\Big)$, $\ell \geq 1$, *we have*

$$\zeta^\star(\mathbf{s}) = \zeta^\sharp\Big(\operatorname*{Cat}_{j=1}^{\ell} \{\overline{2b_j+2}, \{1\}^{c_j-3}, \overline{2a_j+2}\}\Big). \tag{10.30}$$

(**2**-1-**2**-*c*-**2**-1): *For* $\mathbf{s} = \Big(\{2\}^{a_0}, 1, \operatorname*{Cat}_{j=1}^{\ell} \{\{2\}^{b_j}, c_j, \{2\}^{a_j}, 1\}\Big)$, $\ell \geq 0$ *and* $a_0 \geq 1$, *we have*

$$\zeta^\star(\mathbf{s}) = \zeta^\sharp\Big(2a_0+1, \operatorname*{Cat}_{j=1}^{\ell} \{\overline{2b_j+2}, \{1\}^{c_j-3}, \overline{2a_j+2}\}\Big).$$

For example, we have

$$\begin{aligned}\zeta^\star(\{2\}^b, 3, \{2\}^a, 1) &= \zeta^\sharp(\overline{2b+2}, \overline{2a+2}) \\ &= 2\zeta(2a+2b+4) + 4\zeta(\overline{2b+2}, \overline{2a+2}),\end{aligned} \tag{10.31}$$

and

$$\begin{aligned}\zeta^\star(\{2\}^a, 1, \{2\}^b, 3, \{2\}^c, 1) &= \zeta^\sharp(2a+1, \overline{2b+2}, \overline{2c+2}) \\ &= 2\zeta(2(a+b+c)+5) + 4\zeta(2a+1, 2b+2c+4) \\ &\quad + 4\zeta(\overline{2a+2b+3}, \overline{2c+2}) + 8\zeta(2a+1, \overline{2b+2}, \overline{2c+2}).\end{aligned}$$

Proposition 10.8.9. *Let $\ell \in \mathbb{N}$ and $a_j, b_j, c_j - 3 \in \mathbb{N}_0$ for all $j \geq 1$.*

(**2**-1-**2**-*c*-**2**). *For* $\mathbf{s} = \Big(\operatorname*{Cat}_{j=1}^{\ell} \{\{2\}^{a_j}, 1, \{2\}^{b_j}, c_j\}, \{2\}^t\Big)$ *with* $a_1 \geq 1$ *and* $t \geq 0$, *we have*

$$\zeta^\star(\mathbf{s}) = -\zeta^\sharp\Big(2a_1+1, \overline{2b_1+2}, \{1\}^{c_1-3}, \operatorname*{Cat}_{j=2}^{\ell} \{\overline{2a_j+2}, \overline{2b_j+2}, \{1\}^{c_j-3}\}, 2t+1\Big).$$

(**2**-*c*-**2**-1-**2**-*c*-**2**). *For* $\mathbf{s} = \Big(\operatorname*{Cat}_{j=1}^{\ell} \{\{2\}^{b_j}, c_j, \{2\}^{a_j}, 1\}, \{2\}^{b_{\ell+1}}, c_{\ell+1}, \{2\}^t\Big)$ *with* $t \geq 0$, *we have*

$$\zeta^\star(\mathbf{s}) = -\zeta^\sharp\Big(\operatorname*{Cat}_{j=1}^{\ell} \{\overline{2b_j+2}, \{1\}^{c_j-3}, \overline{2a_j+2}\}, \overline{2b_{\ell+1}+2}, \{1\}^{c_{\ell+1}-3}, 2t+1\Big).$$

For example, taking $r = 1$ and $t = 0$ in (**2**-1-**2**-*c*-**2**) we get

$$\begin{aligned}\zeta^\star(\{2\}^a, 1, \{2\}^b, 3) &= -\zeta^\sharp(2a+1, \overline{2b+2}, 1) \\ &= -2\zeta(\overline{2a+2b+4}) - 4\zeta(1+2a, \overline{2b+3}) \\ &\quad - 4\zeta(\overline{2a+2b+3}, 1) - 8\zeta(2a+1, \overline{2b+2}, 1).\end{aligned}$$

In (**2**-c-<u>**2**-1-**2**-c</u>-**2**) we get

$$\begin{aligned}
\zeta^\star(\{2\}^{b_1},3,\{2\}^a,1,\{2\}^{b_2},3) &= -\zeta^\sharp(\overline{2b_1+2},\overline{2a+2},\overline{2b_2+2},1)\\
&= -2\zeta(\overline{2(b_1+a+b_2)+7}) - 4\zeta(2b_1+2a+4,\overline{2b_2+3})\\
&\quad - 4\zeta(\overline{2b_1+2},2a+2b_2+5) - 4\zeta(\overline{2b_1+2a+2b_2+6},1)\\
&\quad - 8\zeta(\overline{2b_1+2},\overline{2a+2},\overline{2b_2+3}) - 8\zeta(\overline{2b_1+2},2a+2b_2+4,1)\\
&\quad - 8\zeta(2b_1+2a+4,\overline{2b_2+2},1) - 16\zeta(\overline{2b_1+2},\overline{2a+2},\overline{2b_2+2},1).
\end{aligned}$$

A beautiful corollary is the following formula:

$$\zeta^\star(\{2\}^a,3,\{2\}^b,1)+\zeta^\star(\{2\}^b,3,\{2\}^a,1)=\zeta^\star(\{2\}^{a+1})\zeta^\star(\{2\}^{b+1}). \quad (10.32)$$

The interested reader can verify numerically all the formulas in this section for small values of parameters such as $a_1, a_2, a, b, t \le 3$ using EZ-face (see Sec. 14.1).

10.9 Three-Two-One Formulas

In this section we collect some results concerning $\zeta^\star(\mathbf{s})$ with the condition that every component in $\mathbf{s}$ is no greater than 3.

Lemma 10.9.1. *Let $n_1,\dots,n_\ell \in \mathbb{D}$. Then*

$$\sum_{\sigma\in\mathfrak{S}_\ell} \zeta(n_{\sigma(1)},\dots,n_{\sigma(\ell)}) = \sum_{(\pi_1,\cdots,\pi_q)} (-1)^{\ell-q}\prod_{h=1}^{q}(\sharp\pi_h-1)!\,\zeta\Big(\bigoplus_{j\in\pi_h} n_j\Big) \quad (10.33)$$

where the sum on the right-hand side is taken over all the possible unordered partitions $(\pi_1,\cdots,\pi_q)$ of the set $\{1,\dots,\ell\}$.

Proof. Without loss of generality we assume n_j are all distinct since otherwise we can use $n_j \oplus \varepsilon_j$ and then take limits $\varepsilon_j \to 0$. The left-hand side of Eq. (10.33) can be written as

$$\sum_{\sigma\in\mathfrak{S}_\ell}\sum_{k_1>\cdots>k_\ell\ge1} \frac{\operatorname{sgn}(n_1)^{k_{\sigma(1)}}\cdots\operatorname{sgn}(n_\ell)^{k_{\sigma(\ell)}}}{k_{\sigma(1)}^{|n_1|}\cdots k_{\sigma(\ell)}^{|n_\ell|}}.$$

Let $\mathbf{k}=(k_1,\dots,k_\ell)\in\mathbb{N}^\ell$ and $O(\mathbf{k})=\{\sigma\in\mathfrak{S}_\ell : n_{\sigma(j)}=n_j\ \forall j\}$ its associated isotropy subgroup. Let $\Pi(\mathbf{k})$ be the corresponding partition of $\{1,\dots,\ell\}$. A term like

$$\frac{\operatorname{sgn}(n_1)^{k_{\sigma(1)}}\cdots\operatorname{sgn}(n_\ell)^{k_{\sigma(\ell)}}}{k_{\sigma(1)}^{|n_1|}\cdots k_{\sigma(\ell)}^{|n_\ell|}}$$

appears exactly once on the left-hand side of Eq. (10.33) if all k_j are distinct. Otherwise, it does not appear at all. We now only need to show that for each fixed $\mathbf{k}$,

$$\sum_{(\pi_1,\dots,\pi_q)\succeq\Pi(\mathbf{k})}(-1)^{\ell-q}\prod_{h=1}^{q}(\sharp\pi_h-1)!=\begin{cases}1, \text{ if } \sharp\Pi(k)=k;\\ 0, \text{ otherwise.}\end{cases} \tag{10.34}$$

Indeed, first note that $(-1)^{\ell-q}=1$ if and only if the permutations of cycle type $(\pi_1,\dots,\pi_q)$ are even (check this!). Second, for each π_h of e_h elements, there corresponds $(e_h-1)!$ different cycles of length e_h. So the left-hand side of Eq. (10.34) is the signed sum of the number of even and odd permutations in $O(\mathbf{k})$. This number is clearly 0 unless $\Pi(\mathbf{k})$ is the trivial partition $\{\{1\},\dots,\{\ell\}\}$ in which case the number becomes 1. This concludes the proof of the theorem. □

Corollary 10.9.2. *Let $n_1,\dots,n_\ell\in\mathbb{D}$. If n_j are all even then*

$$\sum_{\sigma\in\mathfrak{S}_\ell}\zeta\big(n_{\sigma(1)},\dots,n_{\sigma(\ell)}\big)\in\pi^{|n_1|+\cdots+|n_\ell|}\mathbb{Q}.$$

Proof. Since $\zeta(\overline{2n})=(2^{1-2n}-1)\zeta(2n)\in\pi^{2n}\mathbb{Q}$ for every positive integer n the corollary follows from Lemma 10.9.1 immediately . □

Theorem 10.9.3. *Let r be a positive integer, and $e_1,\dots,e_{2r+1}$ nonnegative integers.*

(i) *Put $m=e_1+\cdots+e_{2r}$. Then the sum*

$$\sum_{\tau\in\mathfrak{S}_{2r}}\zeta^\star(\{2\}^{e_{\tau(1)}},3,\{2\}^{e_{\tau(2)}},1,\{2\}^{e_{\tau(3)}},\dots,3,\{2\}^{e_{\tau(2r)}},1)$$

is a rational multiple of π^{2m+4r}.

(ii) *Put $m=e_1+\cdots+e_{2r+1}$. Then the sum*

$$\sum_{\tau\in\mathfrak{S}_{2r+1}}\zeta^\star(\{2\}^{e_{\tau(1)}},3,\{2\}^{e_{\tau(2)}},1,\dots,3,\{2\}^{e_{\tau(2n)}},1,\{2\}^{e_{\tau(2r+1)}+1})$$

is a rational multiple of $\pi^{2m+4r+2}$.

Proof. We start with (i) first. When $r=1$ this follows quickly from Eq. (10.31) by the stuffle relation. For general r, let $a_j=e_{2j}$ and $b_j=e_{2j-1}$ for all $j\le r$ and let $A_j=2e_j+2$ for all $j\le 2r$. Then we can apply Eq. (10.30) of Prop. 10.8.8 to the string $\mathbf{s}=(\{2\}^{b_1},3,\{2\}^{a_1},1,\dots,\{2\}^{b_\ell},3,\{2\}^{a_\ell},1)$ and get

$$\zeta^\star(\mathbf{s})=\zeta^\sharp(\overline{A_1},\dots,\overline{A_{2r}})=\sum_{\mathbf{p}\preceq(\overline{A_1},\dots,\overline{A_{2r}})}2^{\ell(\mathbf{p})}\zeta(\mathbf{p}).$$

For $\mathbf{s} = (\{2\}^{e_1}, 3, \{2\}^{e_2}, 1, \dots, \{2\}^{e_{2r-1}}, 3, \{2\}^{e_{2r}}, 1)$ and any permutation $\tau \in \mathfrak{S}_{2r}$, we define

$$\mathbf{s}^\tau = (\{2\}^{e_{\tau(1)}}, 3, \{2\}^{e_{\tau(2)}}, 1, \dots, \{2\}^{e_{\tau(2r-1)}}, 3, \{2\}^{e_{\tau(2r)}}, 1)$$

Let $\mathbf{A} = (\overline{A_1}, \dots, \overline{A_{2r}})$, $\mathbf{A}^\tau = (\overline{A_{\tau(1)}}, \dots, \overline{A_{\tau(2r)}})$, and $P_\ell(2r)$ be the set of all partitions of $[2r] := \{1, 2, \dots, 2r\}$ into ℓ consecutive subsets. If $\lambda = (\lambda_1, \dots, \lambda_\ell) \in P_\ell(2r)$ then we set $\lambda_j(\mathbf{A}^\tau) = (\overline{A_{\tau(i)}})_{i \in \lambda_j}$ so that the concatenation $\bigsqcup_{j=1}^{\ell} \lambda_j(\mathbf{A}^\tau) = \mathbf{A}^\tau$. Because of the permutation we see that

$$\begin{aligned}\sum_{\tau \in \mathfrak{S}_{2r}} \zeta^\star(\mathbf{s}^\tau) &= \sum_{\tau \in \mathfrak{S}_{2r}} \sum_{\ell=1}^{2r} 2^\ell \sum_{\lambda \in P_\ell(2r)} \zeta\big(\oplus \lambda_1(\mathbf{A}^\tau), \dots, \oplus \lambda_\ell(\mathbf{A}^\tau)\big) \\ &= \sum_{\tau \in \mathfrak{S}_{2r}} \sum_{\ell=1}^{2r} \frac{2^\ell}{\ell!} \sum_{\lambda \in P_\ell(2r)} \sum_{g \in \mathfrak{S}_\ell} \zeta\big(\oplus \lambda_{g(1)}(\mathbf{A}^\tau), \dots, \oplus \lambda_{g(\ell)}(\mathbf{A}^\tau)\big),\end{aligned}$$

where $\oplus\mathbf{t}$ is the $\oplus$-sum of all the components of $\mathbf{t}$ for any composition $\mathbf{t}$. Hence Thm. 10.9.3(i) follows readily from the Cor. 10.9.2 since all A_j's are even numbers.

Thm. 10.9.3(ii) follows from Eq. (10.48) in Exercise 10.8 in a similar fashion, so we leave the details to the interested reader. □

Theorem 10.9.4. *For any positive integers n and m, we have*

$$\zeta^\star(\{2\}^m, 1)\zeta^\star(\{2\}^n, 1) = \zeta^\star(\{2\}^m, 1, \{2\}^n, 1) + \zeta^\star(\{2\}^n, 1, \{2\}^m, 1), \tag{10.35}$$

$$\zeta^\star(\{2\}^m, 1)\zeta^\star(\{2\}^n) = \zeta^\star(\{2\}^m, 1, \{2\}^n) + \zeta^\star(\{2\}^{n-1}, 3, \{2\}^m), \tag{10.36}$$

$$\zeta^\star(\{2\}^m)\zeta^\star(\{2\}^n) = \zeta^\star(\{2\}^{m-1}, 3, \{2\}^{n-1}, 1) + \zeta^\star(\{2\}^{n-1}, 3, \{2\}^{m-1}, 1). \tag{10.37}$$

Proof. By Prop. 10.8.3 and Eq. (10.28) we have

$$\begin{aligned}\frac{1}{4}\zeta^\star(\{2\}^m, 1, \{2\}^n, 1) &= \zeta(2m+1, 2n+1) + \frac{1}{2}\zeta(2m+2n+2), \\ \frac{1}{4}\zeta^\star(\{2\}^m, 1) \cdot \zeta^\star(\{2\}^n, 1) &= \zeta(2m+1)\zeta(2n+1) \\ &= \zeta(2m+1, 2n+1) + \zeta(2m+2n+2) + \zeta(2n+1, 2m+1), \\ -\frac{1}{4}\zeta^\star(\{2\}^m, 1, \{2\}^n) &= \zeta(2m+1, \overline{2n}) + \frac{1}{2}\zeta(\overline{2m+2n+1}), \\ -\frac{1}{4}\zeta^\star(\{2\}^m, 1) \cdot \zeta^\star(\{2\}^n) &= \zeta(2m+1)\zeta(\overline{2n}) \\ &= \zeta(2m+1, \overline{2n}) + \zeta(\overline{2m+2n+1}) + \zeta(\overline{2n}, 2m+1),\end{aligned}$$

$$\frac{1}{4}\zeta^\star(\{2\}^m)\cdot\zeta^\star(\{2\}^n) = \zeta(\overline{2m})\zeta(\overline{2n})$$
$$= \zeta(\overline{2m},\overline{2n}) + \zeta(\overline{2m+2n}) + \zeta(\overline{2n},\overline{2m}),$$

by the stuffle relations. On the other hand, by Eq. (10.31) and Eq. (10.29), we get

$$\frac{1}{4}\zeta^\star(\{2\}^{n-1},3,\{2\}^{m-1},1) = \frac{1}{2}\zeta(2m+2n) + \zeta(\overline{2n},\overline{2m}),$$
$$\frac{1}{4}\zeta^\star(\{2\}^{m-1},3,\{2\}^{n-1},1) = \frac{1}{2}\zeta(2m+2n) + \zeta(\overline{2m},\overline{2n}),$$
$$-\frac{1}{4}\zeta^\star(\{2\}^{n-1},3,\{2\}^{m}) = \zeta(\overline{2n},2m+1) + \frac{1}{2}\zeta(\overline{2m+2n+1}).$$

The rest of the proof is straightforward and is thus left to the interested reader. □

Theorem 10.9.5. *Let m and n be two nonnegative integers.*

(i) *We have*

$$(2n+1)\zeta^\star(\{3,1\}^n,2) = \sum_{j+k=n}\zeta^\star(\{3,1\}^j)\zeta^\star(\{2\}^{2k+1}).$$

(ii) *If $n\ge 1$ then we have*

$$\sum_{\substack{e_1+e_2+\cdots+e_{2n}=1\\ e_1,e_2,\dots,e_{2n}\ge 0}}\zeta^\star(\{2\}^{e_1},3,\{2\}^{e_2},1,\dots,\{2\}^{e_{2n-1}},3,\{2\}^{e_{2n}},1)$$
$$= \sum_{j+k=n-1}\zeta^\star(\{3,1\}^j,2)\zeta^\star(\{2\}^{2k+2}).$$

Remark 10.9.6. Note that Thm. 10.9.5(ii) is related to Exercise 10.10 but is not a consequence of it. Also notice there is no 2-string at the end of any MZ⋆V on the left-hand of (ii). Further, the statement in Thm. 10.9.5 is generally false if $\zeta^\star$ is replaced by ζ.

Proof. (i) By removing all the 2's in **2**-3-**2**-1 formula in Prop. 10.8.8 we get

$$\zeta^\star(\{3,1\}^j) = \zeta^\sharp(\{\overline{2}\}^{2j}) = \sum_{\mathbf{p}_{2j}\preceq(\{\overline{2}\}^{2j})} 2^{\ell(\mathbf{p}_{2j})}\zeta(\mathbf{p}_{2j}).$$

All of the components a_j of $\{\overline{2}\circ\cdots\circ\overline{2}\}$ must satisfy the following sign rule:

$$a_j>0 \text{ if and only if } 4|a_j. \tag{10.38}$$

On the other hand, by removing all the 2's in **2**-c-**2**-1-**2** formula (see Exercise 10.8) except the last 2, we have

$$\zeta^\star(\{3,1\}^n,2) = -\sum_{\mathbf{p}_{2n+1}\preceq(\{\overline{2}\}^{2n+1})} 2^{\ell(\mathbf{p}_{2n+1})}\zeta(\mathbf{p}_{2n+1}).$$

Hence by Eq. (10.28) we need to show that

$$(2n+1)\sum_{\mathbf{p}_{2n+1}\preceq(\{\overline{2}\}^{2n+1})} 2^{\ell(\mathbf{p}_{2n+1})}\zeta(\mathbf{p}_{2n+1})$$
$$=\sum_{j=0}^{n}\sum_{\mathbf{p}_{2j}\preceq(\{\overline{2}\}^{2j})} 2^{\ell(\mathbf{p}_{2j})}\zeta(\mathbf{p}_{2j})\cdot 2\zeta(\overline{4(n-j)+2}). \tag{10.39}$$

Suppose a composition $\mathbf{p}_{2n+1}\preceq(\{\overline{2}\}^{2n+1})$ has the form

$$\mathbf{p}_{2n+1}=(a_1,\dots,a_t)\in\mathbb{D}^t,\quad 1\le t\le 2n+1.$$

We now show that there are exactly $2^t(2n+1)$ copies of such term produced by the stuffle product on the right-hand side of Eq. (10.39). Indeed, for each $i=1,\dots,t$ the entry a_i has two cases:

Case (1). $a_i=4b_i>0$. Then for each $k=1,\dots,b_i$ we may produce such a term on the right-hand side of Eq. (10.39) by stuffing

$$2^t\zeta(a_1,\dots,a_{i-1},\overline{4k-2},a_{i+1},\dots,a_t)$$

from $\mathbf{p}_{2j}$ having length t with the term $2\zeta(\overline{4(b_i-k)+2})$ at the right end of Eq. (10.39). Notice no shuffle is possible since $4(n-j)+2$ is not a multiple of 4. Hence these contribute to $2^{t+1}b_i=2^{t-1}a_i$ copies of $\zeta(a_1,\dots,a_t)$.

Case (2). $a_i=\overline{4b_i+2}$. Then for each $k=1,\dots,b_i$ we may produce such a term on the right-hand side of Eq. (10.39) by stuffing

$$2^t\zeta(a_1,\dots,a_{i-1},\overline{4k},a_{i+1},\dots,a_t)$$

from $\mathbf{p}_{2j}$ having length t with the term $2\zeta(\overline{4(b_i-k)+2})$ at the right end of Eq. (10.39). Further, there is exactly one possible shuffle given by

$$2^{t-1}\zeta(a_1,\dots,a_{i-1},a_{i+1},\dots,a_{t+1})\sqcup\!\sqcup\left\{2\zeta(\overline{4b_i+2})\right\},$$

since the index $(a_1,\dots,a_{i-1},a_{i+1},\dots,a_t)$ has only length $t-1$. Altogether these produce $2^{t+1}b_i+2^t=2^{t-1}|a_i|$ copies of $\zeta(a_1,\dots,a_t)$.

By combining Case (1) and Case (2), we see that the right-hand side of Eq. (10.39) produces exactly

$$\sum_{i=1}^{t}2^{t-1}|a_i|=2^{t-1}\cdot|(a_1,\dots,a_t)|=2^t(2n+1)$$

copies of $\zeta(a_1, \dots, a_t)$ since the weight is $4n+2$. This proves (i).

(ii) We use the same analysis as above and see that we need to prove the following identity:

$$\sum_{\mathbf{q}_{2n}} 2^{\ell(\mathbf{q}_{2n})}\zeta(\mathbf{q}_{2n}) = \sum_{j=0}^{n-1} \sum_{\mathbf{p}_{2j+1}\preceq(\{\overline{2}\}^{2j+1})} 2^{\ell(\mathbf{p}_{2j+1})}\zeta(\mathbf{p}_{2j+1}) \cdot 2\zeta(\overline{4(n-j)}), \tag{10.40}$$

where $\mathbf{q}_{2n}$ runs through all indices of the form $A_1 \circ \cdots \circ A_{2n}$ with one of the A_j's (say A_{j_0}) equal to $\overline{4}$ and all the other A_j's equal to $\overline{2}$. For each choice of $2^t\zeta(a_1, \dots, a_{t+1})$ with length $t+1$ from the left-hand side of Eq. (10.40), all but one of the argument components $a_1, \dots, a_{t+1}$ must satisfy the sign rule Eq. (10.38). The only exceptional component, say a_i, must involve a merge with the special entry $A_{j_0} = \overline{4}$. Now there are two cases:

Case (1). $a_i = 4b_i + 2 > 0$. Then for each $k = 0, \dots, b_i - 1$ we may produce such a term on the right-hand side of Eq. (10.40) by stuffing

$$2^t\zeta(a_1, \dots, a_{i-1}, \overline{4k+2}, a_{i+1}, \dots, a_{t+1})$$

from $\mathbf{p}_{2j+1}$ having length $t+1$ with the term $2\zeta(\overline{4(b_i-k)})$ at the right end of (10.40). Notice no shuffle is possible since $4(n-j)$ is a multiple of 4. Hence these contribute to $2^{t+1}b_i$ copies of $\zeta(a_1, \dots, a_{t+1})$. On the left-hand side, such a term must be produced by setting all $2b_i - 1$ consecutive $\circ$'s around $A_{j_0} = \overline{4}$ to $\oplus$:

$$\dots, \underbrace{A_i \oplus A_{i+1} \oplus \cdots \oplus A_{j_0} \oplus \cdots \oplus A_\ell}_{2b_i \text{ entries}}, \dots.$$

But A_{j_0} can be at any one of the $2b_i$ possible positions, thus producing $2^{t+1}b_i$ copies of $\zeta(a_1, \dots, a_{t+1})$ which match exactly the right-hand side of (10.40).

Case (2). $a_i = \overline{4b_i}$. Then for each $k = 1, \dots, b_i - 1$ we may produce such a term on the right-hand side of (10.40) by stuffing

$$2^t\zeta(a_1, \dots, a_{i-1}, 4k, a_{i+1}, \dots, a_{t+1})$$

from $\mathbf{p}_{2j}$ having length $t+1$ with the term $2\zeta(\overline{4(b_i-k)})$ at the right end of Eq. (10.40). Further, there is exactly one possible shuffle given by

$$2^{t-1}\zeta(a_1, \dots, a_{i-1}, a_{i+1}, \dots, a_{t+1}) ш \left\{2\zeta(\overline{4b_i})\right\}.$$

Hence these contribute to $2^{t+1}(b_i - 1) + 2^t = 2^t(2b_i - 1)$ copies of $\zeta(a_1, \dots, a_{t+1})$. Similar to (1), on the left-hand side, such a term must be produced by setting all $2b_i - 2$ consecutive $\circ$'s around A_{j_0} to $\oplus$. And

A_{j_0} can be at any one of the $2b_i - 1$ possible positions, thus producing $2^t(2b_i - 1)$ copies of $\zeta(a_1, \ldots, a_{t+1})$ which match exactly the right-hand side of Eq. (10.40).

This conclude the proof of theorem. □

Next we will prove a more general formula similar to Thm. 10.9.4. First, we introduce some more notation. For $\infty \geq A \geq B \geq 1$, we put

$$S_{-1}(A,B) = \delta_{A,B}A^2, \qquad S_0(A,B) = 1,$$

$$S_j(A,B) = \sum_{A \geq a_1 \geq \cdots \geq a_j \geq B} \frac{1}{a_1^2 \cdots a_j^2} \qquad (j = 1, 2, 3, \ldots),$$

where $\delta_{A,B}$ is the Kronecker symbol. Then we have

$$S_j(A,B) = \sum_{p=B}^{A} \frac{1}{p^2} S_{j-1}(p,B) = \sum_{p=B}^{A} S_{j-1}(A,p)\frac{1}{p^2} \tag{10.41}$$

for any $j \geq 0$. We now prove a lemma by using partial fractions.

Lemma 10.9.7. *For any $j \geq -1$ and $1 \leq p, q \leq \infty$, we have*

$$\sum_{p_0=1}^{p} S_j(p,p_0)\frac{q}{p_0(p_0+q)} = \sum_{q_0=1}^{q} S_j(q,q_0)\frac{p}{q_0(q_0+p)}. \tag{10.42}$$

Here, $\frac{q}{p_0(p_0+q)}$ (resp. $\frac{p}{q_0(q_0+p)}$) means $\frac{1}{p_0}$ (resp. $\frac{1}{q_0}$) if $q = \infty$ (resp. $p = \infty$), and so on.

Proof. We use induction on j. When $j = -1$, both sides are equal to $\frac{pq}{p+q}$. Thus we assume $j \geq 0$. By Eq. (10.41), we can rewrite the left-hand side of Eq. (10.42) as

$$\begin{aligned}\sum_{p_0=1}^{p} S_j(p,p_0)\frac{q}{p_0(p_0+q)} &= \sum_{p_0=1}^{p}\sum_{p_1=p_0}^{p} S_{j-1}(p,p_1)\frac{1}{p_1^2}\left(\frac{1}{p_0} - \frac{1}{p_0+q}\right)\\ &= \sum_{p_1=1}^{p} S_{j-1}(p,p_1)\frac{1}{p_1^2}\sum_{p_0=1}^{p_1}\left(\frac{1}{p_0} - \frac{1}{p_0+q}\right).\end{aligned}$$

By using the identity

$$\begin{aligned}&\sum_{p_0=1}^{p_1}\left(\frac{1}{p_0} - \frac{1}{p_0+q}\right)\\ &\quad= \sum_{p_0=1}^{\infty}\left\{\left(\frac{1}{p_0} - \frac{1}{p_0+q}\right) - \left(\frac{1}{p_0+p_1} - \frac{1}{p_0+p_1+q}\right)\right\}\\ &\quad= \sum_{p_0=1}^{q}\left(\frac{1}{p_0} - \frac{1}{p_0+p_1}\right) = \sum_{q_1=1}^{q}\frac{p_1}{q_1(q_1+p_1)},\end{aligned}$$

we get

$$\sum_{p_1=1}^{p} \frac{S_{j-1}(p,p_1)}{p_1^2} \sum_{p_0=1}^{p_1} \left(\frac{1}{p_0} - \frac{1}{p_0+q} \right) = \sum_{p_1=1}^{p} \frac{S_{j-1}(p,p_1)}{p_1^2} \sum_{q_1=1}^{q} \frac{p_1}{q_1(q_1+p_1)}$$
$$= \sum_{q_1=1}^{q} \frac{1}{q_1^2} \sum_{p_1=1}^{p} \frac{q_1 S_{j-1}(p,p_1)}{p_1(p_1+q_1)}.$$

Finally, using the induction hypothesis and Eq. (10.41) again, we obtain

$$\sum_{q_1=1}^{q} \frac{1}{q_1^2} \sum_{p_1=1}^{p} \frac{q_1 S_{j-1}(p,p_1)}{p_1(p_1+q_1)} = \sum_{q_1=1}^{q} \frac{1}{q_1^2} \sum_{q_0=1}^{q_1} \frac{p S_{j-1}(q_1,q_0)}{q_0(q_0+p)}$$
$$= \sum_{q_0=1}^{q} \frac{p S_j(q,q_0)}{q_0(q_0+p)}.$$

Thus we have shown Eq. (10.42). □

For $n \in \mathbb{N}$, let $\mathbf{j} = (j_1, \ldots, j_n) \in (\mathbb{N}_0)^n$ and $\mathbf{e} = (e_1, \ldots, e_n)$ where $e_j = 1$ or 3. Define

$$\mathscr{C}(\mathbf{j}; \mathbf{e}) := \zeta^\star\left(\{2\}^{j_1}, e_1, \{2\}^{j_2}, e_2, \ldots, e_{n-1}, \{2\}^{j_n}\right). \tag{10.43}$$

For $n = 0$, we simply put $\mathscr{C} = 1$. Note that the right-hand side of Eq. (10.43) diverges if and only if $n \geq 2$, $j_1 = 0$ and $e_1 = 1$. Otherwise, $(\mathbf{j}, \mathbf{e})$ is called an *admissible pair.*

For convenience, we put $j_{a\uparrow b} = (j_a, j_{a+1}, \ldots, j_b)$ and $j_{b\downarrow a} = (j_b, j_{b-1}, \ldots, j_a)$ for any $a \leq b$.

Theorem 10.9.8. *Let n be a positive integer, $\mathbf{j} = (j_1, \ldots, j_n) \in (\mathbb{N}_0)^n$ and $\mathbf{e} = (e_1, \ldots, e_n)$ with $e_j = 1$ or 3 for all j. Assume that both $(\mathbf{j}, \mathbf{e})$ and $(\overleftarrow{\mathbf{j}}, \overleftarrow{\mathbf{e}})$ are admissible pairs. For $k = 0, \ldots, n$, if $e_k = 1$ then we define*

$$X(k) := \mathscr{C}(j_{1\uparrow k}, 0; e_{1\uparrow k}) \cdot \mathscr{C}(j_{n\downarrow(k+1)}; e_{(n-1)\downarrow k})$$

and if $e_k = 3$ then we define

$$X(k) := \mathscr{C}(j_{1\uparrow(k-1)}, j_k + 1; e_{1\uparrow(k-1)}) \cdot \mathscr{C}(j_{n\downarrow(k+2)}, j_{k+1} + 1; e_{(n-1)\downarrow(k+1)}).$$

Then we have

$$\sum_{k=0}^{n} (-1)^k X(k) = 0.$$

Proof. Set $e_0 = e_n = 1$. For $k = 0, \dots, n$, put

$$E(k) = \sum_{\substack{p_0 \ge \cdots \ge p_k \ge 1 \\ q_{n+1} \ge q_n \ge \cdots \ge q_{k+1} \ge 1}} \prod_{\alpha=1}^{k} \frac{S_{j_\alpha}(p_{\alpha-1}, p_\alpha)}{p_\alpha^{e_\alpha}} \prod_{\beta=k+1}^{n} \frac{S_{j_\beta}(q_{\beta+1}, q_\beta)}{q_\beta^{e_{\beta-1}}} \cdot \frac{p_k^{e_k-1} q_{k+1}}{p_k + q_{k+1}},$$

$$F(k) = \sum_{\substack{p_0 \ge \cdots \ge p_k \ge 1 \\ q_{n+1} \ge q_n \ge \cdots \ge q_{k+1} \ge 1}} \prod_{\alpha=1}^{k} \frac{S_{j_\alpha}(p_{\alpha-1}, p_\alpha)}{p_\alpha^{e_\alpha}} \prod_{\beta=k+1}^{n} \frac{S_{j_\beta}(q_{\beta+1}, q_\beta)}{q_\beta^{e_{\beta-1}}} \cdot \frac{p_k q_{k+1}^{e_k-1}}{p_k + q_{k+1}}.$$

Here we allow $p_0 = \infty$ and $q_{n+1} = \infty$. In particular, we have $E(0) = 0$ (resp. $F(n) = 0$) because of the factor $\frac{q_1}{\infty + q_1}$ (resp. $\frac{p_n}{p_n + \infty}$). For general k, we see that $E(k) + F(k) = X(k)$ since

$$\frac{p_k^{e_k-1} q_{k+1}}{p_k + q_{k+1}} + \frac{p_k q_{k+1}^{e_k-1}}{p_k + q_{k+1}} = \begin{cases} 1, & \text{if } e_k = 1; \\ p_k q_{k+1}, & \text{if } e_k = 3. \end{cases}$$

Moreover, when $1 \le k \le n$, one has

$$\sum_{p_k=1}^{p_{k-1}} \frac{S_{j_k}(p_{k-1}, p_k)}{p_k^{e_k}} \frac{p_k^{e_k-1} q_{k+1}}{p_k + q_{k+1}} = \sum_{q_k=1}^{q_{k+1}} \frac{S_{j_k}(q_{k+1}, q_k)}{q_k^{e_{k-1}}} \frac{p_{k-1} q_k^{e_{k-1}-1}}{p_{k-1} + q_k}$$

by Lemma 10.9.7, hence $E(k) = F(k-1)$. Therefore,

$$\begin{aligned} \sum_{k=0}^{n} (-1)^k X(k) &= \sum_{k=0}^{n} (-1)^k \big(E(k) + F(k)\big) \\ &= \sum_{k=1}^{n} (-1)^k F(k-1) + \sum_{k=0}^{n-1} (-1)^k F(k) = 0 \end{aligned}$$

as desired. □

Example 10.9.9. Taking $\mathbf{e} = (3,3)$ or $\mathbf{e} = (3,1)$ in Thm. 10.9.8, one has

$$\begin{aligned} &\zeta^\star(\{2\}^{j_1}, 3, \{2\}^{j_2}, 3, \{2\}^{j_3}, 1) + \zeta^\star(\{2\}^{j_1+1}) \cdot \zeta^\star(\{2\}^{j_3}, 3, \{2\}^{j_2+1}) \\ &\quad = \zeta^\star(\{2\}^{j_1}, 3, \{2\}^{j_2+1}) \cdot \zeta^\star(\{2\}^{j_3+1}) + \zeta^\star(\{2\}^{j_3}, 3, \{2\}^{j_2}, 3, \{2\}^{j_1}, 1) \end{aligned}$$

or

$$\begin{aligned} &\zeta^\star(\{2\}^{j_1}, 3, \{2\}^{j_2}, 1, \{2\}^{j_3}, 1) + \zeta^\star(\{2\}^{j_1+1}) \cdot \zeta^\star(\{2\}^{j_3}, 1, \{2\}^{j_2+1}) \\ &\quad = \zeta^\star(\{2\}^{j_1}, 3, \{2\}^{j_2}, 1) \cdot \zeta^\star(\{2\}^{j_3}, 1) + \zeta^\star(\{2\}^{j_3}, 1, \{2\}^{j_2}, 3, \{2\}^{j_1}, 1), \end{aligned}$$

respectively. Of course, these two relations can by proved by applying Thm. 10.8.1 to each of the MZ*Vs and then using the stuffle relations.

Corollary 10.9.10. *Let $n > 0$ be an integer. Then*

(i) *For any* $a_1, a_2, \ldots, a_n \in \mathbb{N}_0$ *such that* $a_1, a_n \geq 1$, *we have*

$$\sum_{k=0}^{n}(-1)^k\zeta^\star\Big(\big\{\{2\}^{a_j},1\big\}_{j=1}^{k}\Big)\cdot\zeta^\star\Big(\big\{\{2\}^{a_j},1\big\}_{j=n}^{k+1}\Big)=0. \tag{10.44}$$

Here $\zeta^\star\Big(\big\{\{2\}^{a_j},1\big\}_{j=1}^{0}\Big)=\zeta^\star\Big(\big\{\{2\}^{a_j},1\big\}_{j=n}^{n+1}\Big)=1.$

(ii) *For any* $a_1, a_2, \ldots, a_{2n} \in \mathbb{N}_0$, *we have*

$$\begin{aligned}\sum_{k=0}^{n}&\zeta^\star(\{2\}^{a_1},3,\{2\}^{a_2},1,\ldots,3,\{2\}^{a_{2k}},1)\\ &\times\zeta^\star(\{2\}^{a_{2n}},3,\{2\}^{a_{2n-1}},1,\ldots,3,\{2\}^{a_{2k+1}},1)\\ &=\sum_{k=1}^{n}\zeta^\star(\{2\}^{a_1},3,\{2\}^{a_2},1,\ldots,1,\{2\}^{a_{2k-1}+1})\\ &\times\zeta^\star(\{2\}^{a_{2n}},3,\{2\}^{a_{2n-1}},1,\ldots,1,\{2\}^{a_{2k}+1}).\end{aligned} \tag{10.45}$$

Proof. Taking $\mathbf{e}=(\{1\}^n)$ we get

$$\begin{aligned}X(k)&=\mathscr{C}\big(j_{1\uparrow k},0;\{1\}^k\big)\cdot\mathscr{C}\big(j_{n\downarrow(k+1)},0;\{1\}^{n-k}\big)\\ &=\zeta^\star(\{2\}^{j_1},1,\ldots,\{2\}^{j_k},1)\cdot\zeta^\star(\{2\}^{j_n},1,\ldots,\{2\}^{j_{k+1}},1).\end{aligned}$$

Hence Thm. 10.9.8 implies the identity Eq. (10.44). Similarly, for $\mathbf{e}=(\{3,1\}^n)$, we have

$$\begin{aligned}X(k)=&\zeta^\star(\{2\}^{j_1},3,\{2\}^{j_2},1,\ldots,3,\{2\}^{j_k},1)\\ &\times\zeta^\star(\{2\}^{j_{2n}},3,\{2\}^{j_{2n-1}},1,\ldots,3,\{2\}^{j_{k+1}},1)\end{aligned}$$

if k is even, and

$$\begin{aligned}X(k)=&\zeta^\star(\{2\}^{j_1},3,\{2\}^{j_2},1,\ldots,3,\{2\}^{j_{k-1}},1,\{2\}^{j_k+1})\\ &\times\zeta^\star(\{2\}^{j_{2n}},3,\{2\}^{j_{2n-1}},1,\ldots,3,\{2\}^{j_{k+2}},1,\{2\}^{j_{k+1}+1})\end{aligned}$$

if k is odd. This proves Eq. (10.45). □

Proposition 10.9.11. *For positive integers* m *and* n, *we have*

$$\zeta^\star(\{2m\}^n)=\pi^{2mn}\sum_{\substack{\sum_{j=0}^{m-1}n_j=mn\\ n_i\geq 0}}(-1)^{mn}\exp\left(\frac{2\pi i}{m}\sum_{l=0}^{m-1}ln_l\right)\prod_{k=0}^{m-1}\beta(n_k)$$

where $\beta(n)=(2-2^{2n})B_{2n}/(2n)!$.

Proof. Using the infinite product for the sine function, we have

$$\csc \pi x e^{\frac{\pi i}{m}k} = \left(\pi x e^{\frac{\pi i}{m}k} \prod_{n=1}^{\infty}\left(1-\frac{x^2}{n^2}e^{\frac{2\pi i}{m}k}\right)\right)^{-1}.$$

Substituting $k = 0, 1, \ldots, m-1$ and multiplying both sides, we obtain

$$\left(\pi^m x^m \prod_{k=0}^{m-1} e^{\frac{\pi i}{m}k}\right) \prod_{k=0}^{m-1} \csc \pi x e^{\frac{\pi i}{m}k} = \prod_{n=1}^{\infty}\left(1-\frac{x^{2m}}{n^{2m}}\right)^{-1}. \tag{10.46}$$

The right-hand side of Eq. (10.46) equals

$$1+\left(\sum_{n_1>0}\frac{1}{n_1^{2m}}\right)x^{2m}+\cdots = 1+\sum_{n=1}^{\infty}\zeta^\star(\{2m\}^n)x^{2mn}.$$

On the other hand, the left-hand side of Eq. (10.46) equals

$$\begin{aligned}
&\left(\pi^m x^m \prod_{k=0}^{m-1} e^{\frac{\pi i}{m}k}\right) \prod_{k=0}^{m-1} \sum_{n_k=0}^{\infty} (-1)^{n_k-1}\frac{(2^{2n_k}-2)B_{2n_k}}{(2n_k)!} \\
&\quad \times \pi^{2n_k-1}x^{2n_k-1}e^{\frac{\pi i}{m}k(2n_k-1)} \\
&= \prod_{k=0}^{m-1} \sum_{n_k=0}^{\infty} (-1)^{n_k-1}\frac{(2^{2n_k}-2)B_{2n_k}}{(2n_k)!}\pi^{2n_k}x^{2n_k}e^{\frac{2\pi i}{m}kn_k} \\
&= 1+\sum_{n=1}^{\infty}\sum_{\sum_{j=0}^{m-1} n_j=mn} (-1)^{m(n-1)}(\pi x)^{2mn}\exp\left(\frac{2\pi i}{m}\sum_{j=0}^{m-1} jn_j\right)\prod_{j=0}^{m-1}\beta(n_j).
\end{aligned}$$

Comparing coefficients of both sides, we obtain the desired identity. □

10.10 Historical Notes

Even at the early stage of Euler's study of the MZVs he in fact considered not the MZVs but the MZ⋆Vs. Notice sometimes these values were also called non-strict MZVs in some older literature (for e.g., [445]). Instead of Eq. (1.23) Euler actually showed, for all $n \ge 2$

$$2\zeta^\star(n,1) = (n+2)\zeta(n+1) - \sum_{i=1}^{n-2}\zeta(n-i)\zeta(i+1),$$

where $\zeta^\star(n,1)$ is a double zeta star value. See [210].

In [288] Hoffman established a few fundamental results concerning both the MZVs and the MZ⋆Vs. Since then the MZ⋆Vs frequently appear in the literature although they do not seem to enjoy the same fame as their cousins, the MZVs. Most often one can reduce the study of the MZ⋆Vs to that of the MZVs but sometimes the MZ⋆V version is nicer and easier while at other times the opposite is true. To make things both more complicated and more interesting, the symmetrized MZVs (see Chap. 6) are even more elegant than both the MZVs and the MZ⋆Vs in some situations.

The integral expression of the MZ⋆Vs in Thm. 10.1.1 was first stated in Yamamoto's paper [583]. Using this idea, he gave a few very nice relations between the MZ⋆Vs and Arakawa–Kaneko zeta values (see [21]).

In [288] Hoffman applied some combinatorial theory to the MZ⋆V and proved Thm. 10.2.2. Machide [410] then generalized the MZV part of this theorem to regularized MZVs of weight ≤ 4. The sum relation stated as Eq. (10.7) was first conjectured by C. Moen in an unpublished manuscript according to Hoffman [288]. It was initially verified for depth up to three by Moen and by Hoffman for weight up to six. Our proof here closely follows the argument of [288]. Ohno and Wakabayashi [446] proved the cyclic sum relation of Thm. 10.4.2 for the MZ⋆V by using partial fractions. In [534], Tanaka and Wakabayashi used Hoffman's algebraic machinery to prove some more results akin to the cyclic sum formula. Soon afterwards, Ohno *et al.* [449] extended the cyclic sum formula to its q-analog.

In [14] Aoki and Ohno used generating functions to derive the sum formula of the MZ⋆Vs of fixed weight and height, which is the content of Cor. 10.6.5. Their method was improved to handle the case with the depth also fixed in [15]. This is the source of Thm. 10.6.2.

The identity Eq. (10.10) was proved by Zlobin [641]. He also mentioned it was essentially known to Vasilev [559]. Corollary 10.7.6 was first shown by Kaneko and Ohno [332, Thm. 1.3]. Another proof was given by Yamazaki [584]. It was later generalized to arbitrary height by Z. Li [397] whose proof is reproduced here to show Thm. 10.7.3.

Proposition 10.8.3 was discovered by Ohno and Zudilin [448] numerically first. They dubbed the identity the "Two-one formula", which we denote by the $\underline{\mathbf{2}\text{-}\mathbf{1}}$ formula here. In [582] Yamamoto considered some algebraic structures depending on a variable t which reflect the properties of the MZVs and the MZ⋆Vs when $t = 0$ and $t = 1$, respectively. As he pointed out (see [581, Conjecture 4.4]) the validity of the $\underline{\mathbf{2}\text{-}\mathbf{1}}$ formula implies that the MZ⋆V algebra structure of the form $\zeta^\star(\{2\}^{a_1}, 1, \dots, \{2\}^{a_\ell}, 1)$ is reflected by setting $t = 1/2$. As special cases of the "Two-one formula", Eq. (10.28) (resp. Eq. (10.35)) was first proved by Zlobin [639] (resp. by Ohno and Zudilin [448] by partial fractions).

In [241] Glanois studied Conjecture 10.8.7 using the motivic version of the general **2-3-2** formula and the Hopf algebra structure of motivic MZVs and Euler sums. She also obtained many nice results on the colored MZVs (see Chap. 13) using Galois descents.

When all the arguments $n_1,\dots,n_\ell$ are positive Lemma 10.9.1 becomes [288, Thm. 2.2] whose proof is adopted here. Note that the same result is also stated as [368, Prop. 9.4] whose proof is different from that of [288] but also works here.

The "Three-One" type formula in Theorem 10.9.8 is due to Tasaka and Yamamoto [537]. The "Two-three" type formula in Thm. 10.9.3 was first conjectured by Imatomi *et al.* as Conjectures 4.1 and 4.3 of [313]. Some special cases were confirmed by Imatomi *et al.* [313, Thm. 1.1] and also by Tasaka and Yamamoto [537]. Later, Yamamoto [581] presented a more precise version. Here we have given a different and concise proof. In [581, Thm. 1.1] Yamamoto obtained a more precise formula by using partial sums and generating functions:

$$\sum_{\substack{e_0,e_1,\dots,e_{2r}\geq 0\\ e_0+e_1+\cdots+e_{2r}=m}} \zeta^\star(\{2\}^{e_0},3,\{2\}^{e_1},1,\{2\}^{e_2},3,\dots,3,\{2\}^{e_{2r-1}},1,\{2\}^{e_{2r}})$$
$$= \sum_{\substack{2i+k+u=2r\\ j+l+v=m}} (-1)^{j+k}\binom{k+l}{k}\binom{u+v}{u}\binom{2i+j}{j}\frac{\beta'_{k+l}\beta'_{u+v}\pi^{4r+2m}}{(2i+1)(4i+2j+1)!}, \tag{10.47}$$

where $\beta'_n=(-1)^n(2-2^{2n})B_{2n}/(2n)!$. It is possible to modify the proof of Thm. 10.9.3 to give this more quantified version.

The "Three-Two-One" type formulas in Thm. 10.9.5 were first conjectured by Imatomi *et al.* as [313, Conjecture 4.5]. Note that Thm. 10.9.5(i) is the more precise version of the $n=1$ case of Thm. 10.9.3(i). And Thm. 10.9.5(iii) can be written more compactly as

$$\zeta^\star\big(\{2\}\sqcup\{3,1\}^n\big)=\sum_{k=0}^{n}\zeta^\star(\{3,1\}^{n-k},2)\zeta^\star(\{2\}^{2k}),$$

which is the more precise version of the $m=1$ case of Exercise 10.10, a result of Kondo *et al.* [358]. The case $m=0$ case has the following precise formulation by Muneta [429]:

$$\zeta^\star\big(\{3,1\}^n\big)=\sum_{i=0}^{n}\left\{\frac{2}{(4i+2)!}\sum_{\substack{n_0+n_1=2(n-i)\\ n_0,n_1\geq 0}}(-1)^{n_1}\beta(n_0)\beta(n_1)\right\}\pi^{4n},$$

where $\beta(n)=(2-2^{2n})B_{2n}/(2n)!$. Muneta also found a precise form in the case $m=1$ and Prop. 10.9.11. Of course, these are all special cases of

Yamamoto's general formula in Eq. (10.47). Finally, Tasaka and Yamamoto proved Thm. 10.9.4 in [537] while Kondo *et al.* found an MZ⋆V analog of Thm. 5.8.1 in [358].

Exercises

10.1. Suppose $d, k_1, \dots, k_d \in \mathbb{N}$. For any nonnegative integer $i \le k_1 - 2$, prove that

$$\sum_{n_1\ge n_2\ge\cdots\ge n_d\ge n_{d+1}\ge 1,\ n_1\ne n_{d+1}} \frac{1}{(n_1-n_{d+1})n_1^{k_1-i}\dots n_d^{k_d}n_{d+1}^{i}}$$
$$= \sum_{n_1\ge n_2\ge\cdots\ge n_d\ge n_{d+1}\ge 1,\ n_1\ne n_{d+1}} \frac{1}{(n_1-n_{d+1})n_1^{k_1-i-1}\dots n_d^{k_d}n_{d+1}^{i+1}}$$
$$- \zeta^\star(k_1-i, k_2, k_3, \dots, k_d, i+1) + \zeta(|\mathbf{k}|+1)$$

by mimicking the proof of Lemma 5.2.2.

10.2. Show that the cyclic sum relation Eq. (10.9) implies the sum relation Eq. (10.7).

10.3. Prove Eq. (10.14) by using geometric series.

10.4. Use Lemma C.1 to verify that the two functions in Eq. (10.15) are indeed the solutions of the corresponding homogeneous differential equation (10.13). Then verify that Eq. (10.16) gives a solution of Eq. (10.13). Note that the homogeneous equation is a special case of the Riemann differential equations with the singularities at $0, 1, \infty$ (see Sec. C.1).

10.5. Prove the transformation Eq. (10.17) using the linear transformation Eq. (C.5) with $(a,b,c) = (1-\beta, 1-\beta+x, 2-y)$ followed by the linear transformation in Eq. (C.4).

10.6. Prove Prop. 10.8.9 and Prop. 10.8.8 by repeatedly applying Thm. 10.8.1.

10.7. Prove that for $\mathbf{s} = \left(\operatorname*{\mathbf{Cat}}_{j=1}^{\ell} \{\{2\}^{b_j}, c_j, \{2\}^{a_j}, 1\}\right)$, $\ell \ge 1$, we have

$$\zeta^\star(\mathbf{s}) = \zeta^\sharp\Big(\operatorname*{\mathbf{Cat}}_{j=1}^{\ell} \{\overline{2b_j+2},\ \{1\}^{c_j-3},\ \overline{2a_j+2}\}\Big).$$

10.8. Let $t, \ell \in \mathbb{N}$ and $a_j, b_j, c_j - 3 \in \mathbb{N}_0$ for all $j \ge 1$. Prove the following two general formulas.

(<u>**2**-c-**2**-1</u>-**2**). For $\mathbf{s} = \left(\operatorname*{\mathbf{Cat}}_{j=1}^{\ell}\{\{2\}^{b_j}, c_j, \{2\}^{a_j}, 1\}, \{2\}^t\right)$ with $t, \ell \geq 1$, show that

$$\zeta^\star(\mathbf{s}) = -\zeta^\sharp\Big(\operatorname*{\mathbf{Cat}}_{j=1}^{\ell}\{\overline{2b_j+2},\ \{1\}^{c_j-3},\ \overline{2a_j+2}\}, \overline{2t}\Big). \tag{10.48}$$

(**2**-1-<u>**2**-c-**2**-1</u>-**2**): For $\mathbf{s} = \left(\{2\}^{a_0}, 1, \operatorname*{\mathbf{Cat}}_{j=1}^{\ell}\{\{2\}^{b_j}, c_j, \{2\}^{a_j}, 1\}, \{2\}^t\right)$, with $\ell \geq 0$ and $a_0, t \geq 1$, show that

$$\zeta^\star(\mathbf{s}) = -\zeta^\sharp\Big(2a_0+1, \operatorname*{\mathbf{Cat}}_{j=1}^{\ell}\{\overline{2b_j+2},\ \{1\}^{c_j-3},\ \overline{2a_j+2}\}, \overline{2t}\Big). \tag{10.49}$$

Then verify the following special cases:

$$\begin{aligned}\zeta^\star(\{2\}^b, 3, \{2\}^a, 1, \{2\}^t) &= -2\zeta(\overline{2b+2a+2t+4}) - 4\zeta(2b+2a+4, \overline{2t})\\ &\quad -4\zeta(\overline{2b+2}, 2a+2t+2) - 8\zeta(\overline{2b+2}, \overline{2a+2}, \overline{2t}),\end{aligned}$$

and

$$\begin{aligned}&\zeta^\star(\{2\}^{a_1}, 1, \{2\}^b, 3, \{2\}^{a_2}, 1, \{2\}^t)\\ &= -2\zeta(\overline{2a_1+2b+2a_2+2t+5}) - 4\zeta(2a_1+2b+2a_2+5, \overline{2t})\\ &\quad - 4\zeta(\overline{2a_1+2b+3}, 2a_2+2t+2) - 8\zeta(\overline{2a_1+2b+3}, \overline{2a_2+2}, \overline{2t})\\ &\quad - 4\zeta(2a_1+1, \overline{2b+2a_2+2t+4}) - 8\zeta(2a_1+1, 2b+2a_2+4, \overline{2t})\\ &\quad - 8\zeta(2a_1+1, \overline{2b+2}, 2a_2+2t+2) - 16\zeta(2a_1+1, \overline{2b+2}, \overline{2a_2+2}, \overline{2t}).\end{aligned}$$

10.9. For any positive integer n, show that

$$\zeta^\star(\{2\}^{n-1}, 3) = \sum_{l=1}^{n}\zeta^\star(\{2\}^l, 1, \{2\}^{n-l}) - \sum_{l=0}^{n-1}\zeta^\star(\{2\}^l, 3, \{2\}^{n-1-l}).$$

10.10. For all nonnegative integers m and n, show that

$$\zeta^\star(\{2\}^m \sqcup\!\sqcup \{3,1\}^n) \in \pi^{2m+4n}\mathbb{Q}.$$

Chapter 11

q-Analogs of Multiple Zeta Functions

In this chapter we define some q-analogs of the multiple zeta functions and study their analytic properties. We will also consider some special values at non-positive integers. The theory of special values at positive integers will be developed in the next chapter. As before, throughout this chapter let q be a real number with $0 < q < 1$.

11.1 Definitions

First, for positive integer d, we define an auxiliary function of $2d$ complex variables $s_1, \dots, s_d, t_1, \dots, t_d \in \mathbb{C}$:

$$\zeta_q^{\mathbf{t}}[\mathbf{s}] = \zeta_q^{(t_1,\dots,t_d)}[s_1, \dots, s_d] := \sum_{k_1 > \cdots > k_d > 0} \frac{q^{k_1 t_1 + \cdots + k_d t_d}}{[k_1]^{s_1} \cdots [k_d]^{s_d}}. \tag{11.1}$$

Again, $|\mathbf{s}| := s_1 + \cdots + s_d$ is called the *weight* and the number d is called the *depth*, denoted by $\mathrm{dp}(\mathbf{s})$.

Proposition 11.1.1. *The function $\zeta_q^{\mathbf{t}}[\mathbf{s}]$ converges if $\mathrm{Re}(t_1 + \cdots + t_j) > 0$ for all $j = 1, \dots, d$. It can be analytically continued to a meromorphic function over $\mathbb{C}^{2d}$ via the series expansion*

$$\zeta_q^{\mathbf{t}}[\mathbf{s}] = (1-q)^{|\mathbf{s}|} \sum_{r_1,\dots,r_d=0}^{\infty} \prod_{j=1}^{d} \left[\binom{s_j + r_j - 1}{r_j} \frac{q^{(d+1-j)(r_j+t_j)}}{1 - q^{r_1+t_1+\cdots+r_j+t_j}} \right], \tag{11.2}$$

where $|\mathbf{s}| = s_1 + \cdots + s_d$. It has the following (simple) poles: $t_1 + \cdots + t_j \in \mathbb{Z}_{\le 0} + \frac{2\pi i}{\log q}\mathbb{Z}$ for $j = 1, \dots, d$.

Proof. Assume $|\mathrm{Re}(s_j)| < N_j$ and let $\tau_j = \mathrm{Re}(t_j)$ for all $j = 1, \dots, d$. Since

$$1 < [k] = 1 + q + \cdots + q^{k-1} < k, \tag{11.3}$$

we have

$$\begin{aligned}\left|\zeta_q^{\mathbf{t}}[\mathbf{s}]\right| &< \sum_{k_1>\cdots>k_d>0}\prod_{j=1}^{d}(k_j)^{N_j}q^{k_j\tau_j}\\ &= \sum_{m_1,\ldots,m_d=1}^{\infty}\prod_{j=1}^{d}(m_j+\cdots+m_d)^{N_j}q^{(m_j+\cdots+m_d)\tau_j}\\ &< \prod_{j=1}^{d}j^{N_j}\sum_{m_1,\ldots,m_d=1}^{\infty}\prod_{j=1}^{d}m_j^{N_1+\cdots+N_j}q^{(\tau_1+\cdots+\tau_j)m_j}.\end{aligned}$$

By the root test $\sum_{m>0} m^N q^{m\tau}$ converges if $\tau > 0$. This proves the first part of the proposition.

We now prove Eq. (11.2). By the binomial expansion

$$(1-x)^{-s} = \sum_{r=0}^{\infty}\binom{s+r-1}{r}x^r$$

we get

$$\zeta_q^{\mathbf{t}}[\mathbf{s}] = (1-q)^{|\mathbf{s}|}\sum_{k_1>\cdots>k_d>0}\sum_{r_1,\ldots,r_d=0}^{\infty}\left(\prod_{j=1}^{d}\binom{s_j+r_j-1}{r_j}q^{k_j(r_j+t_j)}\right).$$

As $0<q<1$ the series converges absolutely by Stirling's formula so we can change the order of the summations. The proposition follows immediately from the next lemma by taking $x_j = q^{t_j+r_j}$ for $j=1,\ldots,d$. □

Lemma 11.1.2. *Let $x_j \in \mathbb{C}$ such that $|x_j| < 1$ for $j = 1,\ldots,d$. Then*

$$\sum_{k_1>\cdots>k_d>0}\prod_{j=1}^{d}x_j^{k_j} = \prod_{j=1}^{d}\frac{x_1\cdots x_j}{1-x_1\cdots x_j} = \prod_{j=1}^{d}\frac{x_j^{d+1-j}}{1-x_1\cdots x_j}. \tag{11.4}$$

Proof. By re-indexing $k_j = m_1+\cdots+m_j$ we have

$$\begin{aligned}\sum_{k_1>\cdots>k_d>0}\prod_{j=1}^{d}x_j^{k_j} &= \sum_{m_1=1}^{\infty}\cdots\sum_{m_d=1}^{\infty}x_1^{m_1+\cdots+m_d}\cdots x_{d-1}^{m_{d-1}+m_d}x_d^{m_d}\\ &= \prod_{j=1}^{d}\frac{x_1\cdots x_j}{1-x_1\cdots x_j}.\end{aligned}$$

The second equation in the lemma follows immediately. □

Now we define the first version of the q-analog of the multiple zeta functions.

Definition 11.1.3. For all $\mathbf{s} = (s_1, \dots, s_d) \in \mathbb{C}^d$, we set

$$\zeta_q^I[\mathbf{s}] := \zeta_q^{\mathbf{s}-\mathbf{1}}[\mathbf{s}] = \sum_{k_1 > \cdots > k_d \geq 1} \prod_{j=1}^{d} \frac{q^{k_j(s_j-1)}}{[k_j]^{s_j}}. \tag{11.5}$$

The following corollary follows easily from Prop. 11.1.1.

Theorem 11.1.4. *Set*

$$\mathsf{S}_d' := \left\{ (s_1, \dots, s_d) \in \mathbb{C}^d \;\middle|\; \begin{array}{l} s_1 \in 1 + \dfrac{2\pi i}{\log q}\mathbb{Z}, \ \text{or } s_1 \in \mathbb{Z}_{\leq 0} + \dfrac{2\pi i}{\log q}\mathbb{Z}_{\neq 0}, \\ \text{or } s_1 + \cdots + s_j \in \mathbb{Z}_{\leq j} + \dfrac{2\pi i}{\log q}\mathbb{Z}, \ j > 1 \end{array} \right\}.$$

Then $\zeta_q^I[s_1, \dots, s_d]$ defined by Eq. (11.5) can be extended to a meromorphic function with simple poles lying along S_d' and it can be written as

$$\zeta_q^I[\mathbf{s}] = (1-q)^{|\mathbf{s}|} \sum_{r_1, \dots, r_d = 0}^{+\infty} \prod_{j=1}^{d} \left[\binom{s_j + r_j - 1}{r_j} \frac{q^{(d+1-j)(r_j+s_j-1)}}{1 - q^{r_1+s_1+\cdots+r_j+s_j-j}} \right]. \tag{11.6}$$

It is easy to see that as $q \to 1$ the set S_d' becomes the set defined by Eq. (1.15):

$$\mathsf{S}_d = \left\{ (s_1, \dots, s_d) \in \mathbb{C}^d \;\middle|\; \begin{array}{l} s_1 = 1, s_1 + s_2 = 2, 1, -2m \ \forall m \in \mathbb{Z}_{\geq 0}, \\ \text{and } s_1 + \cdots + s_j \in \mathbb{Z}_{\leq j} \ \forall j \geq 3. \end{array} \right\} \tag{11.7}$$

which is exactly the singular set of the multiple zeta function $\zeta(s_1, \dots, s_d)$.

To see the effect of taking different specializations of t_j in $\zeta_q^{\mathbf{t}}[\mathbf{s}]$ we define the shifting operators $\mathcal{S}_j$ $(1 \leq j \leq d)$ on any function $f_q[s_1, \dots, s_d]$ as follows:

$$\mathcal{S}_j f_q[s_1, \dots, s_d] := f_q[s_1, \dots, s_d] + (1-q) f_q[s_1, \dots, s_j - 1, \dots, s_d]. \tag{11.8}$$

It is obvious that these operators commute.

Proposition 11.1.5. *Let $n_1, \dots, n_d \in \mathbb{N}_0$. Then we have*

$$\begin{aligned} \zeta_q^{(s_1-1-n_1, \dots, s_d-1-n_d)}[s_1, \dots, s_d] &= \mathcal{S}_1^{n_1} \circ \cdots \circ \mathcal{S}_d^{n_d} \zeta_q^I[s_1, \dots, s_d] \\ &= \sum_{r_1=0}^{n_1} \cdots \sum_{r_d=0}^{n_d} \left(\prod_{j=1}^{d} \binom{n_j}{r_j} (1-q)^{r_j} \right) \zeta_q^I[s_1 - r_1, \dots, s_d - r_d]. \end{aligned}$$

Proof. We only sketch the proof in the case $n_2 = \cdots = n_d = 0$. The general case is completely similar. Suppose $n_1 = n = 1$. Then

$$\begin{aligned}
&\zeta_q^{(s_1-2,s_2-1,\dots,s_d-1)}[s_1,\dots,s_d] \\
&\quad = \sum_{k_1>\cdots>k_d>0} \frac{q^{k_1(s_1-2)+k_2(s_2-1)+\cdots+k_d(s_d-1)}}{[k_1]^{s_1}\cdots[k_d]^{s_d}} \\
&\quad = \sum_{k_1>\cdots>k_d>0} \frac{q^{k_1(s_1-2)}(1-q^{k_1})+q^{k_1(s_1-1)}}{[k_1]^{s_1}} \cdot \frac{q^{k_2(s_2-1)+\cdots+k_d(s_d-1)}}{[k_2]^{s_2}\cdots[k_d]^{s_d}} \\
&\quad = \mathcal{S}_1\zeta_q^I[s_1,\dots,s_d].
\end{aligned}$$

The rest follows easily by induction on n and some binomial identities. See Exercise 11.1. □

Corollary 11.1.6. *Let n be a positive integer. Then*

$$\zeta_q^{(s-1-n)}[s] = \mathcal{S}^n\zeta_q^I[s] = \sum_{r=0}^{n}\binom{n}{r}(1-q)^r\zeta_q^I[s-r].$$

Observe that one effect of the shifting operator is to bring in more poles. Essentially, $\mathcal{S}^n$ shifts all the poles of $\zeta_q^I[s]$ by n to the right on the complex plane.

11.2 Analytic Continuation

As a golden rule, the correct q-analog of any classical mathematical object should have the property that when $q \to 1$ the q-analog becomes its classical counterpart. Although it is obvious that when $\mathrm{Re}(s_1 + \cdots + s_j) > j$ we have

$$\lim_{q\to 1}\zeta_q^I[s_1,\dots,s_d] = \zeta(s_1,\dots,s_d),$$

it is highly nontrivial to verify this still holds for all $(s_1,\dots,s_d)$ outside the singular set Eq. (1.15) of the analytic continuation of the multiple zeta function $\zeta(s_1,\dots,s_d)$. We will prove this fact in this section.

Similar to the classical case (see Sec. 1.3) the basic idea is to use the classical Euler–Maclaurin summation formula Eq. (1.12). Let $f(x)$ be any (complex-valued) C^∞ function on $[1,\infty)$ and let m and M be two positive integers. Recall that

$$\begin{aligned}
\sum_{n=1}^{m} f(n) = \int_1^m f(x)\,dx + \frac{1}{2}(f(1)+f(m)) + \sum_{r=1}^{M}\frac{B_{r+1}}{(r+1)!}\Big(f^{(r)}(m)-f^{(r)}(1)\Big) \\
-\frac{(-1)^{M+1}}{(M+1)!}\int_1^m \widetilde{B}_{M+1}(x)f^{(M+1)}(x)\,dx \qquad (11.9)
\end{aligned}$$

where the periodic Bernoulli polynomial satisfies

$$\widetilde{B}_k(x) = -k! \sum_{n\in\mathbb{Z}\setminus\{0\}} \frac{e^{2\pi inx}}{(2\pi in)^k}. \tag{11.10}$$

We first consider the depth one case.

Theorem 11.2.1. *The* q*-analog of the Riemann zeta function* $\zeta_q^I[s]$ *can be extended to a meromorphic function on* $\mathbb{C}\setminus \mathsf{S}_1'$ *with simple poles along* S_1'*. Further, for all* $s \in \mathbb{C}\setminus \mathsf{S}_1$

$$\lim_{q\to 1} \zeta_q^I[s] = \zeta(s).$$

Proof. We set

$$F(x) = \frac{q^{x(s-1)}}{(1-q^x)^s}, \quad F'(x) = (\log q)q^{x(s-1)}\frac{s-1+q^x}{(1-q^x)^{s+1}},$$
$$F''(x) = (\log q)^2\frac{s(s+1)-3s(1-q^x)+(1-q^x)^2}{q^{x(1-s)}(1-q^x)^{s+2}}.$$

Taking $f(x) = F(x)$ in Eq. (11.9) and letting $m \to \infty$, we get

$$\sum_{n=1}^{\infty} \frac{q^{n(s-1)}}{(1-q^n)^s} = \int_1^{\infty} \frac{q^{x(s-1)}}{(1-q^x)^s}\,\mathrm{d}x + \frac{1}{2}\cdot\frac{q^{s-1}}{(1-q)^s} - \frac{(\log q)q^{s-1}}{12}\frac{s-1+q}{(1-q)^{s+1}}$$
$$- \frac{(\log q)^2}{2}\int_1^{\infty} \widetilde{B}_2(x)\frac{s(s+1)-3s(1-q^x)+(1-q^x)^2}{q^{x(1-s)}(1-q^x)^{s+2}}\,\mathrm{d}x$$

for $\mathrm{Re}(s) > 1$. Hence (see Exercise 11.2)

$$\zeta_q^I[s] = \frac{q^{s-1}}{s-1}\cdot\frac{q-1}{\log q} + \frac{q^{s-1}}{2} + \frac{q^{s-1}}{12}\frac{\log q}{q-1}(s-1+q)$$
$$- (1-q)^s\frac{(\log q)^2}{2}\int_1^{\infty} \widetilde{B}_2(x)\frac{s(s+1)-3s(1-q^x)+(1-q^x)^2}{q^{x(1-s)}(1-q^x)^{s+2}}\,\mathrm{d}x. \tag{11.11}$$

By Eq. (11.10)

$$\zeta_q^I[s] = \frac{q^{s-1}}{s-1}\cdot\frac{q-1}{\log q} + \frac{q^{s-1}}{2} + \frac{q^{s-1}}{12}\frac{\log q}{q-1}(s-1+q)$$
$$+ \sum_{n\in\mathbb{Z}\setminus\{0\}} \frac{(1-q)^s(\log q)^2}{(2\pi in)^2}\int_1^{\infty} e^{2\pi inx}\frac{s(s+1)-3s(1-q^x)+(1-q^x)^2}{q^{x(1-s)}(1-q^x)^{s+2}}\,\mathrm{d}x.$$

Setting $q^x = u$ we find

$$\zeta_q^I[s] = \frac{q^{s-1}}{s-1}\cdot\frac{q-1}{\log q} + \frac{q^{s-1}}{2} + \frac{q^{s-1}}{12}\frac{\log q}{q-1}(s-1+q)$$
$$- \sum_{n\in\mathbb{Z}\setminus\{0\}} \frac{(1-q)^s \log q}{(2\pi i n)^2} \sum_{\nu=\pm 1,0} a_\nu(s) b_q(s-1+\delta n, -s+\nu), \quad (11.12)$$

where $\delta = 2\pi i/\log q$, $a_{-1}(s) = s(s+1)$, $a_0(s) = -3s$, $a_1(s) = 1$, and

$$b_t(\alpha,\beta) = \int_0^t u^{\alpha-1}(1-u)^{\beta-1}\,du. \quad (11.13)$$

Observe that $\sigma = \mathrm{Re}(s) > 1$ and for all $\nu = -1, 0, 1$

$$\left| b_q(s-1+\delta n, -s+\nu)\right| \le \int_0^q u^{\sigma-2}(1-u)^{-\sigma+\nu-1}\,du.$$

Hence the sum in Eq. (11.12) converges absolutely. For Eq. (11.13), we have (see Exercise 11.3)

$$b_q(\alpha,\beta) = \sum_{r=1}^{M-1} (-1)^{r-1}\frac{(1-\beta)_{r-1}}{(\alpha)_r} q^{\alpha+r-1}(1-q)^{\beta-r}$$
$$+ (-1)^{M-1}\frac{(1-\beta)_{M-1}}{(\alpha)_{M-1}} b_q(\alpha+M-1, \beta-M+1) \quad (11.14)$$

for any $M \ge 2$. Applying this to $b_q(s-1+\delta n, -s-1)$ and using $q^{\delta n} = 1$ we get

$$b_q(s-1+\delta n, -s+\nu) = \sum_{r=1}^{M-1} (-1)^{r-1}\frac{(s+1-\nu)_{r-1}}{(s-1+\delta n)_r} q^{s+r-2}(1-q)^{-s+\nu-r}$$
$$+ (-1)^{M-1}\frac{(s+1-\nu)_{M-1}}{(s-1+\delta n)_{M-1}} b_q(s-2+M+\delta n, -s+\nu+1-M). \quad (11.15)$$

For all $\nu = -1, 0, 1$, this provides the analytic continuation of $b_q(s-1+\delta n, -s+\nu)$ as functions of s into the region $\mathrm{Re}(s) > 2-M$. This implies that

$$\sum_{n\in\mathbb{Z}\setminus\{0\}} \frac{s(s+1)}{(2\pi i n)^2} b_q(s-1+\delta n, -s-1)$$
$$= \sum_{r=1}^{M-1} (-1)^{r-1} \sum_{n\in\mathbb{Z}\setminus\{0\}} \frac{(s)_{r+1}\cdot q^{s+r-2}(1-q)^{-s-1-r}}{(2\pi i n)^2 (s-1+\delta n)_r}$$

$$-(-1)^{M-1}\sum_{n\in\mathbb{Z}\setminus\{0\}}\frac{(s)_{M+1}\cdot\log q}{(2\pi in)^2(s-1+\delta n)_{M-1}}\int_1^\infty\frac{e^{2\pi inx}q^{x(s-2+M)}}{(1-q^x)^{s+M+1}}\,\mathrm{d}x.$$

Observe that

$$\lim_{q\to1}\frac{\log q}{1-q}=-1,\quad \lim_{q\to1}(1-q)^r(s-1+\delta n)_r=(-2\pi i)^r,\quad \lim_{q\to1}\frac{1-q^x}{1-q}=x.$$

Thus for $\mathrm{Re}(s)>2-M$, by Eq. (11.10) (and its special case when $x=1$), we see that

$$\begin{aligned}
&\lim_{q\to1}(1-q)^s\log q\sum_{n\in\mathbb{Z}\setminus\{0\}}\frac{s(s+1)}{(2\pi in)^2}b_q(s-1+\delta n,-s-1)\\
&=\sum_{r=1}^{M-1}\sum_{n\in\mathbb{Z}\setminus\{0\}}\frac{(s)_{r+1}}{(2\pi in)^{r+2}}-\sum_{n\in\mathbb{Z}\setminus\{0\}}\frac{(s)_{M+1}}{(2\pi in)^{M+1}}\int_1^\infty e^{2\pi inx}x^{-s-M-1}\,\mathrm{d}x\\
&=-\sum_{r=1}^{M-1}\frac{B_{r+2}}{(r+2)!}(s)_{r+1}+\frac{(s)_{M+1}}{(M+1)!}\int_1^\infty\widetilde{B}_{M+1}(x)x^{-s-M-1}\,\mathrm{d}x.
\end{aligned}$$

The same procedure applies to the sums involving $b_q(s-1+\delta n,-s)$ and $b_q(s-1+\delta n,-s+1)$. See Exercise 11.4. Consequently, for $\mathrm{Re}(s)>2-M$, we obtain

$$\begin{aligned}
\lim_{q\to1}\zeta_q^I[s]=&\frac{1}{s-1}+\frac{1}{2}+\sum_{r=1}^{M}\frac{B_{r+1}}{(r+1)!}(s)_r\\
&-\frac{(s)_{M+1}}{(M+1)!}\int_1^\infty\widetilde{B}_{M+1}(x)x^{-s-M-1}\,\mathrm{d}x=\zeta(s)
\end{aligned}$$

by Eq. (1.16). This proves the theorem since M is arbitrary . □

Theorem 11.2.2. *The multiple q-zeta function $\zeta_q^I[s_1,\dots,s_d]$ can be extended to a meromorphic function on $\mathbb{C}^d\setminus\mathsf{S}'_d$ with simple poles along S'_d. Further, for all $(s_1,\dots,s_d)\in\mathbb{C}^d\setminus\mathsf{S}_d$*

$$\lim_{q\to1}\zeta_q^I[s_1,\dots,s_d]=\zeta(s_1,\dots,s_d).$$

Proof. We proceed by induction on d. If $d=1$ the theorem is reduced to Thm. 11.2.1. Now we assume $d\geq2$. In definition Eq. (11.5) we replace s_1, k_1, and k_2 by s, n and k, respectively. Taking $f(x)=F(x+k-1)$ and letting $m\to\infty$ in Eq. (1.12) we get

$$\sum_{n>k}\frac{q^{n(s-1)}}{[n]^s}=(1-q)^s\left(-F(k)+\sum_{n=1}^{\infty}f(n)\right)$$

$$
\begin{aligned}
&= (1-q)^s \left(\int_k^{\infty} F(x)\,\mathrm{d}x - \frac{1}{2}F(k) - \frac{1}{12}F'(k) - \frac{1}{2}\int_1^{\infty} \widetilde{B}_2(x) f''(x)\,\mathrm{d}x \right) \\
&= \frac{(q-1)}{(s-1)\log q}\frac{q^{k(s-1)}}{[k]^{s-1}} - \frac{1}{2}\frac{q^{k(s-1)}}{[k]^s} + \frac{1}{12}\frac{\log q}{q-1}\frac{q^{k(s-1)}(s+q^k-1)}{[k]^{s+1}} \\
&\quad - \frac{(1-q)^s(\log q)^2}{2}\int_k^{\infty} \widetilde{B}_2(x) q^{x(s-1)} \frac{s(s+1)-3s(1-q^x)+(1-q^x)^2}{(1-q^x)^{s+2}}\,\mathrm{d}x
\end{aligned}
\tag{11.16}
$$

because $\widetilde{B}_2(x+k-1) = \widetilde{B}_2(x)$ by periodicity. By exactly the same argument as in the proof of depth 1 case, we can obtain from Eq. (11.10) the following expression for the last term involving the integral in Eq. (11.16):

$$
-\sum_{n\in\mathbb{Z}\setminus\{0\}} \frac{(1-q)^s \log q}{(2\pi i n)^2} \sum_{\nu=\pm 1,0} a_\nu(s) b_{q^k}(s-1+\delta n, -s+\nu). \tag{11.17}
$$

For simplicity, write $\zeta[\mathbf{s}] = \zeta_q^I[\mathbf{s}]$ and $\mathcal{S} = \mathcal{S}_1$. Set $\mathbf{s}' = (s_3, \dots, s_d)$ if $d \geq 3$ and $\mathbf{s}' = \emptyset$ if $d = 2$. Replacing q by q^k in Eq. (11.15), plugging it into Eq. (11.17) which is used to replace the last line of Eq. (11.16), and then applying Prop. 11.1.5, we finally get

$$
\begin{aligned}
\zeta[s_1, s_2, \mathbf{s}'] &= \frac{(q-1)}{(s_1-1)\log q}\zeta[s_1+s_2-1, \mathbf{s}'] - \frac{1}{2}\mathcal{S}\zeta[s_1+s_2, \mathbf{s}'] \\
&\quad + \frac{s_1}{12}\frac{\log q}{q-1}\mathcal{S}^2\zeta[s_1+s_2+1, \mathbf{s}'] + \frac{\log q}{12}\mathcal{S}\zeta[s_1+s_2, \mathbf{s}'] - \sum_{\nu=\pm 1,0}(C_\nu + D_\nu),
\end{aligned}
\tag{11.18}
$$

where C_ν and D_ν are the contributions from the sum involving $b_{q^k}(\dots, -s+\nu)$ (note that $s = s_1$). Explicitly they are computed as follows. Write

$$
T(q, s, n, r) = \prod_{j=0}^{r-1} \big(2\pi i n + (s-1+j)\log q\big)^{-1}. \tag{11.19}
$$

Set $M' = M - 1$. Then

$$
C_{-1} = \sum_{r=1}^{M'} \sum_{n\in\mathbb{Z}\setminus\{0\}} \frac{T(q,s,n,r)}{(2\pi i n)^2}\left(\frac{\log q}{q-1}\right)^{r+1} (s)_{r+1} \cdot \mathcal{S}^3\zeta[s_1+s_2+r+1, \mathbf{s}'],
$$

$$
\begin{aligned}
D_{-1} &= -\sum_{k_2 > \cdots > k_d > 0} \frac{q^{k_2(s_2-1)+\cdots+k_d(s_d-1)}}{[k_2]^{s_2}\cdots[k_d]^{s_d}} \sum_{n\in\mathbb{Z}\setminus\{0\}} \frac{T(q,s,n,M')}{(2\pi i n)^2} \times \\
&\quad \times \left(\frac{\log q}{q-1}\right)^{M+1} (s)_{M+1} \cdot \int_{k_2}^{\infty} e^{2\pi i n x} q^{x(s-2+M)} \left(\frac{1-q^x}{1-q}\right)^{-s-M-1} \mathrm{d}x
\end{aligned}
$$

$$=: \sum_{k_3>\cdots>k_d>0} \frac{q^{k_3(s_3-1)+\cdots+k_d(s_d-1)}}{[k_3]^{s_3}\cdots[k_d]^{s_d}} \sum_{k=k_3+1}^{\infty} R(M,q,k,s_2,s),$$

where we replace the index k_2 by k. Similarly,

$$C_0 = 3\log q \sum_{r=1}^{M'} \sum_{n\in\mathbb{Z}\setminus\{0\}} \frac{T(q,s,n,r)}{(2\pi in)^2} \left(\frac{\log q}{q-1}\right)^r (s)_r \cdot \mathcal{S}^2\zeta[s_1+s_2+r,\mathbf{s}'],$$

$$D_0 = -3\log q \sum_{k_2>\cdots>k_d>0} \frac{q^{k_2(s_2-1)+\cdots+k_d(s_d-1)}}{[k_2]^{s_2}\cdots[k_d]^{s_d}} \sum_{n\in\mathbb{Z}\setminus\{0\}} \frac{T(q,s,n,M')}{(2\pi in)^2}$$
$$\times \left(\frac{\log q}{q-1}\right)^M (s)_M \cdot \int_{k_2}^{\infty} e^{2\pi inx} q^{x(s-2+M)} \left(\frac{1-q^x}{1-q}\right)^{-s-M} dx,$$

and

$$C_1 = (\log q)^2 \sum_{r=1}^{M'} \sum_{n\in\mathbb{Z}\setminus\{0\}} \frac{T(q,s,n,r)}{(2\pi in)^2} \left(\frac{\log q}{q-1}\right)^{r-1}$$
$$\times (s)_{r-1} \cdot \mathcal{S}\zeta[s_1+s_2+r-1,\mathbf{s}'],$$

$$D_1 = -(\log q)^2 \sum_{k_2>\cdots>k_d>0} \frac{q^{k_2(s_2-1)+\cdots+k_d(s_d-1)}}{[k_2]^{s_2}\cdots[k_d]^{s_d}} \sum_{n\in\mathbb{Z}\setminus\{0\}} \frac{T(q,s,n,M')}{(2\pi in)^2}$$
$$\times \left(\frac{\log q}{q-1}\right)^{M'} (s)_{M'} \cdot \int_{k_2}^{\infty} e^{2\pi inx} q^{x(s-2+M)} \left(\frac{1-q^x}{1-q}\right)^{-s-M+1} dx.$$

The crucial step next is to control the summations $\sum_{k=k_3+1}^{\infty}$ and show that they converge uniformly with respect to q. When $0<q\le 1/2$ this is clear. The only non-trivial part is when $q\to 1$. So we assume $1/2<q<1$. Note that

$$\lim_{q\to 1} T(q,s,n,r) = \frac{1}{(2\pi in)^r}, \quad \lim_{q\to 1} \frac{\log q}{q-1} = 1.$$

Lemma 11.2.3. *Let $s_1 = s = \sigma + i\tau$. Let $q_0 = \max\{1/2, e^{(6-2\pi)/\tau}\}$ if $\tau>0$ and let $q_0 = 1/2$ if $\tau \le 0$. Then for all $1>q>q_0$ and positive integer k we have*

$$\left|\frac{\log q}{q-1}\right| < 2, \quad \text{and} \quad |T(q,s,n,r)| < \frac{1}{(6n)^r}.$$

Proof. Let $f(q) = 2(1-q)+\log q$. Then $f'(q) = -2+1/q < 0$ whenever $q>q_0$. So $f(q)>f(1)=0$ whenever $1>q>q_0$. This implies $2(1-q) > -\log q$ whence $\log q/(q-1) < 2$.

To bound $T(q,s,n,r)$ we consider each of its factors in definition Eq. (11.19). For each $0 \le j < r$, we have

$$|2\pi in+(s-1+j)\log q|^2=\big((\sigma-1+j)\log q\big)^2+(2\pi n+\tau\log q)^2 \ge (2\pi n+\tau\log q)^2$$

which is independent j. If $\tau \le 0$ then clearly $|2\pi in+(s-1+j)\log q| > 6n$. If $\tau > 0$ then it follows from $q > e^{(6-2\pi)\tau}$ that

$$2\pi n+\tau\log q > 2\pi n + 6 - 2\pi \ge 6n,$$

as desired. □

Next we want to bound the integral terms in D_{-1}. Let $|\sigma_1| < N$ and $|\sigma_2| < N'$ for some positive integers N and N'. Fix an arbitrary $x > k$ and a positive integer

$$M > 16 + 2N + 6\sum_{j=2}^{d}(N_j+1). \tag{11.20}$$

Then

$$q^{-k(M/6-N'-1)}\left|q^{k(s'-1)}q^{x(s-2+M)}\left(\frac{1-q^x}{1-q}\right)^{-s-M-1}\right| < q^{x(M-N-2)-kM/6}\left(\frac{1-q}{1-q^x}\right)^{M-N+1}.$$

Denote by $g(q)$ the right-hand side of the above inequality. It can be proved that for $1/2 < q < 1$, $g(q)$ is increasing as a function of q so that (see Exercise 11.5)

$$g(q) \le \lim_{q\to 1} g(q) = \frac{1}{x^{M-N+1}}. \tag{11.21}$$

We now can bound the innermost sum of D_{-1}. From Lemma 11.2.3 and Eq. (11.21) we have (if $d=2$ then take $k_3=0$)

$$\begin{aligned}
&\sum_{k=1+k_3}^{\infty} |R(M,q,k,s',s)| \\
&\quad < \sum_{k=1+k_3}^{\infty}(2k)^{N'}\frac{2\zeta(M+1)}{4\pi^2 6^{M-1}}2^{M+1}(N)_{M+1}q^{k(M/6-N'-1)}\int_k^\infty \frac{dx}{x^{M-N+1}} \\
&\quad < \frac{(M+1)!}{M-N}\binom{M+N}{M+1}\sum_{k=1+k_3}^{\infty}\frac{q^{k(M/6-N'-1)}}{k^{M-N-N'}} \\
&\quad < (M+1)!(2M)^{M+1}\sum_{k=1+k_3}^{\infty} q^{k(M/6-N'-1)}
\end{aligned}$$

since $2^{N'+M+2}\zeta(M+1) < 4\pi^2 6^{M-1}$ and $M - N > 2$. Therefore by Eq. (11.3)

$$\big|D_{-1}\big| < (2M)^{2M} \sum_{k_3>\cdots>k_d} \left(\prod_{j=3}^{d} k_j^{N_j} q^{k_j(-N_j-1)}\right) q^{k_3(M/6-N_2-1)} \qquad (11.22)$$

which converges by the condition Eq. (11.20) on M (see the proof of Prop. 11.1.1).

Exactly the same argument applies to the integral terms in D_0 and D_1 (see Exercise 11.7). These convergence results imply two things. First we can show by induction on d that Eq. (11.18) gives rise to an analytic continuation of $\zeta[s_1,\dots,s_d]$ as a meromorphic function on $\mathbb{C}^d \setminus \mathsf{S}'_d$. Second, also by induction on d, we now can conclude that it's legitimate to take the limit $q \to 1$ inside the sums of C_ν and D_ν to get (note that $\lim_{q\to1} \mathcal{S}^n\zeta[\mathbf{s}] = \zeta(\mathbf{s})$ for any $\mathbf{s} \in \mathbb{C}^{d-1} \setminus \mathsf{S}'_{d-1}$ and any positive integer n)

$$\begin{aligned}
&\lim_{q\to1} \zeta[s_1, s_2, \mathbf{s}'] \\
&\quad= \frac{1}{s-1}\zeta(s_1+s_2-1,\mathbf{s}') - \frac{1}{2}\zeta(s_1+s_2,\mathbf{s}'] + \frac{s}{12}\zeta(s_1+s_2+1,\mathbf{s}') \\
&\qquad - \sum_{r=1}^{M-1} \sum_{n\in\mathbb{Z}\setminus\{0\}} \frac{1}{(2\pi i n)^{r+2}} (s_1)_{r+1} \cdot \zeta(s_1+s_2+r+1,\mathbf{s}') \\
&\qquad + \sum_{k_2>\cdots>k_d} \frac{1}{k_2^{s_2}\cdots k_d^{s_d}} \sum_{n\in\mathbb{Z}\setminus\{0\}} \frac{(s)_{M+1}}{(2\pi i n)^{M+1}} \int_{k_2}^{\infty} e^{2\pi i n x} x^{-s_1-M-1}\,\mathrm{d}x \\
&\quad= \sum_{r=0}^{M+1} \frac{B_r}{r!}(s_1)_{r-1} \cdot \zeta(s_1+s_2+r-1, s_3,\dots,s_d) \\
&\qquad - \frac{1}{(M+1)!} \sum_{k_2>\cdots>k_d>0} \frac{1}{k_2^{s_2}\cdots k_d^{s_d}} \int_{k_2}^{\infty} \widetilde{B}_{M+1}(x) \frac{(s)_{M+1}}{x^{s_1+M+1}}\,\mathrm{d}x
\end{aligned}$$

by Eq. (11.10) and its specialization with $x=1$

$$\sum_{n\in\mathbb{Z}\setminus\{0\}} \frac{1}{(2\pi i n)^{r+2}} = -\frac{\widetilde{B}_{r+2}(1)}{(r+2)!} = -\frac{B_{r+2}}{(r+2)!}.$$

The theorem now mostly follows from Thm. 1.3.2 (by replacing M by $M+1$ there). The poles at $s_1 = m - \frac{2\pi i}{\log q} n$ are given by the first term in formula Eq. (11.18) when $m = 1$ and $n = 0$ and by the terms $T(q, s_d, n, r)$ as defined in Eq. (11.19) if $m \le 1$ and $n \ne 0$. The location of the other poles are obtained by induction using those poles of the q-Riemann zeta function presented in Thm. 11.2.1 for the initial step. This completes the proof of the theorem. □

11.3 Special Values at Non-Positive Integers

In this section, we will consider some special values of $\zeta_q^I[s_1, \dots, s_d]$ at non-positive integers.

Theorem 11.3.1. *For any nonnegative integer m, the limiting value $\zeta_q^I[-m] = \lim_{s\to -m} \zeta_q^I[s]$ exists and is given explicitly by*

$$\zeta_q^I[-m] = (1-q)^{-m}\left\{\sum_{r=0}^{m}(-1)^r\binom{m}{r}\frac{1}{q^{m+1-r}-1} + \frac{(-1)^{m+1}}{(m+1)\log q}\right\}.$$

Moreover, for all $m \in \mathbb{N}$, we have

$$\lim_{q\to 1}\zeta_q^I[-m] = -\frac{B_{m+1}}{m+1}. \tag{11.23}$$

Proof. Note the terms with $r \geq m+2$ in the sum Eq. (11.6) vanishes when $s \to -m$ since $\binom{-m+r-1}{r} = 0$ and $1 - q^{-m+r-1} \neq 0$. On the other hand, for $r = m+1$ we have $\lim_{s\to -m}(s+m)/(1-q^{s+m}) = -1/\log q$ and hence

$$\lim_{s\to -m}\binom{s+m}{m+1}\frac{q^{s+m}}{1-q^{s+m}} = \frac{(-1)^m m!}{(m+1)!}\left(-\frac{1}{\log q}\right) = \frac{(-1)^{m+1}}{(m+1)\log q}.$$

To prove Eq. (11.23) we apply Thm. 11.2.1 to get

$$\lim_{q\to 1}\zeta_q^I[-m] = \zeta(-m) = -\frac{B_{m+1}}{m+1}.$$

This completes the proof of the theorem. □

Remark 11.3.2. Thanks to the theorem, we can regard $m\zeta_q^I[1-m]$ as the q-analog of the Bernoulli number B_m.

In general, for integers $n_1, \dots, n_d$ we can define

$$\zeta_q^I[n_1,\dots,n_d] = \lim_{s_d\to n_d}\cdots\lim_{s_1\to n_1}\zeta_q^I[s_1,\dots,s_d],$$
$$\zeta_q^R[n_1,\dots,n_d] = \lim_{s_1\to n_1}\cdots\lim_{s_d\to n_d}\zeta_q^I[s_1,\dots,s_d]$$

if the limits exist. Similarly we can define

$$\zeta(n_1,\dots,n_d) = \lim_{s_d\to n_d}\cdots\lim_{s_1\to n_1}\zeta(s_1,\dots,s_d),$$
$$\zeta^R(n_1,\dots,n_d) = \lim_{s_1\to n_1}\cdots\lim_{s_d\to n_d}\zeta(s_1,\dots,s_d)$$

if the limits exist. Interesting phenomena occur already in the case $d = 2$ and these can be generalized to arbitrary depth. By Thm. 11.1.4 we get

$$\begin{aligned}
&\zeta_q^I[s_1, s_2] \\
&= \sum_{r_1,r_2=0}^{+\infty} \binom{s_1+r_1-1}{r_1}\binom{s_2+r_2-1}{r_2} \frac{(1-q)^{s_1+s_2} q^{2s_1+2r_1+s_2+r_2-3}}{(1-q^{s_1+r_1-1})(1-q^{s_1+r_1+s_2+r_2-2})} \\
&= (1-q)^{s_1+s_2}\left\{\frac{q^{2s_1+s_2-3}}{(1-q^{s_1-1})(1-q^{s_1+s_2-2})} + \frac{s_2 q^{2s_1+s_2-2}}{(1-q^{s_1-1})(1-q^{s_1+s_2-1})}\right. \\
&\quad + \frac{s_1 q^{2s_1+s_2-1}}{(1-q^{s_1})(1-q^{s_1+s_2-1})} + \frac{s_1 s_2 q^{2s_1+s_2}}{(1-q^{s_1})(1-q^{s_1+s_2})} \\
&\quad \left. + \frac{s_2(s_2+1) q^{2s_1+s_2-1}}{2(1-q^{s_1-1})(1-q^{s_1+s_2})} + \frac{s_1(s_1+1) q^{2s_1+s_2+1}}{2(1-q^{s_1+1})(1-q^{s_1+s_2})} + \cdots\right\}. \quad (11.24)
\end{aligned}$$

It is straightforward to get

$$\begin{aligned}
\zeta_q^I[0,0] &= \frac{1}{(q^2-1)(q-1)} - \frac{3}{2(q-1)\log q} + \frac{1}{\log^2 q}, \\
\zeta_q^R[0,0] &= \frac{1}{(q^2-1)(q-1)} - \frac{1}{(q-1)\log q} + \frac{q}{2(q-1)\log q}.
\end{aligned}$$

Taking limit we can find

$$\lim_{q\to 1} \zeta_q^I[0,0] = \frac{1}{3}, \quad \lim_{q\to 1} \zeta_q^R(0,0) = \frac{5}{12}.$$

This is consistent with the behavior of the double zeta functions near $(0,0)$ (see Exercise 1.6). It has the following asymptotic expansion:

$$\zeta(s_1, s_2) = \frac{5s_1 + 4s_2}{12(s_1+s_2)} + R(s_1, s_2)$$

where $R(s_1, s_2)$ is analytic at $(0,0)$ and $\lim_{(s_1,s_2)\to(0,0)} R(s_1, s_2) = 0$.

Let n, k be two nonnegative integers, and $m = k - n - 2$. We now consider the double zeta function around $(s_1, s_2) = (-n, -m)$ which has the following expression by Thm. 1.3.2(ii):

$$\begin{aligned}
&\frac{-1}{n+1}\zeta(s_1+s_2-1) - \frac{1}{2}\zeta(s_1+s_2) - \sum_{r=2}^{n+1} \frac{B_r}{r}\binom{n}{r-1}\zeta(s_1+s_2+r-1) \\
&\qquad + \alpha(m)(-1)^n \frac{B_k}{k!} n!(k-n-2)!(s_1+n)\zeta(s_1+s_2+k-1) \quad (11.25)
\end{aligned}$$

where $\alpha(m) = 0$ if $m \le -1$ and $\alpha(m) = 1$ if $m \ge 0$. Note that the last term is zero when computing

$$\zeta(-n, -m) = \lim_{s_2\to -m} \lim_{s_1\to -n} \zeta(s_1, s_2)$$

while it has possibly nontrivial contribution for $\zeta^R(-n, -m)$, since

$$\lim_{s_1\to -n} \lim_{s_2\to -m} (s_1+n)\zeta(s_1+s_2+k-1) = 1.$$

Table 11.1: Poles and indeterminacy of the double zeta function, $a_k = B_{k-1}/(2k-2)$ and $b_{n,k} = \binom{n}{k-1} B_k/k$.

$k,\ n$	pole, residue	indeterminacy $\zeta=\zeta^R(-n,n+2-k)$
$k=0$	$-1/(n+1)$	NO
$k=1$	$-1/2$	NO
$2\nmid k, 3\le k\le n+1$	NO	$a_k + \sum_{r=k-1}^{n+1} b_{n,r}\zeta(r+1-k)$
$2\nmid k, k=n+2$	NO	$2a_k$
$2\nmid k, k>n+2$	NO	a_k
$2\mid k, 2\le k\le n+1$	$(-1)^{k+1}b_{n,k}$	NO

When $m\ge 0$ and k is even the values of $\zeta(-n,-m)$ and $\zeta^R(-n,-m)$ are usually different:

$$\zeta(-n,-m) =\frac{B_k}{k(n+1)} + \sum_{r=1}^{n+1} \frac{B_r}{r}\binom{n}{r-1}\frac{B_{k-r}}{k-r}, \tag{11.26}$$

$$\zeta^R(-n,-m) =\zeta(-n,-m) + (-1)^n\frac{B_k}{k!}n!(k-n-2)!. \tag{11.27}$$

Note that the term corresponding to $r=1$ is non-zero if and only if $k=2$ (and $n=0$). From this observation we again recover that

$$\zeta(0,0)=\frac{1}{3},\quad \zeta^R(0,0)=\frac{5}{12}. \tag{11.28}$$

We now consider the double q-zeta function.

Theorem 11.3.3. *Let k, n be two nonnegative integers, and $m=k-n-2$. If $m\le -1$ then the double q-zeta function $\zeta_q^I[s_1,s_2]$ has a pole at $(-n,-m)$ with residue given by:*

$$\frac{\operatorname*{Res}_{(s_1,s_2)=(-n,-m)} \zeta_q^I[s_1,s_2]}{-(1-q)^{2-k}(\log q)^{-1}} = \begin{cases} \displaystyle\sum_{r=0}^{k}(-1)^r\binom{n+1-r}{k-r}\binom{n}{r}\Big/(q^{n+1-r}-1) & \text{if } k\le n,\\ \displaystyle\sum_{r=0}^{n}(-1)^r\binom{n}{r}\Big/(q^{n+1-r}-1) - \frac{(-1)^n}{(n+1)\log q} & \text{if } k=n+1. \end{cases}$$

Proof. This follows directly from Eq. (11.24). ☐

Corollary 11.3.4. *Let n be a nonnegative integer. Then*

$$\operatorname*{Res}_{(s_1,s_2)=(-n,1)} \zeta_q^I[s_1,s_2] = \frac{q-1}{\log q}\zeta_q^I[-n]. \tag{11.29}$$

Setting $B_1^* = 1/2$ and $B_n^* = B_n$ for all $n > 1$, then for all $n \geq 0$ we have

$$\lim_{q\to 1} \zeta_q^I[-n] = -\frac{B_{n+1}^*}{n+1} = \operatorname*{Res}_{(s_1,s_2)=(-n,1)} \zeta(s_1,s_2). \tag{11.30}$$

Proof. Equation Eq. (11.29) follows from the case $m = -1$ in the above theorem and Thm. 11.3.1. The first equality in Eq. (11.30) is Eq. (11.23) and the second equality follows from Table 11.1. □

Corollary 11.3.5. *Let k, n be two nonnegative integers, $m = k - n - 2 \leq -2$. Then*

$$\lim_{q\to 1} \operatorname*{Res}_{(s_1,s_2)=(-n,-m)} \zeta_q^I[s_1,s_2] = \operatorname*{Res}_{(s_1,s_2)=(-n,-m)} \zeta(s_1,s_2). \tag{11.31}$$

Proof. By Thm. 11.3.3 and Table 11.1 we only need to prove

$$\lim_{q\to 1} \sum_{r=0}^{k} \frac{(q-1)^{1-k}(-1)^r}{q^{n+1-r}-1}\binom{n+1-r}{k-r}\binom{n}{r} \tag{11.32}$$

$$= \begin{cases} \dfrac{(-1)^k}{n+1} & \text{if } k=0, \\ \dfrac{B_k}{k}\dbinom{n}{k-1} & \text{if } 1 \leq k \leq n. \end{cases} \tag{11.33}$$

First, by the generating function of the Bernoulli numbers

$$\frac{1}{q^{n+1-r}-1} = \frac{1}{e^{(n+1-r)\log q}-1} = \sum_{l=0}^{\infty} \frac{B_l}{l!}\big((n+1-r)\log q\big)^{l-1}.$$

Plugging this into the left-hand side of Eq. (11.32), replacing $1-q$ by $-\log q$, and changing the order of the summations we get

$$(\log q)^{1-k} \sum_{r=0}^{k} (-1)^r \binom{n+1-r}{k-r}\binom{n}{r}/(q^{n+1-r}-1) \tag{11.34}$$

$$= \sum_{l=0}^{\infty} \frac{B_l}{l!}(\log q)^{l-k} \sum_{r=0}^{k} (-1)^r \binom{n+1-r}{k-r}\binom{n}{r}(n+1-r)^{l-1}. \tag{11.35}$$

Then the inner sum over r is the coefficient of x^k of the following polynomial

$$\begin{aligned}
f_l(x) &= \sum_{k=0}^{n+1}\sum_{r=0}^{k}(-1)^r\binom{n+1-r}{k-r}\binom{n}{r}(n+1-r)^{l-1}x^k \\
&= \sum_{r=0}^{n}(-1)^r\binom{n}{r}(n+1-r)^{l-1}\sum_{k=r}^{n+1}\binom{n+1-r}{k-r}x^k \\
&= (x+1)^{n+1}\sum_{r=0}^{n}(-y)^r\binom{n}{r}(n+1-r)^{l-1}, \qquad (11.36)
\end{aligned}$$

where $y = x/(x+1)$. When $l = 0$ this expression becomes

$$\begin{aligned}
f_0(x) &= \frac{(x+1)^{n+1}}{n+1}\sum_{r=0}^{n}(-y)^r\binom{n+1}{r} \\
&= \frac{(x+1)^{n+1}}{n+1}\Big[(1-y)^{n+1} - (-y)^{n+1}\Big] = \frac{1}{n+1}\Big[1-(-x)^{n+1}\Big].
\end{aligned}$$

Note that $k \le n$ we see the coefficient of x^k in $f_0(x)$ is 0 if $k > 0$ and it's $1/(n+1)$ if $k = 0$. If $k = 0$ then only the constant term $-1/(n+1)$ in Eq. (11.34) remains when $q \to 1$ which proves the corollary in this case. So we can assume $l, k > 0$. Then

$$\begin{aligned}
f_l(x) &= \left(z\frac{\mathrm{d}}{\mathrm{d}z}\right)^{l-1}\left\{(x+1)^{n+1}\sum_{r=0}^{n}(-y)^r\binom{n}{r}z^{n+1-r}\right\}\Bigg|_{z=1} \\
&= \left(z\frac{\mathrm{d}}{\mathrm{d}z}\right)^{l-1}\Big\{(x+1)^{n+1}z(z-y)^n\Big\}\Big|_{z=1}.
\end{aligned}$$

Note that highest degree term in $f_l(x)$ is contained in

$$\begin{aligned}
&(x+1)^{n+1}\left(\frac{\mathrm{d}}{\mathrm{d}z}\right)^{l-1}(z-y)^n\Bigg|_{z=1} \\
&\quad = n(n-1)\cdots(n-l+2)(x+1)^{n+1}(1-y)^{n-l+1} \\
&\quad = n(n-1)\cdots(n-l+2)(x+1)^l.
\end{aligned}$$

If $l = 1$ one can easily modify this to get just $x+1$. If $l < k$ then the coefficient of x^k in $f_l(x)$ is 0. If $l = k$ it is equal to

$$n(n-1)\cdots(n-k+2) = (k-1)!\binom{n}{k-1}.$$

The last express is valid even for $k = l = 1$. Thus the range of l in the outer sum of Eq. (11.34) starts from k. Moreover, the first term of Eq. (11.34) is

$$\frac{B_k}{k}\binom{n}{k-1}$$

as desired. This completes the proof of the corollary. □

Proposition 11.3.6. *Let k, n be two nonnegative integers such that $n \geq k$ and k is even. Let $m = k - n - 2$. Then*

$$\operatorname*{Res}_{(s_1,s_2)=(-n,-m)} \zeta_q^I[s_1, s_2] = \frac{-f(q)(q-1)/\log q}{D(q)}, \quad D(q) = \prod_{j=n+1-k}^{n+1} F(q,j)^{\epsilon_j},$$

where $f(q) \in \mathbb{Z}[q]$ is palindromic with leading coefficient $\binom{n}{k}$, $F(q,j) \in \mathbb{Z}[q]$ is a factor of $(q^j - 1)/(q-1)$, $\epsilon_j = 0$ or 1, and $\deg_q D(q) = n + \deg_q f(q)$, such that

$$\lim_{q\to 1} \operatorname*{Res}_{(s_1,s_2)=(-n,-m)} \zeta_q^I[s_1, s_2] = \operatorname*{Res}_{(s_1,s_2)=(-n,-m)} \zeta(s_1, s_2).$$

Proof. The computational proof is left as an exercise for the interested reader. See Exercise 11.6. □

Example 11.3.7. By Thm. 11.3.3 we find with the help of Maple

$$\operatorname*{Res}_{(s_1,s_2)=(-4,4)} \zeta_q^I[s_1, s_2] = \frac{-2q^3(3q^2+4q+3)(q-1)/\log q}{P_1(q,2)P_1(q,3)P_1(q,4)}$$

where $P_a(q, m) = \sum_{j=0}^{m} q^{aj}$. Moreover we can check that

$$\lim_{q\to 1} \operatorname*{Res}_{(s_1,s_2)=(-4,4)} \zeta_q^I[s_1, s_2] = \operatorname*{Res}_{(s_1,s_2)=(-4,4)} \zeta(s_1, s_2) = -\frac{1}{3}$$

by Table 11.1 with $k = 2$ and $n = 4$.

Example 11.3.8. By Thm. 11.3.3, we obtain

$$\operatorname*{Res}_{(s_1,s_2)=(-8,6)} \zeta_q^I[s_1, s_2] = \frac{-14q^5 g(q)(q-1)/\log q}{P_1(q,4)P_1(q,5)P_1(q,6)P_2(q,3)P_3(q,2)}$$

where $g(q)$ is a polynomial in q of degree 14 satisfying

$$q^{14}g(1/q) = g(q) = 5q^{14} + 6q^{13} + 8q^{12} + 7q^{11} - q^{10} - 20q^9 - 30q^8 - 34q^7 - \cdots.$$

Then we can compute with Maple

$$\lim_{q\to 1} \operatorname*{Res}_{(s_1,s_2)=(-8,6)} \zeta_q^I[s_1, s_2] = \operatorname*{Res}_{(s_1,s_2)=(-8,6)} \zeta(s_1, s_2) = \frac{7}{15}$$

by Table 11.1 with $k = 4$ and $n = 8$.

Example 11.3.9. Consider the point $(s_1, s_2) = (-9, 5)$. We have

$$\operatorname*{Res}_{(s_1,s_2)=(-9,5)} \zeta_q^I[s_1, s_2] = \frac{-42q^4 g(q)(q-1)/\log q}{P_1(q,4)P_1(q,6)P_1(q,7)P_2(q,2)P_3(q,2)A(q,4)}$$

where $A(q,m)=\sum_{j=0}^{m}(-1)^j q^j$ and $g(q)$ is a polynomial in q of degree 18 satisfying

$$\begin{aligned}q^{18}&g(1/q)=g(q)\\&=2q^{18}-q^{17}-7q^{15}-11q^{14}-16q^{13}-4q^{12}+9q^{11}+28q^{10}+30q^{9}+\cdots\end{aligned}$$

so that we again have the equality

$$\lim_{q\to1}\operatorname*{Res}_{(s_1,s_2)=(-9,5)}\zeta_q^I[s_1,s_2]=\operatorname*{Res}_{(s_1,s_2)=(-9,5)}\zeta(s_1,s_2)=-\frac{1}{2}$$

by Table 11.1 with $k=6$ and $n=9$.

Theorem 11.3.10. *Let m,n be two nonnegative integers and $k=m+n+2$. Then $\zeta_q^I[s_1,s_2]$ has indeterminacy at $(-n,-m)$ such that*

$$\begin{aligned}\zeta_q^I&[-n,-m](1-q)^{k-2}\\&=\left\{\frac{(-1)^k}{(m+1)(n+1)(\log q)^2}+\sum_{r=0}^{m}\frac{(-1)^{r+n+1}}{(n+1)\log q}\binom{m}{r}\frac{1}{q^{m+1-r}-1}\right.\\&\quad+\sum_{r=0}^{n}\frac{(-1)^{r+m+1}}{\log q}\frac{m!(n+1-r)!}{(k-r)!}\binom{n}{r}\frac{1}{q^{n+1-r}-1}\\&\quad\left.+\sum_{r_1=0}^{m}\sum_{r_2=0}^{n}(-1)^{r_1+r_2}\binom{m}{r_1}\binom{n}{r_2}\frac{1}{q^{n+1-r_2}-1}\frac{1}{q^{k-r_1-r_2}-1}\right\},\end{aligned}$$

and

$$\begin{aligned}\zeta_q^R[-n,-m]=&(1-q)^{2-k}\left\{\sum_{r=0}^{m}\frac{(-1)^{r+n+1}}{(n+1)\log q}\binom{m}{r}\frac{1}{q^{m+1-r}-1}\right.\\&+\sum_{r=0}^{m}\frac{(-1)^{r+n}}{\log q}\binom{k-n-2}{r}\frac{n!(m+1-r)!}{(k-r)!}\frac{q^{m+1-r}}{q^{m+1-r}-1}\\&\left.+\sum_{r_1=0}^{m}\sum_{r_2=0}^{n}(-1)^{r_1+r_2}\binom{m}{r_1}\binom{n}{r_2}\frac{1}{q^{n+1-r_2}-1}\frac{1}{q^{k-r_1-r_2}-1}\right\}.\end{aligned}$$

Proof. This follows from Eq. (11.24) by straightforward computation. □

Similar to Cor. 11.3.4 and Cor. 11.3.5 we have

Corollary 11.3.11. *Let m and n be two nonnegative integers. Then*

$$\lim_{q\to1}\zeta_q^I[-n,-m]=\zeta(-n,-m),\quad \lim_{q\to1}\zeta_q^R[-n,-m]=\zeta^R(-n,-m). \tag{11.37}$$

Proof. Set $k = m+n+2$. We consider $\zeta_q^R[-n,-m]$ first. From Thm. 11.3.10 we get

$$\zeta_q^R[-n,-m] = (q-1)^{-k}\left(\frac{q-1}{\log q}\right)^2 (A+B+C),$$

where

$$A := \sum_{j=0}^{\infty} \frac{B_j}{j!}(\log q)^j \sum_{r=0}^{m} \frac{(-1)^{m+r+1}}{n+1}\binom{m}{r}(m+1-r)^{j-1},$$

$$B := \sum_{i=0}^{\infty} \frac{B_i}{i!}(\log q)^i \sum_{r=0}^{m}(-1)^{r+m+i}\binom{m}{r}\frac{n!(m+1-r)!}{(k-r)!}(m+1-r)^{i-1},$$

$$C = \sum_{i,j=0}^{\infty} \frac{B_iB_j}{i!j!}(\log q)^{i+j} \sum_{r_1=0}^{m}\sum_{r_2=0}^{n}(-1)^{k+r_1+r_2}\binom{m}{r_1}\binom{n}{r_2}$$
$$\times (n+1-r_2)^{i-1}(k-r_1-r_2)^{j-1}.$$

We first compute B as follows. Write

$$B = \sum_{i=0}^{\infty} \frac{B_i}{i!}(\log q)^i W_i^{m,n},$$

where

$$W_i^{m,n} := \sum_{r=0}^{m}(-1)^{r+k+n+i}\frac{m!n!}{r!(k-r)!}(m+1-r)^i.$$

If $i=0$ then we can prove by decreasing induction on n that

$$W_0^{m,n} = \sum_{r=0}^{m}(-1)^{r+k+n}\frac{m!n!}{r!(k-r)!} = \frac{1}{k(n+1)}. \tag{11.38}$$

This is trivial if $n = k-2$. Suppose Eq. (11.38) is true for $n \geq 1$ then

$$W_0^{m,n-1} = -\frac{k-n-1}{n}W_0 + \frac{(n-1)!}{(n+1)!} = \frac{1}{kn},$$

as desired. In the rest of the proof, we write $W_i = W_i^{m,n}$.

Similarly, we can compute C as follows. Put

$$C = \sum_{i,j=0}^{\infty} \frac{B_iB_j(\log q)^{i+j}}{i!j!(n+1)} \sum_{r_1=0}^{m}\sum_{r_2=0}^{n}(-1)^{k+r_1+r_2}$$
$$\times \binom{m}{r_1}\binom{n+1}{r_2}(n+1-r_2)^i(k-r_1-r_2)^{j-1}.$$

We now change the upper limit of r_2 from n to $n+1$ in the above. The extra terms correspond to those by setting $i=0$ and $r_2=n+1$, which produce exactly A. Therefore,

$$C=\sum_{i,j=0}^{\infty}\frac{B_iB_j(\log q)^{i+j}}{i!j!}V_{i,j}-A$$

where

$$V_{i,j}:=\sum_{r_1=0}^{m}\sum_{r_2=0}^{n+1}\frac{(-1)^{k+r_1+r_2}}{n+1}\binom{m}{r_1}\binom{n+1}{r_2}(n+1-r_2)^i(k-r_1-r_2)^{j-1}.$$

For $j\geq 1$, we have

$$\begin{aligned}V_{i,j}&=\left(x\frac{\mathrm{d}}{\mathrm{d}x}\right)^{j-1}\left\{\left(y\frac{\mathrm{d}}{\mathrm{d}y}\right)^{i}\left\{\sum_{r_1=0}^{m}\sum_{r_2=0}^{n+1}\frac{(-1)^{k+r_1+r_2}}{n+1}\right.\right.\\&\quad\left.\left.\times\binom{m}{r_1}\binom{n+1}{r_2}y^{n+1-r_2}x^{k-r_1-r_2}\right\}\Big|_{y=1}\right\}\Big|_{x=1}\\&=\frac{(-1)^k}{n+1}\left(x\frac{\mathrm{d}}{\mathrm{d}x}\right)^{j-1}\left\{\left(y\frac{\mathrm{d}}{\mathrm{d}y}\right)^{i}\left\{x(xy-1)^{n+1}(x-1)^{k-n-2}\right\}\Big|_{y=1}\right\}\Big|_{x=1} \qquad (11.39)\\&=\begin{cases}0, & \text{if } i+j<k;\\ \dfrac{(-1)^kn!}{(n+1-i)!}(k-i-1)!, & \text{if } i+j=k, i\leq n+1.\end{cases}\end{aligned}$$

We have used the fact that if $i+j=k$ and $i>n+1$ then $j<k-n-1$ and by exchanging the two operators $x(\mathrm{d}/\mathrm{d}x)$ and $y(\mathrm{d}/\mathrm{d}y)$ we can easily show that Eq. (11.39) is zero. So if $l<k$ the total contribution to the coefficient of $(\log q)^l$ from $V_{i,j}$ with $j>0$ is trivial and if $l=k$ it is equal to

$$\begin{aligned}&\sum_{i=0}^{n+1}B_iB_{k-i}\frac{(-1)^kn!(k-i-1)!}{i!(k-i)!(n+1-i)!}\\&\quad=\begin{cases}\dfrac{B_{k-1}}{2(k-1)} & \text{if } k \text{ is odd,}\\ \dfrac{B_k}{k(n+1)}+\displaystyle\sum_{i=1}^{n+1}\frac{B_iB_{k-i}}{i(k-i)}\binom{n}{i-1} & \text{if } k \text{ is even,}\end{cases} \qquad (11.40)\end{aligned}$$

because $k\geq n+2\geq 2$ and $B_k=0$ if k is odd.

To deal with $V_{i,0}$, note that Eq. (11.39) still makes sense if we interpret the operator $x(\mathrm{d}/\mathrm{d}x)^{-1}$ as follows:

$$\left(x\frac{\mathrm{d}}{\mathrm{d}x}\right)^{-1}\left\{F(x)\right\}\Big|_{x=1}=\int_0^1\frac{F(x)}{x}\mathrm{d}x$$

whenever $F(0) = 0$. Thus

$$V_{i,0} = \frac{(-1)^k}{n+1}\left(y\frac{\mathrm{d}}{\mathrm{d}y}\right)^i \left\{ \int_0^1 (xy-1)^{n+1}(x-1)^m \,\mathrm{d}x \right\}\Bigg|_{y=1}.$$

Therefore if $i = 0$ then we have

$$V_{0,0} = \frac{(-1)^k}{n+1}\int_0^1 (x-1)^{k-1}\,\mathrm{d}x = -\frac{1}{k(n+1)} = -W_0 \tag{11.41}$$

from Eq. (11.38). If $i \geq 1$ then integrating by parts we get

$$\begin{aligned}
&\int_0^1 (xy-1)^{n+1}(x-1)^m \,\mathrm{d}x \\
&= \frac{(xy-1)^{n+2}}{y(n+2)}(x-1)^m\Bigg|_0^1 - \frac{m}{y(n+2)}\int_0^1 (xy-1)^{n+2}(x-1)^{m-1}\,\mathrm{d}x \\
&= \cdots\cdots \\
&= (-1)^{k+1}\left(\frac{1}{y(n+2)} - \frac{m}{y^2(n+2)(n+3)} + \cdots + (-1)^{m-1}\frac{m!(n+1)!}{y^m(m+n+1)!}\right. \\
&\quad \left. +(-1)^m\frac{m!(n+1)!}{y^m(m+n+1)!}\int_0^1 (xy-1)^{m+n+1}\,\mathrm{d}x,\right) \\
&= (-1)^{k+n}\frac{m!(n+1)!}{k!}\frac{(y-1)^k}{y^{m+1}} + \sum_{r=0}^{m}(-1)^{r+k+1}\frac{m!(n+1)!}{y^{r+1}(m-r)!(n+2+r)!}.
\end{aligned}$$

The substitution $r \to m - r$ yields

$$\begin{aligned}
&V_{i,0} \\
&= (-1)^{k+n}\frac{m!n!}{k!}\left(y\frac{\mathrm{d}}{\mathrm{d}y}\right)^i \left\{\frac{(y-1)^k}{y^{m+1}}\right\}\Bigg|_{y=1} - \sum_{r=0}^{m}\frac{(-1)^{r+n}m!n!}{r!(k-r)!}(r-1-m)^i \\
&= \begin{cases} (-1)^{k+1}W_0 & \text{if } 0 < i < k, \\ (-1)^{k+1}W_0 + (-1)^{k+n}n!(k-n-2)! & \text{if } i = k. \end{cases}
\end{aligned} \tag{11.42}$$

Thus when $0 < i < k$ and k is even we have $V_{i,0} = -W_i$. It follows from Eq. (11.40), Eq. (11.41) and Eq. (11.42) that

$$\lim_{q\to 1}\zeta_q^R(-m,-n) = \zeta^R(s_1,s_2)$$

since $B_k = 0$ if $k > 2$ is odd.

Let's turn to prove the first equality in Eq. (11.37). Theorem 11.3.10 implies that

$$\zeta_q^I[-n,-m] = (q-1)^{-k}\left(\frac{q-1}{\log q}\right)^2 (A+C+D+E)$$

where A and C are as above and

$$D := \frac{1}{(m+1)(n+1)}, \qquad E := \sum_{i=0}^{\infty} \frac{B_i}{i!} (\log q)^i U_i,$$

where

$$U_i = U_i^{m,n} := -\sum_{r=0}^{n} (-1)^{r+n} \frac{m!n!}{r!(k-r)!} (n+1-r)^i.$$

Hence

$$U_0^{m,n} = W_0^{n,m} = \frac{-1}{k(m+1)} = \frac{1}{k(n+1)} - D. \tag{11.43}$$

We only need to show that

$$U_i = \begin{cases} W_i, & \text{if } 0 < i < k; \\ W_i - (-1)^n n!(k-n-2)!, & \text{if } i = k. \end{cases} \tag{11.44}$$

Indeed when $i > 0$ we have

$$\begin{aligned} U_i &= (-1)^{n+1} \left(y \frac{\mathrm{d}}{\mathrm{d}y} \right)^i \left\{ \sum_{r=0}^{n} (-1)^r \frac{m!n!}{k!} \binom{k}{r} y^{n+1-r} \right\} \Bigg|_{y=1} \\ &= (-1)^{n+1} \frac{m!n!}{k!} \left(y \frac{\mathrm{d}}{\mathrm{d}y} \right)^i \left\{ y^{n+1-k} (y-1)^k - \sum_{r=n+1}^{k} (-1)^r \binom{k}{r} y^{n+1-r} \right\} \Bigg|_{y=1} \\ &= (-1)^{n+1} \frac{m!n!}{k!} \left(y \frac{\mathrm{d}}{\mathrm{d}y} \right)^i \left\{ \frac{(y-1)^k}{y^{m+1}} \right\} + \sum_{r=n+1}^{k} (-1)^{r+n} \frac{m!n!}{r!(k-r)!} (n+1-r)^i. \end{aligned}$$

Then first term is 0 if $i < k$ and it's $(-1)^{n+1} m!n!$ if $i = k$. When $r = n+1$ the summand in the second term is 0 since $i > 0$. So we can let r range only from $n+2$ to k. Then change the index $r \to k - r$ (and let r run from 0 to m) we can see immediately that the second term is exactly W_i. This proves Eq. (11.44) which together with Eq. (11.43) implies the first equation in Eq. (11.37). We thus finish the proof of the corollary. □

We conclude this section by remarking that by the stuffle relation Eq. (12.22) from Chap. 12 we can also analyze $\zeta_q^I[s_1, s_2]$ at $(-n, n+2-k)$ for any nonnegative integers k and n. For example, it's easy to compute directly that

$$\begin{aligned} \operatorname*{Res}_{(s_1,s_2)=(2,-3)} \zeta_q^I[s_1, s_2] &= -\frac{1}{(1-q)\log q} \sum_{r=0}^{3} (-1)^r \binom{3}{r} (r+1) \frac{q^{r+1}}{1-q^{r+1}} \\ &= \frac{-q(q-1)^2}{(q+1)(q^2+1)(q^2+q+1)\log q}, \end{aligned} \tag{11.45}$$

which can be obtained also by the shuffle relation in Eq. (12.22) and the expression

$$\operatorname*{Res}_{(s_1,s_2)=(-3,2)} \zeta_q^I[s_1,s_2] = \frac{q(q-1)^2}{(q+1)(q^2+1)(q^2+q+1)\log q}$$

by taking $k=n=3$ in Thm. 11.3.3. Thus $(2,-3)$ is a simple pole of the double q-zeta function $\zeta_q^I[s_1,s_2]$. On the other hand the ordinary double zeta function $\zeta(s_1,s_2)$ does not have a pole along $s_1+s_2=-1$. Indeed from Eq. (11.45) we find that

$$\lim_{q\to 1} \operatorname*{Res}_{(s_1,s_2)=(2,-3)} \{\zeta_q^I[s_1,s_2]\} = 0.$$

11.4 Historical Notes

The q-analog of the Riemann zeta function $\zeta^I[s]$ is first systematically studied from both analytic and algebraic points of view in [337]. In particular, the depth one case of Thms. 11.2.1 and 11.3.1 is proved there. A very interesting paper [341] by Kawagoe, Kurokawa and Wakayama investigated this q-Riemann zeta function along the same line but in more depth. In particular, both the behavior of zeros and the "crystal" case ($q\to 0$) were considered in some details and an analog of the Riemann hypothesis was proposed.

The general multiple q-zeta function $\zeta^I[\mathbf{s}]$ of arbitrary depth was treated first by the author in [615], mostly from the analytic point of view. Nonetheless, he also considered the special values at nonnegative integers when depth is two.

In [337], Kawagoe, Kurokawa and Wakayama showed that if $\varphi(s)$ is a meromorphic function of s then the limit $\lim_{q\to 1}\zeta_q^{\varphi(s)}[s]=\zeta(s)$ if and only if $\varphi(s)=s-\nu$ for some $\nu\in\mathbb{N}$. We don't know in depth $d>1$ case what one should expect of $\varphi_j(\mathbf{s})$ if $\lim_{q\to 1}\zeta_q^{(\varphi_1(\mathbf{s}),\dots,\varphi_d(\mathbf{s}))}[\mathbf{s}]=\zeta(\mathbf{s})$.

Exercises

11.1. Prove Prop. 11.1.5 completely by using the mathematical induction and the binomial identity

$$\binom{n}{r}+\binom{n}{r-1}=\binom{n+1}{r}.$$

11.2. Show by an appropriate substitution that

$$\int_1^\infty \frac{q^{x(s-1)}}{(1-q^x)^s}\,\mathrm{d}x = -\frac{q^{s-1}(1-q)^{1-s}}{(s-1)\log q}.$$

11.3. Prove Eq. (11.14) by repeatedly using integration by parts.

11.4. Show that

$$\lim_{q\to 1}(1-q)^s \log q \sum_{n\in\mathbb{Z}\setminus\{0\}} \frac{s}{(2\pi i n)^2} b_q(s-1+\delta n, -s) = 0,$$

$$\lim_{q\to 1}(1-q)^s \log q \sum_{n\in\mathbb{Z}\setminus\{0\}} \frac{1}{(2\pi i n)^2} b_q(s-1+\delta n, -s+1) = 0.$$

11.5. Let $M > 2N+16$ and $x, k > 0$. Prove that

$$g(q) = q^{x(M-N-2)-kM/6}\left(\frac{1-q}{1-q^x}\right)^{M-N+1}$$

is an increasing function for $1/2 < q < 1$.

11.6. Prove Prop. 11.3.6 using Eq. (11.24).

11.7. Show that D_0 and D_1 satisfy the same bound as Eq. (11.22) if $1/2 < q < 1$ and $M > 16 + 2N_1 + 6\sum_{j=2}^{d}(N_j+1)$ where the positive integers $N_j > |\mathrm{Re}(s_j)|$ for all j.

11.8. Compute the residue

$$\operatorname*{Res}_{(s_1,s_2)=(3,-4)} \zeta_q^I[s_1, s_2].$$

Chapter 12

q-Analogs of Multiple Zeta (Star) Values

There are quite a few different ways to define q-analogs of the MZVs and the MZ*Vs. We require that a good analog should not only deform into its classical counterpart when $q \to 1$ but also satisfy the key relations that hold in the classical setting. Therefore, we seek for the q-analogs of the MZVs (q-MZVs for abbreviation) that obey some kind of double shuffle relations (DBSFs). On the other hand, for the q-analogs of the MZ*Vs (q-MZ*Vs for abbreviation), we choose those which enjoy interesting identities similar to the non-q version. In this chapter we present some of these and consider their most important properties.

12.1 Various Definitions of q-MZVs

We have seen from last chapter that a good q-analog of the multiple zeta function is defined by

$$\zeta_q^I[\mathbf{s}] := \zeta_q^{\mathbf{s}-\mathbf{1}}[\mathbf{s}] = \sum_{k_1>\cdots>k_d\geq 1} \frac{q^{k_1(s_1-1)}\dots q^{k_d(s_d-1)}}{[k_1]^{s_1}\dots[k_d]^{s_d}}.$$

We can define two more analogs as follows:

$$\zeta_q^{\mathrm{II}}[\mathbf{s}] := \zeta_q^{\mathbf{s}}[\mathbf{s}] = \sum_{k_1>\cdots>k_d\geq 1} \prod_{j=1}^{d} \frac{q^{k_1 s_1}\dots q^{k_d s_d}}{[k_1]^{s_1}\dots[k_d]^{s_d}}, \tag{12.1}$$

$$\zeta_q^{\mathrm{III}}[\mathbf{s}] := \zeta_q^{(1,0,\dots,0)}[\mathbf{s}] = \sum_{k_1>\cdots>k_d\geq 1} \frac{q^{k_1}}{[k_1]^{s_1}\dots[k_d]^{s_d}}. \tag{12.2}$$

By Prop. 11.1.1 we obtain the following corollary immediately.

Corollary 12.1.1. *Let* $\mathbf{s} = (s_1, \dots, s_d) \in \mathbb{Z}^d$. *Then*

(i) $\zeta_q^I[\mathbf{s}]$ *converges if* $s_1 + \cdots + s_j > j$, *for all* $j = 1, \dots, d$.
(ii) $\zeta_q^{\mathrm{II}}[\mathbf{s}]$ *converges if* $s_1 + \cdots + s_j > 0$, *for all* $j = 1, \dots, d$.
(iii) $\zeta_q^{\mathrm{III}}[\mathbf{s}]$ *always converges.*

In this chapter we will also consider two more special types of the q-MZVs that are more complicated than the above three types but have some advantages by themselves. Both of them have the form

$$\sum_{k_1>\cdots>k_d\geq 1} \frac{P_{s_1}(q^{k_1})\cdots P_{s_d}(q^{k_d})}{[k_1]^{s_1}\cdots[k_d]^{s_d}}$$

where $P_n(t)$ are polynomials of t without constant term. The first class, called the Bachmann–Kühn brackets (mono- and bi-brackets), uses the Eulerian polynomial $\tilde{P}_n(t)$ for $P_n(t)$, where

$$\tilde{P}_n(t) = \sum_{n=0}^{s-1} E_{s,n}t^n,$$

and the Eulerian numbers $E_{s,n}$ are defined by

$$E_{s,n} := \sum_{i=0}^{n}(-1)^i\binom{s+1}{i}(n+1-i)^s.$$

The second class, called Okounkov's q-MZVs and denoted by $\zeta_q^{\mathrm{O}}[\mathbf{s}]$, is defined using the following palindrome polynomials:

$$P_n(t) := t^{n^+} + t^{n^-} \tag{12.3}$$

where, for any $n \in \mathbb{N}$, n^- and n^+ are the two nonnegative integers such that

$$\frac{n-1}{2} \leq n^- \leq \frac{n}{2} \leq n^+ \leq \frac{n+1}{2}.$$

Clearly we have $n^+ + n^- = n$, $n^+ = n^-$ if n is even, and $n^+ = n^- + 1$ if n is odd.

All of the above q-MZVs can be expressed using the most general *type G* q-MZVs $\zeta_q^{(t_1,\dots,t_d)}[s_1,\dots,s_d]$ defined by Eq. (11.1) where $1 \leq t_1 \leq s_1$, $0 \leq t_j \leq s_j$ for all $j \geq 2$. They are all convergent by Prop. 11.1.1.

12.2 Double Shuffle Relations

In this section, we will use a few algebras of words to encode the structure of various analogs of the q-MZVs. Motivated by the iterated Jackson q-integrals we can apply the theory of Rota–Baxter algebras to study the shuffle algebra. This enables us to obtain the DBSFs of these q-MZVs.

12.2.1 *Rota–Baxter Algebras*

We first consider some Rota–Baxter algebras which encode the (q-)shuffle relations for different versions of the q-MZVs.

Definition 12.2.1. Fix an algebra A over a commutative ring R and an element $\lambda \in R$. We call A a *Rota–Baxter R-algebra* and $\mathscr{P}$ a *Rota–Baxter operator* of weight λ if the operator $\mathscr{P}$ satisfies the following Rota–Baxter relation of weight λ:

$$\mathscr{P}(x)\mathscr{P}(y) = \mathscr{P}(\mathscr{P}(x)y) + \mathscr{P}(x\mathscr{P}(y)) + \lambda\mathscr{P}(xy) \quad \forall x, y \in A. \tag{12.4}$$

Recall that for any continuous function $f(x)$ on $[\alpha, \beta]$ the *Jackson q-integral* is defined by

$$\int_\alpha^\beta f(x)\, d_q x := \sum_{k\geq 0} f\big(\alpha + q^k(\beta - \alpha)\big)(q^k - q^{k+1})(\beta - \alpha). \tag{12.5}$$

Taking $\alpha = 0$ and $\beta = t$ in Eq. (12.5) we now set

$$\begin{aligned} \mathbf{J}[f](t) &:= (1-q)\sum_{k\geq 0} f(q^k t)q^k t = (1-q)\sum_{k\geq 0} \mathbf{E}^k[\mathbf{I}\cdot f](t) \\ &= (1-q)\mathbf{P}[\mathbf{I}\cdot f](t), \end{aligned} \tag{12.6}$$

where $\mathbf{I}(t) = t$ is the identity function,

$$\begin{aligned} \mathbf{E}[f](t) &:= \mathbf{E}_q[f](t) := f(qt), \\ \mathbf{P}[f](t) &:= \mathbf{P}_q[f](t) := f(t) + f(qt) + f(q^2 t) + \cdots \end{aligned}$$

are the *q-expanding* and the *(principle) q-summation operators*, respectively. We also need to define the *(remainder) q-summation operator*

$$\mathbf{R}[f](t) := \mathbf{R}_q[f](t) := f(qt) + f(q^2 t) + \cdots = (\mathbf{P}[f] - [f])(t).$$

So, $\mathbf{P}$ is the principle part (*i.e.*, the whole thing) while $\mathbf{R}$ is the remainder (*i.e.*, without the first term). Clearly, $\mathbf{P} = \mathbf{R} + \mathbf{I}$ where, as an operator, $\mathbf{I}[f] = f$.

Let $t\mathbb{Q}[\![t, q]\!]$ be the space of formal series in two variables with $t > 0$. Then $\mathbf{J}$, $\mathbf{E}$, $\mathbf{P}$ and $\mathbf{R}$ are all $\mathbb{Q}[\![q]\!]$-linear endomorphism of $t\mathbb{Q}[\![t, q]\!]$. We can further define the inverse to $\mathbf{P}$ which is called the *q-difference operator*:

$$\mathbf{D} := \mathbf{I} - \mathbf{E}. \tag{12.7}$$

Proposition 12.2.2. *For any $f, g \in t\mathbb{Q}[\![t,q]\!]$, we have*

$$\begin{aligned}
\mathbf{P}[f]\mathbf{P}[g] &= \mathbf{P}\big[\mathbf{P}[f]g\big] + \mathbf{P}\big[f\mathbf{P}[g]\big] - \mathbf{P}[fg], && (12.8)\\
\mathbf{R}[f]\mathbf{R}[g] &= \mathbf{R}\big[\mathbf{R}[f]g\big] + \mathbf{R}\big[f\mathbf{R}[g]\big] + \mathbf{R}[fg], && (12.9)\\
\mathbf{R}[f]\mathbf{P}[g] &= \mathbf{R}\big[\mathbf{R}[f]g\big] + \mathbf{R}\big[f\mathbf{R}[g]\big] + \mathbf{R}[f]g + \mathbf{R}[fg], && (12.10)\\
\mathbf{J}[f]\mathbf{J}[g] &= \mathbf{J}\big[\mathbf{J}[f]g\big] + \mathbf{J}\big[f\mathbf{J}[g]\big] - (1-q)\mathbf{J}\big[\mathbf{I}fg\big], && (12.11)\\
&= \mathbf{J}\big[f\mathbf{J}[g]\big] + q\mathbf{J}\big[\mathbf{J}\big[E[f]\big]g\big], && (12.12)\\
\mathbf{D}[f]\mathbf{D}[g] &= \mathbf{D}[f]g + f\mathbf{D}[g] - \mathbf{D}[fg], && (12.13)\\
\mathbf{D}[f]\mathbf{P}[g] &= \mathbf{D}\big[f\mathbf{P}[g]\big] + \mathbf{D}[f]g - fg, && (12.14)\\
\mathbf{D}[f]\mathbf{R}[g] &= \mathbf{D}\big[f\mathbf{R}[g]\big] + \mathbf{D}[fg] - fg, && (12.15)\\
\mathbf{DP} &= \mathbf{PD} = \mathbf{I}. && (12.16)
\end{aligned}$$

Proof. For the first identity Eq. (12.8), we have

$$\begin{aligned}
\Big(\mathbf{P}[f]\mathbf{P}[g]\Big)(t) &= \sum_{j\ge 0} f(q^j t) \sum_{k\ge 0} g(q^k t)\\
&= \sum_{j\ge k\ge 0} f(q^j t)g(q^k t) + \sum_{k\ge j\ge 0} f(q^j t)g(q^k t) - \sum_{j\ge 0} f(q^j t)g(q^j t)\\
&= \sum_{\ell,k\ge 0} f(q^{\ell+k} t)g(q^k t) + \sum_{\ell,j\ge 0} f(q^j t)g(q^{\ell+j} t) - \mathbf{P}[fg]\\
&= \Big(\mathbf{P}\big[\mathbf{P}[f]g\big] + \mathbf{P}\big[f\mathbf{P}[g]\big] - \mathbf{P}[fg]\Big)(t)
\end{aligned}$$

All the other identities follow from the same type straightforward computation which is left as an exercise (see Exercise12.8). □

By Prop. 12.2.2 we see that $\mathbf{P}$ and $\mathbf{R}$ are both Rota–Baxter operators on $t\mathbb{Q}[\![t,q]\!]$ (of weight -1 and 1, respectively) but $\mathbf{D}$ is not. In fact, $\mathbf{D}$ satisfies the condition Eq. (12.13) of a differential Rota–Baxter operator (see [259]). Moreover, it is *invertible* in the sense that Rota–Baxter operator $\mathbf{P}$ and the differential $\mathbf{D}$ are mutually inverse by Eq. (12.16).

We end this section with an identity which will be used to derive the duality relations among the q-MZVs. For any $n \in \mathbb{N}$, set

$$\mathbf{P}^n = \underbrace{\mathbf{P}\circ\cdots\circ\mathbf{P}}_{n \text{ times}} \quad \text{and} \quad \mathbf{R}^n = \underbrace{\mathbf{R}\circ\cdots\circ\mathbf{R}}_{n \text{ times}}.$$

Theorem 12.2.3. *Let $d \in \mathbb{N}$ and $\alpha_j, \beta_j \in \mathbb{N}$ for all $j = 1, \ldots, d$. Let $\mathbf{y}(t) = \frac{t}{1-t}$. Then we have*

$$\mathbf{R}^{\alpha_1}\mathbf{y}^{\beta_1}\cdots\mathbf{R}^{\alpha_\ell}\mathbf{y}^{\beta_\ell}(t) = \sum_{\substack{j_1\ge\beta_1,\ldots,j_\ell\ge\beta_\ell\\ k_1\ge\alpha_1,\ldots,k_\ell\ge\alpha_\ell}} \prod_{r=1}^{\ell}\left[\binom{j_r-1}{\beta_r-1}\binom{k_r-1}{\alpha_r-1} q^{k_r\sum_{s=r}^{\ell} j_s} t^{j_r}\right]. \tag{12.17}$$

Proof. First we show that

$$\mathbf{R}^{\alpha}(t^j) = \frac{q^{\alpha j}t^j}{(1-q^j)^{\alpha}}. \tag{12.18}$$

Indeed, if $\alpha = 1$ then

$$\mathbf{R}(t^j) = \sum_{k\ge 1} q^{kj}t^j = \frac{q^j t^j}{1-q^j}.$$

So Eq. (12.18) can be proved easily by the induction.

Now we proceed to prove that for any integer $m \ge 0$

$$\mathbf{R}^{\alpha_1}\mathbf{y}^{\beta_1}\cdots\mathbf{R}^{\alpha_\ell}\Big(\mathbf{y}^{\beta_\ell}(t)\cdot t^m\Big) = \sum_{\substack{j_1\ge\beta_1,\ldots,j_\ell\ge\beta_\ell\\ k_1\ge\alpha_1,\ldots,k_\ell\ge\alpha_\ell}} t^m \prod_{r=1}^{\ell}\left[\binom{j_r-1}{\beta_r-1}\binom{k_r-1}{\alpha_r-1} q^{k_r(m+\sum_{s=r}^{\ell} j_s)} t^{j_r}\right]. \tag{12.19}$$

If $\ell = 1$ then we have

$$\begin{aligned}
\mathbf{R}^{\alpha}\Big(\mathbf{y}^{\beta}(t)\cdot t^m\Big) &= \mathbf{R}^{\alpha}\left(\left(\frac{t}{1-t}\right)^{\beta} t^m\right) = \mathbf{R}^{\alpha}\sum_{j\ge 0}\binom{\beta+j-1}{j} t^{m+\beta+j}\\
&= \mathbf{R}^{\alpha}\sum_{j\ge\beta}\binom{j-1}{\beta-1} t^{m+j}\\
&= \sum_{j\ge\beta}\binom{j-1}{\beta-1}\frac{q^{\alpha(m+j)}t^{m+j}}{(1-q^{m+j})^{\alpha}} \qquad \text{(by Eq. (12.18))}\\
&= \sum_{j\ge\beta}\binom{j-1}{\beta-1}\sum_{k\ge 0}\binom{\alpha+k-1}{k} q^{(\alpha+k)(m+j)}t^{m+j}\\
&= \sum_{j\ge\beta}\sum_{k\ge\alpha}\binom{j-1}{\beta-1}\binom{k-1}{\alpha-1} q^{k(m+j)}t^{m+j}.
\end{aligned}$$

This proves Eq. (12.19) when $\ell = 1$. In general

$$\begin{aligned}
&\mathbf{R}^{\alpha_1}\mathbf{y}^{\beta_1}\cdots\mathbf{R}^{\alpha_{\ell-1}}\Big(\mathbf{y}^{\beta_{\ell-1}}(t)\Big(\mathbf{R}^{\alpha_\ell}\Big(\mathbf{y}^{\beta_\ell}(t)\cdot t^m\Big)\Big)\Big)\\
&=\sum_{\substack{j_\ell\ge\beta_\ell\\ k_\ell\ge\alpha_\ell}}\binom{j_\ell-1}{\beta_\ell-1}\binom{k_\ell-1}{\alpha_\ell-1}q^{k_\ell(m+j_\ell)}\mathbf{R}^{\alpha_1}\mathbf{y}^{\beta_1}\cdots\mathbf{R}^{\alpha_{\ell-1}}\Big(\mathbf{y}^{\beta_{\ell-1}}(t)\cdot t^{m+j_\ell}\Big).
\end{aligned}$$

So Eq. (12.19) follows immediately by induction. We can now finish the proof of the theorem by taking $m = 0$. □

Corollary 12.2.4. *Let $d \in \mathbb{N}$ and $\alpha_j, \beta_j \in \mathbb{N}$ for all $j = 1, \dots, d$. Then we have*

$$\mathbf{R}^{\alpha_1}\mathbf{y}^{\beta_1}\cdots\mathbf{R}^{\alpha_\ell}\mathbf{y}^{\beta_\ell}(1) = \mathbf{R}^{\beta_\ell}\mathbf{y}^{\alpha_\ell}\cdots\mathbf{R}^{\beta_1}\mathbf{y}^{\alpha_1}(1). \tag{12.20}$$

Proof. In Eq. (12.17) we use the substitutions $j_r \leftrightarrow k_{\ell+1-r}$ for all $r = 1, \dots, \ell$. Then we have

$$\begin{aligned}
\sum_{r=1}^{\ell}\sum_{s=r}^{\ell} j_s k_r &\longrightarrow \sum_{r=1}^{\ell}\sum_{s=r}^{\ell} j_{\ell+1-r}k_{\ell+1-s} = \sum_{s=1}^{\ell}\sum_{r=1}^{s} j_{\ell+1-r}k_{\ell+1-s}\\
&= \sum_{s=1}^{\ell}\sum_{r=1}^{\ell+1-s} j_{\ell+1-r}k_s = \sum_{s=1}^{\ell}\sum_{r=s}^{\ell} j_r k_s = \sum_{r=1}^{\ell} k_r \sum_{s=r}^{\ell} j_s
\end{aligned}$$

which follows from $s \leftrightarrow \ell+1-s$ followed by $r \leftrightarrow \ell+1-r$ and $r \leftrightarrow s$. This proves the corollary. □

12.2.2 *q*-Analogs of Hoffman Algebras

We have seen that the DBSFs lead to many (and conjecturally all) $\mathbb{Q}$-linear relations among the MZVs. To study similar relations of the q-MZVs we can modify the Hoffman algebra in this new setting. First we consider some algebras which will be used to define the stuffle relations later.

Definition 12.2.5. Set the alphabet $X_\theta := \{a, b, \theta\}$. Denote by $\mathfrak{A}_\theta := \mathbb{Q}\langle a, b, \theta\rangle$ the (weight) filtered noncommutative polynomial $\mathbb{Q}$-algebra of words from X_θ^*. The weights of a, b and θ are all equal to one. Set

$$\gamma := b - \theta, \qquad \mathsf{z}_s := a^{s-1}b, \quad \mathsf{z}'_s := a^{s-1}\theta, \quad s \in \mathbb{Z}.$$

Let $\mathsf{Y}_{\tilde{\mathrm{I}}} := \{\theta\} \cup \{\mathsf{z}_k\}_{k\ge1}$, $\mathsf{Y}_{\mathrm{II}} := \{\mathsf{z}'_k\}_{k\ge0}$, and $\mathsf{Y}_{\mathrm{III}} := \{\mathsf{z}_k\}_{k\in\mathbb{Z}}$. We put a tilde on top of I since we need to consider some kind of regularization due to convergence issues involved in type I q-MZVs. This is realized by the introduction of the letter θ. Again, we use 1 for the empty word and Y_τ^* to denote the set of words generated on Y_τ for any type τ.

Let $\mathfrak{A}^1_{\tilde{\mathrm{I}}}$, $\mathfrak{A}^1_{\mathrm{II}}$ and $\mathfrak{A}_{\mathrm{III}}$ be the subalgebras of $\mathfrak{A}_\theta$ freely generated by the sets $\mathsf{Y}_{\tilde{\mathrm{I}}}$, Y_{II} and $\mathsf{Y}_{\mathrm{III}}$, respectively. Set

$$\mathfrak{A}^1_{\mathrm{III}} = \sum_{k\in\mathbb{Z}} \mathsf{z}'_k \mathfrak{A}_{\mathrm{III}} \not\subset \mathfrak{A}_{\mathrm{III}}.$$

Here, all integer subscripts are allowed in $\mathsf{Y}_{\mathrm{III}}$ because type III q-MZVs converge for all integer arguments. Further, we define the following subalgebras corresponding to the convergent values:

$$\mathfrak{A}^0_I := \mathbb{Q}1 + \sum_{k\geq 2} \mathsf{z}_k \mathfrak{A}^1_{\tilde{\mathrm{I}}}, \qquad \mathfrak{A}^0_{\tilde{\mathrm{I}}} := \mathbb{Q}1 + \theta\mathfrak{A}^1_{\tilde{\mathrm{I}}} + \sum_{k\geq 2} \mathsf{z}_k \mathfrak{A}^1_{\tilde{\mathrm{I}}} \subsetneq \mathfrak{A}^1_{\tilde{\mathrm{I}}},$$

$$\mathfrak{A}^0_{\mathrm{II}} := \mathbb{Q}1 + \sum_{k\geq 1} \mathsf{z}'_k \mathfrak{A}^1_{\mathrm{II}} \subsetneq \mathfrak{A}^1_{\mathrm{II}}, \qquad \mathfrak{A}^0_{\mathrm{III}} := \mathfrak{A}^1_{\mathrm{III}}.$$

For each type τ, the words in $\mathfrak{A}^0_\tau$ are called *type τ-admissible*. It is consistent with Prop. 11.1.1 since we consider only nonnegative compositions $\mathbf{s}$.

To define the stuffle product for type $\tau = \tilde{\mathrm{I}}$ and II, similar to the MZV case, we define a commutative product $[-,-]_\tau$ first:

$$\begin{aligned} &[\mathsf{z}_k, \mathsf{z}_l]_{\tilde{\mathrm{I}}} := \mathsf{z}_{k+l} + \mathsf{z}_{k+l-1}, \quad [\mathsf{z}'_k, \mathsf{z}'_l]_{\mathrm{II}} := \mathsf{z}'_{k+l}, \\ &[\theta, \mathsf{z}_k]_{\tilde{\mathrm{I}}} := \mathsf{z}_{k+1}, \quad [\theta, \theta]_{\tilde{\mathrm{I}}} := \mathsf{z}_2 - \theta, \end{aligned} \tag{12.21}$$

for all $k, l \geq 1$. Now we define the stuffle product $*_\tau$ on $\mathfrak{A}^1_\tau$ inductively as follows. For any words $\mathbf{u}, \mathbf{v} \in \mathfrak{A}^1_\tau$ and letters $\alpha, \beta \in \mathsf{Y}_\tau$, we set $1 *_\tau \mathbf{u} = \mathbf{u} = \mathbf{u} *_\tau 1$ and

$$(\alpha\mathbf{u}) *_\tau (\beta\mathbf{v}) = \alpha(\mathbf{u} *_\tau \beta\mathbf{v}) + \beta(\alpha\mathbf{u} *_\tau \mathbf{v}) + [\alpha, \beta]_\tau(\mathbf{u} * \mathbf{v}). \tag{12.22}$$

Remark 12.2.6. It is not hard to check that for $\tau = \tilde{\mathrm{I}}$ and II, $(\mathfrak{A}^0_\tau, *_\tau) \subset (\mathfrak{A}^1_\tau, *_\tau)$ as subalgebras.

Definition 12.2.7. Define the injective shifting operator $\mathcal{S}_-$ on any word of $\mathfrak{A}^0_{\mathrm{III}}$ by acting on the first letter:

$$\mathcal{S}_-(\mathsf{z}'_n\mathbf{w}) := \mathsf{z}_n\mathbf{w} - \mathsf{z}_{n-1}\mathbf{w} \quad \text{for all } n \in \mathbb{Z} \text{ and } \mathbf{w} \in \mathsf{Y}^*_{\mathrm{III}}. \tag{12.23}$$

For any $k, l \in \mathbb{Z}$ and any $\mathbf{u}, \mathbf{v} \in \mathsf{Y}^*_{\mathrm{III}}$, define the stuffle product $*_{\mathrm{III}}$ by

$$\mathsf{z}'_k\mathbf{u} *_{\mathrm{III}} \mathsf{z}'_l\mathbf{v} = \mathsf{z}'_k\big(\mathbf{u} *_{\mathrm{II}} \mathcal{S}_-(\mathsf{z}'_l\mathbf{v})\big) + \mathsf{z}'_l\big(\mathcal{S}_-(\mathsf{z}'_k\mathbf{u}) *_{\mathrm{II}} \mathbf{v}\big) + (\mathsf{z}'_{k+l} - \mathsf{z}'_{k+l-1})(\mathbf{u} *_{\mathrm{II}} \mathbf{v}).$$

Here $*_{\mathrm{II}}$ is the ordinary stuffle with $[\mathsf{z}_r, \mathsf{z}_s]_{\mathrm{II}} = \mathsf{z}_{r+s}$ for all $r, s \in Z$.

Lemma 12.2.8. *Let $\tau = \tilde{\mathrm{I}}$, II or III. Then the stuffle products $*_\tau$ is well defined. Namely, $\mathbf{u} *_\tau \mathbf{v} \in \mathfrak{A}^1_\tau$ if $\mathbf{u}, \mathbf{v} \in \mathfrak{A}^1_\tau$.*

Proof. The computation is straightforward. See Exercise 12.2. □

Proposition 12.2.9. *Let* $\tau = \tilde{\mathrm{I}}, \mathrm{II}$ *or* III. *Then the stuffle algebras* $(\mathfrak{A}_\tau^1, *_\tau)$ *are all commutative and associative.*

Proof. This follows from the fact that the product $[-,-]_\tau$ are all commutative and associative which can be verified easily. □

We now turn to the shuffle algebra which is an analog of the corresponding algebra for MZVs reflecting the properties of their representations using iterated integrals.

Definition 12.2.10. Let $X_\pi = \{\pi, \delta, y\}$ be an alphabet and X_π^* be the set of words generated by X_π. Define $\mathfrak{A}_\pi := \mathbb{Q}\langle \pi, \delta, y\rangle$ to be the noncommutative polynomial $\mathbb{Q}$-algebra of words of X_π^*. We may embed $\mathfrak{A}_\theta$ defined by Definition 12.2.5 as a subalgebra of $\mathfrak{A}_\pi$ in two different ways: put $\rho = \pi - 1$ and let

$$\begin{aligned}
&(A)\quad a := \pi, \quad a^{-1} := \delta, \quad b := \pi y, \quad \theta = \rho y \quad \Longrightarrow \quad \gamma := y,\\
&(B)\quad a := \rho, \quad a^{-1} := -, \quad b := \pi y, \quad \theta = \rho y \quad \Longrightarrow \quad \gamma := y.
\end{aligned}$$

We denote the image of the embedding by $\mathfrak{A}_\theta^{(A)}$ and $\mathfrak{A}_\theta^{(B)}$, respectively. The dash $-$ for the image of a^{-1} in (B) means it does not matter what image we choose since a^{-1} appears only when we consider type III q-MZVs using (A). We will use embedding (B) for the other types for which a^{-1} will not be utilized essentially because of convergence issues.

12.2.3 *q-Stuffle Relations*

We now apply the results from the preceding section to study the stuffle relations of the q-MZVs. Define the $\mathbb{Q}$-linear realization maps $\mathfrak{z}_q : \mathfrak{A}_\tau^0 \to \mathbb{C}$ $(\tau = \tilde{\mathrm{I}}, \mathrm{II})$ by $\mathfrak{z}_q[1] = 1$ and

$$\mathfrak{z}_q[y_1^\tau \dots y_d^\tau] := \sum_{k_1 > \cdots > k_d > 0} M_{k_1}^\tau(y_1^\tau) \dots M_{k_d}^\tau(y_d^\tau),$$

where $y_1^\tau \dots y_d^\tau \in \mathfrak{A}_\tau^0$ and the $\mathbb{Q}$-linear maps

$$M_k^{\tilde{\mathrm{I}}}(\theta) := \frac{q^k}{(1-q^k)}, \quad M_k^{\tilde{\mathrm{I}}}(\mathbf{z}_s) := \frac{q^{(s-1)k}}{(1-q^k)^s}, \quad M_k^{\mathrm{II}}(\mathbf{z}_s') := \frac{q^{sk}}{(1-q^k)^s}.$$

Note that $M_k^{\tilde{\mathrm{I}}}(\gamma) = M_k^{\tilde{\mathrm{I}}}(\mathbf{z}_1 - \theta) = 1$. For example, we have

$$\mathfrak{z}_q[\mathbf{z}_2\mathbf{z}_5\gamma^2\mathbf{z}_1] = \mathfrak{z}_q^{(1,4,0,0,0)}[2,5,0,0,1], \quad \mathfrak{z}_q[\theta\mathbf{z}_7\theta\mathbf{z}_4] = \mathfrak{z}_q^{(1,6,1,3)}[1,7,1,4],$$

which are not q-MZVs of type I.

For type Ⅲ, we similarly define the $\mathbb{Q}$-linear realization maps $\mathfrak{z}_q : \mathfrak{A}^0_{\text{Ⅲ}} \to \mathbb{C}$ by $\mathfrak{z}_q[1] = 1$ and

$$\mathfrak{z}_q[y_1 \dots y_d] := \sum_{k_1 > \cdots > k_d > 0} M^{1,\text{Ⅲ}}_{k_1}(y_1) M^{\text{Ⅲ}}_{k_2}(y_2) \dots M^{\text{Ⅲ}}_{k_d}(y_d),$$

where $y_1 \dots y_d \in \mathfrak{A}^0_{\text{Ⅲ}}$ and the $\mathbb{Q}$-linear maps

$$M^{1,\text{Ⅲ}}_k(\mathbf{z}'_s) := \frac{q^k}{(1-q^k)^s}, \quad M^{\text{Ⅲ}}_k(\mathbf{z}_s) := \frac{1}{(1-q^k)^s}.$$

Theorem 12.2.11. *Let $\tau = \tilde{\text{I}}$, Ⅱ or Ⅲ. For any $\mathbf{u}_\tau, \mathbf{v}_\tau \in \mathfrak{A}^0_\tau$, we have*

$$\mathfrak{z}_q[\mathbf{u}_\tau *_\tau \mathbf{v}_\tau] = \mathfrak{z}_q[\mathbf{u}_\tau]\mathfrak{z}_q[\mathbf{v}_\tau]. \tag{12.24}$$

Proof. Let's consider type $\tilde{\text{I}}$ first. For a positive integer N, we define the $\mathbb{Q}$-linear map $F_N : \mathfrak{A}^1_I \to \mathbb{C}$ by $F_N(1) = 1$ and

$$F_N(\alpha_1 \dots \alpha_d) = \sum_{N > k_1 > \cdots > k_d > 0} M^{\tilde{\text{I}}}_{k_1}(\alpha_1) \dots M^{\tilde{\text{I}}}_{k_d}(\alpha_d), \tag{12.25}$$

where $\alpha_j \in \mathsf{Y}_{\tilde{\text{I}}}$ for all $j = 1, \dots, d$. Since $F_N(\alpha\mathbf{u}) = \sum_{N>m>0} M^{\tilde{\text{I}}}_m(\alpha) F_m(\mathbf{u})$ for any $\alpha \in \mathsf{Y}_{\tilde{\text{I}}}$ and $\mathbf{u} \in \mathfrak{A}^1_\theta$. we see that $\mathfrak{z}^I_q[\mathbf{u}] = \lim_{N\to\infty} F_N(\mathbf{u})$. Thus we only need to show that

$$F_N(\mathbf{u} * \mathbf{v}) = F_N(\mathbf{u})F_N(\mathbf{v}) \tag{12.26}$$

for all words $\mathbf{u}$, $\mathbf{v}$ in Y_θ not starting with $\mathbf{z}_1$. Indeed, Eq. (12.26) can be proved by using the induction on $|\mathbf{u}| + |\mathbf{v}|$. Note that Eq. (12.26) is trivial if $\mathbf{u}$ or $\mathbf{v}$ is empty word 1. So we assume $\mathbf{u}, \mathbf{v} \in \mathfrak{A}^0_\theta$ and $\alpha, \beta \in \mathsf{Y}_\theta \setminus \{\mathbf{z}_1\}$. Then

$$\begin{aligned} F_N(\alpha\mathbf{u})F_N(\beta\mathbf{v}) &= \sum_{N>m>0} \Big(M^{\tilde{\text{I}}}_m(\alpha)F_m(\mathbf{u})F_m(\beta\mathbf{v}) + M^{\tilde{\text{I}}}_m(\beta)F_m(\alpha\mathbf{u})F_m(\mathbf{v}) \\ &\quad + M^{\tilde{\text{I}}}_m(\alpha)M^{\tilde{\text{I}}}_m(\beta)F_m(\mathbf{u})F_m(\mathbf{v}) \Big) \\ &= \sum_{N>m>0} \Big(M^{\tilde{\text{I}}}_m(\alpha)F_m(\mathbf{u} * \beta\mathbf{v}) + M^{\tilde{\text{I}}}_m(\beta)F_m(\alpha\mathbf{u} * \mathbf{v}) \\ &\quad + M^{\tilde{\text{I}}}_m(\alpha)M^{\tilde{\text{I}}}_m(\beta)F_m(\mathbf{u} * \mathbf{v}) \Big) \end{aligned}$$

by inductive assumption. Hence Eq. (12.26) follows from Eq. (12.21) and

$$M^{\tilde{\text{I}}}_m(\mathbf{z}_k)M^{\tilde{\text{I}}}_m(\mathbf{z}_l) = M^{\tilde{\text{I}}}_m(\mathbf{z}_{k+l}) + M^{\tilde{\text{I}}}_m(\mathbf{z}_{k+l-1}), \tag{12.27}$$

$$M^{\tilde{\text{I}}}_m(\theta)M^{\tilde{\text{I}}}_m(\mathbf{z}_k) = M^{\tilde{\text{I}}}_m(\mathbf{z}_{k+1}), \tag{12.28}$$

$$M_m^{\tilde{\mathrm{I}}}(\theta)^2 = M_m^{\tilde{\mathrm{I}}}(\mathbf{z}_2) - M_m^{\tilde{\mathrm{I}}}(\theta) \tag{12.29}$$

for all $k, l \geq 2$ and $m \geq 1$.

The same argument works for type II and III if we observe that

$$M_m^{\mathrm{II}}(\mathbf{z}'_k)M_m^{\mathrm{II}}(\mathbf{z}'_l) = M_m^{\mathrm{II}}(\mathbf{z}'_{k+l}), \quad M_m^{1,\mathrm{III}}(\mathbf{z}'_k)M_m^{\mathrm{III}}(\mathbf{z}_l) = M_m^{\mathrm{III}}(\mathbf{z}_{k+l} - \mathbf{z}_{k+l-1}),$$
$$M_m^{\mathrm{III}}(\mathbf{z}_k)M_m^{\mathrm{III}}(\mathbf{z}_l) = M_m^{\mathrm{III}}(\mathbf{z}_{k+l}), \quad M_m^{1,\mathrm{III}}(\mathbf{z}'_k)M_m^{1,\mathrm{III}}(\mathbf{z}'_l) = M_m^{1,\mathrm{III}}(\mathbf{z}'_{k+l} - \mathbf{z}'_{k+l-1}),$$

for all $k, l \geq 0$, $m \geq 1$. This completes the proof of the theorem. □

12.2.4 *Iterated Jackson q-Integrals*

As we have seen in section 3.2, the iterated integral representations of the MZVs and its shuffle product structure is the driving force in defining Hoffman's shuffle algebra of words. Set

$$x_0 := x_0(t) = \frac{1}{t}, \quad x_1 := x_1(t) = \frac{1}{1-t}, \quad \mathbf{y} := \mathbf{y}(t) = \frac{t}{1-t}.$$

Recall that for $a = x_0(t)\,dt$ and $b = x_1(t)\,dt$, we can express MZVs by Chen's iterated integrals:

$$\zeta(s_1, \ldots, s_d) = \int_0^1 a^{s_1-1}b \cdots a^{s_d-1}b.$$

In the current q-analog setting, replacing the Riemann integrals by the Jackson q-integrals in Eq. (12.6) one gets the following two generalizations after slight modification. First, for any $(a_1, \ldots, a_d), (b_1, \ldots, b_d) \in (\mathbb{Z}_{\geq 0})^d$, we define

$$\mathfrak{z}_q[\rho^{a_1}\pi^{b_1}y \ldots \rho^{a_d}\pi^{b_d}y; t]$$
$$:= \mathbf{R}^{a_1}\left[\mathbf{P}^{b_1}[\mathbf{y}\mathbf{R}^{a_2}[\mathbf{P}^{b_2}[\mathbf{y} \cdots \mathbf{R}^{a_d}[\mathbf{P}^{b_d}[\mathbf{y}]] \cdots]]]\right](t). \tag{12.30}$$

Theorem 12.2.12. *Let* $\mathbf{s} = (s_1, \ldots, s_d) \in \mathbb{N}^d$ *and* $\mathbf{a} = (a_1, \ldots, a_d) \in (\mathbb{Z}_{\geq 0})^d$. *Put* $w = |\mathbf{s}|$ *and* $\mathbf{w} = \mathbf{w}^{\mathbf{a}}(\mathbf{s}) = \rho^{a_1}\pi^{s_1-a_1}y \ldots \rho^{a_d}\pi^{s_d-a_d}y$. *If* $a_1 + \cdots + a_j > 0$ *for all* $j = 1, \ldots, d$, *then we have*

$$\zeta_q^{\mathbf{a}}[\mathbf{s}] = (1-q)^w \mathfrak{z}_q[\mathbf{w}^{\mathbf{a}}(\mathbf{s}); 1], \quad \mathfrak{z}_q^{\mathbf{a}}[\mathbf{s}] = \mathfrak{z}_q[\mathbf{w}^{\mathbf{a}}(\mathbf{s})] := \mathfrak{z}_q[\mathbf{w}^{\mathbf{a}}(\mathbf{s}); 1]. \tag{12.31}$$

Proof. First we observe three important facts: for any $k \geq 1$ we have

$$\mathbf{P}(t^k) = \sum_{j\geq 0} q^{kj}t^k = \frac{t^k}{1-q^k}, \quad \mathbf{R}(t^k) = \sum_{j \geq 1} q^{kj}t^k = \frac{q^kt^k}{1-q^k}$$

and $\mathbf{D}(t^k) = t^k(1-q^k)$, by the definition of the two summation operators and the difference operator. Repeatedly applying this we get

$$\mathbf{P}^m(t^k) = \sum_{j\geq 0} q^{kj}t^k = \frac{t^k}{(1-q^k)^m} \qquad \forall m \in \mathbb{Z}, \tag{12.32}$$

$$\mathbf{R}^m(t^k) = \sum_{j\geq 1} q^{kj}t^k = \frac{q^{mk}t^k}{(1-q^k)^m} \qquad \forall m \in \mathbb{Z}_{\geq 0}. \tag{12.33}$$

Thus

$$\mathbf{P}\big(\mathbf{y}(t)\cdot t^k\big) = \sum_{j\geq 0} \frac{q^{j(k+1)}t^{k+1}}{1-q^j t} = \sum_{j\geq 0}\sum_{\ell\geq 0} q^{j(k+\ell+1)}t^{k+\ell+1} = \sum_{\ell>k} \frac{t^\ell}{1-q^\ell}.$$

Similarly, we have

$$\mathbf{D}\big(\mathbf{y}(t)\cdot t^k\big) = \frac{t^{k+1}}{1-t} - \frac{q^{k+1}t^{k+1}}{1-qt} = \sum_{\ell\geq 0}(1-q^{k+\ell+1})t^{k+\ell+1} = \sum_{\ell>k}(1-q^\ell)t^\ell,$$

and

$$\mathbf{R}\big(\mathbf{y}(t)\cdot t^k\big) = \sum_{j\geq 1} \frac{q^{j(k+1)}t^{k+1}}{1-q^j t} = \sum_{j\geq 1}\sum_{\ell\geq 0} q^{j(k+\ell+1)}t^{k+\ell+1} = \sum_{\ell>k} \frac{q^\ell t^\ell}{1-q^\ell}.$$

It follows from Eq. (12.32) and Eq. (12.33) that

$$\mathbf{P}^m\big(\mathbf{y}(t)\cdot t^k\big) = \sum_{\ell>k} \frac{t^\ell}{(1-q^\ell)^m} \qquad \forall m \in \mathbb{Z}, \tag{12.34}$$

$$\mathbf{R}^m\big(\mathbf{y}(t)\cdot t^k\big) = \sum_{\ell>k} \frac{q^{m\ell}t^\ell}{(1-q^\ell)^m} \qquad \forall m \in \mathbb{Z}_{\geq 0}. \tag{12.35}$$

We now prove by induction on d that for all $\mathbf{s} = (s_1, \ldots, s_d) \in \mathbb{N}^d$,

$$\mathfrak{z}_q[\mathbf{w}^{\mathbf{a}}(\mathbf{s}); t] = \sum_{k_1>\cdots>k_d>0} \frac{t^{k_1}q^{k_1a_1}\ldots q^{k_da_d}}{(1-q^{k_1})^{s_1}\cdots(1-q^{k_d})^{s_d}}. \tag{12.36}$$

When $d = 1$, *i.e.*, $\mathbf{s} = s$, then by Eqs. (12.34) and (12.33)

$$\mathfrak{z}_q[\mathbf{w}^a(s); t] = \mathbf{R}^a\mathbf{P}^{s-a}[\mathbf{y}](t) = \sum_{k>0} \frac{\mathbf{R}^a\big(t^k\big)}{(1-q^k)^{s-a}} = \sum_{k>0} \frac{q^{ak}t^k}{(1-q^k)^s} = \mathfrak{z}_q^{(a)}[s; t].$$

This proof works even when $s = a$ because of Eq. (12.35) (take $k = 0$ and $m = a$ there).

In general, assume $d \geq 2$ and Eq. (12.36) is true for smaller depths. Then by the inductive assumption

$$\mathfrak{z}_q[\mathbf{w}^{\mathbf{a}}(\mathbf{s}); t] = \mathbf{R}^{a_1}\mathbf{P}^{s_1-a_1}\big[\mathbf{y}\mathbf{R}^{a_2}\mathbf{P}^{s_2-a_2}[\mathbf{y}\cdots\mathbf{R}^{a_d}\mathbf{P}^{s_d-a_d}[\mathbf{y}]\cdots]\big](t)$$

$$
\begin{aligned}
&= \sum_{k_2>\cdots>k_d>0} \frac{\mathbf{R}^{a_1}\mathbf{P}^{s_1-a_1}\big(\mathbf{y}(t)\cdot t^{k_2}\big)q^{k_2a_2}\dots q^{k_da_d}}{(1-q^{k_2})^{s_2}\cdots(1-q^{k_d})^{s_d}} \\
&\overset{(12.34)}{=\!=} \sum_{k_1>\cdots>k_d>0} \frac{\mathbf{R}^{a_1}\big(t^{k_1}\big)q^{k_2a_2}\dots q^{k_da_d}}{(1-q^{k_1})^{s_1-a_1}(1-q^{k_2})^{s_2}\cdots(1-q^{k_d})^{s_d}} \\
&\overset{(12.33)}{=\!=} \sum_{k_1>\cdots>k_d>0} \frac{t^{k_1}q^{k_1a_1}\dots q^{k_da_d}}{(1-q^{k_1})^{s_1}\cdots(1-q^{k_d})^{s_d}}.
\end{aligned}
$$

Again, if $s_1 = a_1$ the proof is still valid. This completes the proof of Eq. (12.36). Setting $t = 1$ we arrive at Eq. (12.31). □

By the change of variables $a_j \to s_j - a_j$ for all $j = 1, \dots, d$, we immediately obtain the next result.

Theorem 12.2.13. *For* $\mathbf{s} = (s_1, \dots, s_d) \in \mathbb{N}^d$ *and* $\mathbf{a} = (a_1, \dots, a_d) \in (\mathbb{Z}_{\geq 0})^d$, *we set* $\mathbf{s}-\mathbf{a} = (s_1 - a_1, \dots, s_d - a_d)$, $w = |\mathbf{s}|$ *and* $\mathbf{w}^{\mathbf{s}-\mathbf{a}}(\mathbf{s}) = \rho^{s_1-a_1}\pi^{a_1}y\dots\rho^{s_d-a_d}\pi^{a_d}y$. *Define*

$$\mathfrak{z}_q[\mathbf{w}^{\mathbf{s}-\mathbf{a}}(\mathbf{s});t] := \mathbf{R}^{s_1-a_1}\big[\mathbf{P}^{a_1}[\mathbf{y}\mathbf{R}^{s_2-a_2}[\mathbf{P}^{a_2}[\mathbf{y}\cdots\mathbf{R}^{s_d-a_d}[\mathbf{P}^{a_d}[\mathbf{y}]]\cdots]]]\big](t).$$

If $s_1 + \cdots + s_j > a_1 + \cdots + a_j$ *for all* $j = 1, \dots, d$, *then we have*

$$
\begin{aligned}
\zeta_q^{\mathbf{s}-\mathbf{a}}[\mathbf{s}] &= (1-q)^w\mathfrak{z}_q[\mathbf{w}^{\mathbf{s}-\mathbf{a}}(\mathbf{s});1], \\
\mathfrak{z}_q^{\mathbf{s}-\mathbf{a}}[\mathbf{s}] &= \mathfrak{z}_q[\mathbf{w}^{\mathbf{s}-\mathbf{a}}(\mathbf{s})] := \mathfrak{z}_q[\mathbf{w}^{\mathbf{s}-\mathbf{a}}(\mathbf{s});1].
\end{aligned} \tag{12.37}
$$

By specializing the preceding two theorems we quickly find the following corollary. For future reference, we will say $\mathbf{w}_\tau$ has the *typical type* τ *form* for each type τ.

Corollary 12.2.14. *For* $\mathbf{s} = (s_1, \dots, s_d) \in \mathbb{N}^d$, *we set*

$$
\begin{aligned}
\mathbf{w}_I &= \rho^{s_1-1}\pi y\dots\rho^{s_d-1}\pi y = \mathbf{z}_{s_1}\dots\mathbf{z}_{s_d} \in \mathfrak{A}_\theta^{(B)} \subset \mathfrak{A}_\pi y \qquad (s_1 \geq 2), \\
\mathbf{w}_{\mathrm{II}} &= \rho^{s_1}y\dots\rho^{s_d}y = \mathbf{z}'_{s_1}\dots\mathbf{z}'_{s_d} \in \mathfrak{A}_\theta^{(B)} \subset \mathfrak{A}_\pi y, \\
\mathbf{w}_{\mathrm{III}} &= \pi^{s_1-1}\rho y\pi^{s_2}y\dots\pi^{s_d}y = \mathbf{z}'_{s_1}\mathbf{z}_{s_2}\dots\mathbf{z}_{s_d} \in \mathfrak{A}_\theta^{(A)} \subset \mathfrak{A}_\pi y,
\end{aligned}
$$

and

$$
\begin{aligned}
\mathfrak{z}_q[\mathbf{w}_I;t] &:= \mathbf{R}^{s_1-1}\big[\mathbf{P}[\mathbf{y}\mathbf{R}^{s_2-1}[\mathbf{P}[\mathbf{y}\cdots\mathbf{R}^{s_d-1}[\mathbf{P}[\mathbf{y}]]\cdots]]]\big](t), \\
\mathfrak{z}_q[\mathbf{w}_{\mathrm{II}};t] &:= \mathbf{R}^{s_1}\big[\mathbf{y}\mathbf{R}^{s_2}[\mathbf{y}\cdots\mathbf{R}^{s_d}[\mathbf{y}]\cdots]\big](t), \\
\mathfrak{z}_q[\mathbf{w}_{\mathrm{III}};t] &:= \mathbf{P}^{s_1-1}\Big[\mathbf{R}\big[\mathbf{y}[\mathbf{P}^{s_2}[\mathbf{y}[\mathbf{P}^{s_3}[\mathbf{y}\cdots\mathbf{P}^{s_d}[\mathbf{y}]\cdots]]]]\big]\Big](t).
\end{aligned}
$$

Then for all the types $\tau = \mathrm{I}, \mathrm{II}$ *and* III, *we have*

$$\zeta_q^\tau[\mathbf{s}] = (1-q)^w\mathfrak{z}_q[\mathbf{w}_\tau;1], \qquad \mathfrak{z}_q^\tau[\mathbf{s}] = \mathfrak{z}_q[\mathbf{w}_\tau] := \mathfrak{z}_q[\mathbf{w}_\tau;1].$$

Moreover, similar results hold for type $\tilde{\mathrm{I}}$ *q-MZVs. We may replace any of the consecutive strings* $\rho^{s_j-1}\pi$ *by a single* ρ *in* $\mathbf{w}_{\tilde{\mathrm{I}}}$, *and replace the corresponding operator string* $\mathbf{P}^{s_j-1}\mathbf{P}$ *by a single* $\mathbf{R}$.

We now apply the above to Okounkov's q-MZVs. Let $\mathbf{s} \in (\mathbb{Z}_{\geq 2})^d$. Then using Eq. (12.3) we define

$$\zeta_q^{\mathrm{O}}[\mathbf{s}] := \sum_{k_1>\cdots>k_d>0} \prod_{j=1}^{d} \frac{q^{k_j^+ s_j} + q^{k_j^- s_j}}{[k_j]^{s_j}}$$
$$= (1-q)^{|\mathbf{s}|} \sum_{k_1>\cdots>k_d>0} \prod_{j=1}^{d} \frac{q^{k_j^+ s_j} + q^{k_j^- s_j}}{(1-q^{k_j})^{s_j}}.$$

Again, its modified form is:

$$\mathfrak{z}_q^{\mathrm{O}}[\mathbf{s}] := \sum_{k_1>\cdots>k_d>0} \prod_{j=1}^{d} \frac{q^{k_j^+ s_j} + q^{k_j^- s_j}}{(1-q^{k_j})^{s_j}}.$$

Corollary 12.2.15. *For* $\mathbf{s} = (s_1, \ldots, s_d) \in \mathbb{N}^d$, *we set*

$$\mathbf{w}_{\mathrm{O}} = (\rho^{s_1^-}\pi^{s_1^+} + \rho^{s_1^+}\pi^{s_1^-})y \ldots (\rho^{s_d^-}\pi^{s_d^+} + \rho^{s_d^+}\pi^{s_d^-})y \in \mathfrak{A}_\theta^{(B)} \subset \mathfrak{A}_\pi y$$

and

$$\mathfrak{z}_q[\mathbf{w}_{\mathrm{O}}; t] = (\mathbf{R}^{s_1^-}\mathbf{P}^{s_1^+} + \mathbf{R}^{s_1^+}\mathbf{P}^{s_1^-})\big[\mathbf{y} \cdots (\mathbf{R}^{s_d^-}\mathbf{P}^{s_d^+} + \mathbf{R}^{s_d^+}\mathbf{P}^{s_d^-})[\mathbf{y}] \cdots\big](t).$$

Then we have

$$\zeta_q^{\mathrm{O}}[\mathbf{s}] = (1-q)^w \mathfrak{z}_q[\mathbf{w}_{\mathrm{O}}; 1], \qquad \mathfrak{z}_q^{\mathrm{O}}[\mathbf{s}] = \mathfrak{z}_q[\mathbf{w}_{\mathrm{O}}] := \mathfrak{z}_q[\mathbf{w}_{\mathrm{O}}; 1].$$

It is possible to obtain the shuffle relations among $\mathfrak{z}_q^{\mathrm{O}}[\mathbf{s}]$-values using Cor. 12.2.15. The stuffle relations among $\mathfrak{z}_q^{\mathrm{O}}[\mathbf{s}]$ is mentioned implicitly in Okounkov's original paper. For our modified version, they can be derived from the following fact. Let $F_n^{\mathrm{O}}(t) = (t^{n^+} + t^{n^-})/(1-t)^n$ for all $n \geq 2$. Then for all $r, s \in \mathbb{Z}_{\geq 2}$, we have

$$F_r^{\mathrm{O}}(t) \cdot F_s^{\mathrm{O}}(t) = \begin{cases} 2F_{r+s}^{\mathrm{O}}(t), & \text{if } r \text{ or } s \text{ is even;} \\ 2F_{r+s}^{\mathrm{O}}(t) + \frac{1}{2}F_{r+s-2}^{\mathrm{O}}(t), & \text{if } r \text{ and } s \text{ are odd.} \end{cases}$$

For example,

$$\mathfrak{z}_q^{\mathrm{O}}[2,3]\mathfrak{z}_q^{\mathrm{O}}[2] = 2\mathfrak{z}_q^{\mathrm{O}}[2,2,3] + \mathfrak{z}_q^{\mathrm{O}}[2,3,2] + 2\mathfrak{z}_q^{\mathrm{O}}[4,3] + 2\mathfrak{z}_q^{\mathrm{O}}[2,5],$$
$$\mathfrak{z}_q^{\mathrm{O}}[2,3]\mathfrak{z}_q^{\mathrm{O}}[3] = 2\mathfrak{z}_q^{\mathrm{O}}[2,3,3] + \mathfrak{z}_q^{\mathrm{O}}[3,2,3] + 2\mathfrak{z}_q^{\mathrm{O}}[5,3] + 2\mathfrak{z}_q^{\mathrm{O}}[2,6] + \frac{1}{2}\mathfrak{z}_q^{\mathrm{O}}[2,4].$$

12.2.5 *q-Shuffle Relations*

In contrast to the MZV case, the q-shuffle product is much more difficult to define than the q-stuffle product. In this section we will use the Rota–Baxter algebra approach to define this for type $\tilde{\mathrm{I}}$, $\mathrm{I\!I}$, and $\mathrm{I\!I\!I}$ q-MZVs.

The *q-shuffle product* on $\mathfrak{A}_\pi$ is defined recursively as follows: for any words $\mathbf{u},\mathbf{v}\in X_\pi^*$ we define $1\sqcup\!\sqcup\,\mathbf{u}=\mathbf{u}\sqcup\!\sqcup\,1=\mathbf{u}$ and

$$\begin{aligned}(y\mathbf{u})\sqcup\!\sqcup\,\mathbf{v}=\mathbf{u}\sqcup\!\sqcup\,(y\mathbf{v})&=y(\mathbf{u}\sqcup\!\sqcup\,\mathbf{v}), &(12.38)\\ \pi\mathbf{u}\sqcup\!\sqcup\,\pi\mathbf{v}&=\pi(\mathbf{u}\sqcup\!\sqcup\,\pi\mathbf{v})+\pi(\pi\mathbf{u}\sqcup\!\sqcup\,\mathbf{v})-\pi(\mathbf{u}\sqcup\!\sqcup\,\mathbf{v}), &(12.39)\\ \delta\mathbf{u}\sqcup\!\sqcup\,\delta\mathbf{v}&=\mathbf{u}\sqcup\!\sqcup\,\delta\mathbf{v}+\delta\mathbf{u}\sqcup\!\sqcup\,\mathbf{v}-\delta(\mathbf{u}\sqcup\!\sqcup\,\mathbf{v}), &(12.40)\\ \delta\mathbf{u}\sqcup\!\sqcup\,\pi\mathbf{v}=\pi\mathbf{v}\sqcup\!\sqcup\,\delta\mathbf{u}&=\delta(\mathbf{u}\sqcup\!\sqcup\,\pi\mathbf{v})+\delta\mathbf{u}\sqcup\!\sqcup\,\mathbf{v}-\mathbf{u}\sqcup\!\sqcup\,\mathbf{v}, &(12.41)\end{aligned}$$

for any words $\mathbf{u},\mathbf{v}\in X_\pi^*$. The first equation reflects the fact that when $\mathbf{y}(t)$ is multiplied in front of either of the two factors in a product, it can be multiplied after taking the product. The other equations formalize Eq. (12.9),Eq. (12.8), Eq. (12.13), Eq. (12.14), and Eq. (12.15), respectively.

Corollary 12.2.16. *For any words $\mathbf{u},\mathbf{v}\in X_\pi^*$, we have*

$$\begin{aligned}\rho\mathbf{u}\sqcup\!\sqcup\,\rho\mathbf{v}&=\rho(\mathbf{u}\sqcup\!\sqcup\,\rho\mathbf{v})+\rho(\rho\mathbf{u}\sqcup\!\sqcup\,\mathbf{v})+\rho(\mathbf{u}\sqcup\!\sqcup\,\mathbf{v}), &(12.42)\\ \rho\mathbf{u}\sqcup\!\sqcup\,\pi\mathbf{v}&=\pi\mathbf{v}\sqcup\!\sqcup\,\rho\mathbf{u}\\ &=\rho(\rho\mathbf{u}\sqcup\!\sqcup\,\mathbf{v})+\rho(\mathbf{u}\sqcup\!\sqcup\,\rho\mathbf{v})+\rho\mathbf{u}\sqcup\!\sqcup\,\mathbf{v}+\rho(\mathbf{u}\sqcup\!\sqcup\,\mathbf{v}), &(12.43)\\ \delta\mathbf{u}\sqcup\!\sqcup\,\rho\mathbf{v}&=\rho\mathbf{v}\sqcup\!\sqcup\,\delta\mathbf{u}\\ &=\delta(\mathbf{u}\sqcup\!\sqcup\,\pi\mathbf{v})-\mathbf{u}\sqcup\!\sqcup\,\mathbf{v}=\delta(\mathbf{u}\sqcup\!\sqcup\,\rho\mathbf{v})+\delta(\mathbf{u}\sqcup\!\sqcup\,\mathbf{v})-\mathbf{u}\sqcup\!\sqcup\,\mathbf{v}. &(12.44)\end{aligned}$$

Proof. These follow easily from Eq. (12.38)–Eq. (12.41) and the relation $\rho=\pi-1$. □

Corollary 12.2.17. *For $j=1,2$, let $X_\theta^{(j)}$ and $X_\theta^{(j),*}$ be the embedding of X_θ and X_θ^* into X_π^*, respectively, by Definition 12.2.10. For any $\alpha,\beta\in X_\theta^{(j)}$ and $\mathbf{u},\mathbf{v}\in X_\theta^{(j),*}$, we have $1\sqcup\!\sqcup\,\mathbf{u}=\mathbf{u}\sqcup\!\sqcup\,1=\mathbf{u}$ and*

$$\alpha\mathbf{u}\sqcup\!\sqcup\,\beta\mathbf{v}=\alpha(\mathbf{u}\sqcup\!\sqcup\,\beta\mathbf{v})+\beta(\alpha\mathbf{u}\sqcup\!\sqcup\,\mathbf{v})+[\alpha,\beta]_j(\mathbf{u}\sqcup\!\sqcup\,\mathbf{v}),$$

where $[\alpha,\beta]_j$ is determined by $[a,b]_1=[b,a]_1=-b$, $[a,b]_2=[b,a]_2=0$ and

$$[a,a]_j=(-1)^j a,\quad [b,b]_j=-b\gamma,\quad [\alpha,\gamma]_j=[\gamma,\alpha]_j=-\alpha\gamma. \tag{12.45}$$

Proof. All of these identities follow from straightforward computation using Eqs. (12.38)–(12.44). For example,

$$\begin{aligned} b\mathbf{u} \sqcup\!\sqcup\, b\mathbf{v} &= \pi y\mathbf{u} \sqcup\!\sqcup\, \pi y\mathbf{v} \\ &= \pi(y\mathbf{u} \sqcup\!\sqcup\, \pi y\mathbf{v}) + \pi(\pi y\mathbf{u} \sqcup\!\sqcup\, y\mathbf{v}) - \pi(y\mathbf{u} \sqcup\!\sqcup\, y\mathbf{v}) \\ &= \pi y(\mathbf{u} \sqcup\!\sqcup\, \pi y\mathbf{v}) + \pi y(\pi y\mathbf{u} \sqcup\!\sqcup\, \mathbf{v}) - \pi yy(\mathbf{u} \sqcup\!\sqcup\, \mathbf{v}) \\ &= b(\mathbf{u} \sqcup\!\sqcup\, b\mathbf{v}) + b(b\mathbf{u} \sqcup\!\sqcup\, \mathbf{v}) - b\gamma(\mathbf{u} \sqcup\!\sqcup\, \mathbf{v}). \end{aligned} \tag{12.46}$$

Similarly,

$$\begin{aligned} \theta\mathbf{u} \sqcup\!\sqcup\, \theta\mathbf{v} &= \rho y\mathbf{u} \sqcup\!\sqcup\, \rho y\mathbf{v} \\ &= \rho(y\mathbf{u} \sqcup\!\sqcup\, \rho y\mathbf{v}) + \rho(\rho y\mathbf{u} \sqcup\!\sqcup\, y\mathbf{v}) + \rho(y\mathbf{u} \sqcup\!\sqcup\, y\mathbf{v}) \\ &= \rho y(\mathbf{u} \sqcup\!\sqcup\, \rho y\mathbf{v}) + \rho y(\rho y\mathbf{u} \sqcup\!\sqcup\, \mathbf{v}) + \rho yy(\mathbf{u} \sqcup\!\sqcup\, \mathbf{v}) \\ &= \theta(\mathbf{u} \sqcup\!\sqcup\, \theta\mathbf{v}) + \theta(\theta\mathbf{u} \sqcup\!\sqcup\, \mathbf{v}) + \theta\gamma(\mathbf{u} \sqcup\!\sqcup\, \mathbf{v}). \end{aligned} \tag{12.47}$$

The rest of the proof is left to the interested reader. □

Proposition 12.2.18. *The algebra* $(\mathfrak{A}_\pi, \sqcup\!\sqcup)$ *is commutative and associative.*

Proof. Proving commutativity is routine and is thus left to the interested reader as an exercise. The associativity relation $(\mathbf{u} \sqcup\!\sqcup\, \mathbf{v}) \sqcup\!\sqcup\, \mathbf{w} = \mathbf{u} \sqcup\!\sqcup\, (\mathbf{v} \sqcup\!\sqcup\, \mathbf{w})$ is much harder. We now prove it by induction on $|\mathbf{u}| + |\mathbf{v}| + |\mathbf{w}|$.

Without loss of generality we write $\mathbf{u} = \alpha\mathbf{a}$, $\mathbf{v} = \beta\mathbf{b}$, and $\mathbf{w} = \gamma\mathbf{c}$ where $\alpha, \beta, \gamma \in \{\pi, \delta, y\}$ and $\mathbf{a}, \mathbf{b}, \mathbf{c} \in X_\pi^*$. The 27 different cases can all be reduced to the following 5 cases by commutativity:

(1) **One of the letters α, β or γ is a y.** By inductive assumption and Eq. (12.38):

$$\begin{aligned} (y\mathbf{a} \sqcup\!\sqcup\, \mathbf{v}) \sqcup\!\sqcup\, \mathbf{u} &= \big(y(\mathbf{a} \sqcup\!\sqcup\, \mathbf{v})\big) \sqcup\!\sqcup\, \mathbf{u} = y\big((\mathbf{a} \sqcup\!\sqcup\, \mathbf{v}) \sqcup\!\sqcup\, \mathbf{u}\big) \\ &= y\big(\mathbf{a} \sqcup\!\sqcup\, (\mathbf{v} \sqcup\!\sqcup\, \mathbf{u})\big) = y\mathbf{a} \sqcup\!\sqcup\, (\mathbf{v} \sqcup\!\sqcup\, \mathbf{u}). \end{aligned}$$

Hence associativity in this case is proved.

(2) **One π and two δ's.** By commutativity we may assume $\alpha = \beta = \delta$ and $\gamma = \pi$. By Eq. (12.40) and Eq. (12.41)

$$\begin{aligned} &(\delta\mathbf{a} \sqcup\!\sqcup\, \delta\mathbf{b}) \sqcup\!\sqcup\, \pi\mathbf{c} \\ &= \big(\mathbf{a} \sqcup\!\sqcup\, \delta\mathbf{b} + \delta\mathbf{a} \sqcup\!\sqcup\, \mathbf{b} - \delta(\mathbf{a} \sqcup\!\sqcup\, \mathbf{b})\big) \sqcup\!\sqcup\, \pi\mathbf{c} \\ &= \mathbf{a} \sqcup\!\sqcup\, \delta\mathbf{b} \sqcup\!\sqcup\, \pi\mathbf{c} + \mathbf{b} \sqcup\!\sqcup\, \delta\mathbf{a} \sqcup\!\sqcup\, \pi\mathbf{c} - \delta(\mathbf{a} \sqcup\!\sqcup\, \mathbf{b}) \sqcup\!\sqcup\, \pi\mathbf{c} \\ &= \mathbf{a} \sqcup\!\sqcup\, \delta(\mathbf{b} \sqcup\!\sqcup\, \pi\mathbf{c}) + \mathbf{a} \sqcup\!\sqcup\, \delta\mathbf{b} \sqcup\!\sqcup\, \mathbf{c} - \mathbf{a} \sqcup\!\sqcup\, \mathbf{b} \sqcup\!\sqcup\, \mathbf{c} + \mathbf{b} \sqcup\!\sqcup\, \delta(\mathbf{a} \sqcup\!\sqcup\, \pi\mathbf{c}) \\ &\quad + \mathbf{b} \sqcup\!\sqcup\, \delta\mathbf{a} \sqcup\!\sqcup\, \mathbf{c} - \mathbf{b} \sqcup\!\sqcup\, \mathbf{a} \sqcup\!\sqcup\, \mathbf{c} - \delta(\mathbf{a} \sqcup\!\sqcup\, \mathbf{b} \sqcup\!\sqcup\, \pi\mathbf{c}) - \delta(\mathbf{a} \sqcup\!\sqcup\, \mathbf{b}) \sqcup\!\sqcup\, \mathbf{c} + \mathbf{a} \sqcup\!\sqcup\, \mathbf{b} \sqcup\!\sqcup\, \mathbf{c} \end{aligned}$$

$$= \underbrace{\mathbf{a} ш \delta(\mathbf{b} ш \pi\mathbf{c})}_{1} + \underbrace{\mathbf{a} ш \delta\mathbf{b} ш \mathbf{c}}_{2} - \underbrace{\mathbf{a} ш \mathbf{b} ш \mathbf{c}}_{3} + \underbrace{\mathbf{b} ш \delta(\mathbf{a} ш \pi\mathbf{c})}_{4}$$
$$+ \underbrace{\mathbf{b} ш \delta\mathbf{a} ш \mathbf{c}}_{5} - \underbrace{\delta(\mathbf{a} ш \mathbf{b} ш \pi\mathbf{c})}_{6} - \underbrace{\delta(\mathbf{a} ш \mathbf{b}) ш \mathbf{c}}_{7}.$$

On the other hand,

$$\begin{aligned}
&\delta\mathbf{a} ш (\delta\mathbf{b} ш \pi\mathbf{c})\\
&\quad = \delta\mathbf{a} ш \big(\delta(\mathbf{b} ш \pi\mathbf{c}) + \delta\mathbf{b} ш \mathbf{c} - \mathbf{b} ш \mathbf{c}\big)\\
&\quad = \delta\mathbf{a} ш \delta(\mathbf{b} ш \pi\mathbf{c}) + \delta\mathbf{a} ш \delta\mathbf{b} ш \mathbf{c} - \delta\mathbf{a} ш \mathbf{b} ш \mathbf{c}\\
&\quad = \mathbf{a} ш \delta(\mathbf{b} ш \pi\mathbf{c}) + \delta\mathbf{a} ш \pi\mathbf{c} ш \mathbf{b} - \delta(\mathbf{a} ш \mathbf{b} ш \pi\mathbf{c}) + \mathbf{a} ш \delta\mathbf{b} ш \mathbf{c}\\
&\qquad + \delta\mathbf{a} ш \mathbf{b} ш \mathbf{c} - \delta(\mathbf{a} ш \mathbf{b}) ш \mathbf{c} - \delta\mathbf{a} ш \mathbf{b} ш \mathbf{c}\\
&\quad = \underbrace{\mathbf{a} ш \delta(\mathbf{b} ш \pi\mathbf{c})}_{1} + \underbrace{\delta(\mathbf{a} ш \pi\mathbf{c}) ш \mathbf{b}}_{4} + \underbrace{\delta\mathbf{a} ш \mathbf{c} ш \mathbf{b}}_{5} - \underbrace{\mathbf{a} ш \mathbf{c} ш \mathbf{b}}_{3}\\
&\qquad - \underbrace{\delta(\mathbf{a} ш \mathbf{b} ш \pi\mathbf{c})}_{6} + \underbrace{\mathbf{a} ш \delta\mathbf{b} ш \mathbf{c}}_{2} - \underbrace{\delta(\mathbf{a} ш \mathbf{b}) ш \mathbf{c}}_{7}.
\end{aligned}$$

Hence associativity in case (2) is proved.

(3) **Two π's and one δ.** By commutativity we may assume $\alpha = \beta = \pi$ and $\gamma = \delta$. By Eq. (12.39) and Eq. (12.41)

$$\begin{aligned}
&(\pi\mathbf{a} ш \pi\mathbf{b}) ш \delta\mathbf{c}\\
&= \pi(\mathbf{a} ш \pi\mathbf{b}) ш \delta\mathbf{c} + \pi(\pi\mathbf{a} ш \mathbf{b}) ш \delta\mathbf{c} - \pi(\mathbf{a} ш \mathbf{b}) ш \delta\mathbf{c}\\
&= \delta\big(\mathbf{c} ш \pi(\mathbf{a} ш \pi\mathbf{b})\big) + \delta\mathbf{c} ш \mathbf{a} ш \pi\mathbf{b} - \mathbf{a} ш \pi\mathbf{b} ш \mathbf{c}\\
&\quad + \delta\big(\mathbf{c} ш \pi(\mathbf{b} ш \pi\mathbf{a})\big) + \delta\mathbf{c} ш \mathbf{b} ш \pi\mathbf{a} - \mathbf{b} ш \pi\mathbf{a} ш \mathbf{c}\\
&\quad - \delta\big((\mathbf{c} ш \pi(\mathbf{a} ш \mathbf{b})\big) - \delta\mathbf{c} ш \mathbf{a} ш \mathbf{b} + \mathbf{c} ш \mathbf{a} ш \mathbf{b}\\
&= \delta\big(\mathbf{c} ш \pi(\mathbf{a} ш \pi\mathbf{b})\big) + \mathbf{a} ш \delta(\mathbf{c} ш \pi\mathbf{b}) + \mathbf{a} ш \delta\mathbf{c} ш \mathbf{b} - \mathbf{a} ш \mathbf{c} ш \mathbf{b}\\
&\quad - \mathbf{a} ш \pi\mathbf{b} ш \mathbf{c} + \delta\big(\mathbf{c} ш \pi(\mathbf{b} ш \pi\mathbf{a})\big) + \mathbf{b} ш \delta(\mathbf{c} ш \pi\mathbf{a}) + \mathbf{b} ш \delta\mathbf{c} ш \mathbf{a}\\
&\quad - \mathbf{b} ш \mathbf{c} ш \mathbf{a} - \mathbf{b} ш \pi\mathbf{a} ш \mathbf{c} - \delta\big((\mathbf{c} ш \pi(\mathbf{a} ш \mathbf{b})\big) - \delta\mathbf{c} ш \mathbf{a} ш \mathbf{b} + \mathbf{c} ш \mathbf{a} ш \mathbf{b}.\\
&= \underbrace{\delta\big(\mathbf{c} ш \pi(\mathbf{a} ш \pi\mathbf{b})\big)}_{1} + \underbrace{\mathbf{a} ш \delta(\mathbf{c} ш \pi\mathbf{b})}_{2} - \underbrace{\mathbf{a} ш \mathbf{c} ш \mathbf{b}}_{3}\\
&\quad - \underbrace{\mathbf{a} ш \pi\mathbf{b} ш \mathbf{c}}_{4} + \underbrace{\delta\big(\mathbf{c} ш \pi(\mathbf{b} ш \pi\mathbf{a})\big)}_{5} + \underbrace{\mathbf{b} ш \delta(\mathbf{c} ш \pi\mathbf{a})}_{6}\\
&\quad + \underbrace{\mathbf{b} ш \delta\mathbf{c} ш \mathbf{a}}_{7} - \underbrace{\mathbf{b} ш \pi\mathbf{a} ш \mathbf{c}}_{8} - \underbrace{\delta\big(\mathbf{c} ш \pi(\mathbf{a} ш \mathbf{b})\big)}_{9}.
\end{aligned}$$

On the other hand:

$$\pi\mathbf{a} ш (\pi\mathbf{b} ш \delta\mathbf{c})$$

$$
\begin{aligned}
&= \pi\mathbf{a} ⧢ \big(\delta(\mathbf{c} ⧢ \pi\mathbf{b}) + \delta\mathbf{c} ⧢ \mathbf{b} - \mathbf{c} ⧢ \mathbf{b}\big)\\
&= \delta(\mathbf{c} ⧢ \pi\mathbf{b} ⧢ \pi\mathbf{a}) + \delta(\mathbf{c} ⧢ \pi\mathbf{b}) ⧢ \mathbf{a} - \mathbf{c} ⧢ \pi\mathbf{b} ⧢ \mathbf{a}\\
&\quad + \delta(\mathbf{c} ⧢ \pi\mathbf{a}) ⧢ \mathbf{b} + \delta\mathbf{c} ⧢ \mathbf{a} ⧢ \mathbf{b} - \mathbf{a} ⧢ \mathbf{c} ⧢ \mathbf{b} - \pi\mathbf{a} ⧢ \mathbf{c} ⧢ \mathbf{b}\\
&= \underbrace{\delta\big((\mathbf{c} ⧢ \pi(\mathbf{b} ⧢ \pi\mathbf{a})\big)}_{5} + \underbrace{\delta\big(\mathbf{c} ⧢ \pi(\pi\mathbf{b} ⧢ \mathbf{a})\big)}_{1} - \underbrace{\delta\big(\mathbf{c} ⧢ \pi(\mathbf{b} ⧢ \mathbf{a})\big)}_{9}\\
&\quad + \underbrace{\delta(\mathbf{c} ⧢ \pi\mathbf{b}) ⧢ \mathbf{a}}_{2} - \underbrace{\mathbf{c} ⧢ \pi\mathbf{b} ⧢ \mathbf{a}}_{4} + \underbrace{\delta(\mathbf{c} ⧢ \pi\mathbf{a}) ⧢ \mathbf{b}}_{6}\\
&\quad + \underbrace{\delta\mathbf{c} ⧢ \mathbf{a} ⧢ \mathbf{b}}_{7} - \underbrace{\mathbf{a} ⧢ \mathbf{c} ⧢ \mathbf{b}}_{3} - \underbrace{\pi\mathbf{a} ⧢ \mathbf{c} ⧢ \mathbf{b}}_{8}.
\end{aligned}
$$

Hence associativity in case (3) is proved.

(4) $\alpha = \beta = \gamma = \delta$. By inductive assumption, commutativity, and Eq. (12.40)

$$
\begin{aligned}
&(\delta\mathbf{a} ⧢ \delta\mathbf{b}) ⧢ \delta\mathbf{c}\\
&= \big(\mathbf{a} ⧢ \delta\mathbf{b} + \delta\mathbf{a} ⧢ \mathbf{b} - \delta(\mathbf{a} ⧢ \mathbf{b})\big) ⧢ \delta\mathbf{c}\\
&= \mathbf{a} ⧢ \delta\mathbf{b} ⧢ \delta\mathbf{c} + \mathbf{b} ⧢ \delta\mathbf{a} ⧢ \delta\mathbf{c} - \delta(\mathbf{a} ⧢ \mathbf{b}) ⧢ \delta\mathbf{c}\\
&= \mathbf{a} ⧢ \big(\mathbf{b} ⧢ \delta\mathbf{c} + \delta\mathbf{b} ⧢ \mathbf{c} - \delta(\mathbf{b} ⧢ \mathbf{c})\big) - \mathbf{a} ⧢ \mathbf{b} ⧢ \delta\mathbf{c}\\
&\quad + \mathbf{b} ⧢ \big(\mathbf{a} ⧢ \delta\mathbf{c} + \delta\mathbf{a} ⧢ \mathbf{c} - \delta(\mathbf{a} ⧢ \mathbf{c})\big) - \delta(\mathbf{a} ⧢ \mathbf{b}) ⧢ \mathbf{c} + \delta(\mathbf{a} ⧢ \mathbf{b} ⧢ \mathbf{c})\\
&\quad = \underbrace{\mathbf{a} ⧢ \delta\mathbf{b} ⧢ \mathbf{c}}_{1} - \underbrace{\mathbf{a} ⧢ \delta(\mathbf{b} ⧢ \mathbf{c})}_{2} + \underbrace{\mathbf{b} ⧢ \mathbf{a} ⧢ \delta\mathbf{c}}_{3} + \underbrace{\mathbf{b} ⧢ \delta\mathbf{a} ⧢ \mathbf{c}}_{4}\\
&\quad - \underbrace{\mathbf{b} ⧢ \delta(\mathbf{a} ⧢ \mathbf{c})}_{5} - \underbrace{\delta(\mathbf{a} ⧢ \mathbf{b}) ⧢ \mathbf{c}}_{6} + \underbrace{\delta(\mathbf{a} ⧢ \mathbf{b} ⧢ \mathbf{c})}_{7}.
\end{aligned}
$$

On the other hand,

$$
\begin{aligned}
&\delta\mathbf{a} ⧢ (\delta\mathbf{b} ⧢ \delta\mathbf{c})\\
&= \delta\mathbf{a} ⧢ \big(\mathbf{b} ⧢ \delta\mathbf{c} + \delta\mathbf{b} ⧢ \mathbf{c} - \delta(\mathbf{b} ⧢ \mathbf{c})\big)\\
&= \delta\mathbf{a} ⧢ \delta\mathbf{c} ⧢ \mathbf{b} + \delta\mathbf{a} ⧢ \delta\mathbf{b} ⧢ \mathbf{c} - \delta\mathbf{a} ⧢ \delta(\mathbf{b} ⧢ \mathbf{c})\\
&= \mathbf{a} ⧢ \delta\mathbf{c} ⧢ \mathbf{b} + \delta\mathbf{a} ⧢ \mathbf{c} ⧢ \mathbf{b} - \delta(\mathbf{a} ⧢ \mathbf{c}) ⧢ \mathbf{b} + \mathbf{a} ⧢ \delta\mathbf{b} ⧢ \mathbf{c} + \delta\mathbf{a} ⧢ \mathbf{b} ⧢ \mathbf{c}\\
&\quad - \delta(\mathbf{a} ⧢ \mathbf{b}) ⧢ \mathbf{c} - \mathbf{a} ⧢ \delta(\mathbf{b} ⧢ \mathbf{c}) - \delta\mathbf{a} ⧢ \mathbf{b} ⧢ \mathbf{c} + \delta(\mathbf{a} ⧢ \mathbf{b} ⧢ \mathbf{c}).\\
&= \underbrace{\mathbf{a} ⧢ \delta\mathbf{c} ⧢ \mathbf{b}}_{3} - \underbrace{\delta(\mathbf{a} ⧢ \mathbf{c}) ⧢ \mathbf{b}}_{5} + \underbrace{\mathbf{a} ⧢ \delta\mathbf{b} ⧢ \mathbf{c}}_{1} + \underbrace{\delta\mathbf{a} ⧢ \mathbf{b} ⧢ \mathbf{c}}_{4}\\
&\quad - \underbrace{\delta(\mathbf{a} ⧢ \mathbf{b}) ⧢ \mathbf{c}}_{6} - \underbrace{\mathbf{a} ⧢ \delta(\mathbf{b} ⧢ \mathbf{c})}_{2} + \underbrace{\delta(\mathbf{a} ⧢ \mathbf{b} ⧢ \mathbf{c})}_{7}.
\end{aligned}
$$

Hence associativity in case (4) is proved.

(5) $\alpha = \beta = \gamma = \pi$. This is the ordinary stuffle case. We leave it as an exercise.

This concludes the proof of the proposition. $\square$

Corollary 12.2.19. *For $j = 1$ or 2, the algebras $(\mathfrak{A}_\theta^{(j)}, ш)$ are commutative and associative.*

Proof. This follows immediately from Prop. 12.2.18 since $(\mathfrak{A}_\theta^{(j)}, ш)$ are subalgebras of $(\mathfrak{A}_\pi, ш)$ if ш for $\mathfrak{A}_\theta^{(j)}$ is defined as in Cor. 12.2.17. $\square$

Our next theorem shows that we may use the shuffle algebra structure defined above to describe the shuffle relations among different types of the q-MZVs. Before doing so, we need to show that for each type the shuffle product really makes sense.

Proposition 12.2.20. *Embed $\mathfrak{A}_{\tilde{\mathrm{I}}}^0, \mathfrak{A}_{\mathrm{II}}^0 \subset \mathfrak{A}_\theta^{(B)}$ and $\mathfrak{A}_{\mathrm{III}}^0 \subset \mathfrak{A}_\theta^{(A)}$. Then for each type τ, if the two words $\mathbf{u}, \mathbf{v} \in \mathfrak{A}_\tau^0$ have the typical type τ form listed in Cor.* 12.2.14 *then there is an algorithm to express $\mathbf{u} ш \mathbf{v}$ using only those words in the same form.*

Proof. For type $\tilde{\mathrm{I}}$, we prove in general that for $k, l \geq 1$ and $\mathbf{u}, \mathbf{v} \in \mathfrak{A}_{\tilde{I}}^1$

$$\mathbf{z}_k\mathbf{u} ш \mathbf{z}_l\mathbf{v} \in \sum_{j \geq \min(k,l)} \mathbf{z}_j \mathfrak{A}_\theta^1. \tag{12.48}$$

If $k = l = 1$ Eq. (12.48) follows from $b = \mathbf{z}_1$ and $\alpha(b, b) = -b\gamma$. If $k = 1$ and $l \geq 2$, we can prove Eq. (12.48) by induction on l using

$$b\mathbf{u} ш \mathbf{z}_l\mathbf{v} = b(\mathbf{u} ш \mathbf{z}_l\mathbf{v}) + a(b\mathbf{u} ш \mathbf{z}_{l-1}\mathbf{v})$$

since $a\mathbf{z}_j = \mathbf{z}_{j+1}$ for all $j \geq 1$. Now we assume $k, l \geq 2$. Then

$$\begin{aligned}\mathbf{z}_k\mathbf{u} ш \mathbf{z}_l\mathbf{v} &= a(\mathbf{z}_{k-1}\mathbf{u} ш \mathbf{z}_l\mathbf{v} + \mathbf{z}_k\mathbf{u} ш \mathbf{z}_{l-1}\mathbf{v} + \mathbf{z}_{k-1}\mathbf{u} ш \mathbf{z}_{l-1}\mathbf{v}) \\ &\in \sum_{j \geq \min(k,l)} a\mathbf{z}_{j-1}\mathfrak{A}_{\tilde{\mathrm{I}}}^1\end{aligned}$$

by induction assumption. Thus Eq. (12.48) holds. Finally we prove that

$$\theta\mathbf{u} ш \mathbf{z}_k\mathbf{v}, \theta\mathbf{u} ш \theta\mathbf{v} \in \mathfrak{A}_{\tilde{\mathrm{I}}}^0$$

for all $\mathbf{u}, \mathbf{v} \in \mathfrak{A}_{\tilde{\mathrm{I}}}^1$ and $k \geq 2$. This follows easily from Eq. (12.48) and

$$\begin{aligned}\theta\mathbf{u} ш \mathbf{z}_k\mathbf{v} &= \theta(\mathbf{u} ш \mathbf{z}_k\mathbf{v}) + a(b\mathbf{u} ш \mathbf{z}_{k-1}\mathbf{v}), \\ \theta\mathbf{u} ш \theta\mathbf{v} &= \theta\big(\mathbf{u} ш \theta\mathbf{v} + \theta\mathbf{u} ш \mathbf{v} - \gamma(\mathbf{u} ш \mathbf{v})\big).\end{aligned} \tag{12.49}$$

Type Ⅱ is in fact the easiest since we can restrict ourselves to use only Eq. (12.38) and Eq. (12.42) to compute the shuffle and therefore π never comes into the picture. Clearly all such words must start with ρ and end with y.

For type Ⅲ, let's assume $\mathbf{u} = \pi^{s_1-1}\rho y\pi^{s_2}y\dots\pi^{s_d}y$ and $\mathbf{v} = \pi^{t_1-1}\rho y\pi^{t_2}y\dots\pi^{t_d}y$. If we use the definition Eq. (12.39) repeatedly then in each word appearing in $\mathbf{u} ш \mathbf{v}$ the first ρ always appears before all the y's. Such a word can be written in the form $\pi^s\rho^r y\cdots$ for some $s\in\mathbb{Z}$ and $r\geq 1$ (notice that if ρ and π are commutative). Now we can rewrite this as $\pi^s(\pi-1)^{r-1}\rho y\cdots$ and replace all the ρ's after the first y by $\pi-1$. This produces a word of typical type Ⅲ form.

We have completed the proof of the proposition. □

Theorem 12.2.21. *Embed* $\mathfrak{A}^0_{\tilde{\mathrm{I}}}, \mathfrak{A}^0_{\mathrm{II}} \subset \mathfrak{A}^{(B)}_\theta$ *and* $\mathfrak{A}^0_{\mathrm{III}} \subset \mathfrak{A}^{(A)}_\theta$. *Then for each type* τ *and for any* $\mathbf{u}_\tau, \mathbf{v}_\tau \in \mathfrak{A}^0_\tau$, *we have*

$$\mathfrak{z}_q[\mathbf{u}_\tau]\mathfrak{z}_q[\mathbf{v}_\tau] = \mathfrak{z}_q[\mathbf{u}_\tau ш \mathbf{v}_\tau]. \tag{12.50}$$

Proof. For each type τ, we observe that $\mathfrak{z}_q[\mathbf{w}_\tau;t]$ satisfies Eq. (12.50) because of the identities in Prop. 12.2.2. Then the theorem follows from the fact that $\mathfrak{z}_q[\mathbf{w}_\tau] = \mathfrak{z}_q[\mathbf{w}_\tau;1]$ for any word $\mathbf{w}_\tau \in \mathfrak{A}^0_\tau$ by Cor. 12.2.14. □

12.3 Duality Relations

The DBSFs do not contain all linear relations among the various types of the q-MZVs. Some of the missing ones can be proved by the following "duality" relations because of their similarity to the duality relations for the ordinary MZVs.

Theorem 12.3.1. *For a positive integer* k, *set*

$$\varphi_k := (-1)^k\left(\sum_{j=2}^{k}(-1)^j\mathbf{z}_j - \theta\right),$$

where $\varphi_1 = \theta = \rho y \in \mathfrak{A}^{(B)}_\theta$. *Let* $\ell\in\mathbb{N}$ *and* $\alpha_j,\beta_j\in\mathbb{Z}_{\geq 0}$ *for all* $j=1,\dots,\ell$. *Then we have*

$$\zeta_q^{\tilde{\mathrm{I}}}[\varphi_{\alpha_1+1}\gamma^{\beta_1}\cdots\varphi_{\alpha_\ell+1}\gamma^{\beta_\ell}] = \zeta_q^{\tilde{\mathrm{I}}}[\varphi_{\beta_\ell+1}\gamma^{\alpha_\ell}\cdots\varphi_{\beta_1+1}\gamma^{\alpha_1}]. \tag{12.51}$$

Proof. Note that $\gamma = y$, $\mathbf{z}_j = \rho^{j-1}\pi y$ and $\theta = \rho y$ with the embedding

$\mathfrak{A}^0_{\tilde{\mathrm{I}}} \subset \mathfrak{A}^{(B)}_\theta$. Since $\pi = \rho + 1$, for all $k \geq 1$, we have

$$\begin{aligned}\varphi_k =&(-1)^k \left(\sum_{j=2}^{k} (-1)^j \rho^{j-1}(\rho+1)y - \rho y \right) \\ =&(-1)^k \left(\sum_{j=2}^{k} (-1)^j \rho^j y + \sum_{j=2}^{k} (-1)^j \rho^{j-1} y - \rho y \right) = \rho^k y.\end{aligned} \tag{12.52}$$

Thus the theorem follows from Cor. 12.2.4 and Cor. 12.2.14. □

We see that in the expression of φ_k the letter θ appears, however, q-MZVs of the form such as $\zeta_q^I[\theta\gamma\mathbf{z}_2\gamma] = \zeta_q^I[\rho y^2 \rho^2 \pi y]$ is not really defined. In fact, it should be denoted by $\zeta_q^{\tilde{\mathrm{I}}}[\theta\gamma\mathbf{z}_2\gamma] = \zeta_q^{(1,0,1,0)}[1,0,2,0]$ (and such values always converge by Prop. 11.1.1 because of the leading 1 in the auxiliary variable $\mathbf{t}$). But, suitable $\mathbb{Q}$-linear combinations of Eq. (12.51) may lead to identities in which only $\mathbf{z}_k$'s appear. Then all terms can be written as honest ζ_q^I-values. This explains the use of two admissible structures $\mathfrak{A}^0_I$ and $\mathfrak{A}^0_{\tilde{\mathrm{I}}}$.

Similar relations for type II q-MZVs have the most aesthetic appeal and is the primary reason why we prefer to call it by "duality".

Theorem 12.3.2. *Let $\ell \in \mathbb{N}$ and $\alpha_j, \beta_j \in \mathbb{N}$ for all $j = 1, \ldots, \ell$. Then we have*

$$\zeta_q^{\mathrm{II}}[\rho^{\alpha_1} y^{\beta_1} \cdots \rho^{\alpha_\ell} y^{\beta_\ell}] = \zeta_q^{\mathrm{II}}[\rho^{\beta_\ell} y^{\alpha_\ell} \cdots \rho^{\beta_1} y^{\alpha_1}].$$

Proof. This follows from Cor. 12.2.4 and Cor. 12.2.14 immediately. □

Of course we may apply the same idea to type III q-MZVs.

Theorem 12.3.3. *Let $\ell \in \mathbb{N}$ and $\alpha_j, \beta_j \in \mathbb{N}$ for all $j = 1, \ldots, \ell$. Then we have*

$$\begin{aligned}&\zeta_q^{\mathrm{III}}[(\pi-1)^{\alpha_1-1} \rho y^{\beta_1} (\pi-1)^{\alpha_2} y^{\beta_2} \cdots (\pi-1)^{\alpha_\ell} y^{\beta_\ell}] \\ &\qquad = \zeta_q^{\mathrm{III}}[(\pi-1)^{\beta_\ell - 1} \rho y^{\alpha_\ell} (\pi-1)^{\beta_{\ell-1}} y^{\alpha_{\ell-1}} \cdots (\pi-1)^{\beta_1} y^{\alpha_1}].\end{aligned}$$

Proof. Since $\rho = \pi - 1$ the theorem follows readily from Cor. 12.2.4 and Cor. 12.2.14. □

12.4 The P-R and R-P Relations

We observe that there is often more than one way to express a q-MZV using words because of the relation $\pi = \rho + 1$. For example, replacing π by $\rho + 1$ we have

$$\pi^2 \rho y = \pi \rho^2 y + \pi \rho y = \rho^3 y + 2\rho^2 y + \rho y$$

which yields immediately the relations

$$\mathfrak{z}_q^{(1)}[3] = \mathfrak{z}_q^{(2)}[3] + \mathfrak{z}_q^{(1)}[2] = \mathfrak{z}_q^{(3)}[3] + 2\mathfrak{z}_q^{(2)}[2] + \mathfrak{z}_q^{(1)}[1].$$

We call all such relations **P-R** relations (pronounced as "P to R relations"). On the level of the q-MZVs this is a consequence of the following identity

$$\frac{q^{nt}}{(1-q^n)^s} = \frac{q^{nt} - q^{(n+1)t} + q^{(n+1)t}}{(1-q^n)^s} = \frac{q^{nt}}{(1-q^n)^{s-1}} + \frac{q^{(n+1)t}}{(1-q^n)^s}.$$

Now we consider the general type G q-MZVs $\zeta_q^{(t_1,\dots,t_d)}[s_1,\dots,s_d]$ defined by Eq. (11.1) where $1 \le t_1 \le s_1$, $0 \le t_j \le s_j$ for all $j \ge 2$. Repeatedly applications of the **P-R** relation shows that all these values can be reduced to type II q-MZVs since in the word expression we only need to use ρ and y.

On the other hand, if we use ρ by $\pi - 1$ repeatedly we may convert all type G q-MZVs $\zeta_q^{(t_1,\dots,t_d)}[s_1,\dots,s_d]$ to type III values. We call these relations **R-P** relations (pronounced as "R to P relations"). For example, from

$$\rho^3 y = (\pi^2 - 2\pi + 1)\rho y$$

we get

$$\zeta_q^{\mathrm{II}}[3] = \zeta_q^{\mathrm{III}}[3] - 2\zeta_q^{\mathrm{III}}[2] + \zeta_q^{\mathrm{III}}[1].$$

12.5 General Type G q-MZVs

All of the q-MZVs of type $\tilde{\mathrm{I}}$, II and III considered earlier in the chapter are some special forms of the q-MZVs $\zeta_q^{(t_1,\dots,t_d)}[s_1,\dots,s_d]$ defined by Eq. (11.1) where $1 \le t_1 \le s_1$, $0 \le t_j \le s_j$ for all $j \ge 2$, all of which are convergent by Prop. 11.1.1. Let $\mathfrak{A}_{\mathrm{G}}^0$ be the set of all such values which we call *(admissible) type G q-MZVs*. Similar to the above three types, we may use words to encode these values according to Thm. 12.2.12 by setting $a_j = t_j$ there. Namely, setting

$$\mathbf{w}^{\mathbf{t}}(\mathbf{s}) = \rho^{t_1}\pi^{s_1-t_1}y \cdots \rho^{t_d}\pi^{s_d-t_d}y \in X_\pi^*,$$

we have

$$\mathfrak{z}_q^{(t_1,\dots,t_d)}[s_1,\dots,s_d] = \mathfrak{z}_q[\mathbf{w}^{\mathbf{t}}(\mathbf{s})]$$

where the right-hand side is defined by Eq. (12.30). The shuffle product structure is reflected by $(X_\pi^*, ш)$ where the ш is defined by Eqs. (12.38), (12.39), (12.42) and (12.43).

Proposition 12.5.1. *For all* $\mathbf{u}, \mathbf{v} \in \mathfrak{A}_{\mathrm{G}}^0$, *we have* $\mathbf{u} ш \mathbf{v} \in \mathfrak{A}_{\mathrm{G}}^0$.

Proof. Note that admissible words in $\mathfrak{A}_{\mathrm{G}}^0$ must end with y and have at least one ρ before the first y. Moreover, the converse is also true. This is rather straightforward if we use the **P-R** relations repeatedly to get rid of all the π's.

Now, by using the definition of ш it is not hard to see that $\mathbf{u} ш \mathbf{v}$ ends with y and has at least one ρ before the first y if both $\mathbf{u}$ and $\mathbf{v}$ are admissible. So $\mathbf{u} ш \mathbf{v} \in \mathfrak{A}_{\mathrm{G}}^0$ and the proposition is proved. □

To define the stuffle product, we set

$$\mathsf{Y}_{\mathrm{G}} := \{\mathsf{z}_{t,s} \mid t, s \in \mathbb{Z}_{\geq 0}, t \leq s\},$$

and let $\mathfrak{A}_{\mathrm{G}}$ be the noncommutative polynomial $\mathbb{Q}$-algebra of words of $\mathsf{Y}_{\mathrm{G}}^*$ built on the alphabet Y_{G}. Define the type G-admissible words as those in

$$\mathfrak{A}_{\mathrm{G}}^0 := \bigcup_{1\leq t\leq s} \mathsf{z}_{t,s}\mathfrak{A}_{\mathrm{G}}.$$

We can regard $\mathfrak{A}_{\mathrm{G}}$ as a subalgebra of X_π^* by setting $\mathsf{z}_{t,s} = \rho^t\pi^{s-t}y$. Then stuffle product $*_{\mathrm{G}}$ on $\mathfrak{A}_{\mathrm{G}}^0$ can be defined inductively as follows. For any words $\mathbf{u}, \mathbf{v} \in \mathfrak{A}_{\mathrm{G}}^0$ and letters $\mathsf{z}_{t,s}, \mathsf{z}_{t',s'} \in \mathsf{Y}_{\mathrm{G}}$ with $1 \leq t \leq s$ and $1 \leq t' \leq s'$, we set $1 *_{\mathrm{G}} \mathbf{u} = \mathbf{u} = \mathbf{u} *_{\mathrm{G}} 1$ and

$$(\mathsf{z}_{t,s}\mathbf{u}) *_{\mathrm{G}} (\mathsf{z}_{t',s'}\mathbf{v}) = \mathsf{z}_{t,s}(\mathbf{u} *_{\mathrm{G}} \mathsf{z}_{t',s'}\mathbf{v}) + \mathsf{z}_{t',s'}(\mathsf{z}_{t,s}\mathbf{u} *_{\mathrm{G}} \mathbf{v}) + \mathsf{z}_{t+t',s+s'}(\mathbf{u} *_{\mathrm{G}} \mathbf{v}).$$

It is easy to show that $(\mathfrak{A}_{\mathrm{G}}^0, *_{\mathrm{G}})$ is a commutative and associative algebra.

We leave the proof of the following theorems to the interested reader. The first result clearly provides the DBSFs of type G q-MZVs.

Theorem 12.5.2. *For any* $\mathbf{u}, \mathbf{v} \in \mathfrak{A}_{\mathrm{G}}^0 \subset X_\pi^*$, *we have*

$$\mathfrak{z}_q[\mathbf{u} *_{\mathrm{G}} \mathbf{v}] = \mathfrak{z}_q[\mathbf{u} ш \mathbf{v}] = \mathfrak{z}_q[\mathbf{u}]\mathfrak{z}_q[\mathbf{v}]. \tag{12.53}$$

The duality relations are given in the cleanest form by Thm. 12.3.3 which can be translated into the following.

Theorem 12.5.3. *Let $\ell \in \mathbb{N}$ and $\alpha_j, \beta_j \in \mathbb{N}$ for all $j = 1, \dots, \ell$. For $\mathbf{s} = (\alpha_1, 0^{\beta_1 - 1}, \alpha_2, 0^{\beta_2-1}, \dots, \alpha_\ell, 0^{\beta_\ell - 1})$, we define its †-dual by*

$$\mathbf{s}^\dagger = (\beta_\ell, 0^{\alpha_\ell - 1}, \beta_{\ell-1}, 0^{\alpha_{\ell-1}-1}, \dots, \beta_1, 0^{\alpha_1 - 1}).$$

Then we have

$$\zeta_q^{\mathbf{s}}[\mathbf{s}] = \zeta_q^{\mathbf{s}^\dagger}[\mathbf{s}^\dagger].$$

By numerical computation with the help of some computer algebra system one can show that for all of the type $\tilde{\mathrm{I}}$, $\mathrm{I\!I}$ and $\mathrm{I\!I\!I}$ q-MZVs duality relations are necessary to generate some $\mathbb{Q}$-linear relations among q-MZVs that are missed by the DBSFs, at least when the weight is large enough. However, the combination of all the DBSFs and dualities are often not exhaustive yet. Sometimes, this difficulty can be overcome by increasing the weight and depth. But this seems to fail in some other cases, for example, for type $\tilde{\mathrm{I}}$ q-MZVs of weight four.

We can improve the above situation by considering the more general type G values. The advantage is that we have the new **P-R** relations which provide a lot of relations between type G q-MZVs, much more than the DBSFs and the duality combined in quantity (but may not provide new and linearly independent ones). The disadvantage is that there are too many type G values so that even when the weight is five the current computer power is too week to produce all the necessary relations. However, by using **P-R** relations all type G values can be converted to $\mathbb{Q}$-linear combinations of type $\mathrm{I\!I}$ values which can be handled by computers much easier.

12.6 Application to Okounkov's Conjecture

Using Cor. 12.2.15 we may regard Okounkov's q-MZVs as $\mathbb{Q}$-linear combinations of the q-MZVs $\mathfrak{z}_q^{\mathbf{t}}[\mathbf{s}]$ for suitable auxiliary variable $\mathbf{t}$. Further by using the **P-R** relations we may further reduce this to type $\mathrm{I\!I}$ q-MZVs where we don't need the letter π.

Table 12.1: Dimension of type O q-MZVs, proved rigorously for $w \le 6$ and numerically for $w \le 12$. All Re $= \mathsf{DS}^{\mathrm{O}}_{\le w} \cup \mathsf{DU}^{\mathrm{O}}_{\le w}$.

w	2	3	4	5	6	7	8	9	10	11	12
$\sharp(\mathsf{W})^{\mathrm{O}}_{\leqslant w}$	1	2	4	7	12	20	33	54	88	143	232
$\dim \mathsf{Z}^{\mathrm{O}}_{\leqslant w}$	1	2	4	7	11	18	27	42	63	95	142
dim All	0	0	0	0	1	2	6	12	25	48	90

Applying the same idea as above it is possible to verify part (ii) of the following Okounkov's Conjecture, at least when the weight is small.

Conjecture 12.6.1. *Let $q\mathsf{Z}^{\mathrm{O}}$ be the $\mathbb{Q}$-algebra generated by $\mathfrak{z}_q^{\mathrm{O}}[\mathbf{s}]$, where $\mathbf{s} \in (\mathbb{N}_{\geq 2})^d$ and $d \geq 1$. Then*

(i) *The algebra $q\mathsf{Z}^{\mathrm{O}}$ is spanned by $\mathfrak{z}_q^{\mathrm{O}}[\mathbf{s}]$ with $2 \leq s_i \leq 5$, $\mathbb{Z}/2$-graded by weight, and stable under the operator $\mathfrak{D}$ that increases the weight by 2.*
(ii) *Let $q\mathsf{Z}^{\mathrm{O}}_{\leq w}$ be the $\mathbb{Q}$-vector space generated by $\mathfrak{z}_q^{\mathrm{O}}[\mathbf{s}]$, $|\mathbf{s}| \leq w$. Then*

$$\sum_{w=0}^{\infty} t^w \dim q\mathsf{Z}^{\mathrm{O}}_{\leq w} = \frac{1}{1-t-t^2+t^6+t^8-t^{13}} - \frac{1}{1-t}$$
$$= t^2+2t^3+4t^4+7t^5+11t^6+18t^7+27t^8+42t^9+63t^{10}+95t^{11}+142t^{12}+O(t^{13}).$$

For example, one can verify all of the following $\mathbb{Q}$-linearly independent relations in the lower weight cases up to q^{100}. One can also rigorously prove the first identity given by Eq. (12.54) involving only weight four and six values by using the relations we have found for type II q-MZVs ($\mathfrak{z} = \mathfrak{z}^{\mathrm{O}}$):

$$\begin{aligned}
4\mathfrak{z}[6] &= \mathfrak{z}[2,2] + 12\mathfrak{z}[3,3] - 6\mathfrak{z}[4,2], \qquad (12.54)\\
4\mathfrak{z}[7] &= \mathfrak{z}[2,3] + \mathfrak{z}[3,2] + 8\mathfrak{z}[3,4] + 6\mathfrak{z}[4,3] - 4\mathfrak{z}[5,2],\\
\mathfrak{z}[8] &= \mathfrak{z}[2,4] - \mathfrak{z}[6] + 2\mathfrak{z}[3,3] + 6\mathfrak{z}[4,4],\\
9\mathfrak{z}[8] &= \mathfrak{z}[6] - 6\mathfrak{z}[3,3] + 3\mathfrak{z}[4,2] + 20\mathfrak{z}[3,5] + 16\mathfrak{z}[5,3] - 10\mathfrak{z}[6,2],\\
\mathfrak{z}[8] &= 2\mathfrak{z}[2,6] - \mathfrak{z}[6] + 2\mathfrak{z}[3,3] + 4\mathfrak{z}[3,5] - 16\mathfrak{z}[5,3]\\
&\quad - 6\mathfrak{z}[2,3,3] + 3\mathfrak{z}[2,4,2] - 6\mathfrak{z}[3,2,3] - 3\mathfrak{z}[4,2,2],\\
4\mathfrak{z}[3,6] &= \mathfrak{z}[2,5] + 4\mathfrak{z}[5,2] + 3\mathfrak{z}[3,4] + 6\mathfrak{z}[4,5] + 8\mathfrak{z}[5,4] + 2\mathfrak{z}[7,2],\\
8\mathfrak{z}[9] &= \mathfrak{z}[3,4] - 5\mathfrak{z}[2,5] - 8\mathfrak{z}[5,2] - 30\mathfrak{z}[4,5] - 2\mathfrak{z}[4,3] - 36\mathfrak{z}[5,4] - 10\mathfrak{z}[6,3],\\
6\mathfrak{z}[4,2] &= 10\mathfrak{z}[6] + 42\mathfrak{z}[8] - 60\mathfrak{z}[2,6] - 12\mathfrak{z}[3,3] - 120\mathfrak{z}[3,5] + 312\mathfrak{z}[5,3]\\
&\quad - 15\mathfrak{z}[2,2,2] + 180\mathfrak{z}[2,3,3] - 90\mathfrak{z}[2,4,2] + 180\mathfrak{z}[3,2,3] + 60\mathfrak{z}[3,3,2],\\
72\mathfrak{z}[9] &= 62\mathfrak{z}[5,2] + 40\mathfrak{z}[2,5] - 4\mathfrak{z}[3,4] + 40\mathfrak{z}[3,6] - 2\mathfrak{z}[4,3] + 240\mathfrak{z}[4,5]\\
&\quad + 264\mathfrak{z}[5,4] - 5\mathfrak{z}[2,2,3] - 60\mathfrak{z}[3,3,3] - 30\mathfrak{z}[4,2,3],\\
16\mathfrak{z}[9] &= 2\mathfrak{z}[3,4] - 10\mathfrak{z}[2,5] - 12\mathfrak{z}[2,7] - 8\mathfrak{z}[5,2] - 60\mathfrak{z}[4,5] - 24\mathfrak{z}[5,4]\\
&\quad + 4\mathfrak{z}[2,3,2] + 4\mathfrak{z}[3,2,2] + 3\mathfrak{z}[2,2,3] + 24\mathfrak{z}[2,3,4] + 18\mathfrak{z}[2,4,3]\\
&\quad + 12\mathfrak{z}[3,3,3] - 12\mathfrak{z}[2,5,2] + 24\mathfrak{z}[3,2,4] + 6\mathfrak{z}[4,3,2],\\
64\mathfrak{z}[9] &= 40\mathfrak{z}[2,5] + 20\mathfrak{z}[2,7] - 8\mathfrak{z}[3,4] + 44\mathfrak{z}[5,2] + 240\mathfrak{z}[4,5] + 168\mathfrak{z}[5,4]\\
&\quad + 20\mathfrak{z}[3,6] - 4\mathfrak{z}[4,3] - 5\mathfrak{z}[2,3,2] - 5\mathfrak{z}[2,2,3] - 40\mathfrak{z}[2,3,4] - 30\mathfrak{z}[2,4,3]
\end{aligned}$$

$$+\,20\mathfrak{z}[2,5,2]-5\mathfrak{z}[3,2,2]-40\mathfrak{z}[3,2,4]-100\mathfrak{z}[3,3,3]+10\mathfrak{z}[3,4,2],$$

$$\begin{aligned}56\mathfrak{z}[9] =\,& 30\mathfrak{z}[2,5]+20\mathfrak{z}[2,7]+26\mathfrak{z}[5,2]-\mathfrak{z}[3,4]+180\mathfrak{z}[4,5]+112\mathfrak{z}[5,4]\\ &+40\mathfrak{z}[3,6]-6\mathfrak{z}[4,3]-5\mathfrak{z}[2,2,3]-5\mathfrak{z}[2,3,2]-5\mathfrak{z}[3,2,2]-40\mathfrak{z}[2,3,4]\\ &+20\mathfrak{z}[5,2,2]-40\mathfrak{z}[3,2,4]-30\mathfrak{z}[2,4,3]+20\mathfrak{z}[2,5,2]-140\mathfrak{z}[3,3,3].\end{aligned}$$

Therefore, Conjecture 12.6.1 is proved rigorously up to weight 6 (inclusive), and verified numerically up to weight 12 (inclusive). The list of relations for weight 10 to 12 is too long to be presented here.

12.7 A q-Analog of Drinfeld Associator

We have seen that the MZVs can be regarded as the coefficients of the Drinfeld associator related to the KZ equation given by Eq. (4.4) whose q-analog has the following form:

$$\frac{1}{t}\frac{G(t)-G(qt)}{1-q}=\left(\frac{\pi}{t}+\frac{\pi y}{1-t}\right)G(t). \tag{12.55}$$

Note that

$$\lim_{q\to 1}\frac{1}{t}\frac{G(t)-G(qt)}{1-q}=G'(t).$$

Theorem 12.7.1. *The solution to Eq.* (12.55) *is given by*

$$G(t)=\sum_{\mathbf{w}\in X_\pi^*}\mathfrak{Z}_q^{Ш}[\mathbf{w};t]\mathbf{w}\in t\mathbb{Q}[\![t,q]\!]\langle\!\langle X_\pi^*\rangle\!\rangle. \tag{12.56}$$

where $\mathfrak{Z}_q^{Ш}[\pi^{s_1}y\ldots\pi^{s_d}y;t]=(1-q)^{|\mathbf{s}|}\mathfrak{z}_q^{Ш}[\mathbf{s};t]$ *as defined in Eq.* (12.2).

Proof. Let's consider $\mathfrak{z}_q^{Ш}[\mathbf{s};qt]$. There are two cases. If $s_1=1$ then by writing $q^{k_1}t^{k_1}=t^{k_1}-(1-q^{k_1})t^{k_1}$ we have

$$\begin{aligned}\mathfrak{z}_q^{Ш}[\mathbf{s};qt] &= \sum_{k_1>\cdots>k_d>0}\frac{q^{k_1}t^{k_1}}{(1-q^{k_1})(1-q^{k_1})^{s_2}\cdots(1-q^{k_d})^{s_d}}\\ &=\mathfrak{z}_q^{Ш}[\mathbf{s};t]-\sum_{k_2>\cdots>k_d>0}\sum_{k_1>k_2}\frac{t^{k_1}}{(1-q^{k_2})^{s_2}\cdots(1-q^{k_d})^{s_d}}\\ &=\mathfrak{z}_q^{Ш}[\mathbf{s};t]-\frac{t}{1-t}\sum_{k_2>\cdots>k_d>0}\frac{t^{k_2}}{(1-q^{k_2})^{s_2}\cdots(1-q^{k_d})^{s_d}}\\ &=\mathfrak{z}_q^{Ш}[\mathbf{s};t]-\frac{t}{1-t}\mathfrak{z}_q^{Ш}[s_2,\ldots,s_d;t].\end{aligned}$$

Similar but simpler computation shows that if $s_1>1$ then

$$\mathfrak{z}_q^{Ш}[\mathbf{s};qt]=\mathfrak{z}_q^{Ш}[\mathbf{s};t]-\mathfrak{z}_q^{Ш}[s_1-1,s_2,\ldots,s_d;t].$$

Therefore,

$$G(qt) = G(t) - (1-q)\left(\pi + \pi y\frac{t}{1-t}\right)G(t).$$

This is obviously the same as Eq. (12.55). The proof is now complete. □

Definition 12.7.2. We define the q-analog of the Drinfeld associator as

$$\Phi_q := \sum_{\mathbf{w}\in X_\pi^*} \mathfrak{z}_q^{ш}[\mathbf{w};q]\mathbf{w} \in t\mathbb{Q}[\![t,q]\!]\langle\langle X_\pi^*\rangle\rangle. \tag{12.57}$$

12.8 q-MZVs by Bachmann and Kühn

12.8.1 *Mono-Brackets*

It is well known that the Eisenstein series of weight $s \geq 4$

$$G_s(\tau) := \frac{1}{2}\sum_{(m,n)\in\mathbb{Z}^2\setminus\{(0,0)\}} \frac{1}{(m+n\tau)^s} = \zeta(s) + \frac{(2\pi i)^s}{(s-1)!}\sum_{n=1}^{\infty}\sigma_{s-1}(n)q^n, \tag{12.58}$$

where $\tau \in \mathbb{H}$ (the upper half-plane), $q = e^{2\pi i\tau}$, and $\sigma_k(n) = \sum_{d|n} d^k$ is the divisor sum function. To define the multiple Eisenstein series, we first generalize $\sigma_k(n)$ to the multiple divisor sums as follows:

$$\sigma_{r_1,\dots,r_l}(n) = \sum_{\substack{u_1v_1+\cdots+u_lv_l=n\\ u_1>\cdots>u_l\geq 1}} v_1^{r_1}\dots v_l^{r_l}. \tag{12.59}$$

Definition 12.8.1. For any $s_1,\dots,s_l \in \mathbb{N}$, the generating function for the multiple divisor sum $\sigma_{s_1-1,\dots,s_l-1}$ is defined by the formal power series

$$[s_1,\dots,s_l] := \frac{1}{(s_1-1)!\dots(s_l-1)!}\sum_{n\geq 1}\sigma_{s_1-1,\dots,s_l-1}(n)q^n \ \in \mathbb{Q}[\![q]\!],$$

which is called a *mono-bracket*. Here we call $|\mathbf{s}|$ the *weight* and l the *depth*. For convenience, we set $[\emptyset] = 1$.

It is straightforward to see that

$$[s_1,\dots,s_l] = \frac{1}{(s_1-1)!\cdots(s_l-1)!}\sum_{\substack{m_1,\dots,m_l\geq 1\\ d_1,\dots,d_l\geq 1}} d_1^{s_1-1}\cdots d_l^{s_l-1} \times q^{(m_1+\cdots+m_l)d_1+(m_2+\cdots+m_l)d_2+\cdots+m_ld_l}. \tag{12.60}$$

By the Prop. 12.8.3, we will see that the mono-bracket $[\mathbf{s}]$ also provides a q-analog of MZV $\zeta(\mathbf{s})$. First, we define the sth modified Eulerian polynomial by

$$\frac{P_s(z)}{(1-z)^{s+1}} = \left(z\frac{\mathrm{d}}{\mathrm{d}z}\right)^s \frac{z}{1-z} = \sum_{k=1}^{\infty} k^s z^k \tag{12.61}$$

where $P_s(z) = z\tilde{P}_s(z)$. We need to find its special value $P_s(1) = \tilde{P}_s(1)$ even though we do not need the full strength of the following lemma.

Lemma 12.8.2. *For any $s \in \mathbb{N}$, we have*

$$\tilde{P}_s(z) = \sum_{n=0}^{s-1} E_{s,n} z^n,$$

where the Eulerian numbers $E_{s,n}$ are defined by

$$E_{s,n} = \sum_{i=0}^{n} (-1)^i \binom{s+1}{i} (n+1-i)^s.$$

Moreover, $P_s(1) = \tilde{P}_s(1) = s!$.

Proof. Define $\mathcal{P}_s(z) = \sum_{n=0}^{s-1} E_{s,n} z^n$. Then $\mathcal{P}_1(z) = 1$. We will show that $\tilde{P}_1(z) = \mathcal{P}_s(z)$. By the definition of Eq. (12.61) we see that $\tilde{P}_0(z) = \tilde{P}_1(z) = 1$. Now it is not hard to verify that $\mathcal{P}_s(z)$ satisfy the recurrence

$$\mathcal{P}_{s+1}(z) = \mathcal{P}_s(z)(1+sz) + z(1-z)\mathcal{P}_s'(z),$$

the same as that of $\tilde{P}_s(z)$, yielding $\tilde{P}_s(z) = \mathcal{P}_s(z)$. This recurrence further implies that $\tilde{P}_s(1) = s!$, as desired. □

Proposition 12.8.3. *For any $\mathbf{s} \in \mathbb{N}^l$ with $s_1 \geq 2$, we have*

$$\lim_{q\to 1^-} (1-q)^{|\mathbf{s}|}[\mathbf{s}] = \zeta(\mathbf{s}) := \sum_{n_1 > \cdots > n_l \geq 1} \frac{1}{n_1^{s_1} \cdots n_l^{s_l}}. \tag{12.62}$$

Proof. Clear by Lemma 12.8.2. □

Definition 12.8.4. We define a *normalized polylog* by

$$\mathrm{Ln}_s(z) := \frac{\mathrm{Li}_{1-s}(z)}{\Gamma(s)},$$

where for $z \in \mathbb{C}$, $|z| < 1$ the polylog $\mathrm{Li}_s(z)$ is extended to $s \in \mathbb{C}$:

$$\mathrm{Li}_s(z) = \sum_{n\geq 1} \frac{z^n}{n^s}.$$

The normalized polylog extends to an entire function in s and to a holomorphic function in z where $|z| < 1$.

Proposition 12.8.5. *For $q \in \mathbb{C}$ with $|q| < 1$ and for all $s_1, \ldots, s_l \in \mathbb{N}$, we can write the mono-bracket as*

$$[s_1, \ldots, s_l] = \frac{1}{(s_1-1)! \ldots (s_l-1)!} \sum_{n_1 > \cdots > n_l > 0} \prod_{j=1}^{l} \frac{P_{s_j-1}(q^{n_j})}{(1-q^{n_j})^{s_j}}$$

$$= \sum_{n_1 > \cdots > n_l \geq 1} \mathrm{Ln}_{s_1}(q^{n_1}) \ldots \mathrm{Ln}_{s_l}(q^{n_l})$$

where $P_k(t)$ is the kth modified Eulerian polynomial.

Proof. By the definition of Eq. (12.61), we have

$$\sum_{n_1 > \cdots > n_l > 0} \prod_{j=1}^{l} \frac{P_{s_j-1}(q^{n_j})}{(1-q^{n_j})^{s_j}} = \sum_{n_1 > \cdots > n_l > 0} \prod_{j=1}^{l} \sum_{v_j \geq 1} v_j^{s_j-1} q^{v_j n_j}$$

$$= \sum_{n \geq 1} \sigma_{s_1-1, \ldots, s_l-1}(n) q^n.$$

This yields the first equality in the proposition. The second follows from Definition 12.8.4. □

Corollary 12.8.6. *If $s_1 \geq 2$ then*

$$\lim_{q \to 1^-} (1-q)^{|\mathbf{s}|}[\mathbf{s}] = \zeta(\mathbf{s}).$$

Proof. This follows from $P_s(1) = s!$ by Lemma 12.8.2 immediately. □

Definition 12.8.7. Let MB to be the $\mathbb{Q}$-vector space generated by $[\emptyset] = 1 \in \mathbb{Q}[\![q]\!]$ and all the mono-brackets $\Big\{[s_1, \ldots, s_l] : l \in \mathbb{N}, s_1, \ldots, s_l \in \mathbb{N}\Big\}$.

In order to study the product structure of the mono-brackets we need the following lemma.

Lemma 12.8.8. *For $a, b \in \mathbb{N}$, we have*

$$\mathrm{Ln}_a(z) \cdot \mathrm{Ln}_b(z) = \sum_{j=1}^{a} \lambda_{a,b}^{j} \, \mathrm{Ln}_j(z) + \sum_{j=1}^{b} \lambda_{b,a}^{j} \, \mathrm{Ln}_j(z) + \mathrm{Ln}_{a+b}(z),$$

where

$$\lambda_{a,b}^{j} = (-1)^{b-1} \binom{a+b-j-1}{a-j} \frac{B_{a+b-j}}{(a+b-j)!} \in \mathbb{Q}. \tag{12.63}$$

Proof. Set

$$L(X) := \sum_{k\geq 1} \mathrm{Ln}_k(z) X^{k-1} = \sum_{k\geq 1}\sum_{n\geq 1} \frac{n^{k-1} z^n}{(k-1)!} X^{k-1} = \sum_{n\geq 1} e^{nX} z^n = \frac{e^X z}{1-e^X z}.$$

Then

$$\begin{aligned} L(X)\cdot L(Y) &= \frac{1}{e^{X-Y}-1} L(X) + \frac{1}{e^{Y-X}-1} L(Y) \\ &= \sum_{n\geq 1} \frac{B_n}{n!} \Big((X-Y)^{n-1} L(X) + (Y-X)^{n-1} L(Y) \Big) + \frac{L(X)-L(Y)}{X-Y}. \end{aligned}$$

We can prove the lemma by comparing the coefficient of $X^{a-1}Y^{b-1}$. □

It is straightforward to see that the product of two mono-brackets can be found by using Prop. 12.8.5 and Lemma 12.8.8. Furthermore, similar to MZVs, mono-brackets together with their products can be formalized by using some algebra of words.

Definition 12.8.9. Let alphabet $A_{\mathsf{MB}} := \{z_j : j \in \mathbb{N}\}$, A^*_{MB} the set of words over A_{MB}, and $\mathbb{Q}\langle A_{\mathsf{MB}}\rangle$ the noncommutative polynomial algebra over $\mathbb{Q}$ generated by A^*_{MB}. Define $\diamond : A^2_{\mathsf{MB}} \to \mathbb{Q}\langle A_{\mathsf{MB}}\rangle$ by

$$z_a \diamond z_b = z_{a+b} + \sum_{j=1}^{a} \lambda^j_{a,b} z_j + \sum_{j=1}^{b} \lambda^j_{b,a} z_j, \quad \forall a,b \in \mathbb{N}.$$

For all $x, y \in A_{\mathsf{MB}}$ and $\mathbf{u}, \mathbf{v}, \mathbf{w} \in A^*_{\mathsf{MB}}$, we define a multiplication on $\mathbb{Q}\langle A_{\mathsf{MB}}\rangle$ recursively by $1 \star \mathbf{w} = \mathbf{w} \star 1 = \mathbf{w}$ and

$$x\mathbf{u} \star y\mathbf{v} := x(\mathbf{u} \star y\mathbf{v}) + y(x\mathbf{u} \star \mathbf{v}) + (x \diamond y)(\mathbf{u} \star \mathbf{v}).$$

We define the *weight* of $z_{s_1} \dots z_{s_l}$ by $s_1 + \cdots + s_l$ and the *depth* by l.

Proposition 12.8.10. *The algebra* $(\mathbb{Q}\langle A^*_{\mathsf{MB}}\rangle, \star)$ *is a commutative and associative* $\mathbb{Q}$*-algebra.*

Proof. The proof is routine by using induction on the sum of depths of the words involved. For example, see the proof of [296, Thm. 2.1]. □

Proposition 12.8.11. *Define a* $\mathbb{Q}$*-linear map* $[\ \ ] : (\mathbb{Q}\langle A^*_{\mathsf{MB}}\rangle, \star) \to (\mathsf{MB}, \cdot)$ *so that* $[z_{s_1} \dots z_{s_l}] := [s_1, \dots, s_l]$*. Then we have*

$$[\mathbf{u} \star \mathbf{v}] = [\mathbf{u}] \cdot [\mathbf{v}], \quad \forall \mathbf{u}, \mathbf{v} \in A^*_{\mathsf{MB}}.$$

Thus $[\ \ ]$ *is a* $\mathbb{Q}$*-algebra homomorphism.*

Proof. This follows by the same argument as in the MZV case by Lemma 12.8.8 and Prop. 12.8.5. □

Example 12.8.12. For any $r, s, t \in \mathbb{N}$, we have

$$[s] \cdot [t] = [s,t] + [t,s] + [s+t] + \sum_{j=1}^{s} \lambda_{s,t}^{j}[j] + \sum_{j=1}^{t} \lambda_{t,s}^{j}[j]$$

and

$$\begin{aligned}[r] \cdot [s,t] &= [r,s,t] + [s,r,t] + [s,t,r] + [s,r+t] + [r+s,t] \\ &\quad + \sum_{j=1}^{r} \lambda_{r,t}^{j}[s,j] + \sum_{j=1}^{t} \lambda_{t,r}^{j}[s,j] + \sum_{j=1}^{r} \lambda_{r,s}^{j}[j,t] + \sum_{j=1}^{s} \lambda_{s,r}^{j}[j,t].\end{aligned}$$

Considering the analogy between the MZVs and the mono-brackets we see that a missing link is the shuffle product structure. We will turn to this when we study the bi-brackets which are generalizations of mono-brackets.

We know an admissible MZV $\zeta(s_1, \dots, s_l)$ must have $s_1 \geq 2$ so we define $q\mathsf{MZ}$ to be the subset of MB generated by the mono-brackets $[s_1, \dots, s_l]$ with $s_1 \geq 2$. We expect that mono-brackets $[1, \dots]$ should behave quite differently from those in $q\mathsf{MZ}$ which is reflected in the next result.

Proposition 12.8.13. *The mono-bracket* $[1]$ *is transcendental over* $q\mathsf{MZ}$.

Proof. By Cor. 12.8.6 we see that if $s_1 \geq 2$ then near $q = 1$ we have $[\mathbf{s}] \sim \zeta(\mathbf{s})(1-q)^{-|\mathbf{s}|}$. We now claim that

$$[1] \asymp -\frac{\log(1-q)}{1-q}. \tag{12.64}$$

Indeed, by the definition

$$[1] = \sum_{n=1}^{\infty} \sigma_0(n) q^n = \sum_{m,n=1}^{\infty} q^{mn} = \sum_{n=1}^{\infty} \frac{q^n}{1-q^n}.$$

Thus $[n] < n$ implies that

$$[1] = \frac{1}{1-q} \sum_{n=1}^{\infty} \frac{q^n}{[n]} > \frac{1}{1-q} \sum_{n=1}^{\infty} \frac{q^n}{n} = -\frac{\log(1-q)}{1-q}. \tag{12.65}$$

On the other hand, define the sets S_N for $N \geq 1$ by the following. If $N = \lfloor 1 + q + \cdots + q^n \rfloor$ then $n \in S_N$. With this notation,

$$[1] \leq \frac{1}{1-q} \sum_{N=1}^{\lfloor 1/(1-q) \rfloor} \frac{1}{N} \sum_{n \in S_N} q^n.$$

Let $N_* = \min\{n : n \in S_N\}$ and $N^* = \max\{n : n \in S_N\}$. Then

$$\lfloor 1 + q + \cdots + q^{N_*} \rfloor = \lfloor 1 + q + \cdots + q^{N_*} + q^{N_*+1} + \cdots + q^{N^*} \rfloor.$$

Thus $q^{N_*+1} + \cdots + q^{N^*} < 1$. Therefore $\sum_{n \in S_N} q^n = q^{N_*} + q^{N_*+1} + \cdots + q^{N^*} < q^{N_*} + 1 < 2$ and

$$[1] < \frac{2}{1-q} \sum_{N=1}^{\lfloor 1/(1-q) \rfloor} \frac{1}{N} < \frac{2}{1-q} \left(1 + \int_1^{\frac{1}{1-q}} \frac{\mathrm{d}x}{x} \right) = \frac{2(1 - \log(1-q))}{1-q}.$$

Together with Eq. (12.65) this yields Eq. (12.64) which implies that [1] is transcendental over $q\mathsf{MZ}$. □

Proposition 12.8.14. MB *can be regarded as a polynomial ring over* $q\mathsf{MZ}$ *in the variable* $[1]$, *i.e.,* $\mathsf{MB} = q\mathsf{MZ}[[1]]$.

Proof. Let $f = [\{1\}^n, s_1, \ldots, s_{l-n}]$, with $s_1 > 1$. We now prove the proposition by induction on n. The case $n = 0$ is trivial so we assume $n \geq 1$. But by the stuffle product given by Prop. 12.8.11 and the induction

$$[1] \cdot [\{1\}^{n-1}, s_1, \ldots, s_{l-n}] - nf \in q\mathsf{MZ}[[1]].$$

So $f \in q\mathsf{MZ}[[1]]$. This finishes the proof of the proposition. □

12.8.2 *A Derivation*

In this subsection we turn to the operator $\mathfrak{D} := q\frac{\mathrm{d}}{\mathrm{d}q}$ on MB and show it is a derivation and can provide some useful information of MB. For this purpose, we need to consider some finer structures of MB.

Definition 12.8.15. Let $\mathrm{Fil}^{\mathrm{W}}_{\bullet}$ denote the increasing weight filtration on MB and $\mathrm{Fil}^{\mathrm{D}}_{\bullet}$ the increasing depth filtration, namely,

$$\mathrm{Fil}^{\mathrm{W}}_k(\mathsf{MB}) := \left\langle [\mathbf{s}] : \mathbf{s} \in \mathbb{N}^r, r \geq 1, |\mathbf{s}| \leq k \right\rangle_{\mathbb{Q}},$$
$$\mathrm{Fil}^{\mathrm{D}}_l(\mathsf{MB}) := \left\langle [\mathbf{s}] : \mathbf{s} \in \mathbb{N}^r, r \leq l \right\rangle_{\mathbb{Q}}.$$

If we consider both the depth and weight filtrations at the same time we write $\mathrm{Fil}^{\mathrm{W,D}}_{k,l} := \mathrm{Fil}^{\mathrm{W}}_k \mathrm{Fil}^{\mathrm{D}}_l$.

Theorem 12.8.16. *The operator* $\mathfrak{D} = q\frac{\mathrm{d}}{\mathrm{d}q}$ *is a derivation on* MB, *and*

$$\mathfrak{D}\left(\mathrm{Fil}^{\mathrm{W,D}}_{k,l}(\mathsf{MB})\right) \subseteq \mathrm{Fil}^{\mathrm{W,D}}_{k+2,l+1}(\mathsf{MB}).$$

More precisely, let $\oplus$ be componentwise addition and $\mathbf{e}_j = [\dots,0,1,0,\dots]$ where 1 only occurs at the jth component. Then for all $\mathbf{s}=[s_1,\dots,s_l]\in \mathsf{MB}$ we have (setting $s_{l+1}=1$ and $\tilde{\mathbf{s}}=[s_1,\dots,s_{l+1}]$)

$$\begin{aligned}\mathfrak{D}[s_1,\dots,s_l] = [2]\cdot[s_1,\dots,s_l] - \sum_{j=1}^{l+1} s_j(\tilde{\mathbf{s}}+\mathbf{e}_j) + \sum_{k=1}^{l}(l-k+1)s_k(\mathbf{s}+\mathbf{e}_k)\\ -\sum_{j=1}^{l}\sum_{k=1}^{j-1} s_k \sum_{\alpha+\beta=s_j+1}([s_1,\dots,s_{j-1},\alpha,\beta,s_{j+1},\dots,s_l]+\mathbf{e}_k)\\ -\sum_{j=1}^{l}(\alpha-1)\sum_{\alpha+\beta=s_j+2}[s_1,\dots,s_{j-1},\alpha,\beta,s_{j+1},\dots,s_l].\end{aligned}$$

Proof. For any fixed $l\in\mathbb{N}$, define the generating series

$$\begin{aligned}T(X_1,\dots,X_l) &:= \sum_{s_1,\dots,s_l\ge 1}[s_1,\dots,s_l]\prod_{j=1}^{l} X_j^{s_j-1}\\ &= \sum_{n_1,\dots,n_l\ge 1}\prod_{j=1}^{l}\frac{e^{n_jX_j}q^{n_1+\dots+n_j}}{1-q^{n_1+\dots+n_j}}.\end{aligned}$$

Then we have

$$\begin{aligned}&T(X)\cdot T(Y_1,\dots,Y_l)\\ &= \sum_{m,n_1,\dots,n_l\ge 1} e^{mX+n_1Y_1+\dots+n_lY_l}\frac{q^m}{1-q^m}\frac{q^{n_1}}{1-q^{n_1}}\cdots\frac{q^{n_1+\dots+n_l}}{1-q^{n_1+\dots+n_l}}\\ &= T(X+Y_1,\dots,X+Y_l,X)+\sum_{j=1}^{l}T(X+Y_1,\dots,X+Y_j,Y_j,\dots,Y_l)\\ &\quad + R_l - \sum_{j=1}^{l}T(X+Y_1,\dots,X+Y_j,Y_{j+1},\dots,Y_l), \qquad (12.66)\end{aligned}$$

where the last line comes from those terms when $m=n_1+\dots+n_j$ for some j so that

$$R_l = \sum_{j=1}^{l}\left(\sum_{n_1,\dots,n_l\ge 1} e^{\sum_{k=1}^{j}n_k(X+Y_k)+\sum_{k=j+1}^{l}n_kY_k}\prod_{i=1}^{l}\frac{q^{n_1+\dots+n_i}}{(1+q^{n_1+\dots+n_i})^{\delta_{i,j}+1}}\right)$$

since

$$\left(\frac{q^n}{1-q^n}\right)^2 = \frac{q^n}{(1-q^n)^2}-\frac{q^n}{1-q^n}.$$

The first line of Eq. (12.66) comes from the terms whose indices satisfy $m > n_1 + \cdots + n_l$ (put $j = l+1$) or $n_1 + \cdots + n_{j-1} < m < n_1 + \cdots + n_j$ for some j. Setting $m = n_1 + \cdots + n_{j-1} + m'$ and $n_j = m' + n'_j$ for these terms we see the sum is now over $m', n_1, \ldots, n'_j, \ldots, n_l$ which then gives $T(X+Y_1, \ldots, X+Y_j, Y_j, \ldots, Y_l)$.

Define $D(f) := \left(\frac{\mathrm{d}}{\mathrm{d}X} f\right)\big|_{X=0}$. Then

$$D(R_l) = \sum_{j=1}^{l} \sum_{n_1,\ldots,n_l \geq 1} (n_1 + \cdots + n_j) e^{\sum_{k=1}^{l} n_k Y_k} \prod_{i=1}^{l} \frac{q^{n_1+\cdots+n_i}}{(1+q^{n_1+\cdots+n_i})^{\delta_{i,j}+1}},$$

and this is exactly $\mathfrak{D}\, T(Y_1, \ldots, Y_l)$ which can be seen by induction on l and the Leibniz rule. The explicit expression of $\mathfrak{D}$ is a straightforward computation and is thus left to the interested reader. This finishes the proof of the theorem. □

Corollary 12.8.17. *The space* $q\mathsf{MZ}$ *is closed under* $\mathfrak{D}$.

Proof. These follow from straightforward computation. □

Example 12.8.18. We know that the Eulerian polynomials $P_1(t) = t$ and $P_2(t) = t(t+1)$. Therefore we have

$$\frac{t^2}{(1-t)^3} = \frac{P_2(t)}{2(1-t)^3} - \frac{1}{2}\frac{P_1(t)}{(1-t)^2}.$$

This quickly leads to

$$\mathfrak{z}_q^I[3] = [3] - \frac{1}{2}[2] = [2,1] - [2] + \mathfrak{D}[1]$$

using the first identity in Exercise 12.20.

A surprising property of the derivation $\mathfrak{D}$ is that it can provide many $\mathbb{Q}$-linear relations between ordinary MZVs. To see this we define for any $\kappa \geq 1$ the following spaces

$$\mathsf{F}_\kappa := \Big\{ \sum_{n\geq 1} a_n q^n \in \mathbb{R}[\![q]\!] : a_n = O(n^{\kappa-1}) \Big\},$$

$$\mathsf{F}'_\kappa := \Big\{ \sum_{n\geq 1} a_n q^n \in \mathbb{R}[\![q]\!] : a_n = O(n^{\kappa-1}\log^M(n)) \text{ for some } M > 0 \Big\},$$

$$\mathsf{F}_{<\kappa} := \Big\{ \sum_{n\geq 1} a_n q^n \in \mathbb{R}[\![q]\!] : \exists \varepsilon > 0 \text{ with } a_n = O(n^{\kappa-1-\varepsilon}) \Big\}.$$

Lemma 12.8.19. *Let* $\kappa > 1$. *Then*

(i) *All of* $\mathsf{F}_{<\kappa}$, F'_κ *and* F_κ *are* $\mathbb{R}$ *vector spaces.*

(ii) *We have* $\mathsf{F}_{\kappa-1} \subset \mathsf{F}_{<\kappa} \subset \mathsf{F}_{\kappa}$.
(iii) *Let* $r, s \geq 1$. *Then*

$$\mathsf{F}_{<r} \cdot \mathsf{F}_{<s} \subset \mathsf{F}_{<r+s}, \quad \mathsf{F}_{<r} \cdot \mathsf{F}_{s} \subset \mathsf{F}_{<r+s}, \quad \mathsf{F}_{r} \cdot \mathsf{F}_{s} \subset \mathsf{F}_{r+s},$$
$$\mathsf{F}_{<r} \cdot \mathsf{F}'_{s} \subset \mathsf{F}_{<r+s}, \quad \mathsf{F}_{r} \cdot \mathsf{F}'_{s} \subset \mathsf{F}'_{r+s}, \quad \mathsf{F}'_{r} \cdot \mathsf{F}'_{s} \subset \mathsf{F}'_{r+s}.$$

Proof. (i) and (ii) are straightforward. For (iii), let $f = \sum_{n\geq 1} a_n q^n \in \mathsf{F}_r$, $g = \sum_{n\geq 1} b_n q^n \in \mathsf{F}_s$. Assume $fg = \sum_{n\geq 1} c_n q^n$. Then by the definition

$$c_n = \sum_{n_1+n_2=n} a_{n_1} b_{n_2} = \sum_{n_1+n_2=n} O(n_1^{r-1})O(n_2^{s-1}) = O(n^{r+s-1}).$$

Thus $fg \in \mathsf{F}_{r+s}$. The other five cases can be proved similarly. □

For $\kappa > 1$, define the map $Z_\kappa : \mathbb{R}[\![q]\!] \to \mathbb{R}$ by

$$Z_\kappa\Big(\sum_{n\geq 1} a_n q^n\Big) = \lim_{q\to 1^-} (1-q)^\kappa \sum_{n\geq 1} a_n q^n.$$

Proposition 12.8.20. *We have*

(i) Z_κ *is a linear map from* F_κ *to* $\mathbb{R}$
(ii) $\mathsf{F}_{<\kappa} \subset \ker Z_\kappa$.
(iii) $\mathfrak{D}\,\mathsf{F}_{<\kappa-1} \subset \ker(Z_\kappa)$.

Proof. By [152, Cor. 1.2], if $-s \notin \mathbb{N}$ and $e^{-2\pi} < q < 1$ then we have

$$\mathrm{Li}_{-s}(q) = \Gamma(1+s)(-\log q)^{-s-1} + \sum_{n=0}^{\infty} \frac{\zeta(-s-n)}{n!} (\log q)^n,$$

where $\zeta(-s-n)$ is the analytic continuation of the Riemann zeta function. Hence for $\kappa > 1$ and $\varepsilon \geq 0$

$$\begin{aligned}
&\lim_{q\to 1^-} (1-q)^\kappa \sum_{n\geq 1} n^{\kappa-1-\varepsilon} q^n \\
&= \lim_{q\to 1^-} (1-q)^\kappa \,\mathrm{Li}_{\varepsilon+1-\kappa}(q) \\
&= \lim_{q\to 1^-} (1-q)^\kappa \left(\Gamma(\kappa-\varepsilon)(-\log q)^{\varepsilon-\kappa} + \sum_{n=0}^{\infty} \frac{\zeta(\varepsilon-\kappa-n)}{n!} (\log q)^n \right) \\
&= \Gamma(\kappa-\varepsilon) \lim_{q\to 1^-} (-\log q)^\varepsilon \left(\frac{q-1}{\log q} \right)^\kappa \\
&= \begin{cases} \Gamma(\kappa), & \text{if } \varepsilon = 0; \\ 0, & \text{if } \varepsilon > 0. \end{cases}
\end{aligned}$$

Now assume that for any $\varepsilon \geq 0$ we have $f = \sum_{n\geq 1} a_n q^n$ with $a_n = O(n^{\kappa-1-\varepsilon})$, *i.e.* $f \in \mathsf{F}_\kappa$ for $\varepsilon = 0$ and $f \in \mathsf{F}_{<\kappa}$ for $\varepsilon > 0$, then the calculation above gives

$$|Z_\kappa(f)| \ll Z_\kappa\Big(\sum_{n\geq 1} n^{\kappa-1-\varepsilon} q^n\Big) = \begin{cases} C\cdot\Gamma(\kappa), & \text{if } \varepsilon = 0;\\ 0, & \text{if } \varepsilon > 0.\end{cases}$$

Hence (i) and (ii) are proved.

For (iii), note that $\mathfrak{D}\sum_{n\geq 1} a_n q^n = \sum_{n\geq 1} n a_n q^n$. So by (i) we get $\mathfrak{D}\left(\mathsf{F}_{<\kappa-1}\right) \subset \mathsf{F}_{<\kappa} \subset \ker(Z_\kappa)$. □

The brackets $[s_1,\dots,s_l]$ can be considered as elements in the spaces we studied above.

Proposition 12.8.21. *Let* $\mathbf{s} = (s_1,\dots,s_l) \in \mathbb{N}^l$. *Then*

(i) *If all* $s_1,\dots,s_l > 2$, *then* $[s_1,\dots,s_l] \in \mathsf{F}_{|\mathbf{s}|}$.
(ii) *We have* $[s_1,\dots,s_l] \in \mathsf{F}'_{<|\mathbf{s}|} \subset \mathsf{F}_{<|\mathbf{s}|+1}$.
(iii) *We have*

$$[s_1,\dots,s_l] \subset \ker(Z_{|\mathbf{s}|+1}) \quad \textit{and} \quad \mathfrak{D}[s_1,\dots,s_l] \subset \ker(Z_{|\mathbf{s}|+2}).$$

Proof. If $s > 2$ the divisor sum

$$\sigma_{s-1}(n) < \sum_{d=1}^{n}\left(\frac{n}{d}\right)^{s-1} < \zeta(s-1)n^{s-1}$$

and therefore $[s] \in \mathsf{F}_s$. Then by Lemma 12.8.19(iii) we have

$$\sum_{n\geq 1} a_{s_1,\dots,s_l}(n) q^n := [s_1]\dots[s_l] \in \mathsf{F}_{s_1+\dots+s_l}.$$

Hence

$$\begin{aligned}\sigma_{s_1-1,\dots,s_l-1}(n) &= \sum_{\substack{u_1v_1+\dots+u_lv_l=n\\ u_1>\dots>u_l\geq 1}} v_1^{s_1-1}\dots v_l^{s_l-1}\\ &\leq \sum_{u_1v_1+\dots+u_lv_l=n} v_1^{s_1-1}\dots v_l^{s_l-1} = a_{s_1,\dots,s_l}(n),\end{aligned}$$

which implies $[s_1,\dots,s_l] \in \mathsf{F}_{s_1+\dots+s_l}$.

In order to show (ii) we can use the same argument as above except that we have $\sigma_0(n) = O(\log(n))$ and $\sigma_1(n) = O(n\log(n))$ so that $[1] \in \mathsf{F}'_1$ and $[2] \in \mathsf{F}'_2$. By Lemma 12.8.19(iii) we thus get $[s_1,\dots,s_l] \in \mathsf{F}'_{|\mathbf{s}|} \subset \mathsf{F}_{<|\mathbf{s}|+1}$ for $s_1,\dots,s_l \geq 1$.

Finally, (iii) is an easy consequence of Prop. 12.8.20(ii) and (iii). □

Since $\mathsf{MB} = q\mathsf{MZ}[[1]]$ by Prop. 12.8.14 we can define a map

$$Z_k^{alg} : \mathrm{Fil}_k^{\mathrm{W}}(\mathsf{MB}) \longrightarrow \mathbb{R}[T],$$
$$\sum_{j=0}^{k} g_j[1]^{k-j} \longmapsto \sum_{j=0}^{k} Z_j(g_j)T^{k-j},$$

where $g_j \in \mathrm{Fil}_j^{\mathrm{W}}(q\mathsf{MZ})$.

Proposition 12.8.22. *For all $f \in \mathrm{Fil}_k^{\mathrm{W}}(\mathsf{MB})$, we have $Z_{k+2}^{alg}(\mathfrak{D}\, f) = 0$.*

Proof. An element in $\mathrm{Fil}_k^{\mathrm{W}}(\mathsf{MB})$ can be written as $\sum_{j=0}^{k} g_j[1]^{k-j}$ with $g_j \in \mathrm{Fil}_j^{\mathrm{W}}(q\mathsf{MZ})$. The map Z_k^{alg} is linear, it therefore suffices to prove the statement for a $f \in \mathrm{Fil}_k^{\mathrm{W}}(\mathsf{MB})$ of the form $f = g_j[1]^{k-j}$. Now

$$\mathfrak{D}\, f = \mathfrak{D}\, g_j \cdot [1]^{k-j} + (k-j) g_j\, \mathfrak{D}[1] \cdot [1]^{k-j-1}.$$

Since $\mathfrak{D}[1] = [3] + \frac{1}{2}[2] - [2,1] \in q\mathsf{MZ}$ (see Exercise 12.20) and $\mathfrak{D}\, g_j \in q\mathsf{MZ}$ since $\mathfrak{D}$ is closed on $q\mathsf{MZ}$ by Cor. 12.8.17. The map Z_{k+2}^{alg} is therefore given on $\mathfrak{D}\, f$ by

$$Z_{k+2}^{alg}(\mathfrak{D}\, f) = Z_{j+2}\left(\mathfrak{D}\, g_j\right) T^{k-j} + (k+j) Z_{j+3}\left(g_j\, \mathfrak{D}[1]\right) T^{k-j-1}.$$

We get $Z_{j+3}\left(g_j\, \mathfrak{D}[1]\right) = Z_j(g_j) \cdot Z_3(\mathfrak{D}[1])$ and by Prop. 12.8.21 we obtain $Z_3\left(\mathfrak{D}[1]\right) = Z_{j+2}\left(\mathfrak{D}\, g_j\right) = 0$ from which the statement follows. □

12.8.3 *Multiple Eisenstein Series*

We have seen in § 5.9.3 that the Eisenstein series play crucial roles in the study of modular forms. From Eq. (12.58) and the definition of the mono-brackets we get

$$G_s(\tau) = \zeta(s) + (2\pi i)^s [s].$$

As we have the multiple variable version of both $\zeta(s)$ and $[s]$ it is natural to generalize $G_s(\tau)$ as follows.

Definition 12.8.23. For any $m, c \in \mathbb{Z}$ and $\tau \in \mathbb{H}$ (the upper half-plane), we write $m\tau + c \succ 0$ if $m \geq 1$ or $m = 0$ and $c \geq 1$ and $m\tau + c \succ n\tau + d$ if $m\tau + c - (n\tau + d) \succ 0$. For any $\mathbf{a} = (c_1, \ldots, c_d) \in (\mathbb{Z}/N\mathbb{Z})^d$ and $\mathbf{s} = (s_1, \ldots, s_d) \in \mathbb{N}^d$, we define the multiple Eisenstein series by

$$G_{\mathbf{s}}(\tau) := \sum_{\substack{m_1\tau + c_1 \succ \cdots \succ m_d\tau + c_d \succ 0 \\ m_j, c_j \in \mathbb{Z},\ \forall j}} \frac{1}{(m_1\tau + c_1)^{s_1} \cdots (m_d\tau + c_d)^{s_d}}. \tag{12.67}$$

We will call d the *depth* and the sum $s_1 + \cdots + s_d$ the *weight*.

Remark 12.8.24. It is not too hard to show that the series in Eq. (12.67) converges absolutely when $s_1 \geq 3$ and $s_j \geq 2$ for all $j \geq 2$. It converges conditionally if $s_j \geq 2$ for all $j \geq 1$. See [28, Prop. 2.2].

By the definition, multiple Eisenstein series clearly satisfy the stuffle relation. For example,

$$G_2(\tau)G_3(\tau) = G_{2,3}(\tau) + G_{3,2}(\tau) + G_5(\tau).$$

12.8.4 *Bi-Brackets and Double Shuffle Relations*

To find the DBSF for the brackets, Bachmann introduced the *bi-brackets* which are generalizations of the mono-brackets. For any $\mathbf{s} = (s_1, \dots, s_l), \mathbf{r} = (r_1, \dots, r_l) \in \mathbb{N}^l$, we modify his definition by

$$\begin{aligned}\begin{bmatrix}\mathbf{s}\\ \mathbf{r}\end{bmatrix} &:= \prod_{j=1}^{l} \sum_{\substack{n_1 > \cdots > n_l \geq 1\\ d_1,\dots,d_l \geq 1}} \prod_{j=1}^{l} \frac{n_j^{r_j-1}}{(r_j-1)!} \frac{\mathrm{d}_j^{s_j-1}}{(s_j-1)!} q^{n_j d_j} \\ &= \frac{1}{\prod_{j=1}^{l}(r_j-1)!(s_j-1)!} \sum_{n_1 > \cdots > n_l \geq 1} \prod_{j=1}^{l} \frac{n_j^{r_j-1} P_{s_1-1}(q^{n_j})}{(1-q^{n_j})^{s_j}}. \end{aligned} \tag{12.68}$$

Here we call $|\mathbf{s}| + |\mathbf{r}|$ the *weight* and l the *depth*. Let BB be $\mathbb{Q}$-vector space generated by $\left[\begin{smallmatrix}\emptyset\\ \emptyset\end{smallmatrix}\right] = 1 \in \mathbb{Q}[\![q]\!]$ and all the bi-brackets.

As before, to investigate the product structures of BB we resort to an algebra of words.

Definition 12.8.25. Let alphabet $A_{\mathsf{BB}} = \{z_{s,r} : s, r \in \mathbb{N}\}$, A_{BB}^* the set of words over A_{BB}, and $\mathbb{Q}\langle A_{\mathsf{BB}}\rangle$ the noncommutative polynomial algebra over $\mathbb{Q}$ generated by A_{BB}^* (including the empty word 1). Define $\diamond : A_{\mathsf{BB}}^2 \to \mathbb{Q}\langle A_{\mathsf{BB}}\rangle$ by

$$\begin{aligned} z_{s_1,r_1} \diamond z_{s_2,r_2} &:= \binom{r_1+r_2-2}{r_1-1} \\ &\times \left(z_{s_1+s_2,r_1+r_2-1} + \sum_{j=1}^{s_1} \lambda_{s_1,s_2}^j z_{j,r_1+r_2-1} + \sum_{j=1}^{s_2} \lambda_{s_2,s_1}^j z_{j,r_1+r_2-1} \right), \end{aligned}$$

where λ's are defined by Eq. (12.63). For all $x, y \in A_{\mathsf{BB}}$ and $\mathbf{u}, \mathbf{v}, \mathbf{w} \in A_{\mathsf{BB}}^*$, we define a multiplication on $\mathbb{Q}\langle A_{\mathsf{BB}}\rangle$ recursively by $1 \star \mathbf{w} = \mathbf{w} \star 1 = \mathbf{w}$ and

$$x\mathbf{u} \star y\mathbf{v} := x(\mathbf{u} \star y\mathbf{v}) + y(x\mathbf{u} \star \mathbf{v}) + (x \diamond y)(\mathbf{u} \star \mathbf{v}). \tag{12.69}$$

Proposition 12.8.26. *For all $x, y \in A_{\mathsf{BB}}$ and $\mathbf{u}, \mathbf{v} \in A_{\mathsf{BB}}^*$, we have*

$$\mathbf{u}x \star \mathbf{v}y := (\mathbf{u} \star \mathbf{v}y)x + (\mathbf{u}x \star \mathbf{v})y + (\mathbf{u} \star \mathbf{v})(x \diamond y).$$

Proof. This is straightforward. See Exercise 12.15. □

It is clear that every mono-bracket $[\mathbf{s}]$ is equal to the bi-bracket $\left[\begin{smallmatrix}\mathbf{s}\\ \mathbf{1}\end{smallmatrix}\right]$. It is expected that the opposite also holds.

Conjecture 12.8.27. *Every bi-bracket can be expressed as a* $\mathbb{Q}$*-linear combination of mono-bracket.*

The following two results are similar to those for MB so we leave the proofs to the interested reader.

Proposition 12.8.28. *The* $\mathbb{Q}$*-algebra* $(\mathbb{Q}\langle A_{\mathsf{BB}}\rangle, \star)$ *is commutative and associative.*

Proposition 12.8.29. *The* $\mathbb{Q}$*-linear map*

$$\begin{aligned}[\,\cdot\,]\colon \mathbb{Q}\langle A_{\mathsf{BB}}\rangle &\longrightarrow \mathsf{BB}\\ z_{s_1,r_1}\cdots z_{s_l,r_l} &\longmapsto \begin{bmatrix} s_1,\dots,s_l\\ r_1,\dots,r_l\end{bmatrix}\end{aligned}$$

satisfies $[w \star v] = [w]\cdot[v]$*, so that it is a* $\mathbb{Q}$*-algebra homomorphism.*

To define the shuffle product, we need to define a new class of objects which are closely related to bi-brackets.

Definition 12.8.30. Let $\mathbf{s} = (s_1,\dots,s_l), \mathbf{r} = (r_1,\dots,r_l) \in \mathbb{N}^l$. The *multiple q-zeta brackets* are defined by

$$\begin{aligned}\mathfrak{z}_q\begin{bmatrix}\mathbf{s}\\ \mathbf{r}\end{bmatrix} &:= \sum_{\substack{m_1,\dots,m_l>0\\ d_1,\dots,d_l>0}} \prod_{j=1}^{l} \frac{m_j^{r_j-1} d_j^{s_j-1} q^{(m_j+\cdots+m_l)d_j}}{(r_j-1)!(s_j-1)!}\\ &= \sum_{m_1,\dots,m_l>0} \prod_{j=1}^{l} \frac{m_j^{r_j-1} P_{s_j-1}(q^{m_j+\cdots+m_l})}{(r_j-1)!(s_j-1)!(1-q^{m_j+\cdots+m_l})^{s_j}}.\end{aligned} \tag{12.70}$$

Clearly we have

$$\begin{bmatrix}s_1\\ r_1\end{bmatrix} = \mathfrak{z}_q\begin{bmatrix}s_1\\ r_1\end{bmatrix} \quad \text{and} \quad [\mathbf{s}] = \begin{bmatrix}\mathbf{s}\\ \mathbf{1}\end{bmatrix} = \mathfrak{z}_q\begin{bmatrix}\mathbf{s}\\ \mathbf{1}\end{bmatrix}.$$

In general, every bi-bracket can be expressed using the multiple q-zeta brackets, and vice versa.

Proposition 12.8.31. *Let* $\mathbf{s} = (s_1, \ldots, s_l), \mathbf{r} = (r_1, \ldots, r_l) \in \mathbb{N}^l$. *Then we have*

$$\begin{bmatrix}\mathbf{s}\\ \mathbf{r}\end{bmatrix} = \sum_{j_2=1}^{r_1+r_2-1} \binom{j_2-1}{r_2-1} \sum_{j_3=1}^{j_2+r_3-1} \binom{j_3-1}{r_3-1} \cdots \sum_{j_l=1}^{j_{l-1}+r_l-1} \binom{j_l-1}{r_l-1}$$
$$\times \mathfrak{Z}_q\begin{bmatrix} s_1, & s_2, & \ldots, & s_{l-1}, & s_l \\ r_1+r_2-j_2, & j_2+r_3-j_3, & \ldots, & j_{l-1}+r_l-j_l, & j_l\end{bmatrix}$$

and

$$\mathfrak{Z}_q\begin{bmatrix}\mathbf{s}\\ \mathbf{r}\end{bmatrix} = \sum_{i_1=1}^{r_1}\sum_{i_2=1}^{r_2}\cdots\sum_{i_{l-1}=1}^{r_{l-1}} (-1)^{r_1+\cdots+r_{l-1}-i_1-\cdots-i_{l-1}}$$
$$\times \binom{r_1-i_1+i_2-1}{r_1-i_1} \cdots \binom{r_{l-2}-i_{l-2}+i_{l-1}-1}{r_{l-2}-i_{l-2}} \binom{r_{l-1}-i_{l-1}+r_l-1}{r_{l-1}-i_{l-1}}$$
$$\times \begin{bmatrix} s_1, & s_2, & \ldots, & s_{l-1}, & s_l \\ i_1, & r_1-i_1+i_2, & \ldots, & r_{l-2}-i_{l-2}+i_{l-1}, & r_{l-1}-i_{l-1}+r_l\end{bmatrix}.$$

Proof. The equations follow easily from the identities

$$\frac{(m+n)^{r_1-1}}{(r_1-1)!}\frac{n^{r_2-1}}{(r_2-1)!} = \sum_{j=1}^{r_1+r_2-1}\binom{j-1}{r_2-1}\frac{m^{r_1+r_2-j-1}}{(r_1+r_2-j-1)!}\frac{n^{j-1}}{(j-1)!}$$

and

$$\frac{(n-m)^{r-1}}{(r-1)!} = \sum_{i=1}^{r}(-1)^{r+i}\frac{n^{i-1}}{(i-1)!}\frac{m^{r-i}}{(r-i)!}.$$

We leave the details to the interested reader. □

The next result provides a kind of duality for bi-brackets which is quite similar to Thm. 12.2.3 in nature.

Proposition 12.8.32. *Let* $\mathbf{s} = (s_1, \ldots, s_l), \mathbf{r} = (r_1, \ldots, r_l) \in \mathbb{N}^l$. *Then we have*

$$\mathfrak{Z}_q\begin{bmatrix}\mathbf{s}\\ \mathbf{r}\end{bmatrix} = \mathfrak{Z}_q\begin{bmatrix}\overleftarrow{\mathbf{r}}\\ \overleftarrow{\mathbf{s}}\end{bmatrix}.$$

Proof. This follows from the identity

$$\sum_{i=1}^{l} d_i \sum_{j=i}^{l} m_j = \sum_{i=1}^{l} d_i' \sum_{j=i}^{l} m_j'$$

where $d_i' = m_{l+1-i}$ and $m_j' = d_{l+1-j}$. □

If we define a $\mathbb{Q}$-linear map

$$\begin{aligned} \mathfrak{z}_q\colon \mathbb{Q}\langle A_{\mathsf{BB}}\rangle &\longrightarrow \mathsf{BB} \\ z_{s_1,r_1}\cdots z_{s_l,r_l} &\longmapsto \mathfrak{z}_q\begin{bmatrix} s_1,\dots,s_l \\ r_1,\dots,r_l \end{bmatrix} \end{aligned}$$

then the duality can be also written as

$$\mathfrak{z}_q[\mathbf{w}] = \mathfrak{z}_q[\tau\mathbf{w}] \quad \forall \mathbf{w}\in A_{\mathsf{BB}}^*,$$

where τ is the anti-automorphism on $\mathbb{Q}\langle A_{\mathsf{BB}}\rangle$ such that $\tau(z_{s_1,r_1}\cdots z_{s_l,r_l}) = z_{r_l,s_l}\cdots z_{r_1,s_1}$. Moreover, Prop. 12.8.31 provides an isomorphism φ of $\mathbb{Q}\langle A_{\mathsf{BB}}\rangle$ so that

$$[\mathbf{w}] = \mathfrak{z}_q[\varphi\mathbf{w}] \quad \text{and} \quad \mathfrak{z}_q[\mathbf{w}] = [\varphi^{-1}\mathbf{w}] \qquad \forall \mathbf{w}\in A_{\mathsf{BB}}^*.$$

However, φ acts trivially on A_{MB}^* which is regarded as a subset of A_{BB}^* by identifying z_s with $z_{s,1}$.

Definition 12.8.33. For all $\mathbf{u},\mathbf{v}\in\mathbb{Q}\langle A_{\mathsf{BB}}\rangle$, we define the shuffle product

$$\mathbf{u}\sqcup\!\sqcup\,\mathbf{v} := \varphi^{-1}\tau(\tau\varphi\mathbf{u} \star \tau\varphi\mathbf{v}). \tag{12.71}$$

Theorem 12.8.34. *For all* $\mathbf{u},\mathbf{v}\in\mathbb{Q}\langle A_{\mathsf{BB}}\rangle$, *we have*

$$[\mathbf{u}\sqcup\!\sqcup\,\mathbf{v}] = [\mathbf{u}]\cdot[\mathbf{v}].$$

Hence $[-]$ *provides a homomorphism of the* $\mathbb{Q}$*-algebras* $(\mathbb{Q}\langle Z\rangle,\sqcup\!\sqcup) \longrightarrow (\mathsf{BB},\cdot)$.

Proof. This follows quickly from Prop. 12.8.29 and Prop. 12.8.32. □

To see that $\sqcup\!\sqcup$ corresponds to the shuffle product in MZV setting when $q\to 1^-$ we define an alphabet of three letters $A=\{a,b,c\}$ and regard $A_{\mathsf{BB}}\subset A^*$ by setting $z_{s,r}=a^{s-1}bc^{r-1}$ for all $r,s\in\mathbb{N}$. Clearly A_{BB}^* can be identified with the subset of A^* of words not starting with c, not ending with a and not containing the subword ac. The action of τ on this subset becomes the anti-automorphism that exchanges a and c but fixes b.

Theorem 12.8.35. *Modulo the lower weights, we have*

$$x\mathbf{u}\sqcup\!\sqcup\, y\mathbf{v} \equiv x(\mathbf{u}\sqcup\!\sqcup\, y\mathbf{v}) + y(x\mathbf{u}\sqcup\!\sqcup\,\mathbf{v}) \tag{12.72}$$

for arbitrary words $\mathbf{u},\mathbf{v}\in\mathbb{Q}1+\mathbb{Q}\langle a,b\rangle b$ *and* $x,y\in\{a,b\}$.

Proof. By Prop. 12.8.26, modulo lower weights, we have

$$\begin{aligned}\mathbf{u}z_{s_1,r_1} \bigstar \mathbf{v}z_{s_2,r_2} &\equiv (\mathbf{u} \bigstar \mathbf{v}z_{s_2,r_2})z_{s_1,r_1} + (\mathbf{u}z_{s_1,r_1} \bigstar \mathbf{v})z_{s_2,r_2},\\ \mathbf{u}z_{s_1,r_1+1} \bigstar \mathbf{v}z_{s_2,r_2} &\equiv (\mathbf{u}z_{s_1,r_1} \bigstar \mathbf{v}z_{s_2,r_2})c + (\mathbf{u}z_{s_1,r_1+1} \bigstar \mathbf{v})z_{s_2,r_2},\\ \mathbf{u}z_{s_1,r_1+1} \bigstar \mathbf{v}z_{s_2,r_2+1} &\equiv (\mathbf{u}z_{s_1,r_1} \bigstar \mathbf{v}z_{s_2,r_2+1})c + (\mathbf{u}z_{s_1,r_1+1} \bigstar \mathbf{v}z_{s_2,r_2})c.\end{aligned}$$

Hence

$$\mathbf{u}x \bigstar \mathbf{u}y \equiv (\mathbf{u} \bigstar \mathbf{v}y)x + (\mathbf{u}x \bigstar \mathbf{v})y \tag{12.73}$$

for any $\mathbf{u}, \mathbf{v} \in \mathbb{Q}\langle A_{\mathsf{BB}}\rangle$ and $x, y \in A_{\mathsf{BB}} \cup \{c\}$.

Since φ acts trivially on $\mathbb{Q}\langle A_{\mathsf{MB}}\rangle$, applying $\tau\varphi$ to both sides of Eq. (12.71) we get

$$\tau(\mathbf{u} \sqcup\!\sqcup \mathbf{v}) \equiv \tau\mathbf{u} \bigstar \tau\mathbf{v} \quad \forall \mathbf{u}, \mathbf{v} \in \mathbb{Q}\langle A_{\mathsf{MB}}\rangle.$$

By Eq. (12.73) we have

$$\begin{aligned}\tau(x\mathbf{u} \sqcup\!\sqcup y\mathbf{v}) &\equiv \tau(x\mathbf{u}) \bigstar \tau(y\mathbf{v}) \equiv (\tau\mathbf{u})\tau x \bigstar (\tau\mathbf{v})\tau y\\ &\equiv (\tau\mathbf{u} \bigstar (\tau\mathbf{v})\tau y)\tau x + ((\tau\mathbf{u})\tau x \bigstar \tau\mathbf{v})\tau y\\ &\equiv (\tau\mathbf{u} \bigstar \tau(y\mathbf{v}))\tau x + (\tau(x\mathbf{u}) \bigstar \tau\mathbf{v})\tau y\\ &\equiv (\tau(\mathbf{u} \sqcup\!\sqcup y\mathbf{v}))\tau x + (\tau(x\mathbf{u} \sqcup\!\sqcup \mathbf{v}))\tau y\\ &\equiv \tau(x(\mathbf{u} \sqcup\!\sqcup y\mathbf{v}) + y(x\mathbf{u} \sqcup\!\sqcup \mathbf{v})),\end{aligned}$$

which implies the desired result. □

12.9 Some Applications of MZ⋆V Binomial Identities

For convenience, we define more generally the q-analog of Euler (star) sums. For $\mathbf{s} = (s_1, s_2, \ldots, s_d) \in \mathbb{D}^d$ (see Eq. (7.1.1) for the definition of $\mathbb{D}$) we set

$$\zeta_q^{\mathbf{t}}[\mathbf{s}] := \sum_{k_1 > \cdots > k_d \geq 1} \prod_{j=1}^{d} \frac{\operatorname{sgn}(s_j)^{k_j} q^{k_j t_j}}{[k_j]^{|s_j|}}$$

and

$$\zeta_q^{\star,\mathbf{t}}[\mathbf{s}] := \sum_{k_1 \geq \cdots \geq k_d \geq 1} \prod_{j=1}^{d} \frac{\operatorname{sgn}(s_j)^{k_j} q^{k_j t_j}}{[k_j]^{|s_j|}}.$$

As we will only use $\zeta_q^{\star,\mathbf{1}}[\mathbf{s}]$ for the star-type we will simply write $\zeta^\star[\mathbf{s}] = \zeta_q^{\star,\mathbf{1}}[\mathbf{s}]$ in the rest of this section.

12.9.1 *Mollified Companion of q-MZVs and q-MZ⋆Vs*

Adopting the notation from Chap. 9, we define the *mollified companion* of $\zeta_q^{\mathbf{1}}[\mathbf{s}]$ associated with the admissible triple of mollifiers $(\mathbf{s};\mathbf{t};\mathbf{r})$ to be

$$\mathfrak{Z}[\mathbf{s};\mathbf{t};\mathbf{r}] := \mathfrak{Z}[s_1,\dots,s_m;t_1,\dots,t_m;r_1,\dots,r_m] = \sum_{k_1>\cdots>k_m\geq 1}\prod_{j=1}^{m}\frac{\operatorname{sgn}(s_j)^{k_j}q^{t_jk_j+Q(r_j,k_j)}(1+q^{k_j})}{[k_j]^{|s_j|}}.$$

If $m=0$, we put $\zeta^\star[\emptyset]=\mathfrak{Z}[\emptyset;\emptyset;\emptyset]=1$. Throughout this chapter the triples of mollifiers $(\mathbf{s};\mathbf{t};\mathbf{r})$ are chosen in such a way that the above multiple series always converge.

Although our primary goal is to prove q-MZ⋆V identities, in this chapter we will always work with the binomial identities for the q-MHS first. In order to obtain the corresponding q-MZ⋆V identities we need the next result.

Lemma 12.9.1. *Let $0<q<1$, $c,c_1,c_2\in\mathbb{R}$, $c>0$, and let R_k be a sequence of real numbers satisfying $|R_k|<k^{c_1}q^{c_2k}$ for all $k=1,2,\dots$. Then*

$$\lim_{n\to\infty}\sum_{k=1}^{n}q^{ck^2}\left(1-\frac{\begin{bmatrix}n\\k\end{bmatrix}}{\begin{bmatrix}n+k\\k\end{bmatrix}}\right)R_k=0.$$

Proof. For $1\leq k<n/2$ and n sufficiently large,

$$\begin{aligned}1-\frac{\begin{bmatrix}n\\k\end{bmatrix}}{\begin{bmatrix}n+k\\k\end{bmatrix}} &= 1-\frac{(q;q)_n^2}{(q;q)_{n-k}(q;q)_{n+k}}\\ &= 1-\frac{(1-q^{n-k+1})(1-q^{n-k+2})\cdots(1-q^n)}{(1-q^{n+1})(1-q^{n+2})\cdots(1-q^{n+k})}=O(q^{c_3n}),\end{aligned}$$

where c_3 is some positive constant independent of n. Therefore, we have

$$\begin{aligned}\left|\sum_{k=1}^{n/2}q^{ck^2}\left(1-\frac{\begin{bmatrix}n\\k\end{bmatrix}}{\begin{bmatrix}n+k\\k\end{bmatrix}}\right)R_k\right| &< O(q^{c_3n})\sum_{k=1}^{n/2}q^{ck^2}|R_k|\\ &< O(q^{c_3n})\sum_{k=1}^{n/2}k^{c_2}q^{ck^2+c_1k}.\end{aligned}\tag{12.74}$$

On the other side, for $n/2\leq k\leq n$, we can apply the trivial inequality

$$0<1-\frac{\begin{bmatrix}n\\k\end{bmatrix}}{\begin{bmatrix}n+k\\k\end{bmatrix}}=1-\frac{(1-q^{n-k+1})(1-q^{n-k+2})\cdots(1-q^n)}{(1-q^{n+1})(1-q^{n+2})\cdots(1-q^{n+k})}<1$$

to obtain

$$\left| \sum_{k=n/2}^{n} q^{ck^2} \left(1 - \frac{\begin{bmatrix} n \\ k \end{bmatrix}}{\begin{bmatrix} n+k \\ k \end{bmatrix}} \right) R_k \right| < q^{cn^2/4} \sum_{k=n/2}^{n} |R_k| < n^{c_4} q^{cn^2/4+c_5 n}, \quad (12.75)$$

where c_4, c_5 are some real constants independent of n. We now finish the proof by taking $n \to \infty$ using Eq. (12.74) and Eq. (12.75). □

For any admissible triple of mollifiers $(\mathbf{s}; \mathbf{t}; \mathbf{r})$, we define

$$\mathfrak{Z}^\sharp[\mathbf{s}; \mathbf{t}; \mathbf{r}] = \sum_{\underline{\mathbf{p}} \preceq (\mathbf{s}; \mathbf{t}; \mathbf{r})} \mathfrak{Z}[\underline{\mathbf{p}}].$$

Using the above lemma we may take limit $n \to \infty$ in Thm. 9.3.1 and Thm. 9.3.2 to obtain identities for q-MZ*V.

Theorem 12.9.2. *Let $n, c \in \mathbb{N}$ and $c \geq 2$. Then*

$$\zeta^\star[c] = -\mathfrak{Z}^\sharp[\overline{2}, \{1\}^{c-2}; 1, \{0\}^{c-2}; 1, \{\theta\}^{c-2}]. \quad (12.76)$$

Theorem 12.9.3. *Let $n \in \mathbb{N}$ and $\mathbf{s} = (s_1, \dots, s_d) \in \mathbb{N}^d$. Set $\delta(\mathbf{s}) = 1$ if $s_d = 1$ and $\delta(\mathbf{s}) = -1$ if $s_d > 1$. Suppose there is a triple of mollifiers $\underline{\boldsymbol{\lambda}} = (\lambda_1, \dots, \lambda_m; \widetilde{\lambda}_1, \dots, \widetilde{\lambda}_m; \widetilde{\widetilde{\lambda}}_1, \dots, \widetilde{\widetilde{\lambda}}_m)$ satisfying*

$$\widetilde{\widetilde{\lambda}}_1 \boxplus \cdots \boxplus \widetilde{\widetilde{\lambda}}_j \in \{1, 2\} \quad \forall j \geq 1,$$

such that there is an expansion of the form

$$\zeta^\star[\mathbf{s}] = \delta(\mathbf{s}) \mathfrak{Z}^\sharp[\underline{\boldsymbol{\lambda}}].$$

Then for any integers $c \geq 1$, we have

$$\zeta^\star[c, \mathbf{s}] = \delta(\mathbf{s}) \mathfrak{Z}^\sharp[\Pi_c(\underline{\boldsymbol{\lambda}})],$$

where $\Pi_c(\underline{\boldsymbol{\lambda}})$ are defined the same way as in Thm. 9.3.2.

Using the same notation as on page 290 we get the following general result concerning the q-MZ*Vs.

Theorem 12.9.4. *Let $\mathbf{s} = (s_1, \dots, s_\ell) \in \mathbb{N}^\ell$. If $s_1 > 1$ then we have*

$$\zeta^\star[\mathbf{s}] = \delta(\mathbf{s}) \mathfrak{Z}^\sharp[\mathbf{m}_1(\mathbf{s}); \widetilde{\mathbf{m}}_1(\mathbf{s}); \widetilde{\widetilde{\mathbf{m}}}_1(\mathbf{s})], \quad \text{where} \quad \delta(\mathbf{s}) = \begin{cases} 1, & \text{if } s_\ell = 1; \\ -1, & \text{if } s_\ell \geq 2. \end{cases}$$

By letting $q \to 1$ in Thm. 12.9.4 we can finally prove Thm. 10.8.1 which gives the corresponding general rule for the classical MZ*V in Chap. 10. Of course, to guarantee convergence we need to restrict $a \geq 1$ there.

12.9.2 Applications to MZVs and MZ*Vs

By applying Lemma 12.9.1 to Cor. 9.4.3 we immediately get

Corollary 12.9.5. *With the same notation as in Cor.* 9.4.3, *we have*

$$\zeta^\star[\mathbf{s}] = -\mathfrak{Z}^\sharp[\Pi(\mathbf{s})].$$

In particular, if $c_1 = c_2 = \ldots = c_\ell = 3$, *then*

$$\zeta^\star[\mathbf{s}] = \sum_{(\mathbf{p},\widetilde{\mathbf{p}})} \sum_{k_1>\cdots>k_m\ge 1} (-1)^{k_1-1} q^{\binom{k_1}{2}} \prod_{j=1}^{m} \frac{q^{\widetilde{p}_j k_j}(1+q^{k_j})}{[k_j]^{p_j}},$$

where $(\mathbf{p};\widetilde{\mathbf{p}}) \preceq \Big(2a_1+2, \mathop{\mathbf{Cat}}_{j=2}^{\ell}\{2a_j+3\}, 2a_{\ell+1}+1; \mathop{\mathbf{Cat}}_{j=1}^{\ell}\{a_j+1\}, a_{\ell+1}\Big)$.

The following result is a corollary of Exercise 9.3 and 9.4.

Corollary 12.9.6. *Let* $\mathbf{s} = (s_1, \ldots, s_d) \in \mathbb{N}^d$ *and* $s_1 > 1$. *With the same notation as in Exercise* 9.3 *and* 9.4, *we have*

$$\zeta^\star[\mathbf{s}] = \delta(s)\mathfrak{Z}^\sharp[\Pi(\mathbf{s})], \quad \text{where} \quad \delta(\mathbf{s}) = \begin{cases} 1, & \text{if } s_d = 1; \\ -1, & \text{if } s_d \ge 2. \end{cases}$$

For example, taking $\ell = 1$ and $c_1 = 3$, we get

$$\begin{aligned} &\zeta^\star[\{2\}^b, 3, \{2\}^a, 1] = \mathfrak{Z}^\sharp[\overline{2b+2}, \overline{2a+2}; b+1, a+1; 1, 1], \\ &\zeta^\star[\{2\}^{a_0}, 1, \{2\}^b, 3, \{2\}^{a_1}, 1] \\ &\quad = \mathfrak{Z}^\sharp[2a_0+1, \overline{2b+2}, \overline{2a_1+2}; a_0+1, b+1, a_1+1; 2, -1, 1], \qquad (12.77) \\ &\zeta^\star[\{2\}^b, 3, \{2\}^{a_1}, 1, \{2\}^{a_2}] \\ &\quad = -\mathfrak{Z}^\sharp[\overline{2b+2}, \overline{2a_1+2}, \overline{2a_2}; b+1, a_1+1, a_2; 1, 1, -1]. \end{aligned}$$

We can also obtain the following as the q-analog of Eq. (10.32).

Corollary 12.9.7. *Let* a, b *be two nonnegative integers. Then*

$$\begin{aligned} &\zeta^\star[\{2\}^a, 3, \{2\}^b, 1] + \zeta^\star[\{2\}^b, 3, \{2\}^a, 1] \\ &\quad = \zeta^\star[\{2\}^{a+1}]\zeta^\star[\{2\}^{b+1}] + (1-q)\mathfrak{Z}[2a+2b+3; a+b+2; 2]. \end{aligned}$$

Proof. Taking $n \to \infty$ and $c = 2$ in Eq. (9.15), we get, by Lemma 12.9.1,

$$\zeta^\star[\{2\}^a] = \mathfrak{Z}[2a; a; 1] = \sum_{k=1}^{\infty} (-1)^{k-1} \frac{q^{\binom{k}{2}+ak}(1+q^k)}{[k]^{2a}}.$$

Thus

$$\begin{aligned}
&\zeta^\star[\{2\}^a,3,\{2\}^b,1]+\zeta^\star[\{2\}^b,3,\{2\}^a,1]-\zeta^\star[\{2\}^{a+1}]\zeta^\star[\{2\}^{b+1}]\\
&\quad=\sum_{k=1}^{\infty}\frac{q^{(a+b+2)k+Q(2,k)}\big(2(1+q^k)-(1+q^k)^2\big)}{[k]^{2a+2b+4}}\\
&\quad=\sum_{k=1}^{\infty}\frac{q^{(a+b+2)k+Q(2,k)}(1+q^k)(1-q^k)}{[k]^{2a+2b+4}}\\
&\quad=(1-q)\mathfrak{z}[2a+2b+3;a+b+2;2]
\end{aligned}$$

as desired. □

As a non-trivial example of Thm. 12.9.3 we may attach a string of type $(2^a,1)$ to the front of the already treated type $(\{2\}^b,1,\{2\}^c,3,\{2\}^d,1)$ given by Eq. (12.77) and get the following q-MZ*V identity: for any nonnegative integers a,b,c,d

$$\begin{aligned}
&\zeta^\star[\{2\}^a,1,\{2\}^b,1,\{2\}^c,3,\{2\}^d,1]\\
&\quad=\mathfrak{z}^\sharp[2a+1,2b+1,\overline{2c+2},\overline{2d+2};a+1,b+1,c+1,d+1;2,0,-1,1].
\end{aligned}$$

Taking $q\to 1$ in the above, we discover the classical MZ*V identity:

$$\zeta^\star(\{2\}^a,1,\{2\}^b,1,\{2\}^c,3,\{2\}^d,1)=\zeta^\sharp(2a+1,2b+1,\overline{2c+2},\overline{2d+2}).$$

By applying Lemma 12.9.1 to Cor. 9.4.3 we immediately get the following result.

Theorem 12.9.8. *Suppose $\ell\in\mathbb{N}_0$. Let* $\mathbf{s}=\Big(\operatorname*{Cat}_{j=1}^{\ell}\big\{\{2\}^{a_j},c_j\big\},\{2\}^{a_{\ell+1}}\Big)$ *with $a_j,c_j\in\mathbb{N}_0$ and $c_j\ge 3$ for all $j\ge 1$. Then*

$$\zeta^\star[\mathbf{s}]=-\mathfrak{z}^\sharp[\Pi(\mathbf{s})],$$

where $\Pi(\mathbf{s})=\Big(\overline{2a_1+2},\{1\}^{c_1-3},\operatorname*{Cat}_{j=2}^{\ell}\big\{2a_j+3,\{1\}^{c_j-3}\big\},2a_{\ell+1}+1;$
$\operatorname*{Cat}_{j=1}^{\ell}\big\{a_j+1,\{0\}^{c_j-3}\big\},a_{\ell+1};\,1,\operatorname*{Cat}_{j=1}^{\ell}\{\theta\}^{c_j-2}\Big).$

Proof. Clear. □

12.10 Historical Notes

There is a long history of hunting for q-analogs of number theoretical objects in general. Concerning the multiple zeta values, Kaneko *et al.* first studied in detail the analytic properties and some special values of the type Iq-analog of the Riemann zeta functions in [337] in early 2000, which was generalized to multiple variable setting by the author [615] (first version was posted on arXiv in April 2003). In [615], he considered both the analytical and algebraic properties of type Iq-analog of the multiple zeta functions. He then defined their stuffle and shuffle products using the iterated Jackson q-integrals although only very few DBSFs could be found as the shuffle is quite complicated. In 2004, Bradley independently studied type Iq-MZVs in [81] by concentrating on the generalization of classical identities of the MZVs to this q-analog setting, and later, Okuda and Takeyama also looked at some of the relations among this type of the q-MZVs in [456].

Before the study of type Iq-MZVs, slightly different definitions of the q-MZVs were already presented in [499, 643]. Additionally, it is not hard to see that Schlesinger's version diverges when $|q| < 1$ but can converge if $|q| > 1$. In fact, for $\mathbf{s} = (s_1, \dots, s_d) \in \mathbb{Z}^d$

$$\mathfrak{z}_{q^{-1}}^{(0,\dots,0)}[\mathbf{s}] = (-1)^{|\mathbf{s}|}\mathfrak{z}_q^{(\mathbf{s})}[\mathbf{s}] = (-1)^{|\mathbf{s}|}\mathfrak{z}_q^{\mathrm{II}}[\mathbf{s}].$$

So it suffices to consider type II first defined in [643] in order to understand Schlesinger's q-MZVs.

The analytic properties of type Iq-MZVs are treated in [615] which also includes the stuffle relation and the shifting principle. However, the general shuffle structure was not defined until Takeyama's work in Ref. [530], from which our definition for $*_{\bar{\mathrm{I}}}$ is adopted. Unfortunately, his approach to the shuffle relations relies on some auxiliary multiple polylog functions and consequently it is very hard to see why these relations should hold.

The situation looks much better with the appearance of the joint work [120] by Castillo Medina *et al.* who generalized Chen's iterated integrals to Jackson's iterated q-integrals to study type III q-MZVs by using Rota–Baxter algebra technique, which is modified and generalized in Sec. 12.2. Further study was carried out in [515] by Singer along this line. This was later modified to treat all the general q-MZVs in [632].

In 1958, influenced by K.-T. Chen's systematic investigation the iterated integrals, Ree [481] first considered algebras in which the product is expressed in terms of shuffles. Later, G. Baxter [41] constructed algebras with an operator $\mathcal{P}$ satisfying Eq. (12.4) with $\lambda = 1$. In general, Eq. (12.4) was defined and popularized by Rota and his school. See [487–489], or the concise survey by Guo [257] and Chap. 3 of his book [258]. In [179] Ebrahimi-Fard and Guo considered MZVs and their generalizations from the point of view of Rota–Baxter algebras and derived some relations on MZVs from those in Rota–Baxter algebras.

Linear operators satisfying Eq. (12.4) in the context of Lie algebras were introduced in the 1980s and were linked to the solutions of the classical Yang–Baxter equation (after C. Yang and R. Baxter). Along this line, Connes and Kreimer applied the Hopf algebra technique and developed a renormalization theory in perturbative quantum field theory (see [147, 148]) which led to some similar works on renormalizations of the MZVs in [264] and their q-analogs in [618]. There are some further generalizations and some other approaches to the renormalization problem of the MZVs, for example, see [176, 180–183, 413, 414]. In [184], we showed that there is a group acting on the space of all renormalizations of MZVs transitively. Note that the Jackson q-integral was first defined in [315]. However, his difference operator in [316] is different from ours in this chapter.

In [530], Takeyama also noticed the set of type Iq-MZVs needs to be enlarged to type $\tilde{\mathrm{I}}$ q-MZVs which are a kind of "regularized" q-MZVs in the sense that one needs to consider some convergent versions of the q-MZVs when $s_1 = 1$ by modifying the auxiliary variables of $\mathbf{t}$. But for these type $\tilde{\mathrm{I}}$ q-MZVs themselves, the DBSFs are insufficient to provide all the $\mathbb{Q}$-linear relations and a certain "Resummation Identity" defined by Takeyama is required. In this book, we adopt the term "duality" due to its similarity to the duality relations of the ordinary MZVs.

Concerning the individual results, Thm. 12.2.11 is parallel to [120, Prop. 9] and includes [530, Thm. 1] as a special type $\tilde{\mathrm{I}}$ case. Proposition 12.2.18 is [120, Thm. 7] while Corollary 12.2.19 generalizes [530, Prop. 1]. The case for type $\tilde{\mathrm{I}}$ of Prop. 12.2.20 was established by Takeyama as [530, Prop. 2.4]. Note that Thm. 12.2.21 generalizes [530, Thm. 2] but it does not contain [120, Thm. 7] since our word representation of type $\mathbb{III}$ q-MZVs in this book is different from that of [120]. Theorem 12.3.1 was called by Takeyama as the "Resummation Identity" [530, Thm. 4].

Turning to the more complicated q-MZVs, Okounkov stated Conjecture 12.6.1 in [454] after guessing that his version of the q-MZVs is related to some q-series invariants of the Hilbert schemes parameterizing 0-dimensional subschemes of finite lengths on a nonsingular quasi-projective surface. This conjectured relation to Hilbert schemes has been proved by Qin and Yu in [476] when the surface is abelian or, in general, modulo lower weight terms. Our definition of Okounkov's q-MZV here is equal to his original q-MZVs up to a suitable 2-power. Some special cases of part (ii) of Okounkov's Conjecture 12.6.1 on the stability of the operator $\mathfrak{D}$ was proved by Bachmann and Kühn [27].

The double shuffle relations of the double Eisenstein series were first considered by Gangl *et al.* in [233]. The level two case was studied by Kaneko and Tasaka [334]. Yuan and the author extended this to arbitrary levels in [590].

The general multiple Eisenstein series first appeared in the joint works of Bachmann and Kühn [24, 26] via their mono-brackets. The stuffle relations are obvious. However, it is much harder to define the shuffle product structure since no integral expressions for the multiple Eisenstein series are available. Bachmann and Tasaka partially developed such a theory by using the Hopf algebra structure of the (generic) iterated integrals introduced by Goncharov. We refer the interested reader to their original paper [28] for details. The multiple divisor functions used in the definition of the multiple Eisenstein series are generalized to arbitrary levels in [591], through which the MZVs are extended to Hurwitz-type values and their DBSFs are studied.

To help to determine the DBSFs of the mono-brackets, Bachmann introduced the bi-brackets in [25]. Then Zudilin [644] constructed the word algebras that encode the DBSFs of the bi-brackets. In particular, his duality in Proposition 12.8.32 is exactly the partition duality of Bachmann. However, we still need to understand how to extend this to the mono-bracket, one way of which is to solve Conjecture 12.8.27 of Bachmann.

All the results in Sec. 12.9 are consequences of the corresponding results of the q-MH*S obtained in Chap. 9. They are based on the joint work in [280], which generalizes [277], the author's work [628], and the joint work [402] for the ordinary MHS, MZVs and MZ*Vs. In particular, the mollified companions essentially already appeared in [277] and [280]. The key is Lemma 12.9.1 obtained in [277] which generalizes a similar result in the classical non-q setting appearing first as [402, Lemma 4.2].

Finally, some other families of the MZV identities we encountered in Chap. 5 have been extended to the q-setting. Due to space limitation, we refer the interested reader to the Refs. [79–81, 119, 169, 277, 280, 396, 449, 456, 615].

Exercises

12.1. There are many interesting relations among type Iq-MZVs. The following is the q-analog of the cyclic sum formula (cf. Thm. 5.2.3). Let

$\mathbf{s} = (s_1, \dots, s_d) \in \mathbb{N}^d$, $s_1 \geq 2$. Then

$$\sum_{i=1}^{d} \zeta_q^I[s_i+1, s_{i+1}, \dots, s_d, s_1, \dots, s_{i-1}] = \sum_{i=1}^{d}\sum_{j=0}^{s_i-2} \zeta_q^I[s_i-j, s_{i+1}, \dots, s_d, s_1, \dots, s_{i-1}, j+1].$$

12.2. Show that the stuffle products $*_\tau$ is well defined for $\tau = \tilde{\mathrm{I}}$, II or III.

12.3. The following result gives the q-analog of Ohno's Relations (cf. Thm. 5.5.9). Let $\mathbf{s} = (s_1, \dots, s_d) \in \mathbb{N}^d$, $s_1 \geq 2$ and its dual $\mathbf{s}^* = (s'_1, \dots, s'_{d'})$ (see Eq. (5.1)). Then for any nonnegative integer l

$$\sum_{\substack{e_1+\cdots+e_d=l\\ e_i\geq 0}} \zeta_q^I[s_1+e_1, \dots, s_d+e_d] = \sum_{\substack{e'_1+\cdots+e'_{d'}=l\\ e'_i\geq 0}} \zeta_q^I[k'_1+e'_1, \dots, k'_{d'}+e'_{d'}].$$

12.4. Prove the following q-analog of Euler's decomposition formula. For all $a, b \in \mathbb{N}$,

$$\begin{aligned}\mathfrak{z}_q^I[s]\mathfrak{z}_q^I[t] &= \sum_{a=0}^{s-1}\sum_{b=0}^{s-1-a} \binom{a+t-1}{t-1}\binom{t-1}{b} \mathfrak{z}_q^I[t+a, s-a-b] \\ &+ \sum_{a=0}^{t-1}\sum_{b=0}^{t-1-a} \binom{a+s-1}{s-1}\binom{s-1}{b} \mathfrak{z}_q^I[s+a, t-a-b] \\ &- \sum_{j=1}^{\min(s,t)} \frac{(s+t-j-1)!}{(s-j)!\,(t-j)!} \cdot \frac{1}{(j-1)!} \frac{1}{(1-q)^{s+t-j}} \varphi_q[s+t-j],\end{aligned}$$

where φ_q is defined by

$$\varphi_q[n] = \sum_{k=1}^{\infty} (k-1) \frac{q^{(n-1)k}}{[k]^n}. \tag{12.78}$$

12.5. Verify Eqs. (12.27), (12.28) and (12.29) by direct computation.

12.6. Equation (12.12) is commonly considered to be the q-analog for the classical integration by parts rule:

$$R(f)R(g) = R\big(fR(g)\big) + R\big(R(f)g\big), \tag{12.79}$$

which can be seen as dual to Leibniz rule for usual derivations. For any fixed constant a, show that Eq. (12.79) holds if we set $R(f)(t) = \int_a^t f(x)\,\mathrm{d}x$.

12.7. Prove that $\lim_{q\to 1} J_q[f](t) = \int_0^t f(x)\,\mathrm{d}x$.

12.8. Prove Prop. 12.2.2 by direct computation.

12.9. Prove Cor. 12.2.17.

12.10. Show that the shifting operator defined by Eq. (12.23) is injective.

12.11. Prove the following q-analog of Euler's decomposition formula. For all $a, b \in \mathbb{N}$,

$$\begin{aligned}\mathfrak{z}_q^{\text{Ш}}[a]\mathfrak{z}_q^{\text{Ш}}[b] = & \sum_{l=0}^{a-1}\sum_{k=0}^{a-1-l}(-1)^k\binom{l+b-1}{b-1}\binom{b}{k}\mathfrak{z}_q^{\text{Ш}}[b+l, a-l-k] \\ & + \sum_{l=0}^{b-1}\sum_{k=0}^{\min(a,b-1-l)}(-1)^k\binom{l+a-1}{a-1}\binom{a}{k}\mathfrak{z}_q^{\text{Ш}}[a+l, b-l-k] \\ & - \sum_{k=1}^{a}\beta_{a-k}\mathfrak{z}_q^{\text{Ш}}[a+b-k] + \sum_{j=1}^{a-1}\alpha_{a+b-1-j}\,\mathfrak{D}\,\mathfrak{z}_q^{\text{Ш}}[a+b-1-j],\end{aligned}$$

where $\mathfrak{D} := q\frac{\mathrm{d}}{\mathrm{d}q}$ and

$$\begin{aligned}\alpha_k &= \sum_{j=b}^{k}\frac{(-1)^{a+b-j}}{1-j}\binom{j-1}{j-b, j-a, a+b-j-1}, \\ \beta_j &= (-1)^{a-j}\binom{j+b-1}{j, j+b-a, a-j-1}.\end{aligned}$$

12.12. Prove the commutativity of the q-shuffle product by showing that $\mathbf{u} \sqcup\!\sqcup\, \mathbf{v} = \mathbf{v} \sqcup\!\sqcup\, \mathbf{u}$ for all $\mathbf{u}, \mathbf{v} \in \widetilde{Y}^*$ using induction on the sum $|\mathbf{u}| + |\mathbf{v}|$.

12.13. Show that $(\mathfrak{A}_{\mathrm{G}}^0, *_{\mathrm{G}})$ is a commutative and associative algebra.

12.14. Define another type of the q-MZVs by

$$\mathfrak{z}_q^{(s_1-1, s_2, \dots, s_d)}[s_1, \dots, s_d], \quad s_1 \geq 1, s_2, \dots, s_d \geq 0.$$

Define the shuffle and stuffle structures of these q-MZVs by mimicking the treatment of type Iq-MZVs.

12.15. Prove Prop. 12.8.26 by induction on the sum of the weights of the words involved. In fact, if we change the definition of bi-brackets by using the increasing order on the indices in Eq. (12.68) then this is equivalent to the definition in Eq. (12.69).

12.16. Denote by $\mathsf{Z}^{\tau}_{\leq w}$ the $\mathbb{Q}$-space generated by q-MZVs of type τ corresponding to the type τ-admissible words $\mathsf{W}^{\tau}_{\leq w}$, $\mathsf{DS}^{\tau}_{\leq w}$ the space generated by the DBSFs, and $\mathsf{DU}^{\tau}_{\leq w}$ the space generated by the duality relations. Hence $\mathsf{DU}^{\tau}_{\leq w} \setminus \mathsf{DS}^{\tau}_{\leq w}$ gives the duality relations that are not contained in $\mathsf{DS}^{\tau}_{\leq w}$. Using your favorite computer algebra system to confirm the following data contained in Tables 12.2-12.5.

Table 12.2: Dimension of the q-MZVs of type $\tilde{\mathrm{I}}$.

w	2	3	4	5	6	7
$\sharp(\mathsf{W})^{\tilde{\mathrm{I}}}_{\leqslant w}$	4	12	33	88	232	609
lower bound of $\dim \mathsf{Z}^{\tilde{\mathrm{I}}}_{\leqslant w}$	3	7	14	27	50	91
$\dim \mathsf{DS}^{\tilde{\mathrm{I}}}_{\leqslant w}$	1	4	17	56	171	497
$\dim \left(\mathsf{DU}^{\tilde{\mathrm{I}}}_{\leqslant w} \setminus \mathsf{DS}^{\tilde{\mathrm{I}}}_{\leqslant w}\right)$	0	0	1	2	3	6

Table 12.3: Dimension of the q-MZVs of type I.

w	2	3	4	5	6	7	8	9
$\sharp(\mathsf{W})^{I}_{\leqslant w}$	1	3	7	15	31	63	127	255
lower bound of $\dim \mathsf{Z}^{I}_{\leqslant w}$	1	2	4	7	11	18	27	42
$\dim \mathsf{DS}^{I}_{\leqslant w}$	0	1	3	8	20	45		
$\dim \left(\mathsf{DU}^{I}_{\leqslant w} \setminus \mathsf{DS}^{I}_{\leqslant w}\right)$	0	0	0	0	0	0		

Table 12.4: Dimension of the q-MZVs of type II.

w	1	2	3	4	5	6
$\sharp(\mathsf{W})^{\mathrm{II}}_{\leqslant w}$	1	5	19	69	251	923
lower bound of $\dim \mathsf{Z}^{\mathrm{II}}_{\leqslant w}$	1	3	12	30	73	173
$\dim \mathsf{DS}^{\mathrm{II}}_{\leqslant w}$	0	1	5	28	124	536
$\dim \left(\mathsf{DU}^{\mathrm{II}}_{\leqslant w} \setminus \mathsf{DS}^{\mathrm{II}}_{\leqslant w}\right)$	0	1	2	8	35	127

Table 12.5: Dimension of the q-MZVs of type III.

w	1	2	3	4	5	6
$\sharp(\mathsf{W})^{\mathrm{III}}_{\leqslant w}$	1	5	19	69	251	923
lower bound of $\dim \mathsf{Z}^{\mathrm{III}}_{\leqslant w}$	1	4	12	30	73	173
$\dim \mathsf{DS}^{\mathrm{III}}_{\leqslant w}$	0	1	5	28	124	536
$\dim \left(\mathsf{DU}^{\mathrm{III}}_{\leqslant w} \setminus \mathsf{DS}^{\mathrm{III}}_{\leqslant w}\right)$	0	0	1	1	5	4

12.17. For s_1, s_2 with $s_1 + s_2 > 2$ and $s = s_1 + s_2 - 2$, show that

$$\binom{s}{s_1 - 1} \frac{\mathfrak{D}[s]}{s} = [s_1] \cdot [s_2] + \binom{s}{s_1 - 1} [s+1] - \sum_{a+b=s+2} \left(\binom{a-1}{s_1 - 1} + \binom{a-1}{s_2 - 1} \right) [a, b].$$

12.18. Complete the proof of Prop. 12.8.11.

12.19. For $s_1, s_2 \in \mathbb{N}$, show that

$$\begin{aligned} \mathfrak{D}[s_1, s_2] = {} & [2] \cdot [s_1, s_2] - s_1[s_1+1, s_2, 1] - s_2[s_1, s_2+1, 1] - [s_1, s_2, 2] \\ & + 2s_1[s_1+1, s_2] + s_2[s_1, s_2+1] - \Bigg(\sum_{a+b=s_1+2} (a-1)[a, b, s_2] \\ & + \sum_{a+b=s_2+1} s_1[s_1+1, a, b] + \sum_{a+b=s_2+2} (a-1)[s_1, a, b] \Bigg). \end{aligned}$$

12.20. Show that

$$\begin{aligned} \mathfrak{D}[1] &= [3] + \frac{1}{2}[2] - [2, 1], \\ \mathfrak{D}[2] &= [4] + 2[3] - \frac{1}{6}[2] - 4[3, 1], \\ \mathfrak{D}[2] &= 2[4] + [3] + \frac{1}{6}[2] - 2[2, 2] - 2[3, 1], \\ \mathfrak{D}[1, 1] &= [3, 1] + \frac{3}{2}[2, 1] + \frac{1}{2}[1, 2] + [1, 3] - 2[2, 1, 1] - [1, 2, 1], \\ \mathfrak{D}[1, 2] &= -\frac{1}{6}[1, 2] + 2[1, 3] + [1, 4] + \frac{3}{2}[2, 2] + [3, 2] \\ &\quad - 4[1, 3, 1] - [2, 1, 2] - 2[2, 2, 1]. \end{aligned}$$

12.21. The algebra $q\mathsf{MZ}$ view as a $\mathbb{Q}$-subalgebra of MB is filtered by the weight whose graded w piece is denoted by $\mathrm{gr}_w\, q\mathsf{MZ}$. Let d_w be its dimension over $\mathbb{Q}$. Is the following true?

$$\sum_{w=0}^{\infty} d_w t^w = \frac{1-t^2+t^4}{1-2t^2-2t^3}.$$

Verify some lower weight cases.

12.22. For any real number r, we define $[r]_q := (q^r-1)/(q-1)$. Then for all $|x|, |y| < 1$ we have

$$\sum_{a=0}^{\infty}\sum_{b=0}^{\infty}(-1)^{a+b}[x]_q^{a+1}[x]_q^{b+1}\zeta^I[a+2,\{1\}^b] = 1 - \frac{\Gamma_q(1+x)\Gamma_q(1+y)}{\Gamma_q(1+x+y)}$$

where the q-analog of the Gamma function

$$\Gamma_q(x) := (1-q)^{1-x}\prod_{i=0}^{\infty}\frac{1-q^{i+1}}{1-q^{i+x}}.$$

12.23. Use Exercise 12.22 to establish the following expression of the generating function of the height one q-MZVs:

$$\sum_{a=0}^{\infty}\sum_{b=0}^{\infty}u^{a+1}v^{b+1}\zeta^I[a+2,\{1\}^b]$$

$$= 1 - \exp\left(\sum_{i=2}^{\infty}\frac{1}{i}\Big(u^i+v^i-(u+v+(1-q)uv)^i\Big)\sum_{j=2}^{i}(q-1)^{i-j}\zeta^I[j]\right).$$

Chapter 13

Colored Multiple Zeta Values

In Chap. 2 we have studied mainly the analytic properties of the multiple polylog functions. In the following two chapters we turn to the algebraic side of the special values of these functions.

13.1 Definitions

Recall from Definition 2.3.1 that for any positive integers $s_1, \dots, s_d$, the multiple polylog of complex variables is defined as follows:

$$\mathrm{Li}_{s_1,\dots,s_d}(z_1,\dots,z_d) := \sum_{k_1>\cdots>k_d>0} \frac{z_1^{k_1}\cdots z_d^{k_d}}{k_1^{s_1}\cdots k_d^{s_d}}, \tag{13.1}$$

where $|z_1 \cdots z_j| < 1$ for $j = 1, \dots, d$. It is analytically continued to a multi-valued meromorphic function on $\mathbb{C}^d$ in Sec. 2.4.

Definition 13.1.1. Fix an Nth root of unity $\mu = \mu_N := \exp(2\pi\sqrt{-1}/N)$. A *colored multiple zeta value (CMZV) of level* N is a number of the form

$$L_N(s_1,\dots,s_d|i_1,\dots,i_d) := \mathrm{Li}_{s_1,\dots,s_d}(\mu^{i_1},\dots,\mu^{i_d}). \tag{13.2}$$

We will always identify $(i_1, \dots, i_d)$ with $(i_1, \dots, i_d)$ (mod N). The Nth roots of unity are called the *colors.* As before, d is called the *depth* and $s_1 + \cdots + s_d$ the *weight.*

It is easy to see from Eq. (13.1) that a CMZV converges if and only if $(s_1, \mu^{i_1}) \neq (1, 1)$. Clearly, all CMZVs of level N are automatically of level kN for every positive integer k. For example, when $i_1 = \cdots = i_d = 0$ or $N = 1$ one gets the MZV $\zeta(s_1, \dots, s_d)$. When $N = 2$ one recovers so-called Euler sums which will be treated in section 14.2 and whose partial sums have been studied in Chap. 7. We also have the following integral expression by Eq. (2.8):

$$L_N(s_1,\dots,s_d|i_1,\dots,i_d)=\int_0^1\left(\frac{dt}{t}\right)^{s_1-1}\frac{dt}{a_1-t}\cdots\left(\frac{dt}{t}\right)^{s_d-1}\frac{dt}{a_d-t} \tag{13.3}$$

where $a_l=\prod_{j=1}^{l}\mu^{-i_j}$ for $l=1,\dots,d$.

13.2 Dimension Upper Bound by Deligne and Goncharov

Similar to the MZV case, standard conjectures in arithmetic algebraic geometry imply that all $\mathbb{Q}$-linear relations among CMZVs can be derived from those among CMZVs of the same weight. Let $\mathsf{CMZV}(w,N)$ be the $\mathbb{Q}$-span of all the CMZVs of weight w and level N. Let $d(w,N)$ denote its dimension. In general, it is very difficult to determine $d(w,N)$ because any nontrivial lower bound would be equivalent to some nontrivial irrational/transcendental result which in turn would be a consequence of a variant of Grothendieck's period conjecture. For example, one can show easily that $\mathsf{CMZV}(2,4)=\langle\log^2 2,\pi^2,\pi\sqrt{-1}\log 2,(K-1)\sqrt{-1}\rangle$, where $K=\sum_{n\geq 0}(-1)^n/(2n+1)^2$ is the Catalan's constant. From the above-mentioned Grothendieck's conjecture we know $d(2,4)=4$ (see [161] or [240]) but we don't have an unconditional proof of this equlity yet. Namely, we cannot prove that the four numbers $\log^2 2,\pi^2,\pi\sqrt{-1}\log 2,(K-1)\sqrt{-1}$ are linearly independent over $\mathbb{Q}$. Consequently, a nontrivial lower bound of $d(w,N)$ is hard to come by.

On the other hand, one may obtain upper bounds of $d(w,N)$ by finding as many $\mathbb{Q}$-linear relations in $\mathsf{CMZV}(w,N)$ as possible. As in the cases of the MZVs and the Euler sums the double shuffle relations will play important roles in revealing the relations among CMZVs. If such a relation is produced by multiplying two convergent CMZVs then it is called a *finite double shuffle relation* (finite DBSF) similar to the MZV case. In general, we need to regularize the divergent CMZVs to obtain the *regularized double shuffle relations* (regularized DBSFs). We shall recall this theory in Sec. 13.3.1.

The proof of the following important result of Deligne and Goncharov is far beyond the scope of this book, so we omit it here.

Theorem 13.2.1. *For any weight $w\geq 1$ and level $N\geq 1$, we have $d(w,N)\leq D(w,N)$ where $D(w,N)$ are defined by the formal power series*

$$1+\sum_{w=1}^{\infty}D(w,N)t^w=\begin{cases}(1-t^2-t^3)^{-1}, & \text{if } N=1;\\ (1-t-t^2)^{-1}, & \text{if } N=2;\\ (1-at+bt^2)^{-1}, & \text{if } N\geq 3,\end{cases} \tag{13.4}$$

where $a = a(N) := \varphi(N)/2 + \nu(N)$, $b = b(N) := \nu(N) - 1$, φ *is Euler's totient function and* $\nu(N)$ *is the number of distinct prime factors of* N.

If $N > 2$ then we have

$$\sum_{w=1}^{\infty} D(w,N)t^w = at + (a^2 - b)t^2 + (a^3 - 2ab)t^3 + (a^4 - 3a^2b + b^2)t^4 + \cdots$$

In particular, if p is a prime then for any positive integer n

$$D(w,p^n) = a(p^n)^w = \left(\frac{p^{n-1}(p-1)}{2} + 1\right)^w. \qquad (13.5)$$

We will compare the bound $D(w,N)$ of Deligne and Goncharov to the bound obtained by the standard relations to be defined next.

13.3 Standard Relations

It turns out that in higher level cases, the DBSFs are far from enough to produce all the $\mathbb{Q}$-linear relations between CMZVs. In this section, we list some more relations that can be defined without too much difficulty.

Definition 13.3.1. We call a $\mathbb{Q}$-linear relation between CMZVs *standard*[1] if it can be produced by $\mathbb{Q}$-linear combinations of the following four families of relations: the regularized DBSFs (Sec. 13.3.1), the regularized distribution relations (regularized DISTs) (Sec. 13.3.4), the weight one relations (Sec. 13.3.3), and the lifted relations from the above three types (Sec. 13.3.5). Otherwise, it is called a *non-standard* relation.

It is commonly believed that all linear relations among the MZVs (*i.e.*, level one CMZVs) are consequences of the regularized DBSFs. But when level $N \geq 2$ we don't know, even conjecturally, how to produce all linear relations among the CMZVs. Moreover, when $N = 2$ (*i.e.*, for the Euler sums) we don't know whether there are any inclusion relations among the regularized DBSFs, the regularized DISTs, the doubling relations and the generalized doubling relations (see Sec. 14.2.5).

13.3.1 *Regularized Double Shuffle Relations*

Similar to the cases of the MZVs, we first define some algebras which reflect the behavior of multiplications of CMZVs.

[1]This term was suggested by P. Deligne in a letter to Goncharov and Racinet dated Feb. 25, 2008.

Set

$$\mathtt{a} = \frac{dt}{t}, \qquad \mathtt{b}_i = \frac{\mu^i dt}{1-\mu^i t} \qquad \text{for } i = 0, 1, \ldots, N-1.$$

For every positive integer $k \geq 1$, define the word of length k

$$\mathtt{z}_{k,i} := \mathtt{a}^{k-1}\mathtt{b}_i.$$

Then it is straightforward to verify using Eq. (13.3) that if $(s_1, \mu^{i_1}) \neq (1,1)$ then

$$L_N(s_1, \ldots, s_n | i_1, i_2, \ldots, i_n) = \int_0^1 \mathtt{z}_{s_1,i_1}\mathtt{z}_{s_2,i_1+i_2} \cdots \mathtt{z}_{s_n,i_1+i_2+\cdots+i_n}. \quad (13.6)$$

One can now define an algebra of words as follows (cf. Definition 7.2.1):

Definition 13.3.2. The set of alphabet X consists of $N+1$ letters $\mathtt{a}$ and $\mathtt{b}_i$ for $i \equiv 0, \ldots, N-1 \pmod N$. The *weight* of a word $\mathbf{w}$ (*i.e.*, a monomial in the letters in X denoted by $|\mathbf{w}|$) is the number of letters contained in $\mathbf{w}$, and its *depth*, denoted by $\mathrm{dp}(\mathbf{w})$, is the number of $\mathtt{b}_i$'s contained in $\mathbf{w}$. Define the *Hoffman–Racinet* algebra of level N, denoted by $\mathfrak{A}_N$, to be the (weight) graded noncommutative polynomial $\mathbb{Q}$-algebra generated by words over the alphabet X. Let $\mathfrak{A}_N^0$ be the subalgebra of $\mathfrak{A}_N$ generated by words not beginning with $\mathtt{b}_0$ and not ending with $\mathtt{a}$. The words in $\mathfrak{A}_N^0$ are called *admissible words.*

By Eq. (13.6) every CMZV can be expressed as an iterated integral over the closed interval $[0,1]$ of an admissible word $\mathbf{w}$ in $\mathfrak{A}_N^0$. This is denoted by

$$\text{Ł}(\mathbf{w}) := \int_0^1 \mathbf{w}. \quad (13.7)$$

We also extend Ł to $\mathfrak{A}_N$ by $\mathbb{Q}$-linearity. We remark that the length $\lg(\mathbf{w})$ of $\mathbf{w}$ is equal to the weight of $\text{Ł}(\mathbf{w})$. Therefore in general one has

$$L_N(s_1, \ldots, s_n | i_1, i_2, \ldots, i_n) = \text{Ł}(\mathtt{z}_{s_1,i_1}\mathtt{z}_{s_2,i_1+i_2} \cdots \mathtt{z}_{s_n,i_1+i_2+\cdots+i_n}), \quad (13.8)$$

$$\text{Ł}(\mathtt{z}_{s_1,i_1}\mathtt{z}_{s_2,i_2} \cdots \mathtt{z}_{s_n,i_n}) = L_N(s_1, \ldots, s_n | i_1, i_2 - i_1, \ldots, i_n - i_{n-1}). \quad (13.9)$$

For example, $L_3(1,2,2|1,0,2) = \text{Ł}(\mathtt{z}_{1,1}\mathtt{z}_{2,1}\mathtt{z}_{2,0})$. Therefore, we can apply Chen's theory of the iterated integrals to compute the product of two CMZVs. For example, by Lemma 2.1.2 from Chap. 2, one has

$$\begin{aligned}
L_N(1|1)L_N(2,3|1,2) &= \text{Ł}(\mathtt{z}_{1,1})\,\text{Ł}(\mathtt{z}_{2,1}\mathtt{z}_{3,3}) = \text{Ł}(\mathtt{b}_1 \sqcup\!\sqcup (\mathtt{a}\mathtt{b}_1\mathtt{a}^2\mathtt{b}_3)) \\
&= \text{Ł}(\mathtt{b}_1\mathtt{a}\mathtt{b}_1\mathtt{a}^2\mathtt{b}_3 + 2\mathtt{a}\mathtt{b}_1^2\mathtt{a}^2\mathtt{b}_3 + (\mathtt{a}\mathtt{b}_1)^2\mathtt{a}\mathtt{b}_3 + \mathtt{a}\mathtt{b}_1\mathtt{a}^2\mathtt{b}_1\mathtt{b}_3 + \mathtt{a}\mathtt{b}_1\mathtt{a}^2\mathtt{b}_3\mathtt{b}_1) \\
&= \text{Ł}(\mathtt{z}_{1,1}\mathtt{z}_{2,1}\mathtt{z}_{3,3} + 2\mathtt{z}_{2,1}\mathtt{z}_{1,1}\mathtt{z}_{3,3} + \mathtt{z}_{2,1}^2\mathtt{z}_{2,3} + \mathtt{z}_{2,1}\mathtt{z}_{3,1}\mathtt{z}_{1,3} + \mathtt{z}_{2,1}\mathtt{z}_{3,3}\mathtt{z}_{1,1}) \\
&= L_N(1,2,3|1,0,2) + 2L_N(2,1,3|1,0,2) + L_N(2,2,2|1,0,2) \\
&\quad + L_N(2,3,1|1,0,2) + L_N(2,3,1|1,2,N-2).
\end{aligned}$$

Let $\mathfrak{A}_{N,Ш}$ be the algebra of $\mathfrak{A}_N$ where the multiplication is defined by the shuffle product Ш. Denote the subalgebra $\mathfrak{A}_N^0$ by $\mathfrak{A}_{N,Ш}^0$ when one considers the shuffle product. Then one can easily prove

Proposition 13.3.3. *The map* Ł $: \mathfrak{A}_{N,Ш}^0 \longrightarrow \mathbb{C}$ *is an algebra homomorphism.*

Proof. See Exercise 13.2. □

On the other hand, the CMZVs are known to satisfy the series stuffle relations. For example,

$$L_N(2|5)L_N(3|4) = L_N(2,3|5,4) + L_N(3,2|4,5) + L_N(5|9).$$

To study such relations in general one has the following definition.

Definition 13.3.4. Denote by $\mathfrak{A}_N^1$ the subalgebra of $\mathfrak{A}_N$ which is generated by words $\mathsf{z}_{s,i}$ with $s \in \mathbb{N}$ and $i \equiv 0, \dots, N-1 \pmod N$. Equivalently, $\mathfrak{A}_N^1$ is the subalgebra of $\mathfrak{A}_N$ generated by words not ending with a. For any word $\mathbf{w} = \mathsf{z}_{s_1,i_1}\mathsf{z}_{s_2,i_2}\cdots\mathsf{z}_{s_n,i_n} \in \mathfrak{A}_N^1$ and positive integer j, one defines the exponent shifting operator τ_j by

$$\tau_j(\mathbf{w}) = \mathsf{z}_{s_1,j+i_1}\mathsf{z}_{s_2,j+i_2}\cdots\mathsf{z}_{s_n,j+i_n}.$$

For convenience, on the empty word we adopt the convention that $\tau_j(1) = 1$. We then define another multiplication $*$ on $\mathfrak{A}_N^1$ by requiring that $*$ distribute over addition, that $1 * \mathbf{w} = \mathbf{w} * 1 = \mathbf{w}$ for any word $\mathbf{w}$, and that, for any words $\mathbf{u}, \mathbf{v}$,

$$\begin{aligned} \mathsf{z}_{s,j}\mathbf{u} * \mathsf{z}_{t,k}\mathbf{v} = \mathsf{z}_{s,j}\Big(\tau_j\big(\tau_{-j}(\mathbf{u}) * \mathsf{z}_{t,k}\mathbf{v}\big)\Big) + \mathsf{z}_{t,k}\Big(\tau_k\big(\mathsf{z}_{s,j}\mathbf{u} * \tau_{-k}(\mathbf{v})\big)\Big) \\ + \mathsf{z}_{s+t,j+k}\Big(\tau_{j+k}\big(\tau_{-j}(\mathbf{u}) * \tau_{-k}(\mathbf{v})\big)\Big). \qquad (13.10) \end{aligned}$$

This multiplication is called the *stuffle product.*

If we denote by $\mathfrak{A}_{N,*}^1$ the algebra $(\mathfrak{A}_N^1, *)$ then it is not hard to prove the next proposition.

Proposition 13.3.5. *The polynomial algebra $\mathfrak{A}_{N,*}^1$ is a commutative weight graded $\mathbb{Q}$-algebra.*

Proof. See Exercise 13.3. □

Now we can define the subalgebra $\mathfrak{A}^0_{N,*}$ similar to $\mathfrak{A}^0_{N,ш}$ by replacing the shuffle product by the stuffle product. Then by the induction on the lengths and using the series definition one can quickly check that for any $\mathbf{w}_1, \mathbf{w}_2 \in \mathfrak{A}^0_{N,*}$

$$Ł(\mathbf{w}_1)\,Ł(\mathbf{w}_2) = Ł(\mathbf{w}_1 * \mathbf{w}_2).$$

This implies the following result.

Proposition 13.3.6. *The map* $Ł : \mathfrak{A}^0_{N,*} \longrightarrow \mathbb{C}$ *is an algebra homomorphism.*

Proof. See Exercise 13.3. □

Definition 13.3.7. Let w be an integer such that $w \geq 2$. For nontrivial words $\mathbf{w}_1, \mathbf{w}_2 \in \mathfrak{A}^0_N$ with $|\mathbf{w}_1| + |\mathbf{w}_2| = w$, we say that the equation

$$Ł(\mathbf{w}_1 \sqcup\!\sqcup\, \mathbf{w}_2 - \mathbf{w}_1 * \mathbf{w}_2) = 0 \qquad (13.11)$$

provides a *finite double shuffle relation* (finite DBSF) of weight w.

It is known that even at level one (*i.e.*, the MZV case) these relations are not enough to provide all the relations among the MZVs. For example, the weight of the product of two MZVs is at least 4. So the finite DBSFs cannot imply the well-known identity $\zeta(2,1) = \zeta(3)$. However, it is believed that one can remedy this by considering *regularized double shuffle relation* (regularized DBSF) produced by the following mechanism.

First, combining Prop. 13.3.3 and Prop. 13.3.6 we can easily prove the following algebraic result.

Proposition 13.3.8. *We have two algebra homomorphisms:*

$$Ł_* : (\mathfrak{A}^1_{N,*}, *) \longrightarrow \mathbb{C}[T], \quad \textit{and} \quad Ł_{ш} : (\mathfrak{A}^1_{N,ш}, ш) \longrightarrow \mathbb{C}[T]$$

which are uniquely determined by the properties that they both extend the evaluation map $Ł : \mathfrak{A}^0_N \longrightarrow \mathbb{C}$ *by sending* $\mathsf{b}_0 = \mathsf{z}_{1,1}$ *to* T.

Second, in order to establish the crucial relation between $Ł_*$ and $Ł_{ш}$ one can adopt the machinery similar to that of the MZVs developed in section 3.2 as follows. For any $(\mathbf{s}|\mathbf{i}) = (s_1, \dots, s_n | i_1, \dots, i_n)$ where i_j are integers and s_j are positive integers, let the image of the corresponding words in $\mathfrak{A}^1_N$ under $Ł_*$ and $Ł_{ш}$ be denoted by $Ł_*(\mathbf{s}|\mathbf{i}; T)$ and $Ł_{ш}(\mathbf{s}|\mathbf{i}; T)$ respectively.

Theorem 13.3.9. *Define a* $\mathbb{C}$*-linear map* $\rho : \mathbb{C}[T] \to \mathbb{C}[T]$ *by*

$$\rho(e^{Tu}) = \exp\left(\sum_{n=2}^{\infty} \frac{(-1)^n}{n}\zeta(n)u^n\right) e^{Tu}, \qquad |u| < 1.$$

Then for any index set $(\mathbf{s}|\mathbf{i})$ *one has*

$$\text{Ł}_{\sqcup\!\sqcup}(\mathbf{s}|\mathbf{i};T) = \rho\big(\text{Ł}_*(\mathbf{s}|\mathbf{q}(\mathbf{i});T)\big), \tag{13.12}$$

where $\mathbf{q}(i_1,\dots,i_n) = (i_1, i_2 - i_1, \dots, i_n - i_{n-1})$.

Thm. 13.3.9 is a generalization of Thm. 3.3.21 to the higher level CMZV cases, but the proof is much more complicated, which will be given in the next section using some advanced results from Lie algebra.

Definition 13.3.10. Let w be a positive integer such that $w \geq 2$. Let $(\mathbf{s}|\mathbf{i})$ be any index set with the weight of $\mathbf{s}$ equal to w. Then every weight w CMZV relation produced by Eq. (13.12) is called a *regularized double shuffle relation* (regularized DBSF) of weight w. This is obtained by formally setting $T = 0$ in Eq. (13.12).

The above steps can be easily transformed to computer codes which have been used in some Maple programs (see Appendix D.1). For example, one gets by the stuffle product

$$\begin{aligned} TL_N(2|3) &= \text{Ł}_*(1|0;T)\,\text{Ł}_*(2|3;T) = \text{Ł}_*(\mathsf{z}_{1,0} * \mathsf{z}_{2,3}) \\ &= \text{Ł}_*(1,2|0,3;T) + \text{Ł}_*(2,1|3,3;T) + \text{Ł}_*(3|3;T), \end{aligned}$$

while using shuffle product one has

$$\begin{aligned} TL_N(2|3) &= \text{Ł}_{\sqcup\!\sqcup}(1|0;T)\,\text{Ł}_{\sqcup\!\sqcup}(2|3;T) = \text{Ł}_{\sqcup\!\sqcup}(\mathsf{z}_{1,0} \sqcup\!\sqcup \mathsf{z}_{2,3}) = \text{Ł}_{\sqcup\!\sqcup}(\mathsf{b}_0 \sqcup\!\sqcup \mathsf{ab}_3) \\ &= \text{Ł}_{\sqcup\!\sqcup}(1,2|0,3;T) + \text{Ł}_{\sqcup\!\sqcup}(2,1|0,3;T) + \text{Ł}_{\sqcup\!\sqcup}(2,1|3,0;T). \end{aligned}$$

We thus obtains the following regularized DBSF if we compare the above two expressions using Thm. 13.3.9:

$$L_N(2,1|3,0) + L_N(3|3) = L_N(2,1|3,N-3) + L_N(2,1|0,3).$$

13.3.2 *Proof of Regularized Double Shuffle Relations*

When $N = 1$ most results in this subsection have been proved in Chap. 3 already. But the proofs of these results for general levels N contained here are often different because more advanced tools are needed.

Recall that we have defined the alphabet $\mathsf{X} = \mathsf{X}_N = \{\mathsf{a}, \mathsf{b}_j : j \in \mathbb{Z}/N\mathbb{Z}\}$ and chosen $\mu = \mu_N = \exp(2\pi\sqrt{-1}/N)$. Let $\Gamma_N := \{\mu^j : j \in \mathbb{Z}/N\mathbb{Z}\}$ be the set of Nth roots of unity. For convenience, we also use the letters x_ξ, $\xi \in \Gamma_N \cup \{0\}$, such that $\mathsf{x}_0 = \mathsf{a}$ and $\mathsf{x}_{\mu^j} = \mathsf{b}_j$. Thus $\mathsf{X} = \{\mathsf{x}_\xi : \xi \in \Gamma_N \cup \{0\}\}$.

We now denote by Y the subset of X^* of all the words $\mathsf{y}_{k,\xi} = \mathsf{x}_0^{k-1}\mathsf{x}_\xi$ for $k \in \mathbb{N}$ and $\xi \in \Gamma_N$.

Remark 13.3.11. Observe that $\mathsf{y}_{n,\mu^j} = \mathsf{z}_{n,j}$ used in the preceding subsection. It turns out that the notation $\mathsf{y}_{n,\xi}$ adopted throughout this subsection has some theoretical advantage while it is much easier to use $\mathsf{z}_{n,j}$ in computer aided symbolic computations which have been implicitly used in the examples of the preceding subsection.

Let R be any $\mathbb{Q}$-algebra. The subalgebra of $R\langle \mathsf{X} \rangle$ generated by Y is denoted by $R\langle \mathsf{Y} \rangle$ and is free on Y. As a vector space, it is generated by the words over X not ending with x_0 thus it could be identified with the quotient $R\langle \mathsf{X} \rangle / R\langle \mathsf{X} \rangle \mathsf{x}_0$. The corresponding projection is denoted by π_{Y}. The vector subspace $R\langle \mathsf{X} \rangle_{\rm cv}$ of $R\langle \mathsf{X} \rangle$ generated by words over X that are not ending with x_0 and not starting with x_1 is also a graded algebra. It may be identified with the quotient $R\langle \mathsf{X} \rangle / (\mathsf{x}_1 R\langle \mathsf{X} \rangle + R\langle \mathsf{X} \rangle \mathsf{x}_0)$. We denote by $\pi_{\rm cv}$ the corresponding projection. The algebra $R\langle \mathsf{X} \rangle_{\rm cv}$ is same as the subalgebra of $R\langle \mathsf{Y} \rangle$ generated by words over Y that are not starting with $\mathsf{y}_{1,1} = \mathsf{x}_1$. (Note that the second subscript 1 in $\mathsf{y}_{1,1}$ is an element in Γ_N.) Thus it can be identified with the quotient $R\langle \mathsf{Y} \rangle / \mathsf{y}_{1,1} R\langle \mathsf{Y} \rangle$, which is denoted by $R\langle \mathsf{Y} \rangle_{\rm cv}$. Again its corresponding projection is denoted by $\pi_{\rm cv}$. Similarly one defines $R\langle\langle \mathsf{X} \rangle\rangle_{\rm cv}$ and $R\langle\langle \mathsf{Y} \rangle\rangle_{\rm cv}$ as the completion of $R\langle \mathsf{X} \rangle_{\rm cv}$ and $R\langle \mathsf{Y} \rangle_{\rm cv}$ with respect to the weight grading, respectively.

Remark 13.3.12. The algebras $\mathfrak{A}$, $\mathfrak{A}^0$ and $\mathfrak{A}^1$ in Sec. 13.3.1 are essentially equal to $\mathbb{Q}\langle \mathsf{X} \rangle$, $\mathbb{Q}\langle \mathsf{X} \rangle_{\rm cv}$ and $\mathbb{Q}\langle \mathsf{Y} \rangle$, respectively.

Now we take $(\mathbb{C}\langle \mathsf{X} \rangle, \bullet, \Delta)$ to be the graded bialgebra defined by Example B.3 in Appendix B and $(\mathbb{C}\langle \mathsf{Y} \rangle, \bullet, \Delta_*)$ the graded bialgebra defined by Example B.4. In particular, all x_ξ are primitive elements in $(\mathbb{C}\langle \mathsf{X} \rangle, \bullet, \Delta)$. Similarly, $\mathbb{C}\langle\langle \mathsf{X} \rangle\rangle$, $\mathbb{C}\langle\langle \mathsf{Y} \rangle\rangle$, $\mathbb{C}\langle\langle \mathsf{X} \rangle\rangle_{\rm cv}$ and $\mathbb{C}\langle\langle \mathsf{Y} \rangle\rangle_{\rm cv}$ are all graded bialgebras (see Exercise 13.4).

Recall that for any $a_1, \ldots, a_k \in \mathbb{C}$ and any path $\gamma : [0,1] \to \mathbb{C}$, the iterated integral $I_\gamma(a_1, \ldots, a_k)$ is defined by Eq. (2.14). In particular, for $\xi_1, \ldots, \xi_d \in \Gamma_N \cup \{0\}$, $\xi_1 \neq 1$ and $\xi_d \neq 0$, we have

$$I_{[0,1]}(\xi_1, \ldots, \xi_d) = \int_{1>t_1>\cdots>t_d>0} \bigwedge_{j=1}^{d} \omega_{\xi_j}(t_j),$$

where for all $j = 1, \dots, d$

$$\omega_{\xi_j}(t_j) = \begin{cases} \dfrac{dt_j}{\xi_j^{-1} - t_j}, & \text{if } \xi_j \neq 0; \\ \dfrac{dt_j}{t_j}, & \text{if } \xi_j = 0. \end{cases}$$

Recall that an element x in any coalgebra (C, ϵ, Δ) is called *group-like* (see Definition A.6) if $\Delta(x) = x \otimes x$ and $\epsilon(x) = 1$.

Proposition 13.3.13. *For any path $\gamma \in \mathbb{C}^* \setminus \Gamma_N$, we may define*

$$\mathbf{I}_\gamma := \sum_{d \in \mathbb{N}_0, \xi_1, \dots, \xi_d \in \Gamma_N \cup \{0\}} I_\gamma(\xi_1, \dots, \xi_d) \mathsf{x}_{\xi_1} \cdots \mathsf{x}_{\xi_d},$$

where $I_\gamma(\emptyset) = 1$. Then $\mathbf{I}_\gamma$ is a group-like element in $(\mathbb{C}\langle \mathsf{X} \rangle, \bullet, \Delta)$.

Proof. Let X^* be the set of words over X. For any word $\mathbf{w} = l_1 \dots l_n \in \mathsf{X}^*$ and any subset $J = \{i_1, \dots, i_k\}$ of $\{1, \dots, n\}$ in increasing order, we define $\mathbf{w}|J = l_{i_1} \dots l_{i_n}$. Further set $\mathbf{w}|\emptyset = 1$. Since Δ is an algebra homomorphism we have

$$\begin{aligned} \Delta(\mathbf{w}) = \Delta(l_1) \dots \Delta(l_n) &= \prod_{j=1}^{n} (l_j \otimes 1 + 1 \otimes l_j) \\ &= \sum_{J_1 \cup J_2 = \{1, \dots, n\}, J_1 \cap J_2 = \emptyset} (\mathbf{w}|J_1) \otimes (\mathbf{w}|J_2) = \sum_{\mathbf{u}_1, \mathbf{u}_2 \in \mathsf{X}^*} (\mathbf{w}, \mathbf{u}_1 ш \mathbf{u}_2) \mathbf{u}_1 \otimes \mathbf{u}_2 \end{aligned}$$

by the definition of ш, where $(\mathbf{w}, \mathbf{u}_1 ш \mathbf{u}_2)$ is the coefficient of $\mathbf{w}$ in $\mathbf{u}_1 ш \mathbf{u}_2$. This implies that

$$\begin{aligned} \Delta(\mathbf{I}_\gamma) &= \sum_{\substack{n \in \mathbb{N}_0, \xi_1, \dots, \xi_n \in \Gamma_N \cup \{0\} \\ m \in \mathbb{N}_0, \tau_1, \dots, \tau_m \in \Gamma_N \cup \{0\}}} I_\gamma \Big(\boldsymbol{\xi} ш \boldsymbol{\tau} \Big) \mathsf{x}_{\xi_1} \cdots \mathsf{x}_{\xi_n} \otimes \mathsf{x}_{\tau_1} \cdots \mathsf{x}_{\tau_m} \\ &= \sum_{\substack{n \in \mathbb{N}_0, \xi_1, \dots, \xi_n \in \Gamma_N \cup \{0\} \\ m \in \mathbb{N}_0, \tau_1, \dots, \tau_m \in \Gamma_N \cup \{0\}}} I_\gamma(\boldsymbol{\xi}) I_\gamma(\boldsymbol{\tau}) \mathsf{x}_{\xi_1} \cdots \mathsf{x}_{\xi_n} \otimes \mathsf{x}_{\tau_1} \cdots \mathsf{x}_{\tau_m}, \end{aligned}$$

by Lemma 2.1.2(iv), where $\boldsymbol{\xi} = (\xi_1, \dots, \xi_n)$ and $\boldsymbol{\tau} = (\tau_1, \dots, \tau_m)$. Thus $\Delta(\mathbf{I}_\gamma) = \mathbf{I}_\gamma \otimes \mathbf{I}_\gamma$ as desired. □

For any $0 < \alpha < z < 1$ and $\xi_1, \dots, \xi_n \in \Gamma_N \cup \{0\}$, we can also prove $I_{[\alpha,z]}(\xi_1, \dots, \xi_n)$ is group-like in the following way. Fixing α and regarding $\mathbf{I}_{[\alpha,z]}$ as a function of z, we see that it provides a solution of the differential equation

$$d\mathbf{I} = \Omega \mathbf{I} \quad \text{where} \quad \Omega = \sum_{\xi \in \Gamma_N \cup \{0\}} \omega_\xi(z) \mathsf{x}_\xi.$$

Note that Ω is primitive for the coproduct Δ so both $\Delta\,\mathbf{I}_{[\alpha,z]}$ and $\mathbf{I}_{[\alpha,z]}\otimes\mathbf{I}_{[\alpha,z]}$ are the solutions to the differential equation $d\mathcal{J}=\Delta(\Omega)\mathcal{J}$ satisfying the initial condition $\mathcal{J}(\alpha)=1\otimes 1$. Therefore

$$\Delta\,\mathbf{I}_{[\alpha,z]}=\mathbf{I}_{[\alpha,z]}\otimes\mathbf{I}_{[\alpha,z]}\,.$$

Proposition 13.3.14. *For all $z\in(0,1)$, the series*

$$\mathbf{I}_{[0,z]}:=\lim_{\alpha\to 0}\pi_Y(\mathbf{I}_{[\alpha,z]})=\sum_{\substack{\xi_1,\dots,\xi_d\in\Gamma_N\cup\{0\}\\ \xi_d\neq 0,d\in\mathbb{N}_0}} I_{[0,z]}(\xi_1,\dots,\xi_d)\mathsf{x}_{\xi_1}\cdots\mathsf{x}_{\xi_d} \quad (13.13)$$

is a group-like element in $(\mathbb{C}\langle\langle\mathsf{Y}\rangle\rangle,\Delta)$.

Proof. Regarding α as a variable in $(0,1)$, by Prop. 13.3.13 we see that the series $\pi_Y(\mathbf{I}_{[\alpha,z]})$ is a group-like element in $(\mathbb{C}^{(0,1)}\langle\langle Y\rangle\rangle,\Delta)$ since π_Y is a coalgebra homomorphism. Taking $\alpha\to 0$ we see that the limit $\pi_Y(\mathbf{I}_{[0,z]})$ exists by Prop. 2.3.8. This completes the proof of the proposition since taking limit is also a coalgebra homomorphism. □

For any word $\mathsf{y}_{s_1,\xi_1}\mathsf{y}_{s_2,\xi_2}\cdots\mathsf{y}_{s_d,\xi_d}$ over Y, we define the endomorphism $\mathbf{q}$ of $\mathbb{C}\langle\mathsf{Y}\rangle$ as

$$\mathbf{q}(\mathsf{y}_{s_1,\xi_1}\mathsf{y}_{s_2,\xi_2}\cdots\mathsf{y}_{s_d,\xi_d})=\mathsf{y}_{s_1,\xi_1}\mathsf{y}_{s_2,\xi_2/\xi_1}\cdots\mathsf{y}_{s_d,\xi_d/\xi_{d-1}}. \quad (13.14)$$

It has an inverse defined by

$$\mathbf{p}(\mathsf{y}_{s_1,\xi_1}\mathsf{y}_{s_2,\xi_2}\cdots\mathsf{y}_{s_d,\xi_d})=\mathsf{y}_{s_1,\xi_1}\mathsf{y}_{s_2,\xi_1\xi_2}\cdots\mathsf{y}_{s_d,\xi_1\xi_2\cdots\xi_d}. \quad (13.15)$$

Remark 13.3.15. The map $\mathbf{q}$ originates from the formula

$$I_{[0,1]}(\{0\}^{s_1-1},\mu^{i_1},\dots,\{0\}^{s_d-1},\mu^{i_d})=L_N(s_1,\dots,s_d|i_1,i_2-i_1,\dots,i_d-i_{d-1})$$

where $(s_1,\mu^{i_1})\neq(1,1)$. Its inverse $\mathbf{p}$ gives rise to the operator tog of Eq. (8.24).

Proposition 13.3.16. *For any $z\in(0,1)$, we have the following identity in $\mathbb{C}\langle\langle\mathsf{Y}\rangle\rangle$:*

$$\mathbf{L}_z:=\sum_{\substack{d\in\mathbb{N}_0,\mathbf{s}\in\mathbb{N}^d\\ \xi_1,\dots,\xi_d\in\Gamma_N}}\mathrm{Li}_{\mathbf{s}}(z\xi_1,\xi_2,\dots,\xi_d)\mathsf{y}_{s_1,\xi_1}\mathsf{y}_{s_2,\xi_2}\cdots\mathsf{y}_{s_d,\xi_d}=\mathbf{q}(\mathbf{I}_{[0,z]}).$$

Proof. This is just a restatement of Prop. 2.3.8. □

In the next result, we need to apply the projection π_{cv} to remove the divergent terms from Eq. (13.13).

Proposition 13.3.17. *The series*

$$\mathbf{I}_{\rm cv} := \lim_{z\to 1^-} \pi_{\rm cv}\, \mathbf{I}_{[0,z]} = \sum_{\xi_1,\dots,\xi_d\in\Gamma_N\cup\{0\},\xi_1\neq 1,\xi_d\neq 0} I_{[0,1]}(\xi_1,\dots,\xi_d)\mathsf{x}_{\xi_1}\cdots\mathsf{x}_{\xi_d}$$

is a group-like element in $\mathbb{C}\langle\langle \mathsf{X}\rangle\rangle_{\rm cv}$.

Proof. Prop. 13.3.16 shows that as $z\to 1^-$ the asymptotic expansion of every coefficient of the series $\pi_{\rm cv}(\mathbf{I}_{[0,z]})$ is given by the associated power series (see Eq. (2.9)) of $\big(\mathrm{Li}_{\mathbf{s}}^{(M)}(\boldsymbol{\xi})\big)_{M\geq 1}\in \mathsf{DivLog}_{\mathbb{N}}$ for some $\mathbf{s}\in\mathbb{N}^d$ and $\boldsymbol{\xi}\in(\Gamma_N)^d$ $((s_1,\xi_1)\neq(1,1))$, which converges by Cor. 2.3.10. The proposition now follows from Abel's lemma on the convergence of a series at the boundary of its convergence interval (*i.e.*, Lemma 2.3.7). □

By applying $\pi_{\rm cv}$ we can remove the divergent terms as follows:

$$\lim_{z\to 1^-} \pi_{\rm cv}\, \mathbf{L}_z = \sum_{\substack{d\in\mathbb{N}_0,\mathbf{s}=(s_1,\dots,s_d)\in\mathbb{N}^d\\ \xi_1,\dots,\xi_d\in\Gamma_N,(s_1,\xi_1)\neq(1,1)}} \mathrm{Li}_{\mathbf{s}}(\xi_1,\dots,\xi_d)\mathsf{y}_{s_1,\xi_1}\cdots\mathsf{y}_{s_d,\xi_d}.$$

By straightforward computation we find the following crucial relation between the above two elements.

Proposition 13.3.18. *In* $\mathbb{C}\langle\langle \mathsf{Y}\rangle\rangle_{\rm cv}$ *one has*

$$\mathbf{L}_{\rm cv} := \lim_{z\to 1^-} \pi_{\rm cv}\, \mathbf{L}_z = \mathbf{q}(\mathbf{I}_{\rm cv}).$$

Proof. Noticing that $\mathbf{q}$ and $\pi_{\rm cv}$ commute, by the definition of $\pi_{\rm cv}$ we get

$$\begin{aligned}\mathbf{L}_{\rm cv} &= \lim_{z\to 1^-} \pi_{\rm cv}\, \mathbf{L}_z = \lim_{z\to 1^-} \pi_{\rm cv}\mathbf{q}(\mathbf{I}_{[0,z]}) \quad \text{(by Prop. 13.3.16)}\\ &= \mathbf{q}(\lim_{z\to 1^-} \pi_{\rm cv}\, \mathbf{I}_{[0,z]}) = \mathbf{q}(\mathbf{I}_{\rm cv})\end{aligned}$$

by Prop. 13.3.17. □

Proposition 13.3.19. *The series* $\mathbf{L}_{\rm cv}$ *is a group-like element in* $\mathbb{C}\langle\langle \mathsf{Y}\rangle\rangle_{\rm cv}$.

Proof. First we observe that

$$\mathscr{Y} := \sum_{n\in\mathbb{N}_0,\xi\in\Gamma_N} \mathsf{y}_{n,\xi} \tag{13.16}$$

is a group-like element in $(\mathbb{C}\langle\langle \mathsf{Y}\rangle\rangle, \bullet, \Delta_*)$ (see Exercise 13.5). Next it can be checked easily that for any $\lambda\in\mathbb{C}$, both of the maps $\mathsf{y}_{n,\xi}\mapsto\lambda^n\mathsf{y}_{n,\xi}$ and $\mathsf{y}_{n,\xi}\mapsto\xi\mathsf{y}_{n,\xi}$ are homomorphisms of the coalgebra $(\mathbb{C}\langle\langle \mathsf{Y}\rangle\rangle, \Delta_*)$, meaning they commute with Δ_*. Therefore by applying the first map once and the second m times to the sum in Eq. (13.16) we get a group-like element for Δ_* (see Exercise 13.7):

$$\mathscr{Y}_m(\lambda) := \sum_{n\in\mathbb{N}_0,\xi\in\Gamma_N} \xi^m \lambda^n \mathrm{y}_{n,\xi}.$$

Now for any $M \in \mathbb{N}$, we may define the Mth partial sum of CMZVs by (cf. Eq. (2.13))

$$\mathrm{Li}^{(M)}_{s_1,\dots,s_d}(\xi_1,\dots,\xi_d) := \sum_{M\geq k_1>\cdots>k_d>0} \frac{\xi_1^{k_1}\cdots\xi_d^{k_d}}{k_1^{s_1}\cdots k_d^{s_d}}.$$

Then it is straightforward to check that its generating series

$$\mathbf{L}^{(M)} := \sum_{\substack{d\in\mathbb{N}_0,\xi_1,\dots,\xi_d\in\Gamma_N\\ s_1,\dots,s_d\in\mathbb{N}}} \mathrm{Li}^{(M)}_{s_1,\dots,s_d}(\xi_1,\dots,\xi_d)\mathrm{y}_{s_1,\xi_1}\cdots\mathrm{y}_{s_d,\xi_d} \tag{13.17}$$

can be expressed as (see Exercise 13.8)

$$\mathbf{L}^{(M)} = \mathscr{Y}_M\Big(\frac{1}{M}\Big)\mathscr{Y}_{M-1}\Big(\frac{1}{M-1}\Big)\cdots\mathscr{Y}_2\Big(\frac{1}{2}\Big)\mathscr{Y}_1\Big(\frac{1}{1}\Big). \tag{13.18}$$

Since each factor on the right-hand side is group-like we see that $\mathbf{L}^{(M)}$ must be group-like, too.

We now can check that $\mathrm{y}_{1,1}\mathbb{C}\langle\langle\mathsf{Y}\rangle\rangle$ is a co-ideal for Δ_* since $\Delta_*(\mathrm{y}_{1,1}) = \mathrm{y}_{1,1}\otimes 1 + 1\otimes \mathrm{y}_{1,1}$ by the definition. Therefore, by Prop. A.10 in Appendix A, the coproduct Δ_* descends onto the quotient $\mathbb{C}\langle\langle\mathsf{Y}\rangle\rangle/\mathrm{y}_{1,1}\mathbb{C}\langle\langle\mathsf{Y}\rangle\rangle$ which is identified with $\mathbb{C}\langle\langle\mathsf{Y}\rangle\rangle_{\mathrm{cv}}$. As the projection π_{cv} is clearly a coalgebra homomorphism (*i.e.*, commutes with Δ_*) we see immediately that

$$\mathbf{L}_{\mathrm{cv}} = \lim_{M\to\infty} \pi_{\mathrm{cv}}(\mathbf{L}^{(M)})$$

is a group-like element in $(\mathbb{C}\langle\langle\mathsf{Y}\rangle\rangle,\Delta_*)$. This completes the proof of the proposition. □

The following two lemmas are immediate consequences of Cor. B.11 in Appendix B (see Exercise 13.10).

Lemma 13.3.20. *Every group-like element Φ_{cv} in $(\mathbb{Q}\langle\mathsf{X}\rangle_{\mathrm{cv}},\Delta)$ is the image of some group-like elements in $(\mathbb{Q}\langle\mathsf{X}\rangle,\Delta)$ under π_{cv}. Moreover, any two such elements Φ_1 and Φ_2 in $(\mathbb{Q}\langle\mathsf{X}\rangle_{\mathrm{cv}},\Delta)$ are related by*

$$\Phi_2 = \exp((\lambda_2-\lambda_1)\mathrm{x}_1)\Phi_1\exp((\rho_2-\rho_1)\mathrm{x}_0)$$

where λ_j and ρ_j are the coefficients of x_1 and x_0 of Φ_j $(j=1,2)$, respectively.

Proof. Apply Cor. B.11 twice. □

Lemma 13.3.21. *Any group-like series* $\Phi_{\rm cv}$ *of* $(\mathbb{Q}\langle\langle \mathsf{Y}\rangle\rangle_{\rm cv}, \Delta_*)$ *is the image of a group-like element of* $(\mathbb{Q}\langle\langle \mathsf{Y}\rangle\rangle, \Delta_*)$. *Moreover, any two such elements* Φ_1 *and* Φ_2 *are related by*

$$\Phi_2 = \exp((\lambda_2 - \lambda_1)\mathrm{y}_{1,1})\Phi_1,$$

where λ_1 *and* λ_2 *are the coefficients of* $\mathrm{y}_{1,1}$ *in* Φ_1 *and* Φ_2 *respectively.*

With the above lemma available, we can define

Definition 13.3.22. Let $\mathbf{L}$ (resp. $\mathbf{I}$) be the unique group-like element in $(\mathbb{C}\langle\langle \mathsf{Y}\rangle\rangle, \Delta_*)$ (resp. $(\mathbb{C}\langle\langle \mathsf{X}\rangle\rangle, \Delta)$) such that $\pi_{\rm cv}(\mathbf{L}) = \mathbf{L}_{\rm cv}$ (resp. $\pi_{\rm cv}(\mathbf{I}) = \mathbf{I}_{\rm cv}$) and in which the coefficients of $\mathrm{y}_{1,1}$ is 0 (resp. in which the coefficients of x_0 and x_1 are 0). We also call $\mathbf{L}$ (resp. $\mathbf{I}$) the regularization of $\mathbf{L}_{\rm cv}$ (resp. $\mathbf{I}_{\rm cv}$) .

Remark 13.3.23. The coefficients of $\mathbf{L}$ (resp. $\mathbf{I}$) may be viewed as regularized values of the series $\mathrm{Li}_{s_1,\dots,s_n}(\xi_1,\dots,\xi_n)$ (resp. the iterated integral $I_{[0,1]}(\xi_1,\dots,\xi_n)$). Note that Racinet's choice is different from that given by Ihara, Kaneko and Zagier in which the coefficient of $\mathrm{y}_{1,1}$ is T instead of 0.

In what follows we will derive a relation between the two series $\mathbf{L}$ and $\mathbf{I}$. The main tools are contained in the three subsections of Sec. 2.3.

Suppose $f : k_1 \longrightarrow k_2$ is a $\mathbb{C}$-linear map between two rings over $\mathbb{C}$. If $\Phi \in k_1\langle\langle \mathsf{X}\rangle\rangle$ then we denote by $\bar{f}(\Phi)$ the element in $k_2\langle\langle \mathsf{X}\rangle\rangle$ obtained by applying f on the coefficients of Φ. By $\mathbb{C}$-linearity we see that $\bar{f}$ is a homomorphism of $\mathbb{C}\langle\langle \mathsf{X}\rangle\rangle$-bimodules. Further, if f is an algebra homomorphism and Φ is group-like for Δ then so is $\bar{f}(\Phi)$. The same is true for Y and Δ_*.

By Cor. 2.3.10 all the coefficients of the series $\mathbf{L}^{(M)}$ defined by Eq. (13.17) are in $\mathsf{DivLog}_{\mathbb{N}}$. By Remark 2.3.6 we see that As is an algebra homomorphism, so $\mathbf{L}_t := \overline{\mathrm{As}}(\mathbf{L}^{(M)})$ is group-like in $(\mathbb{C}[t]\langle\langle \mathsf{Y}\rangle\rangle, \Delta_*)$, where we have extended As from $\mathbb{C}^{\mathbb{N}}$ to $\mathbb{C}^{\mathbb{N}}\langle\langle \mathsf{Y}\rangle\rangle$. Clearly, $\pi_{\rm cv}\,\mathbf{L}_t = \mathbf{L}_{\rm cv}$ since all convergent coefficients result in constant functions. Hence we can regard the space $\mathbb{C}\langle\langle \mathsf{Y}\rangle\rangle$ as a subspace of $\mathbb{C}[t]\langle\langle \mathsf{Y}\rangle\rangle$.

Lemma 13.3.24. *In* $\mathbb{C}[t]\langle\langle \mathsf{Y}\rangle\rangle$, *we have*

$$\mathbf{L}_t = \exp\big((t+\gamma)\mathrm{y}_{1,1}\big)\,\mathbf{L}\,. \tag{13.19}$$

Proof. Noticing that the coefficient of $\mathrm{y}_{1,1}$ in $\mathbf{L}_t$ is $\mathrm{As}_{\mathbf{h}} = t+\gamma$ (see Example 2.3.3), we can readily prove the lemma by applying Lemma 13.3.21. □

Lemma 13.3.25. *The group-like element* $\exp(T\mathsf{y}_{1,1})\,\mathbf{L}$ *in* $\mathbb{C}[T]\langle\langle \mathsf{Y}\rangle\rangle$ *is the generating series of* $\text{Ł}_*(\mathbf{s}|\mathbf{i};T)$*. Namely,*

$$\exp(T\mathsf{y}_{1,1})\,\mathbf{L} = \sum_{\substack{d\in\mathbb{N}_0,\mathbf{s}=(s_1,\dots,s_d)\in\mathbb{N}^d\\ \mathbf{i}=(i_1,\dots,i_d)\in(\mathbb{Z}/N\mathbb{Z})^d}} \text{Ł}_*(\mathbf{s}|\mathbf{i};T)\mathsf{y}_{s_1,\mu^{i_1}}\cdots\mathsf{y}_{s_d,\mu^{i_d}}.$$

Proof. By the proof of Thm. 4.1.6 we know the generating series of $\text{Ł}_*(\mathbf{s}|\mathbf{i};T)$ is group-like for Δ_* since $\text{Ł}_*(\mathbf{s}|\mathbf{i};T)$ satisfies the stuffle relations. Then we can apply Lemma 13.3.21 to deduce our result since $\text{Ł}_*(1|0;T)=T$ by Prop. 13.3.8. □

Corollary 13.3.26. *Let* $\mathbf{h}(k)=\left(\text{Li}^{(M)}_{\{1\}^k}(\{1\}^k)\right)_{M\ge1}$*. Then the associated power series* $\text{Fp}(\mathbf{h}(k))=(\text{Li}_1)^k/k!$*, where* $\text{Li}_1(z)=-\log(1-z)$*. Moreover,* $\text{As}_{\mathbf{h}(k)}(t)$ *has degree* k*.*

Proof. By the definition, the coefficient of $\mathsf{y}_{1,1}^k$ in $\mathbf{L}^{(M)}$ is $\text{Li}^{(M)}_{\{1\}^k}(\{1\}^k)$. So the last sentence of the corollary follows immediately from Lemma 13.3.24.

By the definition of Eq. (2.9), the function $\text{Fp}(\mathbf{h}(k))$ is given by the power series

$$f_k(z):=\sum_{n_1>n_2>\cdots>n_k}\frac{z^{n_1}}{n_1n_2\cdots n_k}.$$

A simple differentiation yields the recurrence relation $f_k'(z)=f_{k-1}(z)/(1-z)$ for all $k\ge1$ and $f_0(z)=1$. Now by induction and the fact that $f_k(0)=0$ for all $k>0$, we get immediately that $f(z)=(\text{Li}_1(z))^k/k!$, as desired. □

Proposition 13.3.27. *The coefficient of* x_1^k *in the group-like element* $\mathbf{I}_{[0,z]}$ *is* $(\text{Li}_1(z))^k/k!$*.*

Proof. By the definition, the coefficient is

$$\int_0^z\left(\frac{\mathrm{d}t}{1-t}\right)^k=\int_1^{1-z}\left(-\frac{\mathrm{d}t}{t}\right)^k$$

by substitution $t\to 1-t$. This clearly evaluates to $(\text{Li}_1(z))^k/k!$. □

Proposition 13.3.28. *There exists* $\mathbf{S}(\mathsf{y}_{1,1})\in\mathbb{C}[\![\mathsf{y}_{1,1}]\!]$ *such that*

$$\mathbf{L}=\mathbf{S}(\mathsf{y}_{1,1})\mathbf{q}\pi_Y(\mathbf{I}).$$

Proof. Define $\mathbf{I}_t = \overline{\mathrm{As}}(\mathbf{I}_{[0,z]})$ which is a group-like element of $(\mathbb{C}[t]\langle\langle \mathsf{Y}\rangle\rangle, \Delta)$ so that $\pi_{\mathrm{cv}}\,\mathbf{I}_t = \mathbf{I}_{\mathrm{cv}}$. Since the coefficient of x_1 in $\mathbf{I}_t$ is $\mathrm{As}_{\mathrm{Li}_1}(t) = t$ we conclude from Lemma 13.3.21 that

$$\mathbf{I}_t = \exp(\mathrm{x}_1 t)\pi_Y(\mathbf{I}). \tag{13.20}$$

On the other hand, by Prop. 13.3.16 $\mathbf{L}_z = \mathbf{q}(\mathbf{I}_{[0,z]})$ we find $\overline{\mathrm{cp}}(\mathbf{L}_t) = \mathbf{q}\,\mathbf{I}_t$ by using the commutative diagram on page 19. Since $\overline{\mathrm{cp}}$ is an endomorphism of $\mathbb{C}\langle\langle \mathsf{Y}\rangle\rangle$-modules we get from Eqs. (13.19) and (13.20) that $\mathbf{L} = \overline{\mathrm{cp}}\big(\exp\big((2t+\gamma)\mathrm{y}_{1,1}\big)\big)\mathbf{q}\pi_Y(\mathbf{I})$. Now that all the coefficients of $\mathbf{L}$ and $\mathbf{I}$ are in $\mathbb{C}$ we see that $\overline{\mathrm{cp}}\big(\exp\big((2t+\gamma)\mathrm{y}_{1,1}\big)\big)$ must be in $\mathbb{C}[\![\mathrm{y}_{1,1}]\!]$. This completes the proof of the proposition. □

Lemma 13.3.29. *The group-like element* $\exp\big(T\mathrm{x}_1\big)\,\mathbf{I}$ *in* $\mathbb{C}[T]\langle\langle \mathsf{X}\rangle\rangle$ *is the generating series of* $\text{Ł}_{\sqcup\!\sqcup}(\mathbf{s}|\mathbf{i};T)$. *Namely,*

$$\exp\big(T\mathrm{x}_1\big)\,\mathbf{I} = \sum_{\substack{d\in\mathbb{N}_0,\mathbf{s}=(s_1,\dots,s_d)\in\mathbb{N}^d\\ \mathbf{i}=(i_1,\dots,i_d)\in(\mathbb{Z}/N\mathbb{Z})^d}} \text{Ł}_{\sqcup\!\sqcup}(\mathbf{s}|\mathbf{i};T)\mathrm{x}_0^{s_1-1}\mathrm{x}_{\mu^{i_1}}\cdots \mathrm{x}_0^{s_d-1}\mathrm{x}_{\mu^{i_d}}.$$

Proof. By the proof of Prop. 13.3.13 we know the generating series of $\text{Ł}_{\sqcup\!\sqcup}(\mathbf{s}|\mathbf{i};T)$ is group-like for Δ since $\text{Ł}_{\sqcup\!\sqcup}(\mathbf{s}|\mathbf{i};T)$ satisfies the shuffle relations. Then we can apply Lemma 13.3.20 to deduce our result since $\text{Ł}_{\sqcup\!\sqcup}(1|0;T) = T$ by Prop. 13.3.8. □

Next, we want to determine $\mathbf{S}$ explicitly.

Theorem 13.3.30. *The relation between* $\mathbf{L}$ *and* $\mathbf{I}$ *is given by*

$$\mathbf{L} = \mathbf{S}(\mathrm{y}_{1,1})\cdot \mathbf{q}\pi_{\mathsf{Y}}(\mathbf{I}),$$

where $\mathbf{S}(u) = \exp\left(\sum_{n=2}^{\infty}\frac{(-1)^{n-1}\zeta(n)}{n}u^n\right)$.

Proof. This follows easily from Thm. B.7 in Appendix B. □

Let's recall the linear map defined in Eq. (3.3.17):

$$\begin{aligned}\rho &: \mathbb{R}[T]\to\mathbb{R}[T],\\ \rho(e^{Tu}) &\mapsto A(u)e^{Tu},\end{aligned}$$

where $A(u) = \mathbf{S}(u)^{-1}$. Let $\mathbf{L}' := \exp(T\mathrm{y}_{1,1})\cdot\mathbf{L}$ and $\mathbf{I}' := \exp(T\mathrm{x}_1)\cdot\mathbf{I}$. By Lemma 13.3.21, the coefficients of $\mathrm{y}_{1,1}$ and x_1 in $\mathbf{L}'$ and $\mathbf{I}'$ are both equal to T. Note that $\mathrm{y}_{1,1} = \mathrm{x}_1$. For any $\mathbf{s} = (s_1,\dots,s_n)\in\mathbb{N}^n$ and $\mathbf{i} = (i_1,\dots,i_n)\in(\mathbb{Z}/N\mathbb{Z})^n$, by Lemma 13.3.25 (resp. Lemma 13.3.29)

$\text{Ł}_*(\mathbf{s}|\mathbf{i};T)$ (resp. $\text{Ł}_{⧢}(\mathbf{s}|\mathbf{i};T)$) is the coefficient of $\mathsf{y}_{s_1,\mu^{i_1}}\cdots\mathsf{y}_{s_n,\mu^{i_n}}$ (resp. $\mathsf{x}_0^{s_1-1}\mathsf{x}_{\mu^{i_1}}\cdots\mathsf{x}_0^{s_n-1}\mathsf{x}_{\mu^{i_n}}$, in $\mathbf{L}'$ (resp. $\mathbf{I}'$). Finally we prove the regularized DBSFs as given in Thm. 13.3.9 which can be reformulated as follows.

Theorem 13.3.31. *For any indices $s_1,\dots,s_n\in\mathbb{N}$ and $i_1,\dots,i_n\in\mathbb{Z}/N\mathbb{Z}$, we have*

$$\rho\big(\text{Ł}_*(s_1,\dots,s_n|i_1,i_2-i_1,\dots,i_n-i_{n-1};T)\big)=\text{Ł}_{⧢}(s_1,\dots,s_n|i_1,\dots,i_n;T).$$

Proof. Dividing both sides of the equation in Thm. 13.3.30 by $\mathbf{S}(\mathsf{y}_{1,1})$, we obtain

$$\begin{aligned}A(\mathsf{y}_{1,1})\cdot\mathbf{L}'&=\exp(T\mathsf{y}_{1,1})\cdot A(\mathsf{y}_{1,1})\cdot\mathbf{L}\\&=\exp(T\mathsf{y}_{1,1})\cdot\mathbf{q}\pi_{\mathsf{Y}}(\mathbf{I})\\&=\mathbf{q}\pi_{\mathsf{Y}}(\exp(T\mathsf{x}_1)\,\mathbf{I})=\mathbf{q}\pi_{\mathsf{Y}}(\mathbf{I}').\end{aligned}$$

Hence

$$\rho(\mathbf{L}')=\rho(\exp(T\mathsf{y}_{1,1})\cdot\mathbf{L})=A(\mathsf{y}_{1,1})\cdot\exp(T\mathsf{y}_{1,1})\cdot\mathbf{L}=\mathbf{q}\pi_{\mathsf{Y}}(\mathbf{I}').$$

Finally, comparing the coefficient of $\mathsf{y}_{s_1,\mu^{i_1}}\mathsf{y}_{s_1,\mu^{i_2-i_1}}\cdots\mathsf{y}_{s_n,\mu^{i_n-i_{n-1}}}$ on both sides of the equation yields the theorem quickly. $\square$

13.3.3 *Weight One Relations*

If the level $N>3$ then there are some non-trivial linear relations in $\mathsf{CMZV}(1,N)$ of weight one whose structure is known to us, which we now review briefly.

Fix the Nth root of unity $\mu=\exp(2\pi i/N)$ as above. We know the cyclotomic units can be constructed using $1-\mu^j$ for $1\le j<N$ and we have

$$-\log(1-\mu)=\sum_{k>0}\frac{\mu^k}{k}=\mathrm{Li}_1(\mu).$$

Here we have chosen the branch cut of the logarithm by deleting the non-positive real numbers $(-\infty,0]$ from the complex plane. The next lemma is straightforward.

Lemma 13.3.32. *All of the following relations hold for* Li_1.

(i) *(Symmetry relations). For all $1<j<N/2$,*

$$\mathrm{Li}_1(\mu^j)-\mathrm{Li}_1(\mu^{-j})=\frac{(N-2j)\pi i}{N}=\frac{N-2j}{N-2}\big(\mathrm{Li}_1(\mu)-\mathrm{Li}_1(\mu^{-1})\big).$$

(ii) *(Distribution relations). For any divisor d of N and $1 \le a < d$, we have*

$$\sum_{0\le j<N/d} \mathrm{Li}_1(\mu^{a+dj}) = \mathrm{Li}_1(\mu^{aN/d}).$$

Proof. (i) is straightforward by Euler's formula $e^{i\theta} = \cos\theta + i\sin\theta$ and the double angle formula. (ii) follows from the identity

$$\sum_{j=1}^{N} \mu^{jM} = \begin{cases} N, & \text{if } N|M; \\ 0, & \text{otherwise.} \end{cases}$$

We leave the details as Exercise 13.1. □

Definition 13.3.33. For any level N, one may multiply any relation in Lemma 13.3.32 by any level N CMZV of weight $w-1$ to produce a nontrivial linear relation among some CMZVs of weight w (after using either stuffle or shuffle product to expand the products). We call such a relation a *weight one relation.*

Note that the weight one relations have the following forms in words: for all $w \in \mathfrak{A}^0$

$$\begin{cases} (N-2)\,\text{Ł}(\mathtt{y}_{1,j} * w - \mathtt{y}_{1,-j} * w) = (N-2j)\,\text{Ł}(\mathtt{y}_{1,1} * w - \mathtt{y}_{1,-1} * w), \\ \displaystyle\sum_{0\le j<d} \text{Ł}(\mathtt{y}_{1,a+jN/d} * w) = \text{Ł}(\mathtt{y}_{1,ad} * w). \end{cases} \tag{13.21}$$

where $\mathtt{y}_{n,j} = \mathtt{a}^{n-1}\mathtt{b}_j$ as in Sec. 13.3.1

The proof of the following theorem of Bass [38] (corrected by Ennola [197]) will take us too far afield so we have to refer the interested reader to their original papers. We will not use it in the rest of the book.

Theorem 13.3.34. *All $\mathbb{Q}$-linear relations among $\{\mathrm{Li}_1(\mu^j) : j = 1, \dots, N-1\}$ are consequences of relations I and II.*

13.3.4 *Regularized Distribution Relations*

It is well known that MPLs satisfy the following distribution formula: for all $d \in \mathbb{N}$, $\mathbf{s} \in \mathbb{N}^d$ and $\xi_1, \dots, \xi_n \in \Gamma_N$

$$\mathrm{Li}_{\mathbf{s}}(\xi_1, \dots, \xi_n) = d^{|\mathbf{s}|-n} \sum_{z_j^d=\xi_j, 1\le j\le n} \mathrm{Li}_{\mathbf{s}}(z_1, \dots, z_n). \tag{13.22}$$

When $s_1 = 1$ one has to exclude the divergent case $\xi_1 = 1$. We call these *finite distribution relations* (finite DISTs).

We now consider the regularized version of Eq. (13.22). For any divisor d of N, let $i_d : \Gamma_{N/d} \hookrightarrow \Gamma_N$ and let $\mathpzc{p}^d : \Gamma_N \twoheadrightarrow \Gamma_{N/d}$ be the dth power map. They induce two algebra homomorphisms, respectively:

$$\mathpzc{p}^d_* : \mathbb{Q}\langle \mathsf{X}_N \rangle \longrightarrow \mathbb{Q}\langle \mathsf{X}_{N/d} \rangle$$

$$\mathrm{x}_\xi \longmapsto \begin{cases} d\mathrm{x}_0, & \text{if } \xi = 0, \\ \mathrm{x}_{\xi^d}, & \text{if } \xi \in \Gamma_N, \end{cases}$$

and

$$i^*_d : \mathbb{Q}\langle \mathsf{X}_N \rangle \longrightarrow \mathbb{Q}\langle \mathsf{X}_{N/d} \rangle$$

$$\mathrm{x}_\xi \longmapsto \begin{cases} \mathrm{x}_\xi, & \text{if } \xi \in \Gamma_{N/d} \cup \{0\}, \\ 0, & \text{otherwise.} \end{cases}$$

It is easy to see that both i^*_d and $\mathpzc{p}^d_*$ are Δ-coalgebra homomorphisms such that $i^*_d(\mathbf{I})$ and $\mathpzc{p}^d_*(\mathbf{I})$ have the same image under the map π_{cv}.

Theorem 13.3.35. *For every divisor d of N,*

$$\mathpzc{p}^d_*(\mathbf{I}) = \exp\left(\sum_{\xi^d=1, \xi \neq 1} \mathrm{Li}_1(\xi)\mathrm{x}_1 \right) i^*_d(\mathbf{I}). \tag{13.23}$$

Proof. It is not hard to see that Eq. (13.22) is equivalent to $i^*_d(\mathbf{L}_{\mathrm{cv}}) = \pi_{\mathrm{cv}}\mathpzc{p}^d_*(\mathbf{L})$. Since both i^*_d and $\mathpzc{p}^d_*$ commute with $\mathbf{p}$ and $\mathbf{q}$, we have $i^*_d(\mathbf{I}_{\mathrm{cv}}) = \pi_{\mathrm{cv}}\mathpzc{p}^d_*(\mathbf{I})$. Note that both i^*_d and $\mathpzc{p}^d_*$ are coalgebra homomorphisms with respect to Δ_* we know $i^*_d(\mathbf{I})$ and $\mathpzc{p}^d_*(\mathbf{I})$ are both group-like. Further,

$$\pi_{\mathrm{cv}} i^*_d(\mathbf{I}) = i^*_d(\mathbf{I}_{\mathrm{cv}}) = \pi_{\mathrm{cv}}\mathpzc{p}^d_*(\mathbf{I}).$$

The relation Eq. (13.23) now follows immediately from Lemma 13.3.20. □

Combined with Prop. 13.3.42 the above result provides so-called *regularized distribution relations* (regularized DISTs) which of course include all the finite DISTs of CMZVs given by Eq. (13.22).

However, sometimes finite DISTs are not independent of the other relations. In the next theorem one sees that when the weight $w = 2$ and level N is a prime, all the DISTs in Eq. (13.22), where $\mathrm{x}_j = 1$ for all j, are consequences of the regularized DBSFs of CMZVs of level N.

Theorem 13.3.36. *For any prime p, write $L(i,j) = L_p(1,1|i,j)$ and $D(i) = L_p(2|i)$. Define for $1 \le i,j < p$:*

$$\mathrm{FDT} := -D(0) + p\sum_{j=0}^{p-1} D(j), \qquad \mathrm{RDS}(i) := D(i) + L(i,0) - L(i,-i),$$

$$\mathrm{FDS}(i,j) := D(i+j) + L(i,j) + L(j,i) - L(i,j-i) - L(j,i-j).$$

Then one has

$$\mathrm{FDT} = \sum_{1\le i<p} \mathrm{FDS}(i,i) + 2 \sum_{1\le j<i<p} \mathrm{FDS}(i,j) + 2\sum_{i=1}^{p-1} \mathrm{RDS}(i). \qquad (13.24)$$

Proof. When $p=2$ the second term on the right-hand side of Eq. (13.24) is vacuous. Then it is easy to see that both sides of Eq. (13.24) are equal to $D(0)+2D(1)$.

We now assume $p\ge 3$. Changing the order of summation yields that

$$\begin{aligned}
2\sum_{1\le j<i<p} D(i+j) &= \sum_{i=2}^{p-1}\sum_{j=1}^{i-1} D(i+j) + \sum_{j=1}^{p-2}\sum_{i=j+1}^{p-1} D(i+j)\\
&= \sum_{i=2}^{p-2}\sum_{i\ne j=1}^{p-1} D(i+j) + \sum_{j=1}^{p-2} D(j-1) + \sum_{i=2}^{p-1} D(i+1)\\
&= (p-3)\sum_{j=0}^{p-1} D(j) - \sum_{i=2}^{p-2} D(i) - \sum_{i=1}^{p-1} D(2i) + \sum_{j=1}^{p-2} D(j) + \sum_{j=2}^{p-1} D(j) + 2D(0)\\
&= (p-1)D(0) + (p-3)\sum_{j=1}^{p-1} D(j)
\end{aligned}$$

since $\sum_{j=0}^{p-1} D(i+j) = \sum_{j=0}^{p-1} D(j)$ for all i and $\sum_{i=1}^{p-1} D(2i) = \sum_{i=1}^{p-1} D(i)$. This implies that the dilogarithms on the right-hand side of Eq. (13.24) exactly add up to a finite DIST. Thus one only needs to show that all the double logarithms on the right-hand side of Eq. (13.24) cancel out.

First observe that $L(i,0)$ in $\mathrm{FDS}(i,i)$ and $\mathrm{RDS}(i)$ cancel out each other. Now let us consider the lattice points (i,j) of $\mathbb{Z}^2$ corresponding to $L(i,j)$. The points (i,j) corresponding to $L(i,j)$ with positive signs fill in exactly the area inside the square $[1,p-1]\times[1,p-1]$ (including boundary): $L(i,i)$ in $\mathrm{FDS}(i,i)$ provides the diagonal $y=x$, $\sum_{1\le j<i<p} L(i,j)$ (resp. $\sum_{1\le j<i<p} L(j,i)$) form the lower right (resp. upper left) triangular region. For the negative terms of the double logs, $L(i,-i)$ in $\mathrm{RDS}(i)$ provides the diagonal $x+y=p$, $\sum_{1\le j<i<p} L(i,j-i) = \sum_{i=2}^{p-1}\sum_{j=p+1-i}^{p-1} L(i,j)$ forms the upper right triangular region. Similarly, by changing the order of summation $\sum_{1\le j<i<p} L(j,i-j) = \sum_{i=1}^{p-2}\sum_{j=i+1}^{p-1} L(i,j-i) = \sum_{i=1}^{p-2}\sum_{j=1}^{p-1-i} L(i,j)$ fills the lower left region. □

To conclude this section we remark that numerical evidence up to level $N=169$ supports the following conjecture.

Conjecture 13.3.37. *In weight two, all regularized DISTs are consequences of the weight one relations, the regularized DBSFs and the depth two finite DISTs.*

13.3.5 *Lifted Relations from Lower Weights*

When we deal with the MZVs (resp. Euler sums) we hope that all the linear relations come from the regularized DBSFs (resp. regularized DBSFs and DISTs). Since there are no weight one relations when level $N = 3$ it is natural to ask if the regularized DBSFs and DISTs of weight w are enough to produce all the linear relations among weight w CMZVs at level three. Surprisingly, the answer is no.

The first counterexample is in weight four, *i.e.*, $(w, N) = (4, 3)$. Easy computation shows that there are 144 CMZVs in this case among which there are 239 nontrivial regularized DBSFs of weight four which include 191 finite DBSFs of weight four (see Eqs. (13.11) and (13.12)). Furthermore, it is easy to verify that all the seven regularized DISTs (including four finite DISTs) can be derived from the regularized DBSFs. Using these relations we get 127 independent linear relations among the 144 CMZVs. But we expect to have $d(4,3) \le 16$ by Thm. 13.2.1, so there must be at least one more relation linearly independent from the above. Where else can we find it? The answer is so-called *lifted relations.*

We know that a product of two weight two CMZVs is of weight four. So on each of the five regularized DBSFs (including two finite DBSFs) of weight two in $\mathsf{CMZV}(2,3)$ we can multiply any one of the nine CMZVs of $(w, N) = (2, 3)$ to get a relation in $\mathsf{CMZV}(4,3)$. For instance, we have a finite DBSF

$$Ł(\mathsf{y}_{1,1} * \mathsf{y}_{1,1} - \mathsf{y}_{1,1} ш \mathsf{y}_{1,1}) = L_3(2|2) + 2L_3(1,1|1,1) - L_3(1,1|1,0) = 0.$$

Here, as we mentioned in Remark 13.3.11 earlier, $\mathsf{y}_{1,1} = \mathsf{x}_\mu = b_1$ (since we are using it in a concrete computation with the aid of a computer). Multiplying by $L_3(1,1|1,1) = Ł(\mathsf{y}_{1,1}\mathsf{y}_{1,2})$ we obtain a new relation which is linearly independent from the regularized DBSFs of weight four in $\mathsf{CMZV}(4,3)$:

$$\begin{aligned}\zeta(&\mathsf{y}_{1,1}\mathsf{y}_{1,2} ш (\mathsf{y}_{2,0} + 2\mathsf{y}_{1,1}\mathsf{y}_{1,2} - 2\mathsf{y}_{1,1}\mathsf{y}_{1,0}))\\ &= L_3(1,1,2|1,1,0) + 2L_3(1,2,1|1,1,0) + 2L_3(2,1,1|1,1,0)\\ &\quad + L_3(2,1,1|2,2,1) + 4L_3(\{1\}^4|1,1,2,1) + 8L_3(\{1\}^4|1,0,1,0)\\ &\quad - 6L_3(\{1\}^4|1,0,0,1) - 4L_3(\{1\}^4|1,0,1,2) - 2L_3(\{1\}^4|1,1,2,0) = 0.\end{aligned}$$

Such relations coming from the lower weights are called *lifted relations.* In this way, when $(w, N) = (4, 3)$ we can produce 45 lifted regularized DBSFs from weight two, 58 from weight three. We may also lift regularized DISTs and obtain nine and six relations from weight two and three, respectively. However, all the lifted relations together only produce one new linearly independent relation, as expected. Hence we find totally 128 linearly independent relations among the 144 CMZVs with $(w, N) = (4, 3)$. This implies that $d(4, 3) \leq 16$ which is the same bound obtained by Thm. 13.2.1 and is proved to be exact under a variant of Grothendieck's period conjecture by Deligne (see [161] or [240]).

For levels $N \geq 4$, one may lift not only the regularized DBSFs and DISTs but also the weight one relations. But a moment reflection reveals that the lifted weight one relations are still weight one relations by themselves so one does not really need to consider them after all.

Definition 13.3.38. We call a $\mathbb{Q}$-linear relation among CMZVs *standard* if it can be produced by a $\mathbb{Q}$-linear combination of the following four families of relations: the regularized DBSFs, the regularized DISTs, the lifted relations from the above two types, and the weight one relations. Otherwise, it is called a *non-standard* relation.

For general level N and weight w, there are no strict inclusion relations among the four families of the standard relations. But for specific choice of N and w there may exist some. For an earlier example, see Thm. 13.3.36.

Computation in small weight cases supports the following conjecture.

Conjecture 13.3.39. *Suppose $N = 3$ or 4. Every CMZV of level N is a linear combination of CMZVs of the form $L(\{1\}^w | t_1, \ldots, t_w)$ with $t_j \in \{1, 2\}$. Consequently, the $\mathbb{Q}$-dimension of the CMZVs of weight w and level N is given by $d(w, N) = 2^w$ for all $w \geq 1$.*

Remark 13.3.40. The data in Table 14.1 shows that one cannot produce enough relations by using only the standard relations when $(w, N) = (3, 4)$. In fact, even though one has $d(3, 4) \leq 8$ and $d(4, 4) \leq 16$ by Thm. 13.2.1, one can only show that $d(3, 4) \leq 9$ and $d(4, 4) \leq 21$ by using only the standard relations. However, thanks to the octahedral symmetry of $\mathbb{P}^1 - (\{0, \infty\} \cup \boldsymbol{\mu}_4)$ one can find (presumably all) the non-standard relations in these two cases (see Thm. 14.6.1).

Remark 13.3.41. Let $N = 2, 3, 4$ or 8. Assuming a variant of Grothendieck's period conjecture, Deligne constructed explicitly a set of basis for $\mathsf{CMZV}(w, N)$ in [161] (see also [240]). Under the same assumption, his results would also imply that $d(w, 2)$ is given by the Fibonacci numbers, $d(w,3) = d(w,4) = 2^w$, and $d(w,8) = 3^w$.

13.3.6 *Group-Like Element* I

As in Sec. 13.3.2 let $\mathbf{I}$ be the unique group-like element in $(\mathbb{C}\langle\langle \mathsf{X}\rangle\rangle, \Delta)$ whose coefficients of $\mathtt{x}_0$ and $\mathtt{x}_1$ are 0 such that $\pi_{\rm cv}(\mathbf{I}) = \mathbf{I}_{\rm cv}$. In order to do the numerical computation one needs to determine explicitly the coefficients for $\mathbf{I}$. Put

$$\mathbf{I} = \sum_{k\in\mathbb{N}_0,,\xi_1,\dots,\xi_k\in\Gamma_N\cup\{0\}} C(\xi_1,\dots,\xi_k)\mathtt{x}_{\xi_1}\cdots\mathtt{x}_{\xi_k}. \tag{13.25}$$

Proposition 13.3.42. *Let k, m and n be three nonnegative integers. If $k > 0$ then we assume $\xi_1 \neq 1$ and $\xi_k \neq 0$. Set $(\xi_1,\dots,\xi_k,\{0\}^n) = (\xi_1,\dots,\xi_q)$. Then*

$$C(\{1\}^m,\xi_1,\dots,\xi_k,\{0\}^n)$$
$$= \begin{cases} 0, & \text{if } mn = k = 0; \\ Ł\left(\pi_{\rm cv}(\mathtt{x}_{\xi_1}\cdots\mathtt{x}_{\xi_k})\right), & \text{if } m = n = 0; \\ -\dfrac{1}{m}\displaystyle\sum_{i=1}^{q} C(\{1\}^{m-1},\xi_1,\dots,\xi_i,1,\xi_{i+1},\dots,\xi_q), & \text{if } m > 0; \\ -\dfrac{1}{n}\displaystyle\sum_{i=1}^{k} C(\xi_1,\dots,\xi_{i-1},0,\xi_i,\dots,\xi_k,\{0\}^{n-1}), & \text{if } m = 0, n > 0. \end{cases} \tag{13.26}$$

Here Ł is defined by Eq. (13.7) *after using the identification given by Remark* 13.3.12.

Remark 13.3.43. This proposition provides the recursive relations one may use to compute all the coefficients of $\mathbf{I}$.

Proof. Since $\mathbf{I}$ is group-like one has

$$\Delta\mathbf{I} = \mathbf{I}\otimes\mathbf{I}. \tag{13.27}$$

The first case follows from this immediately since $C(0) = C(1) = 0$. The second case is essentially the definition Eq. (13.7) of Ł. If $m > 0$ then one can compare the coefficient of

$$\mathtt{x}_1 \otimes \mathtt{x}_1^{m-1}\mathtt{x}_{\xi_1}\cdots\mathtt{x}_{\xi_q} \tag{13.28}$$

of the two sides of Eq. (13.27) and find the relation Eq. (13.26). Indeed, the coefficient on the right-hand side comes from only one possible tensor product which leads to the coefficient $C(1)C(\{1\}^{m-1},\xi_1,\dots,\xi_k,\{0\}^n)=0$ since $C(1)=1$ by Definition 13.3.22 of **I**. On the left-hand side, the term Eq. (13.28) can be produced by $\Delta(\mathrm{x}_1^m\mathrm{x}_{\xi_1}\cdots\mathrm{x}_{\xi_q})$ which gives m copies, and produced by $\Delta(\mathrm{x}_1^{m-1}\mathrm{x}_{\xi_1}\cdots\mathrm{x}_{\xi_i}\mathrm{x}_{\xi_1}\mathrm{x}_{\xi_{i+1}}\cdots\mathrm{x}_{\xi_q})$ for $i=1,\dots,q$ each of which gives only one copy.

Finally, if $m=0$ and $n>0$ then one may consider the coefficient of $\mathrm{x}_{\xi_1}\cdots\mathrm{x}_{\xi_k}a^{n-1}\otimes a$ in Eq. (13.27). This finishes the proof of the proposition. □

Theorem 13.3.44. *Let* $\mathbf{s}=(s_1,\dots,s_d)\in\mathbb{N}^d$ *and* $\xi_1,\dots,\xi_d\in\Gamma_N$. *Then*

$$\begin{aligned}&C(\{0\}^{s_1-1},\xi_1,\dots,\{0\}^{s_d-1},\xi_d,\{0\}^k)\\&\quad=(-1)^k\sum_{\substack{t_1+\cdots+t_d=k,\\ \mathbf{t}=(t_1,\dots,t_d)\in(\mathbb{N}_0)^d}}\prod_{j=1}^d\binom{s_j+t_j-1}{t_j}\text{Ł}_{\sqcup\!\sqcup}\left((\mathbf{s}+\mathbf{t})|\mathbf{i};0\right)\end{aligned}$$

where $\mathbf{i}=(i_1,\dots,i_d)$ *such that* $\xi_j=\mu^{i_j}$ *for all* $j=1,\dots,d$.

Proof. By induction and Prop. 13.3.42, it is not hard to reduce the proof of the theorem to the case when $k=0$:

$$C(\{0\}^{s_1-1},\xi_1,\dots,\{0\}^{s_d-1},\xi_d)=\text{Ł}_{\sqcup\!\sqcup}(\mathbf{s}|\mathbf{i};0)$$

which follows from Lemma 13.3.29 by setting $T=0$. □

By the substitution $t\to1-t$ we get

Theorem 13.3.45. *Let* $\mathbf{s}=(s_1,\dots,s_d)\in\mathbb{N}^d$ *and* $\xi_1,\dots,\xi_d\in(\Gamma_N\setminus\{1\})\cup\{0\}$. *Then*

$$\begin{aligned}&C(\{1\}^k,\xi_1,\{1\}^{s_1-1},\dots,\xi_d,\{1\}^{s_d-1})\\&\quad=(-1)^k\sum_{\substack{t_1,\dots,t_d\in\mathbb{N}_0\\ t_1+\cdots+t_d=k}}C\Big(\big\{\xi_j,\{1\}^{s_j+t_j-1}\big\}_{j=1}^d\Big)\prod_{j=1}^d\binom{s_j+t_j-1}{t_j}.\end{aligned}$$

13.4* Generalized Drinfeld Associators

It is possible to generalize the setup in Definition 4.1.2 to the higher levels. For $N\in\mathbb{N}$, let $\Psi_N(\mathrm{x}_0,\mathrm{x}_\xi:\xi\in\Gamma_N)$ be the generalized associator defined as the renormalized holonomy from 0 to 1 of

$$\frac{\mathrm{d}G}{\mathrm{d}z}=\left(\frac{\mathrm{x}_0}{z}+\sum_{\xi\in\Gamma_N}\frac{\mathrm{x}_\xi}{\xi^{-1}-z}\right)G,\tag{13.29}$$

i.e., $\Psi_N = G_1^{-1}G_0$, where G_0, G_1 are the solutions of Eq. (13.29) on the open interval $(0,1)$ such that $G_0(z) \simeq z^{\mathtt{x}_0}$ when $z \to 0^+$, $G_1(z) \simeq (1-z)^{\mathtt{x}_1}$ when $z \to 1^-$ and $\mathtt{x}_0$, $\mathtt{x}_\xi$ ($\xi \in \Gamma_N$) are free variables. When $N = 1$ $\Psi_1(\mathtt{x}_0, \mathtt{x}_1) = \Phi(\mathtt{x}_0, \mathtt{x}_1)$ is the KZ associator defined as in Definition 4.1.2. Similar to $\Phi(\mathtt{x}_0, \mathtt{x}_1)$ it is not very difficult to see that Ψ_N is the generating series for CMZVs.

Theorem 13.4.1. *The generalized Drinfeld associator* $\Psi_N(\mathtt{x}_0, \mathtt{x}_\xi : \xi \in \Gamma_N)$ *is the only element in Hopf algebra* $(\mathfrak{R}, \Delta_{Ш}, \epsilon_{Ш})$ *such that*

(i) $\Delta_{Ш}(\Psi_N) = \Psi_N \otimes \Psi_N$ (i.e. Ψ_N *is group-like),*
(ii) $\Psi_N[\mathtt{x}_0] = \Psi_N[\mathtt{x}_1] = 0$,
(iii) *For all* $\mathbf{w} = \mathtt{x}_0^{s_1-1}\mathtt{x}_{\xi_1} \cdots \mathtt{x}_0^{s_d-1}\mathtt{x}_{\xi_d} \in \mathfrak{A}^0$, *the coefficient of* $\mathbf{w}$ *is*

$$\Psi_N[\mathbf{w}] = (-1)^d \operatorname{Li}_{s_1,\dots,s_d}(\xi_1, \xi_2\xi_1^{-1}, \dots, \xi_d\xi_{d-1}^{-1}).$$

Proof. This follows from Lemma 13.3.20 immediately. □

To have a more thorough understanding of these generalized Drinfeld associators we need to study some complicated Lie algebras as follows. This is an extension of the objects defined in Definitions 4.1.16 to 4.1.21.

Definition 13.4.2. Let $\mathbb{K}$ be a field of characteristic 0. For $n \ge 2$, denote by $\mathfrak{t}_{n,N}$ the completed $\mathbb{K}$-Lie algebra with generators

$$t^{1i}\ (2 \le i \le n), \text{ and } t(\alpha)^{ij}\ (i \ne j, 2 \le i, j \le n, \alpha \in \mathbb{Z}/N\mathbb{Z})$$

subject to the following relations

$$\begin{aligned}
t(\alpha)^{ij} &= t(-\alpha)^{ji}, & [t(\alpha)^{ij}, t(\alpha+\beta)^{ik} + t(\beta)^{jk}] &= 0,\\
[t^{1i}, t(\alpha)^{jk}] &= 0, & [t^{1i}, t^{1j} + \sum_{\gamma \in \mathbb{Z}/N\mathbb{Z}} t(\gamma)^{ij}] &= 0,\\
[t(\alpha)^{ij}, t(\beta)^{kl}] &= 0, & [t(\alpha)^{ij}, t^{1i} + t^{1j} + \sum_{\gamma \in \mathbb{Z}/N\mathbb{Z}} t(\gamma)^{ij}] &= 0,
\end{aligned}$$

for all α, $\beta \in \mathbb{Z}/N\mathbb{Z}$ and all distinct i, j, k, l $(2 \le i, j, k, l \le n)$.

Note that $\mathfrak{t}_{n,1}$ is equal to $\mathfrak{t}_n$ for $n \ge 2$ (see Definition 4.1.16).

For a partially defined map $f : \{1, \dots, m\} \to \{1, \dots, n\}$ such that $f(1) = 1$, the Lie algebra morphism $\mathfrak{t}_{n,N} \to \mathfrak{t}_{m,N}$: $x \mapsto x^f = x^{f^{-1}(1),\dots,f^{-1}(n)}$ is uniquely defined by

$$(t(\alpha)^{ij})^f = \sum_{i' \in f^{-1}(i), j' \in f^{-1}(j)} t(\alpha)^{i'j'} \quad (i \ne j,\ 2 \le i,\ j \le n)$$

and $(2 \leq j \leq n)$,

$$(t^{1j})^f = \sum_{j' \in f^{-1}(j)} t^{1j'} + \frac{1}{2} \sum_{j',j'' \in f^{-1}(j)} \sum_{c \in \mathbb{Z}/N\mathbb{Z}} t(c)^{j'j''} + \sum_{i' \neq 1 \in f^{-1}(1), j' \in f^{-1}(j)} \sum_{c \in \mathbb{Z}/N\mathbb{Z}} t(c)^{i'j'}.$$

Again for a partially defined map $g : \{2, \ldots, m\} \to \{1, \ldots, n\}$, the Lie algebra morphism $\mathfrak{t}_n \to \mathfrak{t}_{m,N}$: $x \mapsto x^g = x^{g^{-1}(1), \ldots, g^{-1}(n)}$ is uniquely defined by

$$(t^{ij})^g = \sum_{i' \in g^{-1}(i), j' \in g^{-1}(j)} t(0)^{i'j'} \qquad (i \neq j,\ 1 \leq i,\ j \leq n).$$

Definition 13.4.3. Let σ be the cyclic permutation so that $\sigma(u_1, \ldots, u_N) = (u_2, \ldots, u_N, u_1)$. For $N \geq 1$, the Lie algebra $\mathfrak{grtm}_{(\bar{1},1)}(N, \mathbb{K})$ is defined to be the set of pairs $(\varphi, \psi) \in \mathfrak{grt}_1(\mathbb{K}) \times \mathfrak{t}^0_{3,N}$ satisfying the *mixed pentagon relation* in $\mathfrak{t}^0_{4,N}$

$$\psi^{1,2,34} + \psi^{12,3,4} = \varphi^{2,3,4} + \psi^{1,23,4} + \psi^{1,2,3}, \tag{13.30}$$

the *octagon relations* in $\mathfrak{t}^0_{3,N}$

$$\psi(\mathsf{a}, \mathbf{b}) - \psi(\mathsf{a}, \sigma(\mathbf{b})) + \psi(\mathsf{c}, \sigma^{-2}(\overleftarrow{\mathbf{b}})) - \psi(\mathsf{c}, \sigma^{-1}(\overleftarrow{\mathbf{b}})) = 0 \tag{13.31}$$

with $\mathbf{b} = (\mathsf{b}_0, \mathsf{b}_1, \ldots, \mathsf{b}_{N-1})$ and $\mathsf{a} + \sum_{\alpha \in \mathbb{Z}/N\mathbb{Z}} \mathsf{b}_\alpha + \mathsf{c} = 0$, the *special derivation condition* in $\mathfrak{t}^0_{3,N}$

$$\sum_{\alpha \in \mathbb{Z}/N\mathbb{Z}} [\psi(\mathsf{a}, \mathbf{b}) - \psi(\mathsf{c}, \sigma^{-1}(\overleftarrow{\mathbf{b}})), \mathsf{c}] + [\psi(\mathsf{a}, \sigma^\alpha(\mathbf{b})), \mathsf{b}_\alpha] = 0 \tag{13.32}$$

and $\psi[b_0] = 0$. The Lie bracket is defined in Exercise 13.17.

Definition 13.4.4. For $N \geq 1$, the group $\mathrm{GRTM}_{(\bar{1},1)}(N, \mathbb{K})$ is defined to be the set of pairs $(g, h) \in \mathrm{GRT}_1(\mathbb{K}) \times \exp \mathfrak{t}^0_{3,N}$ with $h[\mathsf{b}_0] = 0$ satisfying the *mixed pentagon relation* in $\exp \mathfrak{t}^0_{4,N}$

$$h^{1,2,34} h^{12,3,4} = g^{2,3,4} h^{1,23,4} h^{1,2,3}, \tag{13.33}$$

the *octagon relations* in $\exp \mathfrak{t}^0_{3,N}$

$$h(\mathsf{c}, \sigma^{-2}(\overleftarrow{\mathbf{b}})) h(\mathsf{c}, \sigma^{-1}(\overleftarrow{\mathbf{b}}))^{-1} h(\mathsf{a}, \mathbf{b}) = h(\mathsf{a}, \sigma(\mathbf{b})) \tag{13.34}$$

with $\mathsf{a} + \sum_{\alpha \in \mathbb{Z}/N\mathbb{Z}} \mathsf{b}_\alpha + \mathsf{c} = 0$ and the *special action condition* in $\exp \mathfrak{t}^0_{3,N}$

$$\mathsf{a} + \mathrm{Ad}\Big(h^{-1} \cdot h(\mathsf{c}, \sigma^{-1}(\overleftarrow{\mathbf{b}}))\Big)(\mathsf{c}) + \sum_{\alpha \in \mathbb{Z}/N\mathbb{Z}} \mathrm{Ad}(\tau_\alpha h^{-1})(\mathsf{b}_\alpha) = 0 \tag{13.35}$$

where τ_α is the automorphism defined by $\mathsf{a} \mapsto \mathsf{a}$ and $\mathsf{b}_\gamma \mapsto \mathsf{b}_{\gamma+\alpha}$ for all $\gamma \in \mathbb{Z}/N\mathbb{Z}$. The corresponding group multiplication is defined in Exercise 13.18.

Definition 13.4.5. The torsor $\mathrm{Pseudo}_{(\bar{1},1)}(N,\mathbb{K})$ is defined to be the set of pairs $(g,h)\in \mathrm{M}_1(\mathbb{K})\times\exp\mathfrak{t}^0_{3,N}$ with $h[b_0]=0$ satisfying the mixed pentagon equation (13.33) and the following variant of octagon equation in $\exp\mathfrak{t}^0_{3,N}$

$$e^{\frac{\mathsf{b}_1}{2}}h\big(\mathsf{c},\sigma^{-2}(\overleftarrow{\mathbf{b}})\big)e^{\frac{\mathsf{c}}{N}}h\big(\mathsf{c},\sigma^{-1}(\overleftarrow{\mathbf{b}})\big)^{-1}e^{\frac{\mathsf{b}_0}{2}}h\big(\mathsf{a},\mathbf{b}\big)e^{\frac{\mathsf{a}}{N}}=h\big(\mathsf{a},\sigma(\mathbf{b})\big). \tag{13.36}$$

It can be shown that $\mathrm{Pseudo}_{(\bar{1},1)}(N,\mathbb{K})$ is a right $\mathrm{GRTM}_{(\bar{1},1)}(N,\mathbb{K})$-torsor. See Exercise 13.19.

13.5 Historical Notes

Most of the recent research on MPLs have been stimulated by their intimate relations to theoretical physics (the Sec. 14.7 for the relevant history).

In the literature, the CMZVs have several aliases, such as MPL values at roots of unity [617], the MZVs rel. μ_N [164, 240], the cyclotomic MZVs [240, 241], or the multiple L-values [331]. Our adoption of the word "color" to denote the roots of unity is influenced by [51, 52] and has been used by some other authors under similar context (for e.g., [282, 373, 620]). The CMZVs are also intimately related to the cyclotomic harmonic polylogs treated in [2, 4]. Sometimes, the CMZVs are also disguised as the special values of multiple L-functions, for example, see [221, 331, 343, 543, 544].

In previous chapters, we concentrated our attention to the level one case of the CMZVs (*i.e.*, the MZVs). In this chapter, we have turned to the algebraic structures of CMZVs at higher levels, paying more attention to the $\mathbb{Q}$-linear relations between these values. Since the linear dependence of the CMZVs can be very hard transcendental number theory problems we are content with the upper bound of the dimensions for the moment. For the interested reader, a precise formulation of this problem is contained in the statement of a variant of Grothendieck's period conjecture which can be found in [161] or in § 5.27(c) of [164].

From the point of view of Lyndon words and quasi-symmetric functions Bigotte *et al.* have studied CMZVs primarily by using the DBSFs and the monodromy argument (cf. [51, Thm. 5.1]) building on earlier works by Hoang Ngoc and his collaborators [281, 287]. However, when the level $N\geq 2$, these DBSFs are frequently incomplete, as we have seen in this chapter (for level two, see also [62]).

Theorem 13.2.1 was proved by Deligne and Goncharov in their seminal joint paper [164, § 5.25] using the theory of motivic fundamental groups of $\mathbb{P}^1-(\{0,\infty\}\cup\boldsymbol{\mu}_N)$. Their machinery has since become the driving force for a series of new discoveries by many followers such as Brown [102–104] and his student Jarossay [317–319].

It is Kontsevich who first noticed that MZVs can be represented by iterated integrals (see [359]). The regularized DBSFs of the MZVs are also called the EDS (extended DBSFs) in [308]. Their approach was improved to treat the CMZVs by Arakawa and Kaneko [21]. Earlier, Racinet [477] extended the DBSF structure to the CMZVs in the more abstract Lie and Hopf algebra setting. Section 13.3.1 was thus built on the results of [308, 477]. The main result, Thm. 13.3.9, is essentially [477, Cor. 2.24]. The presentation of the standard relations in Sec. 13.3 follows mostly that of [477] except for the lifted relations.

When $N \geq 4$ there exist linear relations among CMZVs of weight one Li_1 by a theorem of Bass [38]. Using the general weight one relations induced by them is one of the key ideas in finding the formula of [164, § 5.25] concerning $d(w, N)$. Further, it follows from the motivic theory of classical polylogs developed by Beilinson and Deligne [44] and the Borel's theorem (see [251, Thm. 2.1]) that the $\mathbb{Q}$-dimension of $\mathsf{CMZV}(1, N)$ is

$$d(1, N) = \dim K_1(\mathbb{Z}[\mu_N][1/N]) \otimes \mathbb{Q} + 1 = \varphi(N)/2 + \nu(N),$$

where φ is Euler's totient function and $\nu(N)$ is the number of distinct prime factors of N.

In [173] Drinfeld first showed that there exists a formal Lie series P with $\mathbb{Q}$-coefficients such that $\exp P(ht^{12}, ht^{23})$ is essentially $\Phi(\mathtt{a}, \mathtt{b})$. More important is that his proof utilizes the properties of a remarkable group called the Grothendieck–Teichmüller group GT. He further showed that there is a canonical homomorphism from the absolute Galois group $\mathrm{Gal}(\overline{\mathbb{Q}}/\mathbb{Q})$ into the groups $\mathrm{GT}(\mathbb{Q}_l)$ for all primes l. A number of results were later generalized to higher level cases in [198, 199].

Enriquez showed in [199] that the special derivation condition (13.32) for ψ implies the octagon equation (13.31). However, in contrast to the level one case where the special derivation relation follows from the duality and the hexagon equations it is unknown if the level N special derivation relations (13.32) and (13.35) can be derived from the rest of the equations (see [199, Remark 7.8] and [202]).

On the other hand, Furusho [221] demonstrated that the mixed pentagon equation (13.33) implies the regularized DBSFs among CMZVs. This improved the corresponding result for the MZVs (see [219]) we encountered earlier as Thm. 4.1.23.

Exercises

13.1. (i) Prove the symmetry relations hold for Li_1, *i.e.*, for all $1 < j < N/2$

$$\mathrm{Li}_1(\mu^j) - \mathrm{Li}_1(\mu^{-j}) = \frac{N-2j}{N-2}\big(\mathrm{Li}_1(\mu) - \mathrm{Li}_1(\mu^{-1})\big).$$

(ii) Complete the proof of the distribution relations: for any divisor d of N and $1 \le a < d$

$$\sum_{0\le j<N/d} \mathrm{Li}_1(\mu^{a+dj}) = \mathrm{Li}_1(\mu^{aN/d}).$$

13.2. Prove Prop. 13.3.3 by mimicking the proof of Prop. 3.2.4.

13.3. Prove Prop. 13.3.5 and Prop. 13.3.6 by mimicking the proof of Prop. 3.2.8.

13.4. Show that $\mathbb{C}\langle\langle \mathsf{X}\rangle\rangle$, $\mathbb{C}\langle\langle \mathsf{Y}\rangle\rangle$, $\mathbb{C}\langle\langle \mathsf{X}\rangle\rangle_{\mathrm{cv}}$ and $\mathbb{C}\langle\langle \mathsf{Y}\rangle\rangle_{\mathrm{cv}}$ are all weight graded bialgebras.

13.5. Show that

$$\mathscr{Y} := \sum_{n\in\mathbb{N}_0,\xi\in\Gamma_N} \mathsf{y}_{n,\xi}$$

is a group-like element in $(\mathbb{C}\langle\langle \mathsf{Y}\rangle\rangle, \bullet, \Delta_*)$.

13.6. Prove that $\mathbf{q}$ defined by Eq. (13.14) and $\mathbf{p}$ defined by Eq. (13.15) are inverses to each other.

13.7. For any $\lambda \in \mathbb{C}$, define the $\mathbb{C}$-linear maps $\phi_\lambda, \psi : \mathbb{C}\langle\langle Y\rangle\rangle \longrightarrow \mathbb{C}\langle\langle Y\rangle\rangle$ by the following:

$$\phi_\lambda(\mathsf{y}_{n,j}) = \lambda^n \mathsf{y}_{n,j}, \qquad \psi(\mathsf{y}_{n,j}) = \mu^j \mathsf{y}_{n,j}.$$

Show that both ϕ_λ and ψ commute with Δ_* defined by Eq. (B.1).

13.8. Prove Eq. (13.18) by comparing the coefficient of $\mathsf{y}_{s_1,\xi_1}\cdots\mathsf{y}_{s_d,\xi_d}$ on both sides.

13.9. Verify that $\mathsf{y}_{1,1}\mathbb{C}\langle\langle \mathsf{Y}\rangle\rangle$ is a co-ideal for Δ_* and that the projection π_{cv} is a coalgebra homomorphism.

13.10. Prove Lemma 13.3.21 by taking $\mathfrak{g}_1 = \mathbb{C}\mathsf{y}_{1,1}$ and applying Cor. B.11.

13.11. Prove the FDT stated in Eq. (13.22). Then show that it is equivalent to $i_d^*(\mathbf{L}_{\mathrm{cv}}) = \pi_{\mathrm{cv}}\,\not{p}_*^d(\mathbf{L})$.

13.12. Prove that both i_d^* and $\not{p}_*^d$ commute with $\mathbf{p}$ and $\mathbf{q}$.

13.13. Prove that both i_d^* and p_*^d are coalgebra homomorphisms with respect to Δ_*.

13.14. Show that there is a natural injection $\mathfrak{t}_{n-1,N} \hookrightarrow \mathfrak{t}_{n,N}$ by $x \mapsto x^{0,\dots,n-1}$.

13.15. Show that the Lie subalgebra $\mathfrak{f}_{n,N}$ of $\mathfrak{t}_{n,N}$ generated by t^{1n} and $t(\alpha)^{in}$ $(2 \le i \le n-1,\ \alpha \in \mathbb{Z}/N\mathbb{Z})$ is free of rank $(n-2)N+1$ and forms an ideal of $\mathfrak{t}_{n,N}$. Further, show that $\mathfrak{t}_{n,N}$ is a semi-direct product of $\mathfrak{f}_{n,N}$ and $\mathfrak{t}_{n-1,N}$.

13.16. Prove that the element $z_{n,N} = \sum_{1\le i<j\le n} t^{ij}$ with $t^{ij} = \sum_{a\in\mathbb{Z}/N\mathbb{Z}} t(\alpha)^{ij}$ $(2 \le i < j \le n)$ is central in $\mathfrak{t}_{n,N}$. Let $\mathfrak{t}^0_{n,N}$ be the Lie subalgebra of $\mathfrak{t}_{n,N}$ with the same generators except for t^{1n} and the same relations as $\mathfrak{t}_{n,N}$. Show that

$$\mathfrak{t}_{n,N} = \mathfrak{t}^0_{n,N} \oplus \mathbb{K} \cdot z_{n,N}.$$

In particular, if $n=3$ then $\mathfrak{t}^0_{3,N}$ is the free Lie algebra $\mathfrak{F}_{N+1}$ of rank $N+1$ with generators $\mathsf{a} := t^{12}$ and $\mathsf{b}_\alpha = t(\alpha)^{23}$ $(\alpha \in \mathbb{Z}/N\mathbb{Z})$.

13.17. Let $U\mathfrak{F}_{N+1} = \mathbb{K}\langle\langle \mathsf{a}, \mathsf{b}_\alpha \rangle\rangle_{\alpha\in\mathbb{Z}/N\mathbb{Z}}$. For any $\mathbb{K}$-algebra homomorphism $\iota : U\mathfrak{F}_{N+1} \to S$, we denote by $\varphi\big(\iota(\mathsf{a}), \iota(\mathsf{b}_0), \dots, \iota(\mathsf{b}_{N-1})\big)$ the image $\iota(\varphi) \in S$. The Lie algebra bracket of $\mathfrak{grtm}_{(\bar{1},1)}(N,\mathbb{K})$ is given by

$$\langle (\varphi_1,\psi_1),(\varphi_2,\psi_2)\rangle = (\langle\varphi_1,\varphi_2\rangle, \langle\psi_1,\psi_2\rangle) \tag{13.37}$$

with

$$\begin{aligned} \langle\varphi_1,\varphi_2\rangle &= [\varphi_1,\varphi_2] + D_{\varphi_2}(\varphi_1) - D_{\varphi_1}(\varphi_2), \\ \langle\psi_1,\psi_2\rangle &= [\psi_1,\psi_2] + \bar{D}_{\psi_2}(\psi_1) - \bar{D}_{\psi_1}(\psi_2), \end{aligned}$$

where $\bar{D}_\psi$ is the derivation of $\mathfrak{t}^0_{3,N}$ defined by

$$\bar{D}_\psi(\mathsf{a}) = [\psi, \mathsf{a}],\ \ \bar{D}_\psi(\mathsf{b}_\alpha) = [\psi - \psi\big(\mathsf{a}, \mathsf{b}_\alpha, \mathsf{b}_{\alpha+1}, \dots, \mathsf{b}_{\alpha-1}\big), \mathsf{b}_\alpha]$$

for $\alpha \in \mathbb{Z}/N\mathbb{Z}$ and

$$\bar{D}_\psi(\mathsf{c}) = [\psi\big(\mathsf{c}, \mathsf{b}_0, \mathsf{b}_{N-1}, \dots, \mathsf{b}_1\big), \mathsf{c}].$$

Show that the Lie bracket is well defined. Furthermore, show that the Lie algebra structure is realized by the embedding

$$\mathfrak{grtm}_{(\bar{1},1)}(N,\mathbb{K}) \hookrightarrow \mathrm{Der}(\mathfrak{t}^0_3) \times \mathrm{Der}(\mathfrak{t}^0_{3,N})$$

sending $(\varphi,\psi) \mapsto (D_\varphi, \bar{D}_\psi)$.

13.18. Show that $\mathrm{GRTM}_{(\bar{1},1)}(N,\mathbb{K})$ forms a group if we define the multiplication by

$$(g_1,h_1)\circ(g_2,h_2)=\Big(g_2(\mathsf{a},\mathsf{b})\cdot g_1\big(\mathsf{a},\mathrm{Ad}(g_2^{-1})(\mathsf{b})\big),h_2\big(\mathsf{a},\mathsf{b}_0,\mathsf{b}_1,\dots,\mathsf{b}_{N-1}\big)\cdot h_1\big(\mathsf{a},\mathrm{Ad}(h_2^{-1})\mathsf{b}_0,\mathrm{Ad}(\tau_1h_2^{-1})\mathsf{b}_1,\dots,\mathrm{Ad}(\tau_{N-1}h_2^{-1})\mathsf{b}_{N-1}\big)\Big). \quad (13.38)$$

Further, show that the group structure is realized by the embedding (with opposite homomorphism)

$$\mathrm{GRTM}_{(\bar{1},1)}(N,\mathbb{K})\hookrightarrow \mathrm{Aut}\,\mathfrak{t}_3^0\times\mathrm{Aut}\,\mathfrak{t}_{3,N}^0$$

sending (g,h) to the automorphism $(A_g,\bar{A}_h)$ where $\bar{A}_h$ is defined by $\mathsf{a}\mapsto\mathsf{a}$ and $\mathsf{b}_\alpha\mapsto\mathrm{Ad}(\tau_\alpha h^{-1})\,(\mathsf{b}_\alpha)$ for $\alpha\in\mathbb{Z}/N\mathbb{Z}$. Finally, prove that its associated Lie algebra is $\mathfrak{grtm}_{(\bar{1},1)}(\mathbb{K})$.

13.19. Show that $\mathrm{Pseudo}_{(\bar{1},1)}(N,\mathbb{K})$ forms a right $\mathrm{GRTM}_{(\bar{1},1)}(N,\mathbb{K})$-torsor by (13.38) with $(g_1,h_1)\in\mathrm{Pseudo}_{(\bar{1},1)}(N,\mathbb{K})$ and $(g_2,h_2)\in\mathrm{GRTM}_{(\bar{1},1)}(N,\mathbb{K})$.

Chapter 14

Colored Multiple Zeta Values at Lower Levels

In this chapter, we concentrate on the special cases of CMZVs when the levels are two, four and six. The level two CMZVs are nothing but the (alternating) Euler sums. We start with a very useful online tool which can be used to check numerically if Euler sum identities are correct and also to detect $\mathbb{Q}$-linear relations between them.

14.1 EZ-Face: An Online Computation Resource

We first recall the definition of the one variable MPLs from Sec. 4.2. Let $\mathbf{s} = (s_1, \dots, s_d) \in \mathbb{N}^d$. Then for all $|z| < 1$, we have

$$\mathrm{L}_{\mathbf{s}}(z) = \int_0^z \left(\frac{dt}{t}\right)^{s_1-1} \frac{dt}{1-t} \cdots \left(\frac{dt}{t}\right)^{s_d-1} \frac{dt}{1-t} = \sum_{k_1>\cdots>k_d\geq 1} \frac{z^{k_1}}{k_1^{s_1}\cdots k_d^{s_d}}. \tag{14.1}$$

For convenience, we set $\lambda_z(\mathbf{s}) = \mathrm{L}_{\mathbf{s}}(1/z)$ for any $|z| \geq 1$. For $\mathbf{s}, \mathbf{r} \in \mathbb{N}^d$ and $1 \leq j \leq d$, define the concatenation products

$$\vec{a}_j = \vec{a}_j(\mathbf{s};\mathbf{r}) := = \{s_j+1, \{1\}^{r_j-1}, \dots, s_d+1, \{1\}^{r_d-1}\},$$

$$\overleftarrow{b}_j = \overleftarrow{b}_j(\mathbf{s};\mathbf{r}) := = \{r_j+1, \{1\}^{s_j-1}, \dots, r_1+1, \{1\}^{s_1-1}\},$$

and $\vec{a}_{d+1} := \overleftarrow{b}_0 := \{\}$.

Theorem 14.1.1. *Suppose $|u|, |v| \geq 1$ such that $1/u + 1/v = 1$. Then we have the Hölder convolution for the MZVs:*

$$\zeta(\vec{a}_1) = \zeta(\overleftarrow{b}_d) = \sum_{j=1}^{d} \left\{ \sum_{k=0}^{s_j} \lambda_u\big(s_j+1-k, \{1\}^{r_j-1}, \vec{a}_{j+1}\big)\lambda_v\big(\{1\}^k, \overleftarrow{b}_{j-1}\big) + \sum_{k=1}^{r_j-1} \lambda_u\big(\{1\}^k, \vec{a}_{j+1}\big)\lambda_v\big(r_j+1-k, \{1\}^{s_j-1}, \overleftarrow{b}_{j-1}\big) \right\} + \lambda_v\big(\overleftarrow{b}_d\big).$$

Proof. In the iterated integral representation of MZV by Eq. (3.1)

$$\zeta(\vec{a}_1) = \int_0^1 \mathtt{a}^{s_1}\mathtt{b}^{r_1}\cdots\mathtt{a}^{s_d}\mathtt{b}^{r_d}, \quad \mathtt{a} = \frac{dt}{t}, \ \mathtt{b} = \frac{dt}{1-t},$$

we may regard the path from 0 to 1 as a composition of the two paths: $\alpha_u : [0, 1/u] \to \mathbb{C}$ and $\beta_u : [1/u, 1] \to \mathbb{C}$. Suppose $|\mathbf{s}| + |\mathbf{r}| = w$ and rewrite $\zeta(\vec{a}_1) = \int_{\beta_u \alpha_u} f_1 \cdots f_w$. By Lemma 2.1.2(ii), we can express this iterated integral as a sum of $w+1$ products

$$\int_{\beta_u \alpha_u} f_1 \cdots f_w = \sum_{j=0}^{w} \int_{\beta_u} f_1 \cdots f_j \int_{\alpha_u} f_{j+1} \cdots f_w.$$

Applying the substitution $t_j \to 1 - t_j$ to the first factor of each term in the above sum we can convert these first factors to iterated integrals over the path α_v since $1 - 1/u = 1/v$. Now the theorem follows immediately. □

Remark 14.1.2. The Hölder convolution has important applications.

(i) Note that $\vec{a}_1$ and $\overleftarrow{b}_d$ are dual to each other by the duality relation Eq. (5.1).
(ii) Take $u = v = 2$. Theorem 14.1.1 provides us with a fast numerical computational tool for MZVs with very high precisions because the λ_2 converges geometrically. This is crucial when we want to find integer relations among the MZVs by using LLL or PSLQ algorithms (see Appendix D.3).
(iii) The Hölder convolution can be extended to compute other special values of MPLs. A free online resource [74], called EZ-face, has been developed jointly by J. Borwein, P. Lisonek and P. Irvine based on this idea, which can be used to handle the Euler sums.

Example 14.1.3. Take $\mathbf{s} = \mathbf{r} = (1)$. Then we have

$$\zeta(2) = 2\lambda_2(2) + \lambda_2(1)^2 = 2\sum_{n\geq 1} \frac{1}{2^n n^2} + \left(\sum_{n\geq 1} \frac{1}{2^n n}\right)^2$$

which gives a rather fast way to compute $\pi^2 = 6\zeta(2)$. In fact, by taking only 28 terms in the sums above we already obtain 8 valid digits after the decimal point: $\pi^2 \approx 9.86960440$. To achieve the same precision, we would have to add almost 1.4 billion terms with the definition $\pi^2 = 6\sum_{n\geq 1} \frac{1}{n^2}$.

Example 14.1.4. Take $\mathbf{s} = (2, 1)$ and $\mathbf{r} = (3, 2)$. Then we have

$$\begin{aligned}
&\zeta(3,1,1,2,1) \\
&\quad = \lambda_2(3,1,1,2,1) + \lambda_2(2,1,1,2,1)\lambda_2(1) + \lambda_2(1,1,1,2,1)\lambda_2(1,1) \\
&\qquad + \lambda_2(1,2,1)\lambda_2(3,1) + \lambda_2(1,1,2,1)\lambda_2(2,1) + \lambda_2(2,1)\lambda_2(4,1) \\
&\qquad + \lambda_2(1,1)\lambda_2(1,4,1) + \lambda_2(1)\lambda_2(2,4,1) + \lambda_2(3,4,1) = \zeta(3,4,1).
\end{aligned}$$

14.2 Euler Sums

In the case of $N = 2$ we have called the CMZVs the Euler sums. These values attracted a lot attention when they first appeared prominently in the computation of some Feynman integrals in the 1990s. Many interesting rational relations have been found among these values and the research in this area is still quite active up to this day.

Previously, we have set $\mathsf{a} = \mathrm{d}t/t$ and $\mathsf{b} = \mathrm{d}t/(1-t)$. For level $n = 2$, we need to introduce $\mathsf{c} := -\mathrm{d}t/(1+t)$, which will be in force throughout the rest of this chapter.

14.2.1 *Dimension Conjectures*

If an Euler sum can be expressed by a linear combination of the products of the Euler sums with lower weights and the Euler sums with lower depth then the Euler sum is called *reducible*. Otherwise, it is called *irreducible*.

Let $E_{n,d}$ (resp. $M_{n,d}$) be the minimum number of the irreducible Euler sums of weight n and depth d, such that every Euler sum (resp. MZVs) of weight n and depth d can be written as a $\mathbb{Q}$-linear combination of these irreducibles, augmented by products of irreducibles of lower weights and depths.

Conjecture 14.2.1. *We have*

$$\prod_{n\geq 3}\prod_{d\geq 1}(1-x^n y^d)^{E_{n,d}} = 1 - \frac{x^3 y}{1-x^2}\frac{1}{1-xy}, \tag{14.2}$$

$$\prod_{n\geq 3}\prod_{d\geq 1}(1-x^n y^d)^{M_{n,d}} = 1 - \frac{x^3 y}{1-x^2},. \tag{14.3}$$

Explicitly, these conjectural numbers can be given in terms of the following quantities[1]:

$$T(m,n) := \frac{1}{m+n}\sum_{d|m,d|n}\mu(d)\binom{(m+n)/d}{m/d}, \tag{14.4}$$

[1]Broadhurst calls these numbers *Euler's triangle* in [83, 84].

where μ is the Möbius function defined by Eq. (3.12). When $n \equiv d \pmod 2$ and $n > d$, one can show that

$$E_{n,d} = T((n-d)/2, d). \tag{14.5}$$

Similarly, when $n \equiv d \pmod 2$ and $n > 3d$, one has (see Exercise 14.1)

$$M_{n,d} = T((n-3d)/2, d). \tag{14.6}$$

In order to find the finer structures of the Euler sums we need to use the Lyndon words on the alphabet of all positive integers with its natural order (see Definition 3.3.3).

Conjecture 14.2.2. *Let $w > d$ be two positive integers. Then one can choose the irreducibles by using $\mathrm{L}(\mathbf{w})$ where $\mathbf{w} = \mathsf{c}$ or $\mathbf{w} = \mathsf{a}^{n_1-1}\mathsf{c}\mathsf{a}^{n_2-1}\mathsf{c}\dots\mathsf{a}^{n_d-1}\mathsf{c}$ subject to the following conditions:*

(i) $w = \sum_{j=1}^{d} n_j$,
(ii) $n_j \geq 2$,
(iii) $(n_d, \dots, n_1)$ *is a Lyndon word.*

For example, the words corresponding to the first few irreducible Euler sums are given below:

$$\mathsf{c},\ \mathsf{ac},\ \mathsf{a}^2\mathsf{c},\ \mathsf{a}^3\mathsf{c},\ \mathsf{a}^4\mathsf{c},\ \mathsf{a}^2\mathsf{cac},\ \mathsf{a}^5\mathsf{c},\dots$$

Using this we can produce the conjectured basis of MDVs in lower weights with following words:

$w = 1 : \mathsf{c}$;

$w = 2 : \mathsf{c}^{ш2},\ \mathsf{ac}$;

$w = 3 : \mathsf{c} ш \mathsf{ac},\ \mathsf{c}^{ш3},\ \mathsf{a}^2\mathsf{c}$;

$w = 4 : \mathsf{c}^{ш2} ш \mathsf{ac},\ \mathsf{c}^{ш4},\ \mathsf{c} ш \mathsf{a}^2\mathsf{c},\ \mathsf{ac} ш \mathsf{ac}$;

$w = 5 : \mathsf{c}^{ш3} ш \mathsf{ac},\ \mathsf{c}^{ш5},\ \mathsf{c}^{ш2} ш \mathsf{a}^2\mathsf{c},\ \mathsf{c} ш \mathsf{ac} ш \mathsf{ac},\ \mathsf{a}^2\mathsf{c} ш \mathsf{ac},\ \mathsf{a}^4\mathsf{c},\ \mathsf{a}^2\mathsf{cac};\dots$

When considering only the linear independence of the Euler sums, Broadhurst conjectured that the $\mathbb{Q}$-dimension d_n of the space ES_n of weight n Euler sum sums is given by the Fibonacci numbers: $d_1 = 1$, $d_2 = 2$, $d_3 = 3$, $d_4 = 5$, $d_5 = 8$, and so on. Zlobin further proposed another set of basis.

Conjecture 14.2.3. *The $\mathbb{Q}$-space ES_w has the following basis:*

$$\Big\{\zeta(\bar{b}_1, b_2, \dots, b_r) : b_j = 1 \text{ or } 2,\ \sum_{j=0}^{r} b_j = w\Big\}.$$

Consequently, $\dim_{\mathbb{Q}} \mathsf{CMZV}(w, 2)$ *are given by the Fibonacci numbers.*

14.2.2 *A Possible Integral Structure*

Our numerical evidence suggests there may exist even subtler **integral** structures. So we propose

Problem 14.2.4. Let n be a positive integer. Are there $\mathbb{Q}$-linearly independent Euler sums of weight n such that every Euler sum of weight n is a $\mathbb{Z}$-linear combination of these sums?

This problem can be compared to the corresponding integral structure of the group of cyclotomic units studied by Conrad [149, 150]. The MZVs seems to have an integral structure over $\mathbb{Z}[1/2]$, at least when the weight is less than 6.

We shall now use the regularized DBSF to compute the relations between the Euler sums of weight two, three and four. Most of the computations in this section are carried out by Maple. The relations can be verified numerically by EZ-face (see [74]) with error smaller than 10^{-45}. These numerical results all give affirmative answers to Problem 14.2.4.

Proposition 14.2.5. *All the weight two Euler sums can be expressed as $\mathbb{Z}$-linear combinations of $\zeta(\bar{2})$ and $\zeta(\bar{1},1)$:*

$$\zeta(2) = -2\zeta(\bar{2}), \quad \zeta(\bar{1},\bar{1}) = \zeta(\bar{2}) + \zeta(\bar{1},1)$$

Proof. It is easy to see from regularized DBSF that

$$\zeta(2) = -2\zeta(\bar{1},\bar{1}) + 2\zeta(\bar{1},1), \quad \zeta(\bar{2}) = -\zeta(\bar{1},1) + \zeta(\bar{1},\bar{1}).$$ □

Remark 14.2.6. From the proposition and a stuffle relation we get

$$2\zeta(\bar{1},1) = 2\zeta(\bar{1},\bar{1}) - 2\zeta(\bar{2}) = \zeta(\bar{1})^2 = \ln(2)^2.$$

Hence it is apparent that $\zeta(2)$ and $\zeta(\bar{1},1)$ are linearly independent which verifies the Main Conjecture 1 in this case.

Proposition 14.2.7. *We can express all weight three Euler sums as $\mathbb{Z}$-linear combinations of $\zeta(\bar{2},1)$, $\zeta(\bar{1},1,1)$ and $\zeta(\bar{1},2)$:*

$$\begin{aligned}
\zeta(3) &= 8\zeta(\bar{2},1),\\
\zeta(\bar{3}) &= -6\zeta(\bar{2},1),\\
\zeta(2,1) &= 8\zeta(\bar{2},1),\\
\zeta(2,\bar{1}) &= 2\zeta(\bar{2},1) - 3\zeta(\bar{1},2),\\
\zeta(\bar{2},\bar{1}) &= 3\zeta(\bar{1},2) - 7\zeta(\bar{2},1),\\
\zeta(\bar{1},\bar{2}) &= -2\zeta(\bar{1},2) + \zeta(\bar{2},1),
\end{aligned}$$

$$\begin{aligned}
\zeta(\bar{1},1,\bar{1}) &= \zeta(\bar{2},1) + \zeta(\bar{1},1,1),\\
\zeta(\bar{1},\bar{1},1) &= \zeta(\bar{1},2) - 5\zeta(\bar{2},1) + \zeta(\bar{1},1,1),\\
\zeta(\bar{1},\bar{1},\bar{1}) &= \zeta(\bar{1},2) + \zeta(\bar{1},1,1).
\end{aligned}$$

Proof. When weight is three, by only finite DBSF we have

$$\begin{aligned}
&\zeta(\bar{1},1,\bar{1}) + 2\zeta(\bar{1},\bar{1},1) + \zeta(\bar{1},\bar{2}) + \zeta(2,1) - 3\zeta(\bar{1},1,1) = 0,\\
&2\zeta(\bar{1},\bar{1},\bar{1}) + \zeta(\bar{1},2) + \zeta(2,\bar{1}) - 2\zeta(\bar{1},1,\bar{1}) = 0,\\
&\zeta(\bar{2},\bar{1}) + \zeta(\bar{1},\bar{2}) + \zeta(3) - 2\zeta(\bar{2},1) - \zeta(\bar{1},2) = 0,\\
&\zeta(\bar{1},2) + \zeta(\bar{3}) - \zeta(\bar{2},\bar{1}) - \zeta(\bar{1},\bar{2}) = 0.
\end{aligned}$$

These are far from enough to prove the proposition. But by regularized DBSF we have five more relations:

$$\begin{aligned}
&\zeta(\bar{3}) + 2\zeta(\bar{2},1) + \zeta(\bar{1},2) + 2\zeta(\bar{1},1,1) - \zeta(2,\bar{1}) + \zeta(\bar{1})\zeta(2) - 2\zeta(\bar{1},\bar{1},1) = 0,\\
&\zeta(\bar{1},1,\bar{1}) - \zeta(\bar{2},1) - \zeta(\bar{1},2) - 2\zeta(\bar{1},1,1) + \zeta(\bar{1},\bar{1},\bar{1}) = 0,\\
&\zeta(\bar{1},\bar{1},1) - \zeta(\bar{2},\bar{1}) - \zeta(\bar{1},\bar{2}) - \zeta(\bar{1},1,\bar{1}) = 0,\\
&\zeta(\bar{2},\bar{1}) - \zeta(\bar{3}) - \zeta(\bar{2},1) + \zeta(2,\bar{1}) = 0,\\
&\zeta(2,1) - \zeta(3) = 0.
\end{aligned}$$

Now the proposition follows from the stuffle relation: $\zeta(\bar{1})\zeta(2) = \zeta(\bar{3}) + \zeta(2,\bar{1}) + \zeta(\bar{1},2)$. □

Remark 14.2.8. By our standard conjectures there should be no further linear relations among $\zeta(\bar{2},1)$, $\zeta(\bar{1},1,1)$ and $\zeta(\bar{1},2)$ which gives $\dim_{\mathbb{Q}} \mathsf{CMZV}(3,2) = 3$. This is easy to see to be equivalence to the linear independence of $\zeta(3)$, $\zeta(\bar{1})\zeta(2)$ and $\zeta(\bar{1},\bar{1},1)$.

The preceding two propositions and the following results show that in weight ≤ 5 both Broadhurst–Zlobin Conjecture is true and our Problem 14.2.4 has an affirmative answer.

Proposition 14.2.9. *All weight four Euler sums are $\mathbb{Z}$-linear combinations of $A = \zeta(\bar{2},1,1)$, $B = \zeta(\bar{2},2)$, $C = \zeta(\bar{1},2,1)$, $D = \zeta(\bar{1},1,2)$, and $E = \zeta(\bar{1},1,1,1)$.*

The numerical data in the next proposition also shows that the $\mathbb{Q}$-basis conjectured by Zlobin can not be chosen as the $\mathbb{Z}$-linear basis in general.

Proposition 14.2.10. *All weight five Euler sums are $\mathbb{Q}$-linear combinations of $\zeta(\bar{1},1,1,1,1)$, $\zeta(\bar{1},1,2,1)$, $\zeta(\bar{2},1,1,1)$, $\zeta(\bar{1},1,1,2)$, $\zeta(\bar{1},2,1,1)$, $\zeta(\bar{2},1,2)$, $\zeta(\bar{2},2,1)$ and $\zeta(\bar{1},2,2)$.*

We put the detailed tables of the above in Appendix E.1.

14.2.3 *A Few Useful Integral Substitutions*

Recall that the Hoffman algebra $\mathfrak{A}_1 = \mathbb{Q}\langle \mathtt{a}, \mathtt{b}\rangle$ (resp. the Hoffman–Racinet algebra $\mathfrak{A}_2 = \mathbb{Q}\langle \mathtt{a}, \mathtt{b}, \mathtt{c}\rangle$ by setting $\mathtt{b} = \mathtt{b}_0$ and $\mathtt{c} = \mathtt{b}_1$ in Definition 13.3.2) is the graded noncommutative polynomial $\mathbb{Q}$-algebra generated by words $\mathsf{X}_1 = \{\mathtt{a}, \mathtt{b}\}$ (resp. $\mathsf{X}_2 = \{\mathtt{a}, \mathtt{b}, \mathtt{c}\}$). Let $\mathfrak{A}_2^0$ be the subalgebra generated by words not beginning with $\mathtt{b}$ and not ending with $\mathtt{a}$. Such words are called *admissible words.*

It is possible to prove that the only nontrivial changes of variable that preserve the interval $[0, 1]$ and send $\mathfrak{A}_1$ into $\mathfrak{A}_2$ are given by

$$\mathbf{w}(\mathtt{a}, \mathtt{b}) \mapsto \overleftarrow{\mathbf{w}}(\mathtt{b}, \mathtt{a}) \qquad \text{if } t \mapsto 1 - t, \tag{14.7}$$

$$\mathbf{w}(\mathtt{a}, \mathtt{b}) \mapsto \mathbf{w}(2\mathtt{a}, \mathtt{b} + \mathtt{c}) \qquad \text{if } t \mapsto t^2, \tag{14.8}$$

$$\mathbf{w}(\mathtt{a}, \mathtt{b}) \mapsto \mathbf{w}(\mathtt{a} + \mathtt{c}, \mathtt{b} - \mathtt{c}) \qquad \text{if } t \mapsto \frac{2t}{1+t}, \tag{14.9}$$

$$\mathbf{w}(\mathtt{a}, \mathtt{b}) \mapsto \mathbf{w}(\mathtt{a} + 2\mathtt{c}, 2\mathtt{b} - 2\mathtt{c}) \qquad \text{if } t \mapsto \frac{4t}{(1+t)^2}, \tag{14.10}$$

and their compositions, where $\overleftarrow{\mathbf{w}}(\mathtt{b}, \mathtt{a})$ denotes the word formed by switching $\mathtt{a}$ and $\mathtt{b}$ and then reversing the order of the letters in the word $\mathbf{w}(\mathtt{a}, \mathtt{b})$. For example,

$$\mathbf{w}(\mathtt{a}, \mathtt{b}) \mapsto \mathbf{w}(2\mathtt{a}, \mathtt{b} + \mathtt{c}) \qquad \text{if } t \mapsto 1 - 2t/(1+t) = (1-t)/(1+t). \tag{14.11}$$

Example 14.2.11. We now show that

$$\zeta(3) = 8\zeta(\bar{2}, 1). \tag{14.12}$$

By direct expansion in $\mathfrak{A}_2$ we obtain

$$2[\mathtt{ab}^2 - 2\mathtt{a}(\mathtt{b} + \mathtt{c})^2] + 8[\mathtt{ab}^2 - (\mathtt{a} + \mathtt{c})(\mathtt{b} - \mathtt{c})^2] + [(\mathtt{a} + 2\mathtt{c})(2\mathtt{b} - 2\mathtt{c})^2 - \mathtt{ab}^2] \\ = \mathtt{ab}^2 - 8\mathtt{ac}^2.$$

Using substitutions Eq. (14.8), Eq. (14.9) and Eq. (14.10) respectively for the three square brackets on the left-hand side, we get

$$\zeta(3) - 8\zeta(\bar{2}, 1) = \mathrm{L}_{\sqcup\!\sqcup}(\mathtt{ab}^2 - 8\mathtt{ac}^2) = 0.$$

14.2.4 *A Family of Identities*

In this subsection we generalize Eq. (14.12) as follows.

Theorem 14.2.12. *For every positive integer n,*

$$\zeta(\{3\}^n) = 8^n\zeta(\{\bar{2}, 1\}^n). \tag{14.13}$$

The key idea of the proof is to find some suitable combinations of the finite DBSFs and DISTs at level two.

In order to encode the stuffle relation we need to define the sub-algebra $\mathfrak{A}_2^1 = \mathbb{Q}\langle \mathsf{Y} \rangle$ of $\mathfrak{A}_2 = \mathbb{Q}\langle \mathsf{X} \rangle$ where Y is the alphabet consisting of the letters $\{\beta_n, \gamma_n : n \in \mathbb{N}\}$, where

$$\beta_n := \mathsf{a}^{n-1}\mathsf{b} \quad \text{and} \quad \gamma_n := \mathsf{a}^{n-1}\mathsf{c}.$$

For any word $\mathbf{w} \in \mathsf{Y}^*$ and positive integer n, define the toggling operator $\tau_{\beta_n}(\mathbf{w}) = \mathbf{w}$, and $\tau_{\gamma_n}(\mathbf{w})$ to be the word with β and γ exchanged. For example, $\tau_{\gamma_1}(\gamma_2\beta_4) = \beta_2\gamma_4$. Then the stuffle product defined in Eq. (13.10) can be rephrased as follows. For any $x, y \in \mathsf{Y}$ and $\mathbf{u}, \mathbf{v} \in \mathsf{Y}^*$,

$$\begin{aligned} x\mathbf{u} * y\mathbf{v} = x\Big(\tau_x\big(\tau_x(\mathbf{u}) * y\mathbf{v}\big)\Big) + y\Big(\tau_y\big(x\mathbf{u} * \tau_y(\mathbf{v})\big)\Big) \\ + [x, y]\Big(\tau_{[x,y]}\big(\tau_x(\mathbf{u}) * \tau_y(\mathbf{v})\big)\Big) \end{aligned} \tag{14.14}$$

where

$$[\beta_m, \beta_n] = [\gamma_m, \gamma_n] = \beta_{m+n}, \quad [\gamma_m, \beta_n] = [\beta_m, \gamma_n] = \gamma_{m+n}.$$

For any $l_1, l_2 \in \{\mathsf{a}, \mathsf{b}, \mathsf{c}\}$, define the $\diamond$-concatenation by setting $l_1 \diamond l_2 = l_1 l_2$ except that

$$\mathsf{b} \diamond \mathsf{b} = \mathsf{bc}, \quad \text{and} \quad \mathsf{c} \diamond \mathsf{c} = \mathsf{cb}. \tag{14.15}$$

For any positive integer i, define the ith cut of a word $l_1 \diamond \cdots \diamond l_m$ ($l_i \in \{\mathsf{a}, \mathsf{b}, \mathsf{c}\}$) to be a pair of words given by

$$\begin{aligned} &\mathrm{Cut}_i(l_1 \diamond l_2 \diamond \cdots \diamond l_m) \\ &\quad := \begin{cases} \big(l_1 \diamond l_2 \diamond \cdots \diamond l_{i-1} \diamond l_i,\ l_{i+1} \diamond l_{i+2} \diamond \cdots \diamond l_m\big), & \text{if } i \text{ is odd,} \\ \big(l_i \diamond l_{i-1} \diamond \cdots \diamond l_2 \diamond l_1,\ l_{i+1} \diamond l_{i+2} \diamond \cdots \diamond l_m\big), & \text{if } i \text{ is even,} \end{cases} \end{aligned}$$

for $i = 0, \cdots, m$. For any two words $\mathbf{w}_1$ and $\mathbf{w}_2$, we define the two bilinear maps $\mathfrak{A}_2^0 \times \mathfrak{A}_2^0 \to \mathfrak{A}_2^0$ by

$$ш(\mathbf{u}, \mathbf{v}) = \mathbf{u} ш \mathbf{v}, \quad \text{and} \quad * (\mathbf{u}, \mathbf{v}) = \mathbf{u} * \mathbf{v} \quad \forall \mathbf{u}, \mathbf{v} \in \mathsf{Y}^*.$$

Let $[r]$ and $\{r\}$ be the integral part and the fractional part of any $r \in \mathbb{R}$, respectively.

Theorem 14.2.13. *For every positive integer n, the following holds in $\mathfrak{A}_2^0$:*

$$\begin{aligned} 2^n(\mathsf{ac}^2\mathsf{ab}^2)^{[n/2]}(\mathsf{ac}^2)^{2\{n/2\}} &= (\mathsf{a}^2(\mathsf{b}+\mathsf{c}))^n \\ &+ \sum_{i=0}^{2n} (-1)^{n-i}\Big(ш \circ \mathrm{Cut}_i\big((\mathsf{c} \diamond \mathsf{d})^{\diamond n}\big) - * \circ \mathrm{Cut}_i\big((\mathsf{c} \diamond \mathsf{d})^{\diamond n}\big)\Big). \end{aligned} \tag{14.16}$$

Here, to compute $Ш \circ \mathrm{Cut}_i\left((\mathtt{c}\diamond\mathtt{d})^{\diamond n}\right)$ *we use the following steps: (i) apply* Cut_i *to* $(\mathtt{c}\diamond\mathtt{d})^{\diamond n}$ *by using the* $\diamond$*-concatenation, (ii) replace the letter* $\mathtt{d}$ *by* $\mathtt{a}(\mathtt{b}+\mathtt{c})$, *(iii) expand the* $\diamond$*-concatenation by the definition Eq.* (14.15), *and (iv) apply* $Ш$. *We can compute* $* \circ \mathrm{Cut}_i\left((\mathtt{c}\diamond\mathtt{d})^{\diamond n}\right)$ *similarly.*

Remark 14.2.14. Observe that $(\mathtt{b}+\mathtt{c})\diamond\mathtt{c} = (\mathtt{b}+\mathtt{c})\diamond\mathtt{b} = \mathtt{bc}+\mathtt{cb}$ and therefore

$$Ł((\mathtt{c}\diamond\mathtt{d})^{\diamond n}) = \sum_{t_1,\dots,t_n\in\{2,\bar{2}\}} \zeta(\bar{1},t_1,\bar{1},t_2,\dots,\bar{1},t_n).$$

The $\diamond$-concatenation appears because neither $\mathtt{c}^2$ nor $\mathtt{b}^2$ can appear in any of the Euler sums in the above sum. We should keep this in mind because the operator Cut_i will lead to some order reversals which also should obey this condition.

To prove Thm. 14.2.13 we need two separate identities involving stuffles and shuffles respectively.

Proposition 14.2.15. *For every positive integer* n,

$$\sum_{i=0}^{2n}(-1)^i * \circ \mathrm{Cut}_i\left((\mathtt{c}\diamond\mathtt{d})^{\diamond n}\right) = (-1)^n(\mathtt{a}^2(\mathtt{b}+\mathtt{c}))^n. \tag{14.17}$$

Proof. We proceed by induction on n. When $n=1$ the left-hand side of Eq. (14.17) is

$$\begin{aligned}\mathtt{cd}-\mathtt{c}*\mathtt{d}+\mathtt{d}\diamond\mathtt{c} &= \gamma_1(\gamma_2+\beta_2)-\gamma_1*(\gamma_2+\beta_2)+\mathtt{a}(\mathtt{bc}+\mathtt{cb})\\ &= -(\gamma_2+\beta_2)\tau\gamma_1-\beta_3-\gamma_3+\gamma_2\beta_1+\beta_2\gamma_1 = -\beta_3-\gamma_3.\end{aligned}$$

This is exactly the right-hand side $-\mathtt{a}^2(\mathtt{b}+\mathtt{c})$. Now assume that identity Eq. (14.17) holds up to $n-1$ for some $n\ge 2$. Set $\gamma=\gamma_1=\mathtt{c}$, $\mathtt{d}:=\beta_2+\gamma_2$ and $\mathtt{d}_3=\beta_3+\gamma_3$. Then

$$\Delta := \mathtt{d}\diamond\gamma = \beta_2\gamma+\gamma_2\beta_1.$$

In the rest of this section we set $\tau=\tau_\gamma$. Note that $\tau(\mathtt{d})=\mathtt{d}$ and $\tau(\Delta)=\Delta$. Hence by the recursive definition of the stuffle product Eq. (14.14)

$$\begin{aligned}&\sum_{i=0}^{2n}(-1)^i * \circ \mathrm{Cut}_i\left((\mathtt{c}\diamond\mathtt{d})^{\diamond n}\right)\\ &\quad= \sum_{j=0}^{n}\Delta^j * (\gamma\mathtt{d})^{\diamond(n-j)} - \sum_{j=1}^{n}\left(\gamma\Delta^{j-1}\right) * \left(\mathtt{d}\diamond(\gamma\mathtt{d})^{n-j}\right)\end{aligned}$$

$$
\begin{aligned}
&= \sum_{j=0}^{n}\Big(\beta_2\gamma\Delta^{j-1}\Big) * (\gamma\mathtt{d})^{\diamond(n-j)} - \sum_{j=1}^{n}\Big(\gamma\Delta^{j-1}\Big) * \Big(\beta_2\gamma\Delta^{n-j}\mathtt{d}\Big) \\
&\quad + \sum_{j=0}^{n}\Big(\gamma_2\beta_1\Delta^{j-1}\Big) * (\gamma\mathtt{d})^{\diamond(n-j)} - \sum_{j=1}^{n}\Big(\gamma\Delta^{j-1}\Big) * \Big(\gamma_2\beta_1\Delta^{n-j}\mathtt{d}\Big) \\
&= \sum_{j=1}^{n-1}\Big\{\beta_2\Big(\big(\gamma\Delta^{j-1}\big) * (\gamma\mathtt{d})^{\diamond(n-j)}\Big) + \gamma\tau\Big(\big(\beta_2\gamma\Delta^{j-1}\big) * \big(\Delta^{n-j-1}\mathtt{d}\big)\Big) \\
&\quad + \gamma_3\tau\Big(\big(\gamma\Delta^{j-1}\big) * \big((\mathtt{d}\diamond\gamma)^{n-j-1}\mathtt{d}\big)\Big)\Big\} + (\gamma\mathtt{d})^{\diamond n} + \Delta^n \\
&\quad - \sum_{j=1}^{n}\Big\{\gamma\tau\Big(\Delta^{j-1} * \big(\beta_2\gamma\Delta^{n-j}\mathtt{d}\big)\Big) + \beta_2\Big(\big(\gamma\Delta^{j-1}\big) * \big(\gamma\Delta^{n-j}\mathtt{d}\big)\Big) \\
&\quad + \gamma_3\tau\Big(\Delta^{j-1} * \big(\gamma(\mathtt{d}\diamond\gamma)^{n-j}\mathtt{d}\big)\Big)\Big\} \\
&\quad + \sum_{j=1}^{n-1}\Big\{\gamma_2\tau\Big(\big(\gamma\Delta^{j-1}\big) * (\gamma\mathtt{d})^{\diamond(n-j)}\Big) + \gamma\tau\Big(\big(\gamma_2\beta_1\Delta^{j-1}\big) * \big(\Delta^{n-j-1}\mathtt{d}\big)\Big) \\
&\quad + \beta_3\Big(\big(\gamma\Delta^{j-1}\big) * \big(\Delta^{n-j-1}\mathtt{d}\big)\Big)\Big\} \\
&\quad - \sum_{j=1}^{n}\Big\{\gamma\tau\Big(\Delta^{j-1} * \big(\gamma_2\beta_1\Delta^{n-j}\mathtt{d}\big)\Big) + \gamma_2\tau\Big(\big(\gamma\Delta^{j-1}\big) * \big(\gamma\Delta^{n-j}\mathtt{d}\big)\Big) \\
&\quad + \gamma_3\tau\Big(\Delta^{j-1} * \big(\gamma\Delta^{n-j}\mathtt{d}\big)\Big)\Big\}.
\end{aligned}
$$

Converting $\beta_2\gamma+\gamma_2\beta_1$ back to Δ and canceling all the terms not involving γ_3 or β_3 we get

$$
\begin{aligned}
&\sum_{i=0}^{2n}(-1)^i * \circ\,\mathrm{Cut}_i\left((\mathtt{c}\diamond\mathtt{d})^{\diamond n}\right) \\
&\qquad = \mathtt{d}_3\left\{\sum_{j=1}^{n-1}\big(\gamma\Delta^{j-1}\big) * \big(\Delta^{n-j-1}\mathtt{d}\big) - \sum_{j=1}^{n}\Delta^{j-1} * \big(\gamma\Delta^{n-j}\mathtt{d}\big)\right\}
\end{aligned}
$$

By induction assumption the expression in the last big curly bracket is $(-1)^n(\mathtt{a}^2(\mathtt{b}+\mathtt{c}))^{n-1}$. This proves the proposition since $\mathtt{d}_3 = \mathtt{a}^2(\mathtt{b}+\mathtt{c})$. $\square$

Proposition 14.2.16. *For every positive integer n,*

$$
\sum_{i=0}^{2n}(-1)^i \sqcup\!\sqcup \circ\,\mathrm{Cut}_i\left((\mathtt{c}\diamond\mathtt{d})^{\diamond n}\right) = (-2)^n(\mathtt{ac}^2\mathtt{ab}^2)^{[n/2]}(\mathtt{ac}^2)^{2\{n/2\}} \tag{14.18}
$$

and

$$\sum_{i=0}^{2n}(-1)^i \text{ш} \circ \mathrm{Cut}_i\big((\mathtt{b}\diamond\mathtt{d})^{\diamond n}\big) = (-2)^n(\mathtt{ab}^2\mathtt{ac}^2)^{[n/2]}(\mathtt{ab}^2)^{2\{n/2\}}. \quad (14.19)$$

Here we set $\mathtt{d}\diamond\mathtt{b} = \mathtt{d}\diamond\mathtt{c} = \mathtt{a}(\mathtt{cb}+\mathtt{bc})$.

Proof. We can proceed by induction on n similarly to that of Prop. 14.2.15. We leave the details to the interested reader. □

Proof of Theorem 14.2.13. This follows easily from Prop. 14.2.15 and Eq. (14.18) of Prop. 14.2.16. □

Proof of Theorem 14.2.12. For any positive integer n, we have

$$\zeta(\{\bar{2},1\}^n) = \text{Ł}\left((\mathtt{ac}^2\mathtt{ab}^2)^{[n/2]}(\mathtt{ac}^2)^{2\{n/2\}}\right),$$

and

$$\zeta(\{3\}^n) = \text{Ł}\left((\mathtt{a}^2\mathtt{b})^n\right).$$

On the other hand, by Eq. (14.8) an integral substitution $t \to t^2$ yields

$$\zeta(\{3\}^n) = 4^n\,\text{Ł}\left((\mathtt{a}^2(\mathtt{b}+\mathtt{c}))^n\right). \quad (14.20)$$

This also follows quickly from the following special case of the DIST Eq. (13.22):

$$\begin{aligned}\text{Ł}\left((\mathtt{a}^2(\mathtt{b}+\mathtt{c}))^n\right) &= \sum_{1\le i_1<\cdots<i_j\le n}\ \sum_{k_1>\cdots>k_n>0}\frac{(-1)^{k_{i_1}+\cdots+k_{i_j}}}{(k_1\cdots k_n)^3}\\ &= \sum_{k_1>\cdots>k_n>0}\frac{(1+(-1)^{k_1})\cdots(1+(-1)^{k_n})}{(k_1\cdots k_n)^3} = \frac{1}{4^n}\zeta(\{3\}^n).\end{aligned}$$

Now multiplying 4^n on both sides of Eq. (14.16) in Thm. 14.2.13 and then applying Ł we can see that Thm. 14.13 follows immediately from Eq. (13.11), Eq. (14.17) and Eq. (14.18). □

Remark 14.2.17. Maple computation shows that Eq. (14.20) cannot be derived from finite DBSF in general.

14.2.5 *Generalized Doubling Relations at Level Two*

The finite DISTs at level two are often called *the doubling relations* in the literature. It is possible to generalize this idea by a proper manipulation of the Euler sum series. What one obtains then should be essentially the regularized DISTs. For example, by using partial fraction expansion Eq. (5.14), we get (assuming $s>1$, $w=s+t$, and $A_i^{s,t}=\binom{s+t-i-1}{t-1}$):

$$
\begin{aligned}
E(s,t) &= \frac{1}{2}(\zeta(s,t)+\zeta(\bar{s},\bar{t}))\\
&= \sum_{m=1}^{\infty}\sum_{n=1}^{\infty}\frac{1}{(m+n)^s\ n^t}\frac{1+(-1)^m}{2}=\sum_{m=1}^{\infty}\sum_{n=1}^{\infty}\frac{1}{(2m+n)^s\ n^t}\\
&= \sum_{m=1}^{\infty}\sum_{n=1}^{\infty}\left[\sum_{i=1}^{s}\frac{A_i^{s,t}}{(2m+2n)^{w-i}\ (2m+n)^i}+\sum_{i=1}^{t}\frac{A_i^{t,s}}{(2m+2n)^{w-i}\ n^i}\right]\\
&= \sum_{i=1}^{s}\frac{1}{2^{w-i}}\sum_{m=1}^{\infty}\sum_{n=1}^{\infty}\frac{A_i^{s,t}}{(m+n)^{w-i}\ (2m+n)^i}+\sum_{i=1}^{t}\frac{A_i^{t,s}}{2^{w-i}}\zeta(w-i,i)\\
&= \sum_{i=1}^{t}\frac{A_i^{t,s}}{2^{w-i}}\zeta(w-i,i)+\sum_{i=1}^{s}\sum_{m=1}^{\infty}\sum_{n=m+1}^{\infty}\frac{A_i^{s,t}2^{i-w}}{(m+n)^i n^{w-i}}\\
&= \sum_{i=1}^{t}\frac{A_i^{t,s}}{2^{w-i}}\zeta(w-i,i)+\sum_{i=1}^{s}\sum_{m=1}^{\infty}\left(\sum_{n=1}^{\infty}-\sum_{n=1}^{m}\right)\frac{A_i^{s,t}2^{i-w}}{(m+n)^i n^{w-i}}\\
&= \sum_{i=1}^{t}\frac{A_i^{t,s}}{2^{w-i}}\zeta(w-i,i)+\sum_{i=1}^{s}\frac{A_i^{s,t}}{2^{w-i}}\zeta(i,w-i)-\sum_{i=1}^{s}\sum_{n=1}^{\infty}\sum_{m=n}^{\infty}\frac{A_i^{s,t}2^{i-w}}{(m+n)^i n^{w-i}}\\
&= \sum_{i=1}^{t}A_i^{t,s}2^{i-w}\zeta(w-i,i)+\sum_{i=1}^{s}A_i^{s,t}2^{i-w}\zeta(i,w-i)\\
&\quad -\sum_{i=1}^{s}\sum_{n=1}^{\infty}\sum_{m=1}^{\infty}\frac{A_i^{s,t}2^{i-w}}{(m+2n)^i n^{w-i}}-\sum_{i=1}^{s}\sum_{n=1}^{\infty}\frac{A_i^{s,t}2^{i-w}}{(2n)^i n^{w-i}}\\
&= \sum_{i=1}^{t}A_i^{t,s}2^{i-w}\zeta(w-i,i)+\sum_{i=1}^{s}A_i^{s,t}2^{i-w}\zeta(i,w-i)\\
&\quad -\sum_{i=1}^{s}\sum_{n=1}^{\infty}\sum_{m=1}^{\infty}\frac{1+(-1)^n}{2}\frac{A_i^{s,t}}{(m+n)^i n^{w-i}}-\sum_{i=1}^{s}\sum_{n=1}^{\infty}\frac{A_i^{s,t}2^{-w}}{n^w}\\
&= \sum_{i=1}^{t}A_i^{t,s}2^{i-w}\zeta(w-i,i)+\sum_{i=1}^{s}A_i^{s,t}2^{i-w}\zeta(i,w-i)
\end{aligned}
$$

$$-\sum_{i=1}^{s} A_i^{s,t}\frac{1}{2}\Big(\zeta(i,w-i)+\zeta(i,\overline{w-i})\Big)-\frac{(w-1)!}{(s-1)!\,t!}2^{-w}\zeta(w).$$

For larger depths, one can obtain similar relations as above. We call these *the generalized doubling relations.*

Conjecture 14.2.18. *The regularized DBSFs, finite DISTs and the generalized doubling relations are sufficient to reduce the Euler sums of a given weight and depth to a minimal set that is in agreement with Conjecture* 14.2.1*, both in weight and in depth.*

14.3 Nonstandard Octahedral Relations at Level Four

In this section, we will show that there are explicit non-standard relations at level four. Such a relation occurs already in weight three:

$$\begin{aligned}5\,\mathrm{Li}_{1,2}(-1,-i)&=46\,\mathrm{Li}_{1,1,1}(i,1,1)-7\,\mathrm{Li}_{1,1,1}(-1,-1,i)\\&\quad-13\,\mathrm{Li}_{1,1,1}(i,i,i)+13\,\mathrm{Li}_{1,2}(-i,i)-\mathrm{Li}_{1,1,1}(-i,-1,1)\\&\quad+25\,\mathrm{Li}_{1,1,1}(-i,1,1)-8\,\mathrm{Li}_{1,1,1}(i,i,-1)+18\,\mathrm{Li}_{2,1}(-i,1).\qquad(14.21)\end{aligned}$$

Note that there are $N-1$ weight one CMZVs of level N and there are $N^2(N+1)^{w-2}$ weight $w\geq 2$ CMZVs of level N. Therefore there are 80 CMZVs of weight three and level four. In order to show the nonstandard relation (14.21), one may proceed as follows. First, compute the following relations:

(1) All the finite DBSFs, *i.e.*, multiplying each CMZV of weight one and each CMZV of weight two: there are 48 such relations.
(2) All the regularized non-finite DBSFs: there are 19 such relations.
(3) All the regularized DISTs: there are 14 such relations.
(4) All the weight one relations: there are 16 such relations.

Second, write down all of these relations as linear equations involving 80 formal variables (*i.e.*, weight three CMZVs at level four). Last, use any computer algebra system to solve the linear system. The upshot is that there are 9 free variables given by the CMZVs appearing in Eq. (14.21). But we know the upper bound of $\dim \mathsf{CMZV}(3,4)$ is 8 according to Thm. 13.2.1. Moreover, with computer aided computation it can be shown that the system with all the above relations together with Eq. (14.21) has only 8 free variable. Therefore, if Eq. (14.21) is valid then it is indeed a *non-standard relation.*

Remark 14.3.1. In fact, the computation is a little simpler if we use the Lie algebra approach. However, this would require us to use the motivic language which is beyond the scope of this book. For the interested reader, please see the author's original paper [617].

To prove Eq. (14.21) we need to study the geometry of the Riemann sphere $\mathbb{P}^1$ punctured at $\{0, \infty, \pm 1, \pm i\}$ which clearly possesses an octahedral symmetry. However, one should use a system of tangential base points stable (up to multiplication by roots of unity) by the octahedral group. Hence one can use the straight line path from 1_0 to $(-2)_1$. We now explain the meaning of this in details.

Recall that in § 13.3.2, we considered the Hoffman–Racinet algebra $\mathbb{Q}\langle\langle \mathtt{x}_0, \mathtt{x}_\xi \rangle\rangle_{\xi\in\Gamma_4}$. We define $\mathtt{x}_\infty = -\sum_{\xi\in\Gamma_4\cup\{0\}} \mathtt{x}_\xi$. Define the $\mathbb{C}$-linear map $\rho : \mathbb{Q}\langle\langle \mathtt{x}_0, \mathtt{x}_\xi \rangle\rangle_{\xi\in\Gamma_4} \longrightarrow \mathbb{Q}\langle\langle \mathtt{x}_0, \mathtt{x}_\xi \rangle\rangle_{\xi\in\Gamma_4}$ such that

$$\rho : \mathtt{x}_0 \to \mathtt{x}_1 \to \mathtt{x}_i \to \mathtt{x}_0, \qquad \mathtt{x}_\infty \to \mathtt{x}_{-1} \to \mathtt{x}_{-i} \to \mathtt{x}_\infty.$$

We also need the Möbius transformation

$$R(z) := \frac{i-z}{i+z}, \qquad \text{and its inverse} \qquad R^{-1}(z) = \frac{i(1-z)}{1+z}.$$

Note that both ρ and R have order 3. The key fact is that for any path $\mathfrak{p}$

$$\int_{R(\mathfrak{p})} \Omega = \rho\left(\int_{\mathfrak{p}} \Omega\right). \tag{14.22}$$

Let $0 < \varepsilon < 1/3$ and C_ε be the path $A_1, \dots, A_6$ in the complex plane shown in Figure 14.1, where A_2, A_4 and A_6 are the quarter circles of radii ε, 2ε and 2ε, respectively, oriented clockwise.

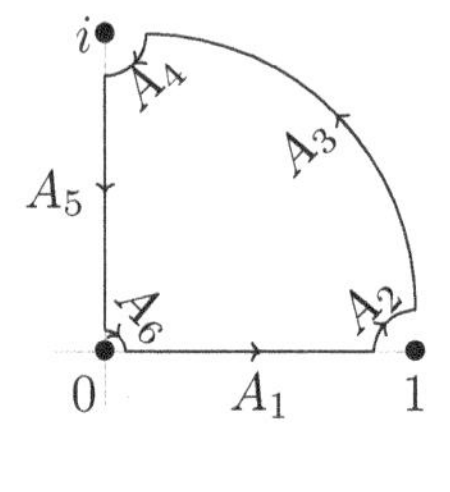

Figure 14.1: The contour of the integral for producing a non-standard relations in weight three and level four.

Let $\Omega = \sum_{\xi\in\Gamma_4\cup\{0\}} \frac{\mathtt{x}_\xi}{z-\xi} dz$. Then by Definition 13.3.22 the regularized iterated integral

$$\mathcal{J} = \sum_{n=0}^{\infty} \mathrm{Reg} \int_0^1 \Omega^{\circ n}$$

is a group-like element in $(\mathbb{C}\langle\langle \mathsf{X}\rangle\rangle, \Delta)$. Since $\int_{\mathfrak{p}} \Omega^{\circ 0} = 1$ for any path $\mathfrak{p}$, we have

$$1 = \int_{C_\varepsilon} \sum_{n=0}^{\infty} \Omega^{\circ n} = \int_{A_6} \sum_{n=0}^{\infty} \Omega^{\circ n} \int_{A_5} \sum_{n=0}^{\infty} \Omega^{\circ n} \cdots \int_{A_1} \sum_{n=0}^{\infty} \Omega^{\circ n}. \tag{14.23}$$

We now compute Eq. (14.23) explicitly. First, by the shuffle property of the iterated integrals

$$\sum_{n=0}^{\infty} \int_{A_6} \Omega^{\circ n} = \sum_{n=0}^{\infty} \int_{A_6} \left(\frac{\mathrm{d}t}{t}\mathsf{x}_0\right)^{\circ n} = \sum_{n=0}^{\infty} \frac{1}{n!} \left(\int_{A_6} \frac{\mathrm{d}t}{t}\mathsf{x}_0\right)^n = e^{(-2\pi i/4)\mathsf{x}_0}$$

Similarly we have

$$\sum_{n=0}^{\infty} \int_{A_2} \Omega^{\circ n} = e^{(-2\pi i/4)\mathsf{x}_1}, \qquad \sum_{n=0}^{\infty} \int_{A_4} \Omega^{\circ n} = e^{(-2\pi i/4)\mathsf{x}_i}.$$

Now A_1 is the straight path $\varepsilon \to 1-\varepsilon$ followed by the straight path $1-\varepsilon \to 1-2\varepsilon$. Thus

$$\begin{aligned}\int_{A_1} \sum_{n=0}^{\infty} \Omega^{\circ n} &= \int_{1-\varepsilon}^{1-2\varepsilon} \sum_{n=0}^{\infty} \Omega^{\circ n} \cdot \int_{\varepsilon}^{1-\varepsilon} \sum_{n=0}^{\infty} \Omega^{\circ n} \\ &= \int_{1-\varepsilon}^{1-2\varepsilon} \sum_{n=0}^{\infty} \left(\frac{\mathrm{d}z}{z-1}\mathsf{x}_1\right)^{\circ n} \cdot \int_{\varepsilon}^{1-\varepsilon} \sum_{n=0}^{\infty} \Omega^{\circ n} + O(\varepsilon).\end{aligned}$$

Regularizing by using Definition 13.3.22 we have

$$\mathrm{Reg} \sum_{n=0}^{\infty} \int_{A_1} \Omega^{\circ n} = e^{\mathsf{x}_1 \log 2} \mathcal{J}.$$

It follows from (14.22) that

$$\begin{aligned}\int_{A_3} \Omega &= \rho\left(\int_{R^2(A_3)} \Omega\right) = \rho\left(\int_{A_1} \Omega\right) + O(\varepsilon), \\ \int_{A_5} \Omega &= \rho^2\left(\int_{R(A_5)} \Omega\right) = \rho^2\left(\int_{A_1} \Omega\right) + O(\varepsilon).\end{aligned}$$

Thus

$$\mathrm{Reg} \sum_{n=0}^{\infty} \int_{A_3} \Omega^{\circ n} = e^{\mathsf{x}_i \ln 2} \rho(\mathcal{J}), \quad \mathrm{Reg} \sum_{n=0}^{\infty} \int_{A_5} \Omega^{\circ n} = e^{\mathsf{x}_0 \ln 2} \rho^2(\mathcal{J}).$$

Remark 14.3.2. Using tangential base points the above computation can be simplified a lot. For example, replacing the straight path $1_0 \to (-1)_1$ by the straight path $1_0 \to (-2)_1$ changes $\mathcal{J}$ to $\mathcal{J}' = e^{\mathsf{x}_1 \log 2} \mathcal{J}$. So the regularized integral I over the path A_1 followed by A_2 gives $I = e^{-2\pi i \mathsf{x}_1/4} \mathcal{J}' = e^{-2\,\mathrm{Li}_1(i)\mathsf{x}_1} \mathcal{J}$.

Hence (14.23) yields $\rho^2(I)\rho(I)I = 1$, i.e.,

$$e^{-2\,\mathrm{Li}_1(i)\mathtt{x}_0}\rho^2(\mathcal{J})e^{-2\,\mathrm{Li}_1(i)\mathtt{x}_i}\rho(\mathcal{J})e^{-2\,\mathrm{Li}_1(i)\mathtt{x}_1}\mathcal{J} = 1. \tag{14.24}$$

Note that all the coefficients in the above equation can be written explicitly as a $\mathbb{Q}$-linear combination of the nine CMZVs appearing in Eq. (14.21). By comparing the coefficient of $\mathtt{x}_2\mathtt{x}_1^2$ in (14.24) one finally arrives at (14.21).

Similarly, in weight four there are five $\mathbb{Q}$-linearly independent non-standard octahedral relations at level four. See the computation in Sec. 14.6.

14.4 Multiple Clausen and Deligne Values at Level Six

In this section, we look at some special forms of CMZVs at level six. Throughout this section, we let $\mu = \mu_6 := (1 + i\sqrt{3})/2$ be a primitive 6th root of unity and put $\theta = \pi/3$.

14.4.1 *Definitions*

Set $\mathsf{a} = \mathrm{d}\log(t)$, $\mathsf{b} = \mathrm{d}\log(1-t)$ and $\boldsymbol{\delta} = \mathrm{d}\log(1-\mu t)$. Recall that we have the evaluation map $\text{Ł} : \mathfrak{A}_6^1 \to \mathbb{C}$, where $\mathfrak{A}_6$ is the level 6 Hoffman–Racinet algebra defined by Definition 13.3.2.

Definition 14.4.1. A *multiple Deligne value* (MDV) is given by $(-1)^d\,\text{Ł}(\mathbf{w})$ for some admissible word $\mathbf{w} \in \{\mathsf{a}, \mathsf{b}, \boldsymbol{\delta}\}^*$ of depth d. Here, the evaluation map Ł is defined by Prop. 13.3.8.

Recall we have defined the one variable MPL $\mathrm{L}_{\mathbf{s}}(z) = \mathrm{Li}_{\mathbf{s}}(z, 1, \dots, 1)$ by Eq. (14.1).

Definition 14.4.2. Let $\mathbf{s} = (s_1, \dots, s_d) \in \mathbb{N}^d$. We set $\lambda(\mathbf{s}) := \mathrm{L}_{\mathbf{s}}(\mu_6)$ and call it a *multiple Clausen value* (MCV). Note that MCVs are complex numbers in general.

Numerical evidence shows that every MDV can be expressed as a $\mathbb{Q}$-linear combination of MCVs of the same weight. See Conjecture 14.4.27.

Lemma 14.4.3. *For all* $\mathbf{s} = (s_1, \dots, s_d) \in \mathbb{N}^d$, *we have*

$$\lambda(s_1 + 1, \dots, s_d) = \int_0^\mu \frac{\mathrm{Li}_{s_1,\dots,s_d}(z)}{z}dz \tag{14.25}$$

$$\lambda(1, s_1, \dots, s_d) = \int_0^\mu \frac{\mathrm{Li}_{s_1,\dots,s_d}(z)}{1-z}dz. \tag{14.26}$$

Proof. By the series definition in Eq. (14.1) we easily see that

$$\frac{\mathrm{d}}{\mathrm{d}z}\,\mathrm{L}_{s_1+1,s_2,\dots,s_n}(z) = \frac{1}{z}\,\mathrm{L}_{s_1,\dots,s_n}(z), \tag{14.27}$$

$$\frac{\mathrm{d}}{\mathrm{d}z}\,\mathrm{L}_{1,s_2,\dots,s_n}(z) = \frac{1}{1-z}\,\mathrm{L}_{s_2,\dots,s_n}(z). \tag{14.28}$$

Integrating these two expressions we obtain Eqs. (14.25) and (14.26), respectively. □

Lemma 14.4.4. *For all $|z| \le 1$ but $z \neq 1$, we have*

$$\mathrm{L}_{\{1\}^n}(z) = \frac{(-\log(1-z))^n}{n!}. \tag{14.29}$$

Proof. Repeated application of (14.28). □

The following formula is straightforward.

Lemma 14.4.5. *For any $\mathbf{s} = (s_1, \dots, s_d) \in \mathbb{N}^d$, we have*

$$\lambda(\mathbf{s}) = (-1)^d\, \text{Ł}(\mathsf{a}^{s_1-1}\boldsymbol{\delta}\cdots\mathsf{a}^{s_d-1}\boldsymbol{\delta}).$$

Hence every MCV is a MDVs.

Definition 14.4.6. Let $\mathbf{s} = (s_1, \dots, s_d) \in \mathbb{N}^d$ and $|\mathbf{s}| = w$. A *multiple Glaishers* (MGl) is the value $\mathrm{mgl}(\mathbf{s}) := \mathrm{Re}\big(i^w \lambda(\mathbf{s})\big)$. On the other hand, we call $\mathrm{mcl}(\mathbf{s}) := \mathrm{Im}\big(i^w \lambda(\mathbf{s})\big)$ a *multiple Clausens* (MCl). We call $|\mathbf{w}|$ their *weight.* Note that both MGl and MCl are real numbers.

By the definition, the MGl at $\mathbf{s}$ has the following expansion

$$\mathrm{mgl}(\mathbf{s}) = \begin{cases} (-1)^{(w+1)/2} \displaystyle\sum_{k_1>\cdots>k_d>0} \frac{\sin(k_1\theta)}{k_1^{s_1}\cdots k_d^{s_d}}, & \text{if } w \text{ is odd;} \\ (-1)^{w/2} \displaystyle\sum_{k_1>\cdots>k_d>0} \frac{\cos(k_1\theta)}{k_1^{s_1}\cdots k_d^{s_d}}, & \text{if } w \text{ is even.} \end{cases}$$

The MCl at $\mathbf{s}$ has the following expansion

$$\mathrm{mcl}(\mathbf{s}) = \begin{cases} (-1)^{(w-1)/2} \displaystyle\sum_{k_1>\cdots>k_d>0} \frac{\cos(k_1\theta)}{k_1^{s_1}\cdots k_d^{s_d}}, & \text{if } w \text{ is odd;} \\ (-1)^{w/2} \displaystyle\sum_{k_1>\cdots>k_d>0} \frac{\sin(k_1\theta)}{k_1^{s_1}\cdots k_d^{s_d}}, & \text{if } w \text{ is even.} \end{cases}$$

In the depth one case, we have the classical results.

Proposition 14.4.7. *For any positive integer n, we have*

$$\mathrm{mcl}(2n+1) = \frac{(-1)^n}{2}(1-2^{-2n})(1-3^{-2n})\zeta(2n+1), \tag{14.30}$$

$$\mathrm{mgl}(n) = -\frac{1}{2}B_n\Big(\frac{1}{6}\Big)\frac{(-2\pi)^n}{n!}. \tag{14.31}$$

Here $B_n(x)$ is the nth Bernoulli polynomial defined by the generating series $te^{xt}/(e^t-1) = \sum_{n\geq 0} B_n(x)t^n/n!$.

Proof. See Exercise 14.5. □

Remark 14.4.8. In depth one case, when $\mathbf{s} = n$ is a positive integer we have the relations

$$\mathrm{mgl}(n) = (-1)^{\lceil n/2\rceil}\mathrm{Gl_n}(\theta) \quad \text{and} \quad \mathrm{mcl}(n) = (-1)^{\lfloor n/2\rfloor}\mathrm{Cl_n}(\theta)$$

using the notation in Lewin's book [392].

Turning to MDVs, let $\overline{\text{Ł}}(\mathbf{w})$ be the complex conjugate of $\text{Ł}(\mathbf{w})$ and extend it $\mathbb{Q}$-linearly to $\mathfrak{A}^1$. Define the *dual* of a word $\mathbf{w}$, denoted by, $\widetilde{\mathbf{w}}$, to be the word obtained by writing $\mathbf{w}$ backwards and then interchanging a and b. For example, $\widetilde{\mathsf{a}^3\mathsf{b}\boldsymbol{\delta}^2} = \boldsymbol{\delta}^2\mathsf{a}\mathsf{b}^3$.

Lemma 14.4.9. *For any admissible word $\mathbf{w} \in \{\mathsf{a},\mathsf{b},\boldsymbol{\delta}\}^*$, we have*

$$\overline{\text{Ł}}(\mathbf{w}) = (-1)^{|\mathbf{w}|}\,\text{Ł}(\widetilde{\mathbf{w}}).$$

Proof. Fix an admissible word $\mathbf{w} \in \{\mathsf{a},\mathsf{b},\boldsymbol{\delta}\}^*$ and let α (resp. β, resp. γ) be the number of appearances of the letter a (resp. b, resp. $\boldsymbol{\delta}$) in $\mathbf{w}$. Since

$$\mathrm{d}\log(1-\overline{\mu}(1-t)) = \mathrm{d}\log(1-\mu t)$$

we get

$$\begin{aligned}\overline{\text{Ł}}(\mathbf{w}) &= (-1)^{\beta+\gamma}\int_0^1 \overline{\mathbf{w}} \xlongequal{t\to 1-t} (-1)^{\alpha+\gamma}\int_1^0 \mathbf{w}(\mathsf{a}\leftrightarrow\mathsf{b}) \\ &= (-1)^{\beta}\int_0^1 \widetilde{\mathbf{w}} = (-1)^{|\mathbf{w}|}\,\text{Ł}(\widetilde{\mathbf{w}})\end{aligned}$$

because $\mathrm{dp}(\widetilde{\mathbf{w}}) = \alpha+\gamma$ and $|\mathbf{w}| = \alpha+\beta+\gamma$. □

Corollary 14.4.10. *All of MCVs, MGl and MCl of weight w are in the $\mathbb{Q}$-vector space MDV_w generated by MDVs of weight w.*

Proof. By Lemmas 14.4.5 and 14.4.9 both $\lambda(\mathbf{s}) + \overline{\lambda(\mathbf{s})}$ and $\lambda(\mathbf{s}) - \overline{\lambda(\mathbf{s})}$ are in MDV. □

Lemma 14.4.11. *For any positive integer n, set $P_n := (-\theta)^n/n!$. Then we have*

$$Ł(\boldsymbol{\delta}^n) = (-1)^n \lambda(\{1\}^n) = i^n P_n.$$

Proof. Observe that by the shuffle relations $n\,Ł(\boldsymbol{\delta}^n) = Ł(\boldsymbol{\delta})\,Ł(\boldsymbol{\delta}^{n-1}) = -i\theta\,Ł(\boldsymbol{\delta}^{n-1})$ since $Ł(\boldsymbol{\delta}) = \log(1-\mu) = -i\theta$. The lemma follows easily by induction. □

Proposition 14.4.12. *For any integer $w \geq 2$, define $G_w := \sum_{w>n>0} \boldsymbol{\delta}^n \mathsf{b} \boldsymbol{\delta}^{w-1-n}$. Then we have*

$$2\mathrm{Re}\,Ł(G_w/i^w) = (w-1)P_w.$$

Proof. By the shuffle relations,

$$Ł(\mathsf{b})\,Ł(\boldsymbol{\delta}^{w-1}) = Ł(\mathsf{b}\boldsymbol{\delta}^{w-1}) + G_w.$$

By the stuffle relations,

$$Ł(\mathsf{b})\,Ł(\boldsymbol{\delta}^{w-1}) = Ł(\mathsf{b}\boldsymbol{\delta}^{w-1}) + (w-1)\,Ł(\boldsymbol{\delta}^w) - \sum_{w>n>0} Ł(\boldsymbol{\delta}^{w-1-n}\mathsf{a}\boldsymbol{\delta}^n).$$

This yields $Ł(G_w) + Ł(\widetilde{G_w}) = (w-1)\,Ł(\boldsymbol{\delta}^w)$. Hence

$$Ł(G_w) + (-1)^w \overline{Ł}(G_w) = (w-1)i^w P_w$$

by Lemma 14.4.9 and 14.4.11. The proposition follows easily. □

14.4.2 *Central Binomial Sums*

In this subsection we consider the following *central binomial sums*:

$$\mathrm{CB}(k) := \sum_{n=1}^{\infty} \frac{1}{n^k \binom{2n}{n}}. \tag{14.32}$$

We will see that these values are can always be expressed as $\mathbb{Q}$-linear combinations of CMZVs at level six (see Thm. 14.4.17). To this end, we define

$$j(s,t) := \int_0^\theta \log^s\left(2\sin\frac{\tau}{2}\right)\tau^t d\tau, \tag{14.33}$$

$$\begin{aligned} r(s,t) :=& \frac{i^{t+1}}{s!t!}\int_0^\theta \log^s(1-e^{i\tau})\tau^t d\tau \\ =& \frac{i^{t+1}}{s!t!}\int_0^\theta \left(\log\left(2\sin\frac{\tau}{2}\right) + \frac{i}{2}(\tau-\pi)\right)^s \tau^t d\tau. \end{aligned} \tag{14.34}$$

Lemma 14.4.13. *For any $s, t \in \mathbb{N}_0$, the real number $j(s,t)$ can be written as a $\mathbb{Q}[i]$-linear combination of $r(c,d)$ with $c \le s$ and $c+d \le s+t$.*

Proof. By the definition $j(0,t) = (-i)^{t+1} t! r(0,t)$. By the binomial expansion we see that

$$r(s,t) = \frac{i^{t+1}}{s!t!} j(s,t) + \sum_{c,d \in \mathbb{N}, c<s, c+d \le s+t} q(c,d) j(c,d)$$

where $q(c,d) \in \mathbb{Q}[i]$. The lemma quickly follows from an induction on s. ☐

Example 14.4.14. We have

$$\begin{aligned} r(1,t) &= \frac{i^{t+1}}{t!}\left(j(1,t) + \frac{i}{2} j(0,t+1) - \frac{i\pi}{2} j(0,t)\right) \\ \Longrightarrow j(1,t) &= \frac{(-i)^{t+1}}{2} t! \Big(2r(1,t) - (t+1) r(0,t+1) + i\pi r(0,t)\Big). \end{aligned}$$

Lemma 14.4.15. *For all positive integers $k \ge 2$, we have*

$$\mathrm{CB}(k) = \frac{(-2)^{k-2}}{(k-2)!} j(k-2,1). \tag{14.35}$$

Proof. Note that Gamma function satisfies $s^{-n}\Gamma(n) = \int_0^1 y^{s-1} \times (-\log y)^{n-1} \mathrm{d}y$. Hence

$$\begin{aligned} \mathrm{CB}(k) &= \sum_{n=1}^{\infty} \frac{1}{\binom{2n}{n} n^k} = \sum_{n=1}^{\infty} \frac{1}{\binom{2n}{n}\Gamma(n)} \int_0^1 (-\log x)^{k-1} x^{n-1} \mathrm{d}x \\ &= \sum_{n=1}^{\infty} \frac{1}{\binom{2n}{n}\Gamma(n)} \int_0^1 (-2\log y)^{k-1} y^{2n-2} (2y) \mathrm{d}y \end{aligned}$$

by substitution $x = y^2$. Using integration by part we get

$$\begin{aligned} \mathrm{CB}(k) &= \frac{(-2)^{k-1}}{(k-2)!} \int_0^1 \frac{(\log y)^{k-2}}{y} \sum_{n=1}^{\infty} \frac{y^{2n}}{n\binom{2n}{n}} \mathrm{d}y \\ &= \frac{(-2)^{k-1}}{(k-2)!} \int_0^1 \frac{(\log y)^{k-2}}{y} \frac{y \sin^{-1}(y/2)}{\sqrt{1-(y/2)^2}} \mathrm{d}y \\ &= \frac{(-2)^{k-2}}{(k-2)!} \int_0^{\theta} \log^{k-2}\left(2\sin\frac{\tau}{2}\right) \tau \mathrm{d}\tau \quad (\text{by } y = 2\sin\frac{\tau}{2}). \end{aligned}$$

Here we have used the Maclaurin series expansion of $\sin^{-1}(t)/\sqrt{1-t^2}$ (see [67, p. 384]). ☐

Lemma 14.4.16. *For all positive integers n and k, we have $r(0,k) = \lambda(\{1\}^{k+1})$ and*

$$(-1)^{k+n} r(n,k) = \zeta(k+2,\{1\}^{n-1}) - \sum_{\ell=0}^{k} (-i)^{\ell} P_{\ell} \lambda(k+2-\ell,\{1\}^{n-1}). \quad (14.36)$$

Proof. The expression of $r(0,k)$ is easy to verify by using the definition Eq. (14.34) and Lemma 14.4.11. By Eq. (14.27) we have

$$\lambda(k+2,\{1\}^{n-1}) = \zeta(k+2,\{1\}^{n-1}) + \int_1^{\mu} \frac{\mathrm{L}_{k+1,\{1\}^{n-1}}(z)}{z} \mathrm{d}z.$$

Applying integration by parts repeatedly we get

$$\lambda(k+2,\{1\}^{n-1}) = \zeta(k+2,\{1\}^{n-1})$$
$$-\sum_{\ell=1}^{k} \frac{(-1)^{\ell} \lambda(k+2-\ell,\{1\}^{n-1}) \log^{\ell}(\mu)}{\ell!} + \frac{(-1)^k}{k!} \int_1^{\mu} \frac{\mathrm{L}_{\{1\}^n}(z) \log^k(z)}{z} \mathrm{d}z.$$

Let $z = e^{i\tau}$. Since $\log(\mu) = i\theta$ by Eq. (14.29) we have

$$\lambda(k+2,\{1\}^{n-1}) = \zeta(k+2,\{1\}^{n-1}) - \sum_{\ell=1}^{k} \frac{(-i\theta)^{\ell} \lambda(k+2-\ell,\{1\}^{n-1})}{\ell!}$$
$$+ \frac{i(-1)^k}{k!n!} \int_0^{\theta} (-\log(1-e^{i\tau}))^n (i\tau)^k \mathrm{d}\tau,$$

which readily reduces to Eq. (14.36). □

Theorem 14.4.17. *For any positive integer $k \geq 2$, the central binomial sum $\mathrm{CB}(k)$ lies in the weight k piece of the $\mathbb{Q}$-algebra generated by the MZVs, MGl and MCl.*

Proof. Note that $\pi = -3\,\mathrm{mgl}(1)$ so the theorem follows from Lemmas 14.4.13, 14.4.15 and 14.4.16. □

14.4.3 *Dualities of Multiple Clausen Values*

We present two different duality results for MCVs in this subsection, depending on whether the first argument is equal to or greater than 1.

Theorem 14.4.18. *For all positive integers $s_1, s_2, \dots, s_n$,*

$$\lambda(1,s_1,\dots,s_n) = \lambda(1)\lambda(s_1,\dots,s_n) - \lambda(2)\lambda(s_1-1,\dots,s_n) + \cdots$$
$$-(-1)^{s_1}\lambda(s_1)\lambda(1,s_2,\dots,s_n) + (-1)^{s_1}\lambda(1,s_1)\lambda(s_2,\dots,s_n) - \cdots$$
$$-(-1)^{|\mathbf{s}|}\lambda(s_n,\dots,s_1)\lambda(1) + (-1)^{|\mathbf{s}|}\lambda(1,s_n,\dots,s_1). \quad (14.37)$$

Proof. This can be proved by repeated integration by parts using Lemma 14.4.3. We leave the details to the interested reader. □

Example 14.4.19. In general, there are $|\mathbf{s}|$ terms on the right-hand side of Eq. (14.37). In weight 7 we have

$$\begin{aligned}\lambda(1,2,1,3) &= \lambda(1)\lambda(2,1,3) - \lambda(2)\lambda(1,1,3) + \lambda(1,2)\lambda(1,3)\\ &\quad - \lambda(1,1,2)\lambda(3) + \lambda(2,1,2)\lambda(2) - \lambda(3,1,2)\lambda(1) + \lambda(1,3,1,2).\end{aligned}$$

Theorem 14.4.20. *For* $\mathbf{s}, \mathbf{r} \in \mathbb{N}^d$ *and* $1 \le j \le d$, *define the compositions*

$$\begin{aligned}\vec{a}_j &= \vec{a}_j(\mathbf{s};\mathbf{r}) := (s_j+1, \{1\}^{r_j-1}, \dots, s_d+1, \{1\}^{r_d-1}),\\ \overleftarrow{b}_j &= \overleftarrow{b}_j(\mathbf{s};\mathbf{r}) := (r_j+1, \{1\}^{s_j-1}, \dots, r_1+1, \{1\}^{s_1-1}),\end{aligned}$$

and $\vec{a}_{d+1} := \overleftarrow{b}_0 := \{\}$. *Then we have*

$$\begin{aligned}\zeta(\vec{a}_1) = \zeta(\overleftarrow{b}_d) &= \sum_{j=1}^{d}\left\{\sum_{k=0}^{s_j}\overline{\lambda(s_j+1-k, \{1\}^{r_j-1}, \vec{a}_{j+1})}\lambda(\{1\}^k, \overleftarrow{b}_{j-1})\right.\\ &\quad \left.+\sum_{k=1}^{r_j-1}\overline{\lambda(\{1\}^k, \vec{a}_{j+1})}\lambda(r_j+1-k, \{1\}^{s_j-1}, \overleftarrow{b}_{j-1})\right\} + \lambda(\overleftarrow{b}_d). \quad (14.38)\end{aligned}$$

Proof. Apply the Hölder convolution as presented in Thm. 14.1.1 with $u = \mu$ and $v = 1-\mu = \overline{\mu}$. Note that in Thm. 14.1.1 $\lambda_v(\mathbf{s}) = \mathrm{Li}_{\mathbf{s}}(\mu) = \lambda(\mathbf{s})$ and $\lambda_u(\mathbf{s}) = \mathrm{Li}_{\mathbf{s}}(\overline{\mu}) = \overline{\lambda(\mathbf{s})}$. □

Example 14.4.21. In general, there are $w+1$ λ-product terms on the right-hand of Eq. (14.38). Take $\mathbf{s} = (2,1)$ and $\mathbf{r} = (3,2)$ in Eq. (14.38). Then the weight $w = 8$ and we have

$$\begin{aligned}\zeta(3,1,1,2,1) &= \overline{\lambda(3,1,1,2,1)} + \overline{\lambda(2,1,1,2,1)}\lambda(1) + \overline{\lambda(1,1,1,2,1)}\lambda(1,1)\\ &\quad + \overline{\lambda(1,2,1)}\lambda(3,1) + \overline{\lambda(1,1,2,1)}\lambda(2,1) + \overline{\lambda(2,1)}\lambda(4,1)\\ &\quad + \overline{\lambda(1,1)}\lambda(1,4,1) + \overline{\lambda(1)}\lambda(2,4,1) + \lambda(3,4,1) = \zeta(3,4,1).\end{aligned}$$

14.4.4 *A* 1-2-1 *Formula of Multiple Glaishers*

Our goal in this subsection is to prove a **1-2-1** type formula for the MGl $\mathrm{mgl}(\{1\}^m, 2, \{1\}^n)$. Recall that the bold-faced **1** stands for a string a 1's.

First, by the definition $\lambda_z(\mathbf{s}) = \mathrm{L}_{\mathbf{s}}(1/z)$ for any $|z| \ge 1$ where the L is the one variable MPL defined by Eq. (14.1).

Proposition 14.4.22. *For all* $|z| \ge 1$, *we have*

$$F_z(x,y) := \sum_{m=0}^{\infty}\sum_{n=0}^{\infty} x^{m+1}y^{n+1}\lambda_z(m+2,\{1\}^n) = 1 - {}_2F_1\left(\begin{matrix} y,-x \\ 1-x \end{matrix}\,\middle|\, \frac{1}{z}\right).$$

Proof. By the definition,

$$\begin{aligned}
&\sum_{m=0}^{\infty}\sum_{n=0}^{\infty} x^{m+1}y^{n+1}\lambda_z(m+2,\{1\}^n) \\
&= y\sum_{m=0}^{\infty} x^{m+1}\sum_{n=0}^{\infty}\frac{1}{(n+1)^{m+2}z^k}\prod_{j=1}^{n}\left(1+\frac{y}{j}\right) = \sum_{m=0}^{\infty} x^{m+1}\sum_{k=1}^{\infty}\frac{(y)_k}{k^{m+1}k!z^k} \\
&= \sum_{k=1}^{\infty}\frac{(y)_k}{k!z^k}\frac{x}{k-x} = -\sum_{k=1}^{\infty}\frac{(y)_k(-x)_k}{k!z^k(1-x)_k} = 1 - {}_2F_1\left(\begin{matrix} y,-x \\ 1-x \end{matrix}\,\middle|\, \frac{1}{z}\right)
\end{aligned}$$

as desired. ☐

Lemma 14.4.23. *Define*

$$\begin{aligned}
M(x,y) &:= F_{\overline{\mu}}(ix,-iy) \\
&= \sum_{m,n\ge 0}(-1)^{n+1}\Big(\mathrm{mgl}(m+2,\{1\}^n) + i\,\mathrm{mcl}(m+2,\{1\}^n)\Big)x^{m+1}y^{n+1}.
\end{aligned}$$

Then we have

$$e^{\theta x}M(x,y) + e^{\theta y}\overline{M(y,x)} + (e^{\theta x}-1)(e^{\theta y}-1) = F_1(ix,-iy). \tag{14.39}$$

Moreover,

$$F_1(x,y) = 1 - \exp\left(\sum_{k\ge 2}\frac{x^k+y^k-(x+y)^k}{k}\zeta(k)\right). \tag{14.40}$$

Proof. Taking $d=1$ in Thm. 14.4.20, we get (see Exercise 14.6)

$$\begin{aligned}
&\zeta(m+2,\{1\}^n) + \frac{(-1)^m(i\theta)^{m+n+2}}{(m+1)!(n+1)!} \\
&= \sum_{r=0}^{m}\frac{(-i\theta)^r}{r!}\lambda(m+2-r,\{1\}^n) + \sum_{r=0}^{n}\frac{(i\theta)^r}{r!}\overline{\lambda(n+2-r,\{1\}^m)}.
\end{aligned}$$

Multiplying the above by $x^{m+1}y^{n+1}$ and summing over all $m,n\ge 0$, we get

$$e^{-ix\theta}F_{\overline{\mu}}(x,y) + e^{iy\theta}\overline{F_{\overline{\mu}}(y,x)} + (e^{-ix\theta}-1)(e^{iy\theta}-1) = F_1(x,y), \tag{14.41}$$

by Prop. 14.4.22. Clearly Eq. (14.41) yields Eq. (14.39). Further, by the definition

$$F_1(x,y) = 1 - {}_2F_1\left(\begin{matrix} y,-x \\ 1-x \end{matrix}\,\middle|\, \frac{1}{z}\right) = 1 - \frac{\Gamma(1-x)\Gamma(1-x)}{\Gamma(1-x-y)}$$

by applying the Gauss summation formula C.2. Finally, Eq. (14.40) can be proved by Eq. (3.16). ☐

Lemma 14.4.24. *For every nonnegative integer n,*

$$\mathrm{mgl}(2,\{1\}^n)=(-1)^{n+1}\frac{\theta^{n+2}}{2(n+2)!}. \tag{14.42}$$

Proof. Multiplying by y/x on both sides of Eq. (14.39)) and then considering the real parts after taking $x\to 0$, we get

$$B(y)+e^{\theta y}C(y)+\theta y(e^{\theta y}-1)=D(y) \tag{14.43}$$

where

$$B(y):=\sum_{n=0}^{\infty}(-1)^{n+1}\,\mathrm{mgl}(2,\{1\}^n)y^{n+2}\quad (m=0),$$

$$C(y):=-\sum_{m=0}^{\infty}\mathrm{mgl}(m+2)y^{m+2}\quad (n=0),$$

$$D(y):=\sum_{m=1}^{\infty}(-1)^{m-1}\zeta(2m)y^{2m}=\frac{\pi y}{e^{2\pi y}-1}-\frac{1}{2}+\frac{\pi y}{2}.$$

In $D(y)$ we have used $(-1)^{m-1}\zeta(2m)=(1/2)(2\pi)^{2m}B_{2m}/(2m)!$ by Euler. Now Eq. (14.31) yields that

$$\begin{aligned}C(y)=&\frac{1}{2}\sum_{m=0}^{\infty}B_{m+2}\Big(\frac{1}{6}\Big)\frac{(-2\pi y)^{m+2}}{(m+2)!}\\ =&\frac{1}{2}\sum_{k=2}^{\infty}B_k\Big(\frac{1}{6}\Big)\frac{(-2\pi y)^k}{k!}\\ =&\frac{-\pi y e^{-\theta y}}{e^{-2\pi y}-1}-\frac{1}{2}\Big(1-B_1\Big(\frac{1}{6}\Big)2\pi y\Big)\\ =&\frac{\pi y e^{2\pi y-\theta y}}{e^{2\pi y}-1}-\frac{1}{2}-\theta y.\end{aligned}$$

After simplification, we see from Eq. (14.43) that

$$B(y)=\frac{e^{\theta y}}{2}-\frac{\theta y}{2}-\frac{1}{2}$$

which is equivalent to Eq. (14.42). □

Finally, we can prove the following **1-2-1** type formula.

Theorem 14.4.25. *For nonnegative integers m and n,*

$$\mathrm{mgl}(\{1\}^m,2,\{1\}^n)=-\frac{1}{2}\cdot\frac{(-\theta)^w}{w!}, \tag{14.44}$$

where $w=m+n+2$.

Proof. By repeated integration by parts, and using $\mathrm{Li}_{\{1\}^m}(\mu) = (i\theta)^n/n!$ and Eq. (14.26), we get

$$\begin{aligned}\lambda(\{1\}^n, 2, \{1\}^m) &= \int_0^\mu \frac{\mathrm{Li}_{\{1\}^{n-1},2,\{1\}^m}(z)}{1-z} dz \\ &= i\theta\lambda(\{1\}^{n-1}, 2, \{1\}^m) + \cdots + (-1)^{n+1}\frac{(i\theta)^n}{n!}\lambda(2, \{1\}^m) \\ &\quad + \int_0^\mu \frac{(\log(1-z))^n}{n!} \cdot \frac{\mathrm{Li}_{\{1\}^{m+1}}(z)}{z} dz \\ &= i\theta\lambda(\{1\}^{n-1}, 2, \{1\}^m) + \cdots + (-1)^{n+1}\frac{(i\theta)^n}{n!}\lambda(2, \{1\}^m) \\ &\quad + (-1)^n \binom{m+n+1}{n} \lambda(2, \{1\}^{m+n})\end{aligned}$$

using Eqs. (14.25) and (14.29). Multiplying by i^{m+n+2} and taking the real parts, we have

$$\begin{aligned}&\mathrm{mgl}(\{1\}^n, 2, \{1\}^m) + \theta\lambda(\{1\}^{n-1}, 2, \{1\}^m) + \cdots + \frac{\theta^n}{n!}\lambda(2, \{1\}^m) \\ &= (-1)^n \binom{m+n+1}{n} \lambda(2, \{1\}^{m+n}).\end{aligned}$$

Using Lemma 14.4.24 and the induction we see that the theorem follows immediately from the binomial identity

$$\sum_{j=0}^{n} (-1)^j \binom{m+n+2}{j} = (-1)^n \binom{m+n+1}{n},$$

which in turn follows from the well-known identity

$$\binom{m+n+2}{j} = \binom{m+n+1}{j} + \binom{m+n+1}{j-1}.$$

We have completed the proof of the theorem. □

Proposition 14.4.26. *For every integer $w \geq 2$, we have*

$$(-1)^w \sum_{j=0}^{w-2} \frac{\theta^j}{j!} \mathrm{mcl}(w-j) = \begin{cases} \mathrm{mcl}(2, \{1\}^{w-2}), & \text{if } 2|w; \\ \mathrm{mcl}(2, \{1\}^{w-2}) + (-1)^{(w-1)/2}\zeta(w), & \text{if } 2 \nmid w. \end{cases}$$

In particular, every MCl of the form $\mathrm{mcl}(2, \{1\}^n)$ *is a* $\mathbb{Q}[\pi]$*-linear combination of depth one MCl and at most one Riemann zeta value.*

Proof. Multiplying by y/x on both sides of Eq. (14.39)) and then considering the imaginary parts after taking $x \to 0$, we get

$$E(y) + e^{\theta y} F(y) = G(y)$$

where

$$E(y) := \sum_{n=0}^{\infty} (-1)^{n+1} \operatorname{mcl}(2, \{1\}^n) y^{n+2} \quad (m = 0),$$

$$F(y) := \sum_{m=0}^{\infty} \operatorname{mcl}(m+2) y^{m+2} \quad (n = 0),$$

$$G(y) := \sum_{m=1}^{\infty} (-1)^{m-1} \zeta(2m+1) y^{2m+1}.$$

The proposition follows immediately. □

14.4.5 *Dimension Conjectures*

Let MGl_w, MCl_w, MCV_w and MDV_w be the $\mathbb{Q}$-vector space generated by weight w MGl, MCl, MCVs and MDVs respectively. First, since $\operatorname{Li}_1(\mu) = i\theta$ and $\operatorname{Li}_{1,1}(\mu) = -\theta^2/2$ we have

$$\operatorname{mgl}(1) = -\theta,\ \operatorname{mcl}(1) = 0,\ \operatorname{mgl}(1,1) = \theta^2/2,\ \operatorname{mcl}(1,1) = 0.$$

Further $\operatorname{mgl}(2) = -\theta^2/4$ by Eq. (14.42) and

$$\operatorname{mcl}(2) = -\frac{\sqrt{3}}{2}\left(1 + \frac{1}{2^2} - \frac{1}{4^2} - \frac{1}{5^2} + \frac{1}{7^2} + \frac{1}{8^2} - \frac{1}{10^2} - \frac{1}{11^2} + \cdots\right) < 0.$$

Thus we know $\mathsf{MGl}_1 = \langle \pi \rangle$, $\mathsf{MCl}_1 = \emptyset$, $\mathsf{MCV}_1 = \langle i\pi \rangle$, $\mathsf{MGl}_2 = \langle \pi^2 \rangle$, $\mathsf{MCl}_2 = \langle \operatorname{mcl}(2) \rangle$, and $\mathsf{MCV}_2 = \langle \pi^2, i \operatorname{mcl}(2) \rangle$. We see easily that MCVs are automatically MDVs. In weight one and two cases, the only possible extra MDV is

$$Ł(\delta \mathsf{b}) = \overline{Ł(\mathsf{a}\delta)} = -\overline{\operatorname{Li}_2(\mu)} = -\frac{1}{4}\theta^2 - i \operatorname{mcl}(2)$$

by Lemma 14.4.9. So $\mathsf{MDV}_1 = \mathsf{MCV}_1$ and $\mathsf{MDV}_2 = \mathsf{MCV}_2$.

The following conjecture is supported by extensive numerical data. In particular, it implies that the dimensions of the MCVs are given by the Fibonacci numbers.

Conjecture 14.4.27. *Set* $\dim \mathsf{MCl}_0 =$ *and* $\dim \mathsf{MGl}_0 = \dim \mathsf{MCV}_0 = 1$. *For any positive integer* $w \geq 2$, *we have*

$$\dim \mathsf{MGl}_w = \dim \mathsf{MGl}_{w-1} + \dim \mathsf{MCl}_{w-2},$$

$$\dim \mathsf{MCl}_w = \dim \mathsf{MCl}_{w-1} + \dim \mathsf{MGl}_{w-2},$$

$$\dim \mathsf{MCV}_w = \dim \mathsf{MCV}_{w-1} + \dim \mathsf{MCV}_{w-2}.$$

Moreover, for all $w \geq 1$, *we have* $\mathsf{MDV}_w = \mathsf{MCV}_w$.

Definition 14.4.28. An *irreducible* MDV is a MDV which is not a $\mathbb{Q}$-linear combination of products of the MDVs of lower depths and weights.

Similar to the situation for the MZVs and the Euler sums, we may define $p_{n,d}$ to be be the minimum number of irreducible MDVs of weight n and depth d, such that a rational basis at this weight and depth is formed from these irreducibles, augmented by products of irreducibles of lower weights and depths. The following conjecture is an analog of Conjecture 14.2.1.

Conjecture 14.4.29. *We have*

$$\prod_{n\geq 2}\prod_{d\geq 1}(1-x^n y^d)^{p_{n,d}} = 1-\frac{x^2 y}{1-x}. \tag{14.45}$$

In order to find the finer structures of MDVs we need to use the Lyndon words on the alphabet of all positive integers with its natural order (see Definition 3.3.3).

Conjecture 14.4.30. *Let $w > d$ be two positive integers. Then one can choose the irreducibles by using* $\mathrm{L}(\mathbf{w})$ *where* $\mathbf{w} = \boldsymbol{\delta}$ *or* $\mathbf{w} = \mathsf{a}^{n_1-1}\boldsymbol{\delta}\mathsf{a}^{n_2-1}\boldsymbol{\delta}\dots\mathsf{a}^{n_d-1}\boldsymbol{\delta}$ *subject to the following conditions:*

(i) $w=\sum_{j=1}^d n_j$,
(ii) $n_j\geq 2$,
(iii) $(n_d,\dots,n_1)$ *is a Lyndon word.*

For example, the words corresponding to the first few irreducible MDVs are given below:

$$\boldsymbol{\delta},\ \mathsf{a}\boldsymbol{\delta},\ \mathsf{a}^2\boldsymbol{\delta},\ \mathsf{a}^3\boldsymbol{\delta},\ \mathsf{a}^4\boldsymbol{\delta},\ \mathsf{a}^2\boldsymbol{\delta}\mathsf{a}\boldsymbol{\delta},\ \mathsf{a}^5\boldsymbol{\delta},\dots$$

Using this we can produce the conjectured basis of MDVs in lower weights with following words:

$w=1:\boldsymbol{\delta}$;

$w=2:\boldsymbol{\delta}^2,\ \mathsf{a}\boldsymbol{\delta}$;

$w=3:\boldsymbol{\delta}\sqcup\!\sqcup\,\mathsf{a}\boldsymbol{\delta},\ \boldsymbol{\delta}^3,\ \mathsf{a}^2\boldsymbol{\delta}$;

$w=4:\boldsymbol{\delta}^2\sqcup\!\sqcup\,\mathsf{a}\boldsymbol{\delta},\ \boldsymbol{\delta}^4,\ \boldsymbol{\delta}\sqcup\!\sqcup\,\mathsf{a}^2\boldsymbol{\delta},\ \mathsf{a}\boldsymbol{\delta}\sqcup\!\sqcup\,\mathsf{a}\boldsymbol{\delta}$;

$w=5:\boldsymbol{\delta}^3\sqcup\!\sqcup\,\mathsf{a}\boldsymbol{\delta},\ \boldsymbol{\delta}^5,\ \boldsymbol{\delta}^2\sqcup\!\sqcup\,\mathsf{a}^2\boldsymbol{\delta},\ \boldsymbol{\delta}\sqcup\!\sqcup\,\mathsf{a}\boldsymbol{\delta}\sqcup\!\sqcup\,\mathsf{a}\boldsymbol{\delta},\ \mathsf{a}^2\boldsymbol{\delta}\sqcup\!\sqcup\,\mathsf{a}\boldsymbol{\delta},\ \mathsf{a}^4\boldsymbol{\delta},\ \mathsf{a}^2\boldsymbol{\delta}\mathsf{a}\boldsymbol{\delta};\dots$

We end this subsection with a conjecture which plays an important role in Broadhurst's discovery of a significant basis in [86] which is different from the one given in Conjecture 14.4.30. It can be used to express numerically the counterterm $P_{7,11}$ of the following 7-loop subdivergence-free Feynman

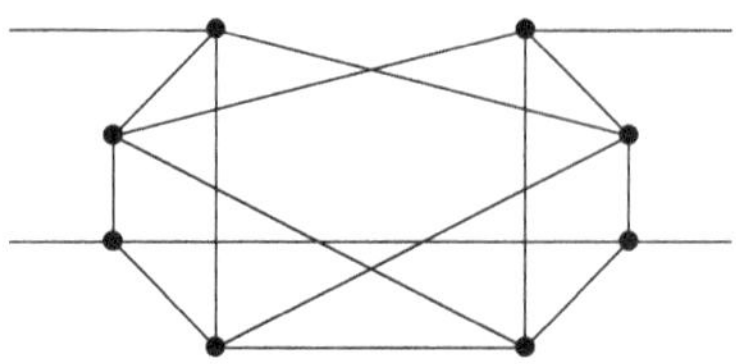

Figure 14.2: The unique 7-loop subdivergence-free Feynman diagram.

diagram in ϕ^4 theory with well-behaved rational coefficients whose denominators don't have large prime factors:

Conjecture 14.4.31. *At odd weight $w > 1$, there exists a* unique $\mathbb{Z}$*-linear reducible combination*

$$\sum_{k=1}^{(w-1)/2} C_{w,k} \mathrm{Im}\, \mathrm{L}\left(\mathsf{a}^{w-2k-1}\boldsymbol{\delta}\mathsf{a}^{2k-1}\mathsf{b}\right) \in \pi^w \mathbb{Q}, \tag{14.46}$$

with $C_{w,1} > 0$ and $\gcd\{C_{w,k} : k = 1, \dots, (w-1)/2\} = 1$. At even weight w, the depth-2 real part $\mathrm{Re}\, \mathrm{L}(\mathsf{a}^{w-2}\boldsymbol{\delta}^2)$ is reducible to MZVs.

14.5* Level p and p^2 Cases for Prime $p \geq 5$

In this optional section, we assume the reader is familiar with the basic facts about the motivic fundamental group laid out in [164]. Let $p \geq 5$ be prime throughout this section. From Thm. 13.2.1 we see that $d(2,p) \leq (p+1)^2/4$. In this section, we will improve this bound in Thm. 14.5.1 by using the motivic setup.

Let $\mathbb{K}$ be the field over $\mathbb{Q}$ generated by a primitive Nth root of unity and $\mathbb{O}$ its ring of integers. Let Γ be the $\mathbb{Q}$-vector subspace of $\mathbb{O}[1/N]^* \otimes \mathbb{Q}$ generated by $1-\xi$ for $1 \neq \xi \in \Gamma_N$. We refer to [164] for the definition of the Tannakian category of mixed Tate motives $\mathrm{MT}(k)_\Gamma$, the definition of the fibre functor $\omega(M) := \bigoplus_n \mathrm{Hom}(\mathbb{Q}(n), \mathrm{Gr}^W_{-2n}(M))$, and of the corresponding motivic fundamental group G. Let $\mathfrak{L}$ be the Lie algebra generated by the generators $\mathtt{x}_\xi, \xi \in \{0, \infty\} \cup \Gamma_N$, with the only relation $\sum \mathtt{x}_\xi = 0$. Set the degree of $\mathtt{x}_\xi$ to be 1. Then $\mathfrak{L}$ is freely generated by the $\mathtt{x}_\xi, \xi \neq \infty$. Let Π be the pro-nilpotent group $\lim_n \exp(\mathfrak{L}/\deg^{>n})$. The completed enveloping algebra of $\mathfrak{L}$ is nothing but $\mathbb{Q}\langle\langle \mathtt{x}_\xi \rangle\rangle_{\xi \in \Gamma_N \cup \{0\}}$ with the coproduct $\Delta : \mathtt{x}_\xi \mapsto \mathtt{x}_\xi \otimes 1 + 1 \otimes \mathtt{x}_\xi$, and Π is the subscheme of the group-like elements under Δ.

Let H_ω be the group of Γ_N-equivariant automorphisms of Π defined in [164, § 5.8]. Set $G_\omega := \omega(G)$, $U_\omega := \ker(G_\omega \to \mathbb{G}_m)$ and $V_\omega := \ker(H_\omega \to$

$\mathbb{G}_m$). There are morphisms of group schemes $G_\omega \to H_\omega$ and $U_\omega \to V_\omega$ compatible with the semi-direct product structure

$$\iota : G_\omega = \mathbb{G}_m \ltimes U_\omega \to H_\omega = \mathbb{G}_m \ltimes V_\omega. \tag{14.47}$$

By [164, § 6.13] one may safely replace $\mathscr{G}_N^{(\ell)}$ by $\mathscr{G}$ throughout [251]. In particular, Goncharov's results can now be used to study the structure of the Galois Lie algebra $\mathscr{G}_{\bullet,\bullet}(\Gamma_N) = \sum_{w\geq 1, l\geq 1} \mathscr{G}_{-w,-l}(\Gamma_N)$ bi-graded by the weight w and depth l. The weight filtration is defined via the action of $\mathbb{G}_m$ and the depth filtration comes from the descending central series of $\ker f_*$, where $f_* : \pi_1(\mathbb{G}_m - \Gamma_N, x) \longrightarrow \pi_1(\mathbb{G}_m)$ for any base point x is induced from the embedding $\mathbb{G}_m - \Gamma_N \hookrightarrow \mathbb{G}_m$.

Theorem 14.5.1. *Let $p \geq 5$ be a prime. Then*

$$d(2,p) \leq \frac{(5p+7)(p+1)}{24}. \tag{14.48}$$

If Grothendieck's period conjecture (see [164, § 5.27(c)]) is true then the equality holds and all the $\mathbb{Q}$-linear relations in $\mathsf{CMZV}(2,p)$ follow from the standard relations.

Proof. Let { , } be the Lie bracket in $\mathscr{G}_{\bullet,\bullet}(\Gamma_N)$. Now in depth 1 one has $\dim \mathscr{G}_{-2,-1}(\Gamma_p) = (p-1)/2$ (see [251, Thm. 2.1]). By [251, Cor. 2.16]

$$\dim \mathscr{G}_{-2,-2}(\Gamma_p) = \frac{(p-1)(p-5)}{12}. \tag{14.49}$$

For any positive integer N, define

$$\beta_N : \bigwedge^2 \mathscr{G}_{-2,-1}(\Gamma_N) \longrightarrow \mathscr{G}_{-2}(\Gamma_N), \quad a \wedge b \longmapsto \{a,b\}. \tag{14.50}$$

Then

$$\dim(\ker \beta_p) = \frac{1}{2}\frac{p-1}{2}\left(\frac{p-1}{2} - 1\right) - \frac{(p-1)(p-5)}{12} = \frac{p^2-1}{24}. \tag{14.51}$$

Let SR be the affine subgroup of Π defined by only those polynomial equations satisfied by the coefficients of $\mathscr{I}$ which are deduced from the standard relations, plus "$2\pi i = 0$", as explained in [164, § 5.22]. This is the group DMRD_0 studied by Racinet in [477, § 3.2, Thm. I]. Its Lie algebra Lie SR is graded by weight and depth, independent of the embedding by Prop. 4.1 of op. cit. and contains $\mathscr{G}$ by [164, § 5.22]. Further, Goncharov shows in [251, § 7.7] that the standard relations provide a complete list of constraints on the diagonal part of the Lie algebra $\mathscr{G}$ in depth ≤ 2, yielding $(\mathrm{Lie}\, SR)_{m,m} = \mathscr{G}_{m,m}$ for $m = -1, -2$. Together with Eq. (14.51) this means in the proof of [164, § 5.25] one can decrease the bound $D(2,p)$ in Thm. 13.2.1 by $(p^2-1)/24$ and arrive at the bound $(5p+7)(p+1)/24$.

Grothendieck's conjecture implies that $\mathscr{G} = \operatorname{Lie} R$ where R is the affine subgroup of Π defined by all the polynomial equations satisfied by the coefficients of $\mathscr{J}$ plus "$2\pi i = 0$". Therefore one obtains both the equality in Eq. (14.48) and the statement about the completeness of the standard relations. This concludes the proof of the theorem. □

Remark 14.5.2. (i) Note that when $p \geq 11$ the vector space $\ker \beta_p$ contains a subspace isomorphic to the space of cusp forms of weight two on $X_1(p)$ which has dimension $(p-5)(p-7)/24$ (see [251, Lemma 2.3 & Thm. 7.8]). So it must contain another piece which has dimension $(p-3)/2$ by (14.51). What is this missing piece?

(ii) When $w > 2$ it's not too hard to improve the bound of $d(w,p)$ given in [164, § 5.24] by the same idea as used in the proof of the theorem (for example, decrease the bound by $(p^2-1)/24$). But they are often not the optimal.

Let's turn to the case of level $N = p^2$. Note that $\varphi(p^2)(\varphi(p^2)-2)/4$ is a trivial upper bound of $\dim \mathscr{G}_{-2,-2}(\Gamma_{p^2})$ (see [251, Thm. 2.1]). If one had any nontrivial upper bound of $\dim \mathscr{G}_{-2,-2}(\Gamma_{p^2})$ one would have a nontrivial lower bound of $\dim(\ker \beta_{p^2})$ where β_{p^2} is defined by (14.50). This would in turn provide an improved upper bound of $d(2,p^2)$ similar to Thm. 14.5.1. However, at present we can only show that

Theorem 14.5.3. *If $p \geq 5$ is a prime, then $\ker \beta_{p^2} \neq 0$ and $d(2,p^2) < p^2(p-1)^2/4$.*

Proof. Fix a primitive p^2th root of unity μ. Put $e(a) = \mathbf{x}_{\mu^a}$ for all integer a. Define

$$g_{k,j} = e(pk+j) + e(p^2-pk-j) + e(pj) + e(p^2-pj)$$

for $0 \leq k < (p-1)/2$, $1 \leq j \leq p-1$, and for $k = (p-1)/2$, $1 \leq j \leq (p-1)/2$. One only needs to prove the following

Claim. Let $h = (p-3)/2$. Then one has

$$\sum_{k=0}^{h}\sum_{l=k}^{h}\sum_{j=2}^{p-2}\{g_{k,1}, g_{l,j}\} + \sum_{k=0}^{h+1}\sum_{j=2}^{h+1}\{g_{k,1}, g_{h+1,j}\}$$
$$+\sum_{k=0}^{h}\sum_{l=k+1}^{h}\sum_{j=2}^{p-2}\{g_{k,p-1}, g_{l,j}\} + \sum_{k=0}^{h}\sum_{j=2}^{h+1}\{g_{k,p-1}, g_{h+1,j}\}$$

$$-\sum_{k=0}^{h}\sum_{l=k}^{h}\sum_{j=2}^{p-2}\{g_{k,j},g_{l,p-1}\}-\sum_{k=0}^{h}\sum_{l=k}^{h}\sum_{j=2}^{p-2}\{g_{k,j},g_{l+1,1}\}=0.$$

There are $h(2h+3)^2 = hp^2$ distinct terms on the left, each with coefficient ± 1.

The proof of the claim is straightforward by a little tedious change of indices and regrouping.

$$-\sum_{k=0}^{h}\sum_{l=k}^{h}\sum_{j=2}^{p-2}\{g_{k,j},g_{l+1,1}\}=\sum_{k=0}^{h}\sum_{l=0}^{k}\sum_{j=2}^{p-2}\{g_{k+1,1},g_{l,j}\}=\sum_{k=1}^{h+1}\sum_{l=0}^{k-1}\sum_{j=2}^{p-2}\{g_{k,1},g_{l,j}\}.$$

Then the expression in the claim becomes

$$\sum_{k=1}^{h}\sum_{l=0}^{h}\sum_{j=2}^{p-2}\{g_{k,1},g_{l,j}\}+\sum_{l=0}^{h}\sum_{j=2}^{p-2}\{g_{0,1},g_{l,j}\}+\sum_{l=0}^{h}\sum_{j=2}^{p-2}\{g_{h+1,1},g_{l,j}\}$$

$$+\sum_{k=0}^{h+1}\sum_{j=2}^{h+1}\{g_{k,1},g_{h+1,j}\}+\sum_{k=0}^{h}\sum_{l=0}^{h}\sum_{j=2}^{p-2}\{g_{k,p-1},g_{l,j}\}+\sum_{k=0}^{h}\sum_{j=2}^{h+1}\{g_{k,p-1},g_{h+1,j}\}$$

$$=\sum_{k=0}^{h+1}\sum_{l=0}^{h}\sum_{j=2}^{p-2}\{g_{k,1},g_{l,j}\}+\sum_{k=0}^{h+1}\sum_{j=2}^{h+1}\{g_{k,1},g_{h+1,j}\}$$

$$+\sum_{k=0}^{h}\sum_{l=0}^{h}\sum_{j=2}^{p-2}\{g_{k,p-1},g_{l,j}\}+\sum_{k=0}^{h}\sum_{j=2}^{h+1}\{g_{k,p-1},g_{h+1,j}\}.$$

By abuse of notation, we use $\{a,b\}=\{e(a),e(b)\}$. By the definition

$$\begin{aligned}
&\{g_{k,1},g_{l,j}\}\\
&=\{pk+1,pl+j\}+\{-pk-1,pl+j\}+\{p,pl+j\}+\{-p,pl+j\}\\
&\quad+\{pk+1,-pl-j\}+\{-pk-1,-pl-j\}+\{p,-pl-j\}\\
&\quad+\{-p,-pl-j\}+\{pk+1,pj\}+\{-pk-1,pj\}+\{p,pj\}+\{-p,pj\}\\
&\quad+\{pk+1,-pj\}+\{-pk-1,-pj\}+\{p,-pj\}+\{-p,-pj\}\\
&=\{pk+1,pl+j\}+\{p(p-k)-1,pl+j\}+\{p,pl+j\}+\{-p,pl+j\}\\
&\quad+\{pk+1,p(p-1-l)+p-j\}+\{p(p-k)-1,p(p-1-l)+p-j\}\\
&\quad+\{p,p(p-1-l)+p-j\}+\{-p,p(p-1-l)+p-j\}+\{pk+1,pj\}\\
&\quad+\{p(p-k)-1,pj\}+\{p,pj\}+\{-p,pj\}+\{pk+1,p(p-j)\}\\
&\quad+\{p(p-k)-1,p(p-j)\}+\{p,p(p-j)\}+\{-p,p(p-j)\}.
\end{aligned}$$

Then

$$\sum_{k=0}^{h+1}\sum_{l=0}^{h}\sum_{j=2}^{p-2}\{g_{k,1},g_{l,j}\}$$

$$=\sum_{k=0}^{h+1}\sum_{l=0}^{h}\sum_{j=2}^{p-2}\{pk+1,pl+j\}+\{p(p-k)-1,pl+j\}$$
$$+\{pk+1,p(p-1-l)+j\}+\{p(p-k)-1,p(p-1-l)+j\}$$
$$+\{p,pl+j\}+\{-p,pl+j\}+\{p,p(p-1-l)+j\}+\{-p,p(p-1-l)+j\}$$
$$+2\{pk+1,pj\}+2\{p(p-k)-1,pj\}+2\{p,pj\}+2\{-p,pj\}$$
$$=\sum_{k=0}^{h+1}\sum_{l=0,l\neq h+1}^{p-1}\sum_{j=2}^{p-2}\{pk+1,pl+j\}+\sum_{k=h+2}^{p}\sum_{l=0,l\neq h+1}^{p-1}\sum_{j=2}^{p-2}\{pk-1,pl+j\}$$
$$+\sum_{k=0}^{h+1}\sum_{l=0,l\neq h+1}^{p-1}\sum_{j=2}^{p-2}\Big(\{p,pl+j\}+\{-p,pl+j\}\Big)+2(h+1)\sum_{k=0}^{h+1}\sum_{j=2}^{p-2}\{pk+1,pj\}$$
$$+2(h+1)\sum_{k=h+2}^{p}\sum_{j=2}^{p-2}\{pk-1,pj\}+2(h+2)(h+1)\sum_{j=2}^{p-2}\Big(\{p,pj\}+\{-p,pj\}\Big).$$

Thus

$$\sum_{k=0}^{h+1}\sum_{l=0}^{h}\sum_{j=2}^{p-2}\{g_{k,1},g_{l,j}\}+\sum_{k=0}^{h+1}\sum_{j=2}^{h+1}\{g_{k,1},g_{h+1,j}\}$$
$$=\sum_{k=0}^{h+1}\sum_{l=0}^{p-1}\sum_{j=2}^{p-2}\{pk+1,pl+j\}+\sum_{k=h+2}^{p}\sum_{l=0}^{p-1}\sum_{j=2}^{p-2}\{pk-1,pl+j\}$$
$$+\frac{p+1}{2}\sum_{l=0}^{p-1}\sum_{j=2}^{p-2}\Big(\{p,pl+j\}+\{-p,pl+j\}\Big)+p\sum_{k=0}^{h+1}\sum_{j=2}^{p-2}\{pk+1,pj\}$$
$$+p\sum_{k=h+2}^{p}\sum_{j=2}^{p-2}\{pk-1,pj\}+\frac{p^2-1}{2}\sum_{j=2}^{p-2}\Big(\{p,pj\}+\{-p,pj\}\Big).$$

Similarly,

$$\sum_{k=0}^{h}\sum_{l=0}^{h}\sum_{j=2}^{p-2}\{g_{k,p-1},g_{l,j}\}+\sum_{k=0}^{h}\sum_{j=2}^{h+1}\{g_{k,p-1},g_{h+1,j}\}$$
$$=\sum_{k=1}^{h+1}\sum_{l=0}^{p-1}\sum_{j=2}^{p-2}\{pk-1,pl+j\}+\sum_{k=h+2}^{p-1}\sum_{l=0}^{p-1}\sum_{j=2}^{p-2}\{pk+1,pl+j\}$$
$$+\frac{p-1}{2}\sum_{l=0}^{p-1}\sum_{j=2}^{p-2}\Big(\{p,pl+j\}+\{-p,pl+j\}\Big)+p\sum_{k=1}^{h+1}\sum_{j=2}^{p-2}\{pk-1,pj\}$$
$$+p\sum_{k=h+2}^{p-1}\sum_{j=2}^{p-2}\{pk+1,pj\}+\frac{(p-1)^2}{2}\sum_{j=2}^{p-2}\Big(\{p,pj\}+\{-p,pj\}\Big).$$

Altogether the expression in the claim is reduced to

$$\sum_{k=0}^{p-1}\sum_{l=0}^{p-1}\sum_{j=2}^{p-2}\{pk+1,pl+j\}+\sum_{k=1}^{p}\sum_{l=0}^{p-1}\sum_{j=2}^{p-2}\{pk-1,pl+j\}$$
$$+p\sum_{l=0}^{p-1}\sum_{j=2}^{p-2}\Big(\{p,pl+j\}+\{-p,pl+j\}\Big)+p\sum_{k=0}^{p-1}\sum_{j=2}^{p-2}\{pk+1,pj\}$$
$$+p\sum_{k=1}^{p}\sum_{j=2}^{p-2}\{pk-1,pj\}+p(p-1)\sum_{j=2}^{p-2}\Big(\{p,pj\}+\{-p,pj\}\Big).$$

This is equal to

$$\sum_{k=0}^{p-1}\sum_{l=0}^{p-1}\sum_{j=2}^{p-2}\Big([pk+1,pl+j]-[p(k+l)+j+1,pl+j]$$
$$+[p(k+l)+j+1,pk+1]\Big)$$
$$\sum_{k=1}^{p}\sum_{l=0}^{p-1}\sum_{j=2}^{p-2}\Big([pk-1,pl+j]-[p(k+l)+j-1,pl+j]$$
$$+[p(k+l)+j-1,pk+1]\Big)$$
$$+p\sum_{l=0}^{p-1}\sum_{j=2}^{p-2}\Big([p,pl+j]-[p(l+1)+j,pl+j]+[p(l+1)+j,p]$$
$$+[-p,pl+j]-[p(l-1)+j,pl+j]+[p(l-1)+j,-p]\Big)$$
$$+p\sum_{k=0}^{p-1}\sum_{j=2}^{p-2}\Big([pk+1,pj]-[p(j+k)+1,pj]+[p(j+k)+1,pk+1]\Big)$$
$$+p\sum_{k=1}^{p}\sum_{j=2}^{p-2}\Big([pk-1,pj]-[p(j+k)-1,pj]+[p(j+k)-1,pk-1]\Big)$$
$$+p(p-1)\sum_{j=2}^{p-2}\Big([p,pj]-[p(j+1),pj]+[p(j+1),p]$$
$$+[-p,pj]-[p(j-1),pj]+[p(j-1),-p]\Big)=0$$

after a number of cancelations. □

14.6 Computational Results in Weight ≤ 5

In this last section we briefly discuss some results in weight three, four and five. Since the computational complexity increases exponentially as the weight increases we cannot do as many cases as we have done in weight two.

Combining the finite DBSFs in weight three to five and the weight one relations Eq. (13.21) and using Maple we have verified that $d(3,1) = 1, d(3,2) \leq 3, d(3,3) \leq 8....$

Table 14.1: Upper bounds of $d(w,N)$ by the standard relations, denoted by $SR(w)$, and those by Thm. 13.2.1, denoted by $D(w)$.

N	1	2	3	4	5	6	7	8	9	10	11
$SR(3)$	1	3	8	9	22	23	50	38	67	70	157
$D(3)$	1	3	8	8	27	21	64	27	64	56	216
$SR(4)$	1	5	16	21	61	69					
$D(4)$	1	5	16	16	81	55					
$SR(5)$	2	8	32								
$D(5)$	2	8	32								

It is possible to do similar computation in other small weight and low level cases. Some of the results are listed in Table 14.1. The values of Deligne and Goncharov's bound $D(w) = D(w,N)$ in Table 14.1 should be compared with the bound $SR(w) = SR(w,N)$ obtained by the standard relations.

Note that $SR(3,4) = D(3,4) + 1$ as we saw in Sec. 14.3. In weight five and level four, there are five non-standard relations (see Exercise 14.8). We now can prove this by using the octahedral symmetry of $\mathbb{P}^1 - (\{0,\infty\} \cup \Gamma_4)$ (see Remark 13.3.40).

Theorem 14.6.1. *If the conjecture in* [164, § 5.27(c)] *is true then all the linear relations among CMZVs of level four and weight three (resp. weight four) are the consequences of the standard relations and the octahedral relation Eq.* (14.21) *(resp. the five octahedral relations Eq.* (14.52)*-Eq.* (14.56)*).*

Proof. We have dealt with the weight three case in Sec. 14.3. Weight 4 case is similar. See also [617, § 3]. □

From the available data in Table 14.1 we can formulate the following conjecture.

Conjecture 14.6.2. *Suppose the level $N = p$ is a prime ≥ 5. Then*

$$d(3,p) \leq \frac{p^3 + 4p^2 + 5p + 14}{12}.$$

Moreover, equality holds if standard relations produce all the linear relations.

The conjecture was formulated under the belief that the upper bound of $d(3,p)$ produced by the standard relations should be a polynomial of p of degree 3. Then one can find the coefficients by the bounds of $d(3,p)$ for $p = 5, 7, 11, 13$ in Table 14.1.

We conclude this chapter with the following conjecture. We call a level N *standard* if $N = 1, 2, 3$ or $N = p^n$ for prime $p \geq 5$. Otherwise, the N is called a *non-standard level.*

Conjecture 14.6.3. (i) *If N is a standard level then the standard relations always provide the sharp bounds of $d(w,N)$, namely, all linear relations can be derived from the standard ones. Further, if $N = p^n$ for prime $p \geq 5$ then the bound $D(w,N)$ in Eq. (13.4) by Deligne and Goncharov can be lowered.*

(ii) *If N is a non-standard level then there exists a positive integer $w_0(N)$ so that at least one non-standard relation exists in $\mathsf{CMZV}(w,N)$ for each $w \geq w_0(N)$. Further, if N is a composite number, then the bound $D(w,N)$ is sharp.*

It is likely that one can take $w_0(4) = w_0(6) = w_0(9) = 3$ and $w_0(N) = 2$ for all the other non-standard levels N.

14.7 Historical Notes

As we mentioned in the historical notes concerning MPLs, the Euler sums and MDVs attracted the attention of physicists in the early 1990s with the advent of high performance computers which enabled them to compute an amazing number of the Feynman integrals numerically with high accuracy. This kind of experimental mathematics directly gives rise to many highly motivating conjectures on these special CMZVs (see [30–32, 66, 154]). With these experiments, D. J. Broadhurst and his collaborators got the first glance at the intimate relations between these objects, see for example, [83–86, 90, 91]. Almost all the dimension conjectures in this chapter belong to them. Some of these have been partially proved by Deligne who showed in [161] that those conjectured dimensions at least provide the upper bounds (see also [240]).

The integral substitutions presented in Sec. 14.2.3 are adopted from [68] in which many different proofs of the identities $\zeta(3) = \zeta(2,1)$ and $\zeta(3) = 8\zeta(\bar{2},1)$ can be found. One of these is the elegant proof given in Example 14.2.11. Notice, however, the significance of the substitution Eq. (14.11) was observed already by Broadhurst (see [85, (127)])

Doubling relations at level two of Sec. 14.2.5 were first investigated in [62] where Blümlein *et al.* collected a lot of data and facts concerning the MZVs and the Euler sums. The generalized doubling relations at level two is explained in details in [62] where Conjecture 14.2.18 can be found on p. 596. Broadhurst pointed out that the relation at level two in [85, (127)] should be related to the doubling relations. Sometimes a MZV that is irreducible over the MZV algebra can be reduced if the Euler sums are allowed. For example, it is unlikely that $\zeta(6,4,1,1)$ is expressed as a $\mathbb{Q}$ linear combination of products of MZVs of depth less than 4. However, it is an element in the $\mathbb{Q}$-algebra generated by depth one and two Euler sums. This phenomenon is called the "pushdown" which is discussed in [31,62,83].

The identity family of Sec. 14.2.5 was first conjectured around 1996 by Borwein, Bradley and Broadhurst [70]. Besides this one, they numerically discovered many other striking identities involving the Euler sums. In the following years a majority of them were rigorously proven except for Eq. (14.13). This identity is remarkable in that it is the only conjectured family of identities relating the Euler sums to MZVs, neither of which have closed evaluations. Several proofs of the case $n = 1$ can be found in [68]. In an email dated Oct. 8, 2006, Prof. J. Borwein informed the author that the only known proof of Eq. (14.13) for $n = 2$ was by computer verification. Within a year, the general case was solved and the proof was given in [622]. Theorem 14.6.1 was proved in [617].

Broadhurst formulated Conjecture 14.2.1 in [83], which provided a great insight into the relations between the MZVs and the Euler sums. Note that Deligne proved the dimensions of ES_n are correct under the assumption of a variant Grothendieck's period conjecture (see [161] or [240]). However, his basis (see Remark 13.3.41) is different from the one given in Conjecture 14.2.3.

The non-standard octahedral relations at level four first appeared in [617,623], where one can also find the improved upper bounds for the prime level p and the prime squared level p^2. To generalize the latter result to arbitrary prime power levels, it is likely the results from [386] will play some key role. Concerning the former result, Okuda discovered some relations in [455, Example 4.2] similar to the octahedral relations. It is possible to verify numerically all the non-standard relations in lower weights. An important tool for such computations of the MPLs, called GiNac, can be found in [562].

It was Broadhurst who first noticed the close connection between Feynman integral and knot theory (see [85]). He and his collaborators carried out extensive computer aided computations in the late 1990s and discovered the detailed structures of the MCl and MGl at level six (see [70, 73]). In particular, the explicit formula for CB(20) is given in [29] as an example of the application of the powerful PSLQ algorithm. The general formula in Sec. 14.4.4 appeared first in [73]. Motivated by some ingenious computations by Panzer [466] and Schnetz [504], Broadhurst also considered the MCVs and posed many conjectures in [86] which we have presented in Sec. 14.4.5. He further explored the relations among some real version of the CMZVs at levels five and seven in [87, 88] which contains many challenging problems (see Exercise 14.9).

Exercises

14.1. Prove the conjectural expression (14.5) and (14.6) assuming Conjecture 14.2.1.

14.2. Prove that the only nontrivial changes of variable that preserve the interval [0, 1] and send $\mathfrak{A}_1$ into $\mathfrak{A}_2$ are given by (14.7) to (14.10) and their compositions.

14.3. By induction and Prop. 13.3.42 reduce the proof of Thm.13.3.44 to the $k = 0$ case.

14.4. Verify the following values of the central binomial sums:

$$\begin{aligned}
S(2) &= \frac{\zeta(2)}{3}, \\
S(3) &= -\frac{2\pi}{3}\,\mathrm{mcl}(2) - \frac{4}{3}\zeta(3), \\
S(4) &= \frac{17}{36}\zeta(4), \\
S(5) &= 2\pi\,\mathrm{mcl}(4) - \frac{19}{3}\zeta(5) + \frac{2}{3}\zeta(3)\zeta(2).
\end{aligned}$$

See p. 25 of [73] for more values of the central binomial sums.

14.5. Show that for all positive integers n

$$\operatorname{mcl}(2n+1) = \frac{(-1)^n}{2}(1-2^{-2n})(1-3^{-2n})\zeta(2n+1),$$
$$\operatorname{mgl}(n) = \frac{(-1)^{n+1}2^{n-1}\pi^n B_n(\frac{1}{6})}{n!}.$$

Here $B_n(x)$ is the nth Bernoulli polynomial defined by the generating series $te^{xt}/(e^t-1) = \sum_{n\geq 0} B_n(x)t^n/n!$.

14.6. Let m, n be two positive integers. Prove that

$$(-1)^{m+n}\zeta(m+1,\{1\}^{n-1}) = (-1)^n P_m P_n + \sum_{r=0}^{m-1} P_r \lambda(m+1-r,\{1\}^{n-1})$$
$$+ \sum_{r=0}^{n-1}(-1)^r P_r \overline{\lambda(n+1-r,\{1\}^{m-1})}.$$

14.7. Use Thm. 14.4.25 to show that

$$\operatorname{mgl}(2,1) = \frac{\pi^3}{324}.$$

Then use the Exercise 14.6 to show

$$\operatorname{mgl}(3,1) = \frac{-23}{19440}\pi^4.$$

14.8. Set $L = L_4$, $1^4 = \{1\}^4$, ... and

$$\begin{aligned}
l_1 &= L(1^4|2,1,0,1), & l_2 &= L(1^4|2,1^2,0), & l_3 &= L(1^4|2,0,3,1),\\
l_4 &= L(1^4|2,0^3), & l_5 &= L(1^4|1,2,0,3), & l_6 &= L(1^4|3^2,0,3),\\
l_7 &= L(1^4|3,1,3,2), & l_8 &= L(1^4|3,0^3), & l_9 &= L(1^4|3,0,1,0),\\
l_{10} &= L(1^4|3,0,1^2), & l_{11} &= L(2,1^2|0,3,0), & l_{12} &= L(3,1|0,3),\\
l_{13} &= L(1^4|2,2,3,0), & l_{14} &= L(2,1^2|3,1^2), & l_{15} &= L(2,1^2|3,0,3),\\
l_{16} &= L(1^2,2|2^3), & l_{17} &= L(1^4|2,0,1,0), & l_{18} &= L(2,1^2|2^2,0),\\
l_{19} &= L(1^4|\{2,0\}^2), & l_{20} &= L(2^2|3,0), & l_{21} &= L(1^4|2^4).
\end{aligned}$$

Check the following five relations in weight three and level four case are $\mathbb{Q}$-linearly independent non-standard relations:

$$\begin{aligned}
0 = &-255608l_1 - 265360l_2 - 219216l_3 - 19306179l_4 - 214008l_5 + 45560l_6\\
&- 148296l_7 - 1117280l_8 - 677152l_9 + 86512l_{10} - 239320l_{11} - 50032l_{12}\\
&- 121008l_{13} - 96944l_{14} + 202328l_{15} - 1178499l_{16} + 98944l_{17}\\
&+ 1565754l_{18} + 23071580l_{19} + 363568l_{20} - 3310177l_{21}, \qquad (14.52)\\
0 = &\ 29752l_1 + 23312l_2 + 10960l_3 + 6123413l_4 + 16440l_5 - 12408l_6\\
&+ 7144l_7 + 58272l_8 + 86976l_9 - 15952l_{10} + 41144l_{11} + 13552l_{12}
\end{aligned}$$

$$\begin{aligned}
&+ 29552l_{13} + 9840l_{14} - 36696l_{15} + 375805l_{16} - 41760l_{17}\\
&- 477366l_{18} - 7196900l_{19} - 62128l_{20} + 1048983l_{21}, \qquad (14.53)\\
0 &= 477444l_1 + 431352l_2 + 268168l_3 + 98404710l_4 + 308964l_5 - 233140l_6\\
&+ 130028l_7 + 1563872l_8 + 1516032l_9 - 296664l_{10} + 702308l_{11} + 190136l_{12}\\
&+ 506440l_{13} + 141592l_{14} - 636468l_{15} + 6027441l_{16} - 701600l_{17}\\
&- 7683609l_{18} - 115803282l_{19} - 1063768l_{20} + 16877562l_{21}, \qquad (14.54)\\
0 &= -5976l_1 + 1776l_2 + 8496l_3 - 2132671l_4 + 3176l_5 + 1752l_6\\
&+ 3832l_7 + 50976l_8 - 2688l_9 + 2320l_{10} - 10264l_{11} - 5808l_{12}\\
&- 6128l_{13} + 2320l_{14} + 8120l_{15} - 132307l_{16} + 13856l_{17}\\
&+ 162614l_{18} + 2487604l_{19} + 12720l_{20} - 368485l_{21}, \qquad (14.55)\\
0 &= -474064l_1 - 405248l_2 - 243520l_3 - 54556373l_4 - 283952l_5 + 84368l_6\\
&- 170640l_7 - 1033056l_8 - 994784l_9 + 174880l_{10} - 540432l_{11} - 156544l_{12}\\
&- 240512l_{13} - 49344l_{14} + 411152l_{15} - 3357683l_{16} + 292256l_{17}\\
&+ 4291792l_{18} + 64572648l_{19} + 743136l_{20} - 9470695l_{21}. \qquad (14.56)
\end{aligned}$$

14.9. Let p be a prime number and $\mu_p = \exp(2\pi i/p)$. Suppose $\left(\begin{smallmatrix} 0 & 1 \\ c & 1 \end{smallmatrix}\right)$ is conjugate in $\mathrm{PGL}(2,\mathbb{C})$ to $\left(\begin{smallmatrix} \mu_p & 0 \\ 0 & 1 \end{smallmatrix}\right)$.

(i) Show that $c = -1/(\mu + \mu^{-1} + 2) \in \mathbb{R}$.

(ii) Define $\gamma(z) = 1/(cz+1)$. Show that the orbit $\{\gamma^i(0) : i \in \mathbb{N}\}$ has p elements:

$$a_0 = 0, a_1 = 1, a_2 = \frac{1}{c+1}, \dots, a_{p-2} = -\frac{1}{c}, a_{p-1} = \infty.$$

(iii) Define the 1-forms $\omega_0 = \mathrm{d}t/t$ and $\omega_j = \mathrm{d}t/(a_j^{-1} - t)$ for $j = 1, \dots, p-2$. Let GR_p be the $\mathbb{Q}$-algebra generated by all the iterated integrals, on the interval $t \in [0,1]$, of the $p-1$ differential forms $\omega_j, j = 1, \dots, p-1$. Define the weight w piece by

$$\mathsf{GR}_{w,p} := \left\{ \int_0^1 \omega_{j_1} \dots \omega_{j_w} : j_1 \neq 1, j_w \neq 0 \right\}.$$

Prove the integrals are all well-defined and then find their expressions by the MPLs.

(iv) Suppose $p = 2n+3$ and $d_w = \dim \mathsf{GR}_{w,p}$. Deligne conjectured [157] that the $d_w \le D_w$ where

$$1 + \sum_{w=1}^{\infty} D_w t^w = \frac{1}{1 - nt - t^2 - t^3}.$$

Can you verify some special cases?

(v)* The algebra GR_p has a obvious shuffle product structure. Is there a kind of "stuffle" product structure?

Chapter 15

Application to Feynman Integrals

In this last chapter we shift our attention to one of the most active research areas in mathematical physics nowadays: the study of the Feynman integrals in the perturbative quantum field theory (QFT). For simplicity, we will consider only the four-dimensional case in Sections 15.1-15.4.

15.1 Why Feynman Integrals?

Generally speaking, a Feynman graph (or Feynman diagram) represents a perturbative contribution to the probability amplitude (given by the corresponding Feynman integral) of a quantum transition from some initial quantum state to some final quantum state. So at high energies this is a valuable tool to obtain predictions of all coupling constants from the QFT. For example, one of the pictures on the book cover (reproduced in Fig. 15.1) represents an electron-positron collision producing a pair of muon and an-

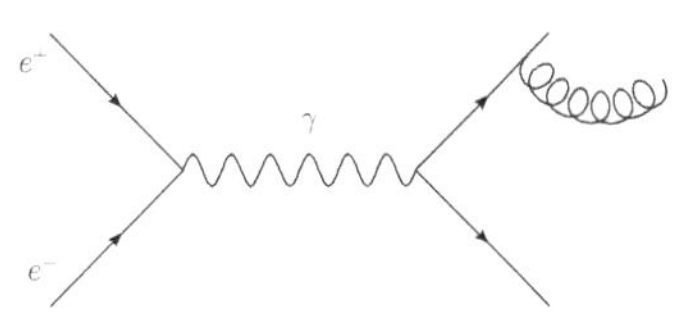

Figure 15.1: A Feynman diagram for electron-positron annihilation.

timuon, or producing some hadrons. Then the Feynman integral can be used to compute the ratio

$$\frac{\sigma(e^+e^- \to \text{hadrons})}{\sigma(e^+e^- \to \mu^+\mu^-)}$$

of the total cross section (*i.e.*, probability) for the production of hadron to the total cross section for the production of $\mu^+\mu^-$ pair. In fact, the calculation using the diagram in Fig. 15.1 already gives pretty good numerical result fitting the experimental data. To reach more accuracy, one needs to consider Z boson and taking quarks (in other words, quantum chromodynamics or QCD) further into account by including not only gluon emission as shown by the diagrams in Fig. 15.1 but also gluon loop corrections which leads the Feynman diagrams with loops shown in Fig. 15.2. More gluons lead to more loops and difference in the order of emissions and absorptions of them corresponds to different configurations of the loops.

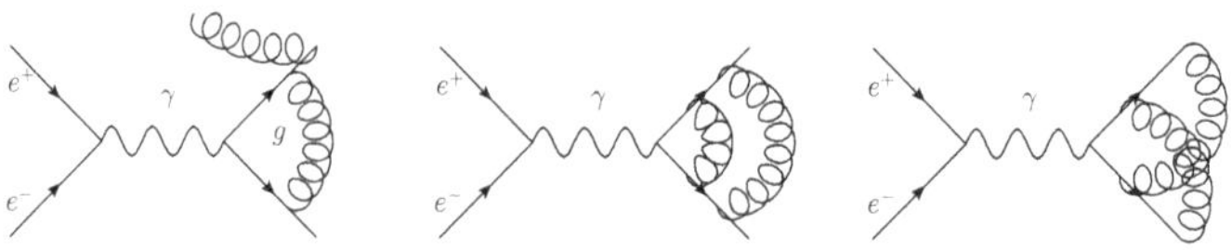

Figure 15.2: Several one- and two-loop Feynman diagrams contributing to the electron-positron annihilation process.

15.2 Feynman Graphs

We first recall some basic definitions in the graph theory.

Definition 15.2.1. A graph $G = (V_G, E_G)$ consists of the following data: V_G is the set of *vertices*, E_G is the set of *edges* whose two ends are vertices in V_G. We usually set $n_G := \sharp(E_G)$. The number of edges incident in a vertex v is called the *degree* (or *valance*) of v. A graph G is in ϕ^4 if the degree of every vertex of G is at most 4.

It is known that the most basic and important Feynman integrals are those associated with graphs in ϕ^4.

If we make G a directed graph and denote by $s(e)$ and $t(e)$ the source and the target of an edge $e \in E_G$ respectively, then the homology complex $H_\bullet(G, \mathbb{Z})$ of G with integral coefficients is defined by

$$0 \longrightarrow \mathbb{Z}[E_G] \stackrel{d}{\longrightarrow} \mathbb{Z}[V_G] \longrightarrow 0$$

where $d(e) = t(e) - s(e)$ and $\mathbb{Z}[E_G]$ sits at degree 0. Let $h_0(G)$ be the number of connected components of G and $h_1(G)$ be the number of independent loops of G. It is straightforward to check that (see Exercise 15.1)

$$H_0(G, \mathbb{Z}) \cong \mathbb{Z}^{h_0(G)}, \qquad H_1(G, \mathbb{Z}) \cong \mathbb{Z}^{h_1(G)}. \tag{15.1}$$

The loop number $h_G := h_1(G)$ is often called the *first Betti number* of G.

Definition 15.2.2. A *Feynman graph*

$$G = (V_G, E_G, E_G^{\text{ext}}, \{m_e\}_{e \in E_G}, \{q_i\}_{i \in E_G^{\text{ext}}})$$

consists of the following: V_G is the set of vertices, E_G is the set of labeled (internal) edges whose two ends are vertices in V_G, E_G^{ext} is the set of external half-edges (or legs), together with the following *kinematics data*:

(i) A particle mass $m_e \in \mathbb{R}$ for each $e \in E_G$,
(ii) An incoming momentum $q_i \in \mathbb{R}^D$ for each $i \in E_G^{\text{ext}}$ such that the *momentum conservation* holds:

$$\sum_{i \in E_G^{\text{ext}}} q_i = 0.$$

It is a convention to draw massive edges (i.e., $m_e \neq 0$) as thickened lines and draw a negative incoming momentum as a positive outgoing momentum. It is also helpful sometimes to think each internal edge as a union of two half-edges. For example, by closing the external legs of the two-point

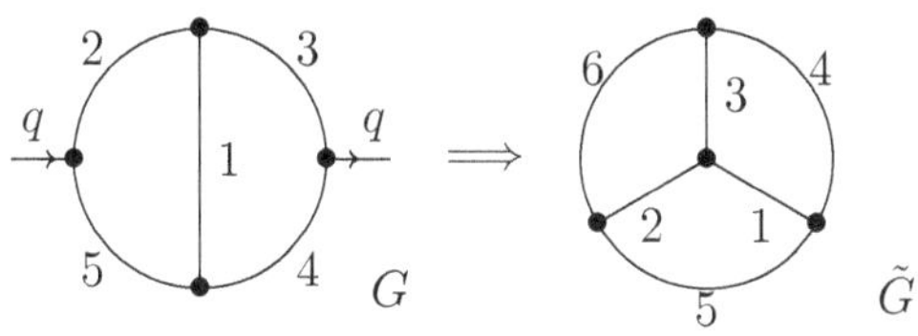

Figure 15.3: Left: 2-point 2-loop massless diagram. Right: wheel with three spokes.

(i.e., two incoming momenta) two-loop graph on the left of Fig. 15.3 we get a three-loop vacuum graph (i.e., no incoming momentum) on the right.

15.3 Feynman Integrals in Momentum Space

We will omit the derivation and the subsequent transformation of the Feynman integrals from the Standard Model of the particle physics into their momentum space form and then to the multi-loop Euclidean parametric form (see [439] for more details).

Suppose a Feynman graph G has m external legs and $n := n_G$ internal edges. For each internal edge e_j, the corresponding propagator in the integrand can be raised to an integer power a_j so that the integral also depends on the numbers $a_1,\dots,a_n$. The associated scalar (*i.e.*, no numerator in the integrand) $h := h_G$-loop integral $I(G)$, called its *period*, is defined by

$$I(G) = \int_{(\mathbb{R}^D)^h} \prod_{r=1}^{h} \frac{\mathrm{d}^D k_r}{i\pi^{\frac{D}{2}}} \prod_{j=1}^{n} \frac{1}{(-p_j^2 + m_j^2)^{a_j}}. \tag{15.2}$$

Here $k_r = (k_{r,1}, \dots, k_{r,D})$ is a D-dimensional vector variable with $\mathrm{d}^D k_r = \mathrm{d}k_{r,1} \dots \mathrm{d}k_{r,D}$, and p^2 means the square of the length of p if p is a D-dimensional momentum vector. The independent (virtual) loop momenta are labeled as $k_1, \dots, k_h$. The momenta flowing through the propagators are then given as a linear combination of the external momenta p and the loop momenta k determined by momentum conservation at every vertex:

$$p_i = \sum_{j=1}^{h} c_{ij} k_j + \sum_{j=1}^{m} d_{ij} q_j, \quad c_{ij}, d_{ij} \in \{-1, 0, 1\}.$$

Notice also by this conservation the number of independent internal momenta must be equal to the loop number h.

15.3.1 *Divergence Types*

Let us first consider the simple example of a scalar two-point one-loop integral with zero external momentum:

$q = 0$ 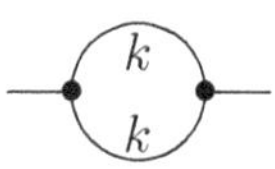

$$= \frac{1}{(2\pi)^4} \int_{\mathbb{R}^4} \frac{\mathrm{d}^4 k}{(k^2)^2} = \frac{1}{8\pi^2} \int_0^\infty \frac{\mathrm{d}r}{r}$$

by using 4-dimensional spherical coordinates. Here $k = (k_1, \dots, k_4)$ is a 4-dimensional vector, $\mathrm{d}^4 k = \mathrm{d}k_1 \cdots \mathrm{d}k_4$, and $k^2 = k_1^2 + \cdots + k_4^2$. This integral diverges when $r \to \infty$ as well as when $r \to 0$. The former divergence is called *ultraviolet (UV) divergence*, the later is called *infrared (IR) divergence*. Another physics term often appears in the literature: if the 4-momentum p and the mass m of a particle satisfies $p^2 = m^2$ then it is said to be *on shell*; otherwise it is said to be *off shell*.

One of the key ideas to handle UV divergence is to use the *dimension regularization*, namely, one can regard $D = 4 - 2\varepsilon$ as a continuous real variable and then expand the integral as a Laurent series of ε. To handle IR divergence coming from massless situation one can take a small mass first and then take the limit when the mass goes to 0.

15.3.2 *A Short Example*

Consider the Feynman graph in Fig. 15.4. The massive Feynman integral

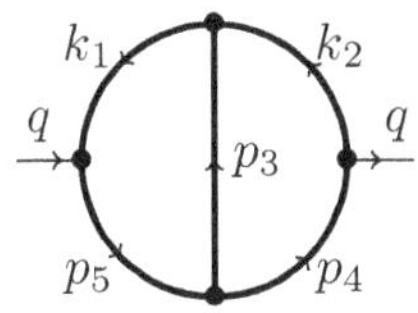

$$p_3 = k_1 - k_2,$$
$$p_4 = q + k_2,$$
$$p_5 = q + k_1.$$

Figure 15.4: The 2-point 2-loop massive diagram.

corresponding to it is:

$$\iint \frac{\mathrm{d}^D k_1 \mathrm{d}^D k_2}{(m_1^2 - k_1^2)^{a_1}(m_2^2 - k_2^2)^{a_2}(m_3^2 - p_3^2)^{a_3}(m_4^2 - p_4^2)^{a_4}(m_5^2 - p_5^2)^{a_5}}, \tag{15.3}$$

where q is the incoming momentum entering from the left and $D = 4 - 2\varepsilon$ is the number of dimensions. If the graph is massless (*i.e.*, $m_j = 0$ for all $j = 1, \dots, 5$) then it is proved [50] that MZVs are sufficient for the Laurent expansion of the integral in Eq. (15.3) if all powers of the propagators are of the form $a_j = n_j + c_j\varepsilon$ with $n_j \in \mathbb{N}$ and $c_j \in \mathbb{R}_{\geq 0}$.

15.3.3 *Massive and Massless Integrals*

When allowing the Feynman integrals to depend on kinematical invariants and particle masses, we know MPLs are not sufficient to express the result. Here the first cases show up already in the massive sunrise graphs $\ominus$ (see [57]). Because our main interest in this book is on MPLs and MZVs, we will consider only massless Feynman integrals in the rest of this chapter if we do not point out otherwise.

15.4 Graphical Functions in Position Space

In this section we consider loop integrals associated with ladder graphs and some generalizations. These integrals can be expressed using single-valued polylogs satisfying certain differential equations. Since these equations reduce the loop level by one, they can be solved iteratively in the loop order.

We begin with the box ladder graphs in Fig. 15.5. Let $p = p_1 + p_4 + \sum_{j=1}^{h} k_j$ and $y_j = k_1 + \cdots + k_j$ for $j = 1, \dots, h$. Then the period of the h-loop box ladder with incoming momenta $p_1, \dots, p_4$ is given by the period integral (up to some prefactor depending on p_j's)

$$I(\mathrm{B}_h(p_1, p_2, p_3, p_4)) = \frac{1}{\pi^{2h}} \int \frac{(p_1 + p_4)^2}{(y_h + p_1 + p_4)^2} \prod_{i=1}^{h} \frac{(p_3 + p_4)^2 \mathrm{d}^4 k_j}{k_j^2 (y_j + p_1)^2 (y_j - p_2)^2}.$$

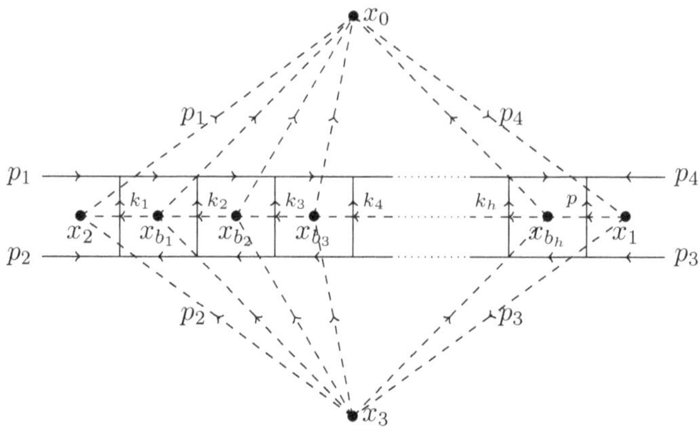

Figure 15.5: The box ladder graph and its dual graph.

Let $x_{ij} = x_j - x_i$ be the vector from x_i to x_j where x's are points in 4-dimensional space such that $p_1 = x_{20}$, $p_2 = x_{32}$, $p_3 = x_{13}$, $p_4 = x_{01}$ (note that $\sum p_j = 0$). Then by the change of variables $k_1 = x_{b_1,2}$ and $k_j = x_{b_j,b_{j-1}}$ for $j = 2, \dots, h$ we get

$$I(\mathrm{B}_h(p_1, p_2, p_3, p_4)) = \frac{x_{12}^2}{\pi^{2h}} \int \frac{1}{x_{b_1,2}^2} \prod_{j=1}^{h-1} \left(\frac{x_{03}^2 \cdot \mathrm{d}^4 x_{b_j}}{x_{b_j,b_{j+1}}^2 x_{b_j,0}^2 x_{b_j,3}^2} \right) \frac{x_{03}^2 \cdot \mathrm{d}^4 x_{b_h}}{x_{b_h,0}^2 x_{b_h,1}^2 x_{b_h,3}^2}.$$

Since the integral is conformally covariant we can take $x_3 \to \infty$ and find that there is some function $\mathscr{B}^{(h)}$ such that (see Eq. (15.9))

$$\begin{aligned} I(\mathrm{B}_h(p_1, p_2, p_3, p_4)) &= \frac{x_{12}^2}{\pi^{2h}} \int \frac{1}{x_{b_1,2}^2} \prod_{i=1}^{h-1} \left(\frac{\mathrm{d}^4 x_{b_i}}{x_{b_i,b_{i+1}}^2 x_{b_i,0}^2} \right) \frac{\mathrm{d}^4 x_{b_h}}{x_{b_h,1}^2 x_{b_h,0}^2} \\ &= \mathscr{B}^{(h)} \left(\frac{x_{02}^2}{x_{12}^2}, \frac{x_{01}^2}{x_{12}^2} \right). \end{aligned}$$

15.4.1 *Sequential Graphs and Graphical Functions*

We now generalize the (planar dual of the) box ladder graphs to the following setting.

Definition 15.4.1. Let $\mathbf{w}$ be a word in $\{0, 1, 2\}^*$. We denote by $\lg(\mathbf{w})$ the number of letters in $\mathbf{w}$. The *sequential graph* $G_{\mathbf{w}}$ is the graph with two distinguished vertices 0 and 1 and a horizontal chain of vertices that connect to either 0 or 1, or both. We can obtain the word $\mathbf{w}$ by recording the connections for each vertex from left to right sequentially: 0 (resp. 1) for a single connection to 0 (resp. 1), or 2 for a double connection to both 0 and 1.

Example 15.4.2. The left graph in Fig. 15.6 is the sequential graph $G_{20111202}$ if z and its edge are both removed. The right one corresponds to the dual graph[1] of the zig-zag graph Z_5 if $z \to 0$.

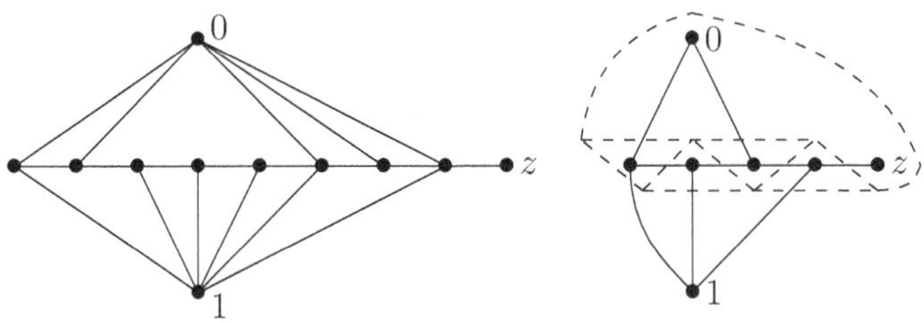

Figure 15.6: Left: the sequential graph $G_{20111202}$. Right: the zig-zag Z_5 (dashed graph) and its dual G_{2101} (solid graph, with edge $0 \to 1$ removed).

Given a word $\mathbf{w} = a_1 \dots a_{h-1}$ with letters $a_i \in \{0, 1\}$, we can define the generalized h-loop (planar dual of) three-point ladder integral by

$$I_{\mathbf{w}}(x_0, x_1, x_2) = \frac{1}{\pi^{2h}} \int \frac{1}{x_{b_1,2}^2} \prod_{i=1}^{h-1} \left(\frac{d^4 x_{b_i}}{x_{b_i,b_{i+1}}^2 x_{b_i,a_i}^2} \right) \frac{d^4 x_{b_h}}{x_{b_h,0}^2 x_{b_h,1}^2}. \tag{15.4}$$

Here the integration vertices are labeled as $x_{b_1}, \dots, x_{b_h}$ while x_{ij}^2 denotes the square distance $(x_i - x_j)^2$ in 4-dimensional Euclidean space. We now show that these integrals can be expressed in terms of classical polylogs. In particular we will derive the formula for $\mathscr{B}^{(h)}$ in terms the single-valued polylogs.

Definition 15.4.3. Let vertex 0 correspond to the origin in $\mathbb{R}^4$ and $1 = (1, 0, 0, 0) \in \mathbb{R}^4$. Let $z \neq 0, 1$ be a complex variable embedded in $\mathbb{R}^4$ as $(\mathrm{Re}(z), \mathrm{Im}(z), 0, 0)$. Given a graph G with three distinguished vertices $0, 1, z$ with position space Feynman rules, we define its *graphical function*

$$f_G(z) := \left(\prod_{v \in V_G \setminus \{0,1,z\}} \int_{\mathbb{R}^4} \frac{d^4 x_v}{\pi^2} \right) \frac{1}{\prod_{e \in E_G} Q_e} \tag{15.5}$$

[1]Strictly speaking, in the dual graph 0 and 1 should be joint by a solid curve. But this doesn't change its Feynman integral.

where $Q_e = \|x - y\|^2$ for every edge e connecting the two positions x and y in $\mathbb{R}^4$. If $G = G_{\mathbf{w}}$ is a sequential graph with $\mathbf{w} \in \{0,1,2\}^*$ then we write $f_{\mathbf{w}}$ for $f_{G_{\mathbf{w}}}$.

Remark 15.4.4. By definition, $\|z-1\|^2$ in $\mathbb{R}^4$ is equal to $(z-1)(\bar{z}-1) \in \mathbb{R}$. Similarly, $\|z\|^2 = z\bar{z}$.

Example 15.4.5. Let I be the graph $\underset{0}{\bullet}\!-\!\underset{z}{\bullet}\!-\!\underset{1}{\bullet}$. Then we have.

$$f_I(z) = \frac{1}{Q_{0,z}Q_{1,z}} = \frac{1}{z\bar{z}(z-1)(\bar{z}-1)}. \tag{15.6}$$

In general, we can compute the period integrals for the box ladder graphs B_n, the wheel graphs WS_n (see Thm. 15.4.19), and the zig-zag graphs Z_n (see Thm. 15.4.20), by using some sequential graphical functions:

$$I(\mathrm{B}_n) = f_{2\{0\}^{n-1}}(1),$$
$$I(\mathrm{Z}_{2n+3}) = f_{2\{01\}^n1}(0) = f_{2\{01\}^n0}(1) = f_{2\{10\}^n1}(0) = f_{2\{10\}^n0}(1), \tag{15.7}$$
$$I(\mathrm{Z}_{2n+4}) = f_{2\{01\}^{n+1}}(0) = f_{2\{01\}^n00}(1) = f_{2\{10\}^n11}(0) = f_{2\{10\}^{n+1}}(1). \tag{15.8}$$

In particular, by the right picture in Fig. 15.6 we obtain $I(\mathrm{Z}_5) = f_{2101}(0)$.

Example 15.4.6. The dual of Fig. 15.5 for the box ladder graph is the sequential graph $G_{2\{0\}^h}$ if we set $x_0 = 0$, $x_2 = 1$ and $x_1 = z$. Thus $p_4^2 = x_{01}^2 = \|z\|^2$, $p_1^2 = x_{02}^2 = 1$, and $(p_1+p_4)^2 = (p_2+p_3)^2 = x_{12}^2 = \|z-1\|^2$.

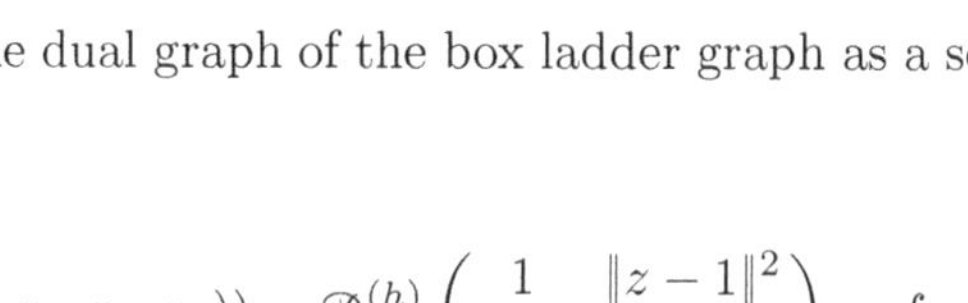

Figure 15.7: The dual graph of the box ladder graph as a sequential graph.

So we have

$$I(\mathrm{B}_h(p_1,p_2,p_3,p_4)) = \mathscr{B}^{(h)}\left(\frac{1}{\|z\|^2}, \frac{\|z-1\|^2}{\|z\|^2}\right) = f_{2\{0\}^{h-1}}(z).$$

Therefore period of the box ladder graph is given by

$$I(\mathrm{B}_h) = f_{2\{0\}^{h-1}}(1). \tag{15.9}$$

15.4.2 *Properties of Graphical Functions*

Recall that a complex variable function $f(z)$ is called *real analytic* in an open set $D \subseteq \mathbb{C}$ if $f(z)$ is real valued and is infinitely differentiable with respect to both ∂_z and $\partial_{\bar{z}}$. More precisely, such a function should be written as $f(z, \bar{z})$, where $f(x, y) \in C^\infty$ satisfies $f(x, y) = f(\bar{y}, \bar{x})$.

Definition 15.4.7. A single-valued real analytic function $f : \mathbb{C}\backslash\{0, 1\} \to \mathbb{R}$ is in $\mathsf{C}_{1,1}$ if for all $\epsilon > 0$

(i) $f(z) = o(|z|^{-1-\epsilon})$ for $z \to 0$, *i.e.*, $\lim_{z\to 0} f(z)|z|^{1+\epsilon} = 0$, and
(ii) $f(z) = o(|z|^{-1+\epsilon})$ for $z \to \infty$, *i.e.*, $\lim_{z\to\infty} f(z)|z|^{1-\epsilon} = 0$.

Proposition 15.4.8. *Let G be a graph with three distinguished vertices $0, 1, z$. Then its graphical function $f_G(z) : \mathbb{C} \setminus \{0, 1\} \to \mathbb{R}$ satisfies*

(i) $f_G(z) = f_G(\bar{z})$,
(ii) f_G *is a single-valued,*
(iii) $f_{\mathbf{w}} \in \mathsf{C}_{1,1}$ *for every sequential graph* $G_{\mathbf{w}}$,
(iv) $f_{\mathbf{w}}$ *is real analytic in* $\mathbb{C} \setminus \{0, 1\}$ *for every sequential graph* $G_{\mathbf{w}}$.

Proof. (i) This follows from the change of coordinates $x_v = (x_v^1, x_v^2, x_v^3, x_v^4) \mapsto (x_v^1, -x_v^2, x_v^3, x_v^4)$.

(ii) An infinitesimal ε-deformation of the denominator factors avoiding all singularities shows that the monodromy for f_G is always trivial. So (ii) follows by taking the limit $\varepsilon \to 0$.

The proofs of (iii) and (iv) are long and very technical so we refer the interested reader to Schnetz's original paper where more general results can be found in [505, Prop. 3.14 and Prop. 3.15]. □

Definition 15.4.9. Let G be a graph with three marked vertices $0, 1, z$. We can produce a new graph $G^\sharp$ by appending an edge to the vertex z and relabel the new vertex by z. See Fig. 15.8.

Figure 15.8: Graph $G^\sharp$ is obtained from G by appending an edge to z.

In what follows we will assume that both G and $G^\sharp$ have well-defined graphical functions.

Proposition 15.4.10. *The following identity holds*

$$\mathfrak{D}_{z,\bar z} f_{G^\sharp}(z) := \frac{1}{z-\bar z}\,\partial_z\partial_{\bar z}\,(z-\bar z) f_{G^\sharp}(z) = -f_G(z). \tag{15.10}$$

Proof. By using D-dimensional spherical coordinates $(r,\phi^1,\dots,\phi^{D-1})$ we see that, as a function in D-dimensional space

$$\begin{aligned} h(z) :=& \Delta_z^{(D)} \int \frac{F(y)}{\|y-z\|^2} \mathrm{d}^D y \\ =& 2(4-D)\int \frac{F(y)}{\|y-z\|^4}\mathrm{d}^D y = 2(4-D)(F*g)(z) \end{aligned} \tag{15.11}$$

where $*$ stands for the convolution and $g(x)=\|x\|^{-4}$ has Fourier transform given by

$$\hat g(\xi) = \int \frac{e^{2\pi i x\cdot\xi}}{\|x\|^4}\mathrm{d}^D x = \frac{\pi^{4-D/2}}{\|\xi\|^{D-4}}\Gamma(D/2-2).$$

Thus by dimension regularization $D=4-2\varepsilon$ we get from Eq. (15.11)

$$\begin{aligned} \hat h(\xi) = \hat F(\xi)\hat g(\xi) =& 2(4-D)\frac{\pi^{4-D/2}}{\|\xi\|^{D-4}}\Gamma(D/2-2)\hat F(\xi) \\ =& 4\varepsilon\pi^{2+\varepsilon}\|\xi\|^{2\varepsilon}\Gamma(-\varepsilon)\hat F(\xi) = -4\pi^{2+\varepsilon}\|\xi\|^{2\varepsilon}\Gamma(1-\varepsilon)\hat F(\xi). \end{aligned}$$

Taking $\varepsilon\to 0$ we obtain

$$\Delta^{(4)}\int \frac{F(y)}{\|y-z\|^2}\mathrm{d}^4 y = -4\pi^2 F(z).$$

Thus, for $F(y)=f_G(y)$, we have

$$\Delta^{(4)} f_{G^\sharp}(z) = -4 f_G(z). \tag{15.12}$$

Let $e_1=(1,0,0,0)$ in $f_{G^\sharp}$ then $f_{G^\sharp}(z)$ depends only on r and ϕ^1 in spherical coordinates. Hence the Laplacian becomes (see Exercise 15.4)

$$\Delta^{(4)} f_{G^\sharp}(r,\phi^1) = \left(\partial_r^2 + \frac{3}{r}\partial_r + \frac{1}{r^2}\partial_{\phi^1}^2 + \frac{2\cos\phi^1}{r^2\sin\phi^1}\partial_{\phi^1}\right) f_{G^\sharp}(r,\phi^1)$$

which coincides with the following operator in 2-dimensional polar coordinates (r,ϕ^1):

$$\frac{1}{r\sin\phi^1}\Delta_z^{(2)}\circ M(r\sin\phi^1) = \partial_r^2 + \frac{3}{r}\partial_r + \frac{1}{r^2}\partial_{\phi^1}^2 + \frac{2\cos\phi^1}{r^2\sin\phi^1}\partial_{\phi^1},$$

where $M(r\sin\phi^1)$ just means multiplication by $r\sin\phi^1$ (see Exercise 15.5). By transition to $\mathbb{C}$ the proposition follows immediately from Eq. (15.12) since $\Delta^{(2)}=4\partial_z\partial_{\bar z}$ and $z-\bar z=2ir\sin\phi^1$. $\square$

Recall that the important properties of SVMPLs are given by Thm. 4.3.2. Although they are not defined at 0, 1 and ∞ we can define their regularized values at these points. In particular, the SVMZVs are just the regularized values of SVMPLs at 1 (see Definition 4.4.3). Also recall from Sec. 4.4 we have defined the sets $\mathsf{O} = \mathbb{Q}\left[z, \frac{1}{z}, \frac{1}{z-1}\right]$, $\overline{\mathsf{O}} = \mathbb{Q}\left[\bar{z}, \frac{1}{\bar{z}}, \frac{1}{\bar{z}-1}\right]$. and the two bi-differential $(\partial_z, \partial_{\bar{z}})$-algebras $\mathsf{A} := \mathsf{O}\overline{\mathsf{O}}\mathsf{Ls}$ and $\mathsf{A}^{\text{sv}} := \mathsf{O}\overline{\mathsf{O}}\mathsf{Ls}^{\text{sv}}$, where Ls (resp. Ls^{sv}) is the shuffle-algebra spanned by SVMPLs over $\mathbb{C}$ (resp. over $\mathsf{MZ}_{\mathbb{Z}}^{\text{sv}}$, the set of SVMZVs).

Definition 15.4.11. Let

$$\mathsf{B} := \left\{ \frac{f(z)}{z-\bar{z}} : f \in \mathsf{A}, f(z) = -f(\bar{z}) \right\}. \tag{15.13}$$

Similarly, $\mathsf{B}^0, \mathsf{B}^{0,\text{sv}}$ are defined with $f \in \mathsf{Ls}, f \in \mathsf{Ls}^{\text{sv}}$ in Eq. (15.13), respectively. Let $\mathsf{B}_n^\bullet$ be the (total) weight n piece of $\mathsf{B}^\bullet$.

The differential operator in equation (15.10) maps graphical functions in B into B. By integration in A the differential operator can be inverted in B, namely, we can produce $f_{G^\sharp}$ from f_G.

Theorem 15.4.12. *Assume the graphical function $f_{G^\sharp}(z)$ is real analytic in $\mathbb{C}\setminus\{0,1\}$ and for $z \in \mathbb{R}^4$, $f_{G^\sharp}(z), f_{G^\sharp}(1-z) \in \mathsf{C}_{1,1}$. Moreover, let*

$$f_G(z) = \sum_{a,b\in\{0,1\}} \frac{g_{a,b}(z)}{(z-a)(\bar{z}-b)(z-\bar{z})} \in \mathsf{B} \tag{15.14}$$

for some SVMPL $g_{a,b}$ in Ls. Then

$$f_{G^\sharp}(z) = \mathfrak{I}_{z,\bar{z}} f_G(z) := \frac{-1}{2(z-\bar{z})} \left(\int_0 \mathrm{d}z \int_0 \mathrm{d}\bar{z} + \int_0 \mathrm{d}\bar{z} \int_0 \mathrm{d}z \right) (z-\bar{z}) f_G(z). \tag{15.15}$$

The function $f_{G^\sharp}$ is the unique solution of Eq. (15.10) *in* B^0.

Proof. It is easy to verify that the right-hand side of Eq. (15.15), denoted by $F(z)$, satisfies Eq. (15.10). So we see that

$$(z-\bar{z})\big(f_{G^\sharp}(z) - F(z)\big) \in \ker(\partial_z \partial_{\bar{z}}). \tag{15.16}$$

Our goal is to prove that $F(z) = f_{G^\sharp}(z)$.

Observe that $\ker(\partial_z\partial_{\bar{z}})$ is a sum of a holomorphic h and an anti-holomorphic $\overline{h}$ (*i.e.*, $\partial_{\bar{z}} h = \partial_z \overline{h} = 0$). By Prop. 15.4.8(i) both f_G and $f_{G^\sharp}$ are invariant under $z \leftrightarrow \bar{z}$ so that

$$(z-\bar{z})\big(f_{G^\sharp}(z) - F(z)\big) = h(z) - h(\bar{z}).$$

Since $f_{G^\sharp}$ and F are real analytic on $\mathbb{C}\setminus\{0,1\}$ we get

$$h'(z) = \partial_z(z-\bar{z})(f_{G^\sharp}(z) - F(z)).$$

Now, h' is a single-valued real analytic function on $\mathbb{C}\setminus\{0,1\}$ with $\partial_{\bar{z}}h' = 0$. Hence h' is holomorphic on $\mathbb{C}\setminus\{0,1\}$. We now claim that

$$h'(z) = c/z + d/(z-1) \quad \text{for some constants } c, d \in \mathbb{C}. \tag{15.17}$$

Indeed, by assumption $f_{G^\sharp}(z) \in \mathsf{C}_{1,1}$ we have $f_{G^\sharp}(z) = o(|z|^{-1-\epsilon})$ for $z \to 0$ so that $\partial_z(z-\bar{z})f_{G^\sharp}(z) = o(|z|^{-1-\epsilon})$. By Eq. (15.14), Prop. 4.4.9 and Prop. 4.4.12 we have $(z-\bar{z})F(z) \in \mathsf{Ls}$. Hence $(z-\bar{z})F(z) = o(|z|^{-\epsilon})$ by Exercise 4.24 and $\partial_z(z-\bar{z})F(z) = o(|z|^{-1-\epsilon})$. Thus $h'(z) = o(|z|^{-1-\epsilon})$ and $z^2h'(z) = o(|z|^{1-\epsilon})$. In particular $z^2h'(z)$ is continuous at $z=0$ and by Riemann's Theorem on Removable Singularities $z^2h'(z)$ is holomorphic at 0. It vanishes at 0 and therefore $zh'(z)$ is holomorphic at 0.

For $z=1$ the above argument can be applied to $f_{G^\sharp}(1-z)$ and $f_G(1-z)$ so that we get that $(z-1)h'(z)$ is holomorphic at 1. Hence

$$h'(z) = c/z + d/(z-1) + H(z)$$

where $H(z)$ is holomorphic on all of $\mathbb{C}$. To show $H(z) = 0$, notice that $(z-\bar{z})F(z) \in \mathsf{Ls}$ yields $\partial_z(z-\bar{z})F(z) = o(|z|^{-1+\epsilon})$ as $z \to \infty$ by Exercise 4.25. By assumption $f_{G^\sharp}(z) \in \mathsf{C}_{1,1}$ so $\partial_z(z-\bar{z})f_{G^\sharp}(z) = o(|z|^{-1+\epsilon})$. Thus $\lim_{z\to\infty} h'(z)|z|^{1-\epsilon} = 0$. But this implies $\lim_{z\to\infty} H(z) = 0$ so $H(z)$ is a bounded holomorphic function on $\mathbb{C}$. Therefore $H(z)$ is a constant equal to $\lim_{z\to\infty} H(z) = 0$. The Claim Eq. (15.17) is now proved. Consequently,

$$f_{G^\sharp}(z) - F(z) = \frac{c(\log(z) - \log(\bar{z})) + d(\log(z-1) - \log(\bar{z}-1))}{z-\bar{z}}.$$

But $f_{G^\sharp}(z) - F(z)$ is single-valued so we must have $c = d = 0$, namely, $f_{G^\sharp} = F$. Further, $f_{G^\sharp} \in \mathsf{B}^0$ follows from Eqs. (15.14), (15.15), Props. 4.4.9 and 4.4.12.

To prove the uniqueness in the last sentence, assume $f \in \mathsf{B}^0$ is another solution of Eq. (15.10). Since the kernel of $\partial_z\partial_{\bar{z}}$ in Ls is the constant function, $f_{G^\sharp}(z) = f(z) + c/(z-\bar{z})$ by Eq. (15.16). So $c = 0$ by the invariance of $f_{G^\sharp}$ and f under $z \leftrightarrow \bar{z}$.

We now have completed the proof of the theorem. □

Remark 15.4.13. It is clear that the differential operator $\mathfrak{D}_{z,\bar{z}}$ in Eq. (15.10) can be rewritten as

$$\mathfrak{D}_{z,\bar{z}} := \frac{1}{z-\bar{z}}\partial_z\partial_{\bar{z}}(z-\bar{z}) = \partial_z\partial_{\bar{z}} - \frac{\partial_z - \partial_{\bar{z}}}{z-\bar{z}}.$$

This is sometimes useful in the computation because by the theorem if f_G is known and we can find a solution in B^0 of Eq. (15.10) (by whatever method) then it must be the unique solution of Eq. (15.15) as well.

Example 15.4.14. Consider the following graphs:

$I :=$ 

and $I^\sharp =$ 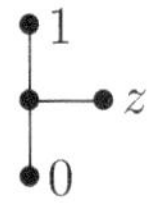

By Example 15.4.5 we have $f_I(z) = 1/\big(z\bar{z}(z-1)(\bar{z}-1)\big)$. Thus $f_{I^\sharp} = f_2$ is the sequential function of the word 2. We can check trivially that

$$\mathfrak{D}_{z,\bar{z}}\left(\frac{4i\,\mathscr{L}i_2(z)}{z-\bar{z}}\right) = -\frac{1}{z-\bar{z}}\left(\frac{1}{\bar{z}(z-1)} - \frac{1}{z(\bar{z}-1)}\right) = f_I(z)$$

where $\mathscr{L}i_2(z)$ is the single-valued Bloch–Wigner dilog function defined by Eq. (2.45). So we get

$$f_2(z) = \frac{4i\,\mathscr{L}i_2(z)}{z-\bar{z}} \in \mathsf{B}^0 \tag{15.18}$$

by Remark 15.4.13. In fact, it is not difficulty to use the theorem to integrate and obtain directly that $f_2(z)$ is equal to

$$\frac{\mathrm{L}_{10}(z) + \mathrm{L}_1(z)\,\mathrm{L}_0(\bar{z}) + \mathrm{L}_{01}(\bar{z}) - \mathrm{L}_{10}(\bar{z}) - \mathrm{L}_1(\bar{z})\,\mathrm{L}_0(z) - \mathrm{L}_{01}(z)}{z-\bar{z}} \tag{15.19}$$

which, by the identities in Example 4.2.13, also implies Eq. (15.18).

Lemma 15.4.15. *Let* $\mathbf{w}$ *be a word in* $\{0,1,2\}^*$ *that begins with* 2. *Then* $f_{\mathbf{w}}(e_1 - z)$ *is the sequential function of the word with* 0*'s and* 1*'s swapped.*

Proof. Note that for the graphical function defined on $\mathbb{R}^4$ we interpret $1-z$ as $e_1 - z$ for the unit vector $e_1 \in \mathbb{R}^4$. Denote by $\nu_{x,y}$ the number of edges connecting two vertices x and y. Assume

$$f_{\mathbf{w}}(z) = f_{0,1,z,\infty} = \int \frac{f^0_{0,1,z,\infty}}{\|z\|^{2\nu_{0,z}}\|e_1 - z\|^{2\nu_{1,z}}} \prod_{a=0,1}\ \prod_{\substack{x:\text{internal}\\ \nu(x,a)=1}} \frac{\mathrm{d}^4 x}{\|x-a\|^2}, \tag{15.20}$$

where f^0 only involves propagators between internal vertices (i.e., those $\neq 0,1,z$). A change of variables $x_i \mapsto 1 - x_i$ leaves the integration measure invariant (since we are in even dimensions) and also keeps all the propagators between internal vertices intact so we get $f^0_{0,1,1-z,\infty} = f^0_{1,0,z,\infty}$. A propagator from an internal vertex x_i to an external vertex a is mapped to a propagator from x_i to $1-a$. Thus we have

$$\begin{aligned} f_{\mathbf{w}}(e_1 - z) &= f_{0,1,1-z,\infty} \\ &= \int \frac{f^0_{1,0,z,\infty}}{\|e_1 - z\|^{2\nu_{0,z}}\|-z\|^{2\nu_{1,z}}} \prod_{a=0,1}\ \prod_{\substack{x:\text{internal}\\ \nu(x,1-a)=1}} \frac{\mathrm{d}^4 x}{\|x-a\|^2} = f_{\tau(\mathbf{w})}(z) \end{aligned}$$

where τ exchanges 0 and 1, as desired. ☐

Proposition 15.4.16. *In four dimensions, the sequential functions $f_{\mathbf{w}}$, where $\mathbf{w} \in \{0,1,2\}^*$ begins with 2, provide the unique solutions in B^0 of*

$$\mathfrak{D}_{z,\bar{z}} f_{\mathbf{w}a}(z) = \begin{cases} \dfrac{-1}{(z-a)(\bar{z}-a)} f_{\mathbf{w}}(z), & \text{if } a = 0, 1; \\ \dfrac{-1}{z\bar{z}(z-1)(\bar{z}-1)} f_{\mathbf{w}}(z), & \text{if } a = 2 \end{cases} \tag{15.21}$$

with the initial condition given by Eq. (15.18).

Proof. By Lemma 15.4.15, Prop. 15.4.8(iii) and (iv) both $f_{\mathbf{w}}$ and $f_{\mathbf{w}}(e_1 - z)$ are real analytic on $\mathbb{C} \setminus \{0,1\}$ and in $\mathsf{C}_{1,1}$. After adding an edge from a to z we get a graph G with $f_G(z) = f_{\mathbf{w}}/\|z-a\|^2$ (see the first two pictures in Fig. 15.9). If adding both edges from 0 to z and from 1 to z we would get a graph G with $f_G(z) = f_{\mathbf{w}}/(\|z\|\|z-1\|)^2$. On the other hand, by the definition of the sequential graphs, appending an edge to z to $G_{\mathbf{w}}$ we immediately get the graph $G_{\mathbf{w}a}$ (see the last picture in Fig. 15.9). Equation (15.14) is obvious for $a = 0, 1$. For $a = 2$, Eq. (15.14) follows

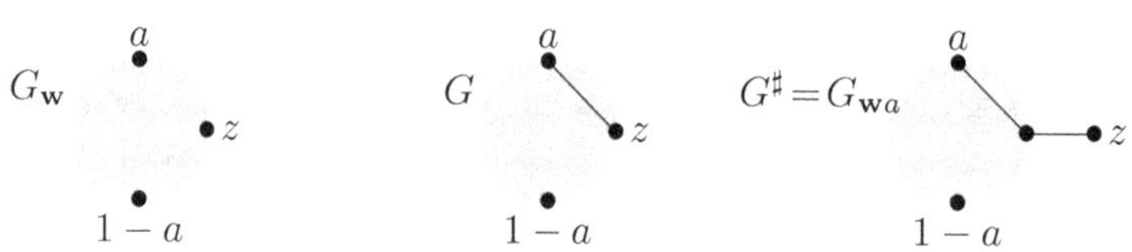

Figure 15.9: Adding edge a to z and applying $\sharp$ to $G_{\mathbf{w}}$ produces $G_{\mathbf{w}a}$.

from $1/(z(z-1)) = 1/(z-1) - 1/z$ and $1/(\bar{z}(\bar{z}-1)) = 1/(\bar{z}-1) - 1/\bar{z}$. The proposition now follows from Thm. 15.4.12. □

The solution of the recursive differential equation (15.21) can be given explicitly.

Proposition 15.4.17. *Let $\mathbf{w} \in \{0,1,2\}^*$ that begins with 2. Then the sequential function $f_{\mathbf{w}}$ is recursively given by Eq.* (15.18) *and*

$$f_{\mathbf{w}a}(z) = \begin{cases} \mathfrak{I}_{z,\bar{z}} \left(\dfrac{(z-\bar{z}) f_{\mathbf{w}}(z)}{(z-a)(\bar{z}-a)} \right), & \text{if } a = 0, 1; \\ \mathfrak{I}_{z,\bar{z}} \left(\dfrac{(z-\bar{z}) f_{\mathbf{w}}(z)}{z\bar{z}(z-1)(\bar{z}-1)} \right), & \text{if } a = 2. \end{cases} \tag{15.22}$$

Moreover, $f_{\mathbf{w}} \in \mathsf{B}^{0,\mathrm{sv}}_{2\lg(\mathbf{w})}$.

Proof. Equation (15.22) is clear from Thm. 15.4.12 and Prop. 15.4.16. The nontrivial task is to prove that the coefficients of $f_{\mathbf{w}}$ are in $\mathsf{MZ}^{\text{sv}}_{\mathbb{Z}}$. We proceed by induction on $\lg(\mathbf{w})$. The length 1 case $\mathbf{w} = 2$ is Eq. (15.18). Assuming the proposition holds for $\mathbf{w}$ we now prove the case for $\mathbf{w}a$ where $a \in \{0, 1, 2\}$.

Case (i) $a \in \{0, 1\}$. Assume $b \in \{0, 1\}$. Then

$$Q_b(z) := \int_0 \frac{(z - \bar{z}) f_{\mathbf{w}}(z) \mathrm{d}\bar{z}}{\bar{z} - b} \in \mathsf{Ls}^{\text{sv}}_{2\lg(\mathbf{w})+1}$$

by Prop. 4.4.12. Thus

$$R_{a,b}(z) := \int_0 \frac{Q_b(z) \mathrm{d}z}{z - a} \in \mathsf{Ls}^{0,\text{sv}}_{2\lg(\mathbf{w})+2} \tag{15.23}$$

by Prop. 4.4.9. For $\int_0 \mathrm{d}\bar{z} \int_0 \mathrm{d}z$, since

$$\overline{Q_a(z)} = \int_0 \frac{(\bar{z} - z) f_{\mathbf{w}}(\bar{z}) \mathrm{d}z}{z - a} = -\int_0 \frac{(z - \bar{z}) f_{\mathbf{w}}(z) \mathrm{d}z}{z - a}$$

we get by setting $P(z) = (z - \bar{z}) f_{\mathbf{w}}(z)$ in Exercise 4.26

$$\begin{aligned}
&\int_0 \mathrm{d}\bar{z} \int_0 \mathrm{d}z \left(\frac{(z - \bar{z}) f_{\mathbf{w}}(z)}{(z - a)(\bar{z} - b)} \right) \\
&\quad = R_{a,b}(z) - \left(Q_b(a) + \overline{Q_a(b)}\right) \mathscr{L}_1(z) \in \mathsf{Ls}^{0,\text{sv}}_{2\lg(\mathbf{w})+2}
\end{aligned}$$

since $\overline{Q_a(b)} = Q_a(b) \in \mathsf{MZ}^{\text{sv}}_{\mathbb{Z}, 2\lg(\mathbf{w})+1}$ for $a, b \in \{0, 1\}$. Combining with Eq. (15.23) we have

$$\mathfrak{I}_{z,\bar{z}} \left(\frac{(z - \bar{z}) f_{\mathbf{w}}(z)}{(z - a)(\bar{z} - b)} \right) = \frac{-1}{z - \bar{z}} \left(R_{a,b}(z) - \frac{1}{2} (Q_b(a) + Q_a(b)) \mathscr{L}_1(z) \right). \tag{15.24}$$

Consequently when $a = b \in \{0, 1\}$ we get $f_{\mathbf{w}a} \in \mathsf{B}^{0,\text{sv}}_{2\lg(\mathbf{w})+2}$.

Case (ii) $a = 2$. By partial fractions we get

$$\frac{1}{z\bar{z}(z - 1)(\bar{z} - 1)} = \left(\frac{1}{z - 1} - \frac{1}{z} \right) \left(\frac{1}{\bar{z} - 1} - \frac{1}{\bar{z}} \right) = \sum_{a,b=0}^{1} \frac{(-1)^{a+b}}{(z - a)(\bar{z} - b)}.$$

Then Eq. (15.24) yields

$$f_{\mathbf{w}2} = \frac{-1}{z - \bar{z}} \sum_{a,b=0}^{1} (-1)^{a+b} (R_{a,b}(z) - Q_a(b) \mathscr{L}_1(z)) \in \mathsf{B}^{0,\text{sv}}_{2\lg(\mathbf{w})+2},$$

as desired. □

Corollary 15.4.18. *For every word* $\mathbf{w} \in 2\{0, 1, 2\}^*2$, *the graph period* $I(G_{\mathbf{w}})$ *of the sequential graph* $G_{\mathbf{w}}$ *lies in* $\mathsf{MZ}^{\text{sv}}_{\mathbb{Z}, 2|\mathbf{w}|-1}$.

Proof. Assume $\mathbf{w} = \mathbf{u}2$. Then we have $I(G_{\mathbf{u}2}) = f_{\mathbf{u}1}(0)$ with $f_{\mathbf{u}1}(z) = g(z)/(z-\bar{z})$ and $g \in \mathsf{Ls}^{\mathrm{sv}}_{2\lg(\mathbf{w})}$ by Prop. 15.4.17. But by setting $\mathbf{w} = \mathbf{u}$ in the proof of Prop. 15.4.17 we get

$$Q_1(z) := \int_0 \frac{(z-\bar{z})f_{\mathbf{u}}(z)\mathrm{d}\bar{z}}{\bar{z}-1}, \quad R_{1,1}(z) := \int_0 \frac{Q_1(z)\mathrm{d}z}{z-1}$$

and from Eq. (15.24)

$$g(z) = (z-\bar{z})f_{\mathbf{u}1}(z) = Q_1(1)\mathscr{L}_1(z) - \int_0 \frac{Q_1(z)\mathrm{d}z}{z-1}.$$

By Example 4.3.3 we know $\mathscr{L}_1(z) = \log(1-z) + \log(1-\bar{z})$. Thus

$$\partial_z g(z) = \frac{Q_1(1)}{z-1} - \frac{Q_1(z)}{z-1}.$$

From Exercise 15.6 we see that $I(G_{\mathbf{u}2}) = \frac{1}{2}(\partial_z g(0) - \partial_{\bar{z}} g(0)) = Q_1(0) - Q_1(1) \in \mathsf{MZ}^{\mathrm{sv}}_{\mathbb{Z}}$ since $g(z) = -g(\bar{z})$. $\square$

We are now ready to compute the period of the box ladder graph and the *wheel graph* WS_{n+1} *with* $n+1$ *spokes*, $n \geq 1$. See Fig. 15.10.

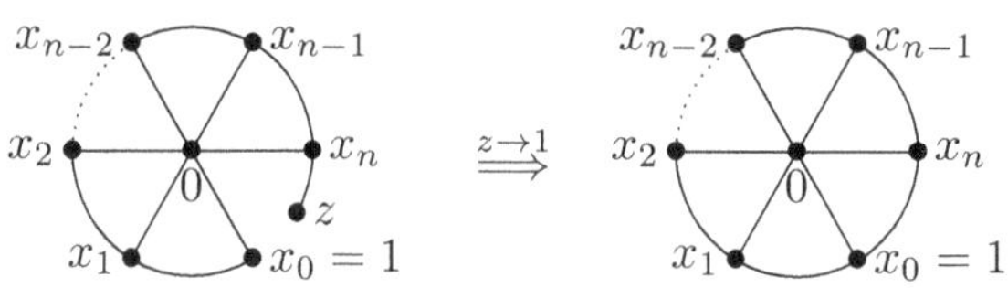

Figure 15.10: The Wheel Graph is the limit of $G_{2\{0\}^{n-1}}$ when $z \to 1$.

Theorem 15.4.19. *For any $n \geq 2$, we have*

$$I(\mathrm{WS}_{n+1}) = I(\mathrm{B}_n) = \binom{2n}{n}\zeta(2n-1).$$

Proof. We only need to compute the graphical function $f_{2\{0\}^{n-1}}(z)$ and then take the limit $z \to 1$. The main idea is repeated integration starting from the initial case f_2 which is given by Eq. (15.18), or, by Eq. (15.19) and Example 4.2.13

$$f_2(z) = \frac{\mathscr{L}_{10}(z) - \mathscr{L}_{01}(z)}{z-\bar{z}}. \tag{15.25}$$

Note that an integration of $\mathscr{L}_{\mathbf{w}}$ with $\int_0 \mathrm{d}z/z$ or $\int_0 \mathrm{d}\bar{z}/\bar{z}$ is equivalent to prepending or appending one 0, respectively, according to Eq. (4.58) (notice that Lemma 4.3.1 says that $\mathbf{y}_1 - \mathbf{x}_1$ contains at least two $\mathbf{x}_1$'s). Hence applying Prop. 15.4.17 with $a = 0$ exactly $(n-1)$-times we obtain

$$f_{2\{0\}^{n-1}}(z) = (-1)^n \frac{\mathscr{L}_{\{0\}^n 1\{0\}^{n-1}}(z) - \mathscr{L}_{\{0\}^{n-1}1\{0\}^n}(z)}{z - \bar{z}}.$$

If g is a single-valued function which vanishes at $z = 1$, we can take $z = 1+i\epsilon$ and use L'Hôpital's Rule to obtain

$$\lim_{z\to 0} \frac{g(z)}{z - \bar{z}} = \frac{1}{2}(\partial_z g(1) - \partial_{\bar{z}} g(1)).$$

Applying this to $f_{2\{0\}^{n-1}}(z)$ and using $\mathscr{L}_{\{0\}^{n-1}1\{0\}^n}(\bar{z}) = \mathscr{L}_{\{0\}^n 1\{0\}^{n-1}}(z)$ by Thm. 4.3.2(ii), we obtain

$$\begin{aligned} f_{2\{0\}^{n-1}}(1) &= \frac{(-1)^n}{2}(\partial_z - \partial_{\bar{z}})\Big(\mathscr{L}_{\{0\}^n 1\{0\}^{n-1}}(z) - \mathscr{L}_{\{0\}^{n-1}1\{0\}^n}(z)\Big)\Big|_{z=1} \\ &= (-1)^n \partial_z \Big(\mathscr{L}_{\{0\}^n 1\{0\}^{n-1}}(z) - \mathscr{L}_{\{0\}^{n-1}1\{0\}^n}(z)\Big)\Big|_{z=1}. \end{aligned} \tag{15.26}$$

It follows from Exercise 4.22 that

$$I(\mathrm{WS}_{n+1}) = f_{2\{0\}^{n-1}}(1) = \binom{2n}{n}\zeta(2n-1). \tag{15.27}$$

By combining Eqs. (15.9) and (15.27) we also get

$$I(\mathrm{B}_n) = \binom{2n}{n}\zeta(2n-1).$$

This completes the proof of the theorem. □

Theorem 15.4.20. *For any $n \ge 3$, the period of the zig-zag graph Z_n is given by*

$$I(\mathrm{Z}_n) = \frac{4\cdot(2n-2)!}{n!(n-1)!}\Big(1 - \frac{1-(-1)^n}{2^{2n-3}}\Big)\zeta(2n-3). \tag{15.28}$$

Proof. By Eqs. (15.7) and (15.8) we only need to compute $I(\mathrm{Z}_{2n+3}) = f_{2\{10\}^n 1}(0)$ and $I(\mathrm{Z}_{2n+2}) = f_{2\{01\}^n}(0)$. In what follows, we will use the notation and a few key results from Sec. 4.5.

For all alternating words $\mathbf{w} \in \mathsf{H}^0$, we set $\mathbf{v} = \overleftarrow{\mathbf{w}}\mathbf{x}_1\mathbf{x}_0\mathbf{w}$. From Thms. 4.5.6 and 4.5.7 it is not hard to see that the functions $\tilde{f}_{2,\mathbf{w}}$ defined by

$$\tilde{f}_{2,\mathbf{w}}(z) = (-1)^{\lg(\mathbf{w})} \begin{cases} \dfrac{F_{\mathbf{v}}(z) - F_{\overleftarrow{\mathbf{v}}}(z)}{z - \bar{z}} & \text{if } \mathbf{w} = \mathbf{x}_0\mathbf{u}, \\ \dfrac{\hat{F}_{\mathbf{v}}(z) - \hat{F}_{\overleftarrow{\mathbf{v}}}(z)}{z - \bar{z}} & \text{if } \mathbf{w} = \mathbf{x}_1\mathbf{u}, \end{cases} \tag{15.29}$$

are single-valued, real-valued and satisfy the system of differential equations

$$\mathfrak{D}_{z,\bar z}\tilde f_{2,\mathbf{w}\mathrm{x}_a}(z) = \frac{-1}{(z-a)(\bar z-a)}\tilde f_{2,\mathbf{w}}(z), \quad a\in\{0,1\}. \tag{15.30}$$

Moreover, if $\mathbf{w}=1$ is the empty word then going back to the $\{0,1\}^*$-word indexing for the MPLs we get

$$F_{\mathrm{x}_0\mathrm{x}_1}(z) = \hat F_{\mathrm{x}_0\mathrm{x}_1}(z) = \mathrm{L}_{01}(z) + \mathrm{L}_0(z)\,\mathrm{L}_1(\bar z) + \mathrm{L}_{01}(\bar z),$$
$$F_{\mathrm{x}_1\mathrm{x}_0}(z) = \hat F_{\mathrm{x}_1\mathrm{x}_0}(z) = \mathrm{L}_{10}(z) + \mathrm{L}_1(z)\,\mathrm{L}_0(\bar z) + \mathrm{L}_{10}(\bar z),$$

since the coefficients of x_0, x_1, $\mathrm{x}_0\mathrm{x}_1$ and $\mathrm{x}_1\mathrm{x}_0$ all vanish in the series S_X and $\hat{\mathsf{S}}_\mathsf{X}$ by Definitions 4.5.3 and 4.5.3. Thus

$$\begin{aligned}\tilde f_2(z) &= \frac{F_{\mathrm{x}_1\mathrm{x}_0}(z) - F_{\mathrm{x}_0\mathrm{x}_1}(z)}{z-\bar z}\\ &= \frac{\mathrm{L}_{10}(z) + \mathrm{L}_1(z)\,\mathrm{L}_0(\bar z) + \mathrm{L}_{01}(\bar z) - \mathrm{L}_{10}(\bar z) - \mathrm{L}_1(\bar z)\,\mathrm{L}_0(z) - \mathrm{L}_{01}(z)}{z-\bar z}\\ &= f_2(z) = \frac{4i\,\mathscr{L}i_2(z)}{(z-\bar z)}\end{aligned}$$

by (15.19). By the uniqueness in Thm. 15.4.12 we see that

$$\tilde f_{2,\mathbf{w}}(z) = f_{2\,\mathscr{b}(\mathbf{w})}(z)$$

where $\mathscr{b} : \{\mathrm{x}_0,\mathrm{x}_1\}^* \to \{0,1\}^*$ is the homomorphism of the word algebras such that $\mathscr{b}(\mathrm{x}_a) = a$ for $a=0$ and 1.

Let us determine the regular value of $f_{2\mathbf{w}}(z)$ at $z=0$ for $\mathbf{w} = \{10\}^n1$ or $\{01\}^n$ (see Definition 4.4.1). Similar to the computation in Eq. (15.26) we can take $z=i\epsilon$ and use L'Hôpital's Rule to compute $\tilde f_{2\mathbf{w}}(0) = \mathrm{reg}_{z=0}\,\tilde f_{2\mathbf{w}}(z)$. Setting $\mathbf{v} = \overleftarrow{\mathbf{w}}\mathrm{x}_1\mathrm{x}_0\mathbf{w} = \mathrm{x}_1\mathbf{u}\,\mathrm{x}_1$, we thus obtain by Thms. 4.5.6 and 4.5.7

$$\tilde f_{2\mathbf{w}}(0) = \begin{cases} F_{\mathbf{u}\,\mathrm{x}_1}(0) - F_{\mathrm{x}_1\mathbf{u}}(0) & \text{if } \mathbf{w} = (\mathrm{x}_0\mathrm{x}_1)^n;\\ \hat F_{\mathrm{x}_1\mathbf{u}}(0) - \hat F_{\mathbf{u}\,\mathrm{x}_1}(0) & \text{if } \mathbf{w} = (\mathrm{x}_1\mathrm{x}_0)^n\mathrm{x}_1,\end{cases}$$

where we have used the fact that $F_{\mathrm{x}_1\overleftarrow{\mathbf{u}}}(z) = F_{\mathbf{u}\,\mathrm{x}_1}(\bar z)$ and $F_{\overleftarrow{\mathbf{u}}\,\mathrm{x}_1}(z) = F_{\mathrm{x}_1\mathbf{u}}(\bar z)$. To calculate the regularized values above using the definitions in Eqs. (4.62) and (4.67), we note that $\mathrm{reg}_{z=0}\,\mathrm{L}_\mathbf{t} = 0$ for all nonempty words $\mathbf{t}$, so that by Lemma 4.4.2

$$\tilde f_{2\mathbf{w}}(0) = \begin{cases} \mathsf{S}_{\mathbf{u}\,\mathrm{x}_1} - \mathsf{S}_{\mathrm{x}_1\mathbf{u}} & \text{if } \mathbf{w} = (\mathrm{x}_0\mathrm{x}_1)^n;\\ \hat{\mathsf{S}}_{\mathrm{x}_1\,\mathbf{u}} - \hat{\mathsf{S}}_{\mathbf{u}\,\mathrm{x}_1} & \text{if } \mathbf{w} = (\mathrm{x}_1\mathrm{x}_0)^n\mathrm{x}_1.\end{cases}$$

Now, in the even weight case since $\mathbf{u} = (\mathrm{x}_0\mathrm{x}_1)^n\mathrm{x}_0(\mathrm{x}_0\mathrm{x}_1)^{n-1}\mathrm{x}_0$ we find that $\mathsf{S}_{\mathrm{x}_1\mathbf{u}}$ and $\mathsf{S}_{\mathbf{u}\,\mathrm{x}_1}$ are summands in $\mathsf{S}^1_{1,0}$ and $\mathsf{S}^1_{0,1}$, respectively. Hence by Eq. (4.59) we have

$$\begin{aligned} I(\mathrm{Z}_{2n+2}) &= \mathsf{S}^1_{1,0}(n,n) - \mathsf{S}^1_{1,0}(n+1,n-1) \\ &= 4\left[\binom{4n}{2n} - \binom{4n}{2n+2}\right]\zeta(4n+1) = \frac{4\cdot(4n+2)!}{(2n+2)!(2n+1)!}\zeta(4n+1) \end{aligned}$$

by Eq. (4.59). On the other hand, for Z_{2n+3}, we see that $\hat{\mathsf{S}}_{\mathrm{x}_1\mathbf{u}}$ and $\hat{\mathsf{S}}_{\mathbf{u}\,\mathrm{x}_1}$ are summands in $\hat{\mathsf{S}}^1_{1,0}$ and $\hat{\mathsf{S}}^1_{0,1}$, respectively. Thus we get

$$\begin{aligned} I(\mathrm{Z}_{2n+3}) &= \hat{\mathsf{S}}^1_{1,0}(n,n+1) - \hat{\mathsf{S}}^1_{1,0}(n+1,n) \\ &= 4\left(1 - \frac{1}{2^{4n+2}}\right)\frac{(4n+4)!}{(2n+3)!(2n+2)!}\zeta(4n+3) \end{aligned}$$

by Eq. (4.60). Finally, straightforward simplification finishes the proof of the zig-zag theorem. □

15.5 Feynman Integrals in Parametric Form

The essential difficulty in evaluating Feynman integrals lies in the complexity of the corresponding Feynman graphs. To clarify this we now define some polynomials associated with these graphs. Later this will be used to express Feynman integrals.

15.5.1 *Graph Polynomials in Edge Variables*

Definition 15.5.1. Suppose G is a Feynman graph as in Definition 15.2.2. Let $\mathscr{T}_1$ be the set of the spanning trees of G. Note that to get a spanning tree we need to remove exactly h_G edges, one from each loop. Cutting one more line of a spanning tree leads to two disconnected trees (T_1, T_2), or a 2-tree (2-*forest* in graph theory). Let $\mathscr{T}_2$ be the set of all such pairs. To each $e \in E_G$ we associate a "Schwinger parameter" α_e. The *first* and *second Symanzik polynomials* are respectively defined by

$$\Psi_G := \sum_{T\in\mathscr{T}_1}\prod_{e\notin T}\alpha_e, \quad \Phi_G(\mathbf{q}) := \sum_{(T_1,T_2)\in\mathscr{T}_2}\left(\prod_{e\notin(T_1,T_2)}\alpha_e\right)\left(q^{T_1}\right)^2 \quad (15.31)$$

where $q^{T_1} = -q^{T_2}$ is the total momentum into T_1. Finally, we define

$$\Xi_G(\mathbf{m},\mathbf{q}) := \Phi_G(\mathbf{q}) + \left(\sum_{e\in E_G} m_e^2\alpha_e\right)\Psi_G. \quad (15.32)$$

The following results are straightforward.

Lemma 15.5.2. *The first (resp. second) Symanzik polynomial is homogeneous of degree h_G (resp. h_G+1) in the edge variables $\{\alpha_e : e \in E_G\}$.*

Example 15.5.3. Let G be the Feynman graph in Figure 15.11. Suppose

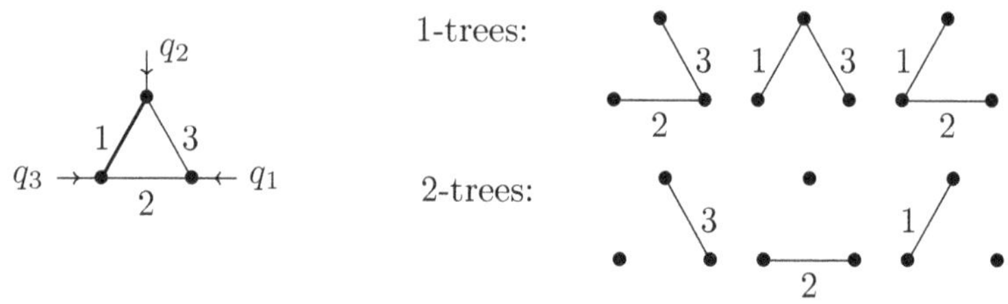

Figure 15.11: A 3-point 1-loop massive diagram and its spanning trees.

$m_1 \neq 0$ (denoted by a thickened line) and $m_2 = m_3 = 0$. We have

$$\Psi_G = \alpha_1 + \alpha_2 + \alpha_3, \quad \Phi_G(\mathbf{q}) = q_1^2\alpha_2\alpha_3 + q_2^2\alpha_1\alpha_3 + q_3^2\alpha_1\alpha_2.$$

Hence

$$\Xi_G(\mathbf{m},\mathbf{q}) = q_1^2\alpha_2\alpha_3 + q_2^2\alpha_1\alpha_3 + q_3^2\alpha_1\alpha_2 + m_1^2\alpha_1(\alpha_1+\alpha_2+\alpha_3).$$

15.5.2 *Feynman Integrals in Projective Space*

Let G be a Feynman graph as in Definition 15.2.2. Let n_G be the number of edges of G as before. Using Schwinger parameter (edge variables) we can define the Feynman integral associated with G by the integral in projective space $\mathbb{P}^{n_G-1}$:

$$I_G(\mathbf{m},\mathbf{q},D) := \Gamma\Big(n_G - h_G\frac{D}{2}\Big)\int_\sigma f_G(\mathbf{m},\mathbf{q},D)\cdot\Omega_G \tag{15.33}$$

where

$$f_G(\mathbf{m},\mathbf{q},D) = \frac{1}{\Psi_G^{D/2}}\left(\frac{\Psi_G}{\Xi_G(\mathbf{m},\mathbf{q})}\right)^{n_G-h_G\frac{D}{2}},$$

$$\Omega_G = \sum_{j=1}^{n_G}(-1)^j\alpha_j d\alpha_1\cdots\widehat{d\alpha_j}\cdots d\alpha_{n_G},$$

$$\sigma = \Big\{(\alpha_1:\cdots:\alpha_{n_G}):\ \alpha_j\in\mathbb{R}_{\ge 0}\Big\}\subset\mathbb{P}^{n_G-1}(\mathbb{R}).$$

Note that $f_G(\mathbf{m},\mathbf{q},D)$ is homogeneous of degree $-n_G$ so that by Euler's Homogeneous Function Theorem

$$\left(n_G+\sum_{e\in E_G}\alpha_e\frac{\partial}{\partial\alpha_e}\right)f_G(\mathbf{m},\mathbf{q},D)=0$$

so $d(f_G(\mathbf{m},\mathbf{q},D)\cdot\Omega_G)=0$. Therefore $I_G(\mathbf{m},\mathbf{q},D)$ is well defined. One can always take $\alpha_j=1$ and change the integral into an ordinary affine integral.

Example 15.5.4. Let G be the Feynman graph in Figure 15.11. Suppose further $m_1=m_2=m_3=0$ but $q_1^2+q_2^2+q_3^2\neq 0$. We have

$$I_G:=I_G(\mathbf{m},\mathbf{q},4)=\int_\sigma\frac{\Omega_G}{(\alpha_1+\alpha_2+\alpha_3)(q_1^2\alpha_2\alpha_3+q_2^2\alpha_1\alpha_3+q_3^2\alpha_1\alpha_2)}.$$

If one of the incoming momenta vanishes then the integral becomes trivial (see Exercise 15.2). So we now assume $q_j^2\neq 0$ for $j=1,2,3$. By the change of variables $\alpha_j\to q_j^2\alpha_j$ we get

$$I_G=\int_\sigma\frac{\Omega_G}{(q_1^2\alpha_1+q_2^2\alpha_2+q_3^2\alpha_3)(\alpha_2\alpha_3+\alpha_1\alpha_3+\alpha_1\alpha_2)}.$$

First we assume

$$q_1^2t^2+(q_3^2-q_1^2-q_2^2)t+q_2^2=0 \tag{15.34}$$

has two distinct real roots z and $\bar{z}$. Then we get

$$\begin{aligned}q_1^2I_G&=\int_\sigma\frac{\Omega_G}{(\alpha_1+z\bar{z}\alpha_2+(1-z)(1-\bar{z})\alpha_3)(\alpha_2\alpha_3+\alpha_1\alpha_3+\alpha_1\alpha_2)}\\&=\int_{\substack{(\alpha_1:\alpha_2)\in\mathbb{P}^1\\ \alpha_1,\alpha_2\geq 0}}\frac{\Omega_G}{(\alpha_1+z\alpha_2)(\alpha_1+\bar{z}\alpha_2)}\log\frac{(\alpha_1+\alpha_2)(\alpha_1+z\bar{z}\alpha_2)}{(1-z)(1-\bar{z})\alpha_1\alpha_2}\end{aligned}$$

by integrating over $\alpha_3\in(0,\infty)$ using the partial fraction expansion:

$$\frac{1}{(at+b)(ct+d)}=\det\begin{vmatrix}a&b\\c&d\end{vmatrix}^{-1}\left(\frac{1}{t+b/a}-\frac{1}{t+d/c}\right)$$

where

$$\begin{bmatrix}a&b\\c&d\end{bmatrix}=\begin{bmatrix}\alpha_1+\alpha_2&\alpha_1\alpha_2\\(1-z)(1-\bar{z})&\alpha_1+z\bar{z}\alpha_2\end{bmatrix}.$$

By messy computation (see Exercise 15.3) we finally obtain

$$I_G=\frac{1}{q_1^2(z-\bar{z})}\left(2\,\mathrm{Li}_2(z)-2\,\mathrm{Li}_2(\bar{z})+\log(z\bar{z})\log\frac{1-z}{1-\bar{z}}\right)=\frac{2\mathscr{L}i_2(z)}{q_1^2\mathrm{Im}(z)}$$

where $\mathscr{L}i_2(z)=\mathrm{Im}\left(\mathrm{Li}_2(z)\right)+\arg(1-z)\log|z|$ is the single-valued Bloch–Wigner dilog function (2.45).

If Eq. (15.34) has real roots, then similarly we can see that I_G can be expressed in terms of dilogs and logarithms.

The following result can be proved by complicated reduction process which we refer the interested reader to [156]. The motivic explanation is given by [56, Prop. 8.3].

Theorem 15.5.5. *Every one-loop Feynman integral can be expressed in terms of dilogs and logarithms evaluated at algebraic functions of the kinematic invariants (momenta and masses).*

15.6 Historical Notes

In particle physics, Z boson was found in 1983 at CERN using proton anti-proton collisions. However, some very important measurements of the Z boson were made in the 1980s (before its discovery) and 1990s by electron-positron experiments at CERN's Large Electron-Positron Collider.

For an amazing application of the Feynman graphs in particle physics, the interested reader is strongly recommended to read the fascinating story about the precise computation of the muon g factor by Hayes' essay [274] in the *American Scientist.* The final computation step of the three-loop order contributions can be found in [376]. The four-loop order computation is still on-going.

The structure of ϕ^4 periods was first studied systematically by Broadhurst and Kreimer in 1995 [89] (see also [90]). Usually, four equivalent ways can be used to define a graph period: position space, momentum space, parametric space, and dual parametric space. Hence different methods have been utilized: Mellin-Barnes integrals, hypergeometric functions, differential equation, difference equations and so on. To apply techniques from algebraic geometry one often uses the Feynman (Schwinger) parametric form.

The use of differential equations in the study of loop integrals has a long history (see [519]). The equations considered in Sec. 15.4 are for off-shell integrals and are second-order inhomogeneous equations. Similar equations for on-shell integrals can be found in [175].

According to [504], the Feynman integral corresponding to the Feynman diagram $K_{3,4}$ in Fig. 15.12 is given by the period of the its completion $K_{4,4}$ (denoted by $P_{6,4}$ in loc. cit.):

$$\frac{16704}{5}\zeta(8)-2592\zeta(5)\zeta(3)-\frac{6912}{5}\zeta(5,3).$$

Note that this graph is the simplest graph that does not have Hamiltonian circles.

Until 2013 only eight ϕ^4 periods have been calculated: The trivial period $1\ (=P(\mathrm{Z}_1))$, the wheels with three spokes [141] and with four spokes [140]

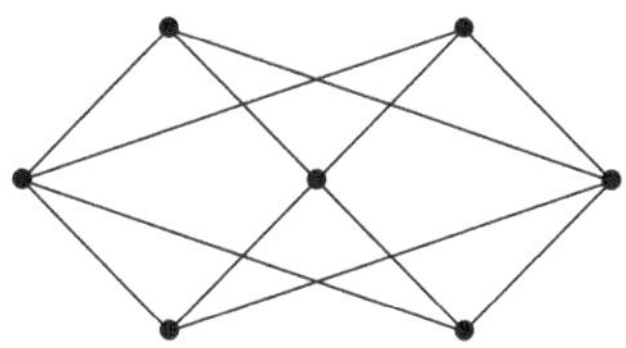

Figure 15.12: The first irreducible double zeta $\zeta(5,3)$ appears in this Feynman diagram.

which are Z_3 and Z_4, respectively. The period of the zig-zag graph Z_5 with 5 (see Figure 15.6) loops was calculated by Kazakov in 1983 [344], while Z_6 by Broadhurst in 1985 [82] and Ussyukina in 1991 [554]. Before 2013, the only calculated non zig-zag periods were $G(3,1,0)$ in [89] which is $P_{6,2}$ in [504], $P_{6,3}$ by Panzer, and the bipartite graph $K_{4,4}$ in [503] which is $P_{6,4}$ in [504]. By exact numerical methods twenty-four more periods were determined up to eight loops [504], all of which are MZVs.

In the mid 1990s it was realized that there is a close connection between perturbative QFT/QCD and knot theory. Kreimer [364] described a way to connect the Feynman integrals with UV-divergences to knot invariants where the knots are extracted from the topology of the corresponding Feynman diagrams.

The sector decomposition method (see [65] and its references) was developed to compute numerically rather complicated multi-scale Feynman integrals. But this method typically produces a large number of convergent parametric integrals which are difficult to analyze precisely and rigorously by analytical tools in mathematics. Moreover, it often produces spurious intermediate quantities which are canceled out at the end. In [467, 468], based on the works of Brown and his collaborators [63, 99, 100, 109], Panzer improved on an approach to computing these integrals by applying an appropriate sequence of partial integrations which ultimately render all Feynman parameter integrations in terms of CMZVs when the integrands are linearly reducible (see [99, p. 4] for relevant details).

The computation of the periods of wheel graphs and the zig-zag graphs in Sec. 15.4.1 by using sequential graphs and graphical functions follows mostly that of Schnetz [505] and Brown and Schnetz [108]. Sequential graphs were also independently considered by Drummond [174].

The period equation of the boxed ladder graph in Exercise 15.7 was first derived by N. Ussyukina and A. Davydychev in [555].

As this chapter is the only "applied" content of this book, the reader with little physics background may found Weinzierl's introductory lecture notes [570] very helpful. On the other hand, for those who are thirsty for more mathematical theory on Feynman graphs, they are suggested to read the articles [91, 107, 466].

Exercises

15.1. Show that the homological groups of a graph are given by (15.1).

15.2. Show that

$$\int_0^\infty \int_0^\infty \frac{d\alpha_1 d\alpha_2}{(\alpha_1+\alpha_2+q^2)(\alpha_1+\alpha_2+\alpha_1\alpha_2)}$$

can be expressed using dilogarithms evaluated at some algebraic functions of q^2.

15.3. We can use the following steps to compute I_G in Example 15.5.4.

(i) Let $\alpha_2 = 1$. We get

$$q_1^2 I_G = \int_0^\infty \frac{d\alpha}{(\alpha+z)(\alpha+\bar{z})} \log \frac{(\alpha+1)(\alpha+z\bar{z})}{(1-z)(1-\bar{z})\alpha}.$$

Let $f(z,\bar{z})$ be the above integral over $(0,1)$. Then

$$q_1^2 I_G = f(z,\bar{z}) - z\bar{z} f(z^{-1},\bar{z}^{-1}).$$

(ii) By partial fractions

$$(z-\bar{z}) f(z,\bar{z}) = g(\bar{z},z) - g(z,\bar{z})$$

where

$$g(z,\bar{z}) = \int_0^1 \frac{d\alpha}{\alpha+z} \log \frac{(\alpha+1)(\alpha+z\bar{z})}{(1-z)(1-\bar{z})\alpha}.$$

(iii) Evaluate $g(z,\bar{z})$ by using the formula

$$\int_0^x \frac{\log(t+b)}{t+a}\,dt = \mathrm{Li}_2\left(\frac{a+x}{a-b}\right) - \mathrm{Li}_2\left(\frac{a}{a-b}\right) + \log(b-a)\log\frac{a+x}{a}.$$

(iv) Simplify the above results by using the five-term relation a few times:

$$\mathrm{Li}_2(xy) = \mathrm{Li}_2(x) + \mathrm{Li}_2(y) + \mathrm{Li}_2\left(\frac{xy-x}{1-x}\right) + \mathrm{Li}_2\left(\frac{xy-y}{1-y}\right) + \frac{1}{2}\log^2\left(\frac{1-x}{1-y}\right).$$

15.4. Suppose the D-dimensional spherical coordinates are given by

$$\begin{aligned} x^1 &= r\cos\phi^1, \\ x^2 &= r\sin\phi^1\cos\phi^2, \\ &\vdots \\ x^{D-1} &= r\sin\phi^1\cdots\sin\phi^{D-2}\cos\phi^{D-1}, \\ x^D &= r\sin\phi^1\cdots\sin\phi^{D-2}\sin\phi^{D-1}, \end{aligned}$$

where $r\in[0,\infty)$, $\phi^1,\ldots,\phi^{D-2}\in[0,\pi]$, and $\phi^{D-1}\in[0,2\pi]$. Show that the Laplacian is

$$\begin{aligned} \Delta^{(D)} &:= \sum_{k=1}^{D}\partial_{x^k}^2 \\ &= \frac{1}{r^{D-1}}\partial_r r^{D-1}\partial_r + \frac{1}{r^2(\sin\phi^1)^{D-2}}\partial_{\phi^1}(\sin\phi^1)^{D-2}\partial_{\phi^1} \\ &\quad + \frac{1}{(r\sin\phi^1)^2(\sin\phi^2)^{D-3}}\partial_{\phi^2}(\sin\phi^2)^{D-3}\partial_{\phi^2} \\ &\quad + \ldots \\ &\quad + \frac{1}{(r\sin\phi^1\cdots\sin\phi^{D-3})^2\sin\phi^{D-2}}\partial_{\phi^{D-2}}\sin\phi^{D-2}\partial_{\phi^{D-2}} \\ &\quad + \frac{1}{(r\sin\phi^1\cdots\sin\phi^{D-2})^2}\partial_{\phi^{D-1}}^2. \end{aligned}$$

15.5. By using Exercise 15.4 show that

$$\frac{1}{r\sin\phi^1}\Delta^{(2)}\circ M(r\sin\phi^1) = \partial_r^2 + \frac{3}{r}\partial_r + \frac{1}{r^2}\partial_{\phi^1}^2 + \frac{2\cos\phi^1}{r^2\sin\phi^1}\partial_{\phi^1}.$$

15.6. Assume $f_{\mathbf{v}1}(z) = g(z)/(z-\bar{z})$. By taking $z = i\epsilon$ and using L'Hôpital's Rule show that $I(G_{\mathbf{v}2}) = \frac{1}{2}(\partial_z g(0) - \partial_{\bar{z}} g(0))$.

15.7. By using Exercise 4.22 show that

$$f_{2\{0\}^{n-1}}(z) = \sum_{k=n}^{2n}(-1)^k\binom{k}{n}\frac{(\log z\bar{z})^{2n-k}}{(2n-k)!}\frac{\mathrm{Li}_k(z)-\mathrm{Li}_k(\bar{z})}{z-\bar{z}}.$$

Appendix A

Key Concepts of Hopf Algebras

In this appendix we provide a quick review of the basic definitions and properties of Hopf algebras that we have used in this book.

Definition A.1. A *coalgebra* (C, Δ, ϵ) over a field $\mathbb{K}$ is a $\mathbb{K}$-vector space C with two $\mathbb{K}$-linear maps $\Delta : C \to C \otimes C$ (called *coproduct*) and $\epsilon : C \to \mathbb{K}$ (called *counit*) such that the following two diagrams commutate:

(i) Coassociativity

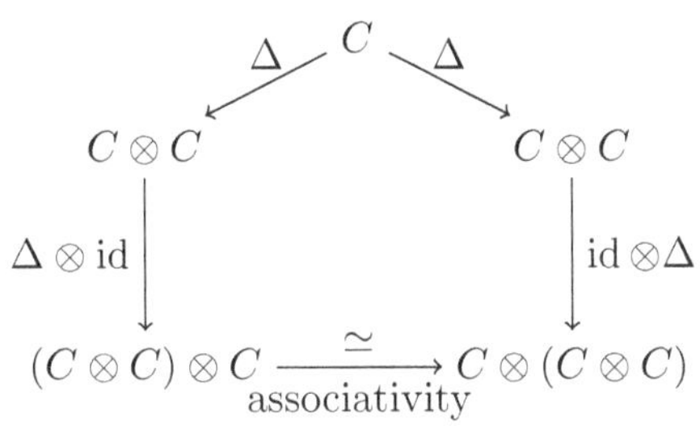

(ii) ϵ and Δ are compatible.

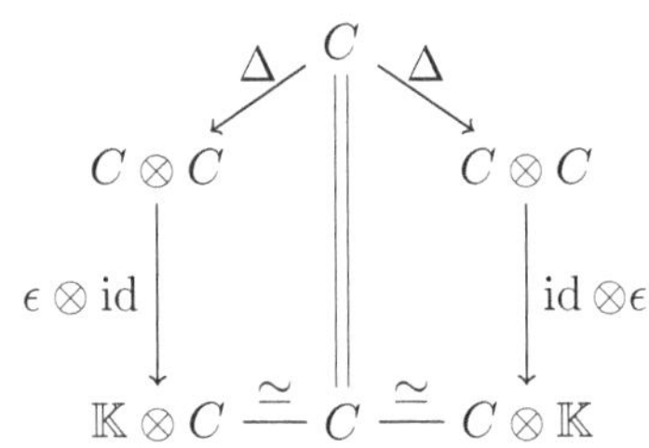

Dual to the concept of commutativity, we say C is *cocommutative* if $\tau \circ \Delta = \Delta$ where $\tau(x \otimes y) = y \otimes x$ for all $x, y \in C$.

Example A.2. Let $\mathbb{K}[x]$ be the polynomial ring over $\mathbb{K}$. We can define

$$\Delta(1) = 1 \otimes 1, \ \Delta(x) = 1 \otimes x + x \otimes 1 \quad \text{and} \quad \epsilon(x^n) = \delta_{0,n}$$

where $\delta_{0,n}$ is the Kronecker symbol. Then it is not hard to see that

$$\Delta(x^n) = \sum_{j=0}^{n} \binom{n}{j} x^j \otimes x^{n-j}$$

and that $(\mathbb{K}[x], \Delta, \epsilon)$ is a coalgebra.

Definition A.3. A *bialgebra* $(B, \mu, \iota, \Delta, \epsilon)$ over a field $\mathbb{K}$ is a $\mathbb{K}$-vector space B with four $\mathbb{K}$-linear maps $\mu : B \otimes B \to B$ (the *product*, also denoted by $\bullet$ sometimes), $\iota : \mathbb{K} \to B$ (the *unit*, or *identity*), $\Delta : B \to B \otimes B$ (the coproduct) and $\epsilon : B \to \mathbb{K}$ (the counit), such that (B, μ, ι) is an associative algebra and (C, Δ, ϵ) is a coassociative coalgebra satisfying the following four compatibility conditions:

(iii) μ and Δ

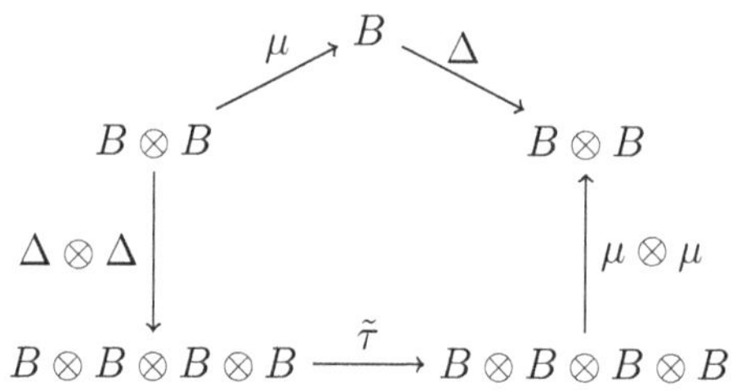

(iv) ι and ϵ

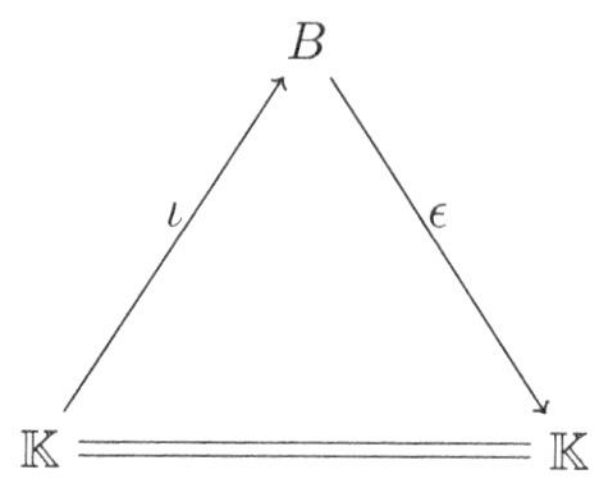

where $\tilde{\tau} = \mathrm{id} \otimes \tau \otimes \mathrm{id}$ and

(v) μ and ϵ

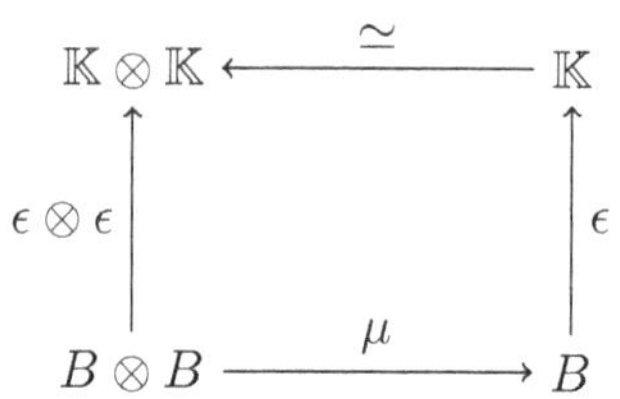

(vi) ι and Δ

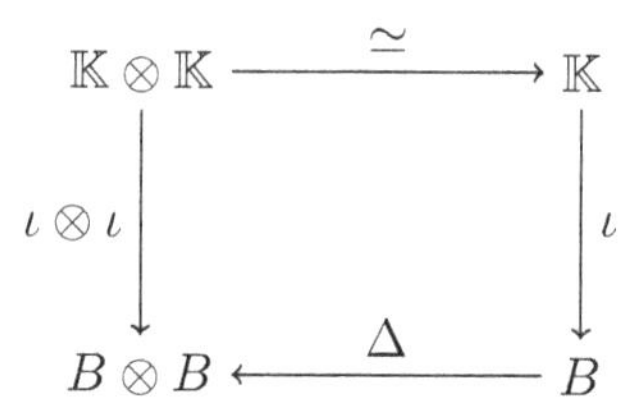

Condition (iii) implies the unit ι is a coalgebra homomorphism and condition (vi) means the counit ϵ is an algebra homomorphism. We say B is *commutative* if it is commutative as an algebra. We say B is *cocommutative* if it is cocommutative as a coalgebra.

Definition A.4. A *Hopf algebra* is a bialgebra H over a field $\mathbb{K}$ together with a $\mathbb{K}$-linear anti-homomorphism $S : H \to H$ (called the *antipode*) such that (vii) S is compatible with all the other four structure maps, *i.e.*, the following diagram commutes:

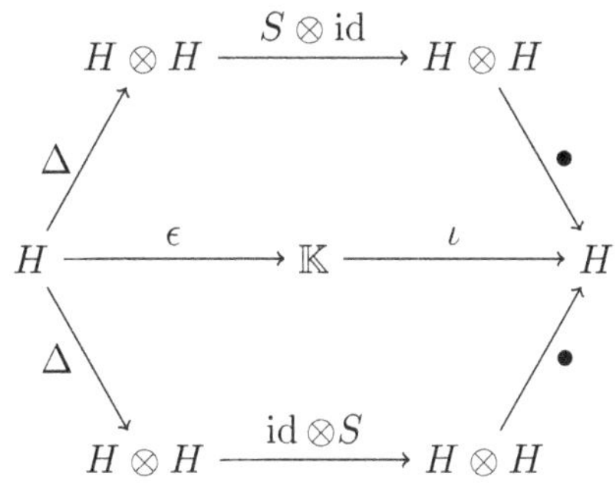

Example A.5. The simplest nontrivial example of a Hopf algebra is given by the group algebra

$$k[G] := \left\{ \sum_{j=1}^{n} a_j g_j : n \in \mathbb{N}, a_j \in \mathbb{K}, g_j \in G \ \forall j = 1, \cdots, n \right\}$$

for any group G and field $\mathbb{K}$. The algebra structure is straightforward. For the coalgebra structure, we define $\Delta(g) = g \otimes g$, $\epsilon(g) = 1$ and $S(g) = g^{-1}$ for all $g \in G$. It is easy to check all the Hopf algebra axioms are satisfied.

Because of this example, we have the following definition.

Definition A.6. An element x in any coalgebra (C, ϵ, Δ) is called *group-like* (or *diagonal*) if $\Delta(x) = x \otimes x$ and $\epsilon(x) = 1$.

Example A.7. In the study of Lie algebras $\mathfrak{g}$ over a field $\mathbb{K}$ (see Appendix B), we can give the universal enveloping algebra $U\mathfrak{g}$ a natural Hopf algebra structure by imposing some conditions on the structure maps as follows. For all $x \in \mathfrak{g}$, we set $\Delta(x) = x \otimes 1 + 1 \otimes x$, $\epsilon(x) = 0$ and $S(x) = -x$, where 1 is the unit in $\mathbb{K}$.

Definition A.8. An element x in any bialgebra B over a field $\mathbb{K}$ with coproduct Δ is called *primitive* if $\Delta(x) = x \otimes 1 + 1 \otimes x$, where 1 is the unit in $\mathbb{K}$.

Definition A.9. A subset I of a coalgebra (C, Δ, ϵ) is called a *co-ideal* if $I \subseteq \ker(\epsilon)$ and $\Delta(I) \subseteq I \otimes C + C \otimes I$.

Proposition A.10. If I is a co-ideal of a coalgebra (C, Δ, ϵ) then the quotient space C/I has a natural coalgebra structure.

Proof. It is straightforward to check that

$$\big((\Delta\otimes\mathrm{id})\circ\Delta\big)(I)\subseteq I\otimes C\otimes C+C\otimes I\otimes C+C\otimes C\otimes I,$$
$$\big((\mathrm{id}\otimes\Delta)\circ\Delta\big)(I)\subseteq I\otimes C\otimes C+C\otimes I\otimes C+C\otimes C\otimes I,$$

which vanishes in $(C/I)\otimes(C/I)\otimes(C/I)$. Similar,

$$\big((\epsilon\otimes\mathrm{id})\circ\Delta\big)(I)\subseteq I,\quad \big((\mathrm{id}\otimes\epsilon)\circ\Delta\big)(I)\subseteq I,$$

which vanishes in C/I. The proposition follows easily from these facts. □

We now consider bialgebras with some grading.

Definition A.11. A bialgebra B over a field $\mathbb{K}$ is *graded* if $B=\oplus_{n\geq 0}B_n$ where the elements in B_n all have weight n such that

(1) $\dim_k B_n<0$ as a $\mathbb{K}$-vector space for all $n\geq 0$
(2) The product $B_m\bullet B_n\subseteq B_{m+n}$ for all $m,n\geq 0$
(3) The coproduct $\Delta(B_n)\subseteq\oplus_{j=0}^{n}B_j\otimes B_{n-j}$ for all $n\geq 0$

In this book we always consider *connected* graded bialgebras, *i.e.*, $\dim_k B_0=1$.

Example A.12. Let Γ_N be the set of Nth roots of unity. Let $\mathsf{X}=\mathsf{X}_N$ be an alphabet with $N+1$ letters $\{x_0,x_\xi:\xi\in\Gamma_N\}$ and X^* the set of words over X. All the letters in X have weight one. The empty word 1 has weight zero. Let $\mathbb{Q}\langle\mathsf{X}\rangle$ be the $\mathbb{Q}$-linear vector space generated by 1 and X^*. It becomes a graded Hopf algebra when we define the product by concatenation, the unit by $\iota(\alpha)=\alpha 1$ for all $\alpha\in\mathbb{Q}$, the coproduct by $\Delta(x)=1\otimes x+x\otimes 1$ for all $x\in\mathsf{X}$, the counit by $\epsilon(\mathbf{w})=0$ for all nontrivial word $\mathbf{w}$ and $\epsilon(1)=1$, and the antipode by $S(1)=1$ and $S(x)=-x$ for all $x\in\mathsf{X}$.

For any field $\mathbb{K}$ containing $\mathbb{Q}$, we may define $\mathbb{K}\langle\mathsf{X}\rangle$ easily by extension of scalars.

Theorem A.13. For any word $\mathbf{w}\in\mathsf{X}^*$, we have

$$S(\mathbf{w})=(-1)^{|\mathbf{w}|}\overleftarrow{\mathbf{w}},$$

where $\overleftarrow{\mathbf{w}}$ is the reversal of $\mathbf{w}$.

Proof. Clear by anti-homomorphism. □

Example A.14. Another important example of Hopf algebra is the set of quasi-symmetric functions QSym_2 defined in Definition 7.3.2. We can make QSym_2 a commutative Hopf algebra by setting its coproduct Δ by

$$\Delta(M_{s_1,\dots,s_d}) = \sum_{j=0}^{d} M_{s_1,\dots,s_j} \otimes M_{s_{j+1},\dots,s_d},$$

and counit ε with $\varepsilon(u) = 0$ for all elements u of positive degree. Its antipode S is an involution of QSym_2, *i.e.*, an automorphism with $S^2 = \mathrm{id}$.

Theorem A.15. The antipode S of QSym_2 is given by the followings: for every composition $\mathbf{s}$ of signed numbers,

(i) $S(M_{\mathbf{s}}) = (-1)^{\mathrm{dp}(\mathbf{s})} E_{\overleftarrow{\mathbf{s}}}$, where $E_{\mathbf{t}} = \sum_{\mathbf{s} \preceq \mathbf{t}} M_{\mathbf{s}}$ and $\mathbf{t} \preceq \mathbf{s}$ means $\mathbf{t}$ can be obtained from $\mathbf{s}$ by combining some of its parts using $\oplus$.
(ii) $S(M_{\mathbf{s}}) = (-1)^{\mathrm{dp}(\mathbf{s})} E_{\overleftarrow{\mathbf{s}}} = \sum_{\bigsqcup_{j=1}^{r} \mathbf{s}_j = \mathbf{s}} (-1)^r M_{\mathbf{s}_1} M_{\mathbf{s}_2} \cdots M_{\mathbf{s}_r}$, where $\bigsqcup_{j=1}^{r} \mathbf{s}_j$ is the concatenation of $\mathbf{s}_1$ to $\mathbf{s}_r$.
(iii) $(-1)^{\mathrm{dp}(\mathbf{s})} M_{\overleftarrow{\mathbf{s}}} = \sum_{\bigsqcup_{j=1}^{r} \mathbf{s}_j = \mathbf{s}} (-1)^r E_{\mathbf{s}_1} E_{\mathbf{s}_2} \cdots E_{\mathbf{s}_r}$.

Proof. Assume first $\mathbf{s}$ is a composition of positive integers. Then part (i) is given by [188, Prop. 3.4] or [412, Cor. 2.3]. Part (ii) follow from the general theory of Hopf algebras (see [188, p. 7]). Applying antipode S to (ii) and use (i) we can derive (iii) quickly. A careful reading reveals that all the proofs are still valid when all the components of $\mathbf{s}$ are signed number. $\square$

Appendix B

Some Useful Results from Lie Algebras

In this appendix we give a very concise review of the basic results of Lie algebras that we need in this book.

A *Lie algebra* is an algebra $(\mathfrak{g}, [\cdot,\cdot])$ where $\mathfrak{g}$ is a vector field over some field $\mathbb{K}$ together with a $\mathbb{K}$-bilinear binary operation $[\cdot,\cdot]$, called *Lie bracket*, that satisfies $[x, x] = 0$ for all $x \in \mathfrak{g}$ and the following *Jacobi identity*:

$$[x, [y, z]] + [z, [x, y]] + [y, [z, x]] = 0 \quad \forall x, y, z \in \mathfrak{g}.$$

Note that $[x, x] = 0$ implies the skew-symmetry (or anti-commutativity) $[x, y] = -[y, x]$ for all $x \in \mathfrak{g}$.

Theorem B.1. If B is a bi-algebra (see Appendix A), then the set of primitive elements form a Lie algebra with the usual commutator bracket $[x, y] = xy - yx$.

Proof. We only need to show $[x, y]$ is primitive if both x and y are primitive. Indeed,

$$\begin{aligned}
\Delta([x, y]) &= \Delta(x)\Delta(y) - \Delta(y)\Delta(x) \\
&= (x \otimes 1 + 1 \otimes x)(y \otimes 1 + 1 \otimes y) - (y \otimes 1 + 1 \otimes y)(x \otimes 1 + 1 \otimes x) \\
&= xy \otimes 1 + 1 \otimes xy - yx \otimes 1 - 1 \otimes yx \\
&= [x, y] \otimes 1 + 1 \otimes [x, y],
\end{aligned}$$

as desired. □

Example B.2. Let A be an associative algebra over the field $\mathbb{K}$. Then A becomes a Lie algebra over $\mathbb{K}$, denoted by A_L, if we define the Lie bracket by

$$[a, b] = ab - ba \quad \forall a, b \in A.$$

Let $\mathfrak{g}$ be any Lie algebra over $\mathbb{K}$. Given a unital associative $\mathbb{K}$-algebra U and a Lie algebra homomorphism: $h : \mathfrak{g} \longrightarrow U_L$ (notation as in Example B.2), we say that U is the *universal enveloping algebra* of $\mathfrak{g}$, denoted by $U\mathfrak{g}$, if it satisfies the following universal property: for any unital associative $\mathbb{K}$-algebra A and Lie algebra homomorphism $f : \mathfrak{g} \longrightarrow A_L$ there exists a unique unital algebra homomorphism $g : U \longrightarrow A$ such that the following diagram commutes:

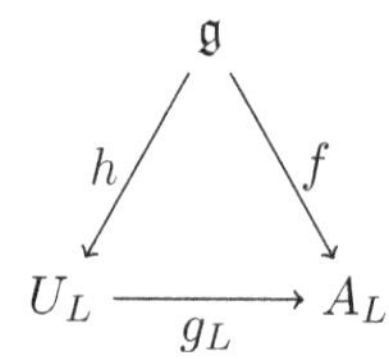

A standard way to construct the universal enveloping algebra of $\mathfrak{g}$ is to take $T\mathfrak{g}/I$ where $T\mathfrak{g}$ is the tensor algebra of $\mathfrak{g}$ and I is the two-sided ideal generated by $a \otimes b - b \otimes a - [a, b]$ for all $a, b \in \mathfrak{g}$. The universal enveloping algebra $U\mathfrak{g}$ has a natural way to become a Hopf algebra (see Example A.7).

Example B.3. For any set X and a $\mathbb{Q}$-algebra R, let $R\langle\mathsf{X}\rangle$ (resp. $\mathfrak{Lib}_R(\mathsf{X})$) be the free[1] noncommutative associative algebra (resp. the free Lie algebra constructed as in Example B.2) on X with coefficients in R, where the product is given by concatenation. Note that $R\langle\mathsf{X}\rangle$ can be viewed as the universal enveloping algebra of $\mathfrak{Lib}_k(\mathsf{X})$. When all elements in X are given weight one $R\langle\mathsf{X}\rangle$ has a graded Hopf algebra structure as defined in Example A.12. We denote by $R\langle\langle\mathsf{X}\rangle\rangle$ the set of all formal series on X over the field $\mathbb{K}$.

The above example and the next one have played important roles in the study of regularized double shuffle relations of multiple polylog values.

Example B.4. Let N be a positive integer and $\mu = \exp(2\pi\sqrt{-1}/N)$. Let $\Gamma_N := \{\mu^j : j \in \mathbb{Z}/N\mathbb{Z}\}$ be the set of Nth roots of unity. Denote by Y the infinite alphabet consisting of letters $\mathsf{y}_{n,\xi}$ for $n \in \mathbb{N}$ and $\xi \in \Gamma_N$, and Y^* the set of the words over Y. For any fixed $n \in \mathbb{N}$ and $\xi \in \boldsymbol{\mu}_N$, the letters $\mathsf{y}_{n,\xi}$ has colored weight (n, ξ). The empty word 1 has weight zero. We extend the definition of $\mathsf{y}_{n,\xi}$ to the case where $n = 0$ by setting

$$\mathsf{y}_{0,\xi} = \begin{cases} 1, & \text{if } \xi = 1; \\ 0, & \text{if } \xi \neq 1. \end{cases}$$

[1] $\mathfrak{Lib}$ comes from the French word *libre* for free.

Let $\mathbb{K}\langle\langle \mathsf{Y}\rangle\rangle$ be the graded noncommutative associative algebra on Y with coefficients in $\mathbb{K}$, where the product is given by concatenation. Put on $\mathbb{K}\langle\langle \mathsf{Y}\rangle\rangle$ a grading using *colored weights* as follows:

$$\mathbb{K}\langle\langle \mathsf{Y}\rangle\rangle = \bigoplus_{\substack{n\in\mathbb{N}_0\\ \xi\in\boldsymbol{\mu}}} \mathbb{K}B_{n,\xi}, \quad \text{where } B_{n,\xi} = \sum_{\substack{d\in\mathbb{N},\ \xi_1\xi_2\cdots\xi_d=\xi\\ s_1+\cdots+s_d=n}} \mathsf{y}_{s_1,\xi_1}\cdots\mathsf{y}_{s_d,\xi_d}.$$

Similar to Example A.12 we can equip $\mathbb{K}\langle\langle \mathsf{Y}\rangle\rangle$ with a graded bialgebra structure when we define the unit by $\iota(\alpha) = \alpha 1$ for all $\alpha \in \mathbb{K}$, the coproduct by

$$\Delta_*(\mathsf{y}_{n,\xi}) = \sum_{k+\ell=n,\lambda\nu=\xi} \mathsf{y}_{k,\lambda}\otimes \mathsf{y}_{\ell,\nu} \quad \forall n\in\mathbb{N}_0, \xi\in\boldsymbol{\mu}, \tag{B.1}$$

the counit by $\epsilon(\mathbf{w}) = 0$ for all nontrivial word $\mathbf{w}$ and $\epsilon(1) = 1$.

Remark B.5. It is crucial to understand that the graded bialgebra $(R\langle \mathsf{Y}\rangle, \Delta_*)$ is not the universal enveloping algebra of the free Lie algebra $\mathfrak{Lib}_k(\mathsf{Y})$.

Put

$$\mathcal{Y} := \sum_{n\in\mathbb{N}_0,\xi\in\boldsymbol{\mu}} \mathsf{y}_{n,\xi}.$$

Since it is group-like for Δ_* (see Exercise 13.5) we know $\log(\mathcal{Y})$ is primitive by [482, Thm. 3.2]. Let $u_{n,\xi}$ be the component of $\log(\mathcal{Y})$ in $B_{n,\xi}$ (*i.e.*, the colored weight (n,ξ) piece). It is obvious that $u_{0,\xi} = 0$ for all $\xi\in\boldsymbol{\mu}$. Let

$$U = \{u_{n,\xi} : n\in\mathbb{N}, \xi\in\boldsymbol{\mu}\}.$$

Then all the elements in U are primitive since $\log(\mathcal{Y})$ is primitive and Δ_* respects the colored weight grading. So by Example A.12 we see that $(\mathbb{Q}\langle U\rangle, \bullet, \Delta)$ is the universal enveloping algebra of the free Lie algebra $\mathfrak{Lib}_{\mathbb{Q}}(U)$.

Proposition B.6. The associative $\mathbb{Q}$-algebra $\mathbb{Q}\langle\langle \mathsf{Y}\rangle\rangle$ is freely generated by U. Moreover, as bialgebras

$$(\mathbb{Q}\langle\langle \mathsf{Y}\rangle\rangle, \bullet, \Delta_*) \cong (\mathbb{Q}\langle U\rangle, \bullet, \Delta)$$

Proof. Since $-\sum_{n\in\mathbb{N},\xi\in\boldsymbol{\mu}} \mathsf{y}_{n,\xi} = 1-\mathcal{Y}$, by the definition

$$\sum_{n\in\mathbb{N},\xi\in\boldsymbol{\mu}} u_{n,\xi} = \log(\mathcal{Y}) = \sum_{j=1}^{\infty}\frac{(-1)^{j-1}}{j}\Bigg(\sum_{n\in\mathbb{N},\xi\in\boldsymbol{\mu}} \mathsf{y}_{n,\xi}\Bigg)^j.$$

By considering the depth filtration we see immediately that

$$u_{n,\xi} = \mathsf{y}_{n,\xi} + (\text{terms with depth} > 1) \tag{B.2}$$

and more generally,

$$u_{s_1,\xi_1} \cdots u_{s_d,\xi_d} = \mathsf{y}_{s_1,\xi_1} \cdots \mathsf{y}_{s_d,\xi_d} + (\text{terms with depth} > d).$$

This yields the proposition directly. □

We now take $\mathsf{X} = \mathsf{X}_N$ as in Example A.12 and use the notation in Example B.3. Further we regard Y as a subset of X^* by defining $\mathsf{y}_{n,\xi} = x_0^{n-1}x_\xi$ and set $\pi_{\mathsf{Y}} : R\langle\langle \mathsf{X}\rangle\rangle \to R\langle\langle \mathsf{Y}\rangle\rangle$ by sending all words ending with x_0 to 0. Let

$$\mathbf{q}(\mathsf{y}_{s_1,\xi_1}\mathsf{y}_{s_2,\xi_2}\cdots\mathsf{y}_{s_d,\xi_d}) = \mathsf{y}_{s_1,\xi_1}\mathsf{y}_{s_2,\xi_2\xi_1^{-1}}\cdots\mathsf{y}_{s_d,\xi_d\xi_{d-1}^{-1}}.$$

Then we have

Theorem B.7. Let $\mathbb{K}$ be a field extension of $\mathbb{Q}$. Let Φ be a group-like element in $(R\langle\langle \mathsf{X}\rangle\rangle, \Delta)$ whose coefficient of x_1 is 0. Then

(i) $\log(\Phi)$ is a primitive element.

(ii) $\overleftarrow{\Phi}(-x_0, -x_1) = \Phi(x_0, x_1)^{-1}$ where $\leftarrow$ means the reversal of all the words in the coefficients.

(iii) Suppose $S \in \mathbb{K}[\![t]\!]$ such that

$$\Phi_* := S(\mathsf{y}_{1,1}) \cdot \mathbf{q}\pi_{\mathsf{Y}}(\Phi) \tag{B.3}$$

in $R\langle\langle \mathsf{Y}\rangle\rangle$ is diagonal for Δ_* whose coefficient of $\mathsf{y}_{1,1}$ is 0. Then we have

$$S(t) = \exp\left(\sum_{n=2}^{\infty} \frac{(-1)^{n-1}}{n}(\Phi|x_0^{n-1}x_1)t^n\right).$$

Proof. (i) follows easily from [482, Thm. 3.2].

(ii) follows from the formula for the antipode (see Thm. A.13) of the Hopf algebra $(R\langle\langle \mathsf{X}\rangle\rangle, \Delta)$ (*i.e.*, signed reversal of words) and [482, Prop. 1.10].

For (iii), regard the bialgebra $(k[\![t]\!], \Delta)$ as the completion (with respect to the natural grading) of the universal enveloping algebra of the free Lie-algebra $\{\alpha t : \alpha \in \mathbb{K}\}$ with $\Delta(t) = 1 \otimes t + t \otimes 1$. Then

$$\begin{aligned} \pi_t : (R\langle\langle \mathsf{X}\rangle\rangle, \Delta) &\longrightarrow (k[\![t]\!], \Delta) \\ x_\alpha &\longmapsto \delta_{\alpha,1}t \end{aligned}$$

is a coalgebra homomorphism. Clearly it descends to $(R\langle\langle \mathsf{Y}\rangle\rangle, \Delta) \to (k[\![t]\!], \Delta)$ so that $\pi_t(\mathsf{y}_{1,1}) = t$ and $\pi_t(\mathsf{y}_{n,\xi}) = 0$ if $(n, \xi) \neq (1, 1)$.

By assumption, Φ is group-like so $\pi_t(\Phi)$ is group-like and therefore $\pi_t(\Phi) = \exp(\alpha t)$ for some $\alpha \in \mathbb{K}$. But the coefficient of x_1 is 0 which is mapped to that of t in $\pi_t(\Phi)$ so we must have $\alpha = 0$. Hence by Eq. (B.3)

$$\pi_t(\Phi_*) = \pi_t(S(\mathsf{y}_{1,1}))\pi_t(\mathbf{q}\pi_{\mathsf{Y}}(\Phi)) = S(t)\pi_{\mathsf{Y}}\pi_t(\Phi) = S(t)$$

since $\mathbf{q}$ fixes all depth 1 elements of $R\langle\langle \mathsf{Y}\rangle\rangle$.

On the other hand, by Prop. B.6 we see that

$$\Phi_* = \exp\Big(\sum_{n\in\mathbb{N},\xi\in\boldsymbol{\mu}} \alpha_{n,\xi} u_{n,\xi} + (\text{terms of depth} > 1)\Big) \tag{B.4}$$

where all the larger depth terms comes from the Lie brackets. But $\mathbb{K}[\![t]\!]$ is commutative so that π_t sends all Lie brackets to 0. Hence

$$\log S(t) = \log \pi_t(\Phi_*) = \sum_{n\in\mathbb{N},\xi\in\boldsymbol{\mu}} \alpha_{n,\xi}\pi_t(u_{n,\xi}).$$

If $\boldsymbol{\mu} \ni \xi = \xi_1\xi_2\dots\xi_d \neq 1$ then there must be $\xi_j \neq 1$ for some j. Hence if the color weight $\xi \neq 1$ then $\pi_t(u_{n,\xi}) = 0$. Thus we get

$$\sum_{n\in\mathbb{N}} \pi_t(u_{n,1}) = \pi_t\Big(\sum_{n\in\mathbb{N},\xi\in\boldsymbol{\mu}} u_{n,\xi}\Big) \overset{\text{defn}}{=\!=} \pi_t \log\Big(\sum_{n\in\mathbb{N},\xi\in\boldsymbol{\mu}} \mathsf{y}_{n,\xi}\Big) = \log(1+t).$$

This implies that $\pi_t(u_{n,1}) = (-1)^{n-1}t^n/n$ and therefore

$$\log S(t) = \log \pi_t(\Phi_*) = \sum_{n\in\mathbb{N}} \frac{(-1)^{n-1}}{n}\alpha_{n,1}t^n.$$

Combining Eq. (B.2) and Eq. (B.4) we see that

$$\Phi_* = 1 + \sum_{n\in\mathbb{N},\xi\in\boldsymbol{\mu}} \alpha_{n,\xi}\mathsf{y}_{n,\xi} + (\text{terms of depth} > 1).$$

Thus by Eq. (B.3) we get $\alpha_{n,1} = (\Phi|x_0^{n-1}x_1)$ if $n \geq 2$. Finally, $\alpha_{1,1} = 0$ by the assumption of Φ_*. This completes the proof of the theorem. □

One of the fundamental results of Lie algebras is the following Poincaré–Birkhoff–Witt Theorem (PBW Theorem). From the definition of universal enveloping algebra we know that there is a canonical map $h : \mathfrak{g} \to U\mathfrak{g}$.

Theorem B.8. Let $\mathfrak{g}$ be a Lie algebra over $\mathbb{K}$ and X a totally ordered basis of $\mathfrak{g}$. A canonical monomial over X is a finite sequence $(x_1, x_2, \dots, x_n)$ of elements of X which is non-decreasing in the order $\leq$, that is, $x_1 \leq x_2 \leq \cdots \leq x_n$. Extend h to all canonical monomials as follows: If $(x_1, x_2, \dots, x_n)$ is a canonical monomial, let

$$h(x_1, x_2, \dots, x_n) = h(x_1) \cdot h(x_2) \cdots h(x_n).$$

Then h is injective on the set of the canonical monomials and its range forms a basis of the $\mathbb{K}$-vector space $U\mathfrak{g}$.

The PBW Theorem has the following immediate corollary.

Corollary B.9. If $\mathfrak{g}$ is a Lie algebra over a field then the canonical map $\mathfrak{g} \to U\mathfrak{g}$ is injective. In particular, any Lie algebra over a field is isomorphic to a Lie subalgebra of an associative algebra.

The PBW Theorem has a coalgebra version when the base field contains $\mathbb{Q}$.

Theorem B.10. Let $\mathfrak{g}_1$ and $\mathfrak{g}_2$ be two sub-Lie algebras of $\mathfrak{g}$ over $\mathbb{K}$ containing $\mathbb{Q}$ such that $\mathfrak{g} = \mathfrak{g}_1 \oplus \mathfrak{g}_2$ as vector spaces. Let $\widehat{U\mathfrak{g}}$ denote the completion of $U\mathfrak{g}$. Then the map

$$\begin{aligned} \text{PBW}: \widehat{U\mathfrak{g}_1} \otimes \widehat{U\mathfrak{g}_2} &\longrightarrow \widehat{U\mathfrak{g}} \\ a \otimes b &\longmapsto ab \end{aligned}$$

is a coalgebra isomorphism.

Keeping the same notation, Thm. B.10 has the following corollary.

Corollary B.11. Let $\pi : \widehat{U\mathfrak{g}} \to \widehat{U\mathfrak{g}}/\mathfrak{g}_1\widehat{U\mathfrak{g}}$ be the projection. Then every diagonal element of $\widehat{U\mathfrak{g}}/\mathfrak{g}_1\widehat{U\mathfrak{g}}$ is the image of some element in $\exp(\mathfrak{g})$ by π. Moreover, if $f_1, f_2 \in \exp(\mathfrak{g})$ with $\pi(f_1) = \pi(f_2)$ then $f_1 = \exp(\lambda)f_2$ for some $\lambda \in \mathfrak{g}_1$.

Proof. Let ϵ be the counit of $\widehat{U\mathfrak{g}_1}$. The map $\epsilon \otimes \mathrm{id} : \widehat{U\mathfrak{g}_1} \otimes \widehat{U\mathfrak{g}_2} \to \otimes\widehat{U\mathfrak{g}_2}$ is a coalgebra homomorphism whose kernel is given by $\mathfrak{g}_1\widehat{U\mathfrak{g}_1} \otimes \widehat{U\mathfrak{g}_2} \simeq \mathfrak{g}_1\widehat{U\mathfrak{g}}$ by Thm. B.10. Thus the coalgebra quotient $\widehat{U\mathfrak{g}}/\mathfrak{g}_1\widehat{U\mathfrak{g}} \simeq \widehat{U\mathfrak{g}_2}$. Therefore we have the following coalgebra commutative diagram:

$$\begin{array}{ccc} \widehat{U\mathfrak{g}_1} \otimes \widehat{U\mathfrak{g}_2} & \xrightarrow{\ \text{PBW}\ } & \widehat{U\mathfrak{g}} \\ \epsilon\otimes\mathrm{id} \downarrow \uparrow 1\otimes\mathrm{id} & & \downarrow \pi \\ \widehat{U\mathfrak{g}_2} & \xrightarrow{\ \simeq\ } & \widehat{U\mathfrak{g}}/\mathfrak{g}_1\widehat{U\mathfrak{g}} \end{array}$$

The corollary now follows from the fact that a diagonal element of $\widehat{U\mathfrak{g}}$ must have the form $\exp(\lambda)$ form some $\lambda \in \mathfrak{g}$ by [482, Thm. 3.2]. ◻

Another deep result from Lie algebra is the following theorem whose proof is too long to be presented here. See the proof of Prop. 5.7 of [173].

Definition B.12. A *Lie monomial* in the Lie algebra $\mathbb{K}\langle x, y\rangle$ freely generated by x and y is an expression formed from x and y by using only brackets, but not addition. A *Lie polynomial* over $\mathbb{K}$ is a $\mathbb{K}$-linear combination of Lie monomials.

For instance, $[[x, y], [x, [[x, y], x]]]$ and $[x, [x, \ldots, [x, y] \ldots]]$ are both Lie monomials.

Theorem B.13. Let $\mathfrak{g}$ be a Lie algebra containing two generators x and y. Suppose $s(x, y) = s(y, x)$ is a Lie polynomial in $U\mathfrak{g}$ (hence $\Delta s = 1 \otimes s + s \otimes 1$). If

$$s(x, y) - s(x, y + z) + s(x + y, z) - s(y, z) = 0$$

for any generators $x, y, z \in \mathfrak{g}$ then $s = 0$.

Appendix C

Basics of Hypergeometric Functions

Given any n parameters $b_1, \ldots, b_n$ which are not negative integers, m parameters $a_1, \ldots, a_m$, and a complex variable t, the (generalized) hypergeometric function is defined by

$$\begin{aligned} &{}_mF_n(a_1, \ldots, a_m; b_1, \ldots, b_n; t) \\ &= {}_mF_n\left(\begin{matrix} a_1, \ldots, a_m \\ b_1, \ldots, b_n \end{matrix}; t\right) = \sum_{k=0}^{\infty} \frac{(a_1)_k (a_2)_k \cdots (a_m)_k}{(b_1)_k (b_2)_k \cdots (b_n)_k} \frac{t^k}{k!}, \end{aligned} \tag{C.1}$$

where we have used the Pochhammer symbols defined by Eq. (1.13). In this Appendix, we collect some results on hypergeometric functions which we have used in the book. The interested reader is strongly urged to read the old but authoritative classic Ref. [34] by Bailey and Chapter 15 of the comprehensive handbook [6] for more detailed information.

C.1 Solutions of Some Ordinary Differential Equations

The classical hypergeometric function usually refers to the ${}_2F_1$-series. The first influential result concerning these series was discovered by Gauss (see § 15.1.20 of Ref. [6]). He showed that

$${}_2F_1\left(\begin{matrix} a, b \\ c \end{matrix} \middle| 1\right) = \frac{\Gamma(c)\Gamma(c-a-b)}{\Gamma(c-a)\Gamma(c-b)}, \quad \mathrm{Re}(c) > \mathrm{Re}(a+b), \tag{C.2}$$

which is now called the Gauss summation formula/theorem. These series are important in many branches of mathematics, mainly because they satisfy some ubiquitous differential equations. For example, we have the following important result.

Lemma C.1. The hypergeometric function ${}_2F_1(a, b; c; t)$ provides a solution of the following homogeneous differential equation:

$$t(1-t)\frac{d^2 f}{dt^2} + \Big(c - (1+a+b)t\Big)\frac{df}{dt} - abf = 0. \tag{C.3}$$

Proof. This can be proved in a straightforward manner using the definition. □

It turns out that a more general type second order ordinary differential equations, called the Riemann (or Papperitz) differential equations, can always be solved using ${}_2F_1$-series. These equations have the following general form

$$\frac{d^2 f}{dt^2} + \left[\frac{1-\alpha-\alpha'}{t-a} + \frac{1-\beta-\beta'}{t-b} + \frac{1-\gamma-\gamma'}{t-c}\right]\frac{df}{dt}$$
$$+ \left[\frac{\alpha\alpha' g(a,b,c)}{t-a} + \frac{\beta\beta' g(b,c,a)}{t-b} + \frac{\gamma\gamma' g(c,a,b)}{t-c}\right]\frac{f}{(t-a)(t-b)(t-c)} = 0,$$

where $g(a,b,c) = (a-b)(a-c)$, subject to the condition

$$\alpha + \alpha' + \beta + \beta' + \gamma + \gamma' = 1.$$

Note that these equations have regular singular points at a, b and c. It has the general solution

$$f(t) = \left(\frac{t-a}{t-b}\right)^{\alpha}\left(\frac{t-c}{t-b}\right)^{\gamma} {}_2F_1\left(\begin{matrix}\alpha+\beta+\gamma, \alpha+\beta'+\gamma \\ 1+\alpha-\alpha'\end{matrix}\middle| \frac{(t-a)(c-b)}{(t-b)(c-a)}\right).$$

C.2 Transformation Formulas

There are many functional relations among the hypergeometric functions which are usually called the transformation formulas. In this section we collect some of those that we have used in this book.

C.2.1 *Linear Transformations*

Combining the integral representation

$${}_2F_1\left(\begin{matrix}a, b \\ c\end{matrix}\middle| t\right) = \frac{\Gamma(c)}{\Gamma(b)\Gamma(c-b)}\int_0^1 v^{b-1}(1-v)^{c-b-1}(1-vt)^{-a}\, dv$$

where $\mathrm{Re}(c) > \mathrm{Re}(b) > 0$, with some Mellin-Barnes type integrals we have

$${}_2F_1\left(\begin{matrix}a, b \\ c\end{matrix}\middle| 1-t\right) = (1-t)^{c-a-b}{}_2F_1\left(\begin{matrix}c-a, c-b \\ c\end{matrix}\middle| t\right), \tag{C.4}$$

and

$${}_2F_1\left(\begin{matrix}a, b \\ c\end{matrix}\middle| 1-t\right) = \frac{\Gamma(c)\Gamma(c-a-b)}{\Gamma(c-a)\Gamma(c-b)}{}_2F_1\left(\begin{matrix}a, b \\ a+b-c+1\end{matrix}\middle| t\right)$$
$$+ t^{c-a-b}\frac{\Gamma(c)\Gamma(a+b-c)}{\Gamma(a)\Gamma(b)}{}_2F_1\left(\begin{matrix}c-a, c-b \\ a+b-c+1\end{matrix}\middle| t\right), \tag{C.5}$$

for all t with $|\arg(t)| < \pi$.

C.2.2 *Some Transformations of* $_3F_2$

Let $\alpha_{i,j} = \alpha_i + \alpha_j$, $\alpha_{i,j,k} = \alpha_i + \alpha_j + \alpha_k$ and similarly for β. The first transformation formula of the $_3F_2$-series is (see § 3.8, Eq. (1) of Ref. [34])

$$\begin{aligned}{}_3F_2\left(\begin{matrix}\alpha_1,\alpha_2,\alpha_3\\ \beta_1,\beta_2\end{matrix}\right) &= \frac{\Gamma(\beta_1)\Gamma(\beta_1-\alpha_{1,2})}{\Gamma(\beta_1-\alpha_1)\Gamma(\beta_1-\alpha_2)}\,{}_3F_2\left(\begin{matrix}\alpha_1,\alpha_2,\beta_2-\alpha_3\\ \alpha_{1,2}-\beta_1+1,\beta_2\end{matrix}\right)\\ &\quad+\frac{\Gamma(\beta_1)\Gamma(\beta_2)\Gamma(\alpha_{1,2}-\beta_1)\Gamma(\beta_{1,2}-\alpha_{1,2,3})}{\Gamma(\alpha_1)\Gamma(\alpha_2)\Gamma(\beta_2-\alpha_3)\Gamma(\beta_{1,2}-\alpha_{1,2})}\\ &\quad\times{}_3F_2\left(\begin{matrix}\beta_1-\alpha_1,\beta_1-\alpha_2,\beta_{1,2}-\alpha_{1,2,3}\\ \beta_1-\alpha_{1,2}+1,\beta_{1,2}-\alpha_{1,2}\end{matrix}\right), \qquad \text{(C.6)}\end{aligned}$$

provided that $\mathrm{Re}(\beta_{1,2}-\alpha_{1,2,3})>0$ and $\mathrm{Re}(\alpha_3-\beta_1+1)>0$. The second one is (see Ex. 7 on p. 98 of Ref. [34])

$${}_3F_2\left(\begin{matrix}\alpha_1,\alpha_2,\alpha_3\\ \beta_1,\beta_2\end{matrix}\right) = \frac{\Gamma(\beta_2)\Gamma(\beta_{1,2}-\alpha_{1,2,3})}{\Gamma(\beta_2-\alpha_3)\Gamma(\beta_{1,2}-\alpha_{1,2})}\,{}_3F_2\left(\begin{matrix}\beta_1-\alpha_1,\beta_1-\alpha_2,\alpha_3\\ \beta_1,\beta_{1,2}-\alpha_{1,2}\end{matrix}\right), \qquad \text{(C.7)}$$

provided that $\mathrm{Re}(\beta_{1,2}-\alpha_{1,2,3})>0$ and $\mathrm{Re}(\beta_2-\alpha_3)>0$.

We now use the above two formulas to prove the following result (due to Z. Li in Ref. [397]) used in the proof of Thm. 10.7.3.

Proposition C.2. Let $a, b, c \in \mathbb{C}$ with sufficiently small real parts. Then

$$\begin{aligned}{}_3F_2\left(\begin{matrix}a,b,c\\ a+b,1+c\end{matrix}\right) &= -\frac{\Gamma(a+b)\Gamma(1+c)\Gamma(1+c-a-b)}{\Gamma(a)\Gamma(b)\Gamma(1+c-a)\Gamma(1+c-b)}\sum_{n=1}^{\infty}\frac{(a)_n(1-b)_n}{n(1+c-b)_n n!}\\ &\quad+\frac{\Gamma(a+b)\Gamma(1+c)\Gamma(1+c-a-b)}{\Gamma(a)\Gamma(b)\Gamma(1+c-a)\Gamma(1+c-b)}(\psi(1+c-b)-\psi(a)-\psi(b)-\gamma).\end{aligned}$$

Proof. Taking a parameter ε with $|\varepsilon|$ sufficient small, we have

$${}_3F_2\left(\begin{matrix}a,b,c\\ a+b,1+c\end{matrix}\right) = \lim_{\varepsilon\to 0}{}_3F_2\left(\begin{matrix}a,b,c\\ a+b+\varepsilon,1+c-\varepsilon\end{matrix}\right).$$

Now we consider the series ${}_3F_2\left(\begin{smallmatrix}a,b,c\\ a+b+\varepsilon,1+c-\varepsilon\end{smallmatrix}\right)$. Applying Eq. (C.6), we get

$${}_3F_2\left(\begin{matrix}a,b,c\\ a+b+\varepsilon,1+c-\varepsilon\end{matrix}\right) = \frac{\Gamma(a+b+\varepsilon)\Gamma(\varepsilon)}{\Gamma(a+\varepsilon)\Gamma(b+\varepsilon)}\,{}_3F_2\left(\begin{matrix}a,b,1-\varepsilon\\ 1-\varepsilon,1+c-\varepsilon\end{matrix}\right) \qquad \text{(C.8)}$$

$$+\frac{\Gamma(a+b+\varepsilon)\Gamma(1+c-\varepsilon)\Gamma(-\varepsilon)}{\Gamma(a)\Gamma(b)\Gamma(1-\varepsilon)\Gamma(1+c)}\,{}_3F_2\left(\begin{matrix}a+\varepsilon,b+\varepsilon,1\\ 1+\varepsilon,1+c\end{matrix}\right). \qquad \text{(C.9)}$$

Applying Eq. (C.2) to the ${}_3F_2$-series in Eq. (C.8), we get

$$
\begin{aligned}
{}_3F_2\left(\begin{matrix} a, b, 1-\varepsilon \\ 1-\varepsilon, 1+c-\varepsilon \end{matrix}\right) &= {}_2F_1\left(\begin{matrix} a, b \\ 1+c-\varepsilon \end{matrix}\right) \\
&= \frac{\Gamma(1+c-\varepsilon)\Gamma(1+c-\varepsilon-a-b)}{\Gamma(1+c-\varepsilon-a)\Gamma(1+c-\varepsilon-b)}
\end{aligned}
$$

for $\mathrm{Re}(1+c-\varepsilon-a-b)>0$. Applying Eq. (C.7) to the ${}_3F_2$-series in Eq. (C.9), we have

$$
\begin{aligned}
&{}_3F_2\left(\begin{matrix} a, b, c \\ a+b+\varepsilon, 1+c-\varepsilon \end{matrix}\right) \\
&= \frac{\Gamma(1+\varepsilon)\Gamma(a+b+\varepsilon)\Gamma(1+c-\varepsilon)\Gamma(1+c-a-b-\varepsilon)}{\varepsilon\Gamma(a+\varepsilon)\Gamma(b+\varepsilon)\Gamma(1+c-a-\varepsilon)\Gamma(1+c-b-\varepsilon)} \\
&\quad - \frac{\Gamma(a+b+\varepsilon)\Gamma(1+c-\varepsilon)\Gamma(1+c-a-b-\varepsilon)}{\varepsilon\Gamma(a)\Gamma(b)\Gamma(1+c-a-\varepsilon)\Gamma(1+c-b)} {}_3F_2\left(\begin{matrix} \varepsilon, 1-b, a+\varepsilon \\ 1+\varepsilon, 1+c-b \end{matrix}\right) \\
&= \frac{1}{\varepsilon}\left(\frac{\Gamma(1+\varepsilon)\Gamma(a+b+\varepsilon)\Gamma(1+c-\varepsilon)\Gamma(1+c-a-b-\varepsilon)}{\Gamma(a+\varepsilon)\Gamma(b+\varepsilon)\Gamma(1+c-a-\varepsilon)\Gamma(1+c-b-\varepsilon)}\right. \\
&\quad \left. - \frac{\Gamma(a+b+\varepsilon)\Gamma(1+c-\varepsilon)\Gamma(1+c-a-b-\varepsilon)}{\Gamma(a)\Gamma(b)\Gamma(1+c-a-\varepsilon)\Gamma(1+c-b)}\right) \\
&\quad - \frac{\Gamma(a+b+\varepsilon)\Gamma(1+c-\varepsilon)\Gamma(1+c-a-b-\varepsilon)}{\Gamma(a)\Gamma(b)\Gamma(1+c-a-\varepsilon)\Gamma(1+c-b)} \sum_{n=1}^{\infty} \frac{(a+\varepsilon)_n(1-b)_n}{(n+\varepsilon)(1+c-b)_n n!}.
\end{aligned}
$$

by expanding the ${}_3F_2$-series on the right. Using L'Hôpital's rule and the fact that $\psi(1)=-\gamma$ we can now finish the proof of the proposition by taking $\varepsilon \to 0$. □

The following result is used in the proof of Thm. 5.7.4.

Proposition C.3. Suppose $\mathrm{Re}(\alpha-2\beta)>-2$. Then

$$
\begin{aligned}
&{}_3F_2\left(\begin{matrix} 1, \alpha, \beta \\ 1+\alpha, 2+\alpha-\beta \end{matrix}; 1\right) \\
&= \frac{\alpha(1+\alpha-\beta)}{\beta-1}\left[\psi(\alpha)-\psi(2+\alpha-2\beta)-\psi\Big(\frac{\alpha+1}{2}\Big)+\psi\Big(\frac{\alpha+3}{2}-\beta\Big)\right].
\end{aligned}
\tag{C.10}
$$

Proof. Taking $m=1$ and $k=l=0$ in (4) of Ref. [383], we get

$$
\begin{aligned}
{}_3F_2\left(\begin{matrix} 1, \alpha, \beta \\ 1+\alpha, 2+\alpha-\beta \end{matrix}; 1\right) = \frac{-\alpha(1+\alpha-\beta)}{2(\beta-1)}&[\psi((\alpha+1)/2)-\psi(\alpha/2) \\
&+\psi(\alpha/2-\beta+1)-\psi((\alpha+1)/2-\beta+1)].
\end{aligned}
$$

An application of the duplication relation (1.6) yields the proof of the proposition. □

Appendix D

Sample Computer Codes

In this appendix, we provide the pseudo-codes for a few key computer programs we have used in the book. The readers who like to explore by themselves should be able to modify these easily using their favorite computer algebra systems.

D.1 Double Shuffle Relations

Both the shuffle and the stuffle relations can be computed using the recursive definitions. Note that we need to use the result from Exercise 3.1: if $a = \mathrm{dp}(\mathbf{u})$ and $b = \mathrm{dp}(\mathbf{v})$ then $\sharp(\mathbf{u} ш \mathbf{v}) = \binom{a+b}{a}$ and $\sharp(\mathbf{u} * \mathbf{v}) = D(a,b)$ is the Delannoy number.

(i) **The stuffle program.**

```
stuffle:=proc(s, t)
    d =depth(s); l =depth(t);
    if d = 0 or l = 0 then
        RETURN([s,t]);
    else
        s' = [s_2, ..., s_d]; t' = [t_2, ..., t_l];
        a =stuffle(s', t); b =stuffle(s, t'); c =stuffle(s', t');
        RETURN([s_1, a] + [t_1, b] + [s_1 + t_1, c]);
    end if;
end proc:
```

(ii) **The shuffle program.**

We can modify program (i) by removing the term **c** from the output.

(iii) Other variations of the double shuffle relations such as the q-analogs can be obtained by further modifying the above programs.

D.2 Multiple Harmonic Sums

We use the formula

$$H_n(s_1, \dots, s_d) = \sum_{k=d}^{n} \frac{H_k(s_2, \dots, s_d)}{k^{s_1}}.$$

(i) **The MHS program.**

```
H:=proc(s, n)
    d =depth(s);
    S_0 := 1; # S_k stands for the partial sums.
    for i to d do
        S_i := 0; # Initialize the partial sums
    end do;
    for k from 0 to n - d do
        for i from 1 to d do
                S_i := S_i + S_{i-1}/(k+i)^{t_i};
        end do;
    end do;
    RETURN(S_d);
end proc:
```

(ii) **The alternating MHS program and the q-analogs.** These can be written down by modifying (i).

D.3 Finite Multiple Zeta Values

In Chapter 8, we found many linear relations between FMZVs and FESs experimentally and formed a few conjectures based on these numerical results. The following pseudo-code gives the idea to detect such relations. The key idea is to use the Chinese Remainder's Theorem and the PSLQ algorithm [29, 212].

Suppose $\mathbf{s}_1, \dots, \mathbf{s}_k$ are k different compositions of positive integers (or signed numbers for AMHS). We want to know whether there are rational numbers $c_1, \dots, c_k$ such that

$$c_1 H_{p-1}(\mathbf{s}_1) + \cdots + c_k H_{p-1}(\mathbf{s}_k) \equiv 0 \pmod{p}$$

for all sufficiently large primes p.

Take a large number M, say $M = 1000$. Suppose $p_1, \dots, p_M$ are M sufficiently large primes. By the Chinese Remainder's Theorem there are positive integers $A_1, \dots, A_k$ (independent of $p_1, \dots, p_M$) such that

$$H_{p_j-1}(\mathbf{s}_1) \equiv A_1, \dots, H_{p_j-1}(\mathbf{s}_k) \equiv A_k \pmod{p_j}$$

for all $j = 1, \ldots, M$. This can be found by the command `chrem` in Maple and `ChineseRemainder` in Mathematica. Assume $P = \prod_{j=1}^{M}$. Then we can use PSLQ[1] algorithm to find rational numbers $c_0', c_1', \ldots, c_k' \in \mathbb{Q}$ such that

$$c_0' P + c_1' A_1 + \cdots + c_k' A_k = 0.$$

This algorithm has been implement in Maple simply as `PQSL`. Mathematica users can consult [153]. Or, one can use the `LLL` algorithm in PARI/GP.

Now, if all of $c_0', c_1', \ldots, c_k'$ stay relatively small and stabilize for all integers M beyond a certain large value, then we can assume that we have found a linear relation

$$c_1 \zeta_{\mathscr{A}_1}(\mathbf{s}_1) + \cdots + c_k \zeta_{\mathscr{A}_1}(\mathbf{s}_k) = 0.$$

The interested reader should have no trouble modifying the above to treat higher superbities, *i.e.*, the super congruences.

[1]The name "PSLQ" comes from a combination of partial sums of squares vector and LQ (lower-diagonal-orthogonal) matrix factorizations.

Appendix E

Tables of Special Values

E.1 Euler Sums of Lower Weights

We list all the $\mathbb{Q}$-linear independent relations obtained by the DBSFs between the Euler sums of weight < 6. Most of the computations are carried out by Maple. We have checked the consistency of these relations with the many known ones and verified numerically all these identities by EZ-face [74] with error smaller than 10^{-50}.

All weight four Euler sums are $\mathbb{Z}$-linear combinations of $A = \zeta(\bar{2},1,1)$, $B = \zeta(\bar{2},2)$, $C = \zeta(\bar{1},2,1)$, $D = \zeta(\bar{1},1,2)$, and $E = \zeta(\bar{1},1,1,1)$:

$$\begin{aligned}
\zeta(4) &= 64A + 16B, \\
\zeta(\bar{4}) &= -56A - 14B, \\
\zeta(3,1) &= 16A + 4B, \\
\zeta(3,\bar{1}) &= 118A + 19B + 14C, \\
\zeta(2,2) &= 48A + 12B, \\
\zeta(\bar{3},1) &= 10A + 2B, \\
\zeta(\bar{3},\bar{1}) &= -140A - 24B - 14C, \\
\zeta(2,\bar{2}) &= -24A - 7B, \\
\zeta(\bar{2},\bar{2}) &= -12A - 3B, \\
\zeta(\bar{1},3) &= -38A - 5B - 6C, \\
\zeta(\bar{1},\bar{3}) &= 58A + 8B + 8C, \\
\zeta(2,1,1) &= 64A + 16B, \\
\zeta(2,1,\bar{1}) &= 16A + 2B + 6C + 3D, \\
\zeta(2,\bar{1},1) &= 22A + 3B + C - 3D, \\
\zeta(2,\bar{1},\bar{1}) &= 100A + 13B + 9C - 6D, \\
\zeta(\bar{2},1,\bar{1}) &= 91A + 14B + 8C - 3D, \\
\zeta(\bar{2},\bar{1},1) &= -161A - 26B - 15C + 3D, \\
\zeta(\bar{2},\bar{1},\bar{1}) &= -101A - 14B - 9C + 6D, \\
\zeta(\bar{1},2,\bar{1}) &= -102A - 14B - 8C + 6D, \\
\zeta(\bar{1},\bar{2},1) &= 69A + 11B + 8C, \\
\zeta(\bar{1},\bar{2},\bar{1}) &= 63A + 8B + 3C - 6D,
\end{aligned}$$

$$
\begin{aligned}
\zeta(\bar{1},1,\bar{2}) &= 21A + 3B + C - 2D,\\
\zeta(\bar{1},\bar{1},\bar{2}) &= A + 2B + D,\\
\zeta(\bar{1},1,1,\bar{1}) &= A + E,\\
\zeta(\bar{1},1,\bar{1},1) &= 11A + 2B + C + E,\\
\zeta(\bar{1},1,\bar{1},\bar{1}) &= C + E,\\
\zeta(\bar{1},\bar{1},1,1) &= -83A - 16B - 5C + D + E,\\
\zeta(\bar{1},\bar{1},1,\bar{1}) &= -38A - 5B - 5C + D + E,\\
\zeta(\bar{1},\bar{1},\bar{1},1) &= D + E,\\
\zeta(\bar{1},\bar{1},\bar{1},\bar{1}) &= A + B + D + E.
\end{aligned}
$$

All weight five Euler sums are $\mathbb{Z}$-linear combinations of the following

$$
\begin{aligned}
&A = \zeta(\bar{1},\bar{1},\bar{1},2), \quad B = \zeta(\bar{2},1,\bar{1},\bar{1}), C = \zeta(\bar{1},1,\bar{1},\bar{2}), \quad D = \zeta(\bar{2},1,1,1),\\
&E = \zeta(\bar{1},\bar{1},\bar{1},1,1), F = \zeta(2,2,\bar{1}), \quad G = \zeta(\bar{1},1,\bar{1},1,\bar{1}), H = \zeta(\bar{1},\bar{1},\bar{1},\bar{1},\bar{1}).
\end{aligned}
$$

$$
\begin{aligned}
\zeta(5) &= -13504A + 1856B - 1344C + 26880D - 18752E - 640F - 31552G + 50304H,\\
\zeta(\bar{5}) &= 12660A - 1740B + 1260C - 25200D + 17580E + 600F + 29580G - 47160H,\\
\zeta(4,1) &= -9808A + 1344B - 944C + 19632D - 13648E - 464F - 22848G + 36496H,\\
\zeta(4,\bar{1}) &= -14918A + 2044B - 1434C + 29862D - 20758E - 704F - 34748G + 55506H,\\
\zeta(\bar{4},1) &= 3638A - 498B + 346C - 7296D + 5066E + 172F + 8466G - 13532H,\\
\zeta(\bar{4},\bar{1}) &= 19862A - 2722B + 1914C - 39744D + 27634E + 938F + 46274G - 73908H,\\
\zeta(3,2) &= 22672A - 3104B + 2160C - 45456D + 31568E + 1072F + 52768G - 84336H,\\
\zeta(3,\bar{2}) &= 4562A - 626B + 446C - 9108D + 6342E + 216F + 10642G - 16984H,\\
\zeta(\bar{3},2) &= -6552A + 898B - 632C + 13110D - 9116E - 310F - 15266G + 24382H,\\
\zeta(\bar{3},\bar{2}) &= -17848A + 2444B - 1704C + 35772D - 24848E - 844F - 41548G + 66396H,\\
\zeta(2,3) &= -26368A + 3616B - 2560C + 52704D - 36672E - 1248F - 61472G + 98144H,\\
\zeta(2,\bar{3}) &= 6792A - 934B + 680C - 13506D + 9428E + 322F + 15878G - 25306H,\\
\zeta(\bar{2},\bar{3}) &= 24902A - 3412B + 2394C - 49854D + 34654E + 1178F + 58004G - 92658H,\\
\zeta(\bar{2},3) &= -8622A + 1182B - 834C + 17244D - 11994E - 408F - 20094G + 32088H,\\
\zeta(\bar{1},4) &= 5266A - 720B + 494C - 10582D + 7338E + 248F + 12240G - 19578H,\\
\zeta(\bar{1},\bar{4}) &= -8990A + 1230B - 850C + 18044D - 12522E - 424F - 20910G + 33432H,\\
\zeta(3,1,1) &= -9808A + 1344B - 944C + 19632D - 13648E - 464F - 22848G + 36496H,\\
\zeta(3,1,\bar{1}) &= -5314A + 725B - 500C + 10677D - 7402E - 250F - 12339G + 19741H,\\
\zeta(3,\bar{1},1) &= -2257A + 312B - 225C + 4489D - 3137E - 108F - 5290G + 8427H,\\
\zeta(3,\bar{1},\bar{1}) &= -7299A + 1005B - 713C + 14566D - 10151E - 347F - 17057G + 27208H,\\
\zeta(\bar{3},1,1) &= 4482A - 614B + 430C - 8974D + 6238E + 212F + 10438G - 16676H,\\
\zeta(\bar{3},1,\bar{1}) &= 9570A - 1308B + 908C - 19204D + 13328E + 452F + 22250G - 35578H,\\
\zeta(\bar{3},\bar{1},1) &= 12462A - 1710B + 1204C - 24924D + 17338E + 590F + 29056G - 46394H,\\
\zeta(\bar{3},\bar{1},\bar{1}) &= -4288A + 582B - 396C + 8646D - 5978E - 201F - 9922G + 15900H,\\
\zeta(2,2,1) &= 22672A - 3104B + 2160C - 45456D + 31568E + 1072F + 52768G - 84336H,\\
\zeta(2,\bar{2},1) &= 3025A - 414B + 287C - 6065D + 4213E + 143F + 7038G - 11251H,\\
\zeta(2,\bar{2},\bar{1}) &= 6421A - 881B + 627C - 12818D + 8927E + 303F + 14977G - 23904H,\\
\zeta(2,1,2) &= -26368A + 3616B - 2560C + 52704D - 36672E - 1248F - 61472G + 98144H,\\
\zeta(2,\bar{1},2) &= 2206A - 302B + 210C - 4428D + 3074E + 104F + 5134G - 8208H,\\
\zeta(2,1,\bar{2}) &= 7958A - 1093B + 786C - 15861D + 11056E + 377F + 18581G - 29637H,\\
\zeta(2,\bar{1},\bar{2}) &= 23513A - 3221B + 2255C - 47094D + 32727E + 1113F + 54757G - 87484H,\\
\zeta(\bar{2},2,1) &= -12813A + 1755B - 1227C + 25664D - 17835E - 606F - 29835G + 47670H,\\
\zeta(\bar{2},2,\bar{1}) &= -20468A + 2804B - 1964C + 40988D - 28488E - 968F - 47668G + 76156H,\\
\zeta(\bar{2},\bar{2},1) &= -5477A + 750B - 523C + 10977D - 7625E - 259F - 12750G + 20375H,\\
\zeta(\bar{2},\bar{2},\bar{1}) &= 12308A - 1686B + 1180C - 24654D + 17132E + 582F + 28662G - 45794H,\\
\zeta(\bar{2},1,2) &= 12622A - 1729B + 1210C - 25281D + 17568E + 597F + 29393G - 46961H,\\
\zeta(\bar{2},1,\bar{2}) &= -3065A + 420B - 295C + 6135D - 4265E - 145F - 7140G + 11405H,\\
\zeta(\bar{2},\bar{1},2) &= -14047A + 1923B - 1337C + 28170D - 19561E - 665F - 32691G + 52252H,\\
\zeta(\bar{2},\bar{1},\bar{2}) &= -9411A + 1290B - 909C + 18831D - 13095E - 445F - 21930G + 35025H,
\end{aligned}
$$

$$
\begin{aligned}
\zeta(\bar{1},3,1) &= 123A - 17B + 13C - 242D + 171E + 6F + 289G - 460H,\\
\zeta(\bar{1},3,\bar{1}) &= -11820A + 1614B - 1120C + 23726D - 16460E - 557F - 27466G + 43926H,\\
\zeta(\bar{1},\bar{3},1) &= 6380A - 874B + 612C - 12776D + 8880E + 302F + 14858G - 23738H,\\
\zeta(\bar{1},\bar{3},\bar{1}) &= 12610A - 1722B + 1194C - 25312D + 17560E + 594F + 29302G - 46862H,\\
\zeta(\bar{1},2,2) &= -190A + 26B - 18C + 384D - 266E - 9F - 442G + 708H,\\
\zeta(\bar{1},2,\bar{2}) &= 13726A - 1880B + 1314C - 27494D + 19106E + 649F + 31960G - 51066H,\\
\zeta(\bar{1},\bar{2},2) &= -13631A + 1867B - 1305C + 27302D - 18973E - 644F - 31739G + 50712H,\\
\zeta(\bar{1},\bar{2},\bar{2}) &= 599A - 82B + 57C - 1203D + 835E + 28F + 1394G - 2229H,\\
\zeta(\bar{1},1,3) &= -3186A + 435B - 300C + 6399D - 4438E - 150F - 7401G + 11839H,\\
\zeta(\bar{1},1,\bar{3}) &= -2732A + 376B - 268C + 5448D - 3798E - 130F - 6384G + 10182H,\\
\zeta(\bar{1},\bar{1},3) &= 20431A - 2799B + 1969C - 40888D + 28427E + 966F + 47591G - 76018H,\\
\zeta(\bar{1},\bar{1},\bar{3}) &= -7808A + 1070B - 758C + 15608D - 10858E - 369F - 18196G + 29054H,\\
\zeta(2,1,1,1) &= -13504A + 1856B - 1344C + 26880D - 18752E - 640F - 31552G + 50304H,\\
\zeta(2,1,1,\bar{1}) &= -11109A + 1518B - 1044C + 22320D - 15477E - 523F - 25812G + 41289H,\\
\zeta(2,1,\bar{1},1) &= 1174A - 158B + 101C - 2395D + 1642E + 54F + 2691G - 4333H,\\
\zeta(2,1,\bar{1},\bar{1}) &= 14927A - 2044B + 1431C - 29899D + 20773E + 705F + 34745G - 55518H,\\
\zeta(2,\bar{1},1,1) &= -2712A + 371B - 258C + 5439D - 3777E - 128F - 6306G + 10083H,\\
\zeta(2,\bar{1},1,\bar{1}) &= -14828A + 2030B - 1419C + 29709D - 20641E - 701F - 34517G + 55158H,\\
\zeta(2,\bar{1},\bar{1},1) &= -7120A + 977B - 681C + 14262D - 9911E - 337F - 16585G + 26496H,\\
\zeta(2,\bar{1},\bar{1},\bar{1}) &= 11204A - 1534B + 1074C - 22440D + 15595E + 530F + 26096G - 41691H,\\
\zeta(\bar{2},\bar{1},1,1) &= 8717A - 1197B + 847C - 17415D + 12122E + 412F + 20334G - 32456H,\\
\zeta(\bar{2},\bar{1},1,\bar{1}) &= -8511A + 1162B - 806C + 17085D - 11852E - 401F - 19775G + 31627H,\\
\zeta(\bar{2},\bar{1},\bar{1},1) &= 3432A - 470B + 327C - 6882D + 4779E + 162F + 7980G - 12759H,\\
\zeta(\bar{2},\bar{1},\bar{1},\bar{1}) &= -652A + 89B - 66C + 1296D - 905E - 31F - 1531G + 2436H,\\
\zeta(\bar{2},1,\bar{1},1) &= 8659A - 1183B + 822C - 17376D + 12059E + 409F + 20134G - 32193H,\\
\zeta(\bar{2},1,1,\bar{1}) &= 3571A - 490B + 344C - 7145D + 4969E + 169F + 8322G - 13291H,\\
\zeta(\bar{1},2,1,1) &= 190A - 26B + 18C - 384D + 265E + 9F + 442G - 707H,\\
\zeta(\bar{1},2,1,\bar{1}) &= 190A - 25B + 18C - 385D + 265E + 9F + 442G - 707H,\\
\zeta(\bar{1},2,\bar{1},1) &= -27776A + 3801B - 2654C + 55667D - 38668E - 1313F - 64648G + 103316H,\\
\zeta(\bar{1},2,\bar{1},\bar{1}) &= -19006A + 2604B - 1828C + 38048D - 26452E - 900F - 44288G + 70740H,\\
\zeta(\bar{1},\bar{2},1,1) &= -2407A + 330B - 233C + 4812D - 3347E - 113F - 5610G + 8957H,\\
\zeta(\bar{1},\bar{2},1,\bar{1}) &= -202A + 32B - 34C + 353D - 274E - 12F - 533G + 807H,\\
\zeta(\bar{1},\bar{2},\bar{1},1) &= 6507A - 893B + 631C - 13009D + 9054E + 309F + 15184G - 24238H,\\
\zeta(\bar{1},\bar{2},\bar{1},\bar{1}) &= 31628A - 4333B + 3038C - 63328D + 44021E + 1497F + 73681G - 117702H,\\
\zeta(\bar{1},1,2,1) &= -122A + 17B - 13C + 242D - 170E - 6F - 288G + 458H,\\
\zeta(\bar{1},1,2,\bar{1}) &= 3310A - 453B + 314C - 6641D + 4609E + 156F + 7692G - 12301H,\\
\zeta(\bar{1},1,\bar{2},\bar{1}) &= 6195A - 850B + 600C - 12383D + 8619E + 294F + 14454G - 23073H,\\
\zeta(\bar{1},1,\bar{2},1) &= 2888A - 394B + 272C - 5803D + 4023E + 136F + 6706G - 10729H,\\
\zeta(\bar{1},\bar{1},2,1) &= -7433A + 1019B - 711C + 14888D - 10348E - 352F - 17315G + 27663H,\\
\zeta(\bar{1},\bar{1},2,\bar{1}) &= 6793A - 932B + 658C - 13586D + 9453E + 322F + 15848G - 25301H,\\
\zeta(\bar{1},\bar{1},\bar{2},1) &= -313A + 43B - 30C + 626D - 437E - 15F - 730G + 1167H,\\
\zeta(\bar{1},\bar{1},\bar{2},\bar{1}) &= -18914A + 2592B - 1822C + 37855D - 26321E - 895F - 44073G + 70394H,\\
\zeta(\bar{1},1,1,2) &= 191A - 26B + 18C - 384D + 267E + 9F + 442G - 709H,\\
\zeta(\bar{1},1,1,\bar{2}) &= -2521A + 345B - 240C + 5054D - 3510E - 119F - 5864G + 9374H,\\
\zeta(\bar{1},1,\bar{1},2) &= 13126A - 1798B + 1257C - 26291D + 18271E + 621F + 30567G - 48838H,\\
\zeta(\bar{1},\bar{1},1,2) &= 13312A - 1826B + 1295C - 26595D + 18512E + 630F + 31043G - 49555H,\\
\zeta(\bar{1},\bar{1},1,\bar{2}) &= -4812A + 661B - 475C + 9593D - 6687E - 228F - 11237G + 17924H,\\
\zeta(\bar{1},\bar{1},\bar{1},\bar{2}) &= -13127A + 1798B - 1258C + 26291D - 18271E - 621F - 30565G + 48836H,\\
\zeta(\bar{1},1,1,1,1) &= -191A + 26B - 18C - 442G + 384D - 266E - 9F + 709H,\\
\zeta(\bar{1},1,1,1,\bar{1}) &= -191A + 26B - 18C + 385D - 266E - 9F - 442G + 709H,\\
\zeta(\bar{1},1,1,\bar{1},1) &= 4481A - 614B + 430C - 8973D + 6237E + 212F + 10438G - 16674H,\\
\zeta(\bar{1},1,1,\bar{1},\bar{1}) &= -A - E + 2H,\\
\zeta(\bar{1},1,\bar{1},1,1) &= -4693A + 643B - 451C + 9395D - 6531E - 222F - 10930G + 17462H,\\
\zeta(\bar{1},1,\bar{1},\bar{1},1) &= -313A + 43B - 31C - 730G + 626D - 436E - 15F + 1167H,\\
\zeta(\bar{1},1,\bar{1},\bar{1},\bar{1}) &= -13126A + 1798B - 1258C + 26291D - 18271E - 621F - 30565G + 48837H,\\
\zeta(\bar{1},\bar{1},1,1,1) &= 7496A - 1031B + 747C - 14915D + 10408E + 355F + 17522G - 27929H,\\
\zeta(\bar{1},\bar{1},1,1,\bar{1}) &= 2081A - 285B + 194C - 4183D + 2901E + 98F + 4840G - 7740H,\\
\zeta(\bar{1},\bar{1},1,\bar{1},1) &= -3308A + 452B - 313C + 6641D - 4607E - 156F - 7689G + 12297H,\\
\zeta(\bar{1},\bar{1},1,\bar{1},\bar{1}) &= -12121A + 1660B - 1164C + 24269D - 16868E - 573F - 28225G + 45094H,\\
\zeta(\bar{1},\bar{1},\bar{1},1,\bar{1}) &= 12622A - 1729B + 1210C - 25280D + 17569E + 597F + 29393G - 46961H,\\
\zeta(\bar{1},\bar{1},\bar{1},\bar{1},1) &= 10552A - 1445B + 1008C - 21144D + 14690E + 499F + 24565G - 39254H.
\end{aligned}
$$

E.2 Finite Multiple Zeta Values

We now list the FMZVs in lower weights. Note that by the reversal relation in fact we only need to list non-decreasing entries using lexicographic order. Moreover, by the duality relation given by Eq. (8.33) we only need to list entries of depth no greater than $(w+1)/2$. In odd weights all palindromic entries vanish by reversal relation. Furthermore, we omit all depth one and two elements, and depth 3 odd weight elements because we may use Thm. 8.5.12 and Thm. 8.5.13 to compute them fairly quickly.

Table E.1: FMZVs of superbity 1 in weight 5, 6 and 7.

$\zeta_{\mathscr{A}}(1,1,1,2)$	$-5\beta_5$	$\zeta_{\mathscr{A}}(1,1,2,1)$	$10\beta_5$	$\zeta_{\mathscr{A}}(1,2,3)$	$3\beta_3^2$
$\zeta_{\mathscr{A}}(1,3,2)$	$-\frac{9}{2}\beta_3^2$	$\zeta_{\mathscr{A}}(1,4,1)$	$3\beta_3^2$	$\zeta_{\mathscr{A}}(1,1,4)$	$-\frac{3}{2}\beta_3^2$
$\zeta_{\mathscr{A}}(2,1,3)$	$\frac{3}{2}\beta_3^2$	$\zeta_{\mathscr{A}}(1,1,2,2)$	$3\beta_3^2$	$\zeta_{\mathscr{A}}(1,1,1,3)$	$-\frac{3}{2}\beta_3^2$
$\zeta_{\mathscr{A}}(1,1,3,1)$	$\frac{3}{2}\beta_3^2$	$\zeta_{\mathscr{A}}(1,2,1,2)$	$\frac{9}{2}\beta_3^2$	$\zeta_{\mathscr{A}}(1,2,2,1)$	$-\beta_3^2$
$\zeta_{\mathscr{A}}(2,1,1,2)$	$-6\beta_3^2$			$\zeta_{\mathscr{A}}(1,1,1,4)$	$-27\beta_7'$
$\zeta_{\mathscr{A}}(1,1,2,3)$	$69\beta_7'$	$\zeta_{\mathscr{A}}(1,1,3,2)$	$-27\beta_7'$	$\zeta_{\mathscr{A}}(1,1,4,1)$	$33\beta_7'$
$\zeta_{\mathscr{A}}(1,2,1,3)$	$-27\beta_7'$	$\zeta_{\mathscr{A}}(1,2,2,2)$	$-27\beta_7'$	$\zeta_{\mathscr{A}}(1,2,3,1)$	$-63\beta_7'$
$\zeta_{\mathscr{A}}(1,3,1,2)$	$-9\beta_7'$	$\zeta_{\mathscr{A}}(2,1,1,3)$	$33\beta_7'$	$\zeta_{\mathscr{A}}(2,1,2,2)$	$-63\beta_7'$

Table E.2: FMZVs of superbity 1 in weight 8, $E = \zeta_{\mathscr{A}}(1,1,6)$ and $F = \beta_3\beta_5$.

$\zeta_{\mathscr{A}}(1,2,5)$	$E+F$	$\zeta_{\mathscr{A}}(1,3,4)$	$-\frac{25}{2}E-\frac{9}{2}F$
$\zeta_{\mathscr{A}}(1,4,3)$	$10E+4F$	$\zeta_{\mathscr{A}}(1,5,2)$	$\frac{5}{2}E-\frac{1}{2}F$
$\zeta_{\mathscr{A}}(1,6,1)$	$-2E$	$\zeta_{\mathscr{A}}(2,1,5)$	$-\frac{7}{2}E-\frac{1}{2}F$
$\zeta_{\mathscr{A}}(2,2,4)$	$10E+2F$	$\zeta_{\mathscr{A}}(2,3,3)$	$10E+2F$
$\zeta_{\mathscr{A}}(2,4,2)$	$-20E-4F$	$\zeta_{\mathscr{A}}(3,1,4)$	$\frac{5}{2}E+\frac{1}{2}F$
$\zeta_{\mathscr{A}}(3,2,3)$	$-20E-4F$	$\zeta_{\mathscr{A}}(4,1,3)$	$\frac{5}{2}E+\frac{1}{2}F$
$\zeta_{\mathscr{A}}(1,1,1,5)$	$\frac{3}{4}E-\frac{1}{4}F$	$\zeta_{\mathscr{A}}(1,1,2,4)$	$\frac{3}{4}E+\frac{5}{4}F$
$\zeta_{\mathscr{A}}(1,1,3,3)$	$-\frac{21}{2}E-3F$	$\zeta_{\mathscr{A}}(1,1,4,2)$	$\frac{33}{4}E+\frac{9}{4}F$
$\zeta_{\mathscr{A}}(1,1,5,1)$	$-\frac{3}{4}E+\frac{1}{4}F$	$\zeta_{\mathscr{A}}(1,3,1,3)$	0
$\zeta_{\mathscr{A}}(1,3,2,2)$	$\frac{3}{2}F$	$\zeta_{\mathscr{A}}(1,3,3,1)$	$\frac{27}{2}E+\frac{9}{2}F$
$\zeta_{\mathscr{A}}(2,2,1,3)$	$-\frac{9}{4}F-\frac{45}{4}E$	$\zeta_{\mathscr{A}}(3,1,1,3)$	$\frac{15}{2}E+\frac{3}{2}F$
$\zeta_{\mathscr{A}}(1,2,1,4)$	$-\frac{9}{2}E-F$	$\zeta_{\mathscr{A}}(1,2,2,3)$	$\frac{9}{2}E-\frac{3}{2}F$
$\zeta_{\mathscr{A}}(1,2,3,2)$	$-\frac{27}{4}E-\frac{1}{4}F$		

Table E.3: FMZVs of superbity 1 in weight 9, depth 4 and 5.

$\zeta_{\mathscr{A}}(1,1,1,6)$	$-\frac{1889}{72}\beta_9 - \frac{1}{2}\beta_3^3$	$\zeta_{\mathscr{A}}(1,1,2,5)$	$\frac{5551}{72}\beta_9 + \beta_3^3$
$\zeta_{\mathscr{A}}(1,1,3,4)$	$-\frac{4697}{72}\beta_9 - \frac{1}{2}\beta_3^3$	$\zeta_{\mathscr{A}}(1,1,4,3)$	$\frac{2609}{36}\beta_9 + \beta_3^3$
$\zeta_{\mathscr{A}}(1,1,5,2)$	$-\frac{821}{36}\beta_9 - \frac{5}{2}\beta_3^3$	$\zeta_{\mathscr{A}}(1,1,6,1)$	$\frac{521}{24}\beta_9 + \frac{1}{18}\beta_3^3$
$\zeta_{\mathscr{A}}(1,2,1,5)$	$-\frac{209}{72}\beta_9 + \beta_3^3$	$\zeta_{\mathscr{A}}(1,2,2,4)$	$-\frac{3857}{72}\beta_9 - \frac{2}{27}\beta_3^3$
$\zeta_{\mathscr{A}}(1,2,3,3)$	$-\frac{1591}{36}\beta_9 - \frac{2}{27}\beta_3^3$	$\zeta_{\mathscr{A}}(1,2,4,2)$	$-\frac{401}{36}\beta_9 + \frac{5}{27}\beta_3^3$
$\zeta_{\mathscr{A}}(1,2,5,1)$	$-\frac{1399}{24}\beta_9 - \frac{1}{9}\beta_3^3$	$\zeta_{\mathscr{A}}(1,3,1,4)$	$-\frac{127}{12}\beta_9 - \frac{1}{18}\beta_3^3$
$\zeta_{\mathscr{A}}(1,3,2,3)$	$\frac{292}{3}\beta_9 + \frac{2}{9}\beta_3^3$	$\zeta_{\mathscr{A}}(1,3,3,2)$	$-\beta_9 - \frac{1}{6}\beta_3^3$
$\zeta_{\mathscr{A}}(1,3,4,1)$	$\frac{1567}{18}\beta_9 + \frac{5}{2}\beta_3^3$	$\zeta_{\mathscr{A}}(1,4,1,3)$	$-\frac{251}{9}\beta_9 + \frac{1}{2}\beta_3^3$
$\zeta_{\mathscr{A}}(1,4,2,2)$	$-\frac{122}{3}\beta_9 - \frac{1}{9}\beta_3^3$	$\zeta_{\mathscr{A}}(1,5,1,2)$	$\frac{167}{72}\beta_9 + \frac{2}{27}\beta_3^3$
$\zeta_{\mathscr{A}}(2,1,1,5)$	$\frac{749}{18}\beta_9 + \frac{1}{2}\beta_3^3$	$\zeta_{\mathscr{A}}(2,1,2,4)$	$-\frac{1813}{18}\beta_9 - \beta_3^3$
$\zeta_{\mathscr{A}}(2,1,3,3)$	$\frac{1579}{36}\beta_9 + \frac{1}{2}\beta_3^3$	$\zeta_{\mathscr{A}}(2,1,4,2)$	$-\frac{2071}{36}\beta_9 + \beta_3^3$
$\zeta_{\mathscr{A}}(2,2,1,4)$	$\frac{357}{8}\beta_9$	$\zeta_{\mathscr{A}}(2,2,2,3)$	$54\beta_9$
$\zeta_{\mathscr{A}}(2,2,3,2)$	$84\beta_9$	$\zeta_{\mathscr{A}}(2,3,1,3)$	$\frac{43}{3}\beta_9 - \frac{1}{9}\beta_3^3$
$\zeta_{\mathscr{A}}(2,3,2,2)$	$-84\beta_9$	$\zeta_{\mathscr{A}}(3,1,1,4)$	$-\frac{3647}{72}\beta_9 + \beta_3^3$
$\zeta_{\mathscr{A}}(3,1,2,3)$	$\frac{1567}{18}\beta_9 - \frac{2}{27}\beta_3^3$		

$\zeta_{\mathscr{A}}(1,1,1,1,5)$	$-\frac{3965}{144}\beta_9 - \beta_3^3$	$\zeta_{\mathscr{A}}(1,1,1,2,4)$	$\frac{11107}{144}\beta_9 + 2\beta_3^3$
$\zeta_{\mathscr{A}}(1,1,1,3,3)$	$-\frac{2333}{144}\beta_9 + \frac{1}{2}\beta_3^3$	$\zeta_{\mathscr{A}}(1,1,1,4,2)$	$\frac{1709}{48}\beta_9 - \frac{3}{2}\beta_3^3$
$\zeta_{\mathscr{A}}(1,1,1,5,1)$	$\frac{1481}{72}\beta_9 + 2\beta_3^3$	$\zeta_{\mathscr{A}}(1,1,2,1,4)$	$-\frac{2333}{144}\beta_9 + \frac{1}{2}\beta_3^3$
$\zeta_{\mathscr{A}}(1,1,2,2,3)$	$-\frac{13133}{144}\beta_9 - 4\beta_3^3$	$\zeta_{\mathscr{A}}(1,1,2,3,2)$	$-\frac{1891}{48}\beta_9 + 3\beta_3^3$
$\zeta_{\mathscr{A}}(1,1,2,4,1)$	$-\frac{5239}{72}\beta_9 - 4\beta_3^3$	$\zeta_{\mathscr{A}}(1,1,3,1,3)$	$-\frac{85}{144}\beta_9 + \beta_3^3$
$\zeta_{\mathscr{A}}(1,1,3,2,2)$	$\frac{3485}{48}\beta_9 + 3\beta_3^3$	$\zeta_{\mathscr{A}}(1,1,3,3,1)$	$\frac{1015}{24}\beta_9 - \frac{3}{2}\beta_3^3$
$\zeta_{\mathscr{A}}(1,1,4,1,2)$	$\frac{5185}{144}\beta_9 + \frac{5}{2}\beta_3^3$	$\zeta_{\mathscr{A}}(1,1,4,2,1)$	$-\frac{1865}{36}\beta_9 - \frac{5}{2}\beta_3^3$
$\zeta_{\mathscr{A}}(1,2,1,1,4)$	$\frac{1709}{48}\beta_9 + 3\beta_3^3$	$\zeta_{\mathscr{A}}(1,2,1,2,3)$	$-\frac{1891}{48}\beta_9 - 6\beta_3^3$
$\zeta_{\mathscr{A}}(1,2,1,3,2)$	$\frac{223}{16}\beta_9 + \frac{9}{2}\beta_3^3$	$\zeta_{\mathscr{A}}(1,2,1,4,1)$	$-\frac{131}{8}\beta_9 - \frac{9}{2}\beta_3^3$
$\zeta_{\mathscr{A}}(1,2,2,1,3)$	$\frac{3485}{48}\beta_9 - \frac{3}{2}\beta_3^3$	$\zeta_{\mathscr{A}}(1,2,2,2,2)$	$\frac{255}{16}\beta_9$
$\zeta_{\mathscr{A}}(1,2,2,3,1)$	$-\frac{765}{8}\beta_9 + \frac{1}{2}\beta_3^3$	$\zeta_{\mathscr{A}}(1,2,3,1,2)$	$\frac{595}{16}\beta_9 - \frac{1}{3}\beta_3^3$
$\zeta_{\mathscr{A}}(1,2,4,1,1)$	$\frac{1865}{36}\beta_9 + \frac{5}{2}\beta_3^3$	$\zeta_{\mathscr{A}}(1,3,1,1,3)$	$-\frac{5185}{144}\beta_9 - 2\beta_3^3$
$\zeta_{\mathscr{A}}(1,3,1,2,2)$	$\frac{595}{16}\beta_9 + \frac{9}{2}\beta_3^3$	$\zeta_{\mathscr{A}}(1,3,2,1,2)$	$-\frac{935}{16}\beta_9 + \frac{9}{2}\beta_3^3$
$\zeta_{\mathscr{A}}(1,4,1,1,2)$	$\frac{4633}{144}\beta_9 - 4\beta_3^3$	$\zeta_{\mathscr{A}}(2,1,1,1,4)$	$\frac{1481}{72}\beta_9 - \frac{5}{2}\beta_3^3$
$\zeta_{\mathscr{A}}(2,1,1,2,3)$	$-\frac{5239}{72}\beta_9 + 5\beta_3^3$	$\zeta_{\mathscr{A}}(2,1,1,3,2)$	$-\frac{131}{8}\beta_9 - \frac{1}{3}\beta_3^3$
$\zeta_{\mathscr{A}}(2,1,2,1,3)$	$\frac{1015}{24}\beta_9 + 3\beta_3^3$	$\zeta_{\mathscr{A}}(2,1,2,2,2)$	$\frac{765}{8}\beta_9$
$\zeta_{\mathscr{A}}(2,1,3,2,1)$	$-\frac{595}{16}\beta_9 + \frac{1}{3}\beta_3^3$	$\zeta_{\mathscr{A}}(2,2,1,1,3)$	$-\frac{1865}{36}\beta_9 + 2\beta_3^3$

Table E.4: FMZVs of superbity 2 in weight 5, where $z311 = \zeta_{\mathscr{A}_2}(3,1,1)$.

$\zeta_{\mathscr{A}_2}(1,4)$	$3p\beta_3^2 - 2z311$	$\zeta_{\mathscr{A}_2}(4,1)$	$-3p\beta_3^2 + 2z311$
$\zeta_{\mathscr{A}_2}(1,1,3)$	$-z311$	$\zeta_{\mathscr{A}_2}(2,3)$	$-3p\beta_3^2 + 4z311$
$\zeta_{\mathscr{A}_2}(1,2,2)$	$-\frac{3}{2}p\beta_3^2 + 3z311$	$\zeta_{\mathscr{A}_2}(2,1,2)$	$-3p\beta_3^2$
$\zeta_{\mathscr{A}_2}(3,2)$	$3p\beta_3^2 - 4z311$	$\zeta_{\mathscr{A}_2}(2,2,1)$	$\frac{9}{2}p\beta_3^2 - 3z311$
$\zeta_{\mathscr{A}_2}(1,3,1)$	0		

Table E.5: FMZVs of superbity $\ell = 2$ in weight six, where $z411 = \zeta_{\mathscr{A}_2}(4,1,1) = \zeta_{\mathscr{A}_2}(1,1,4)$.

$\zeta_{\mathscr{A}_2}(6)$	$6\beta_7$	$\zeta_{\mathscr{A}_2}(2,1,3)$	$18\beta_7 - z411$
$\zeta_{\mathscr{A}_2}(5,1)$	$-10\beta_7$	$\zeta_{\mathscr{A}_2}(1,2,3)$	$-\frac{11}{4}\beta_7 - 2z411$
$\zeta_{\mathscr{A}_2}(1,5)$	$4\beta_7$	$\zeta_{\mathscr{A}_2}(1,3,2)$	$-\frac{65}{4}\beta_7 + 3z411$
$\zeta_{\mathscr{A}_2}(2,4)$	$-10\beta_7$	$\zeta_{\mathscr{A}_2}(2,3,1)$	$-\frac{9}{4}\beta_7 + 3z411$
$\zeta_{\mathscr{A}_2}(3,3)$	$-3\beta_7$	$\zeta_{\mathscr{A}_2}(3,1,2)$	$11\beta_7 - z411$
$\zeta_{\mathscr{A}_2}(4,2)$	$4\beta_7$	$\zeta_{\mathscr{A}_2}(3,2,1)$	$\frac{17}{4}\beta_7 - 2z411$
$\zeta_{\mathscr{A}_2}(2,2,2)$	$2\beta_7$	$\zeta_{\mathscr{A}_2}(1,4,1)$	$6\beta_7 - 2z411$

Appendix F

Answers to Some Exercises

Chapter 1.

Exercise 1.4. See [363].

Exercise 1.6. See [605, p. 1278].

Chapter 2.

Exercise 2.3. Differentiate and simplify.

Exercise 2.13. We need to evaluate the integral $\int_0^1 F(x)\,\mathrm{d}x$ where

$$F(x) = \frac{1}{x}\left(\frac{\log((1+x)/2)}{1-x} - \frac{\log((1-x)/2)}{1+x}\right).$$

Using series expansion of log and geometric expansion we can get some double log. Simplifying, we find an anti-derivative of $F(x)$:

$$h(x) = \mathrm{Li}_2\left(\frac{1-x}{2}\right) - \mathrm{Li}_2\left(\frac{1+x}{2}\right) - \mathrm{Li}_2(1-x) - \mathrm{Li}_2(-x) - \log(x)\log(1-x).$$

Indeed, this can checked be directly by differentiation. So

$$\int_0^1 F(x)\,\mathrm{d}x = -\,\mathrm{Li}_2(-1) = -\zeta(\bar{2}) = \frac{1}{2}\zeta(2) = \frac{\pi^2}{12},$$

which agrees with Schwinger's result.

Exercise 2.17. See [606].

Chapter 3.

Exercise 3.3. See [587].

Exercise 3.4. See [289, Theorems 2.1 and 4.2].

Exercise 3.5. See [289, Lemma 2.2].

Exercise 3.6. See [289, Lemma 6.1].

Exercise 3.14. See [308, Cor. 7].

Chapter 4.

Exercise 4.1. Assume $\Phi(\mathrm{x}_0, \mathrm{x}_1) = \Phi(z)$. By the definition, $G_1(z)\Phi(z) = G_0(z)$. Now differentiate both sides and use the KZ differential equation (4.4).

Exercise 4.5. See [477, 2.4.1].

Exercise 4.9. See [36].

Exercise 4.10. The proof of the cases of W_2, W_3, W_4 are almost identical to that of W_1. For example, for W_2, we let $u = (z_3 - z_2)/(z_3 - z_1)$, $v = (z_3 - z_1)/(z_4 - z_1)$ and $W = g(u, v)(z_4 - z_1)^{\hbar T}$ where $T = \sum_{1=i<j\le n} t^{ij}$.. Then $g(u, v)$ satisfies

$$\frac{\partial g}{\partial u} = \hbar\left(\frac{t^{23}}{u} - \frac{t^{13}}{1-u} - \frac{vt^{24}}{1-uv}\right)g(u,v), \tag{F.1}$$

$$\frac{\partial g}{\partial v} = \hbar\left(\frac{t^{12}+t^{13}+t^{23}}{v} - \frac{ut^{24}}{1-uv} - \frac{t^{34}}{1-v}\right)g(u,v). \tag{F.2}$$

Note that $[t^{23}, t^{12} + t^{13} + t^{23}] = 0$ by Lemma 4.1.24 so the above system is integrable. The rest is now straightforward.

The case W_5 is a little different. We let $u = (z_2 - z_1)/(z_4 - z_1)$ and $v = (z_4 - z_3)/(z_4 - z_1)$ and $W = g(u, v)(z_4 - z_1)^{\hbar T}$. Then g satisfies

$$\frac{\partial g}{\partial u} = \hbar\left(\frac{t^{12}}{u} - \frac{t^{23}}{1-u-v} - \frac{t^{24}}{1-u}\right)g(u,v),$$

$$\frac{\partial g}{\partial v} = \hbar\left(\frac{t^{13}}{1-v} + \frac{t^{23}}{1-u-v} - \frac{t^{34}}{v}\right)g(u,v).$$

Note that $[t^{12}, t^{34}] = 0$ by the definition of the embedding of t^{ij} so the above system is integrable. The rest is now easy.

Exercise 4.11. The proofs of the rest of four identities are almost identical to that of $W_1 = W_2(\Phi_h \otimes 1)$. For example, to prove

$$W_2 = W_3(\mathrm{id} \otimes \Delta \otimes \mathrm{id})(\Phi_h),$$

we set

$$V_2 = W_2(z_3 - z_2)^{-\hbar t^{23}},$$

$$V_3 = W_3(z_3 - z_2)^{-\hbar t^{23}}(\mathrm{id} \otimes \Delta \otimes \mathrm{id})(\Phi_h)$$

and so on.

Exercise 4.12. To prove identity $W_4 = W_5 e^{ht^{13}/2}$ can be proved essentially by applying the transposition $1 \leftrightarrow 2$ on the proof of $W_2 = W_3 e^{ht^{23}/2}$.

To prove the last identity $W_6 = W_1 e^{h(t^{13}+t^{23})/2}$ we define $V_6 = W_6(z_2 - z_1)^{-\hbar t^{12}}$ and $V_1 = W_1(z_2-z_1)^{-\hbar t^{12}} e^{h(t^{13}+t^{23})/2}$. We now show that $V_6 = V_1$. We see that both functions satisfy

$$\frac{\partial V}{\partial z_1} = \hbar \sum_{j\neq 1} \frac{t^{1j}}{z_1 - z_j} V - \hbar V \cdot \frac{t^{12}}{z_1 - z_2} = \frac{t^{13}V}{z_1 - z_3} + \frac{[t^{12}, V]}{z_1 - z_2},$$

$$\frac{\partial V}{\partial z_2} = \hbar \sum_{j\neq 2} \frac{t^{2j}}{z_2 - z_j} V - \hbar V \cdot \frac{t^{12}}{z_2 - z_1} = \frac{t^{23}V}{z_2 - z_3} + \frac{[t^{12}, V]}{z_2 - z_1},$$

$$\frac{\partial V}{\partial z_3} = \hbar \sum_{j\neq 3} \frac{t^{ij}}{z_3 - z_j} V.$$

Thus $V_6 - V_1$ is an absolute constant. Moreover, both functions have the same asymptotics as $z_2 \to z_1$ which implies $V_6 = V_1$.

Exercise 4.14. See [254, Prop. 1.2].

Exercise 4.15. Suppose $K(z)$ is any other holomorphic solution of (4.1) on Uw. As L(z) is invertible we set $F(z) = L(z)^{-1}K(z)$. On differentiating the equation $K(z) = L(z)F(z)$, we obtain by (4.1)

$$\sum_{i=0}^{N} \frac{x_i}{z - p_i} K(z) = \sum_{i=0}^{N} \frac{x_i}{z - p_i} L(z)F(z) + L(z)F'(z).$$

Thus $L(z)F'(z) = 0$. So $F'(z) = 0$, and so $F(z)$ is independent of z.

Exercise 4.26. See [505, Lemma 2.23].

Exercise 4.29. See [108, Lemma 3.5-3.9].

Exercise 4.30. See [108, Lemma 4.3-4.5].

Chapter 5.

Exercise 5.3. See [320] or [613]. This has been generalized along several directions in a series of papers [110–112, 320, 425, 510, 566–568, 580, 631].

Exercise 5.5. See [333].

Exercise 5.6. See [304].

Exercise 5.8. See [409].

Exercise 5.15. By use the Maple package in [603] one can find

$$\begin{aligned}
c_1 =& 4(n+1)[20736n^8 - 42624n^7m + 30720n^6m^2 - 5984n^5m^3 \\
& - 2352n^4m^4 + 616n^3m^5 + 136n^2m^6 + 6nm^7 + 233856n^7 \\
& - 395712n^6m + 217680n^5m^2 - 18976n^4m^3 - 14560n^3m^4 + 293nm^6 \\
& + 1036n^2m^5 + 10m^7 + 1102848n^6 - 1498864n^5m + 592128n^4m^2 \\
& + 3560n^3m^3 - 27864n^2m^4 - 379nm^5 + 111m^6 + 2837664n^5 \\
& - 2981696n^4m + 764308n^3m^2 + 64512n^2m^3 - 18768nm^4 - 669m^5 \\
& + 4342320n^4 - 3323100n^3m + 453800n^2m^2 + 67107nm^3 - 3056m^4 \\
& + 4019904n^3 + 84643nm^2 - 2028388n^2m + 18425m^3 + 2172432n^2 \\
& - 598934nm - 9019m^2 + 612576n - 58146m + 65664], \\
c_2 =& (n+2)[13824n^8 - 39168n^7m + 43392n^6m^2 - 21760n^5m^3 \\
& + 3904n^4m^4 + 208n^3m^5 - 48n^2m^6 + 168192n^7 - 393984n^6m \\
& + 341088n^5m^2 - 120960n^4m^3 + 10792n^3m^4 + 1480n^2m^5 - 92nm^6 \\
& - +853440n^6 - 1603136n^5m + 1035776n^4m^2 236424n^3m^3 \\
& + 3280n^2m^4 + 2296nm^5 - 36m^6 + 2348352n^5 - 3397920n^4m \\
& + 1530908n^3m^2 - 185740n^2m^3 - 9629nm^4 + 912m^5 + 3810816n^4 \\
& - 4009996n^3m + 1122686n^2m^2 - 40102nm^3 - 5487m^4 + 3704808n^3 \\
& - 2586126n^2m + 356293nm^2 + 7266m^3 + 2080560n^2 \\
& - 812238nm + 27573m^2 + 602808n - 86928m + 65520].
\end{aligned}$$

Exercise 5.16. See [406, 407, 410].

Exercise 5.17. See [411, Cor. 1.2].

Exercise 5.18. See [195].

Exercise 5.21. Assume $f(x) = \sum_{j=1}^{d} a_{2j-1}x^{2j-1}$ and $g(x) = \sum_{j=0}^{2d} b_j x^j$ with $a_{2d-1} \neq 0$. Then all the coefficients of $g(x)$ are uniquely determined by b_{2d} and a_{2j-1} $(j = 1, \dots, d)$. This in fact also shows that the kernel of the map has dimension 1.

Exercise 5.23. Take $H(X,Y) = X^{k-2} + Y^{k-2}$ and then $H(X,Y) = (X - Y)^{k-2}$.

Chapter 6.

Exercise 6.2. Let $\alpha = s_i$ and $\beta = s_{i+l+1}$. Suppose $z_\alpha u = \mathtt{a}^{s_i-1}\mathtt{b}\cdots\mathtt{a}^{s_1-1}\mathtt{b}$ and $z_\beta v = \mathtt{a}^{s_{i+l+1}-1}\mathtt{b}\cdots\mathtt{a}^{s_d-1}\mathtt{b}$ (recall $z_s = \mathtt{a}^{s-1}\mathtt{b}$). Then

$$\begin{aligned}
&\sum_{j=0}^{l}(-1)^j(z_1^j z_\alpha u) * (z_1^{l-j} z_\beta v)\\
&= z_\alpha(u * z_1^l z_\beta v) + \sum_{j=1}^{l}(-1)^j z_1(z_1^{j-1} z_\alpha u * z_1^{l-j} z_\beta v) + (-1)^l z_\beta(z_1^l z_\alpha u * v)\\
&\quad + \sum_{j=0}^{l-1}(-1)^j z_1(z_1^j z_\alpha u * z_1^{l-j-1} z_\beta v) + z_{\alpha+1}(u * z_1^{l-1} z_\beta v)\\
&\quad + z_{\beta+1}(z_1^{l-1} z_\alpha u * v) + \sum_{j=1}^{l-1}(-1)^j z_2(z_1^{j-1} z_\alpha u * z_1^{l-j-1} z_\beta v)\\
&= z_\alpha(u * z_1^l z_\beta v) + (-1)^l z_\beta(z_1^l z_\alpha u * v) + z_{\alpha+1}(u * z_1^{l-1} z_\beta v)\\
&\quad + z_{\beta+1}(z_1^{l-1} z_\alpha u * v) + \sum_{j=1}^{l-1}(-1)^j z_2(z_1^{j-1} z_\alpha u * z_1^{l-j-1} z_\beta v) \in \mathfrak{A}^0.
\end{aligned}$$

Chapter 7.

Exercise 7.1. $(1,3,2)^\vee = (2,1,2,1)$ and $(4,2,1,3)^\vee = (\{1\}^3, 2, 3, 1, 1)$.

Exercise 7.6. See the proof of Lemma 10.9.1.

Chapter 8.

Exercise 8.6. By Definition in Eq. (8.24)

$$\mathrm{tog}(\mathtt{z}_{\bar{4}}\mathtt{z}_2\mathtt{z}_1\mathtt{z}_{\bar{3}}\mathtt{z}_2\mathtt{z}_{\bar{2}}\mathtt{z}_{\bar{1}}\mathtt{z}_7\mathtt{z}_{\bar{6}}) = \mathtt{z}_{\bar{4}}\mathtt{z}_{\bar{2}}\mathtt{z}_{\bar{1}}\mathtt{z}_3\mathtt{z}_2\mathtt{z}_{\bar{2}}\mathtt{z}_1\mathtt{z}_7\mathtt{z}_{\bar{6}}.$$

By Definition in Eq. (8.25)

$$\mathrm{tog}^{-1}(\mathtt{b}_1\mathtt{a}^2\mathtt{b}_1\mathtt{a}^3\mathtt{b}_0\mathtt{b}_1\mathtt{a}^4\mathtt{b}_0\mathtt{a}^4\mathtt{b}_1\mathtt{a}\mathtt{b}_0) = \mathtt{b}_1\mathtt{a}^2\mathtt{b}_0\mathtt{a}^3\mathtt{b}_1\mathtt{b}_1\mathtt{a}^4\mathtt{b}_1\mathtt{a}^4\mathtt{b}_1\mathtt{a}\mathtt{b}_1.$$

Exercise 8.13. These are part of Conjecture 3.20 in [619]. Note that part (iii) of the conjecture there is trivial by the reversal relation because **s** is palindrome and of odd weight.

Chapter 9.

Exercise 9.12. This is Conjecture 1.1 of [279].

Chapter 10.

Exercise 10.2. See the proof at the end of [446].

Exercise 10.9. Two different proofs can be found: by the derivation relation for the MZ*Vs [307, eq. (12)], or by the Hoffman algebra [537, Prop. 2.6]

Exercise 10.10 See [358].

Chapter 12.

Exercise 12.1. One can prove this by modifying the proof of Thm. 5.2.3 in Sec. 5.2. See [81] or [456].

Exercise 12.3. One may find two different proofs in [81] and [456].

Exercise 12.4. See [79, Theorem 2.1].

Exercise 12.13. By Eq. (12.42)

$$
\begin{aligned}
&(\rho a \shuffle \rho b) \shuffle \rho c \\
&= \rho(\rho a \shuffle b) \shuffle \rho c + \rho(\rho b \shuffle a) \shuffle \rho c - \rho(a \shuffle b) \shuffle \rho c \\
&= \rho\Big(\rho a \shuffle b \shuffle \rho c + \rho(\rho a \shuffle b) \shuffle \mathbf{v} - \rho a \shuffle b \shuffle c + \rho b \shuffle a \shuffle \rho c \\
&\quad + \rho(\rho b \shuffle a) \shuffle c - \rho b \shuffle a \shuffle c - \rho(a \shuffle b) \shuffle c - a \shuffle b \shuffle \rho c + a \shuffle b \shuffle c \Big) \\
&= \rho\Big(\underbrace{b \shuffle \rho(a \shuffle \rho c)}_{1} + \underbrace{b \shuffle \rho(\rho a \shuffle c)}_{2} - \underbrace{b \shuffle \rho(a \shuffle c)}_{3} + \underbrace{\rho(\rho a \shuffle b) \shuffle c}_{4} \\
&\quad - \underbrace{\rho a \shuffle b \shuffle c}_{5} + \underbrace{a \shuffle \rho(b \shuffle \rho c)}_{6} + \underbrace{a \shuffle \rho(\rho b \shuffle c)}_{7} - \underbrace{a \shuffle \rho(b \shuffle c)}_{8} \\
&\quad + \underbrace{\rho(\rho b \shuffle a) \shuffle c}_{9} - \underbrace{\rho b \shuffle a \shuffle c}_{10} - \underbrace{\rho(a \shuffle b) \shuffle c}_{11} - \underbrace{a \shuffle b \shuffle \rho c}_{12} + \underbrace{a \shuffle b \shuffle c}_{13} \Big).
\end{aligned}
$$

On the other hand:

$$
\begin{aligned}
&\rho a \shuffle (\rho b \shuffle \rho c) \\
&= (\rho b \shuffle \rho c) \shuffle \rho a \\
&= \rho\Big(\underbrace{c \shuffle \rho(b \shuffle \rho a)}_{4} + \underbrace{c \shuffle \rho(\rho b \shuffle a)}_{9} - \underbrace{c \shuffle \rho(b \shuffle a)}_{11} + \underbrace{\rho(\rho b \shuffle c) \shuffle a}_{7} \\
&\quad - \underbrace{\rho b \shuffle c \shuffle a}_{10} + \underbrace{b \shuffle \rho(c \shuffle \rho a)}_{2} + \underbrace{b \shuffle \rho(\rho c \shuffle a)}_{1} - \underbrace{b \shuffle \rho(c \shuffle a)}_{3} \\
&\quad + \underbrace{\rho(\rho c \shuffle b) \shuffle a}_{6} - \underbrace{\rho c \shuffle b \shuffle a}_{12} - \underbrace{\rho(b \shuffle c) \shuffle a}_{8} - \underbrace{b \shuffle c \shuffle \rho a}_{5} + \underbrace{b \shuffle c \shuffle a}_{13} \Big).
\end{aligned}
$$

Exercise 12.11. See [119, (2)].

Exercise 12.21. This is a conjecture of Bachmann.

Exercise 12.22. See [81, Thm. 16].

Exercise 12.23. See [81, Thm. 15].

Chapter 13.

Exercise 13.4. See [477, Prop. 2.3.7] and [482].

Chapter 14.

Exercise 14.1. See [84] for Eq. (14.5) and [90] for Eq. (14.6).

Exercise 14.2. Let $F := \{1, t, 1-t, 1+t\}$. Let S be the set of polynomials which is constant multiple of a product of elements from F (repetition allowed). We then need to find all rational functions $r(t) = f(t)/g(t)$ such that $f, g, f+g, f-g \in S$ since $d\log r$, $d\log(1-r)$ and $d\log(1+r)$ all must be linear combinations of $d\log t$, $d\log(1-t)$ and $d\log(1+t)$. Now a lengthy case by case study finishes the proof.

Exercise 14.5. See [392, pp. 198 and 202].

Bibliography

[1] ABEL, N.H. Note sur la fonction $\psi x = x + \frac{x^2}{2^2} + \frac{x^3}{3^2} + \cdots + \frac{x^n}{n^2} + \cdots$ (in French). In *Œuvres complètes de Niels Henrik Abel–Nouvelle édition, Tome II (in French)*, L. Sylow and S. Lie, Ed. Christiania [Oslo]: Grøndahl & Søn., 1881, pp. 189–193.

[2] ABLINGER, J. Discovering and proving infinite binomial sums identities. *arXiv:1507.01703* (2015).

[3] ABLINGER, J. AND BLÜMLEIN, J. Harmonic sums, polylogarithms, special numbers, and their generalizations. In *Computer Algebra in Quantum Field Theory*, C. Schneider and J. Bluemlein, Ed., Texts and Monographs in Symbolic Computation. Springer, 2013, pp. 1–32.

[4] ABLINGER, J., BLÜMLEIN, J. AND SCHNEIDER, C. Harmonic sums and polylogarithms generated by cyclotomic polynomials. *J. Math. Phys. 52 (10)* (2011), 102–301.

[5] ABOUZAHRA, M. AND LEWIN, L. Ladders in the trans-Kummer region. In *[393]*. 1991, pp. 49–68.

[6] ABRAMOWITZ, M. AND STEGUN, I.A. *Handbook of Mathematical Functions with Formulas, Graphs, and Mathematical Tables.* Dover, New York, 1972.

[7] AGOH, T. AND DILCHER, K. Convolution identities and lacunary recurrences for Bernoulli numbers. *J. Number Theory 124* (2007), 105–122.

[8] AKIYAMA, S. AND ISHIKAWA, H. An analytic continuation of multiple L-functions and related zeta-functions. In *Analytic Number Theory*, C. Jia and K. Matsumoto, Ed., Developments in Math. Springer, 2002, pp. 1–16.

[9] AKIYAMA, S. AND TANIGAWA, Y. Multiple zeta values at non-positive integers. *Ramanujan J. 5* (2001), 327–351.

[10] AKIYAMA, S., EGAMI, S. AND TANIGAWA, Y. Analytic continuation of multiple zeta-functions and their values at non-positive integers. *Acta Arith. 98* (2001), 107–116.

[11] ANDERSON, G.W. AND THAKUR, D.S. Multizeta values for $\mathbb{F}_q[t]$, their period interpretation, and relations between them. *Intl. Math. Res. Notices* (2009), 2038–2055.

[12] ANDREWS, G.E. *The Theory of Partitions*, vol. 2 of *Encyclopedia Math. Appl.* Addison-Wesley Publishing Co., London, Amsterdam, Don Mills,

Ontario, Sydney, Tokyo, 1976.

[13] ANDREWS, G.E. q-analogs of the binomial coefficient congruences of Babbage, Wolstenholme and Glaisher. *Discrete Math. 204* (1999), 15–25.

[14] AOKI, T. AND OHNO, Y. Sum relations for multiple zeta values and connection formulas for the Gauss hypergeometric functions. *Publ. Res. Inst. Math. Sci. 41* (2005), 329–337.

[15] AOKI, T., KOMBU, Y. AND OHNO, Y. A generating function for sums of multiple zeta values and its applications. *Proc. Amer. Math. Soc. 136(2)* (2007), 387–395.

[16] AOKI, T., OHNO, Y. AND WAKABAYASHI, N. On generating functions of multiple zeta values and generalized hypergeometric functions. *Manuscripta Math. 134* (2011), 139–155.

[17] AOMOTO, K. Addition theorem of Abel type for hyperlogarithms. *Nagoya Math. J. 88* (1982), 55–71.

[18] AOMOTO, K. Special values of hyperlogarithms and linear difference schemes. *Illinois J. Math. 34(2)* (1990), 191–216.

[19] APÉRY, R. Irrationalité de $\zeta(2)$ et $\zeta(3)$ (in French). *Astérisque 61* (1979), 11–13.

[20] APOSTOL, T.M. AND VU, T.H. Dirichlet series related to the Riemann zeta function. *J. Number Theory 19* (1984), 85–102.

[21] ARAKAWA, T. AND KANEKO, M. Multiple zeta values, poly-Bernoulli numbers, and related zeta functions. *Nagoya Math. J. 153* (1999), 189–209.

[22] ARAKAWA, T., IBUKIYAMA, T. AND KANEKO, M. *Bernoulli Numbers and Zeta Functions, with an appendix by Zagier, D.* Monographs in Math. Springer, 2014.

[23] ATKINSON, F.V. The mean value of the Riemann zeta-function. *Acta Math. 81* (1949), 353–376.

[24] BACHMANN, H. *Multiple zeta-Werte und de Verbindung zu Modulformen durch Multiple Eisensteinreihen (in German), master thesis.* Hamburg Univesity, 2012.

[25] BACHMANN, H. The algebra of bi-brackets and regularized multiple Eisenstein series. *arXiv:1504.08138* (2015).

[26] BACHMANN, H., AND KÜHN, U. The algebra of multiple divisor functions and applications to multiple zeta values. *Ramanujan J.* (2015).

[27] BACHMANN, H. AND KÜHN, U. A short note on a conjecture of Okounkov about a q-analogue of multiple zeta values. *arXiv:1407.6796* (2014).

[28] BACHMANN, H. AND TASAKA, K. The double shuffle relations for multiple Eisenstein series. *arXiv:1502.03408* (2015).

[29] BAILEY, D.H. AND BROADHURST, D.J. Parallel integer relation detection: Techniques and applications. *Math. Comp. 70* (2000), 1719–1736.

[30] BAILEY, D.H., BORWEIN, J.M. AND GIRGENSOHN, R. Experimental evaluation of Euler sums. *Experiment. Math. 3* (1994), 17–30.

[31] BAILEY, D.H., BORWEIN, J.M., BROADHURST, D.J. AND ZUDILIN, W. *Experimental Math. and mathematical physics*, vol. 517 of *Contemp. Math.* Amer. Math. Soc., Providence, RI, 2010.

[32] BAILEY, D.H., BORWEIN, J.M., CALKIN, N., GIRGENSOHN, R., LUKE, R.

AND MOLL, V. *Experimental Math. in Action.* A K Peters, Natick, MA, 2007.

[33] BAILEY, D.H. AND BROADHURST, D.J. A seventeenth-order polylogarithm ladder. *arXiv:9906134* (1999).

[34] BAILEY, W.N. *Generalited Hypergeometric Series.* Cambridge Uniyersity Press, Cambridge, 1935.

[35] BALL, K. AND RIVOAL, T. Irrationalité d'une infinité de valeurs de la fonction zêta aux entiers impairs (in French). *Inv. Math. 146(1)* (2001), 193–207.

[36] BAR-NATAN, D. On associators and the Grothendieck–Teichmüller group I. *Selecta Math. New Series 4* (1998), 183–212.

[37] BARRIERI, R., MIGNACO, J.A. AND REMIDDI, E. Electron Form Factors up to Fourth Order. II. *Il Nuovo Cimento 11(4)* (1972), 865–916.

[38] BASS, H. Generators and relations for cyclotomic units,. *Nagoya Math. J. 27(2)* (1966), 401–407.

[39] BAUMARD, S. Aspects modulaires et elliptiques des relations entre multizêtas (in French). *Number Theory, ⟨NNT:2014PA066100⟩.⟨tel-01017022v2⟩* (2014).

[40] BAUMARD, S. AND SCHNEPS, L. Period polynomial relations between double zeta values. *Ramanujan J. 32 (1)* (2013), 83–100.

[41] BAXTER, G. An analytic problem whose solution follows from a simple algebraic identity. *Pacific J. Math. 10* (1960), 731–742.

[42] BAYAD, A. AND HAMAHATA, Y. Multiple polylogarithms and multi-poly-Bernoulli polynomials. *Funct. Approx. Comment. Math. 46* (2012), 45–61.

[43] BAYAD, M. A generalization of Wolstenholme's Theorem. *Amer. Math. Monthly 104(6)* (1997), 557–560.

[44] BEILINSON, A.A. AND DELIGNE, P. Interprétation motivique de la conjecture de Zagier reliant polylogarithmes et régulateurs (in French). In *Motives*, vol. 55 (part 2) of *Proc. Sym. Pure Math.* Amer. Math. Soc., Providence, RI, 1994, pp. 97–121.

[45] BEILINSON, A.A., GONCHAROV, A.B., SCHECHTMAN, V.V. AND VARCHENKO, A.N. Aomoto dilogarithms, mixed Hodge structures and motivic cohomology of pairs of triangles in the plane. In *Grothendieck Festschrift II*, vol. 87 of *Prog. in Math.* Birkhäuser, Boston, 1991, pp. 78–131.

[46] BEILINSON, A.A., MACPHERSON, R. AND SCHECHTMAN, V.V. Notes on motivic cohomology. *Duke Math. J. 54* (1987), 679–710.

[47] BEILINSON, A.A., VARCHENKO, A.N., GONCHAROV, A.B. AND SCHECHTMAN, V.V. Projective geometry and algebraic K-theory. *Leningrad Math. J 2* (1991), 523–576.

[48] BESSER, A. Finite and p-adic polylogarithms. *Compos. Math. 130(2)* (2002), 215–223.

[49] BESSER, A. AND FURUSHO, H. The double shuffle relations for p-adic multiple zeta values. In *Primes and Knots*, Toshitake Kohno and Masanori Morishita, Ed., vol. 416 of *Contemp. Math.* Amer. Math. Soc., Providence, RI, 2006, pp. 9–29.

[50] BIERENBAUM, I. AND WEINZIERL, S. The massless two-loop two-point function. *Eur. Phys. J. C32* (2003), 67–78.

[51] BIGOTTE, M., JACOB, G., OUSSOUS, N.E. AND PETITOT, M. Generating power series of coloured polylogarithm functions and Drinfel'd associator. In *Computer Math., Chiang Mai*, vol. 8 of *Lecture Notes Ser. Comput.* World Scientific Publishing, River Edge, NJ, 2000, pp. 39–48.

[52] BIGOTTE, M., JACOB, G., OUSSOUS, N.E. AND PETITOT, M. Lyndon words and shuffle algebras for generating the coloured multiple zeta values relations tables. *Theoretical Computer Sci. 273 (1-2)* (2002), 271–282.

[53] BLOCH, S. Application of the dilogarithm function in algebraic K-theory and algebraic geometry. In *Proc. Intl. Symp. Alg. Geom. (Kyoto Univ. Kyoto, 1977)*. Kinokuniya Book Store, Tokyo, 1978, pp. 103–114.

[54] BLOCH, S. Lectures on mixed motives given at Santa Cruz. *Notes online, available at www.math.uchicago.edu/~bloch/publications.html* (1995).

[55] BLOCH, S. Higher Regulators. In *Algebraic K-Theory, and Zeta Functions of Elliptic Curves*, vol. 11 of *CRM Monograph Series*. Amer. Math. Soc., Providence, RI Centre de Recherches Math., 2000.

[56] BLOCH, S. AND KREIMER, D. Feynman amplitudes and Landau singularities for one-loop graphs. *Comm. Number Thy. Phys. 4 (4)* (2010), 709–753.

[57] BLOCH, S. AND VANHOVE, P. The elliptic dilogarithm for the sunset graph. *J. Number Theory 148 (3)* (2015), 328–364.

[58] BLÜMLEIN, J. Harmonic sums and Mellin transforms. *Nucl. Phys. Proc. Suppl. 79* (1999), 166–168.

[59] BLÜMLEIN, J. Reduction of multiple harmonic sums and harmonic polylogarithms. *Nuclear Instruments and Methods in Physics Research Section A: Accelerators, Spectrometers. Detectors and Associated Equipment 534 (1-2)* (2004), 279–283.

[60] BLÜMLEIN, J. Relations between harmonic sums in massless QCD calculations. *Few-Body Systems 36* (2005), 29–34.

[61] BLÜMLEIN, J. AND S. KURTH. Harmonic sums and Mellin transforms up to two-loop order. *Phys. Rev. D60* (1999 014018).

[62] BLÜMLEIN, J., BROADHURST, D.J. AND VERMASEREN, J.A.M. The multiple zeta value data mine. *Comput. Phys. Commun. 181* (2010), 582–625.

[63] BOGNER, C., AND BROWN, F. Symbolic integration and multiple polylogarithms. *PoS(LL2012)053. arXiv:1209.6524*.

[64] BOGNER, C. AND LÜDERS, M. Multiple polylogarithms and linearly reducible Feynman graphs. In *Feynman Amplitudes, Periods and Motives*, L. Alvarez-consul, J. I. Burgos-gil and K. Ebrahimi-Fard, Ed., vol. 648 of *Contemp. Math.* Amer. Math. Soc., Providence, RI, 2015, pp. 11–28.

[65] BOROWKA, S. AND HEINRICH, G. Massive non-planar two-loop four-point integrals with SecDec 2.1. *Comput. Phys. Commun. 184* (2013), 2552.

[66] BORWEIN, J.M. AND BAILEY, D.H. *Math. by Experiment, 2nd edition.* A K Peters Ltd., Natick, MA, 2008.

[67] BORWEIN, J.M. AND BORWEIN, P. *Pi and the AGM.* John Wiley and Sons, Toronto and New York, 1987 (paperback 1998).

[68] Borwein, J.M. and Bradley, D.M. Thirty-two Goldbach variations. *Intl. J. Number Theory 2(1)* (2006), 65–103.

[69] Borwein, J.M. and Girgensohn, R. Evaluation of triple Euler sums. *Electron. J. Combin. 3* (1996), R23.

[70] Borwein, J.M., Bradley, D.M. and Broadhurst, D.J. Evaluations of k-fold Euler/Zagier sums: a compendium of results for arbitrary k. *Electron. J. Combin. 4(2)* (1997), R5.

[71] Borwein, J.M., Bradley, D.M., Broadhurst, D.J. and Lisonek, P. Combinatorial aspects of multiple zeta values. *Electron. J. Combin. 5* (1998), R38.

[72] Borwein, J.M., Bradley, D.M., Broadhurst, D.J. and Lisonek, P. Special values of multidimensional polylogarithms. *Trans. Amer. Math. Soc. 353* (2001), 907–941.

[73] Borwein, J.M., Broadhurst, D.J. and Kamnitzer, J. Central binomial sums, multiple Clausen values, and zeta values. *Experiment. Math. 10 (1)* (2001), 25–34.

[74] Borwein, J.M., Lisonek, P., and Irvine, P. An interface for evaluation of Euler sums at `oldweb.cecm.sfu.ca/cgi-bin/EZFace/zetaform.cgi`.

[75] Bouillot, O. The algebra of Hurwitz multizeta functions. *Comptes Rendus Math. 352 (11)* (2014), 865–869.

[76] Bowman, D.M. and Bailey, D.F. Multiple polylogarithms: a brief survey. In *Proc. Conf. on q-Series with Applications to Combinatorics, Number Theory and Physics*, B.C. Berndt and K. Ono, Ed., vol. 291 of *Contemp. Math.* Amer. Math. Soc., Providence, RI, 2001, pp. 71–92.

[77] Bowman, D.M. and Bradley, D.M. The algebra and combinatorics of shuffles and multiple zeta values. *J. Combin. Theory Ser. A 97(1)* (2002), 43–63.

[78] Boyd, D.W. A p-adic study of the partial sums of the harmonic series. *Experimental Math. 4(3)* (1994), 287–302.

[79] Bradley, D.M. A q-analog of Euler's decomposition formula for the double zeta function. *Int. J. Math. and Math. Sci. 21* (2005), 3453–3458.

[80] Bradley, D.M. Duality for finite multiple harmonic q-series. *Discrete Math. 300* (2005), 44–56.

[81] Bradley, D.M. Multiple q-zeta values. *J. Alg. 283* (2005), 752–798.

[82] Broadhurst, D.J. Massless scalar Feynman diagrams: five loops and beyond. *Open Univ. preprint OUT-4102-18* (1985).

[83] Broadhurst, D.J. Conjectured enumeration of irreducible multiple zeta values from knots and Feynman diagrams. *arXiv:9612012* (1996).

[84] Broadhurst, D.J. On the enumeration of irreducible k-fold Euler sums and their roles in knot theory and field theory. *arXiv:9604128* (1996).

[85] Broadhurst, D.J. Massive 3-loop Feynman diagrams reducible to SC^* primitives of algebras of the sixth root of unity. *European Phys. J. C (Fields) 8* (1999), 311–333.

[86] Broadhurst, D.J. Multiple Deligne values: a data mine with empirically tamed denominators. *arXiv:1409.7204* (2014).

[87] Broadhurst, D.J. Multiple Landen values and the tribonacci numbers.

arXiv:1504.05303 (2015).

[88] Broadhurst, D.J. Tests of conjectures on multiple Watson values. *arXiv:1504.08007* (2015).

[89] Broadhurst, D.J. and Kreimer, D. Knots and numbers in ϕ^4 Theory to 7 loops and beyond. *Int. J. Mod. Phys. C6* (1995), 519–524.

[90] Broadhurst, D.J. and Kreimer, D. Association of multiple zeta values with positive knots via Feynman diagrams up to 9 loops. *Phys. Lett. B 393(3-4)* (1997), 403–412.

[91] Broadhurst, D.J., and Schnetz, O. Algebraic geometry informs perturbative quantum field theory. *PoS(LL2014)078. arXiv:1409.5570.*

[92] Broadhurst, D.J., Gracey, J.A. and Kreimer, D. Beyond the triangle and uniqueness relations: Non-zeta counterterms at large N from positive knots. *Z. Phys. C 75* (1997), 559–574.

[93] Brochier, A. A Kohno–Drinfeld Theorem for the monodromy of cyclotomic KZ connections. *Comm. Math. Phys. 311(1)* (2012), 55–96.

[94] Broedel, J., Mafra, C.R., Matthes, N., and Schlotterer, O. Elliptic multiple zeta values and one-loop superstring amplitudes. *J. High Energy Phys. 07#112 (41 pp)* (2015).

[95] Broedel, J., Matthes, N. and Schlotterer, O. Relations between elliptic multiple zeta values and a special derivation algebra. *arXiv:1507.02254* (2015).

[96] Broedel, J., Schlotterer, O. and Stieberger, S. Polylogarithms, multiple zeta values and superstring amplitudes. *arXiv:1304.7267* (2013).

[97] Brown, F. Single-valued multiple polylogarithms in one variable. *C. R. Acad. Sci. Paris, Ser. I 338* (2004), 527–532.

[98] Brown, F. Single-valued hyperlogarithms and unipotent differential equations. *preprint* (2005).

[99] Brown, F. On the periods of some Feynman integrals. *arXiv:0910.0114* (2009).

[100] Brown, F. The massless higher-loop two-point function. *Commun. Math. Phys. 287* (2009), 925–958.

[101] Brown, F. Multiple zeta values and periods: from moduli spaces to Feynman integrals. In *Combinatorics and physics*, vol. 539 of *Contemp. Math.* Amer. Math. Soc., Providence, RI, Providence, RI, 2011, pp. 27–52.

[102] Brown, F. Mixed Tate motives over Spec($\mathbb{Z}$). *Ann. Math. 175(2)* (2012), 949–976.

[103] Brown, F. On the decomposition of motivic multiple zeta values. In *Galois–Teichmüller theory and Arithmetic Geometry*, vol. 63 of *Adv. Studies Pure Math.* The Math. Soc. of Japan, 2012, pp. 31–58.

[104] Brown, F. Depth-graded motivic multiple zeta values. *arXiv:1301.3053* (2013).

[105] Brown, F. Irrationality proofs for zeta values, moduli spaces and dinner parties. *arXiv:1412.6508* (2014).

[106] Brown, F. and Levin, A. Multiple elliptic polylogarithms. *arXiv:1110.6917* (2011).

[107] Brown, F. and Schnetz, O. Modular forms in quantum field theory.

arXiv:1304.5342 (2013).

[108] Brown, F. and Schnetz, O. Single-valued multiple polylogarithms and a proof of the zig-zag conjecture. *J. Number Theory 148 (3)* (2015), 478–506.

[109] Brown, F., and Yeats, K. Spanning forest polynomials and the transcendental weight of Feynman graphs. *Comm. Math. Phy. 301 (2)* (2011), 357–382.

[110] Cai, T. and Shen, Z. A congruence involving alternating harmonic sums modulo $p^\alpha q^\beta$. *arXiv:1503.03154* (2015).

[111] Cai, T. and Shen, Z. Super congruences involving alternating harmonic sums modulo prime powers. *arXiv:1503.03156* (2015).

[112] Cai, T., Shen, Z. and Jia, L. A congruence involving harmonic sums modulo $p^\alpha q^\beta$. *arXiv:1503.02798* (2015).

[113] Carlitz, L. q-Bernoulli numbers and polynomials. *Duke Math. J. 15* (1948), 987–1000.

[114] Carlitz, L. A degenerate Staudt–Clausen theorem. *Arch. Math. 7* (1956), 28–33.

[115] Carr, S., Gangl, H. and Schneps, L. On the Broadhurst–Kreimer generating series for multiple zeta values. In *Feynman Amplitudes, Periods and Motives*, L. Alvarez-consul, J. I. Burgos-gil and K. Ebrahimi-Fard, Ed., vol. 648 of *Contemp. Math.* Amer. Math. Soc., Providence, RI, 2015, pp. 57–78.

[116] Cartier, P.E. On the structure of free Baxter algebras. *Adv. in Math. 9* (1972), 253–265.

[117] Cartier, P.E. Fonctions polylogarithms, nombres polyzêtas et groups prounipotents (in French). *Astérisque 282* (2002), 137–173.

[118] Cartier, P.E. On the double zeta values. In *Galois–Teichmüller Theory and Arithmetic Geometry*, H. Nakamura et al., Ed., Adv. Studies in Pure Math. 68. Math. Soc. Japan, Tokyo, 2012, pp. 91–119.

[119] Castillo Medina, J., Ebrahimi-Fard, K., and Manchon, D. On Euler's decomposition formula for qMZVs. *Ramanujan J. 37 (2)* (2015), 365–389.

[120] Castillo Medina, J., Ebrahimi-Fard, K. and Manchon, D. Unfolding the double shuffle structure of q-multiple zeta values. *Bull. Aust. Math. Soc. 91 (3)* (2015), 368–388.

[121] Cathelineau, J.L. Remarques sur les différentielles des polylogarithmes uniformes (in French). *Ann. Inst. Fourier (Grenoble) 46(5)* (1996), 1327–1347.

[122] Cattani, E. *Introduction to variations of hodge structure.* Summer school on Hodge theory, ICTP, available at `www.math.umass.edu/~cattani/ICTP/cattani_vhs.pdf`, June 2010.

[123] Chamberland, M. and Dilcher, K. Divisibility properties of a class of binomial sums. *J. of Number Theory 120* (2006), 349–371.

[124] Chang, C.-Y. Linear independence of monomials of multizeta values in positive characteristic. *Compo. Math. 150* (2014), 1789–1808.

[125] Chang, C.-Y., Papanikolas, M.A. and Yu, J. An effective criterion for Eulerian multizeta values in positive characteristic. *arXiv:1411.0124*

(2014).
[126] CHARLTON, S. $\zeta(\{\{2\}^m, 1, \{2\}^m, 3\}^n, \{2\}^m)/\pi^{4n+2m(2n+1)}$ is rational. *J. Number Theory 148 (3)* (2015), 463–477.
[127] CHATZISTAMATIOU, A. On integrality of p-adic iterated integrals. *arXiv:1501.05760* (2015).
[128] CHAVEZ, F., AND DUHR, C. Three-mass triangle integrals and single-valued polylogarithms. *J. High Energy Phy. 114* (November 2012).
[129] CHEN, H.-J. Anderson Thakur polynomials and multizeta values in positive characteristic. *arXiv:1506.06463* (2015).
[130] CHEN, H.-J. On shuffle of double zeta values over $\mathbb{F}_q[t]$. *J. Number Theory 148 (3)* (2015), 153–163.
[131] CHEN, K.-T. Iterated integrals and exponential homomorphisms. *Proc. London Math. Soc. 4(3)* (1954), 502–512.
[132] CHEN, K.-T. Algebras of iterated path integrals and fundamental groups. *Trans. Amer. Math. Soc. 156* (1971), 359–379.
[133] CHEN, K.-T. Iterated path integrals. *Bull. Amer. Math. Soc. 83* (1977), 831–879.
[134] CHEN, K.-T., FOX, R.H. AND LYNDON, R.C. Free differential calculus. IV The quotient groups of the lower central series. *Ann. Math. 68(1)* (1958), 81–95.
[135] CHEN, K.-W. Sums of products of generalized Bernoulli polynomials. *Pacific J. Math. 208(1)* (2003), 39–52.
[136] CHEN, K.-W. Applications of stuffle product of multiple zeta values. *J. Number Theory 153* (2015), 107–116.
[137] CHEN, K.-W. Generalized harmonic numbers and Euler sums. *preprint* (2015).
[138] CHEN, K.-W., CHUNG, C.-L., AND EIE, M. Combinatorial implications of decomposition theorems of multiple zeta values. *J. Combin. Number Theory 7 (2)* (2015).
[139] CHEN, K.-W., CHUNG, C.-L. AND EIE, M. Sum formulas and duality theorems of multiple zeta values. *J. Number Theory 158* (2016), 33–53.
[140] CHETYRKIN, K.G. AND TKACHOV, F.V. Integration by parts: The algorithm to calculate β-functions in 4 Loops. *Nucl. Phys. B192* (1981), 159–204.
[141] CHETYRKIN, K.G., KATAEV, A.L. AND TKACHOV, F.V. New approach to evaluation of multiloop Feynman integrals: The Gegenbauer polynomial x-space technique. *Nucl. Phys. B174* (1980), 345–377.
[142] CHOIE, Y. AND MATSUMOTO, K. Functional equations for double series of Euler type with coefficients. *arXiv:1306.0987* (2013).
[143] COHEN, H. *Number Theory. Vol. II. Analytic and Modern Tools*, vol. 240 of *Graduate Texts in Math.* Springer, New York, 2007.
[144] COHEN, H., LEWIN, L. AND ZAGIER, D. A sixteenth-order polylogarithm ladde. *Experimental Math. 1(1)* (1992), 25–34.
[145] COMBARIZA, G. A few conjectures about the multiple zeta values. *arXiv:1207.1735* (2012).
[146] COMTET, L. *Advanced Combinatorics.* D. Reidel Publ. Company, 1974.

[147] CONNES, A. AND KREIMER, D. Renormalization in quantum field theory and the Riemann–Hilbert problem. I. The Hopf algebra structure of graphs and the main theorem. *Comm. Math. Phys. 210 (1)* (2000), 249–273.

[148] CONNES, A. AND KREIMER, D. Renormalization in quantum field theory and the Riemann–Hilbert problem. II. The β-function, diffeomorphisms and the renormalization group. *Comm. Math. Phys. 216 (1)* (2001), 215–241.

[149] CONRAD, M. Basen vo Moduln mit Anwendung auf Kresenheiten und Stickelbergerelemente (in German). *Ph. D. thesis* (1997).

[150] CONRAD, M. Construction of bases for the group of cyclotomic units. *J. Number Theory 81* (2000), 1–15.

[151] COPPO, M.-A. AND CANDELPERGHER, B. The Arakawa–Kaneko zeta function. *Ramanujan J. 22* (2010), 153–162.

[152] COSTIN, O. AND GAROUFALIDIS, S. Resurgence of the fractional polylogarithms. *Math. Res. Lett. 16(5)* (2009), 817–826.

[153] CRANDALL, R. *Topics in Advanced Scientific Computation.* Springer, 1996.

[154] CRANDALL, R. AND BUHLER, J. On the evaluation of Euler sums. *Experiment Math. 3* (1994), 275–285.

[155] CURRY, C., EBRAHIMI-FARD, K., MALHAM, S. AND WIESE, A. Levy processes and quasi-shuffle algebras. *Stochastics: An International Journal of Probability and Stochastic Processes 86(4)* (2014), 632–642.

[156] DAVYDYCHEV, A.I. AND DELBOURGO, R. A geometrical angle on Feynman integrals. *J. Math. Phys. 39* (1998), 4299–4334.

[157] DELIGNE, P. letters to the Broadhurst on April 22 and 23, in response to [87].

[158] DELIGNE, P. *Equations differentielles á points singuliers réguliers (in French)*, vol. 163 of *Lecture Notes in Math.* Springer-Verlag, Berlin-Heidelberg-New York, 1970.

[159] DELIGNE, P. Le groupe fondamental de la droite projective moins trois points (in French). In *Galois groups over* $\mathbb{Q}$, vol. 16 of *Publ. MSRI.* Springer-Verlag, 1989, pp. 79–297.

[160] DELIGNE, P. Analytic continuation of multiple zeta functions. *Unpublished manuscript* (2005).

[161] DELIGNE, P. Le groupe fondamental de la $\mathbb{G}_m - \boldsymbol{\mu}_N$, pour $N = 2, 3, 4, 6$ ou 8 (in French). *Publ. Math. Inst. Hautes Etudes Sci. 112* (2010), 101–141.

[162] DELIGNE, P. Multizêtas d'après Francis Brown (in French). *Astérisque 352* (2013).

[163] DELIGNE, P. Letter to Spencer Bloch. *private communication* (April 3, 1984).

[164] DELIGNE, P. AND GONCHAROV, A.B. Groupes fondamentaux motiviques de Tate mixte (in French). *Ann. Sci. Ecole Norm. S. 38(1)* (2005), 1–56.

[165] DEVOTO, A. AND DUKE, D.W. Table of integrals and formulae for Feynman diagram calculations. *Riv. Nuovo Cim. 7(6)* (1984), 1–39.

[166] DIAMOND, F. AND SHURMAN, J. *A First Course in Modular Forms, first edtion*, vol. 228 of *Graduate Texts in Math.* Springer-Verlag, 2007.

[167] DILCHER, K. Sums of products of Bernoulli numbers. *J. number theory 60* (1996), 23–41.

[168] DILCHER, K. Determinant expressions for q-harmonic congruences and degenerate Bernoulli numbers. *Electron. J. Combin. 15(1)* (2008), R63.

[169] DILCHER, K., HESSAMI PILEHROOD, KH. AND HESSAMI PILEHROOD, T. On q-analogues of double Euler sums. *J. Math. Anal. Appl. 420* (2014), 979–988.

[170] DING, S., FENG, L. AND LIU, W. A combinatorial identity of multiple zeta values with even arguments. *Electron. J. Combin. 21(2)* (2014), #P2.27.

[171] DIXON, L., DUHR, C. AND PENNINGTON, J. Single-valued harmonic polylogarithms and the multi-Regge limit. *arXiv:1207.0186* (2012).

[172] DRINFELD, V.G. Quasi-Hopf algebras. *Leningrad Math J. 1* (1989), 1419–1457.

[173] DRINFELD, V.G. On quasitriangular quasi-Hopf algebras and a group closely connected with Gal($\mathbb{Q}/Q$). *Leningrad Math. J. 2* (1991), 829–860.

[174] DRUMMOND, J.M. Generalised ladders and single-valued polylogs. *J. High Energy Phys. 92* (2012).

[175] DRUMMOND, J.M., HENN, J.M. AND TRNKA, J. New differential equations for on-shell loop integrals. *JHEP 1104* (2011), 083.

[176] DUCHAMP, G., HOANG NGOC, M., TOLLU, C., BUI, V. AND NGÔ, Q. (Pure) transcendence bases in φ-deformed shuffle bialgebras. *arXiv:1507.01089* (2015).

[177] DUHR, C. Scattering amplitude, Feynman integrals and multiple polylogarithms. In *Feynman Amplitudes, Periods and Motives*, L. Alvarez-consul, J. I. Burgos-gil and K. Ebrahimi-Fard, Ed., vol. 648 of *Contemp. Math.* Amer. Math. Soc., Providence, RI, Providence, RI, 2015, pp. 109–134.

[178] DUPONT, C. The combinatorial Hopf algebra of motivic dissection polylogarithms. *Adv. Math. 264 (10)* (2014), 646–699.

[179] EBRAHIMI-FARD, K. AND GUO, L. Rota–Baxter algebras and multiple zeta values. *Integers 8*, A4 (2008).

[180] EBRAHIMI-FARD, K., GUO, L. AND KREIMER, D. Integrable renormalization I: the ladder case. *J. Math. Phys. 45* (2004), 3758–3769.

[181] EBRAHIMI-FARD, K., GUO, L. AND KREIMER, D. Integrable Renormalization II: the general case. *Ann. Henri Poincaré 6* (2005), 369–395.

[182] EBRAHIMI-FARD, K., MANCHON, D. AND SINGER, J. Renormalisation of q-regularised multiple zeta values. *arXiv:1508.02144* (2015).

[183] EBRAHIMI-FARD, K., MANCHON, D. AND SINGER, J. The Hopf algebra of (q)multiple polylogarithms with nonpositive arguments. *arXiv:1503.02977* (2015).

[184] EBRAHIMI-FARD, K., MANCHON, D., SINGER, J. AND ZHAO, J. Transfer group for renormalizations of multiple zeta values. *preprint* (2015).

[185] ECALLE, J. Les fonctions résurgentes, Vol. 2. Les fonctions résurgentes appliquées à l'itération (in French). *Publ. Math. Orsay 81.06* (1981), # 283 pp.

[186] ECALLE, J. ARI/GARI, la dimorphie et l'arithmétique des multizetas: un premier bilan (in French). *J. Théorie Nombres Bordeaux 15* (2003), 411–478.

[187] EDEN, R.J., LANDSHOFF, P.V., OLIVE, D.I. AND POLKINGHORNE, J.C.

The Analytic S-Matrix, (first edition). Cambridge University Press, 1966.

[188] Ehrenborg, R. On posets and Hopf algebras. *Adv. Math. 119* (1996), 1–25.

[189] Eie, M. A note on Bernoulli numbers and Shintani generalized Bernoulli polynomials. *Trans. Amer. Math. Soc. 348* (1996), 1117–1136.

[190] Eie, M. *Topics in Number Theory*, vol. 2 of *Monographs in Number Theory*. World Scientific Publishing, River Edge, NJ, 2009.

[191] Eie, M. *The Theory of Multiple Zeta Values with Applications in Combinatorics*, vol. 7. World Scientific Publishing, River Edge, NJ, 2013.

[192] Eie, M. and Lai, K.F. On Bernoulli identities and applications. *Rev. Mat. Iberoamericana 14(1)* (1998), 167–213.

[193] Eie, M. and Wei, C.-S. Generalizations of Euler decomposition and their applications. *J. Number Theory 133(8)* (2013), 2475–2495.

[194] Eie, M., Liaw, W. and Ong, Y.L. A restricted sum formula among multiple zeta values. *J. Number Theory 129(4)* (2009), 908–921.

[195] Eie, M., Yang, F-Y. and Ong, Y.L. Some alternating double sum formulae of multiple zeta values. *Comput. Appl. Math. 29* (2010), 375–391.

[196] Elbaz-Vincent, P. and Gangl, H. On Poly(ana)logs I. *Compos. Math. 130(2)* (2002), 161–214.

[197] Ennola, V. On relations between cyclotomic units. *J. Number Theory 4* (1972), 236–247.

[198] Enriquez, B. On some relations between generalized associators. *arXiv:0609440* (2006).

[199] Enriquez, B. Quasi-reflection algebras and cyclotomic associators. *Selecta Math. New Series 13* (2007), 391–463.

[200] Enriquez, B. Analogues elliptiques des nombres multizétas (in French). *arXiv:1301.3042* (2013).

[201] Enriquez, B. Elliptic associators. *Selecta Math. New Series 20* (2014), 491–584.

[202] Enriquez, B. and Furusho, H. Mixed pentagon, octagon and Broadhurst duality equation. *J. Pure Appl. Alg. 216* (2012), 982–995.

[203] Enriquez, B. and Gavarini, F. A formula for the logarithm of the KZ associator. *SIGMA 2* (2006), #80.

[204] Ernst, T. A method for q-calculus. *J. Nonlinear Math. Phys. 10 (4)* (2003), 487–525.

[205] Espie, M., Novelli, J.-C. and Racinet, G. Formal computations about multiple zeta values. In *From combinatorics to dynamical systems*, vol. 3. de Gruyter, Berlin, 2003, pp. 1–16.

[206] Essouabri, D. Singularité de séries de Dirichlet associéees à des polynômes de plusieurs variables et applications en théorie analytique des nombres (in French). *Thése* (1995).

[207] Essouabri, D. Singularité de séries de Dirichlet associées à des polynômes de plusieurs variables et applications en théorie analytique des nombres (in French). *Ann. Inst. Fourier Grenoble 47 (2)* (1997), 429–483.

[208] Essouabri, D., Matsumoto, K. and Tsumura, H. Multiple zeta-functions associated with linear recurrence sequences and the vectorial sum

formula. *Canad. J. Math. 63 (2)* (2011), 241–276.

[209] Eswarathasan, A. and Levine, E. p-Integral harmonic sums. *Discrete Math. 91* (1991), 249–257.

[210] Euler, L. Meditationes circa singulare serierum genus, Novi Comm. Acad. Sci. Petropol., **20** (1776), 140–186. *reprinted in Opera Omnia, Ser. I 15* (1927), 217–267.

[211] Euler, L. Demonstratio insignis theorematis numerici circa unicias potestatum binomialium, Nova Acta Acad. Sci. Petropol. **15** (1799/1802), 33–43. *reprinted in Opera Omnia, Ser. I 16(2)* (1935), 104–116.

[212] Ferguson, H. and Bailey, D.H. A polynomial time numerically stable integer relation algorithm. *RNR Technical Report RNR-91-032;* (jul 1992).

[213] Fomin, S. and Zelevinsky, A. Y-systems and generalized associahedra. *Ann. Math. 158(3)* (2003), 977–1018.

[214] Frenkel, E. and Szenes, A. Thermodynamic Bethe ansatz and dilogarithm identities. I. *Math. Res. Lett. 2(6)* (1995), 677–693.

[215] Furusho, H. The multiple zeta value algebra and the stable derivation algebra. *Publ. Res. Inst. Math. Sci. 39* (2003), 695–720.

[216] Furusho, H. p-adic multiple zeta values I.: p-adic multiple polylogarithms and the p-adic KZ equation. *Inv. Math. 155* (2004), 253–286.

[217] Furusho, H. Multiple zeta values and Grothendick–Teichmüller groups. In *Primes and Knots (Baltimore, 2003)*, T. Kohno and M. Morishita, Ed., vol. 416 of *Contemp. Math.* Amer. Math. Soc., Providence, RI, Providence, RI, 2006, pp. 49–82.

[218] Furusho, H. p-adic multiple zeta values II – tannakian interpretations. *Amer. J. Math. 129* (2007), 1105–1144.

[219] Furusho, H. Pentagon and hexagon equations. *Ann. Math. 171* (2010), 545–556.

[220] Furusho, H. Double shuffle relation for associators. *Ann. Math. 174* (2011), 341–360.

[221] Furusho, H. Geometric interpretation of double shuffle relation for multiple L-values. In *Galois–Teichmüller Theory and Arithmetic Geometry*, H. Nakamura et. Al., Ed., vol. 68 of *Adv. Studies in Pure Math.* Math. Soc. Japan, Tokyo, 2012, pp. 163–187.

[222] Furusho, H. On relations among multiple zeta values obtained in knot theory. *arXiv:1501.06638* (2015).

[223] Furusho, H. and Jafari, A. Algebraic cycles and motivic generic iterated integrals. *Math. Res. Lett. 14 (6)* (2007), 923–942.

[224] Furusho, H. and Jafari, A. Regularization and generalized double shuffle relations for p-adic multiple zeta values. *Compos. Math. 143* (2007), 1089–1107.

[225] Furusho, H., Komori, Y., Matsumoto, K. and Tsumura, H. Desingularization of complex multiple zeta functions. *arXiv:1508.06920* (2015).

[226] Furusho, H., Komori, Y., Matsumoto, K. and Tsumura, H. Fundamentals of p-adic multiple L-functions and evaluation of their special values. *arXiv:1508.07185* (2015).

[227] Gangl, H. Some Computations in weight 4 motivic cohomology. In *Regu-*

lators in Analysis, Geometry and Number Theory, vol. 171 of *Progress in Math.* Birkhäuser, Boston, 2000, pp. 117–125.

[228] Gangl, H. Functional equations of higher logarithms. *Selecta Math. 9* (2003), 361–379.

[229] Gangl, H. Functional equations and ladders for polylogarithms. *Comm. Num. Theory Phys. 7(3)* (2013), 397–410.

[230] Gangl, H. and Müller-Stach, S. Polylogarithmic identities in cubical higher Chow groups. In *Algebraic K-theory (Seattle, WA, 1997)*. Amer. Math. Soc., Providence, RI, 1999, pp. 25–40.

[231] Gangl, H., Goncharov, A.B. and Levin, A. Multiple logarithms, trees and algebraic cycles. In *Frontiers in Number Theory, Physics, and Geometry II*, P.E. Cartier, B. Julia, P. Moussa and P. Vanhove, Ed. Springer, 2007, pp. 759–776.

[232] Gangl, H., Goncharov, A.B. and Levin, A. Multiple polylogarithms, polygons, trees and algebraic cycles. In *Proc. Symp. Pure Math.*, vol. 80.2. Amer. Math. Soc., Providence, RI, 2009, pp. 547–594.

[233] Gangl, H., Kaneko, M. and Zagier, D. Double zeta values and modular forms. In *Automorphic Forms and Zeta Functions*, S. Böcherer, T. Ibukiyama, M. Kaneko and F. Sato, Ed. World Scientific Publishing, River Edge, NJ, 2006, pp. 71–106.

[234] Gardiner, A. Four problems on prime power divisibility. *Amer. Math. Monthly 95* (1988), 926–931.

[235] Gelfand, I.M., Krob, D., Lascoux, A., Leclerc, B., Retakh, V.S. and Thibon, J.-Y. Noncommutative symmetric functions. *Adv. Math. 112* (1995), 218–348.

[236] Gessel, I. Multipartite P-partitions and inner product of skew Schur functions. In *Contemp. Math.*, vol. 34. Amer. Math. Soc., Providence, RI, 1984, pp. 289–301.

[237] Gessel, I. and Reutenauer, C. Counting permutations with given cycle-structure and, descent set. *J. Combin. Theory Ser. A 64* (1993), 189–215.

[238] Glaisher, J.W.L. On the residues of the sums of products of the first $p-1$ numbers, and their powers, to modulus p^2 or p^3. *Quart. J. Math. 31* (1900), 321–353.

[239] Glaisher, J.W.L. On the residues of the sums of the inverse powers of numbers in arithmetical progression. *Quart. J. Math. 32* (1900), 271–288.

[240] Glanois, C. Motivic unipotent fundamental groupoid of $\mathbb{G}_m \setminus \mu_N$ for $N = 2, 3, 4, 6, 8$ and Galois descents. *arXiv:1411.4947* (2014).

[241] Glanois, C. Periods of the motivic fundamental groupoid of. *Ph.D. thesis, Université Pierre et Marie Curie* (2016).

[242] Golden, J., Goncharov, A.B., Spradlin, M., Vergu, C., and Volovich, A. Motivic amplitudes and cluster coordinates. *J. High Energy Phys. 91* (2014).

[243] Golden, J., Paulos, M., Spradlin, M., and Volovich, A. Cluster polylogarithms for scattering amplitudes. *J. Physics A: Math. Theoretical* (November 2014).

[244] Goncharov, A.B. Polylogarithms and motivic Galois groups. In *Motives,*

vol. 55 (part 2) of *Proc. Symp. Pure Math.* Amer. Math. Soc., Providence, RI, 1994, pp. 43–97.

[245] GONCHAROV, A.B. Geometry of configurations, polylogarithms, and motivic cohomology. *Adv. Math. 114* (1995), 197–318.

[246] GONCHAROV, A.B. Polylogarithms in arithmetic and geometry. In *Proc. ICM, Vol. **1** (Zurich, 1994)*. Birkhäuser, Basel, 1995, pp. 374–387.

[247] GONCHAROV, A.B. The double logarithm and Manin's complex for modular curves. *Math. Res. Letters 4* (1997), 617–636.

[248] GONCHAROV, A.B. Mixed elliptic motives. In *London Math. Soc. Lecture Notes*, vol. 254. Cambridge Univ. Press, Cambridge, 1998, pp. 147–221.

[249] GONCHAROV, A.B. Multiple polylogarithms, cyclotomy and modular complexes. *Math. Res. Letters 5* (1998), 497–516.

[250] GONCHAROV, A.B. Multiple ζ-values. In *Galois groups, and geometry of modular varieties, European Congress of Math. (Barcelona, 2000), Vol. **I***, vol. 201 of *Progr. Math.* Birkhäuser, Basel, 2001, pp. 361–392.

[251] GONCHAROV, A.B. The dihedral Lie algebaras and Galois symmetries of $\pi_1(\mathbb{P}^1 - \{0, \infty, \mu_N\})$. *Duke J. Math. 110(3)* (2001), 397–487.

[252] GONCHAROV, A.B. AND ZHAO, J. Grassmannian trilogarithms. *Compos. Math. 127* (2001), 83–108.

[253] GONCHAROV, A.B., SPRADLIN, M., C. VERGU AND VOLOVICH, A. Classical polylogarithms for amplitudes and Wilson loops. *Phys. Rev. Lett. 105* (2010), 151605.

[254] GONZALEZ-LORCA, J. Série de Drinfel'd, monodromie et algèbres de Hecke (in French). *preprint ENS LMENS-98-12* (1998).

[255] GOSS, D. *Basic Structures of Function Field Arithmetic*, vol. 35 of *Ergebnisse der Mathematik und ihrer Grenzgebiete (3) [Results in Math. and Related Areas (3)]*. Springer-Verlag, Berlin, 1996.

[256] GRANVILLE, A. A decomposition of Riemann's zeta-function. In *Analytic Number Theory*, Y. Motohashi, Ed., vol. 247 of *London Math. Soc. Lecture Note Ser.* Cambridge Univ. Press, Cambridge, 1997, pp. 95–101.

[257] GUO, L. What is a Rota–Baxter algebra. *Notice Amer. Math. Soc. 56* (2009), 1436–1437.

[258] GUO, L. *An Introduction to Rota–Baxter Algebra*, vol. 4 of *Surveys of Modern Math.* International Press of Boston, Incorporated, 2012.

[259] GUO, L. AND KEIGHER, W. On differential Rota–Baxter algebras. *J. Pure and Appl. Alg. 212* (2008), 522–540.

[260] GUO, L. AND XIE, B. Weighted sum formula for multiple zeta values. *J. Number Theory 129 (11)* (2009), 2747–2765.

[261] GUO, L. AND XIE, B. The shuffle relation of fractions from multiple zeta values. *Ramanujan J. 25* (2011), 307–317.

[262] GUO, L. AND XIE, B. Explicit double shuffle relations and a generalization of Euler's decomposition formula. *J. Alg. 380* (2013), 46–77.

[263] GUO, L. AND XIE, B. Structure theorems of mixable shuffle algebras and free commutative Rota–Baxter algebras. *Comm. Algebra 41* (2013), 2629–2649.

[264] GUO, L. AND ZHANG, B. Renormalization of multiple zeta values. *J. Alg.*

319(9) (2008), 3770–3809.

[265] Guo, L., Lei, P. and Ma, B. Applications of shuffle product to restricted decomposition formulas for multiple zeta values. *J. Number Theory 144* (2014), 219–233.

[266] Guo, L., Lei, P. and Zhao, J. Families of weighted sum formulas for multiple zeta values. *Intl. J. Number Theory 11 (3)* (2015), 997–1025.

[267] Hain, R. Classical polylogarithms. In *Motives*, vol. 55 (part 2) of *Proc. Symp. Pure Math.* Amer. Math. Soc., Providence, RI, 1994, pp. 3–42.

[268] Hain, R. Classical polylogarithms. *Lectures on the Hodge-de Rham theory of the fundamental group of* $\mathbb{P} - \{0, 1, \infty\}$. *Lecture notes dated March 10* (2005).

[269] Hain, R. and MacPherson, R. Higher logarithms. *Ill. J. Math 34* (1990), 392–475.

[270] Hain, R. and Zucker, S. A guide to unipotent variations of mixed Hodge structure. In *Hodge theory (Sant Cugat, 1985)*, vol. 1246 of *Lecture Notes in Math.* Springer, Berlin–New York, 1987, pp. 92–106.

[271] Hain, R. and Zucker, S. Unipotent variations of mixed Hodge structure. *Inv. Math. 88* (1987), 83–124.

[272] Hamahata, Y. and Masubuchi, H. Special multi-poly-Bernoulli numbers. *J. Integer Seq. 10* (2007), Article 07.4.1.

[273] Hanamura, M. Mixed motives and algebraic cycles. I. *Math. Res. Lett. 2* (1995), 811–821.

[274] Hayes, B. g-ology, The decades-long quest for the elusive physical constant g. *Amer. Scientist 92, May-June (3)* (2004), 212–214.

[275] Hazewinkel, M. The algebra of quasi-symmetric functions is free over the integers. *Adv. Math. 164* (2001), 283–300.

[276] Heisenberg, W. and Euler, H. Folgerungen aus der Diracschen Theorie des Positrons (Translation available: arXiv:physics/0605038). *Zeitschr. Phys. 98* (1936).

[277] Hessami Pilehrood, Kh. and Hessami Pilehrood, T. On q-analogues of two-one formulas for multiple harmonic sums and multiple zeta star values. *Monatsh. Math. 176 (2)* (2015), 275–291.

[278] Hessami Pilehrood, Kh., Hessami Pilehrood, T. and Tauraso, R. New properties of multiple harmonic sums modulo p and p-analogues of Leshchiner's series. *Trans. Amer. Math. Soc. 366 (6)* (2014), 3131–3159.

[279] Hessami Pilehrood, Kh., Hessami Pilehrood, T. and Tauraso, R. Some q-congruences for homogeneous and quasi-homogeneous multiple q-harmonic sums. *arXiv:1406.4022* (2014).

[280] Hessami Pilehrood, Kh., Hessami Pilehrood, T. and Zhao, J. On q-analogues of some families of multiple harmonic sum and multiple zeta star value identities. *arXiv:1307.7985* (2013).

[281] Hoang Ngoc, M. Dirichlet functions of order n and t parameter. *Discrete Math. 180* (1998), 221–241.

[282] Hoang Ngoc, M. Shuffle algebra and differential Galois group of colored polylogarithms. *Nuclear Phys. B135* (2004), 220–224.

[283] Hoang Ngoc, M. *Calcul symbolique non commutatif: Aspects combina-*

toires des fonctions, spéciales et des nombres spéciaux (in French). Presses Académiques Francophones, 2014.
[284] HOANG NGOC, M. AND PETITOT, M. Lyndon words, polylogarithms, and the Riemann ζ function. *Discrete Math. 217* (2000), 273–292.
[285] HOANG NGOC, M., JACOB, G., PETITOT, M. AND N. OUSSOUS. Aspects combinatoires des polylogarithmes et des sommes d'Euler–Zagier (in French). *Sém. Lothar. Combin. 43*, Art. B43e (1999), 29 pp.
[286] HOANG NGOC, M., JACOB, G., PETITOT, M. AND N. OUSSOUS. De l'algèbre des ζ de Riemann multivariées à, l'algèbre des ζ de Hurwitz multivariées (in French). *Sém. Lothar. Combin. 44*, Art. B44i (2000), 21p.
[287] HOANG NGOC, M., PETITOT, M. AND J. VAN DER HOEVEN. Shuffle algebra and polylogarithms. *Discrete Math. 225 (1-3)* (2000), 217–230.
[288] HOFFMAN, M.E. Multiple harmonic series. *Pacific J. Math. 152* (1992), 275–290.
[289] HOFFMAN, M.E. The algebra of multiple harmonic series. *J. Alg. 194* (1997), 477–495.
[290] HOFFMAN, M.E. Quasi-shuffle products. *J. Alg. Combin. 11* (2000), 49–68.
[291] HOFFMAN, M.E. Periods of mirrors and multiple zeta values. *Proc. Amer. Math. Soc. 130* (2002), 971–974.
[292] HOFFMAN, M.E. Quasi-symmetric functions and mod p multiple harmonic sums. *Kyushu J. Math. 69* (2015), 345–366.
[293] HOFFMAN, M.E. The Hopf algebra structure of multiple harmonic sums. *Nuclear Phys. B (Proc. Suppl.) 135* (2004), 214–219.
[294] HOFFMAN, M.E. Algebraic aspects of multiple zeta values. In *Zeta Functions, Topology and Quantum Physics*, T. Aoki et al., Ed., vol. 14 of *Dev. Math.* Springer, New York, 2005, pp. 51–74.
[295] HOFFMAN, M.E. On multiple zeta values of even arguments. *arXiv:1205.7051* (2012).
[296] HOFFMAN, M.E. AND IHARA, K. Quasi-shuffle products revisited. *Max-Planck-Institut für Mathematik Preprint Series 2012 (16)* (2012).
[297] HOFFMAN, M.E. AND MOEN, C. Sums of triple harmonic series. *J. Number Theory 60* (1996), 329–331.
[298] HOFFMAN, M.E. AND OHNO, Y. Relations of multiple zeta values and their algebraic expression. *J. Alg. 262* (2003), 332–347.
[299] HOROZOV, I. Multiple zeta functions and Ideles. *arXiv:0611849v4* (2006).
[300] HOROZOV, I. Multiple Dedekind zeta functions. *J. Reine Angew. Math, to appear* (2011).
[301] HOWARD, F.T. Explicit formulas for degenerate Bernoulli numbers. *Discrete Math. 162* (1996), 175–185.
[302] HUDSON, R. The $\mathbb{Z}_2$-graded sticky shuffle product Hopf algebra. In *Quantum Probability*, M. Bozejko, Ed., vol. 73. Banach Centre Publ., 2006, pp. 237–243.
[303] IGARASHI, M. Cyclic sum of certain parametrized multiple series. *J. Number Theory 131* (2011), 508–518.
[304] IGARASHI, M. A generalization of Ohno's relation for multiple zeta values. *J. Number Theory 132* (2012), 565–578.

[305] IHARA, K. AND OCHIAI, H. Symmetry on linear relations for multiple zeta values. *Nagoya Math. J. 189* (2008), 49–62.

[306] IHARA, K. AND TAKAMUKI, T. The quantum $\mathfrak{g}_2$ invariant and relations of multiple zeta values. *J. Knot Theory Ramifications 10 (7)* (2001), 983–997.

[307] IHARA, K., KAJIKAWA, J., OHNO, Y. AND OKUDA, J. Multiple zeta values vs. multiple zeta-star values. *J. Alg. 332* (2011), 187–208.

[308] IHARA, K., KANEKO, M. AND ZAGIER, D. Derivation and double shuffle relations for multiple zeta values. *Compos. Math. 142* (2006), 307–338.

[309] IHARA, Y. Profinite braid groups, Galois representations and complex multiplications. *Ann. Math. 123* (1986), 43–106.

[310] IHARA, Y. The Galois representation arising from $\mathbb{P}^1 - \{0, 1, \infty\}$ and Tate twists of even degree. In *Galois groups over* $\mathbb{Q}$, vol. 16 of *Publ. MSRI.* Springer-Verlag, 1989, pp. 299–313.

[311] IHARA, Y. On the stable derivation algebra associated with some braid groups. *Israel J. Math. 80 (1-2)* (1992), 135–153.

[312] IMATOMI, K., TANAKA, T. AND KANEKO, M. Multi-poly-Bernoulli numbers and finite multiple zeta values. *J. Integer Sequences 17* (2014), 1–12.

[313] IMATOMI, K., TANAKA, T., TASAKA, K. AND WAKABAYASHI, N. On some combinations of multiple zeta-star values. *Acta Human. Sci. Univ. Sangio Kyotiensis, Natural science series 42* (2013), 1–20.

[314] IRELAND, K. AND ROSEN, M. *A Classical Introduction to Modern Number Theory (second edition)*, vol. 84 of *Graduate Texts in Math.* Spring-Verlag, Berlin, New York, 1990.

[315] JACKSON, F.H. On q-definite integrals. *Quart. J. Pure and Appl. Math. 41* (1908), 193–203.

[316] JACKSON, F.H. On q-functions and a certain difference operator. *Trans. Roy. Soc. Edin. 46* (1908), 253–281.

[317] JAROSSAY, D. Double mélange des multizêtas finis et multizêtas symétrisés (in French). *C. R. Acad. Sci. Paris, Ser. I 352* (2014), 767–771.

[318] JAROSSAY, D. Finite multiple zeta values, formal algebraic relations and the fundamental group of $M_{0,4}$ and $M_{0,5}$ - I. *arXiv:1412.5099* (2014).

[319] JAROSSAY, D. p-adic multiple zeta values and multiple harmonic sums - II: p-adic decompositions of multiple harmonic sums. *arXiv:1501.04893* (2015).

[320] JI, C. A simple proof of a curious congruence by Zhao. *Proc. Amer. Math. Soc. 133* (2005), 3469–3472.

[321] JONQUIÈRE, A. Note sur la série $\sum_{n=1}^{\infty} \frac{x^n}{n^s}$ (in French). *Bull. SocMath. France 17* (1889), 142–152.

[322] JOYNER, S. On an algebraic version of the Knizhnik–Zamolodchikov equation. *Ramanujan J. 28* (2012), 361–384.

[323] JUSKEVIC, A.P. AND WINTER, E. *Leonhard Euler und Christian Goldbach: Briefwechsel 1729–1764 (in German).* Akademie-Verlag, Berlin, 1965.

[324] KAJIKAWA, J. Duality and double shuffle relations of multiple zeta values. *J. Number Theory 121(1)* (2006), 1–6.

[325] KALLEN, G. AND SABRY, A. Fourth order vacuum polarization. *K. Dan. Vidensk. Selsk. Mat.-Fys. Medd. 29* (1955), No. 17.

[326] KANEKO, M. Poly-Bernoulli numbers. *J. Théorie Nombres Bordeaux 9* (1997), 221–228.

[327] KANEKO, M. On an extension of the derivation relation for multiple zeta values. In *The Conference on L-Functions.* World Scientific Publishing, River Edge, NJ, 2007, pp. 89–94.

[328] KANEKO, M. A note on poly-Bernoulli numbers and multiple zeta values. In *Diophantine Analysis and Related Fields: DARF 2007/2008*, T. Komatsu, Ed., vol. 976 of *AIP Conf. Proc.* American Institute of Physics, Melville, NY, 2008, pp. 118–124.

[329] KANEKO, M. Poly-Bernoulli numbers and related zeta functions. In *Algebraic and Analytic Aspects of Zeta Functions and L-Functions*, G. Bhowmik et. al., Ed., vol. 21 of *Math. Soc. Japan Memoirs.* Math. Soc. Japan, 2010, pp. 73–85.

[330] KANEKO, M. Finite multiple zeta values (in Japanese). *RIMS Kôkyûroku Bessatsu, to appear* (2015).

[331] KANEKO, M. AND ARAKAWA, T. On multiple L-values. *J. Math. Soc. Japan 56* (2004), 967–991.

[332] KANEKO, M. AND OHNO, Y. On a kind of duality of multiple zeta-star values. *Internat. J. Number Theory 6(8)* (2010), 1927–1932.

[333] KANEKO, M. AND SAKATA, M. On multiple zeta values of extremal height. *arXiv:1505.01014* (2015).

[334] KANEKO, M. AND TASAKA, K. Double zeta values, double Eisenstein series, and modular forms of level 2. *Math. Ann. 357 (3)* (2013), 1091–1118.

[335] KANEKO, M. AND TSUMURA, H. Multi-poly-Bernoulli numbers and related zeta functions. *arXiv:1503.02156* (2015).

[336] KANEKO, M. AND ZAGIER, D. Finite multiple zeta values. *in preparation.*

[337] KANEKO, M., KUROKAWA, N. AND WAKAYAMA, M. A variation of Euler's approach to values of the Riemann zeta function. *Kyushu J. Math. 57* (2003), 175–192.

[338] KANEKO, M., M. NORO AND TSURUMAKI, K. On a conjecture for the dimension of the space of the multiple zeta values. In *The IMA Volumes in Math. and Appl.*, vol. 148. Software for Algebraic Geometry, 2010, pp. 47–58.

[339] KARPLUS, R. AND KROLL, N.M. Fourth-order corrections in quantum electrodynamics and the magnetic moment of the electron. *Phys. Rev. 76* (1949), 846–847.

[340] KARPLUS, R. AND KROLL, N.M. Fourth-order corrections in quantum electrodynamics and the magnetic moment of the electron. *Phys. Rev. 77* (1950), 536–549.

[341] KAWAGOE, K., WAKAYAMA, M. AND YAMASAKI, Y. q-analogues of the Riemann zeta, the Dirichlet L-functions and a crystal zeta function. *Forum Math. 20* (2008), 1–26.

[342] KAWASHIMA, G. A class of relations among multiple zeta values. *J. Number Theory 129(4)* (2009), 755–788.

[343] KAWASHIMA, G., TANAKA, T. AND WAKABAYASHI, N. Cyclic sum formula for multiple L-values. *J. Alg. 348 (1)* (2011), 336–349.

[344] Kazakov, D. The method of uniqueness, a new powerful technique for multiloop calculations. *Phys. Lett. B133 (6)* (1983), 406–410.

[345] Kelliher, J.P. and Masri, R. Analytic continuation of multiple Hurwitz zeta functions. *Math. Proc. Camb. Phil. Soc. 145* (2008), 605–617.

[346] Kirillov, A.N. Dilogarithm identities. *Prog. Theor. Phys. Suppl. 118* (1995), 61–142.

[347] Klassen, T. and Melzer, E. The thermodynamics of purely elastic scattering theories and conformal perturbation theory. *Nucl. Phys. B 350* (1991), 635–689.

[348] Knauer, J. and Richstein, J. The continuing search for Wieferich primes. *Math. Comp. 74* (2005), 1559–1563.

[349] Kohnen, W. and Zagier, D. Modular forms with rational periods. In *Modular Forms*, R.A. Rankin and E. Horwood, Ed. Chichechester, 1984, pp. 197–249.

[350] Kohno, T. Monodromy representations of braid groups and Yang–Baxter equations. *Ann. Inst. Fourier 37 (4)* (1987), 139–160.

[351] Komori, Y. An integral representation of multiple Hurwitz–Lerch zeta functions and generalized multiple Bernoulli numbers. *Quart. J. Math. 61(4)* (2010), 437–496.

[352] Komori, Y., Matsumoto, K. and Tsumura, H. Functional relations for zeta-functions of root systems. In *Number Theory: Dreaming in Dreams*, T. Aoki, , S. Kanemitsu and J.-Y. Liu, Ed., Proc. 5th China-Japan Seminar. World Scientific Publishing, River Edge, NJ, 2009, pp. 135–183.

[353] Komori, Y., Matsumoto, K., and Tsumura, H. Barnes multiple zeta-functions, Ramanujan's formula, and relevant series involving hyperbolic functions. *J. Ramanujan Math. Soc. 28* (2013), 49–69.

[354] Komori, Y., Matsumoto, K. and Tsumura, H. On Witten multiple zeta-functions associated with semisimple Lie algebras II. *J. Math. Soc. Japan. 62(2)* (2010), 355–394.

[355] Komori, Y., Matsumoto, K. and Tsumura, H. Multiple zeta values and zeta-functions of root systems. *Proc. Japan Acad. Ser. A Math. Sci. 87 (6)* (2011), 103–107.

[356] Komori, Y., Matsumoto, K. and Tsumura, H. On Witten multiple zeta-functions associated with semisimple Lie algebras IV. *Glasgow Math. J. 53* (2011), 185–206.

[357] Komori, Y., Matsumoto, K. and Tsumura, H. On Witten multiple zeta-functions associated with semisimple Lie algebras III. In *Multiple Dirichlet Series, L-functions and Automorphic Forms*, D. Bump, S. Friedberg and D. Goldfeld, Ed., vol. 300 of *Progress in Math.* Birkhäuser, 2012, pp. 215–276.

[358] Kondo, Y., Saito, S. and Tanaka, T. The Bowman-Bradley theorem for multiple zeta-star values. *J. Number Theory 132(9)* (2012), 1984–2002.

[359] Kontsevich, M. Vassiliev's knot invariants. In *I. M. Gel'fand Seminar*, vol. 16(2) of *Adv. Soviet Math.* Amer. Math. Soc., Providence, RI, 1993, pp. 137–150.

[360] Kontsevich, M. The 1/2-logarithm. *reproduced as an appendix [196]*

(1995).
[361] KONTSEVICH, M. AND ZAGIER, D. Periods. In *Math. Unlimited – 2001 and beyond*, Engquist and Schmidt, Ed. Springer, 2001, pp. 771–808.
[362] KOORNWINDER, T.H. On Zeilberger's algorithm and its q-analogue. *J. Comput. Appl. Math. 48* (1993), 91–111.
[363] KRATTENTHALER, C. AND RIVOAL, T. An identity of Andrews, multiple integrals, and very-well-poised hypergeometric series. *Ramanujan J. 13* (2007), 203–219.
[364] KREIMER, D. Knots and divergences. *Phys. Lett. B354 (1?2)* (1995), 117–124.
[365] KREIMER, D. Quantum fields, periods and algebraic geometry. In *Feynman Amplitudes, Periods and Motives*, L. Alvarez-consul, J. I. Burgos-gil, and K. Ebrahimi-Fard, Ed., vol. 648 of *Contemp. Math.* Amer. Math. Soc., Providence, RI, 2015, pp. 11–28.
[366] KUMMER, E.E. Über die Transcendenten, welche aus wiederholten, Integrationen rationaler Formeln entstehen. *J. Reine Angew. Math. 21* (1840), 74–90.
[367] KUROKAWA, N. Multiple zeta functions: an example. In *Zeta Functions in Geometry (Tokyo, 1990)*, vol. 21 of *Adv. Stud. Pure. Math.* Kinokuniya, Tokyo, 1992, pp. 219–226.
[368] KUROKAWA, N., LALIN, M.N. AND OCHIAI, H. Higher Mahler measures and zeta functions. *Acta Arith. 135(3)* (2008), 269–297.
[369] LAGARIAS, J.C. AND LI, W-C. W. The Lerch zeta function I. Zeta integrals. *Forum Math. 24* (2012), 1–48.
[370] LAGARIAS, J.C. AND LI, W-C. W. The Lerch zeta function II. Analytic continuation. *Forum Math. 24* (2012), 49–84.
[371] LAGARIAS, J.C. AND LI, W-C. W. The Lerch zeta function III: Polylogarithms and special values. *arXiv:1506.06161* (2015).
[372] LALIN, M.N. Some examples of Mahler measures as multiple polylogarithms. *J. Number Theory 103* (2003), 85–108.
[373] LALIN, M.N. On a certain combination of colored multizeta values. *J. Ramanujan Math. Soc. 20* (2006), 115–127.
[374] LAMBERT, J. Mémoire sur quelques propriétés remarquables des quantités transcendantes circulaires et logarithmiques (in French). In *Pi: A Source Book (3rd ed.)*, L. Berggren, J. Borwein and P. Borwein, Ed. Springer-Verlag, New York, 2004, pp. 129–140.
[375] LANG, S. *Introduction to Modular Forms (corrected 3rd printing).* Grundlehren der mathematischen Wissenschaften. Springer-Verlag, 2001.
[376] LAPORTA, S. AND REMIDDI, E. The analytical value of the electron $(g-2)$ at order α^3 in QED. *Phys. Lett. B 379* (1996), 283–291.
[377] LAPPO-DANILEVSKY, J.A. Théorie algorithmique des corps de Riemann (in French). *Rec. Math. Moscou 34 (6)* (1927), 113–146.
[378] LAPPO-DANILEVSKY, J.A. *Memoires sur la theorie des systemes des équations différentieles linéaires (in French).* Amer. Math. Soc. Chelsea Publishing, 1953.
[379] LARA RODRIUEZ, J.A. Some conjectures and results about multizeta values

over $\mathbb{F}_q[t]$. *J. Number Theory 130* (2010), 1013–1023.
[380] LARA RODRIUEZ, J.A. Relations between multizeta values in characteristic p. *J. Number Theory 131* (2011), 2081–2099.
[381] LARA RODRIUEZ, J.A. Special relations between multizeta values and parity results. *arXiv:1108.4726* (2011).
[382] LARA RODRIUEZ, J.A., AND THAKUR, D.S. Zeta-like multizeta values for $\mathbb{F}_q[t]$. *Indian J. Pure Appl. Math. 45(5)* (2014), 785–798.
[383] LAVOIE, J., GRONDIN, F., RATHIE, A. AND ARORA, K. Generalizations of Dixon's theorem on the sum of a ${}_3F_2$. *Math. Comp. 62* (1994), 267–272.
[384] LE, T.Q.T. AND MURAKAMI, J. Kontsevich's integral for the Homfly polynomial and relations between values of multiple zeta functions. *Topology Appl. 62* (1995), 193–206.
[385] LE, T.Q.T. AND MURAKAMI, J. Kontsevich's integral for the Kauffman polynomial. *Nagoya Math. J. 142* (1996), 39–65.
[386] LEE, R. AND SCHWERMER, J. Cohomology of arithmetic subgroups of SL_3 at infinity. *J. Reine Angew. Math. 330* (1982), 100–131.
[387] LESHCHINER, D. Some new identities for $\zeta(k)$. *J. Number Theory 13* (1981), 355–362.
[388] LEUDESDORF, C. Some results in the elementary theory of numbers. *Proc. London Math. Soc. 20* (1889), 199–212.
[389] LEURENT, S. AND VOLIN, D. Multiple zeta functions and double wrapping in planar $N = 4$ SYM. *arXiv:1302.1135* (2013).
[390] LEVINE, M. *Mixed Motives.* Amer. Math. Soc., Providence, RI, 1998.
[391] LEWIN, L. *Dilogarithms and associated functions.* Macdonald, London, 1958.
[392] LEWIN, L. *Polylogarithms and Associated Functions.* Elsevier Sci. Publishers, New York, New York, 1981.
[393] LEWIN, L. *Structural properties of polylogarithms*, vol. 37 of *Math. Surverys and Monographs.* Amer. Math. Soc., Providence, RI 1991, 1991.
[394] LI, Z. Sum of multiple zeta values of fixed weight, depth and i-height. *Math. Z. 258* (2008), 133–142.
[395] LI, Z. Higher order shuffle regularization of multiple zeta values. *Proc. Amer. Math. Soc. 138* (2010), 2321–2333.
[396] LI, Z. Sum of multiple q-zeta values. *Proc. Amer. Math. Soc. 138* (2010), 505–516.
[397] LI, Z. On a conjecture of Kaneko and Ohno. *Pacific J. Math. 257* (2012), 419–430.
[398] LI, Z. Another proof of Zagier's evaluation formula of the multiple zeta values $\zeta(2,\dots,2,3,2,\dots,2)$. *Math. Research Lett. 20* (2013), 947–950.
[399] LI, Z. Regularized double shuffle and Ohno–Zagier relations of multiple zeta values. *J. Number Theory 133* (2013), 596–610.
[400] LI, Z. Some identities in the harmonic algebra concerned with multiple zeta values. *Int. J. Number Theory 9* (2013), 783–798.
[401] LINDEMANN, F. Über die Ludolph'sche Zahl. *Sitzungsber. Königl. Preuss. Akad. Wissensch. zu Berlin 2 1882* (1882), 679–682.
[402] LINEBARGER, E. AND ZHAO, J. A family of multiple harmonic sum and

multiple zeta star value identities. *Mathematika 61 (1)* (2015), 63–71.

[403] LIPSTEIN, A.E. AND MASON, L. From dlogs to dilogs; the super Yang–Mills MHV amplitude revisited. *arXiv:1307.1443* (2013).

[404] LOTHAIRE, M. *Combinatorics of Words*, vol. 17 of *Encycl. Math. Appl.* Addison-Wesley, 1983.

[405] MA, D. Period polynomial relations between double zeta values of odd weight. *arXiv:1409.7717* (2014).

[406] MACHIDE, T. Weighted sums with two parameters of multiple zeta values and their formulas. *Int. J. Number Theory 8* (2012), 1903–1921.

[407] MACHIDE, T. Extended double shuffle relations and the generating function of triple zeta values of any fixed weight. *Kyushu Math. J. 67* (2013), 281–307.

[408] MACHIDE, T. Generators for vector spaces consisting of double zeta values with even weight. *J. Number Theory 133* (2013), 2240–2246.

[409] MACHIDE, T. Some restricted sum formulas for double zeta values. *Proc. Japan Acad. Ser. A 89* (2013), 51–54.

[410] MACHIDE, T. Identities involving cyclic sums of regularized multiple zeta values each of depth less than 5. *arXiv:1502.02505* (2015).

[411] MACHIDE, T. Use of the generating funtion to generalize the sum formula for quadruple zeta values. *arXiv:1503.01194* (2015).

[412] MALVENUTO, C. AND REUTENAUER, C. Duality between quasi-symmetric functions and the Solomon descent algebra. *J. Alg. 177* (1994), 967–982.

[413] MANCHON, D. Renormalised multiple zeta values which respect quasi-shuffle relations. In *Combinatorics and physics*, vol. 539 of *Contemp. Math.* Amer. Math. Soc., Providence, RI, 2011, pp. 255–267.

[414] MANCHON, D. AND PAYCHA, S. Nested sums of symbols and renormalised multiple zeta functions. *Int. Math. Res. Notices* (2010), 4628–4697.

[415] MARKETT, C. Triple sums and the Riemann zeta function. *J. Number Theory 48* (1994), 113–132.

[416] MASRI, R. Multiple Dedekind zeta functions and evaluations of extended multiple zeta values. *J. Number Theory 115* (2005), 295–309.

[417] MASRI, R. Multiple zeta values over global function fields. In *Multiple Dirichlet Series, Automorphic Forms and Analytic Number Theory (Bretton Woods, 2005)*, S. Friedberg et al., Ed., vol. 75 of *Proc. Symp. Pure Math.* Amer. Math. Soc., Providence, RI, 2006, pp. 157–175.

[418] MATSUMOTO, K. On the analytic continuation of various multiple zeta functions. In *Number Theory for the Millenium II*, M.A. Bennett et al., Ed. A. K. Peters, 2002, pp. 417–440.

[419] MATSUMOTO, K. The analytic continuation and the asymptotic behaviviour of certain multiple zeta functions I. *J. Number Theory 101* (2003), 223–243.

[420] MATSUMOTO, K. Functional equations for double zeta-functions. *Math. Proc. Camb. Phil. Soc. 136(1)* (2004), 1–7.

[421] MATSUMOTO, K. Analytic properties of multiple zeta-functions in several variables. In *Number Theory: Tradition and Modrnization*, W. Zhang and Y. Tanigawa, Ed., Proc. 3rd China-Japan Seminar. Springer, 2006, pp. 153–

173.

[422] MATSUMOTO, K. AND TSUMURA, H. On Witten multiple zeta-functions associated with semisimple Lie algebras I. *Annales de l'institut Fourier 56 (5)* (2006), 1457–1504.

[423] MATSUMOTO, K., NAKAMURA, T. AND TSUMURA, H. Functional relations and special values of Mordell–Tornheim triple zeta and L-functions. *Proc. Amer. Math. Soc. 136(3)* (2008), 2135–2145.

[424] MATTHES, N. Elliptic double zeta values. *arXiv:1509.08760* (2015).

[425] MCCOY, M., THIELEN, K., WANG, L., AND ZHAO, J. A family of super congruences involving multiple harmonic sums. *Intl. J. Number Theory* (2016).

[426] MESTROVIC, R. Wolstenholme's theorem: its generalizations and extensions in the last hundred and fifty years (1862–2012). *arXiv:1111.3057* (2011).

[427] MOLL, V.H., JIU, L. AND VIGNAT, C. A symbolic approach to multiple zeta values at the negative integers. *arXiv:1503.04257* (2015).

[428] MORTENSON, E. A p-adic supercongruence conjecture of van Hamme. *Proc. Amer. Math. Soc. 136* (2008), 4321–4328.

[429] MUNETA, S. On some explicit evaluations of multiple zeta-star values. *J. Number Theory 128(9)* (2008), 2538–2548.

[430] MUNETA, S. A note on evaluations of multiple zeta values. *Proc. Amer. Math. Soc. 137(3)* (2009), 931–935.

[431] MURAHARA, H. A note on finite real multiple zeta values. *arXiv:1411.1123* (2014).

[432] MURTY, M.R. AND SINHA, K. Multiple Dirichlet series, automorphic forms, and analytic number theory. In *Proc. Symp. Pure Math.*, vol. 75. Amer. Math. Soc., Providence, RI, 2006, pp. 135–156.

[433] NAKAMURA, T. Bernoulli numbers and multiple zeta values. *Proc. Japan Acad. Ser. A 81* (2005), 21–22.

[434] NAKAMURA, T. Restricted and weighted sum formulas for double zeta values of even weight. *Šiauliai Math. Seminar 4* (2009), 151–155.

[435] NAKAMURA, T. Zeros and the universality for the Euler–Zagier–Hurwitz type of multiple zeta-functions,. *Bull. London Math. Soc. 41* (2009), 691–700.

[436] NAKAMURA, T. Non-existence of the Euler product of the Euler–Zagier L- and L-star functions. *preprint* (2015).

[437] NAKAMURA, T., AND PAŃKOWSKI, Ł. Any non-monomial polynomial of the Riemann zeta-function has complex zeros off the critical line. *arXiv:1212.5890* (2012).

[438] NAKAMURA, T. AND TASAKA, K. Remarks on double zeta values of level 2. *J. Number Theory 133* (2013), 48–54.

[439] NAKANISHI, N. *Graph theory and Feynman integrals.* Gordon and Breach Science Publishes, New York, London, Paris, 1971.

[440] NEUMANN,W. AND ZAGIER, D. Volumes of hyperbolic three-manifolds. *Topology 24* (1985), 307–332.

[441] NIELDSEN, N. *Handbuch der Theorie der Gammafunktion, Leipzig.* Internet

Archive, 1906.

[442] OESTERLÉ, J. Polylogarithmes (in French). *Astérisque 216* (1993), 49–67.

[443] OHNO, Y. A generalization of the duality and sum formulas on the multiple zeta values. *J. Number Thery 74* (1999), 39–43.

[444] OHNO, Y. Sum relations for multiple zeta values. In *Zeta Functions, Topology and Quantum Physics*, vol. 14 of *Dev. Math.* Springer, 2005, pp. 131–144.

[445] OHNO, Y. AND OKUDA, J. On the sum formula for the -analogue of non-strict multiple zeta values. *Proc. Amer. Math. Soc. 135* (2007), 3029–3037.

[446] OHNO, Y. AND WAKABAYASHI, N. Cyclic sum of multiple zeta values. *Acta Arith. 123(3)* (2006), 289–295.

[447] OHNO, Y. AND ZAGIER, D. Multiple zeta values of fixed weight, depth, and height. *Indag. Math. 12* (2001), 483–487.

[448] OHNO, Y. AND ZUDILIN, W. Zeta stars. *Commun. Number Theory Phys. 2* (2008), 325–347.

[449] OHNO, Y., OKUDA, J. AND ZUDILIN, W. Cyclic q-MZSV sum. *J. Number Theory 132* (2012), 144–155.

[450] OKAMOTO, T. Generalizations of Apostol–Vu and Mordell–Tornheim multiple zeta functions. *Acta Arith. 140* (2009), 169–187.

[451] OKAMOTO, T. Multiple zeta values related with the zeta-functions of the root system of type B_r. *JP J. Alg. Number Theory Appl. 22* (2011), 1–44.

[452] OKAMOTO, T. Multiple zeta values related with the zeta-function of the root system of type A_2, B_2 and G_2. *Comment. Math. Univ. St. Pauli 61* (2012), 9–27.

[453] OKAMOTO, T. On alternating analogues of the Mordell–Tornheim triple zeta values. *J. Ramanujan Math. Soc. 28(2)* (2013), 247–269.

[454] OKOUNKOV, A. Hilbert schemes and multiple q-zeta values. *Funct. Anal. Appl. 48* (2014), 138–144.

[455] OKUDA, J. Duality formulas of the special values of multiple polylogarithms. *Bull. London Math. Soc. 37* (2005), 230–242.

[456] OKUDA, J. AND TAKEYAMA, Y. On relations for the q-multiple zeta values. *Ramanujan J. 14* (2007), 379–387.

[457] OKUDA, J. AND UENO, K. New approach to Ohno relations for multiple zeta values. *arXiv:0106148* (2001).

[458] OKUDA, J. AND UENO, K. Relations for multiple zeta values and Mellin transforms of multiple polylogarithms. *Publ. Res. Inst. Math. Sci. 40* (2004), 537–564.

[459] OKUDA, J. AND UENO, K. The sum formula for multiple zeta values and connection problem of the formal Knizhnik–Zamolodchikov equation. In *Zeta functions, topology and quantum physics*, vol. 14 of *Dev. Math.* Springer, New York, 2005, pp. 145–170.

[460] ONG, Y.L., EIE, M. AND LIAW, W.-C. Explicit evaluation of triple Euler sums. *Intl. J. Number Theory 4 (3)* (2008), 437–451.

[461] ONG, Y.L., EIE, M. AND WEI, C-S. Explicit evaluations of quadruple Euler sums. *Acta Arith. 144* (2010), 213–230.

[462] ONOZUKA, T. Analytic continuation of multiple zeta-functions and the

asymptotic behavior at non-positive integers. *Funct. Approx. Comment. Math.. 49 (2)* (2013), 331–348.

[463] Ore, A. and Powell, J.L. Three-photon annihilation of an electron-positron pair. *Phys. Rev. 75* (1949), 1696–1699.

[464] Oyama, K. Ohno's relation for finite multiple zeta values. *arXiv:1506.00833* (2015).

[465] Pan, H. A generalization of Wolstenholme's harmonic series congruence. *Rocky Mountain J. Math. 38* (2008), 1263–1269.

[466] Panzer, E. *Feynman integrals and hyperlogarithms, PhD thesis.* Humboldt-Universität zu Berlin, submitted 26th of August 2014. Updated version at arXiv:1506.07243, 2014.

[467] Panzer, E. Feynman integrals via hyperlogarithms. In *Proceedings of "Loops & Legs in Quantum Field Theory", Weimar (Germany), April 27th - May 2nd.*, Proc. Sci. (LL2014)049. 2014.

[468] Panzer, E. On hyperlogarithms and Feynman integrals with divergences and many scales. *JHEP (71)* (2014), 1403.

[469] Parker, D., Scherlis, A., Spradlin, M. and Volovich, A. Hedgehog bases for A_n cluster polylogarithms and an application to six-point amplitudes. *arXiv:1507.01950* (2015).

[470] Petermann, A. Fourth order magnetic moment of the electron. *Helv. Phys. Acta. 30* (1957), 407–408.

[471] Petitot, M. Table des multiple zeta values jusqu'au poids 16, available at `www.lifl.fr/~petitot/recherche/MZV`.

[472] Petkovsek, M., Wilf, H. and Zeilberger, D. *A=B.* A. K. Peters/CRC Press, 1996.

[473] Petrogradsky, V.M. Multiple zeta functions and asymptotic structure of free abelian groups of finite rank. *J. Pure Appl. Alg. 208 (3)* (2007), 1137–1158.

[474] Pickard, W.F. On polylogarithms. *Publ. Math. Debrecen 15* (1968), 33–43.

[475] Poincaré, H. *Oevres*, vol. 2. Paris, 1916.

[476] Qin, Z. and Yu, F. On Okounkov's conjecture connecting Hilbert schemes of points and multiple q-zeta values. *arXiv:1510.00837* (2015).

[477] Racinet, G. Doubles mélanges des polylogarithmes multiples aux racines de l'unité (in French). *Publ. Math. IHES 95* (2002), 185–231.

[478] Rademacher, H. *Topics in Analytic Number Theory.* Springer-Verlag, 1973.

[479] Ramakrishnan, D. On the monodromy of higher logarithms. *Proc. Amer. Math. Soc. 85* (1982), 596–599.

[480] Ramanujan, S. Modular equations and approximations to π. *Quart. J. Math. (Oxford) 45* (1914), 350–372.

[481] Ree, R. Lie elements and an algebra associated with shuffles. *Ann. Math. 68* (1958), 210–220.

[482] Reutenauer, C. *Free Lie Algebras.* Oxford Science Publications, Oxford, 1993.

[483] Riemann, B. Über die Anzahl der Primzahlen unter einer gegebenen

Grösse. *Monatsberichte der Berliner Akademie, 1859, Gesammelte Werke, Teubner, Leipzig, 1892.*

[484] Rivoal, T. La fonction zêta de Riemann prend une infinité de valeurs irrationnelles aux entiers impairs (in French). *C. R. Acad. Sci. Ser. A. Math. 331* (2000), 267–270.

[485] Rolfsen, D. *Knots and Links.* Amer. Math. Soc. Chelsea Publishing, 2003.

[486] Rosen, J. Multiple harmonic sums and Wolstenholme's theorem. *Intl. J. Number Theory 9 (8)* (2013), 2033–2052.

[487] Rota, G.-C. Baxter algebras and combinatorial identities I. *Bull. Amer. Math. Soc. 5* (1969), 325–329.

[488] Rota, G.-C. Baxter operators, an introduction. In *Gian-Carlo Rota on Combinatorics, Introductory papers and commentaries*, J.P.S. Kung, Ed., Contemp. Mathematicians. Birkhäuser Boston, Boston, MA, 1995.

[489] Rota, G.-C. and Smith, D.A. Fluctuation theory and Baxter algebras. In *Symposia Math.*, vol. IX. Academic Press, 1971, pp. 179–201.

[490] Sadaoui, B. Multiple zeta values at the non-positive integers. *Comptes Rendus Math. 352 (12)* (2014), 977–984.

[491] Saito, S. and Tanaka, T. A note on infinite divisibility of zeta distributions. *Appl. Math. Sci. 6 (30)* (2012), 1455–1461.

[492] Saito, S. and Wakabayashi, N. The Bowman-Bradley type theorem for finite multiple zeta values. *Tohoku Math. J., to appear* (2013).

[493] Saito, S. and Wakabayashi, N. Sum formula for finite multiple zeta values. *J. Math. Soc. Japan 67* (2015), 1069–1076.

[494] Saito, S., Tanaka, T. and Wakabayashi, N. Combinatorial remarks on the cyclic sum formula for multiple zeta values. *J. Integer Sequences 14* (2011), Article 11.2.4.

[495] Sakugawa, K. and Seki, S. On functional equations of finite multiple polylogarithms. *arXiv:1509.07653* (2015).

[496] Sankaranarayanan, A. An identity involving Riemann zeta function. *Indian J. Pure Appl. Math. 18(7)* (1987), 794–800.

[497] Sasaki, Y. Multiple zeta values for coordinatewise limits at non-positive integers. *Acta Arith. 136* (2009), 299–317.

[498] Sasaki, Y. Some formulas of multiple zeta values for coordinate-wise limits at non-positive integers. In *New Directions in Value-Distribution Theory of Zeta and L-Functions, (Wuerzburg Conference, Oct 6-10, 2008)*, R. Steuding and J. Steuding, Ed. Shaker Verlag, 2009, pp. 317–325.

[499] Schlesinger, K.-G. Some remarks on q-deformed multiple polylogarithms. *arXiv:math/0111022* (2011).

[500] Schlotterer, O. and Stieberger, S. Motivic multiple zeta values and superstring amplitudes. *J. Phys. A: Math. Theor. 46* (2013), 475401 (37 pp).

[501] Schneps, L. On the Poisson bracket on the free Lie algebra in two generators. *J. Lie Theory 16 (1)* (2006), 19–37.

[502] Schneps, L. ARI, GARI, Zig and Zag: An introduction to Ecalle's theory of multiple zeta values. *arXiv:1507.01534* (2015).

[503] Schnetz, O. Calculation of the ϕ^4 6-loop non-zeta transcendental.

arXiv:9912149 (1999).

[504] SCHNETZ, O. Quantum periods: A census of ϕ^4 transcendentals. *Comm. in Number Theory Phys. 4 (1)* (2010), 1–48.

[505] SCHNETZ, O. Graphical functions and single-valued multiple polylogarithms. *arXiv:1302.6445* (2013).

[506] SCHUTZENBERGER, M.P. On a factorisation of free monoids. *Proc. Amer. Math. Soc. 16* (1965), 21–24.

[507] SCHWINGER, J. Quantum electrodynamics. III. The electromagnetic properties of the electron – radiative corrections to scattering. *Phys. Rev. 79* (1949), 790–817.

[508] SEKINE, C. Partial sums of multiple zeta value series. *Tokyo J. Math. 29 (2)* (2006), 465–474.

[509] SHEN, Z. AND CAI, T. Some identities for multiple Hurwitz-zeta values (in Chinese). *Sci. Sinica Math. 41* (2011), 955–970.

[510] SHEN, Z. AND CAI, T. Congruences for alternating triple harmonic sums. *Acta Mathematica Sinica, Chinese Series 55 (4)* (2012), 737–748.

[511] SHEN, Z. AND CAI, T. Some identities for multiple zeta values. *J. Number Theory 132* (2012), 314–323.

[512] SHI, L.-L. AND PAN, H. A q-analogue of Wolstenholme's harmonic series congruence. *Amer. Math. Monthly 114(6)* (2007), 529–531.

[513] SHIMURA, G. *Elementary Dirichlet Series and Modular Forms.* Springer, New York, 2007.

[514] SILVERMAN, J.H. Wieferich's criterion and the *abc*-conjecture. *J. Number Theory 30* (1988), 226–237.

[515] SINGER, J. On q-analogues of multiple zeta values. *Funct. Approx. Comment. Math. 53*, 1 (2015), 135–165.

[516] SITARAMACHANDRARAO RAO, R. AND DAVIS, B. Some identities involving the Riemann zeta function, II. *Indian J. Pure Appl. Math. 17(10)* (1986), 1175–1186.

[517] SLATER, L. *Generalized Hypergeometric Functions.* Cambridge Univ. Press, Cambridge, 1966.

[518] SLOAN, N.J.A. The On-Line Encyclopedia of Integer Sequences.

[519] SMIRNOV, V.A. *Feynman integral calculus.* Springer, Berlin, 2006.

[520] SOMMERFIELD, C.M. Magnetic dipole moment of the electron. *Phys. Rev. 107* (1957), 328–329.

[521] SOUDÈRES, I. Motivic double shuffle. *Intl. J. Number Theory 6* (2010), 339–370.

[522] SOUDÈRES, I. Cycle complex over the projective line minus three points: toward multiple zeta values cycles. *J. Pure Appl. Alg.* (2016. arXiv:1210.4653).

[523] SOUDÈRES, I. A relative basis for mixed Tate motives over the projective line minus three points. *arXiv:1312.1849* (2013).

[524] SOUDÈRES, I. Multiple zeta value cycles in low weight. In *Feynman Amplitudes, Periods and Motives*, L. Alvarez-consul, J. I. Burgos-gil and K. Ebrahimi-Fard, Ed., vol. 648 of *Contemp. Math.* Amer. Math. Soc., Providence, RI, 2015, pp. 203–248.

[525] STANLEY, R. On the number of reduced decompositions of elements of Coxeter groups. *European J. Combin. 5* (1984), 359–372.
[526] STEENBRINK, J. AND ZUCKER, S. Variation of mixed Hodge structure. I. *Inv. Math. 80* (1985), 489–542.
[527] SUN, Z.-W. Congruences concerning Bernoulli numbers and Bernoulli polynomial. *Disc. Applied Math. 105* (2000), 193–223.
[528] 'T HOOFT, G. AND VELTMAN, M. Scalar one loop integrals. *Nucl. Phys. B 153* (1979), 365–401.
[529] TAKAMUKI, T. The Kontsevich invariant and relations of multiple zeta values. *Kobe J. Math. 16* (1999), 27–43.
[530] TAKEYAMA, Y. The algebra of a q-analogue of multiple harmonic series. *SIGMA 9* (2013), Paper 0601, 15 pp.
[531] TANAKA, T. On the quasi-derivation relation for multiple zeta values. *J. Number Theory 129(9)* (2009), 2021–2034.
[532] TANAKA, T. Algebraic interpretation of Kawashima relation for multiple zeta values. In *Algebraic number theory and related topics 2008*, Res. Inst. Math. Sci. (RIMS), Kyoto. MS Kôkyûroku Bessatsu, B19, 2010, pp. 117–134.
[533] TANAKA, T. Restricted sum formula and derivation relation for multiple zeta values. *arXiv:1303.0398* (2013).
[534] TANAKA, T. AND WAKABAYASHI, N. An algebraic proof of the cyclic sum formula for multiple zeta values. *J. Alg. 323(3)* (2010), 766–778.
[535] TANGEDAL, B. AND YOUNG, P. On p-adic multiple zeta and log gamma functions. *J. Number Theory 131(7)* (2011), 1240–1257.
[536] TASAKA, K. On linear relations among totally odd multiple zeta valuesrelated to period polynomials. *Kyushu J. Math., to appear* (2014).
[537] TASAKA, K. AND YAMAMOTO, S. On some multiple zeta-star values of one-two-three indices. *Intl. J. Num. Theory 9 (5)* (2013), 1171–1184.
[538] TAURASO, R. Congruences involving alternating multiple harmonic sum. *Electron. J. Combin. 17(1)* (2010), R16.
[539] TAURASO, R. AND ZHAO, J. Congruences of alternating multiple harmonic sums. *J. Comb. Number Theory 2(2)* (2010), 129–159.
[540] TERASOMA, T. Mixed Tate motives and multiple zeta values. *Inv. Math. 149* (2002), 339–369.
[541] TERASOMA, T. Geometry of multiple zeta values. In *Proc. ICM (Madrid, 2006)*, vol. II of *European Mathe. Soc., Zürich.* Asociación Int. Congress Math., 2006, pp. 627–635.
[542] TERASOMA, T. Brown–Zagier relation for associators. *arXiv:1301.7474* (2013).
[543] TERHUNE, D. Evaluation of double L-values. *J. Number Theory 105* (2004), 275–301.
[544] TERHUNE, D. Evaluations of a class of double L-values. *Proc. Amer. Math. Soc. 134* (2006), 1881–1889.
[545] THAKUR, D.S. *Function Field Arithmetic.* World Scientific Publishing, River Edge, NJ, 2004.
[546] TODOROV, I. Polylogarithms and multizeta values in massless Feynman

amplitudes. *IHES preprint* (2014), IHES/P/14/10.

[547] TORNHEIM, L. Harmonic double series. *Amer. J. Math. 72* (1950), 303–314.

[548] TRUESDELL, C. On a function which occurs in the theory of the structure of polymers. *Ann. Math. 46* (1945), 144–157.

[549] TSUMURA, H. A note on q-analogues of the Dirichlet series and q-Bernoulli numbers. *J. Number Theory 39 (3)* (1991), 251–256.

[550] TSUMURA, H. Combinatorial relations for Euler–Zagier sums. *Acta Arith. 111* (2004), 27–42.

[551] TSUMURA, H. On alternating analogues of Tornheim's double series II. *Ramanujan J. 18* (2009), 81–90.

[552] TSUNOGAI, H. On ranks of the stable derivation algebra and Deligne's problem. *Proc. Japan Acad. Ser. A Math. Sci. 73* (1997), 29–31.

[553] ULANSKII, E.A. Identities for generalized polylogarithms (Russian), Mat. Zametki **73**(4) (2003), 613–624. English translation available. *Math. Notes 73(3-4)* (2003), 571–581.

[554] USSYUKINA, N. Calculation of multiloop diagrams in high orders of perturbation theory. *Phys. Lett. B267* (1991), 382–388.

[555] USSYUKINA, N. AND DAVYDYCHEV, A. Exact results for three- and four-point ladder diagrams with an arbitrary number of rungs. *Phys. Lett. B305* (1993), 136–143.

[556] VAN DER POORTEN, A. A proof did Euler missed ... Apéry's proof of the irrationality of $\zeta(3)$. An informal report. *The Mathematical Intelligencer 1 (December)* (1979), 195–203.

[557] VAN HAMME, L. Some conjectures concerning partial sums of generalized hypergeometric series. In *p-adic functional analysis (Nijmegen, 1996)*, vol. 192 of *Lecture Notes in Pure and Appl. Math.* Dekker, 1997, pp. 223–236.

[558] VANDIVER, H.S. On development in an arithmetic theory of the Bernoulli and allied numbers. *Scripta Math. 25* (1961), 273–303.

[559] VASILEV, D.V. Some formulas for Riemann zeta-function at integral points. *Vestnik Moskov. Univ. Ser. I, Mat. Mekh.* ***1****(1996), 81–84. English translation available, Moscow Univ. Math. Bull. 51(1)* (1996), 41–43.

[560] VERMASEREN, J.A.M. Harmonic sums, Mellin transforms and integrals. *Internat. J. Modern Phys. A 14(13)* (1999), 2037–2076.

[561] VOEVODSKY, V. Triangulated Category of Motives over a field. In *Cycles, Transfers and Motivic Homology Theories*, V. Voevodsky, A. Suslin and E. Friedlander, Ed., vol. 43 of *Annals of Math. Studies.* Princeton Press, 1999.

[562] VOLLINGA, J. AND WEINZIERL, S. Numerical evaluation of multiple polylogarithms. *Computer Phys. Comm. 167(3)* (2005), 177–194.

[563] WALDSCHMIDT, M. Valeurs zêta multiples. Une introduction. *J. Théor. Nombres Bordeaux 12* (2000), 581–595.

[564] WALDSCHMIDT, M. Multiple polylogarithms: an introduction. In *Number Theory and Discrete Math.*, A.K. Agarwal, B.C. Berndt, C.F. Krattenthaler, G.L. Mullen, K. Ramachandra, M. Waldschmidt, Ed., Trends in Math. Birkhäuser, Boston, 2002, pp. 1–12.

[565] WAN, J. Some notes on weighted sum formulae for double zeta values. In

Number Theory and Related Fields: In Memory of Alf van der Poorten, vol. 43. Springer Proc. in Math. & Stat, 2013, pp. 361–379.

[566] WANG, L. A new congruence on multiple harmonic sums and Bernoulli numbers. *arXiv:1504.03227* (2015).

[567] WANG, L. A new curious congruence involving multiple harmonic sums. *J. Number Theory 154 (9)* (2015), 16–31.

[568] WANG, L. AND CAI, T. A curious congruence modulo prime powers. *J. Number Theory 144* (2014), 15–24.

[569] WANG, Q. Moduli spaces and multiple polylogarithm motives. *Adv. in Math. 206* (2006), 329–357.

[570] WEINZIERL, S. Feynman graphs. *arXiv:1301.6918* (2013).

[571] WHITTAKER, E.T. AND WATSON, G.N. *A Course of Modern Analysis.* Cambridge Univ. Press, Cambridge, 1996.

[572] WILDESHAUS, J. *Realizations of Polylogarithms*, vol. 1650 of *Lecture Notes in Math.* Springer-Verlag, 1997.

[573] WITTEN, E. On quantum gauge theories in two-dimensions. *Commun. Math. Phys. 141(1)* (1991), 153–209.

[574] WOJTKOWIAK, Z. A construction of analogs of the Bloch–Wigner function. *Math. Scandinavica 65* (1989), 140–142.

[575] WOJTKOWIAK, Z. The basic structure of polylogarithmic functional equations. In *Structural Properties of Polylogarithms*, L. Lewin, Ed., vol. 37 of *Math. Surveys and Monographs.* Amer. Math. Soc., Providence, RI, 1992, pp. 205–231.

[576] WOJTKOWIAK, Z. On l-adic iterated integrals. I. Analog of Zagier conjecture. *Nagoya Math. J. 176* (2004), 113–158.

[577] WOJTKOWIAK, Z. On l-adic iterated integrals. II. Functional equations and l-adic iterated polylogarithms. *Nagoya Math. J. 177* (2005), 117–153.

[578] WOLSTENHOLME, J. On certain properties of prime numbers. *Quarterly J. Pure & Appl. Math. 5* (1862), 35–39.

[579] WU, A.C.T. On the analytic properties of the 4-point function in perturbation theory. *Mat. Fys. Medd. Dan. Vid. Selsk. 33(3)* (1961), 1–88.

[580] XIA, B. AND CAI, T. Bernoulli numbers and congruences for harmonic sums. *Intl. J. of Number Theory 6 (4)* (2010), 849–855.

[581] YAMAMOTO, S. Explicit evaluation of certain sums of multiple zeta-star values. *Funct. Approx. Comment. Math. 49 (2)* (2013), 283–289.

[582] YAMAMOTO, S. Interpolation of multiple zeta and zeta-star values. *J. Alg. 385* (2013), 102–114.

[583] YAMAMOTO, S. Multiple zeta-star values and multiple integrals. *arXiv:1405.6499* (2014).

[584] YAMAZAKI, C. On the duality for multiple zeta-star values of height one. *Kyushu J. Math. 64(1)* (2010), 145–152.

[585] YASUDA, S. Finite real multiple zeta values generate the whole space $\mathbb{Z}$. *Intl. J. Number Theory* (2016).

[586] YASUDA, S. $W(F_4)$-symmetry and the linearized double shuffle spaces of depth four. *unpublished manuscript* (Dec. 6, 2013).

[587] YEE, A.J. A new shuffle convolution for multiple zeta values. *J. Alg.ic*

Combin. 21 (2005), 55–69.

[588] Yu, J. Transcendence and special zeta values in characteristic p. *Ann. Math. 134* (1991), 1–23.

[589] Yuan, H., and Zhao, J. New families of weighted sum formulas for multiple zeta values. *Rocky Mountain J. Math., available* (2014).

[590] Yuan, H., and Zhao, J. Double shuffle relations of double zeta values and double Eisenstein series of level N. *J. London Math. Soc. 92 (3)* (2015), 520–546.

[591] Yuan, H., and Zhao, J. Multiple divisor functions and multiple zeta values at level N. *Manuscripta Math.* (2016).

[592] Yuan, H. and Zhao, J. Restricted sum formula of multiple zeta values. *Funct. Approx. Comment. Math 51 (1)* (2014), 111–119.

[593] Zagier, D. Hyperbolic manifolds and special values of Dedekind zeta-functions. *Inv. Math. 83* (1986), 285–301.

[594] Zagier, D. The remarkable dilogarithm. *J. Math. Phys. Sci. 22* (1988), 131–145.

[595] Zagier, D. The Bloch–Wigner–Ramakrishnan polylogarithm function. *Math. Ann. 286* (1990), 613–624.

[596] Zagier, D. Polylogarithms, Dedekind zeta functions and the algebraic K-theory of fields. In *Arithmetic Algebraic Geometry*, vol. 89 of *Prog. in Math.* Birkhäuser, Boston, 1991, pp. 391–430.

[597] Zagier, D. Periods of modular forms, trace of Hecke operators, and multiple zeta values. *RIMS Kokyuroku 843* (1993), 162–170.

[598] Zagier, D. Values of zeta functions and their applications. In *First European Congress of Math. (Paris, 1992)*, A. Joseph et al., Ed., vol. II. Birkhäuser, Basel, 1994, pp. 497–512.

[599] Zagier, D. Quelques conséquences surprenantes de la cohomologie de $\mathrm{SL}(2,\mathbb{Z})$ (in French). In *Leçons de Mathématiques d'aujourd'hui.* Cassini, Paris, 2000, pp. 99–123.

[600] Zagier, D. The dilogarithm function. In *Frontiers In Number Theory, Physics and Geometry II*, P.E. Cartier, B. Julia, P. Moussa and P. Vanhove, Ed. Springer, 2007, pp. 3–65.

[601] Zagier, D. Evaluation of the multiple zeta values $\zeta(2,\dots,2,3,2,\dots,2)$. *Ann. Math. 175* (2012), 977–1000.

[602] Zagier, D. Multiple zêta values (in French). *Lecture notes of a course at College de France* (December 5, 2011).

[603] Zeilberger, D. Maple Package EKHAD. *Computer program implimenting WZ method.*

[604] Zhang, W.P. On the several identities of Riemann zeta-function. *Chinese Sci. Bull. 36(22)* (1991), 1852–1856.

[605] Zhao, J. Analytic continuation of multiple zeta functions. *Proc. Amer. Math. Soc. 128* (1999), 1275–1283.

[606] Zhao, J. Remarks on a Hopf algebra for defining motivic cohomology. *Duke Math. J. 103* (2000), 445–458.

[607] Zhao, J. Motivic complexes of weight three and pairs of simplices in projective 3-space. *Adv. Math. 161* (2001), 141–208.

[608] ZHAO, J. Multiple polylogarithms: analytic continuation, monodromy and variations of mixed Hodge structures. In *Contemporary Trends in Alg. Geometry and Alg. Topology*, S.S. Chern, L. Fu and R. Hain, Ed., vol. 5 of *Nankai Tracts in Math.* World Scientific Publishing, River Edge, NJ, 2002, pp. 167–193.

[609] ZHAO, J. Multiple harmonic series I: generalizations of Wolstenholme's Theorem. *arXiv:0301252. This is the unpublished version of [619]* (2003).

[610] ZHAO, J. Motivic cohomology of pairs of simplices. *Proc. London Math. Soc. 88(3)* (2004), 313–354.

[611] ZHAO, J. Variations of mixed Hodge structures of multiple polylogarithms. *Canad. J. Math. 56(6)* (2004), 1308–1338.

[612] ZHAO, J. Analytic continuation of multiple polylogarithms. *Anal. Math. 33* (2007), 301–323.

[613] ZHAO, J. Bernoulli numbers, Wolstenholme's Theorem and p^5 variations of Lucas' Theorem. *J. Number Theory 123* (2007), 18–26.

[614] ZHAO, J. Goncharov's relations in Bloch's higher Chow group $CH^3(F, 5)$. *J. Number Theory 124* (2007), 1–25.

[615] ZHAO, J. Multiple q-zeta functions and multiple q-polylogarithms. *Ramanujan J. 14(2)* (2007), 189–221.

[616] ZHAO, J. An exotic shuffle relation of multiple zeta values. *Arch. Math. (Basel) 91(5)* (2008), 409–415.

[617] ZHAO, J. Multiple polylogarithm values at roots of unity. *C. R. Acad. Sci. Paris, Ser. I 346* (2008), 1029–1032.

[618] ZHAO, J. Renormalization of multiple q-zeta values. *Acta Math. Sin. (Engl. Ser.) 24(10)* (2008), 1593–1616.

[619] ZHAO, J. Wolstenholme type Theorem for multiple harmonic sums. *Int. J. of Number Theory 4(1)* (2008), 73–106.

[620] ZHAO, J. A note on colored Tornheim's double series. *Integers 10* (2010), A59.

[621] ZHAO, J. Multi-polylogs at twelfth roots of unity and special values of Witten multiple zeta function attached to the exceptional Lie algebra $\mathfrak{g}_2$. *J. Alg. Appl. 9(2)* (2010), 327–337.

[622] ZHAO, J. On a conjecture of Borwein, Bradley and Broadhurst. *J. Reine Angew. Math. 639* (2010), 223–233.

[623] ZHAO, J. Standard relations of multiple polylogarithm values at roots of unity. *Doc. Math. 15* (2010), 1–34.

[624] ZHAO, J. Alternating Euler sums and special values of Witten multiple zeta function attached to $\mathfrak{so}(5)$. *J. Aust. Math. Soc. 89(3)* (2011), 419–430.

[625] ZHAO, J. Mod p structure of alternating and non-alternating multiple harmonic sums. *J. Théor. Nombres Bordeaux 23(1)* (2011), 259–268.

[626] ZHAO, J. Witten volume formulas for semi-simple Lie algebras. *Integers 11A* (2011).

[627] ZHAO, J. Finiteness of p-divisible sets of multiple harmonic sums. *Ann. Sci. Math. Québec. 36* (2012), 419–443.

[628] ZHAO, J. Identity families of multiple harmonic sums and multiple zeta (star) values. *J. Math. Soc. Japan, to appear* (2013), arXiv:1303.2227.

[629] ZHAO, J. On q-analog of Wolstenholme type congruences for multiple harmonic sums. *Integers 13* (2013).

[630] ZHAO, J. Restricted sum formula of alternating Euler sums. *Ramanujan J. 36 (3)* (2013), 375–401.

[631] ZHAO, J. A super congruence involving multiple harmonic sums. *arXiv:1404.3549* (2014).

[632] ZHAO, J. Uniform approach to double shuffle and duality relations of various q-analogs of multiple zeta values via Rota–Baxter algebras. *arXiv:1412.8044* (2014).

[633] ZHAO, J. Finite multiple zeta values and finite Euler sums. *arXiv:1507.04917* (2015).

[634] ZHAO, J. Sum formula of multiple Hurwitz-zeta values. *Forum Mathematicum 27 (2)* (2015), 929–936.

[635] ZHAO, J. AND ZHOU, X. Reducibility of signed cyclic sums of Mordell–Tornheim zeta and L-values. *J. Ramanujan Math. Soc. 26* (2011), 383–414.

[636] ZHAO, J. AND ZHOU, X. Witten multiple zeta values attached to $\mathfrak{sl}(4)$. *Tokyo J. Math. 34* (2011), 135–152.

[637] ZHOU, X. AND CAI, T. A generalization of a curious congruence on harmonic sums. *Proc. of Amer. Math. Soc. 135* (2007), 1329–1333.

[638] ZHOU, X., CAI, T. AND BRADLEY, D.M. Signed q-analogs of Tornheim's double series. *Proc. Amer. Math. Soc. 136* (2008), 2689–2698.

[639] ZLOBIN, S. Generating functions for the values of a multiple zeta function, Vestnik Moskov. Univ. Ser. I Mat. Mekh. **2** (2005), 55–59. English translation available. *Moscow Univ. Math. Bull. 60(2)* (2005), 44–48.

[640] ZLOBIN, S. Special values of generalized polylogarithms. *arXiv:0712.1656* (2007).

[641] ZLOBIN, S. Relations for multiple zeta values. *Math. Notes 84(6)* (2008), 771–782.

[642] ZUDILIN, W. One of the numbers $\zeta(5)$, $\zeta(7)$, $\zeta(9)$, $\zeta(11)$ is irrational. *Russian Math. Surveys 56(4)* (2001), 774–776.

[643] ZUDILIN, W. Algebraic relations for multiple zeta values. *Russian Math. Surveys 58(1)* (2003), 1–29.

[644] ZUDILIN, W. Multiple q-zeta brackets. *Mathematics 3* (2015), 119–130.

List of Abbreviations

AMHS	alternating multiple harmonic sum
CMZV	colored multiple zeta value
DBSF	double shuffle relation
DIST	distribution relation
DTI	diagonal translation invariant (only in Sec. 6.4)
FES	finite Euler sum
FMZV	finite multiple zeta value
MCl	multiple Clausens (singular or plural form)
MCV	multiple Clausen value
MDV	multiple Deligne value
MGl	multiple Glaishers (singular or plural form)
MHS	multiple harmonic sum
MH*S	multiple harmonic star sum
MHS	mixed Hodge structures (only in Chap. 2)
MPL	multiple polylog(arithm)
MZV	multiple zeta value
MZ*V	multiple zeta star value
SVMPL	single-valued single-variable multiple polylog(arithm)
q-MHS	q-analog of multiple harmonic sum
q-MH*S	q-analog of multiple harmonic star sum
q-MZV	q-analog of the multiple zeta value
q-MZ*V	q-analog of the multiple zeta star value

List of Symbols

Index

www.ingramcontent.com/pod-product-compliance
Lightning Source LLC
LaVergne TN
LVHW010636300926
849161LV00005B/955